Smart Innovation, Systems and Technologies

Volume 243

The Smart Innovation, Systems and Technologies book series encompasses the topics of knowledge, intelligence, innovation and sustainability. The aim of the series is to make available a platform for the publication of books on all aspects of single and multi-disciplinary research on these themes in order to make the latest results available in a readily-accessible form. Volumes on interdisciplinary research combining two or more of these areas is particularly sought.

The series covers systems and paradigms that employ knowledge and intelligence in a broad sense. Its scope is systems having embedded knowledge and intelligence, which may be applied to the solution of world problems in industry, the environment and the community. It also focusses on the knowledge-transfer methodologies and innovation strategies employed to make this happen effectively. The combination of intelligent systems tools and a broad range of applications introduces a need for a synergy of disciplines from science, technology, business and the humanities. The series will include conference proceedings, edited collections, monographs, handbooks, reference books, and other relevant types of book in areas of science and technology where smart systems and technologies can offer innovative solutions.

High quality content is an essential feature for all book proposals accepted for the series. It is expected that editors of all accepted volumes will ensure that contributions are subjected to an appropriate level of reviewing process and adhere to KES quality principles.

Indexed by SCOPUS, EI Compendex, INSPEC, WTI Frankfurt eG, zbMATH, Japanese Science and Technology Agency (JST), SCImago, DBLP.

All books published in the series are submitted for consideration in Web of Science.

More information about this series at http://www.springer.com/series/8767

P. Karuppusamy · Isidoros Perikos ·
Fausto Pedro García Márquez
Editors

Ubiquitous Intelligent Systems

Proceedings of ICUIS 2021

Springer

Editors
P. Karuppusamy
Department of EEE
Shree Venkateshwara Hi-Tech Engineering
Erode, Tamil Nadu, India

Fausto Pedro García Márquez
ETSI Industriales de Ciudad Real
University of Castile-La Mancha
Ciudad Real, Spain

Isidoros Perikos
Department of Computer Engineering
and Informatics
University of Patras
Patra, Greece

ISSN 2190-3018 ISSN 2190-3026 (electronic)
Smart Innovation, Systems and Technologies
ISBN 978-981-16-3674-5 ISBN 978-981-16-3675-2 (eBook)
https://doi.org/10.1007/978-981-16-3675-2

This Springer imprint is published by the registered company Springer Nature Singapore Pte Ltd.
The registered company address is: 152 Beach Road, #21-01/04 Gateway East, Singapore 189721, Singapore

We are honored to dedicate the proceedings of 1st ICUIS 2021 to all the participants, organizers and editors of this conference proceedings.

Preface

On behalf of the conference committee, I take this opportunity to welcome you all to the International Conference on Ubiquitous Computing and Intelligent Information Systems [ICUIS 2021]. The conference theme is *Ubiquitous Computing and Communication Systems*, a topic that is gaining significant research attraction form both academia and industries due to its relevance in solving the challenges in real-world applications ranging from smart cities to industries. The recently well-established track record of ubiquitous systems makes ICUIS as an excellent venue for exploring the complex challenges associated with the intelligent systems.

The conference includes different technical sessions by categorizing the computing and communication systems. The main aim of these sessions is to disseminate the state-of-the-art research results and findings and discuss the same with the session chair, who have professional expertise in the same field. In this first conference, totally 336 papers were submitted by the authors from all over the world, and out of these, about 57 papers were selected to present at the conference.

We were really honored and delighted to have prominent guests as keynote speakers and session chairs of the conference event. The grand opening of the conference is with the distinguished keynote speaker: Dr. R. Kanthavel, Professor, Department of Computer Engineering, King Khalid University, Abha, Kingdom of Saudi Arabia.

The success of ICUIS 2021 depends completely on the efforts of the authors, who have taken huge effort in submitting the papers on different varieties of topics. A huge appreciation is also deserved for the technical program committee, internal and external reviewers, and faculty and non-faculty members of the institution, who have invested their significant time and effort for maintaining the international quality for this first conference series. Additionally, we thank Springer publication for their extended publication support.

Erode, India
Patra, Greece
Ciudad Real, Spain

P. Karuppusamy
Isidoros Perikos
Fausto Pedro García Márquez

Acknowledgements

We are deeply grateful to our institution Shree Venkateshwara Hi-Tech Engineering College for sponsoring the first conference series of ICUIS 2021 and would like to acknowledge all members of Advisory Committee and Program Committee for providing excellent guidance. In particular, the organizer and editor of the conference wish to acknowledge the authors for delivering their presentation on ICUIS 2021. Also, the organizers wish to forthrightly acknowledge the timely technical assistance and services provided by reviewers. The efforts of reviewers helped the editors to maintain the high standard of the conference.

The organizers wish to acknowledge Dr. R. Kanthavel, Professor, Department of Computer Engineering, King Khalid University, Abha, Kingdom of Saudi Arabia, for their discussion and cooperation in successfully organizing the keynote session in this conference. The organizers also wish to acknowledge all the participants of the conference amidst the current global pandemic situation.

Organizing this event would not have been possible without the continual effort of the organizing committee members, notably: Dr. P. Karuppusamy, who served as the conference chair and organizing secretary of the conference.

Finally, we thank the Springer publications for their valuable suggestions and technical support throughout the publication process.

Contents

About the Editors

Dr. P. Karuppusamy is working as Professor and Head in Department of Electrical and Electronics Engineering at Shree Venkateshwara Hi-Tech Engineering College, Erode. In 2017, he had completed doctorate in Anna University, Chennai, and in 2007, he had completed his postgraduate in power electronics and drives in Government College of Technology, Coimbatore, India. He has more than 10 years of teaching experience. He has published more than 40 papers in national and international journals and conferences. He has acted as a conference chair in IEEE international conferences and a guest editor in reputed journals. His research area includes modeling of PV arrays and adaptive neuro-fuzzy model for grid-connected photovoltaic system with multilevel inverter.

Dr. Isidoros Perikos completed his Ph.D. in Computer Engineering and Informatics, Computer Engineering and Informatics Department at University of Patras, Greece (2016), and M.Sc. in Computer Science and Technology, Computer Engineering and Informatics Department at University of Patras (2010). He has completed Engineering Diploma (5-year program, M.Eng.) in Computer Engineering and Informatics, Computer Engineering and Informatics Department at University of Patras (2008). His research interest includes Semantic Web and ontology engineering, Web intelligence, natural language processing and understanding, human–computer interaction, and affective computing robotics. He has published in national and international journals and conferences.

Dr. Fausto Pedro García Márquez works at UCLM as Full Professor (Accredited as Full Professor from 2013), Spain, Honorary Senior Research Fellow at Birmingham University, UK, Lecturer at the Postgraduate European Institute, and he has been Senior Manager in Accenture (2013–2014). He obtained his European Ph.D. with a maximum distinction. He has been distingueed with the prices: Advancement Prize for Management Science and Engineering Management Nominated Prize (2018), and he has published more than 150 papers (65% ISI, 30% JCR, and 92% internationals), some recognized as: "Renewable Energy" (as "Best Paper 2014"), "ICMSEM" (as "excellent"), "International Journal of Automation and Computing" and "IMechE

Part F: Journal of Rail and Rapid Transit" (most downloaded), etc. He is an author and an editor of 25 books (Elsevier, Springer, Pearson, Mc-GrawHill, Intech, IGI, Marcombo, AlfaOmega...) and 5 patents. He is Editor of 5 international journals and Committee Member of more than 40 international conferences. He has been Principal Investigator in 4 European Projects, 5 National Projects, and more than 150 projects for universities, companies, etc. His main interest includes maintenance management, renewable energy, transport, advanced analytics, and data science.

Improvement of QoS Parameters of IoT Networks Using Artificial Intelligence

Anjum Sheikh, Sunil Kumar, and Asha Ambhaikar

Abstract The quality of service (QoS) parameters of IoT networks plays an important role in knowing the efficiency of an application. As the number of IoT users and devices is increasing and the number is envisaged to grow fast in the future, it has become extremely important to pay attention toward the QoS parameters for increasing the acceptability of the technology among the people. The IoT networks should be capable of handling devices of diverse nature and at the same time should provide wireless access to all of them. Artificial intelligence (AI) is one of the techniques to improve the QoS and has been used in this paper to know the change in the QoS parameters for networks with a varying number of nodes. The parameters studied in this paper are end-to-end delay, throughput, packet delivery ratio, and jitter and energy consumption. All the values have been calculated for a network with 30, 40, 50, 60, 70, 80, and 90 nodes. A comparison of QoS parameters by using AI and without AI has been explained. The results indicate that most of the parameters showed improvement in the values for all the sizes of network with the application of AI.

1 Introduction

The technological advancements have brought the world at our fingertips. New technologies are arriving continuously to make our life more comfortable. In earlier days, Internet was used to connect two computers for the exchange of information. The rapid development of technologies has enabled a connection between devices or things. Internet of things (IoT) is one such technology that connects any two devices at any time from any part of the world with the help of Internet. The IoT devices

A. Sheikh (✉)
Research Scholar, Kalinga University, Raipur, Chhatisgarh, India

S. Kumar
HoD Electrical and Electronics Engg., Kalinga University, Raipur, Chhatisgarh, India

A. Ambhaikar
Professor and Dean Students Welfare, Kalinga University, Raipur, Chhatisgarh, India

P. Karuppusamy et al. (eds.), *Ubiquitous Intelligent Systems*, Smart Innovation, Systems and Technologies 243, https://doi.org/10.1007/978-981-16-3675-2_1

are equipped with sensors that exchange data with each other for accomplishing the assigned tasks. Due to the heterogeneity of the networks and scalability issues, it is not possible to use the routing algorithms that were used for the computer networks. The features of IoT networks are different as compared to the computer networks, and therefore, it is not possible to work with the traditional routing algorithms that were used by the networks to forward data using Internet. The rapid increase in the number of IoT devices has increased the amount of data that is forwarded continuously. The success of IoT depends on the efficiency of its network to transmit and receive this data at correct time intervals without any changes or losses.

IoT thus includes a large amount of data that is transmitted or received through the networks by the devices. The data on the IoT networks goes through three stages of forwarding, processing, and analyzing. The IoT networks should be able to provide wireless connectivity to the large number of devices without affecting the data being transferred over them. The task of transfer toward the intended devices would be challenging with the increase in the density of the devices over the networks. At the same time, handling the data transfer among the heterogeneous devices poses another challenge on the functioning of IoT networks. It is therefore essential that the networks should be able to forward the correct data toward the destination and at the same time prevent collision or loss of data packets [1]. The dynamic nature of the IoT networks, in which the nodes keep on changing their positions, poses another challenge while maintaining the QoS parameters. Artificial intelligence (AI) is a powerful tool that helps in analyzing the voluminous IoT data [2]. AI is increasingly being used as a tool for improving the existing systems and has been integrated with applications like health care, analysis of data, and security for the development of innovative and higher quality systems [3]. The IoT applications become smarter when combined with AI. This is the reason that many companies working on IoT are merging AI to achieve better operational efficiency. Ai is generally distinguished into two types as narrow AI and general AI. The narrow AI includes intelligent systems that are able to perform certain tasks without being programmed specifically while the general AI is a form of intelligence technique that is able to learn the methods needed for performing the assigned tasks [4].

Some of the QoS parameters studied in this paper are end-to-end delay, throughput, packet delivery ratio, and jitter and energy consumption. All the parameters are equally important for the efficient working of the IoT networks. Section 2 of this paper discusses some of the research works already done in the area of QoS parameters for the IoT networks. Section 3 gives a brief description of the experimental setup, while Sect. 4 explains the changes in the values of QoS parameters by using AI. The comparative results for the networks with 30, 40, 5, 60, 70, 80, and 90 nodes have been studied to know the effect of AI on the values of the QoS parameters. Most of the parameters have shown significant improvement by using AI.

2 Related Research Work

With the development of IoT and increase in its users, it has become essential to develop methods to improve the QoS metrics. This section presents some of the researches that have been done for the improvement of QoS on IoT networks.

The reliability of an IoT network is determined by its capacity to handle continuous transmissions among the devices without affecting the QoS metrics. The authors in [5] have used QoS categories activeness awareness adaptive enhanced distribution channel access (QCAAAE) to improve the efficiency of the uplink access in the networks. The simulation results indicated improved values of throughput and slight increase in the values of delay for video and video services. A multiple quality of service parameter-based routing protocol (MQSPR) in [6] to improve the performance of the network that supports communication systems among the aircraft and ground along with the IoT communication. The algorithm helps to improve the network performance of the aeronautical ad hoc networks by overcoming the challenges of reliable data delivery. The MQSPR improves the packet delivery ratio and at the same time achieves good connectivity.

Backtracking search optimization algorithm (BSOA) in [7] has introduced a QoS provisioning framework (QOPF) to maintain the level of QoS for satisfying the consumers demands while using the latest IoT applications. The service providers sometimes fail to fulfill the requirements of the users by using traditional algorithms for the applications that combine IoT and cloud computing. QOPF approach ensures better utilization of infrastructure to meet the complex demands and at the same time maintain the good values of performance metrics. The simulation results indicate better values for delay, throughput, packet delivery ratio, and jitter as compared to the other algorithms. Limited bandwidth, interference, and multipath reflections are some of the limitations that affect the QoS while working with the radio frequency-based wireless networks. An integrated and visible light communication and positioning (VLCP) system in [8] has been used to provide improved speed of communication while maintaining the QoS requirements.

The growing number of devices and heterogeneity of the networks are the factors that make it difficult to achieve better performance while maintaining optimum values of QoS metrics. The service providers are unable to provide the appropriate network connections to the users due to the different characteristics of devices and large amount of information that has to be exchanged on the IoT networks. A QoS scheduling module for service-oriented IoT has been proposed in [9] deals with the scheduling of heterogeneous IoT networks. The decision model in the network layer has been developed by classifying the network traffic into two types according to the types of services. The first type is the delay or jitter sensitive service that can be used for real-time applications and the best effort (BE) class service for peer to peer applications.

Artificial intelligence is being considered to be one of the suitable methods to maintain and improve the QoS of the IoT networks. One of such algorithms has been discussed in [10] that use AI to improve the overall quality of the system for Internet of vehicles (IoV). This algorithm improves the lifetime of the portable devices. AI system has been used in [11] to detect the kind of network traffic and to inform the network controller about the actions to be taken for guaranteeing the QoS. Using AI for the given software-defined network (SDN) reduces the jitter in the network, thereby reducing the transmission losses. This improvement in the performance of network was observed due to the capability of AI to accurately detect the network traffic. A multilayer neural network (MNN) for long short-term memory (LSTM) learning, based on machine learning (a subset of AI), has been used in [12] for optimization of resources to achieve QoS. The proposed model helps in obtaining better bandwidth and energy utilization and at the same time maintaining suitable values of QoS metrics for the given IoT environment.

The authentication and encryption methods used for security of IoT networks consume a lot of energy that may reduce the network lifetime. An algorithm based on AI for adaptive security proposed in [13] uses extended Kalman filtering (EKF) that estimates the energy requirements of the various available security methods and then selects the method that provides good protection but at the same time does not exhaust the energy of IoT network. This method has improved the security and also provides better values of throughput and network lifetime. An AI-based energy-efficient model to ensure better spectrum utilization has been presented in [14] to overcome the problems of less throughput and increased delay observed in the real-time applications. The network gateways use energy detection technique to sense the available channels and forward it to the cognitive engine. The cognitive engine helps to select the channels and divides the time slots for each channel to ensure delivery of data on time. This model works well for the resource-constrained IoT devices and obtains better values of packet delivery ratio, delay, and throughput. Imbalance of load in controllers and switches results into poor values of QoS. To deal with this issue, a software-defined IoT model based on AI has been discussed in [15] that improves QoS by classifying network traffic and then constructs a network topology that would help in efficient routing of data over the network. This method reduces the latency time and packet loss and thus obtains increased throughput. Machine learning, subset of AI has been used in [16] to increase the network lifetime by increasing the energy efficiency of the routing algorithm. It uses clustering-based method to select the cluster heads on the basis of their residual energy. Along with network lifetime, this method helps in obtaining better packet delivery ratio and reduction in the transmission delay.

3 Simulation Setup

The simulations for the experiment have been done using network simulator -2 and FEDORA 7. A topology of 300 length and width has been used, and the nodes are mobile that keep on changing the position. The routing protocol used in AODV with packet size of 1000. The codes used for the calculation of QoS parameters have been done using AI and without AI.

For all the simulations, the first step is to specify whether the QoS metrics has to be calculated using AI or without it. After the option has been selected, the next step includes specifying the number of communications to be performed in that iteration. In this paper, all the values of QoS metrics have been studied and evaluated for 5 as well as 10 communications. The next step is to specify source node, sink node, and packet priority node for each communication. If we consider the case for 5 communications, the sink, source, and packet priority will have to specify for 5 events. The steps involved in the execution are given in Fig. 1, and the average values for the QoS parameters are given in Table 1. All the steps have been repeated for 5 and 10 communications using 30, 40, 50, 60, 70, 80, and 90 nodes.

```
# Do you want to apply AI and machine based routing (1-yes)?
# enter number of Communications: 2
# enter source sensor for event number 1
# enter sink receiver for event number 1
# enter packet priority for event number 1
# enter source sensor for event number 2
# enter sink receiver for event number 2
# enter packet priority for event number 2
```

Fig. 1 Steps for execution

Table 1 Average values of QoS parameters for 5 and 10 communications

Nodes	End-to-end delay		Energy consumption		Jitter		Packet delivery ratio		Throughput	
	AI	Without AI	AI	Without AI	AI	Without AI	AI	Without AI	AI	Without AI
30	254	856	7.745	59.504	121	36	98.01	99.68	830.41	773.07
40	253.5	769	4.6225	42.79	54.5	69.5	98.79	99.77	837.15	791.06
50	254	501	3.092	32.88	112	94.5	99.06	99.63	837.75	793.19
60	181	617	2.496	27.38	29.5	47	99.19	99.7	835.72	795.04
70	187	691	1.748	21.51	21	132	99.43	99.74	845.21	792.91
80	156.5	671.5	1.608	20.31	64	75	99.61	99.65	836.87	793.76
90	150.5	626.5	1.2	17.63	65.5	102	99.11	99.76	845.63	797.63

4 Results and Discussions

This section shows the variation in the values for QoS parameters for 5 and 10 communications with respect to networks with different number of nodes. The values for the QoS parameter have been calculated without using AI and then with AI and the values have been compared. The charts given below show the changes in the metrics by which the effect of AI on the performance of networks can be studied.

4.1 End-to-End Delay

The first parameter studied in this section is an end-to-end delay which is the time required for the data packets to travel from source to the destination. The value of end-to-end delay should be less which means that it is preferable to have lower values of delay. Large values of delay indicate that more time is required for the data to reach its destination device. The value of end-to-end delay is dependent on the density of nodes. As the number of nodes in the given network increases, the distance among the nodes will reduce, and therefore, the end-to-end delay will also be reduced. Figure 2 shows the comparison of end-to-end delay for completing 5 communications over the network by using AI and without using AI, while Fig. 3 shows the comparison for 10 communications. Both the graphs show that the values of delay are very less for the networks by using AI. The values of end-to-end delay have been reduced by 53.47% by using AI.

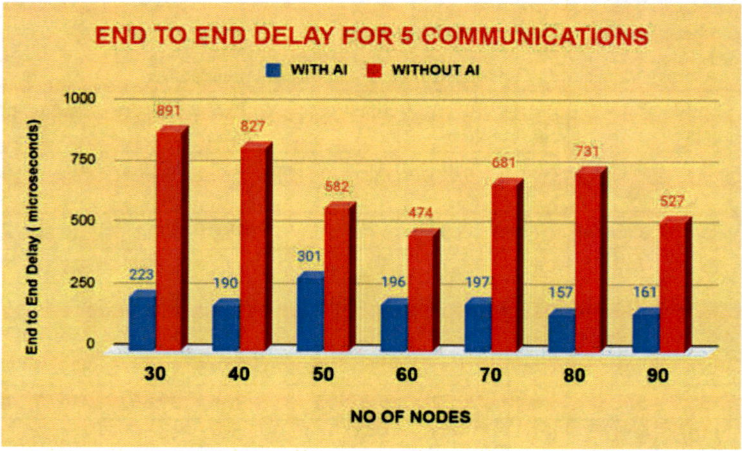

Fig. 2 End-to-end delay for 5 communications

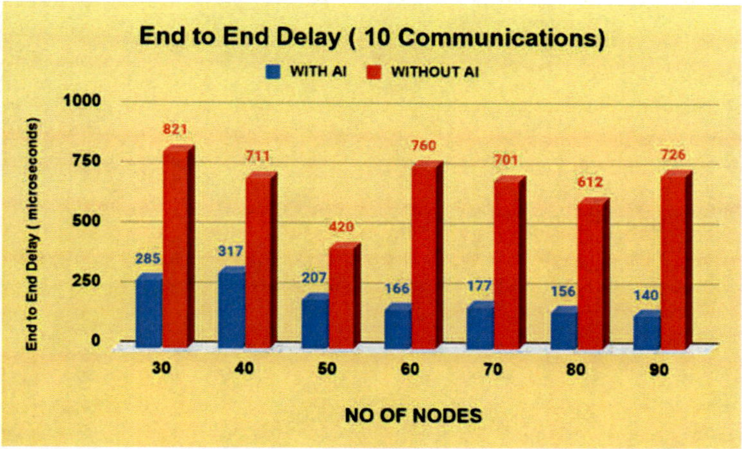

Fig. 3 End-to-end delay for 10 communications

4.2 *Energy Consumption*

The next QoS parameter studied in this section is energy consumption of the network. The IoT devices are battery operated and it is advisable to use the algorithms that consume less energy so that the network can be active for longer time duration. The routing path used by nodes will be able to transmit the data efficiently if all the nodes are working properly. Energy dissipation by the nodes can discharge the batteries due to which the nodes cannot work and the path will be broken resulting in loss of data over the network. It is therefore necessary that the routing techniques or algorithm used for the IoT nodes should consume less energy to increase the lifetime of the networks. Energy consumption is directly proportional to the distance among the nodes. When the number of nodes in the network is increased, the distance among the nodes decreases, and hence, a reduction in energy consumption can be observed for the networks with more number of nodes. Figure 4 and 5 shows the comparison for the energy consumption of the networks for 5 and 10 communications, respectively. The energy consumption has been reduced by 81.23% with the help of AI.

4.3 *Jitter*

Inconsistency in the delay of data packets traveling over the routing path causes jitter. The data packets routed toward the destination may use different paths due to which they do not reach the destination in sequence. Higher values of jitter can be disturbing for the real-time applications and can be the reason for unreliable and distorted communications. In IoT networks, the jitter increases with the increase in

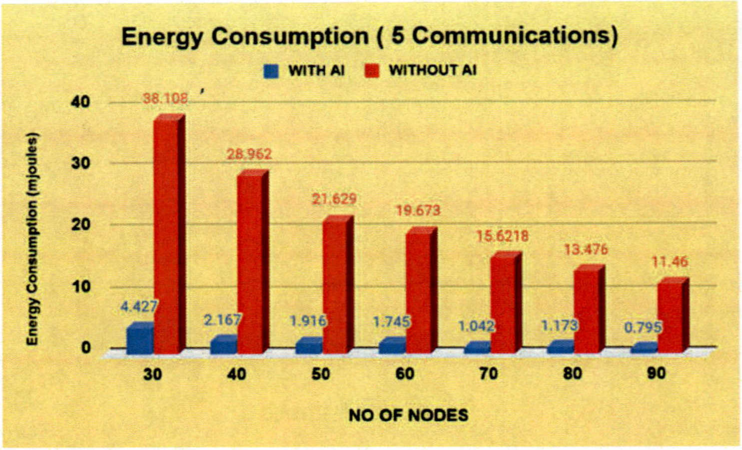

Fig. 4 Energy consumption for 5 communications

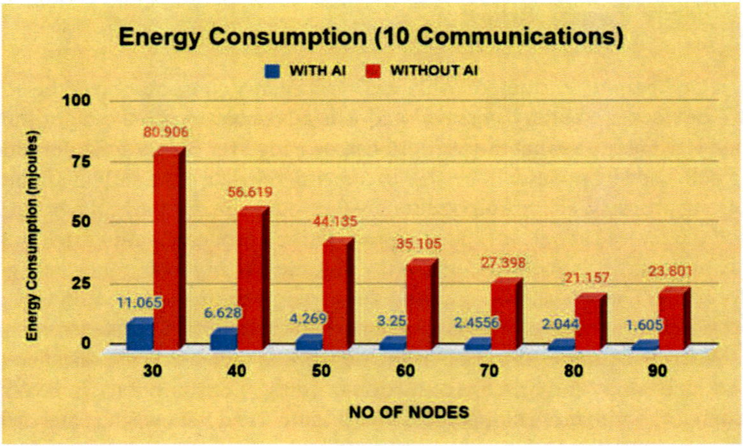

Fig. 5 Energy consumption for 10 communications

number of data packets on the routing path. With the increase in the number of IoT devices, dealing with jitter is another important issue to ensure better services to the IoT consumers. Figures 6 and 7 show the values of jitter in the network for 5 and 10 communications, respectively. The values of jitter are higher in some of the cases with using AI for network with 30, 40, and 80 nodes for completing five communications and for the network with 30 nodes while completing 10 communications. In the rest of the communications, the value of jitter is higher for the networks when AI is not used. The value of jitter has reduced averagely by 8.6% by using AI.

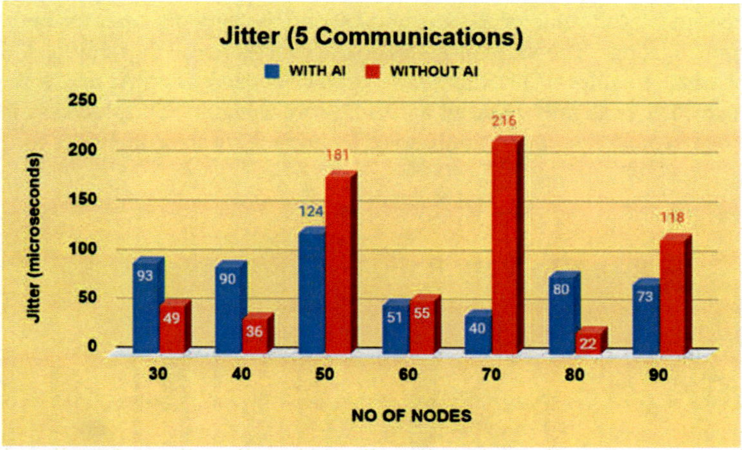

Fig. 6 Jitter for 5 communications

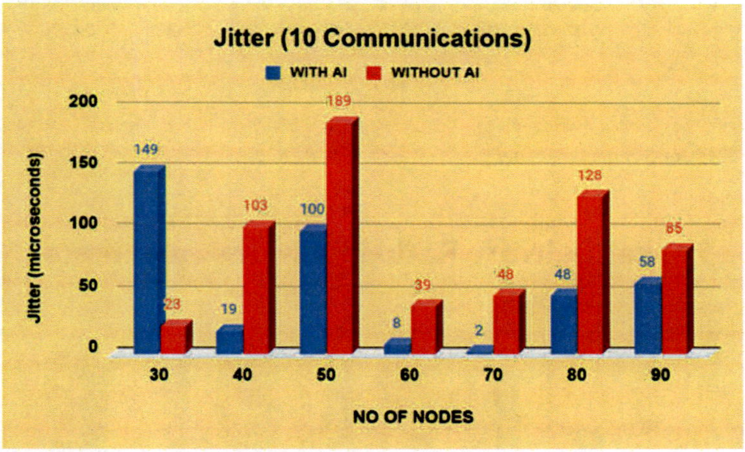

Fig. 7 Jitter for 10 communications

4.4 Packet Delivery Ratio

Packet delivery ratio (PDR) is another very important performance metric that helps in knowing the reliability of the network in transmitting information. PDR is the ratio of the number of data packets received at the sink node to the number of data packets that were actually transmitted by the source. It is desirable to have large values of PDR as it indicates that more amount of information is reaching its destination. Less value of PDR is an indication that data packets are being lost in the routing path. In this experiment, the value of PDR is good for all the networks and for both 5 and 10 communications. The value of PDR is more than 98 for all the iterations. But the comparisons of values of PDR that have been calculated using AI and without AI show that PDR is slightly reduced by 0.342% by using AI. This decrease in PDR can be due to more jitter that was observed by using AI in some instances for Figs. 8 and 9.

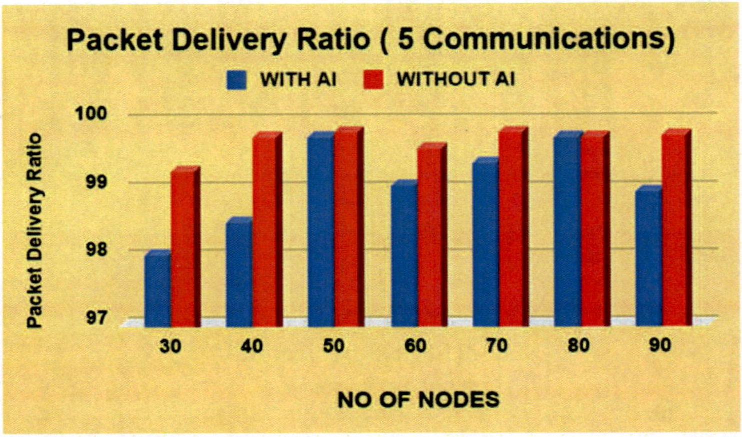

Fig. 8 PDR for 5 communications

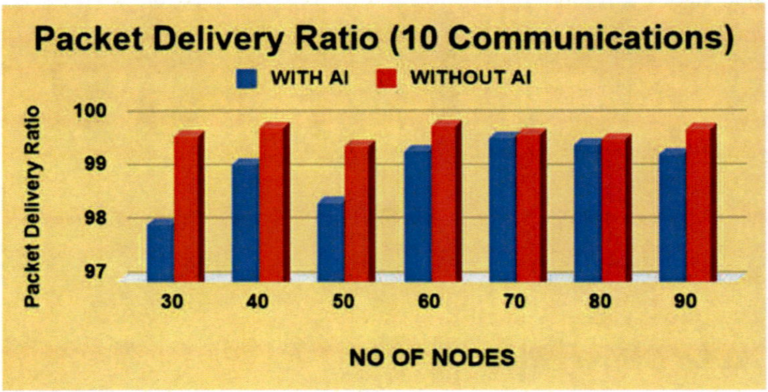

Fig. 9 PDR for 10 communications

4.5 Throughput

Throughput is a very important QoS metric as its value indicates the amount of data bits that have been successfully transmitted during the given time unit. Large values of throughput mean more amount of data could be transferred using the algorithm on the given route. The values of throughput have to be maintained keeping in view the application and kind of routing algorithm as energy consumption increases with the amount of data transfer. In some of the applications, it is preferred to have low values of throughput to preserve energy level of the network for a longer time. Figures 10 and 11 show the values of throughput obtained for 5 and 10 communications, respectively. It can be seen that the values of throughput have improved by 2.91% by using AI.

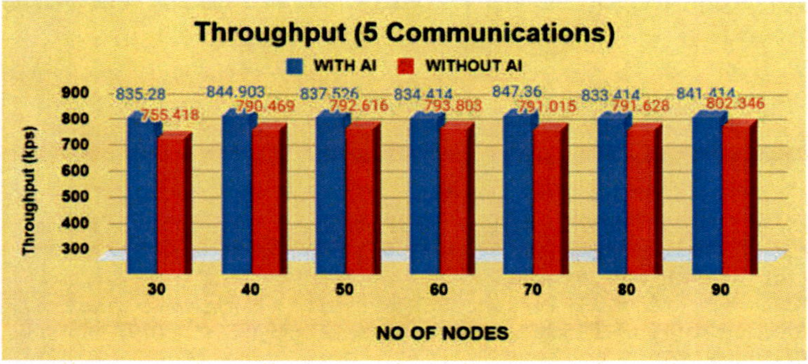

Fig. 10 Throughput for 5 communications

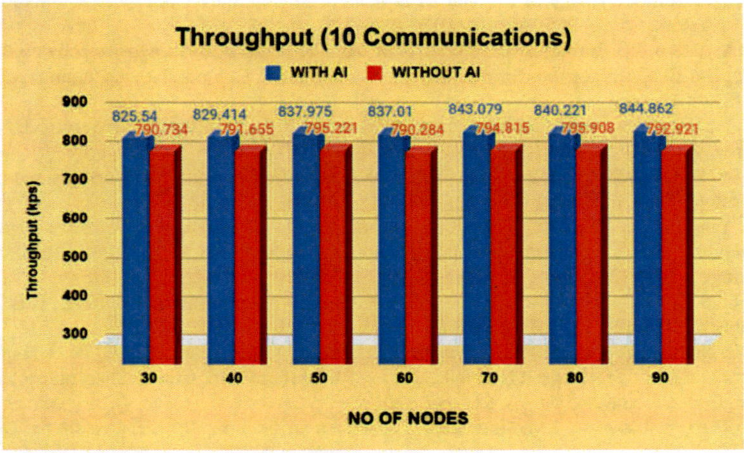

Fig. 11 Throughput for 10 communications

5 Conclusion

The variation of QoS parameters by using AI for the given networks has been studied for 5 and 10 communications. The values of QoS parameters have been calculated by using AI and without AI for the different network sizes. A comparison of some of the important parameters, end-to-end delay, energy consumption, throughput, jitter, and PDR has been given for the networks. Most of the parameters have improved by using AI. The improvements observed in the values of QoS metrics are, reduction in end-to-end delay by 53.47%, energy consumption by 81.23%, and Jitter by 8.61%. The throughput improved by 2.91% but PDR has reduced by 0.342%. It can be seen that all the other parameters except PDR showed improvement in their values. The integration of IoT and AI can therefore be beneficial to meet the requirements of transferring data in less time and less energy and at the same time maintain the quality of data.

References

1. H. Song, J. Bai, Y. Yi, J. Wu, L. Liu, Artificial intelligence enabled internet of things: network architecture and spectrum access. IEEE Comput. Intell. Mag. **15**(1), 44–51 (2020). https://doi.org/10.1109/MCI.2019.2954643
2. S.K. Singh, S. Rathore, J.H. Park, BlockIoT intelligence: a blockchain-enabled intelligent IoT architecture with artificial intelligence. Fut. Gener. Comput. Syst. (2019). https://doi.org/10.1016/j.future.2019.09.002
3. A.A. Osuwa, E.B. Ekhoragbon, L.T. Fat, Application of artificial intelligence in internet of things. In: Proceedings of the 2017 9th international conference on computational intelligence and communication networks (CICN), Girne, Cyprus, 16–17 Sept 2017, pp. 169–173
4. S.G. Tzafestas, Synergy of IoT and AI in modern society: the robotics and automation case. Rob. Autom. Eng. J. **3**(5), 00118–00132 (2018)
5. M.A. Salem, I.F. Tarrad, M.I. Youssef, S.M. Abd El-Kader, Qos categories activeness-aware adaptive EDCA algorithm for dense IoT networks. Int. J. Comput. Netw. Commun. (IJCNC) **11**(3) (2019)
6. Q. Luo, J. Wang, Multiple QoS parameters based routing for civil aeronautical Ad Hoc networks. IEEE Internet Things J. **4**(3), 804–814 (2017)
7. M.M. Badawy, Z.H. Ali, H.A. Ali, QoS provisioning framework for service-oriented internet of things (IoT), in *Cluster Computing* (Springer 2019), pp. 575–591
8. H. Yang, W.-D. Zhong, C. Chen, A. Alphones, P. Du, QoS-driven optimized design based integrated visible light communication and positioning for indoor IoT networks. IEEE Internet Things J. 1–15 (2019)
9. L. Li, S. Li, S. Zhao, QoS-aware scheduling of services-oriented internet of things. IEEE Trans. Industr. Inf. **10**(2), 1497–1505 (2014)
10. A.H. Sodhro, Z. Luo, G.H. Sodhro, M. Muzammal, J. Rodrigues, V.H.C. de Albuquerque, Artificial intelligence based QoS optimization for multimedia communication in IoV systems. Fut. Gener. Comput. Syst. **95**, 687–680 (2019)
11. A. Rego, A. Canovas, J.M. Jimenez, J. Lloret, An artificial intelligence system for QoS and QoE guarantee in IoT using software defined networks. IEEE Access **6**, 31580–31598 (2018)
12. R.C. Bhaddurgatte, B.P. Vijaya Kumar, S.M. Kusuma, Machine learning and prediction-based resource management in IoT considering Qos. Int. J. Recent Technol. Eng. (IJRTE) **8**(2), 687–694 (2019). ISSN: 2277-3878

13. B. Mao, Y. Kawamoto, N. Kato, AI-based joint optimization of QoS and security for 6G energy harvesting internet of things. IEEE Internet Things J. **7**(8), 7032–7042 (2020)
14. W. Yao, F. Khan, M. Ahmad, N. Shah, I. ur Rahman, A. Yahya, A. ur Rehman, Artificial intelligence-based load optimization in cognitive internet of things, in *Neural Computing and Applications* (Springer, 2020), pp. 16179–16181
15. M. Begovic, S. Causevic, B. Memic, A. Haskovic, AI-aided traffic differentiated QoS routing and dynamic offloading in distributed fragmentation optimized SDN-IoT. Int. J. Eng. Res. Technol. **13**(8), 1880–1895 (2020). ISSN 0974-3154
16. K. Li, H. Huang, X. Gao, F. Wu, G. Chen, QLEC: a machine-learning-based energy-efficient clustering algorithm to prolong network lifespan for IoT in high-dimensional space, in *ICPP* (2019) ACM ISBN 978-1-4503-6295-5/19/08

Early Diagnosis of Alzheimer's Disease Using ACO Optimized Deep CNN Classifier

Simarjeet Singh and Rekh Ram Janghel

Abstract Alzheimer's disease (AD) is a human brain disease that remains as a common cause of dementia, which occurs mainly in middle-aged or grown-up individuals. AD results in cognitive decline and memory loss. AD is caused by the decomposition of plaques around the nerves of brains or around the brain cells, where the brain cells get neurofibrillary tangled and result in various instability and mental illness. AD is a chronic and irreversible disease; the reasons and disease identification are still not known, but research says it can be identified during the early stages. In spite of that fact, this research work has proposed a computer-aided Alzheimer's classification method that will classify the class of an image either in normal class or demented class. The method uses the hybrid strategy of ant colony optimization (ACO) and feed forward convolutional neural network (CNN or ConvNet); however, identifying the architecture of CNN requires lots of expertise and is time-consuming. Henceforth, this research work has used the bio-inspired optimization strategy, which will identify the optimal combination of hyper-parameters, i.e. it recommends the configuration for the CNN model, and with that, configuration of hyper-parameters with the CNN model is trained with the training dataset, and CNN performs feature extraction alongside classification for arranging the gatherings possibly, when the model undergoes validation, where the performance metric of the model is evaluated and to identify whether the validating images are falling in the category of normal class (i.e. non-demented) or Alzheimer's class (i.e. demented class) with good results or not, the classification error is measured during this phase and is backpropagated to ACO optimizer, iteratively ACO is used to minimize the classification error by tuning the hyper-parameters, and after few iterations, a CNN architecture with optimal hyper-parameters combinations is obtained to result in least classification errors. The method was applied to the Alzheimer's Disease Neuroimaging Initiative (ADNI) dataset, which constitutes the fMRI images of Alzheimer's affected patients and resulted in developing efficient and state-of-the-art method for the classification of Alzheimer's disease. The proposed method performance metrics were recorded

S. Singh (✉) · R. R. Janghel
National Institute of Technology Raipur, Raipur, India

R. R. Janghel
e-mail: rrjanghel.it@nitrr.ac.in

P. Karuppusamy et al. (eds.), *Ubiquitous Intelligent Systems*, Smart Innovation, Systems and Technologies 243, https://doi.org/10.1007/978-981-16-3675-2_2

as 98.67% as accuracy, 97.63% as sensitivity and 99.02% as specificity. The results were seen noticeably better than other proposed methodologies.

1 Introduction

Alzheimer's disease (AD) is a neurological brain disease [1] that causes perpetual harm to synapses related to the ability of thinking and remembrance [2]. The cognitive decay brought about by this issue is at last prompts dementia, where the disease starts with gentle decay of protein decompositions around nerve cells and results as a neurodegenerative kind of dementia [3]. Diagnosing Alzheimer's disease requires expertise and clinical evaluations, persistent history, smaller than expected mini-mental state assessment exam score (MMSE), also physical and neurobiological exams [4]. There are various modalities that clearly depict the brain structure; among them, resting-state functional magnetic resonance imaging (rs-fMRI) [5] is a modality that gives non-obtrusive methods for estimating practical brain structure and changes in the brain [6]. AD is mostly observed in the grown-ups and as per the Alzheimer Disease International Survey in 2015, there were roughly 46.8 million people in the world who were having dementia and 22.9 million groups were in Asia, and the numbers are expected to turn twice in the next 20 years. Plenty of computer-aided mechanisms for Alzheimer's classification and early detection have been proposed by using machine learning and deep learning. Deep learning (DL) is the super subset of artificial intelligence and subset of machine learning, whose functionality as well as the structure is similar to the organization of human brain [7], where it is used for image classification, text or voice recognition, etc. Further, it is composed of a large number of hidden layers that help in modelling the features and updating the probabilities for obtaining the overall results. Deep learning models are capable of extracting thousands of features from a set of input data and helps in making the prediction of a new data with high percentage of accuracy.

In Fig. 1, the architecture of DL is composed of a large number of hidden layers,

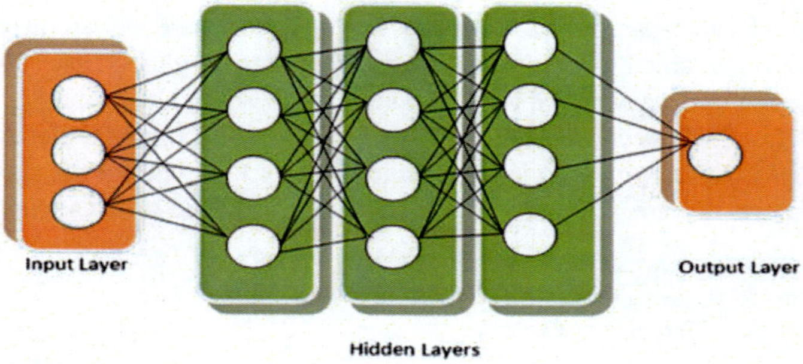

Fig. 1 Deep learning architecture

when compared to the shallow learning neural networks, where DL has been discovered to be particularly effective in recognizing the patterns present in datasets. A robust artificial intelligence algorithm that enables computational models that require multiple processing layers for learning [8]. For example, DL can group Alzheimer's disease and will support analysts and clinicians in diagnosing the brain disease with greater efficiency. This research work incorporates the convolutional neural networks, which is a feed forward network [9] and is widely used in the field of image recognition.

In this research, ADNI dataset which contains the fMRI images of Alzheimer's disease patients undergoes conversion from 3D image to comma-separated value (CSV) dataset, which included a total of 2652 rows and 4097 columns data entries of two classes, i.e. normal and Alzheimer's class. Further, pre-processing the dataset using some traditional missing data handling to fill the data entries with some valid values (used mean for filling the missing values), standard scalar for normalizing the data and principal component analysis (PCA) for feature reduction mechanism, reduces the number of features in the given dataset to avoid the model from overfitting and getting better performance results. Among various defined hyper-parameters combination architecture of CNN, the optimal hyper-parameter values which resulted in maximum performance result and least error was defined as the architecture having: 3 convolution layers, 3 max-pooling layers, utilizing ReLu as normalization layer between two convolution layers, followed by 1 flatten layer and 3 dense layers with final layer of fully connected (dense) connected to the sigmoid activation function. The weights and filters used by the CNN architecture can be seen in Sect. 3.3, the optimization parameters evaluated by ant colony optimization (ACO) was taken into consideration on two parameters, i.e. optimizers and learning rates. The optimizer and learning rate combination that produced the best results with the CNN's hyper-parameters combination was "Adam" as an optimizer and "0.01" as a learning rate. CNN architecture was cascaded in parallel with ACO which uses three different functions, ant solutions construct, pheromone update, daemon action [10], where fitness value is assessed and various global subsets are generated as long as convergence is met. The algorithm undergoes until any termination condition is satisfied and returns the value with best optimal combinations of hyper-parameters which will result in best performance metrics over without optimized methods.

This paper is coordinated as follows: the background knowledge related to Alzheimer's and deep learning is in Sect. 1 followed by including the works done in the past for the early diagnosis of Alzheimer's in Sect. 2. Proposed methodology, dataset description, pre-processing applied in this work, ConvNet and ACO are depicted in Sects. 3, and 4 is formed with all the experimental results and discussions. Section 5 presents the Conclusion, and finally Acknowledgment.

2 Literature Review

Over the recent years, distinct methodologies have been proposed; several researchers have applied diverse deep learning algorithms and bio-inspired optimization procedures alongside the hybrid of both techniques for the early detection and diagnosis of Alzheimer's disease. A touch of papers have been depicted underneath:

Rishu Garg et al. proposed a method that is used to enhance the learnability of classifiers by some simple data pre-processings, i.e. grayscale conversion, selective clipping of dataset in fMRI scans, and the model achieved a classification accuracy of 97.52% [11]. Wang et al. [12] put forward three new variations of feed forward neural networks, which consists of IABAP-FNN, ABC-SPSO-FNN and HPA-FNN, which utilized the combination of CNN and artificial bee colony (ABC). The research accomplished an accuracy of 99.45% for abnormal brain detection. Zhang et al. [13] built up a novel artificial intelligence model that can make classification naturally from brain MRI images. The strategies involved in this research were as accordingly: first, the brain images were handled, including skull stripping and spatial standardization. Second, one hub cut was chosen from the volumetric picture, and fixed wavelet entropy (SWE) was done to remove the surface highlights. Third, a solitary hidden layer neural network organization was utilized as the classifier and the model recorded an accuracy of 92.71% for the detection of Alzheimer's disease. Rekh Ram Janghel et al. proposed a unique method to increase the performance of CNN architecture by applying some pre-processing in the dataset before sending the dataset to extract features, the method has achieved an average accuracy of about 99.45% on fMRI data [14]. Khagi et al. [15] proposed a method which performed classification of Alzheimer's disease based on transfer learning [TL] from various pre-trained CNN models and one scratch model, wherein the scratch model achieved greater accuracy of about 53.69% among various models, but it can be highly improved by tuning the parameters of scratch CNN model. Khvostikov et al. [16] developed a model which used 3D-CNN using s-MRI images along with other modalities of brain images of Alzheimer's patients for Alzheimer's disease classification and gained an accuracy of 96.7%.

3 Proposed Methodology

The proposed methodology can be seen in Fig. 2, where the Alzheimer's dataset has undergone pre-processing, and the missing values are handled using the mean of following features value, forwarded with the feature reduction mechanism carried out by principal component analysis (PCA). Further, the dataset is split into training set and validation set with the ratio of 80-20% of total dataset, and training dataset is fed to the feed forward neural network which is connected in parallel with ant colony optimizer block which finds out the optimal hyper-parameter combination (optimizer and learning rate) for CNN model, and iteratively, the model trains the training

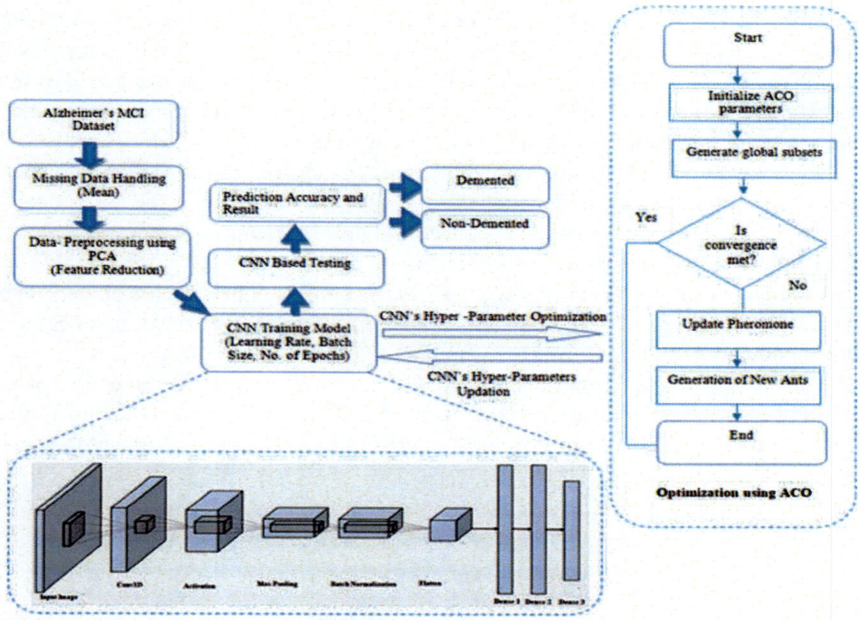

Fig. 2 The proposed methodology of ACO-optimized CNN

dataset until the convergence is met or the number of iterations is reached or we get the minimum error-rate for the training dataset. The convolutional neural network used in this research was built in scratch with a total of 10 layers (3 convolution, 3 max-pooling, 1 flatten, 3 fully connected layer), also described in detail in Sect. 3.3. The output layer of fully connected layer is connected with the sigmoid activation function, and with this combination, model is fit. Later when the model is trained, the validation phase is carried out in which the performance metric of the model is evaluated, and the classification is carried out depicting the model to predict the following data among two classes, i.e. Alzheimer's class with label encoded as 0, or normal class label encoded as 1 with higher accuracy and minimum resource utilization.

3.1 Dataset and Environment

Data which is utilized in this research was assembled from Open Access Series of Imaging Studies (OASIS) [17] where the neuroimaging datasets are uninhibitedly accessible to the researchers. The dataset consists of total 3689 images divided into two different classes, i.e. Alzheimer's class (demented) and normal class (non-demented) which contains in particular 1915 images and 1775 images, respectively.

The early prediction and diagnosis is the main target as there is no such cure available for Alzheimer's disease. So, depicting the class of Alzheimer is very important as it will give idea to the clinicians to give the treatment for the particular phase of the disease and to minimize the risk and maximizing the cure required for the patients. For the latest information, you can visit http://www.adni-info.org [18] (Fig. 3).

The total number of classification of persons whether demented or non-demented can be understood by the below graphs:

On the basis of age of patients, we can find out the count of demented patients, the below graphs can make us understand more clearly.

From Figs. 4 and 5, we can identify that there is a higher concentration of demented patients with the age 60–80. In addition, a demented person's survival rate declines when compared to a non-demented person.

The proposed ACO optimized deep CNN classifier was operated on 64-bit Windows Operating System, with Intel Core i5-8265U CPU at 1.60 GHz, with Intel UHD Graphics 620. And for developing proposed model, the coding environment

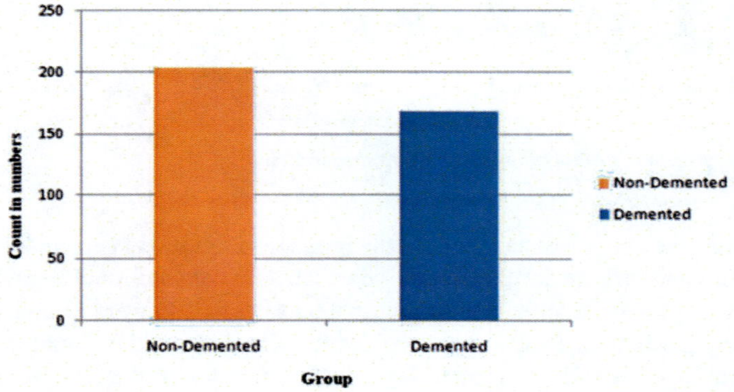

Fig. 3 The count of total patients whether demented or non-demented

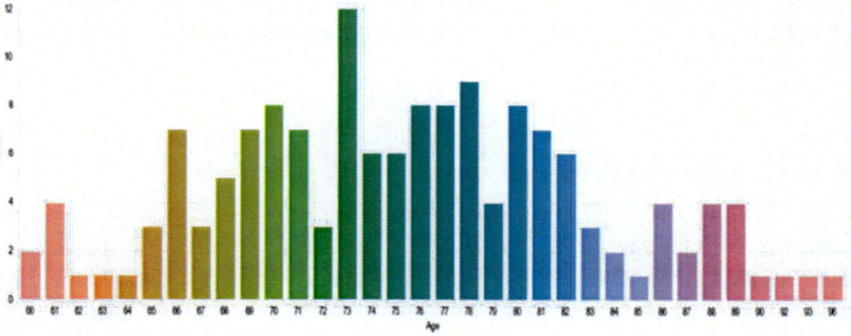

Fig. 4 The count of patients with respect to their ages

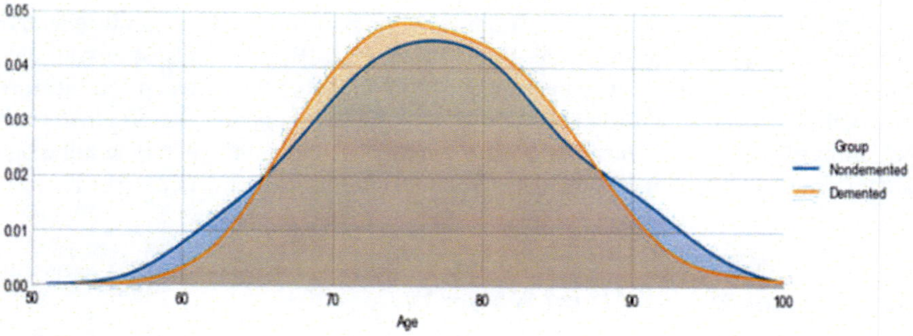

Fig. 5 The comparison of concentrations and the age of survivals of demented and non-demented patients

was taken as Jupyter Notebook version 6.0.3 of Anaconda Navigator utilizing Python 3.7.9 along with tensorflow 2.4.1, numpy 1.19.5 and pandas 1.0.5.

3.2 Missing Data Handling and PCA

When the data is collected from some source, then there is the possibility that data may contain some inconsistency in the form of redundant data or some missing values, and this result in one-sided estimations and hides the actual capability of the system [19]. Thus in order to handle such inconsistency, certain calculations and manipulations are performed which may help us in getting the enhanced results.

There are various mechanisms to handle this issue, like *Missing completely at random* (*MCAR*), *Missing at Random* (*MAR*), *Missing not at Random* (*MNAR*), *Mean Imputation or List-wise Deletion* [19].

These are some statistical approaches that can let us handle minimizing the loss and yield better performance metrics results, but we have incorporated with the mean strategy where missing value is evaluated with the existing values mean calculations. Thus, method helps in minimizing the errors and enhances our performance. These methods usually work when the data is linear.

Principal Component Analysis (*PCA*) is a dimension reduction mechanism that is used to reduce the features of large datasets. The major advantage of using PCA is that it extracts useful information and de-correlates the variables based on the extracted information [20]. The main property of PCA is that number of principal components (PCs) are generated which are basically the linear combination of original variables, and the weight vector which is likewise the eigenvector that fulfils the property of principal of least squares [21]. We have used X as the input matrix, k is the number of variables which is 1497 and t is the number of observations which is 4096.

PCA algorithms undergo five different steps: firstly, the standardization method is performed. This is done by subtracting each value with mean and dividing the standard deviation for each value of each variable. In second step, covariance matrix is evaluated, and in step three, the Eigen vectors and Eigen values of co-variance matrix is computed to identify the principal components (PCs). In step four, feature vector is calculated which helps in identifying which components to keep or to discard which has lesser significance and forms a matrix of remaining vectors and the last step, i.e. step five is to recast the data over principal component (PCs) axes [22].

3.3 Convolutional Neural Network (ConvNet)

The convolutional neural network design is composed of various hidden layers. A portion of the hidden layers is portrayed underneath:

Convolution Layer: The goal of this layer is to register the output of nodes that are inter-connected with the nearby areas in input. It figures a cross-multiplication between the weights along with the input, and a convoluted pixel matrix is created which is fed as in input to next layer [23].

Convolution layer applies various numbers of filters that cycle small nearby parts of the input [24] where these filters are duplicated along the entire input space [25].

The convolution equation can be represented by:

$$S(i, j) = (I * K)(i, j) = \sum_m \sum_n I(m, n)K(i - m, j - n) \quad (1)$$

where I is the image (in pixel format), K is the filter used in convolution process, m and n are row and column of image, respectively [26].

Normalization Layer: This layer's objective is to perform element-wise activation function max $(0, x)$ which is additionally named as the ReLu feature extraction layer resulting in rectified feature map [27].

$$f(x) = \max(0, x) \quad (2)$$

Pooling Layer: A pooling layer or pooling filter's, i.e. a max or min filter depending on the criteria is opted, objective is to perform down sampling operation. Max-pooling can assist with keeping the most required features for recognizing an image. Through max-pooling, the features become more minimal and proficient from lower layer to higher layers. And later, this layer generates pooled feature map as its output which is further flattened and fed as an input to next layer [16].

Fully Connected Layer: This layer will evaluate the class scores [16], where classification after feature extraction takes place, i.e. the classification of image in output layer takes place [28].

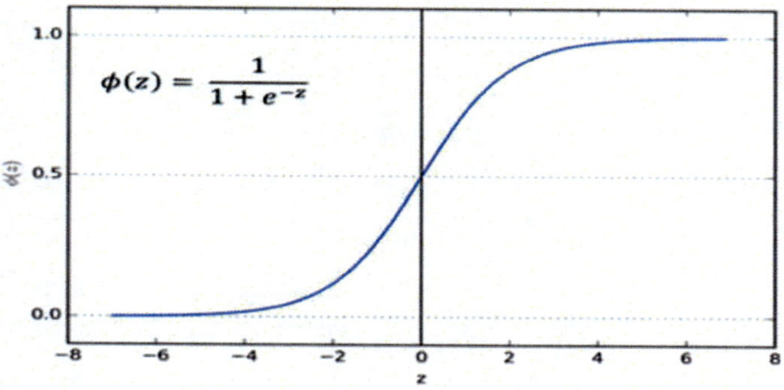

Fig. 6 Sigmoid activation function

Sigmoid Activation Function: Activation functions are significant in neural networks to comprehend the complex patterns. Activation functions convert the input signals of neurons into an artificial neural network to output signals. Generally, there are two different sorts of activation functions contingent upon their conduct and behaviour: linear as well as non-linear activation functions. Because of the nature of non-linearity and computational simplicity in neural networks, sigmoid is the most commonly used activation function [29]. Sigmoid additionally named as logistic function is a non-linear activation function which [30] has a bend-like S-shape. At the point when we make a model and whose objective is to foresee the probability as an output, then sigmoid function is utilized on the grounds that it exists (0–1). These activation functions can make a neural network stall out during the training time and the excellence of this activation functions is that the value never arrives at zero nor it exceeds one. The large negative numbers will in general tend to zero and large positive numbers will tend towards one (Fig. 6).

The architecture of CNN model utilized in this research work can be understood by understanding the number of layers as depicted in Fig. 7.

3.4 Ant Colony Optimization (ACO)

Swarm intelligence is a new approach of solving problems [31] that is motivation-inspired from the behaviour of insects and animals [32]. ACO is a meta-heuristic probabilistic method for taking care of computational issues which were proposed by Dorigo et al. [33]. ACO is an algorithm which is a part of ant colony algorithms family in bio-inspired swarm-based algorithms.

ACO considers artificial systems which take motivations from the conduct of real ant colonies which are utilized to solve different computational optimization problems. Ants use stigmergic communication by means of pheromone trails since they do not have some other method of doing communication, as they are practically

```
_________________________________________________________________
Layer (type)                 Output Shape              Param #
=================================================================
conv2d_4 (Conv2D)            (None, 148, 148, 32)      896
_________________________________________________________________
activation_7 (Activation)    (None, 148, 148, 32)      0
_________________________________________________________________
max_pooling2d_4 (MaxPooling2 (None, 74, 74, 32)        0
_________________________________________________________________
conv2d_5 (Conv2D)            (None, 72, 72, 32)        9248
_________________________________________________________________
activation_8 (Activation)    (None, 72, 72, 32)        0
_________________________________________________________________
max_pooling2d_5 (MaxPooling2 (None, 36, 36, 32)        0
_________________________________________________________________
conv2d_6 (Conv2D)            (None, 34, 34, 64)        18496
_________________________________________________________________
activation_9 (Activation)    (None, 34, 34, 64)        0
_________________________________________________________________
max_pooling2d_6 (MaxPooling2 (None, 17, 17, 64)        0
_________________________________________________________________
flatten_2 (Flatten)          (None, 18496)             0
_________________________________________________________________
dense_4 (Dense)              (None, 64)                1183808
_________________________________________________________________
activation_10 (Activation)   (None, 64)                0
_________________________________________________________________
dense_5 (Dense)              (None, 32)                2080
_________________________________________________________________
activation_11 (Activation)   (None, 32)                0
_________________________________________________________________
dropout_2 (Dropout)          (None, 32)                0
_________________________________________________________________
dense_6 (Dense)              (None, 1)                 33
_________________________________________________________________
activation_12 (Activation)   (None, 1)                 0
=================================================================
Total params: 1,214,561
Trainable params: 1,214,561
Non-trainable params: 0
_________________________________________________________________
```

Fig. 7 Convolutional neural network architecture utilized in this research work

visually impaired and cannot do the complex task alone [33]. They depend on the marvels of swarm intelligence for the endurance so as to accumulate some food.

They first move in random directions; when they get food, they set down pheromone along their ways which goes about as a communication.

Medium among ants and they additionally discovered the shortest way from their position to the position of food.

The operator's values of ACO are Pheromone Update and measure, trail evaporations.

Control Parameters of ACO are Number of Ants, Pheromone evaporation rate, iterations, what's more, measure of Reinforcement.

ACO is also called as the autocatalytic positive feedback algorithm. Initially, it was characterized for traveling salesman problem; however, later on, it started getting applied to the hard optimizations problems. The primary ant optimization was named ant system, and till now, different expansions of ant colony optimization algorithms have occurred to be specific they are Elitist Ant System (EAS), MMAS, ACS and ACO with fuzzy. There is a wide scope of applications where ACO can be forced; some of them are graph colouring, classification problem in data mining, shortest path problem, travelling salesman problem and more. Even before the start of search process, the equal amount of pheromone is assigned in all directions [10]. When an ant '*k*' is at a node '*i*', then ant uses pheromone trail to compute the probability of choosing '*j*', a next node. An ant will move from node '*i*' to node '*j*' with probability given as:

$$p_{i,j} = \frac{(\tau_{i,j})^{\alpha}.(\eta_{i,j})^{\beta}}{\sum(\tau_{i,j})^{\alpha}.(\eta_{i,j})^{\beta}} \tag{3}$$

where $\tau_{i,j}$ is the amount of pheromone on edge i, j; α is the parameter to Influence $(\tau_{i,j})$; $\eta_{i,j}$ is the desirability of edge i, j; β is the parameter to Influence $(\eta_{i,j})$.

And the amount of pheromone is updated using the equation:

$$\tau_{i,j} = (1 - \rho).\tau_{i,j} + \Delta\tau_{i,j} \tag{4}$$

where ρ is the rate of pheromone evaporation; and $\Delta\tau_{i,j}$ is the amount of pheromone deposited.

Algorithm 1 Ant Colony Optimization

Step 1: Initialization

Determine the population of ants.
Set the initial pheromone intensities for each ant.
Set the ACO parameters such as of: α, β, η

Step 2: Evaluation of Ants Selected Subsets.
Step 3: Check the stopping criteria if satisfied move to Step 7 else continue.
Step 4: Pheromone Updating.
Step 5: Generation of new ants and create new feature subset.
Step 6: Remove the previous ants and evaluate the path using Probability rule.
Step 7: Evaluation of Accuracy of final subsets.

Here, the major objective of using ACO as an optimizer is to optimize the hyper-parameters of convolutional neural network model to get optimal combinations of hyper-parameters, which will result in enhancing the performance metrics of the model more effectively and to build an effective neural network. *The Hyper-parameters related to Convolutional Neural Network can be of any type*:

Some Common Hyper-Parameters of CNN

I. *Number of Convolution Layers*
II. *Number of kernels in each Convolution Layer*
III. *Activation Function in each Convolution Layer*
IV. *No. of Dense Layer*
V. *Batch Size*
VI. *Learning Rate*
VII. *Number of Neurons in each layers: Convolution, Max-Pooling, Dense*
VIII. *Learning Rule*
IX. *Optimizers*

In our work, the position of the food source encodes a possible hyper-parameters combination that represents the new CNN design, and ant behaviour can aid in the search for the best food source positions (hyper-parameters) via fitness evaluation. Initially, we select N number of ants, then we initialize the matrix of pheromone deposited, then using pheromone matrix, Ants will start exploring some paths, with probability equation, i.e. using Eq. 3, ants will decide which city to go (resulting in some combination of hyper-parameters), the ants will keep on going city to city according to the above choosing rule until all cities are visited (all hyper-parameters combinations are generated), then based on the amount of pheromone deposited on the paths some ants are selected (some hyper-parameters are selected), then wait for the pheromone to get evaporated and new ants are generated and the process goes on (to search for the optimal combinations of hyper-parameters) until any termination condition or convergence is met.

Hyper-Parameters Optimized in This Work Are

Learning Rate $\in$ *{0.0001,0.001,0.01,0.1...}*
Optimizers $\in$ *{SGD, Adam, RMSprop, AdaGrad, AdaDelta...}*

4 Result and Discussion

4.1 Results

In this segment of paper proposed, we will be depicting the experimental results over different parameters of feed forward network and hybrid models, i.e. the performance metrics of ConvNet, along with ConvNet and ACO.

Table 1 represents the different mathematical performance metrics such as accuracy, precision, recall/sensitivity, specificity and error rate for the proposed work. And, we can record when batch size is 6, learning rate is 0.01 and the size ratio is

Table 1 The performance metrics with the variation of batch sizes when learning rate is 0.01 and size ratio is 80-20

Batch-size	Accuracy	Error-rate	Recall/Sensitivity	Specificity
2	93.33	6.67	92.67	94.66
4	94.67	5.333	93.75	95.78
6	**98.67**	**1.33**	**97.63**	**99.02**
8	94.00	6.00	92.88	95.70
10	91.86	8.14	90.96	93.30

80-20; we get maximum accuracy of 98.67%, specificity with 99.02% and sensitivity with 97.63% (Fig. 8).

Table 2 depicts the performance of two models: one is CNN without optimization strategy and another is optimized CNN when learning rate is 0.01.

Table 3 and Fig. 9 represent the accuracy with the variations of batch sizes using different learning rates 0.0001, 0.001, 0.01 and size ratio 80-20 and up to 200 epochs, we can note that the maximum accuracy is found when batch size is 6 and learning rate is 0.01, the accuracy is 98.67%.

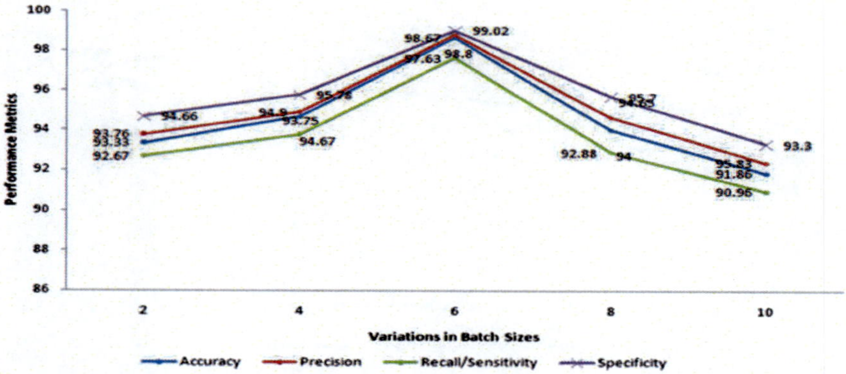

Fig. 8 The graph of different performance metrics with the variations in batch size when learning rate is 0.01 and size ratio is 80-20

Table 2 Performance metric of with and without optimized CNN

Number of epochs	CNN		CNN + ACO	
	Accuracy	Error-rate	Accuracy	Error-rate
50	93.09	6.91	96.00	4.0
100	96.30	3.70	97.33	2.67
150	96.88	3.12	98.00	2.0
200	**97.33**	**2.67**	**98.67**	**1.33**

Table 3 Accuracy of hybridized ACO + CNN using variation in batch sizes with different learning rates 0.0001, 0.001, 0.01

Batch size	Learning rate	Accuracy	Error-rate
2	0.0001	89.33	10.67
4	0.0001	93.63	6.37
6	0.0001	94.67	5.33
8	0.0001	97.67	2.33
10	0.0001	95.89	4.11
2	0.001	96.00	4.00
4	0.001	97.33	2.67
6	0.001	98.00	2.00
8	0.001	97.70	2.30
10	0.01	95.83	4.17
2	0.01	93.33	6.67
4	0.01	94.67	5.33
6	**0.01**	**98.67**	**1.33**
8	0.01	94.00	6.00
10	0.01	91.86	8.14

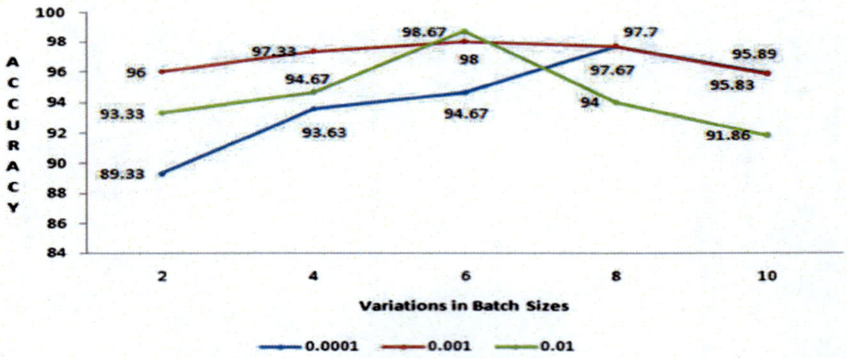

Fig. 9 The accuracy of model using different learning rates

4.2 Discussion

Table 4 is the comparative analysis table in which it depicts that the proposed methodology gained an accuracy of 98.67%, specificity of 99.02% and sensitivity of 97.63 using the hybrid mechanism of CNN optimized with ACO which is a better and efficient approach than many existing methodologies.

Table 4 The comparative analysis of accuracies with already proposed methodologies

S. No.	Author name	Techniques used	Accuracy (%)
1	Ji et al. [34]	ConvNet using MRI	97.65
2	Ratna et al. [35]	Deep belief Network	91.76
3	Behesti et al. [36]	Histogram + SVM	84.07
4	**Proposed**	**CNN + ACO**	**98.67**

5 Conclusion

In this research, we developed an ant colony optimized convolutional neural network as a hybrid methodology for the classification of Alzheimer's disease into two different classes, i.e. Alzheimer's class or normal class.

All the experimental data have been taken from the ADNI. The dataset is of image type which contains 3689 images divided into classes; the dataset was converted to CSV dataset, and the methodology begins with pre-processing the dataset using the missing data handling and feature reduction mechanism (PCA). The training data is fed to the neural network where the model is fit from the training dataset which is connected in parallel with the ant colony optimization which finds the optimal hyper-parameter combination for the feed forward neural network, and this phenomenon is performed iteratively during training phase followed by validating the neural network by performing classification tests on the model which resulted in the following performance metrics, i.e. 98.67% as accuracy, 99.02% as specificity and 97.63% as sensitivity. Future works may incorporate clinical data being taken into contemplations with different other hybrid methods consolidating all the more better results and improving accuracies as well as other parametric dynamics that are the basis for experimental analysis.

Acknowledgements This work is upheld and supported by SEED grant project of National Institute of Technology Raipur. The authors appreciate the help of Dr. Rekh Ram Janghel, Assistant Professor (Information Technology Department) at National Institute of Technology-Raipur. We thank sir for his consistent support and Guidance. Further, sir consistently helped in each conceivable way, and permitted me to finish the undertaking in the right direction.

References

1. T. Altaf, S.M. Anwar, N. Gul, M.N. Majeed, M. Majid, Multi-class Alzheimer's disease classification using image and clinical features. Biomed. Signal Process. Control **43**, 64–74 (2018). https://doi.org/10.1016/j.bspc.2018.02.019

2. A. Farooq, S. Anwar, M. Awais, S. Rehman, A deep CNN based multi-class classification of Alzheimer's disease using MRI, in *IST 2017-IEEE International Conference on Imaging Systems and Techniques Proceedings* (2017), pp. 1–6. http://doi.org/10.1109/IST.2017.8261460
3. S. Sarraf, G. Tofighi, Classification of Alzheimer's disease structural MRI data by deep learning convolutional neural networks (2016), pp. 1–14 [Online]. Available: http://arxiv.org/abs/1607.06583
4. R.R. Janghel, Deep-learning-based classification and diagnosis of Alzheimer's disease. https://www.igi-global.com/viewtitlesample.aspx?id=237939&ptid=228600&t=deep-learning-based+classification+and+diagnosis+of+alzheimer%27s+disease. Accessed Dec 12, 2020
5. F. Saeed, Towards quantifying psychiatric diagnosis using machine learning algorithms and big fMRI data. Big Data Anal. **3**(1), 18–20 (2018). https://doi.org/10.1186/s41044-018-0033-0
6. S. Sarraf, G. Tofighi, Deep learning-based pipeline to recognize Alzheimer's disease using fMRI data, in *FTC 2016—Proceedings Future Technologies Conference* (2017), pp. 816–820. http://doi.org/10.1109/FTC.2016.7821697
7. K.L. Hua, C.H. Hsu, S.C. Hidayati, W.H. Cheng, Y.J. Chen, Computer-aided classification of lung nodules on computed tomography images via deep learning technique. Onco Targets Ther. **8**, 2015–2022 (2015). https://doi.org/10.2147/OTT.S80733
8. Y. Lecun, Y. Bengio, G. Hinton, Deep learning. Nature **521**(7553), 436–444 (2015). https://doi.org/10.1038/nature14539
9. S. Shukla, R.K. Chaurasiya, Emotion analysis through EEG and peripheral physiological signals using KNN classifier, vol. 30 (2019)
10. S. Binitha, S.S. Sathya, A survey of bio inspired optimization algorithms. Int. J. Soft. Comput. Eng. (IJSCE) **2**(2) (2012)
11. R. Garg, R.R. Janghel, Y. Rathore, Enhancing learnability of classification algorithms using simple data pre-processing in fMRI scans of Alzheimer's disease (2019)
12. S. Wang et al., Feed-forward neural network optimized by hybridization of PSO and ABC for abnormal brain detection. Int. J. Imaging Syst. Technol. **25**(2), 153–164 (2015). https://doi.org/10.1002/ima.22132
13. Y. Zhang et al., Multivariate approach for Alzheimer's disease detection using stationary wavelet entropy and predator-prey particle swarm optimization. J. Alzheimer's Dis. **65**(3), 855–869 (2018). https://doi.org/10.3233/JAD-170069
14. R.R. Janghel, Y.K. Rathore, Deep convolution neural network based system for early diagnosis of Alzheimer's disease. Irbm **1**, 1–10 (2020). https://doi.org/10.1016/j.irbm.2020.06.006
15. B. Khagi, C.G. Lee, G.R. Kwon, Alzheimer's disease classification from brain MRI based on transfer learning from CNN, in *BMEiCON 2018—11th Biomedical Engineering International Conference* (2019), pp. 1–4. http://doi.org/10.1109/BMEiCON.2018.8609974
16. A. Khvostikov, K. Aderghal, J. Benois-Pineau, A. Krylov, G. Catheline, 3D CNN-based classification using sMRI and MD-DTI images for Alzheimer disease studies. [Online]. Available: https://ida.loni.usc.edu
17. D.S. Marcus, A.F. Fotenos, J.G. Csernansky, J.C. Morris, R.L. Buckner, Open access series of imaging studies: longitudinal MRI data in non-demented and demented older adults. J. Cogn. Neurosci. **22**(12), 2677–2684 (2010). https://doi.org/10.1162/jocn.2009.21407
18. J. Escudero, E. Ifeachor, J.P. Zajicek, C. Green, J. Shearer, S. Pearson, Machine learning-based method for personalized and cost-effective detection of Alzheimer's disease. IEEE Trans. Biomed. Eng. **60**(1), 164–168 (2013). https://doi.org/10.1109/TBME.2012.2212278
19. S. KumarPandey, R. RamJanghel, A survey on missing information strategies and imputation methods in healthcare, in *2018 8th International Conference on Cloud Computing, Data Science and Engineering (Confluence)* (2018), pp. 299–304
20. R.R. Janghel, A. Shukla, C.P. Rathore, K. Verma, S. Rathore, A comparison of soft computing models for Parkinson's disease diagnosis using voice and gait features. Netw. Model Anal. Health Inform. Bioinform **6**(6) (2017). http://doi.org/10.1007/s13721-017-0155-8

21. E. Alickovic, J. Kevric, A. Subasi, Performance evaluation of empirical mode decomposition, discrete wavelet transform, and wavelet packed decomposition for automated epileptic seizure detection and prediction. Biomed. Signal Process. Control **39**, 94–102 (2018). https://doi.org/10.1016/j.bspc.2017.07.022
22. S. Wold, K. Esbensen, P. Geladi, Chemom. Intell. Lab. Syst. **2**(1–3), 37–52 (1987) [Online]. Available: http://files.isec.pt/DOCUMENTOS/SERVICOS/BIBLIO/Documentos%20de%20acesso%20remoto/Principal%20components%20analysis.pdf
23. J. Wu, Introduction to convolutional neural networks (2017)
24. M. Imani, E. Pakizeh, M.M. Pedram, H.R. Arabnia, Improving MAX-MIN ant system performance with the aid of ART2-based twin removal method, in *Proceedings 9th IEEE International Conference on Cognitive Informatics, ICCI 2010* (2010), pp. 186–193. http://doi.org/10.1109/COGINF.2010.5599744
25. O. Abdel-Hamid, A.R. Mohamed, H. Jiang, G. Penn, Applying convolutional neural networks concepts to hybrid NN-HMM model for speech recognition, in *ICASSP, IEEE International Conference on Acoustics, Speech, and Signal Processing* (2012), pp. 4277–4280. http://doi.org/10.1109/ICASSP.2012.6288864
26. S.K. Pandey, R.R. Janghel, Recent deep learning techniques, challenges and its applications for medical healthcare system: a review. Neural Process. Lett. **50**(2), 1907–1935 (2019). https://doi.org/10.1007/s11063-018-09976-2
27. W. Jung, D. Jung, B. Kim, S. Lee, W. Rhee, J.H. Ahn, Restructuring batch normalization to accelerate CNN training. July 2018. Accessed: Dec 12, 2020. [Online]. Available: http://arxiv.org/abs/1807.01702
28. S.K. Pandey, R.R. Janghel, Automatic detection of arrhythmia from imbalanced ECG database using CNN model with SMOTE. Australas. Phys. Eng. Sci. Med. **42**(4), 1129–1139 (2019). https://doi.org/10.1007/s13246-019-00815-9
29. J. Han, C. Moraga, The influence of the sigmoid function parameters on the speed of backpropagation learning, in *Lecture Notes in Computer Science* (including subseries *Lecture Notes in Artificial Intelligence and Lecture Notes in Bioinformatics*), vol 930 (1995), pp. 195–201. http://doi.org/10.1007/3-540-59497-3_175
30. M. Liu, D. Zhang, D. Shen, Hierarchical fusion of features and classifier decisions for Alzheimer's disease diagnosis. Hum. Brain Mapp. **35**(4), 1305–1319 (2014). https://doi.org/10.1002/hbm.22254
31. R.S. Parpinelli, H.S. Lopes, New inspirations in swarm intelligence: a survey. Int. J. Bio-Inspired Comput. **3**(1), 1–16 (2011). https://doi.org/10.1504/IJBIC.2011.038700
32. C. Blum, M. López-Ibáñez, Ant colony optimization. Intell. Syst. (2016). http://doi.org/10.4249/scholarpedia.1461
33. M. Dorigo, V. Maniezzo, A. Colorni, Ant system: optimization by a colony of cooperating agents. IEEE Trans. Syst. Man Cybern. Part B **26**(1), 29–41 (1996). http://doi.org/10.1109/3477.484436
34. H. Ji, Z. Liu, W.Q. Yan, R. Klette, Early diagnosis of Alzheimer's disease using deep learning, in *Proceedings of the 2nd International Conference on Control and Computer Vision—ICCCV 2019*, June 2019, pp. 87–91, http://doi.org/10.1145/3341016.3341024.
35. M. Ratna, W. Ito, H. Nurul, F. Moh, Structural MRI classification for Alzheimer's (2017), pp. 37–42
36. I. Beheshti, N. Maikusa, H. Matsuda, H. Demirel, G. Anbarjafari, Histogram-based feature extraction from individual gray matter similarity-matrix for Alzheimer's disease classification. J. Alzheimer's Dis. **55**(4), 1571–1582 (2017). https://doi.org/10.3233/JAD-160850

Performance Evaluation of Throughput and End-to-End Delay Using an Optimized Cluster Based Data Forwarding (OCDF) Protocol

Shaik Mazhar Hussain, Kamaludin Mohamad Yusof, and Shaik Ashfaq Hussain

Abstract V2X communications are defined as the communication between vehicles and various elements of the intelligent transportation system (ITS). Two potential technologies of V2X communication are cellular and dedicated short-range communication (DSRC). Each of the technologies have their own limitations. DSRC offers low latency which are vital for vehicle safety applications. However, due to limited spectrum and short range, the performance of DSRC under high vehicle density scenarios degrades. Cellular network offers larger coverage range, high data rates, and high bandwidth. However, it suffers from higher latencies due to long transmission time intervals. Hence, there is a need to integrate DSRC and LTE as a heterogeneous solution to enhance the performance of vehicular networks in urban environments. In this paper, we have proposed a novel optimized cluster-based data forwarding (OCDF) protocol with an intelligent radio interface selection scheme to overcome the issues related with network performance when DSRC and cellular networks used alone. To evaluate the proposed protocol, three traffic applications were considered—safety services, bandwidth services, and voice services. Appropriate radio interface will be selected by determining packet loss ratio (PLR) levels. A minimum threshold value of PLR will be set by which radio interface can be selected intelligently. The proposed approach is compared with the existing approaches, and the performance of the throughput and end-to-end delay is evaluated using NS-3 simulation tool. Result shows that the performance of throughput and end-to-end delay is improved in comparison to the existing approaches under urban environments.

Supported by organization x.

S. M. Hussain (✉) · K. M. Yusof · S. A. Hussain
Department of Communications Engineering & Advanced Telecommunication Technology (ATT), Faculty of Engineering, School of Electrical Engineering, Universiti Teknologi Malaysia, Johor Bahru, Malaysia
e-mail: mazhar@graduate.utm.my

K. M. Yusof
e-mail: kamalmy@fke.utm.my

P. Karuppusamy et al. (eds.), *Ubiquitous Intelligent Systems*, Smart Innovation, Systems and Technologies 243, https://doi.org/10.1007/978-981-16-3675-2_3

1 Introduction

The past ten years have shown sharp incline in the development of mobile technologies as well as an increased interest of researchers in the field of intelligent systems. This has enabled ease of data interchange and better data management and transfer. The rapid development is only expected to increase further as the years progress, and also to be widely implemented in vehicles in the future, giving rise to a new form of travel for people. Vehicular systems are sensitive as human lives hang in the balance, so the structure of such systems needs to be smart and strong. Adhoc networks formed by vehicles are called as VANETs. VANETs are considered as the key components of intelligent transportation systems (ITS). Vehicular ad hoc networks (VANETs) are characterized as objects that are connected freely, capable of random motion, and wireless communication. Such are the features that VANETs inherit from the mobile ad hoc networks (MANETs). VANETs also alludes to the setup, wherein vehicles are seen as intelligent objects. It is essentially the creation of a wireless network of vehicles instead of mobile devices as is with MANETs. They have smart communication—the intelligent transportation system (ITS), a relatively newer concept which describes the exchange of data between the vehicles, a variety of sensors and the infrastructure in place. This concept gives way to numerous applications for the vehicles' drivers, including but not limited to, driver safety, assistance, and multiple features related to information and entertainment. The Internet of things (IoT) is a concept that has not yet been defined very specifically and concisely as it encompasses all types of systems. Generally, it can be said to be a network comprising of objects with sensors. These sensors gather data from the environment and their system, and then the data is distributed through the Internet. Applying the concept of IoT to vehicular systems produces the notion of Internet of vehicles (IoV). This is an integrated network that enables data collection and information sharing for vehicles, their surroundings, and the road systems. It also allows for treatment and computing of data, alongside the secure sharing of data over different platforms. The collected data aids in performing efficient monitoring of vehicles, vehicle control. There are many different aspects that come up for discussion regarding VANETs: the architecture, communication domains, wireless access technologies, and applications. There are different levels to VANETs: the first level involving the devices used to collect the data from the environment, the second layer which includes the different wireless communication networks, leading to seamless networking. And then the last layer, the third layer, which contains analytic, processing and storage tools, for data processing and analytics and the subsequent decision-making. The communication system of VANETs is designed to facilitate vehicles in communicating with one another and the road infrastructure (provides the vehicle with updated information). There are three possible combinations of communication with the vehicle as the epicenter

1. V2V communication—each vehicle can contact its neighboring vehicle directly, which means there is no infrastructure involved. This form of communication is used for safety purposes.

2. V2R communication—This is the data exchange between a vehicle and roadside units, for example, traffic lights and warning signs.
3. V2I communication—The vehicle(s) can connect to the Internet infrastructure and benefit from the wide range of available services.

Further development in IoV systems led to the advent of three more connection types:

1. V2P/V2H—This refers to the communication between the vehicle and the personal devices of people such as the driver, passengers, and pedestrians. This can lead to the use of other services such as playing music, viewing files from the personal device.
2. Vehicle-to-Sensor (V2S)—This type of connection allows the vehicle to keep a check on its own conditions, like position, speed, and oil level among others.
3. Vehicle-to-Everything (V2X)—Since the system uses IoT, the vehicle is able to communicate and share data with anything that pertains to its surroundings. This leads to the communication network to no longer be limited, but rather vast and widespread (with the help of IoV) and opens a whole new area of vehicular communication. However, there is a form of communication where the vehicle can communicate with the first three of the above possible communication combinations, for optimum usefulness. This is hybrid communication, where the vehicle is enabled to communicate with the roadside units, other vehicles, and the infrastructure. The form of connection depends on the distance between both points: Is there a direct line for communication or not? IoV vehicles are intelligent in the sense that they have complex inner systems that employ the use of multiple sensors and other devices that allow for detection of various elements, such as other vehicles and road infrastructure. Data is collected from the surroundings, and embedded communication services will communicate the data to other units or the Internet and a 'vehicular operating system' is essential for the processing of all the collected data. There are two technological tools used in VANETs—WAVE and CALM, whereas IoV uses broad range of wireless technologies including cellular networks. Ad hoc architecture forms ad hoc networks using wireless networks and on-road vehicles, and these networks can work without the need for external infrastructure support. There are two communication standards supported by these:
4. WAVE—This standard uses dedicated short-range communication as a communication technology. The coverage range of DSRC is 1 km and supports data rate of up to 27 Mbps.
5. CALM—This includes a combination of different wireless technologies like GPRS-2.5G, UMTS-3G, wireless in 60 GHz band, and infrared communication systems. Additionally, IoV also supports Bluetooth, ZigBee, 4G LTE, and WiMAX.

IoV system enables the vehicles to multitask, employing different and several facilities provided by the Internet, functioning as consumers and suppliers simultaneously. This makes IoV a mixed system, with client-server system and peer-2-peer system. Client-server system allows vehicles to obtain the servers' services—infrastructure, vehicles, and roadside units, to name some. The peer-2-peer system enables the vehicles to interact with each other and execute a myriad tasks: video streaming, playing music, sharing and downloading files, and many others. In this sense, the cloud platform is also a greatly useful tool as it allows the execution of many taxing tasks and the management of different applications and functions by the IoV, all at the same time. The cloud platform augments the data processing of road data collected in real time, while also applying AI to allow for smart client services and making intelligent decisions. The IoV cloud platform has been segregated into three main layers:

1. Cloud services—Includes all the services of the cloud that enable applications like networking, computing, cooperation, and data storage.
2. Application Servers—The IoV system has intelligent functionality which includes: congestion management, entertainment application, and road safety among others. There are two engines for processing: the internal one and the external one. The internal engine includes applications relating to big data: storage, processing and analyzing, and the cloud platform's basic serves implement the above. The external engine is further divided into two units, the unit that performs information gathering (tasked with data collection) and the data diffusion unit, which delivers the services to the clients.
3. Information Consumer and Producer—There are many intelligent devices that are a part of the overall IoV system. These devices receive the data and information gathered and processed earlier. The devices are tasked with data collection from the vehicular surroundings, and the data collected by these is immensely useful for establishments related to production, repair and servicing of Internet-based automobiles.

VANETs are increasingly shorthanded in terms of data processing, computing and storage due to the absence of cloud technologies. The layered architectural model of the VANETs is based on six layers and is detailed below.

1. The Access Layer—this layer has two further sub layers:
2. Network and Transport Layer—a dependable communication system is required from the routing protocols, which ensures the users of the network will not face connection issues at any point, especially during data transfer. Different protocols are defined for VANETs, namely UDP, TCP, and others. There are also a myriad communication paradigms supported by the network, like unicast, multicast, speed based routing, broadcast, and others.
3. Security layer—this protects against firewall and unauthorized access. It has different modules for guaranteeing, authentication, authorization, identification, hardware security, to name a few.
4. Facilities layer—this layer is intended to be used for information presentation to the users using hardware and human-machine interface. It codes and decodes messages in accordance to the language in use.
5. Application layer—this layer has all the different features that the VANETs system offers.

6. Management layer—involves management of networks, VANETs features, legacy system protection, communication services, etc. As for the model architecture of IoV, that is defined and explained below. It is based on five instead of six layers and allows interaction and connection of all the components of the network itself and the data dispersion elements.
7. User Interaction Layer—the different elements and devices of the communication aspect of IoV are present in this layer, such as the sensors of the vehicle, smart phones, cellular infrastructure, etc. This layer is curated and aimed at collecting data from the vehicle's surroundings and to convert the data to EM form and secure it.
8. Coordination Layer—this is the second layer, and it includes the myriad of heterogeneous networks: WAVE, 4G/LTE, satellite, and, most importantly, Wi-Fi. This layer conducts data treatment by collecting data from all the networks and then turning them into a structure that is uniform and readable by all the other succeeding networks.
9. Processing and Analysis—It is the central layer. It constitutes following tasks: storage, processing, and analysis of the data that has been sent in by the coordination layer.
10. Application layer—This is the fourth layer of the IoV architecture, and it comprises of the intelligent applications and features of IoV, like the safety applications, entertainment features, parking, and fuel indications, to name a few. The layer provides the vehicle's users with the above services (and more) based on the analysis of the collected data and the decisions made by the third layer.
11. Business Layer—this is the fifth and final layer of the IoV architecture. It makes action plans and strategies for new and improved business models. These depend on the features used by the users and the statistical data collected and analyzed from the same. Hence, this section also involves the decision-making and relating to the economic aspects of the services offered by the system and the employment of resources.

1.1 Cooperative Intelligent Transportation Systems (C-ITS)

Intelligent [1] transportation system (ITS) provides intelligent services to various types of transport and traffic management and facilitate users with effective transport networks. ITS is defined as systems in which various communication and information technologies are applied in the area of road transportation to improve the efficiency. Intelligent transportation technologies include vehicle navigation, traffic signal controller, and integrating live data from various sources. Several wireless communication technologies have been proposed for ITS such as for short-range communications using IEEE802.11p protocols or WAVE /DSRC. WiMAX, GSM or 3G for long-range communications. The main purpose of ITS is to maximize traffic efficiency by minimizing traffic problems by enriching users by providing real-time information and enhancing safety and comfort. ITS is mainly composed of three

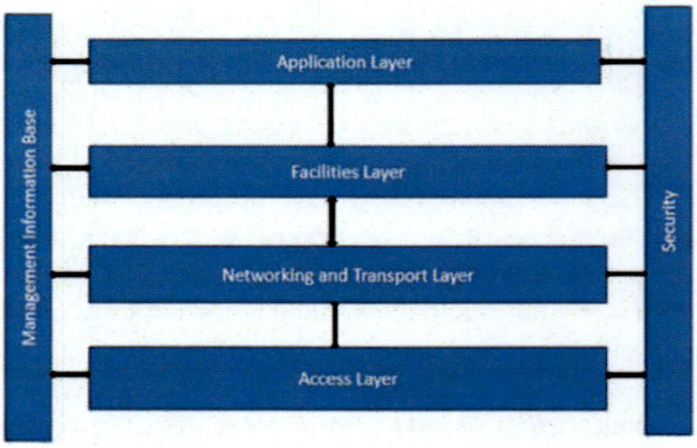

Fig. 1 C-ITS architecture [1]

major functions: data collection, analytics, controlling , coordinating, and decision-making. A complete list of standards and protocols focusing on C-ITS is available in ISO-21217. It provides complete information on global standardization focusing C-ITS. It serves as a guide for designers and developers. In this paper, we will be discussing C-ITS standards specified by European Telecommunications Standards Institute (ETSI), published in 2014 specified in ISO-21217.

1. The horizontal layers include—This layer includes access layer, networking and transport layer, facilities and application Layer.
2. The vertical layer include: Management layer and security layer.

C-ITS is a subset of standards for ITS. ITS aims on improving

1. Safety—crash avoidance, obstacle detection, emergency call
2. Efficiency—navigation, lane access control, speed limits
3. Comfort—telematics and infotainment services
4. Sustainability-C-ITS supports Wi-Fi and cellular networks.

Figure 1 shows the C-ITS architecture.

1. Application Layer—This layer provides services such as road safety, traffic efficiency, and other applications.
2. Facilities Layer—It provides services such as CAM and DENM.
3. Security Layer—This layer provides services such as authentication of the sender of a broadcast message used for information dissemination and secure session establishment and maintenance.
4. Access Layer—This layer provides access to all kinds of cellular access technologies, other technologies such as infrared, millimeter wave (ultra wideband communications), and vehicular Wi-Fi optical light communications.
5. Network and Transport Layer—This layer comprises protocols for ensuring secure end-to-end data delivery.
6. Management Layer—This layer is responsible for configuring ITS station.

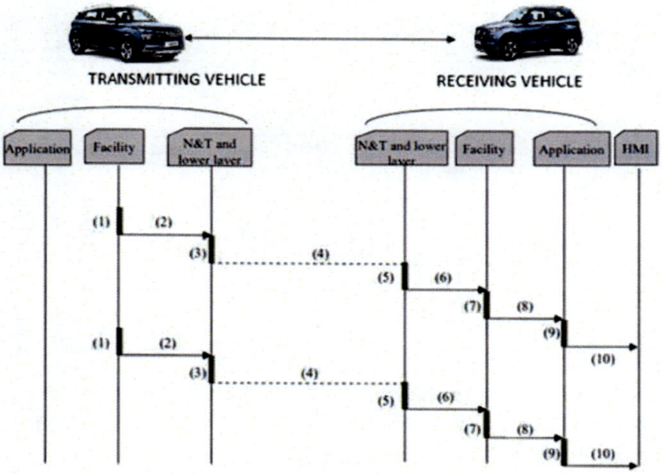

Fig. 2 CAM transmission data flow [1]

There are two types of messages specified by ETSI for C-ITS:

1. Cooperative Awareness Message (CAM)
2. Decentralized Environmental Alert Message (DEAM).

1.2 Cooperative Awareness Message (CAM)

Cooperative awareness messages (CAMs) create awareness among vehicles about road network. There are four use cases which falls under the category of CAM—vehicle emergency warning, slow vehicle indication, intersection collision warning, and indication of motorcycle approaching. CA basic service is responsible for generation and transmission of CAM by implementing CAM protocol (Fig. 2).

1.2.1 CAM Transmission Data Flow

1.2.2 Transmission

1. The facilities layer collects the necessary data from the relevant facilities and constructs the CAM according to the format specified in ETSI EN 302 637-2
2. Network and transport layer receives the CAM with the required transmission parameters. The basic transport protocol (BTP) is responsible for multiplexing messages from the facilities layer to the networking and transport layer
3. CAM is broadcasted.

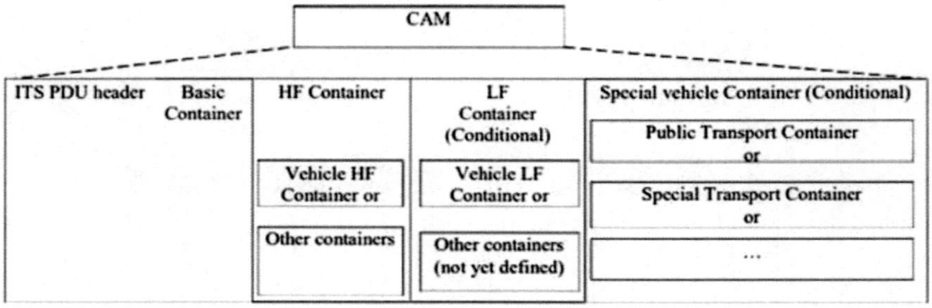

Fig. 3 CAM format [1]

1.2.3 Reception

1. CAM is received by the receiver vehicle
2. CAM is given to the facilities layer for processing and dispatches the information to the application layer
3. Received CAM information will be processed at application layer and provides the necessary warning to the driver.

1.3 CAM FORMAT

1. ITS PDU HEADER—This section contains protocol version, type of message, and address of the sender.
2. BASIC CONTAINER—This section contains type of the station and position.
3. HIGH FREQUENCY CONTAINER—This section contains information about vehicle heading, speed, and acceleration.
4. LOW FREQUENCY CONTAINER—Path history and vehicle role.
5. SPECIAL VEHICLE CONTAINER—Public transport, dangerous goods, and road works.
6. CAM period is given as Tmin = 100 ms and Tmax = 1 s (Fig. 3).

1.4 De-centralized Environmental Alert Message (DEAM)

DEAM is for event-driven safety information which alerts road users of detected event. The exchange of DENM among vehicles is operated by DENM protocol.

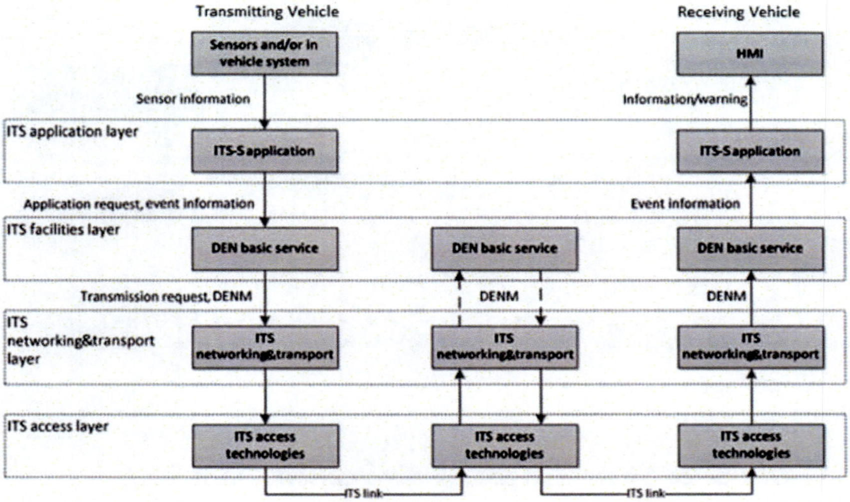

Fig. 4 DENM data flow [1]

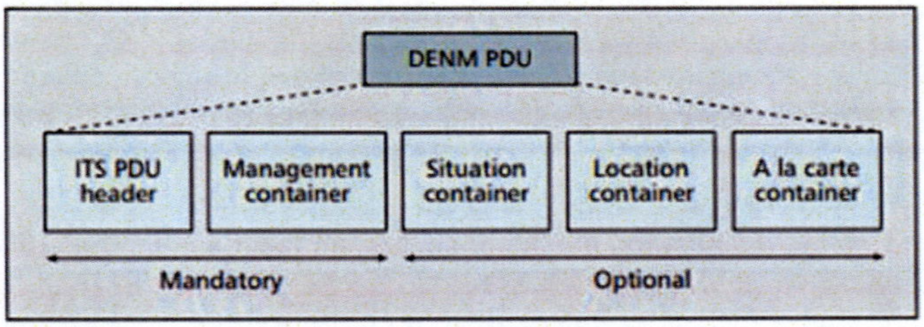

Fig. 5 DENM format [1]

1.4.1 DENM Protocol

Vehicle transmits DENM to neighboring vehicles upon detection of any event. DENM messages are initiated at the application layer and remain active till the event exists. The messages are terminated once the event is terminated. Vehicle on receiving DENM will process and alerts the users about the specific event (Figs. 4 and 5).

1. Management Container—Action identifier, detection time, and event position
2. Situation Container—Predefined code is assigned for causing and related events
3. Location Container—Event speed, heading
4. A la carte container—Lane position, road works.

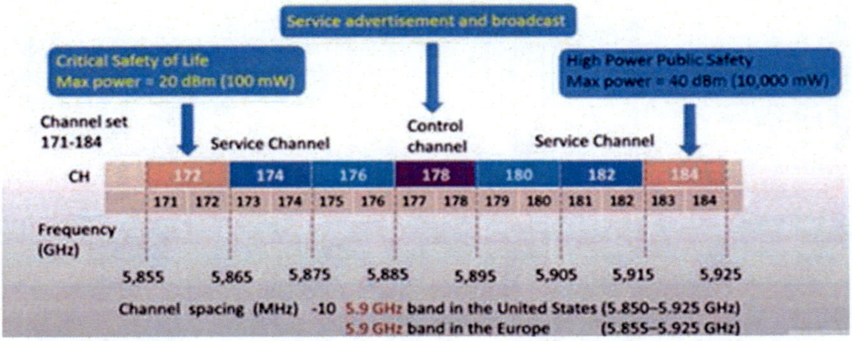

Fig. 6 DSRC spectrum [1]

1.5 Dedicated Short-Range Communication (DSRC)

DSRC is a wireless communication technology providing service for both vehicle to vehicle (V2V) and vehicle to infrastructure communications allocating 75 MHz spectrum in the 5.9 GHz band. Figure 6 shows the DSRC spectrum band.

DSRC has seven channels out of which 1 is dedicated to service channels and the remaining 6 are dedicated to service channels. Channel 178 (CH178) is dedicated for control channel, CH 172, CH174, CH176, CH180, CH182, CH184 are dedicated for service channels. Channel 172 is dedicated for critical safety of life like accidence avoidance. CH184 is dedicated for public safety applications like road intersection collision avoidance. DSRC is comprised of on-board units (OBU) and road-side units (RSU). Figure 7 shows the block diagram of DSRC components. It comprises of GPS for determining vehicle position, internal sensors for collecting data from the surroundings, and computer for processing the data and DSRC radio for broadcasting the information at an angle of 360° using omni-directional antenna.

Fig. 7 DSRC infrastructure [1]

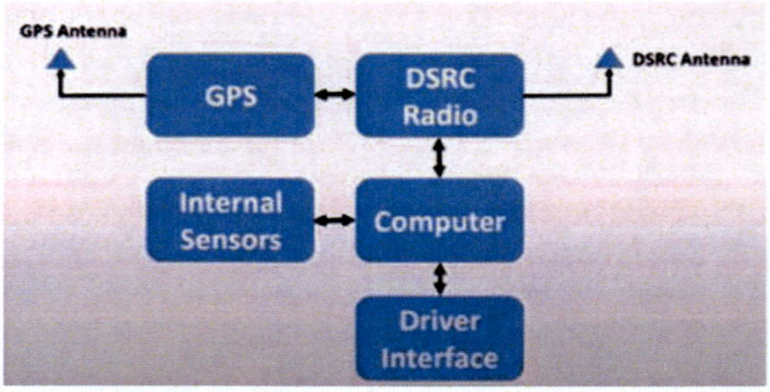

2 Existing Works

In [2], the paper is mainly focused on one application that is intersection collision avoidance. To do this, firstly the problem of DSRC is highlighted. LTE is proposed for transmitting collaborative awareness message (CAM). A cluster-based architecture is proposed where Wi-Fi is used for cluster formation and LTE is used for transmission of CAM packets. The algorithm is a light weight as it is focused only one application based on which clusters are formed. It is shown in the results that when DSRC is used alone and LTE is used alone, the average delay is comparatively high with heterogeneous architecture. The main drawback of this paper is the source of transmission is considered as LTE whose latency is varied from 1.5 to 3 s and might not be suitable for the applications where critical safety of life is a major concern. In [3], an advanced AODV protocol is proposed. The authors have designed algorithms for cluster head selection (CH), gateway selection (GW), and packet forwarding and junction services. Comparative analysis is done between normal AODV and Advance AODV Protocol (AAP). The paper did not addressed the real-time issues with high vehicle density and also the impact on PDR and latency which are considered as very crucial parameters in vehicular environment. The paper has proposed hybrid architecture called vehicular multihop algorithm for stable clustering (VMaSC-LTE) integrating DSRC-based multihop clustering and the long-term evolution (LTE) with the aim of attaining high data packet delivery ratio (PDR) and low delay while keeping the cellular architecture usage at a minimum level. Cluster head (CH) selection is based on average relative speed with respect to the neighboring vehicles. The performance metrics considered in this paper are data packet delivery ratio, delay, control overhead, and clustering stability. In this paper, IEEE 802.11p–LTE hybrid architecture is proposed where vehicles form multihop clustered topology in each direction of the road. The average relative speed is considered as a clustering metric. The paper does not investigated the use of proposed algorithm in urban scenarios. Several hybrid architectures were proposed recently to exploit both DSRC and cellular technologies. In [4–6], the authors have proposed hybrid architectures for more efficient clustering. Authors in [5] have demonstrated the use of cellular communication signaling in hybrid architectures. Authors in [6] demonstrate the use of centralized architecture to minimize the clustering overhead. Authors in [7] have proposed a new protocol based on efficient path selection for connecting more time to connect to the network for services such as Internet access and driver information services. Authors in [8–10] proposed a novel cluster based hybrid architecture for dissemination of messages where the goal was to minimize the number of cluster heads (CHs) communicating with the cellular network which in turn reduces the cost of cellular architecture and handoff occurrences at the base station. The motive of efficient clustering is to reduce the CH, minimize the overhead and stabilize clusters. It is observed that none of the hybrid architectures performed any stability analysis. Also in [8], the delay performance of message dissemination is not considered. In contrast, authors in [9, 10] provided the delay performance but failed to show the effect of overheads and clustering stability. None of the hybrid architectures compared their performance with DSRC-based alternative routing mechanisms such as flooding and cluster-based routing. Several literature articles are available based on vehicle clustering which mainly focused on network performance metrics in high-density vehicular networks [11]. In [12], cluster-based directional routing protocol is proposed for dense networks where mainly the clustering metric is considered as direction to select the cluster head for forwarding packets. The proposed protocol is compared with AODV and GPSR protocols. The proposed protocol found to be superior than the existing protocols in terms of packet delivery ratio and minimal latency. However, the impact of high vehicle density on PDR and latency is not shown. Only the impact of distance on overhead packets, packet delivery ratio, number of hops, and latency is evaluated. In [13], a cluster-based multichannel communications scheme is proposed. This protocol not only supports safety messages, but also non-safety messages such as multimedia and data applications. The protocol integrates clustering both contention free and contention-based MAC protocols. The schemes use contention free MAC within a cluster and contention-based MAC among cluster head vehicles to guarantee reliable delivery of messages. A theoretical model was developed to investigate the delay of safety messages transmitted by cluster head vehicles. A contention window size is derived to balance the tradeoff between delay of safety messages and the successful rate of delivery of safety messages. From the simulation results, it is shown that the proposed protocol worked efficiently to support non-real time traffic under high way traffic scenarios and guarantees real time delivery of safety messages. Another clustering approach is proposed in [14] based on distributed adaptive clustering algorithm based on revised group mobility metric and spatial dependency. In this paper, the clustering is based on reactive approach where the clustering is triggered if and only if the cluster head lost its connection to the cluster or the cluster member cannot connect to the cluster. In [12, 15–17], periodic re-clustering is adapted where the clustering procedures are periodically executed. In [13, 14, 18, 19] reactive clustering mechanism is proposed where the cluster will be triggered only if cluster head loses its connection with its cluster or cluster member cannot join its cluster. All the above-mentioned mechanisms are based on cluster merging where the clusters are activated if the distance between two neighboring cluster heads is below a certain threshold or the duration of cluster head connection time is greater than the predetermined value. The drawback of cluster merging is overheads. Hence, there is a need to limit the cluster size and hop count. Authors in [20, 21] presented the concept of merging clusters in which the connection time of cluster heads is greater than the predetermined value. One of the major findings from the above clustering approaches is none of them focused on sparse networks where network disconnectivity is a major concern.

3 Proposed Approach

We propose an optimal cluster based data forwarding (OCDF) protocol as a heterogeneous IoV solution. In OCDF, first we introduce an improved beetle swarm optimization (IBSO) algorithm for optimal cluster head (CH) selection and clustering. The cluster member, i.e., vehicular node forward data to owned CH in a cluster, and then it should be forward to the corresponding radio access unit (RAU)/Base station. CH transmits the packet to the destination either through DSRC or LTE. The proposed protocol is developed at the network and transport layer of ITS standard. Secondly, new congestion control technique using intelligent radio interface selection algorithm (ERIS) is proposed at service layer. Each vehicle is assumed to be equipped with DSRC and LTE terminals for transmission and reception of packets. The radio interfaces are selected by determining the packet loss ratio (PLR) levels. Both the interfaces are capable of transmitting and receiving the information. Three use cases Safety services, Bandwidth service and Infotainment services are considered for evaluation of algorithm. Packet loss ratio (PLR) levels will be monitored and assessed intelligently. Threshold values of PLR will be set for DSRC, and LTE clustering approach is applied. No of CAM transmissions over LTE network and DSRC network will be reduced. Hence, control packets containing PLR levels will be sent over LTE networks and DSRC networks parallely. Switching time of LTE and DSRC is avoided. Consequently, reducing vertical handover delays. Figure 8 shows the proposed framework.

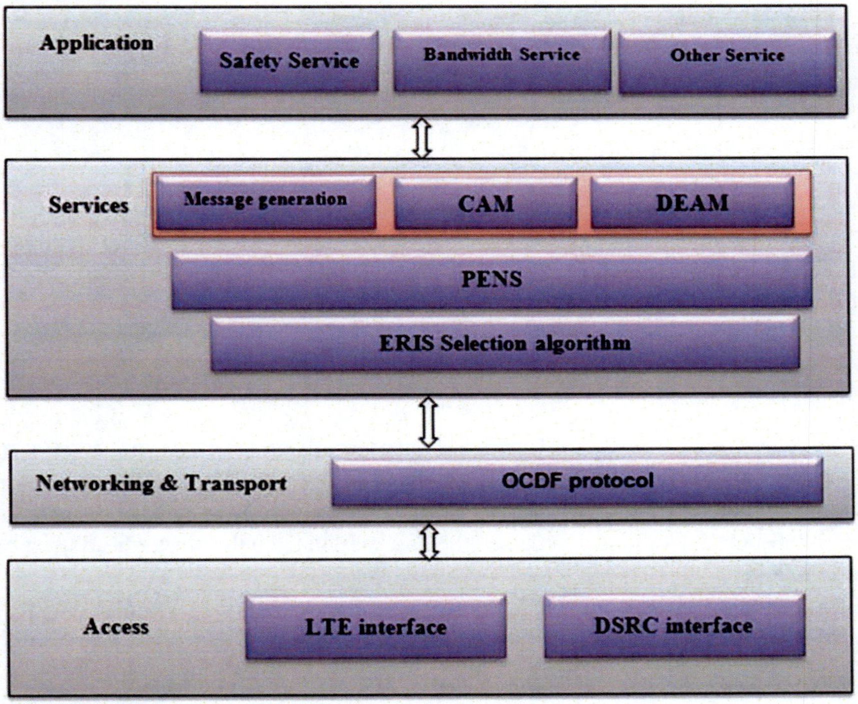

Fig. 8 Proposed framework

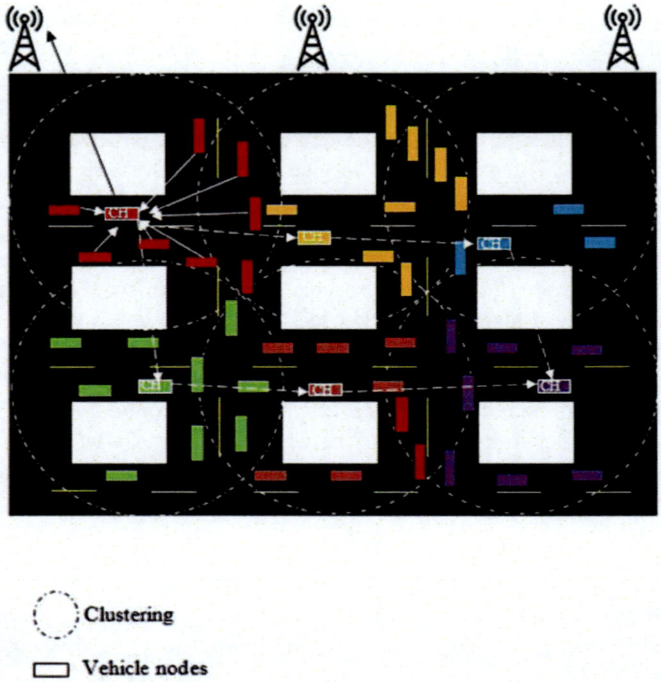

Fig. 9 Network model of OCDF protocol

Network model of Proposed OCDF protocol is shown in Fig. 9. An improved beetle swarm optimization (IBSO) algorithm is used to create energy-aware clusters through optimal selection of the cluster head. The main purpose of IBSO algorithm is to address the issues related to packet delivery ratio, latency, end-to-end delay, and throughput. In OCDF protocol, vehicular nodes are clustered and cluster head (CH) is selected for every cluster. The CH selection is based on base station, cluster distance, and energy parameter. The vehicular node estimates the distance of neighboring nodes by receiving the signal strength. Let us assume five cluster heads CH_i = CH_1,CH_2,CH_3,CH_4,CH_5 forming five clusters (C_1,C_2,C_3,C_4,C_5). The selection of cluster heads is based on average distance and energy. The average distance of each intra-cluster vehicular node and the base station from the cluster head must be minimized.

$$\text{Minimum}(\text{VH}_1) = 1/\text{k}\sum_{i=1}^{n} \text{d}(\text{VN}_\text{j}, \text{CH}_\text{i}) + \text{d}(\text{CH}_\text{i}, \text{BS})$$

where m—number of vehicular nodes in the coverage region;

n—Number of CH to be selected;

$1/k \sum_{i=1}^{n} d(VN_j, CH_i)$—Average distance between CH and Vehicular nodes;

$d(CH_i, BS)$—Average distance between base station and cluster head The next factor for the selection of optimal cluster head is residual energy of all CH must be maximized.

The average power for all VNs determined as,

$$EN_A = \sum_{i=1}^{x} EN_i/X$$

X is the number of active VNs, EN_i residual energy for VN_i, Each vehicular node in the cluster region joins to the corresponding CH for cluster formation. It depends on the weight of the cluster head and is calculated as below

$$W(VN_j, CH_i) = \alpha E_r es\ (CH_i)/d(VN_j, CH_i)d(CH_i, BS)$$

where $E_i es\ (CH_i)$ represents the residual energy of CH. VNs link to the CH through advanced residual energy

$$1 \div d(VN_j, BS)$$

Represents mutual of distance among VN in addition to CH. The VN joins to the adjacent CH in its communication range

$$1 \div d(CH_i, BS)$$

Defines the reciprocal of distance among CH as well as RAU. The VN is link to the CH, which is closer to the base station BS.

VN is link to the CH, which is closer to the base station BS. α Represents a stable value

During this formation of clusters, every VN calculates this weight esteem utilizing the above condition. At that point, the VN joins to the CH with the most noteworthy weight value.

4 Simulation Environment and Findings

It is assumed that vehicles will be equipped with both DSRC and LTE radio interfaces in access layer and are able to transmit data via both. Around 600 vehicles will be considered in urban environment (Table 1).

The proposed approach optimized clustering data forwarding protocol (OCDF) is compared with the existing approaches—long-range Wi-Fi, WAVE, 4G LTE, and heterogeneous architecture. From the simulation results, it is observed that proposed protocol outperforms the existing approaches in terms of throughput and delay as shown in Figs. 15 and 16 (Figs. 10, 11, 12, 13 and 14).

Table 1 Simulation setup

Parameters	Values
Simulator	NS3
Wireless technologies	Long-range Wi-Fi, 4G LTE and WAVE Hetero arch
Frequency ranges	2.4 GHz, 700–2570 MHz, 5.9 GHz
Simulation time	200 s
Number of vehicle nodes/speed	600/varying
Antenna	Omni directional
Traffic application	Voice, video and safety messages
Packet size/data rate	500B/100 kb
Road	Urban scenario
Proposed protocol	OCDF
Channel	Wireless channel
Video size	IKB

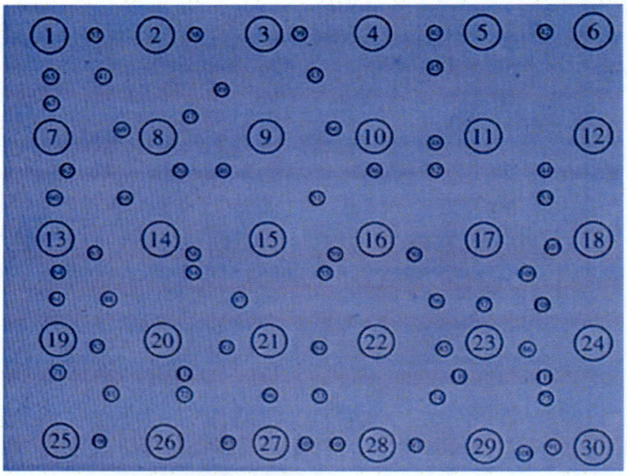

Fig. 10 Road scenario

Figure 15 compares throughput of 4G LTE, LR Wi-Fi, WAVE, HETRO, and proposed method for vehicle nodes at varying speeds. As shown in Fig. 3, the proposed method is significantly better than the existing approaches. There is approximately 150% increase in throughput using OCDF as compared to 4GLTE, 114% increase as compared to WAVE, 66.66% increase as compared to LR Wi-Fi, 36.36% increase as compared to HETRO. The higher throughput is achieved due to less packet drops with the proposed technique. However, there is a slight decrease in the throughput as the vehicle node density increases and also due to increase in vehicle speed. This degradation of throughput is due to couple of reasons such as unsuccessful handovers and inappropriate selection of target networks at very high vehicle speeds.

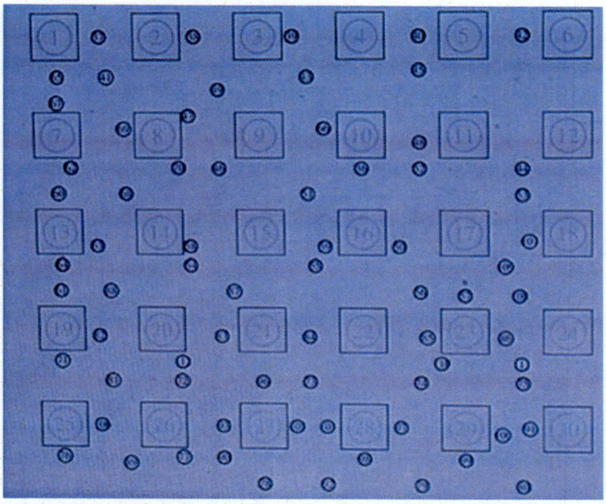

Fig. 11 Vehicle generation

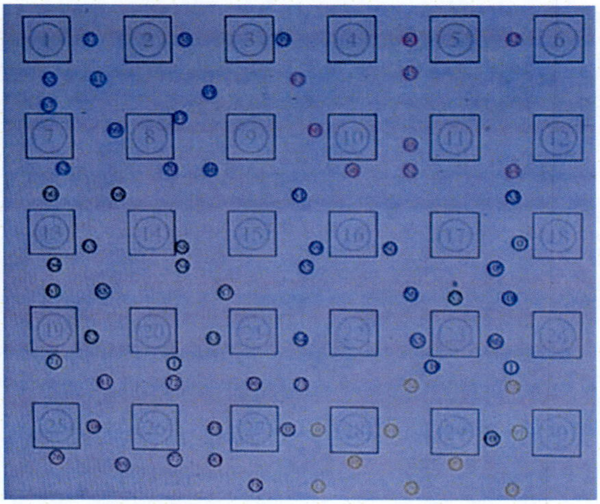

Fig. 12 Node deployment and cluster formation (shown in different colors)

Figure 16 shows the delay experienced by the vehicle on-board unit when running delay sensitive applications. Higher delay significantly affects the QoS performance of delay sensitive applications. In our work, our focus was mainly to reduce the network delay which can be achieved by selecting the network well before approaching the access points. From the results obtained, our proposed radio access selection method reduces the delay drastically in comparison to the existing approaches (almost reduced to 2 ms) when used for delay sensitive applications as shown in Fig. 16.

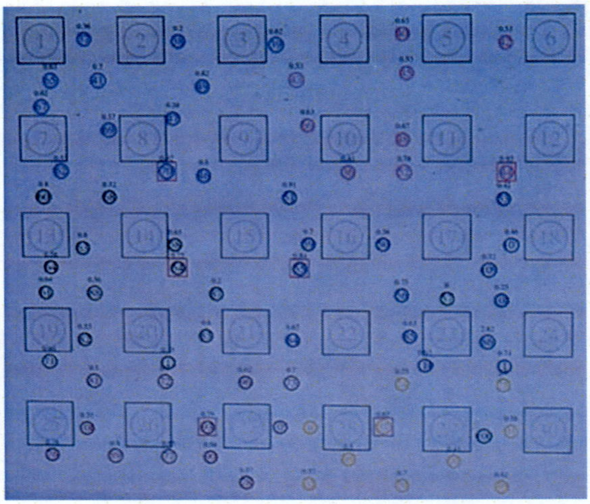

Fig. 13 Cluster head selection (in square brackets)

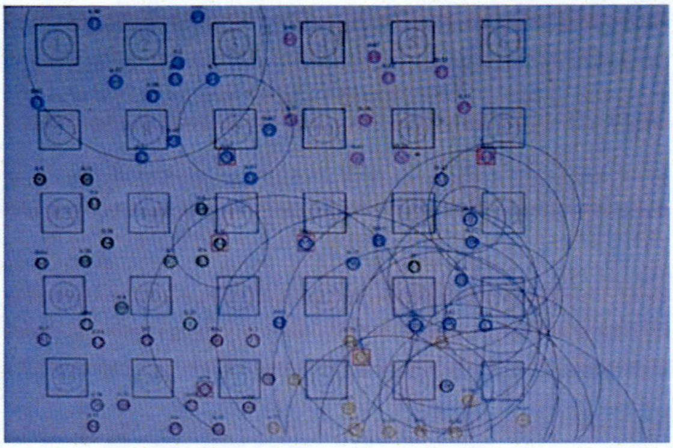

Fig. 14 Data transmission

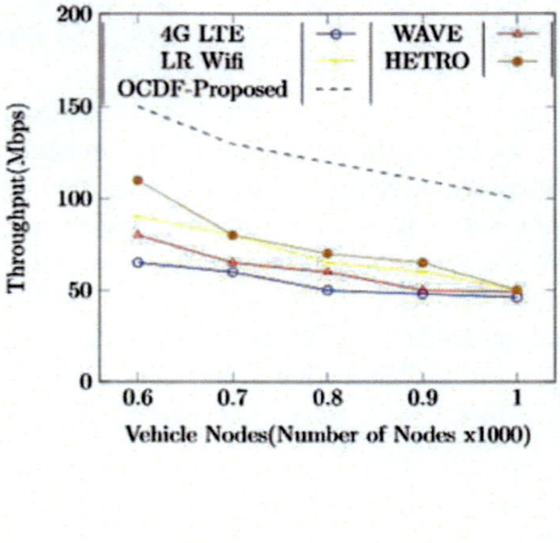

Fig. 15 Vehicle nodes versus throughput

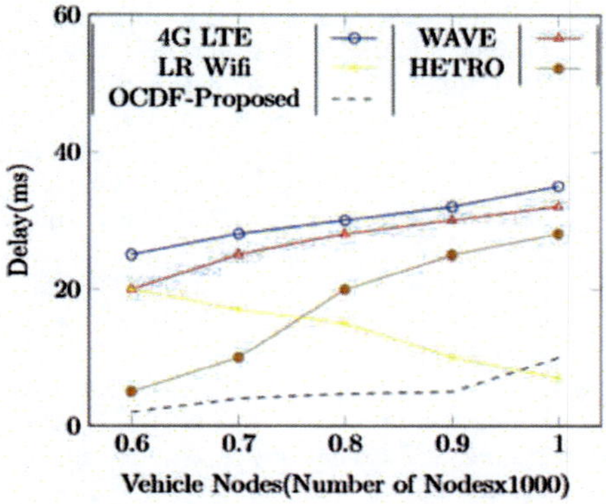

Fig. 16 Vehicle nodes versus delay

5 Conclusion

In our research work, we have proposed heterogeneous solution using OCDF protocol for data dissemination and intelligently selecting radio technology. IBSO algorithm is a meta heuristic algorithm which gives very competitive results with good robustness and running speeds in comparison to the current popular optimization algorithms. it also exhibits higher performance and can handle multi-objective optimization problems more efficiently. In our work, we have considered three use cases—voice, video, and safety message services. The objective is to compare and investigate the performance of throughput and delay. A comparative analysis is done with the existing approaches long-range Wi-Fi, WAVE, and 4G LTE. In our research work, we have integrated 4G LTE and DSRC to enhance vehicular network performance. In the future work, the algorithm for reducing handover delays yet to be incorporated to avoid packet losses and to get more effective results.

References

1. T. Mai, R. Jiang, E. Chung, A Cooperative Intelligent Transport Systems (C-ITS)-based lane-changing advisory for weaving sections. J. Adv. Transp. **50**(5), 752–768 (2016)
2. L.C. Tung, J. Mena, M. Gerla, C. Sommer, A cluster based architecture for intersection collision avoidance using heterogeneous networks, in 2013 12th Annual Mediterranean Ad Hoc Networking Workshop (MED-HOC-NET) (IEEE, 2013), pp. 82–88
3. K. Namdev, P. Singh, Clustering in vehicular ad hoc network for efficient communication. Int. J. Comput. Appl. 115(11) (2015)
4. S. Ucar, S.C. Ergen, O. Ozkasap, Multihop-cluster-based IEEE 802.11 p and LTE hybrid architecture for VANET safety message dissemination. IEEE Trans. Veh. Technol. 65(4), 2621–2636 (2015)
5. I. Lequerica, P.M. Ruiz, V. Cabrera, Improvement of vehicular communications by using 3G capabilities to disseminate control information. IEEE Netw. **24**(1), 32–38 (2010)

6. G. Remy, S.M. Senouci, F. Jan, Y. Gourhant, LTE4V2X: LTE for a centralized VANET organization, in 2011 IEEE Global Telecommunications Conference-GLOBECOM 2011 (IEEE, 2011), pp. 1–6
7. A. Benslimane, S. Barghi, C. Assi, An efficient routing protocol for connecting vehicular networks to the Internet. Pervasive Mob. Comput. **7**(1), 98–113 (2011)
8. T. Taleb, A. Benslimane, Design guidelines for a network architecture integrating VANET with 3G & beyond networks, in 2010 IEEE Global Telecommunications Conference GLOBECOM 2010 (IEEE, 2010), pp. 1–5
9. A. Benslimane, T. Taleb, R. Sivaraj, Dynamic clustering-based adaptive mobile gateway management in integrated VANET-3G heterogeneous wireless networks. IEEE J. Sel. Areas Commun. **29**(3), 559–570 (2011)
10. R. Sivaraj, A.K. Gopalakrishna, M.G. Chandra, P. Balamuralidhar, QoS-enabled group communication in integrated VANET-LTE heterogeneous wireless networks, in 2011 IEEE 7th International Conference on Wireless and Mobile Computing, Networking and Communications (WiMob) (IEEE, 2011), pp. 17–24
11. R.S. Bali, N. Kumar, J.J. Rodrigues, Clustering in vehicular ad hoc networks: taxonomy, challenges and solutions. Veh. Commun. **1**(3), 134–152 (2014)
12. T. Song, W. Xia, T. Song, L. Shen, A cluster-based directional routing protocol in VANET, in 2010 IEEE 12th International Conference on Communication Technology (IEEE, 2010), pp. 1172–1175
13. H. Su, X. Zhang, Clustering-based multichannel MAC protocols for QoS provisionings over vehicular ad hoc networks. IEEE Trans. Veh. Technol. **56**(6), 3309–3323 (2007)
14. Y. Zhang, J.M. Ng, C.P. Low, A distributed group mobility adaptive clustering algorithm for mobile ad hoc networks. Comput. Commun. **32**(1), 189–202 (2009)
15. D. Zhang, H. Ge, T. Zhang, Y.Y. Cui, X. Liu, G. Mao, New multi-hop clustering algorithm for vehicular ad hoc networks. IEEE Trans. Intell. Transp. Syst. **20**(4), 1517–1530 (2018)
16. A. Daeinabi, A.G.P. Rahbar, A. Khademzadeh, VWCA: an efficient clustering algorithm in vehicular ad hoc networks. J. Netw. Comput. Appl. **34**(1), 207–222 (2011)
17. G. Wolny, Modified DMAC clustering algorithm for VANETs, in 2008 Third International Conference on Systems and Networks Communications (IEEE, 2008), pp. 268–273
18. Z.Y. Rawashdeh, S.M. Mahmud, A novel algorithm to form stable clusters in vehicular ad hoc networks on highways. Eurasip J. Wirel. Commun. Netw. **2012**(1), 1–13 (2012)
19. Z. Wang, L. Liu, M. Zhou, N. Ansari, A position-based clustering technique for ad hoc intervehicle communication. IEEE Trans. Syst. Man Cybern. Part C (Appl. Rev.) 38(2), 201–208 (2008)
20. R. Neelaveni, Performance enhancement and security assistance for VANET using cloud computing. J. Trends Comput. Sci. Smart Technol. (TCSST) **1**(01), 39–50 (2019)
21. D. Sivaganesan, Efficient routing protocol with collision avoidance in vehicular networks. J. Ubiquitous Comput. Commun. Technol. (UCCT) **1**(02), 76–86 (2019)

Three Level Synthesis of Biometrics for Secured Authorization System with Hybrid Optimization

R. Sindhuja and S. Srinivasan

Abstract Biometric modalities are used in wide variety of applications such as banking, safety lockers, payment gateways, and lot more. Currently, security systems have greatly improved in all aspects and especially in the area of biometrics and its applications. A recent study reveals that nearly 88% of human recognition system works with the concept of biometric authorization. Most of the time, human biometrics are used for identifying the individuals. Three major biometrics of human's merely not duplicable and those are the human face, human iris, and human fingerprint. This paper attempts to fuse all the aforementioned human biometrics as a multimodal system and it creates an ultra-secured system that can able to identify individuals with less error rate. Apart from the fusion of three human biometrics, this paper implements the optimization technique for all the three human biometrics. The main attribute of this paper is to verify the multimodal output.

1 Introduction

Biometrics are meant by the measurement of human biological substances with various units. Each and every individual varies in their biological values, hence measurements of those values also vary respectably. With the help of the above statement, an attempt was made in the late eighteenth century in France. Biometric was used for the anthropological technique of anthropometry to law enforcement with help of a biometrics researcher. Biometrics are used to identified or recognize an individual with their physical biometric substances because of this reason only the biometrics recognition system has a huge market in the world. They are retina, finger vein, ıris, finger print, palm print, hand geometry, ear, face, sweat pore, lips, DNA, odour, vascular imaging, and brainwave [1].

The above-mentioned are the physical-biological substances used to be identified or recognize an individual. Apart from physical-biological substances, the

R. Sindhuja (✉) · S. Srinivasan
Department of Electronics and Instrumentation and Engineering, Annamalai University, Chidambaram, India

P. Karuppusamy et al. (eds.), *Ubiquitous Intelligent Systems*, Smart Innovation, Systems and Technologies 243, https://doi.org/10.1007/978-981-16-3675-2_4

behavioural characteristics of humans are also used to identify or recognize an individual. Gait analysis, keystroke dynamics, signature, voice ID, mouse use characteristics, and cognitive biometrics are the behavioural characteristics in biometrics. In the last decade, authentication was a huge factor in terms of security systems [2].

Many techniques and methods were followed such as a token, smart keys, smart cards (RF-ID card) secret password and dozens of personal questionnaires regarding the birth date, first school, and a lot more. In the beginning stage, to identify the individuals a unique token was provided, smart keys are still better with an inclusive of few electronic parts, which can able detect the individuals with the help of sensors [3].

The smart cards are very portable for usage and also very effective compared to the smart keys and the smart cards working with the help of the integrated chip. In the later stage, the passwords only characters (alphabets) then slowly numbers and special symbols were included in the passwords. PIN is the another authentication method, and it is easy to use in our day-to-day lives. The above-mentioned are the techniques and methods, which were ruling the security system past ten and more years because these techniques have good authentication detecting skill. As equal to its merits, it has multiple disadvantages as follows,

- Duplication of the passwords and secret PINs can be prepared easily.
- The users of the authentication systems can forget the passwords and digital keys.
- All the above-mentioned methods can be hacked or stolen at any point in time.
- High possibility of spoofing.
- High-maintenance cost.

The above-mentioned are the highlighted demerits of the past authentication system, but all these can be overcome by using the biometric authentication system. The biometric system has different advantage such as security that cannot be hacked, the accuracy of access, accountability, convenient without any mantel stress, scalability with a different application, non-memorizable, trustworthy, reducing the access time [4] and finally, it can be implemented with a one-time investment (reduction in cost).

In this paper, a multimodal biometric authentication system is implemented to attain maximum security in its application. The multimodal is a collection of unimodal biometrics techniques such as the face, fingerprint, and iris, and the accuracy, sensitivity, and specificity of all unimodal biometrics are compared [1].

2 Literature Review

Damousis and Argyropoulos were working closely on a multimodal biometrics system and they performed biometrics fusion using couples of algorithms such as Gaussian mixture models (GMM), artificial neural networks (ANN), fuzzy expert systems (FES), and support vector machines (SVM). The output was validated with a prompt database that was considered as a bench mark and all unimodal biometrics

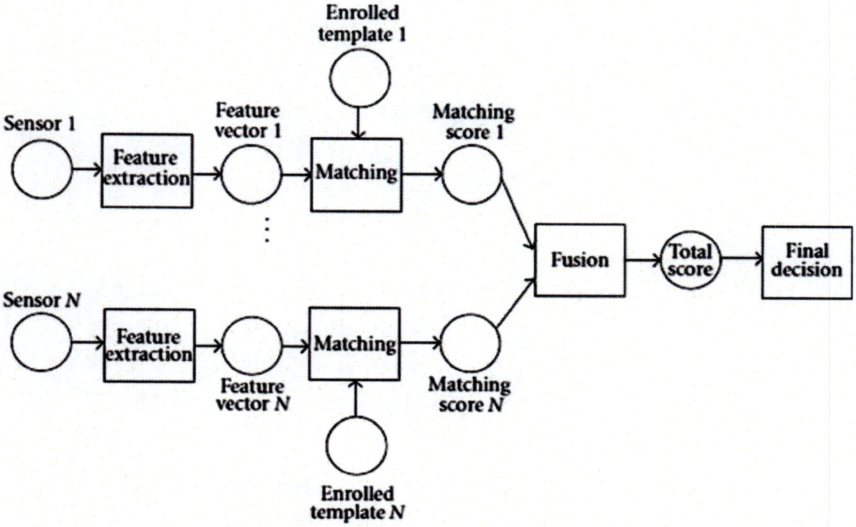

Fig. 1 Multimodal biometrics system with N-sensors

were pipelined separately and match with one another before the output get fused together. Based on the total score, the final decision was made for identifying the individual and expert 6 with EER 1.09% (Fig. 1) [5].

Sujatha and Chilambuchelvan have performed research that overcomes the disadvantage of the unimodal such as distinctiveness, spoof attacks, noise in sensed data, intra-class variations, and non-universality. They also address the false rejection rate and false acceptance rate from the biometrics system and this system have the capacity to enhance accuracy and equal error rate [6].

Houda and Touahria were completed a research that investigates the comparative performance from three different approaches for multimodal recognition of combined fingerprints and iris. They have concentrated in the matching score and decision-making levels and they also suggested that fuzzy logic mimics the human reasoning in a soft. Both were occupied the database of iris and face from CASIA and FVC 2004. Figure 2 represents the overall ideology of the researchers for biometrics multimodal system [7].

Experimental results achieved best compromise between FRR and FAR (0% FAR and 0.05% FRR) with accuracy 99.975% and EER equal to 0.038 and matching time equal to 0.1754 s [8].

The term "multimodal biometric" refers to multiple biometric traits used together at a specific level of fusion to recognize persons. The "multibiometrics" includes either the use of multiple algorithms, also called classifiers at enrolment matching stages for the same biometric trait, or the use of multiple sensors of the same biometric trait like using different instruments to capture the biometric details, or using multiple

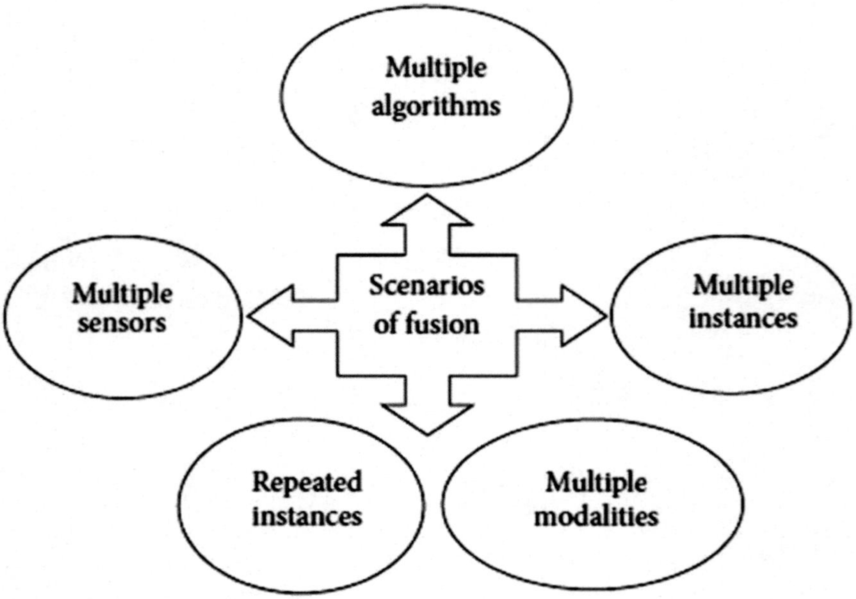

Fig. 2 Combined multimodal system

instances of the same biometric trait like the use of fingerprints of three fingers, or finally using repeated instances like repeated impressions of one finger.

Chia and Dzati both have performed three types of fusion which involve score level fusion, feature level fusion and decision level fusion. Researchers high concentrated on the multibiometric system that deals with one or more physical information of human or behavioural habits used to identify an individual. They have used speech signal and lip-reading for final decision making with the help of AND and OR logic [10]. For extracting, the speech features MFCC (Mel Frequency Ceptral Coefficient) was used, and for visual feature ROI (region of interest) was used which derived from lip-reading data set is used as visual features. Finally, for classifier SVM (support vector machine) is used to discriminate the dataset [11].

Anil and Subramoniam were performed a research multimodal verification authentication systems constructed on machine learning algorithm is inculcated. They achieved performances with face plus palm print feature level fusion is 91.52% and decision level fusion is 91.63%, face plus ear recognition is 96.8% and verification is 97.1%, and finally face plus finger plus iris produces recognition is 78.5%.

3 Materials and Methods

Purely this research is based on the image processing methodology and the medical image is used to extract information to detect diseases and disorders, monitor the physical conditions or treatment conditions, and a lot more.

The three-level fusion of biometrics is implemented in this research paper as a multimodal system. Three unimodal biometrics are coupled to create a single-multimodal biometrics system that can able to recognize and identify the particular individual from the known dataset (Fig. 3).

The biometrics database of face, iris, and fingerprint were taken from different organizations. All the face, iris, and fingerprint datasets were taken separately for

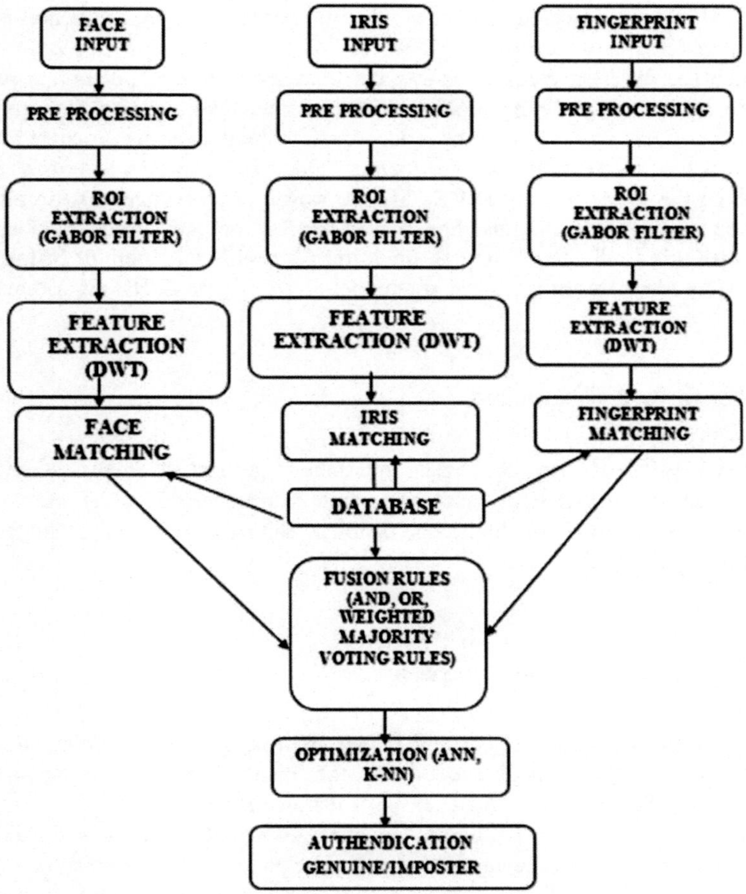

Fig. 3 Flow chart of proposed architecture

processing as unimodal threads and the results fused together to produce a valid output. The output is retrieved from an optimized system that considered all three results of fingerprint, iris, and face values for identifying an individual from a known dataset of people. Figure 3 demonstrates the three-level synthesis multimodal biometrics system for detecting or recognize an individual from a known dataset. This system uses the image processing technique for identifying the similarity and differences between bio-images.

Primarily, all three bio-images are subjected to the image pre-processing stage and this stage helps the system to reduce the error rate in the detection of individuals and improve the overall efficiency. In the pre-processing stages, the raw image or original image from the database can be resized, colour conversion and noise reduced. The resizing is very important because all input images should be in the same size which helps the system to improve competence. The input image can be any coloured image, which has to be changed into a greyscale image for reducing the noises from the raw images.

There are different types of noises that may exist in an input image such as Gaussian noise, salt-and-pepper noise, shot noise, quantization noise, film grain, and a lot more. These noises have to be reduced before the images get processed further to extract the information from the images. The noise-free feed to next stage for extracting the features using DWT (discrete wavelet transformation) and all three types of images (face, iris, and fingerprint) will be compared to the database and the decision is made after fusing all three images results with help of fusion rules. Finally, the fusion result is classified with help of ANN and K-NN algorithms [13].

4 Result and Discussion

The entire system has been catalogued into three stages such as stage 1, stage 2, and stage 3, and in the first stage input images are pre-processed, in the second stage feature extraction and in the final stage fusion of all three (face, iris, and fingerprint) results in declaring the decision.

4.1 First Stage: Pre-processing

Figure 4 represents the first level of image processing and here colour image is converted into the greyscale image to reduce the noise and also contrast enhancement is performed. This process is made easy for feature extraction.

Figure 5 explains the histogram equalization which distributes the contrast throughout the input face image. This method helps to reduce the noise rate, whilst segmenting the image for feature extraction. In Fig. 4i, the contrast not evenly speared, but Fig. 4iii shows the after the histogram equalization that makes the contrast of the image evenly speared throughout the image.

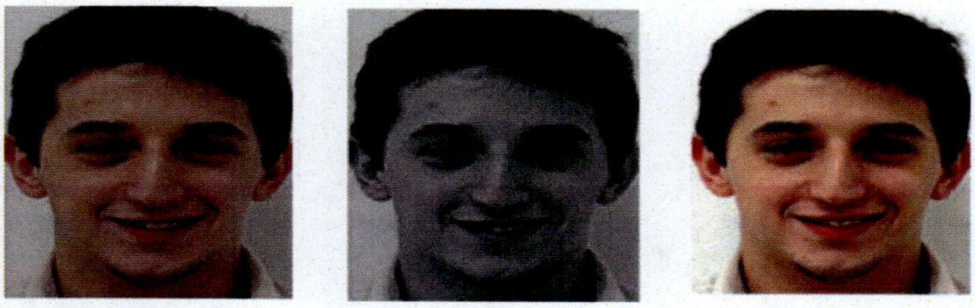

i. Colour image ii. Grey scale image iii. Contrast enhanced image

Fig. 4 Image pre-processing for face image

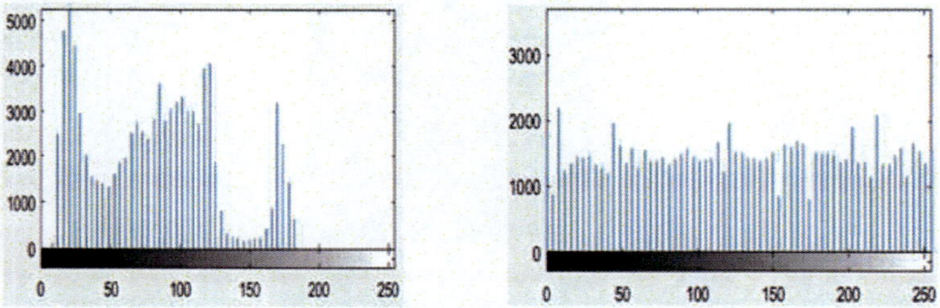

i. .Histogram plot of Grey Scale image ii. Histogram equalization

Fig. 5 Image histogram of face

Figure 6 shows different image processing techniques to enhance the image quality and these help in feature extraction. Especially in Fig. 6iii shows the output of Binarized image that convert the grayscale image into a binary image and here all black and white pixels are converted into zeros and ones.

Figure 6iv is the output of the Canny edge detection that detects the edges in the image and removes the unwanted textures, details, and noises using a Gaussian filter, and also it smoothens the image. The Ridge thinning method is used to thinner the fingerprint between each track's ridge, hence the uniqueness of the fingerprint can be found easily (Fig. 6v). Figure 6vi shows the output of Minutiae Marking helps enhance image quality without any information losses [5].

Same kind of pre-processing was applied to the iris image with some additional techniques such as Edge detection and Hough Circle. These techniques were helpful to segments the features from the iris image for identifying the individuals. All different outputs are displayed in Fig. 7.

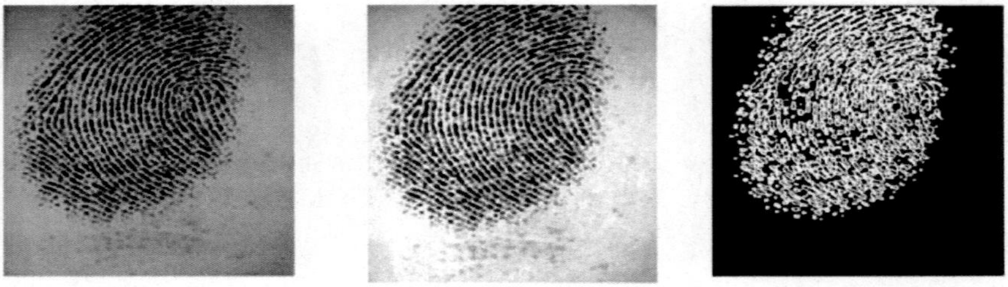

i. Fingerprint input image ii. Contrast Enhanced image iii. Binarized image

iv. Canny edge detection v. Ridge thinning vi. Minutiae Marking

Fig. 6 Image pre-processing for fingerprint image

i. Iris input image ii. Edge detection iii. Contrast Enhanced image

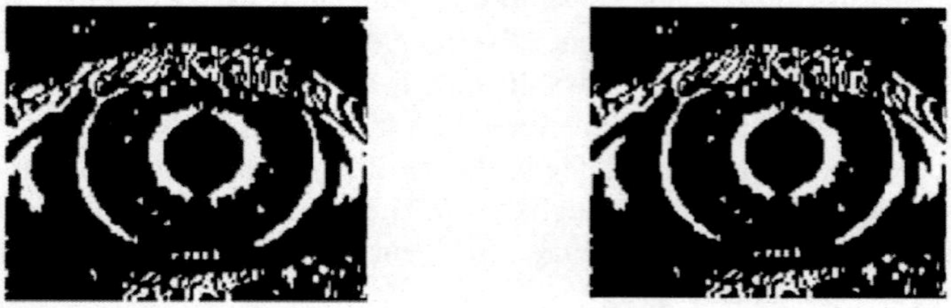

iv. Binarized image v. Hough Circle

Fig. 7 Image pre-processing for iris image

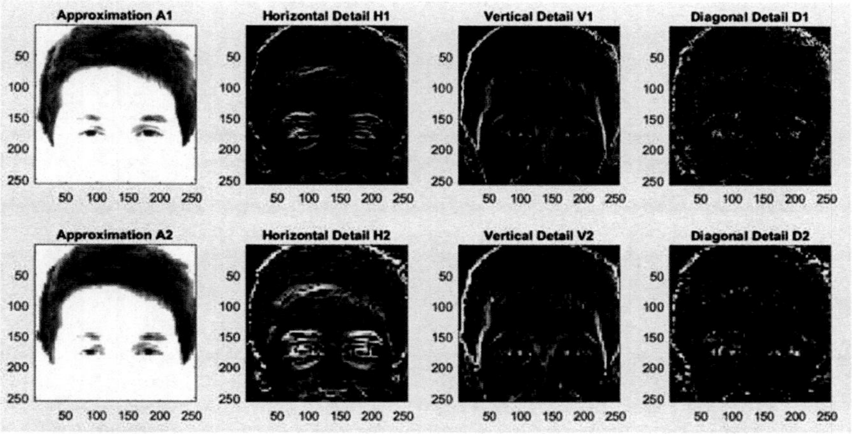

Fig. 8 Second level DWT decomposition of face input image

4.2 Second Stage: Feature Extraction

The feature extraction is completely depend on discrete wavelet transformation (DWT) decomposition image processing technique. Figure 8 displays different levels of DWT decomposition of the pre-processed input face images. Here both vertical, horizontal, and diagonal decompositions were performed and approximation finalize the feature from vertical, horizontal, and diagonal decompositions.

Table 1 displays different attributes of twenty different face images and it contains nine different field which were calculated from DWT decomposition step.

Figure 9 shows different levels of decomposition steps initially started with roes wise and then column-wise and finally different combination of comparison was performed between input image plus horizontally and vertically and diagonally.

Figure 10 shows how a pre-processed image decomposed using DWT technique, and here three levels of decomposition were performed between high-frequency bands and low-frequency bands such as HH1, HL1, and LH1 then again HH2, HL2, and LH2, and so on (Fig. 11).

Table 2 displays different attributes of twenty different fingerprint images, and it contains nine different field which were calculated from DWT decomposition steps (Fig. 12).

Table 3 displays different attributes of twenty different iris images, and it contains nine different field which were calculated from DWT decomposition steps.

Here, the face, iris, and fingerprint were combination together as a fusion using an hybrid technique, and Table 4 is showing the measurement of decision fusion with hybrid technique. Table 4 compares three decision rules such as AND rule, OR rule, and weighted majority voting. Overall performances of weighted majority voting are efficient and it has been displayed in Fig. 13.

Table 1 Extracted features from LH_2 sub band (features of 20 face image samples)

Features/samples	Auto correlation	Dissimilarity	Energy	Entropy	Homogeneity	Maximum probability	Average	Variance
Sample 1	55.08702	19.77753	0.439568	0.952209	1.577999	0.309954	14.68666	31.7473
Sample 2	50.83038	11.43288	0.478327	0.971927	1.379777	0.34067	10.57076	30.28957
Sample 3	51.26759	10.77455	0.34525	0.972992	1.808056	0.334266	11.80746	30.60432
Sample 4	50.81655	10.59283	0.34762	0.922145	1.322368	0.36015	10.47752	30.11174
Sample 5	53.32464	16.54089	0.34446	0.997235	1.633868	0.1471	13.71482	31.39928
Sample 6	53.63489	16.32207	0.313264	0.940797	1.93371	0.386265	13.88391	31.06924
Sample 7	52.43525	11.77372	0.361871	0.971142	1.147664	0.352819	12.1051	31.38129
Sample 8	50.88579	19.4232	0.330576	0.96824	1.129618	0.362371	10.43878	30.79406
Sample 9	56.8732	17.24692	0.345126	0.977188	1.786858	0.836445	16.76059	31.85522
Sample 10	56.79676	15.47395	0.38777	0.935653	1.204024	0.317432	16.79804	31.99281
Sample 11	54.74281	18.97996	0.332217	0.986996	1.910248	0.390832	14.93694	31.43804
Sample 12	57.07464	10.52082	0.396605	0.928736	1.618537	0.383657	17.78074	31.83588
Sample 13	55.72302	16.64261	0.328957	0.953736	1.71531	0.308917	15.80158	32.53408
Sample 14	50.81655	10.2517	0.394807	0.922145	1.322368	0.301505	10.24775	30.11174
Sample 15	56.55576	111.19	0.386106	0.953405	1.015553	0.399227	16.31602	31.67693
Sample 16	56.64658	12.89521	0.316547	0.941262	1.635221	0.341982	16.6044	31.60252
Sample 17	58.94155	16.06217	0.45054	0.974056	1.975637	0.371879	18.54732	31.11421
Sample 18	58.84487	15.84738	0.458993	0.992236	1.220845	0.32696	18.7122	31.98606
Sample 19	58.51079	18.35392	0.461466	0.922246	1.906661	0.354378	18.40749	31.52068
Sample 20	57.6214	15.35631	0.480576	0.97246	1.699526	0.312493	17.50884	31.11601

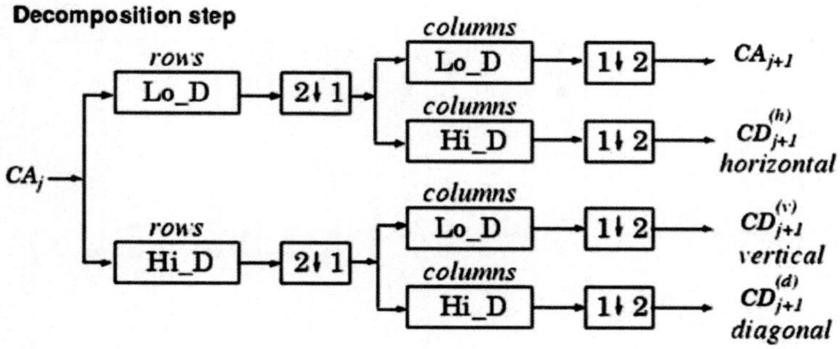

Fig. 9 DWT filter bank

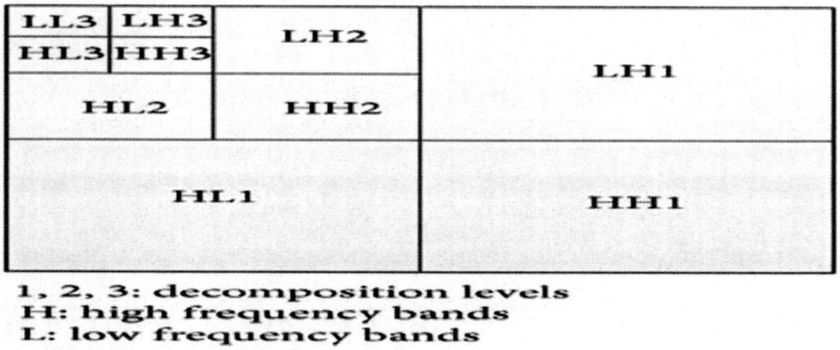

Fig. 10 Wavelet decomposition levels

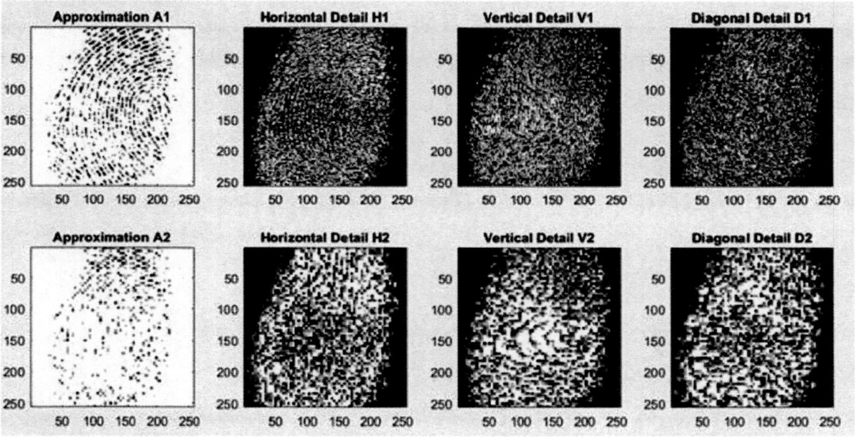

Fig. 11 Second level DWT decomposition of finger print input image

Table 2 Extracted features from LH_2 Sub band (features of 20 fingerprint image samples)

Features/samples	Auto correlation	Dissimilarity	Energy	Entropy	Homogeneity	Maximum probability	Average	Variance
Sample 1	15.073	1.69592102	0.78801	1.155265	0.613278	0.56126	6.446435	60.4096
Sample 2	16.958976	2.02719081	0.746601	1.243059	0.330215	0.489262	7.123143	68.26335
Sample 3	17.168444	1.94389964	0.757013	1.231915	0.336065	0.496514	7.10491	68.59135
Sample 4	14.763	1.74575	0.781781	1.159994	0.380924	0.559857	6.41625	59.57933
Sample 5	14.9375	1.76712201	0.77911	1.167132	0.376514	0.554373	6.47165	60.26247
Sample 6	13.996333	1.88508333	0.764365	1.173046	0.374652	0.554333	6.35425	57.61841
Sample 7	16.744063	1.8765625	0.76543	1.214719	0.346327	0.511801	6.958229	66.85214
Sample 8	16.363779	1.91684534	0.760394	1.216626	0.345937	0.512722	6.905053	65.74682
Sample 9	14.806793	2.25168413	0.718539	1.246982	0.330321	0.494916	6.819486	62.07091
Sample 10	17.994957	1.87402501	0.765747	1.229765	0.335773	0.492267	7.234232	71.08033
Sample 11	17.424971	1.59265155	0.800919	1.16892	0.369732	0.537045	6.888723	67.93888
Sample 12	17.071343	1.80768362	0.77404	1.206547	0.350196	0.515352	6.977386	67.65565
Sample 13	15.793485	1.73680044	0.7829	1.174934	0.370362	0.544637	6.638286	63.02166
Sample 14	14.099155	1.69084119	0.788645	1.137162	0.395317	0.577367	6.226022	57.09772
Sample 15	15.369213	1.71921296	0.785098	1.164784	0.376958	0.553604	6.530324	61.51096
Sample 16	16.730997	2.1897096	0.726286	1.265107	0.318143	0.472243	7.198885	68.25783
Sample 17	17.242242	1.74901661	0.781373	1.197765	0.354632	0.520089	6.969733	67.97975
Sample 18	13.202528	2.15217461	0.730978	1.203925	0.356821	0.533017	6.385586	56.20317
Sample 19	13.135839	1.75433268	0.780708	1.131486	0.400311	0.584595	6.061334	54.13028
Sample 20	15.073	1.69592102	0.78801	1.155265	0.613278	0.56126	6.446435	60.4096

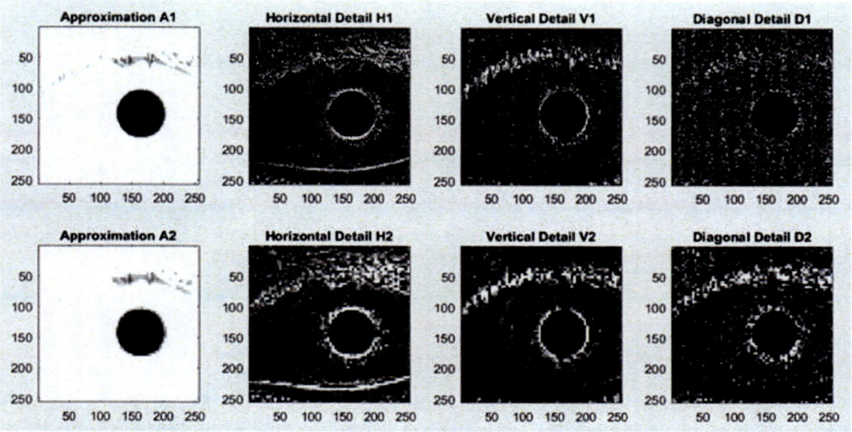

Fig. 12 Second level DWT Decomposition of Iris input image

Table 5 shows the details of the performance comparison of neural network classifier with different authentication methods. The accuracy, sensitivity, and specificity for fingerprint authentication, iris authentication, and face authentication were displayed individually, but when fingerprint authentication, iris authentication, and face authentication were fused together with help of neural network classifier as a single-authentication module then overall synergy and efficiency increases and all accuracy, sensitivity, and specificity were increased by 4 points.

Figure 14 displays the graphical representation of the fusion authentication of fingerprint, iris, and face authentications. The blue, carroty, and grey represent the accuracy, sensitivity, and specificity fields.

Fingerprint authentication, iris authentication, and face authentication were fused together with help of k nearest neighbour (K-NN) classifier as a single-authentication module (Table 6; Fig. 15).

Results of neural network classifier was increased by only 4%, but K-NN the classifier increases the overall efficiency by 6%. Hence K-NN the classifier was considered as one of the best classifiers for fusion of different biometrics samples for authentication which can able to identify any individuals from the known database.

Figure 16 represents the performance comparison of the Neural network classifier and K-NN classifier and all attributes of the K-NN classifier is higher than the neural network classifier. The accuracy increased by 2.37%, sensitivity increased by 2.1%, and specificity increased by 2.4%.

Table 3 Extracted features from LH_2 sub band (features of 20 iris image samples)

Features/sample s	Auto correlation	Dissimilarity	Energy	Entropy	Homogeneity	Maximum probability	Average	Variance
Sample 1	35.0261	29.3778	0.236781	0.885221	2.331858	0.1431	34.99969	11.48775
Sample 2	30.43938	21.54936	0.267783	0.871927	2.31438	0.155408	30.46057	10.74429
Sample 3	31.31591	20.05251	0.270459	0.870407	2.285808	0.157334	31.32081	10.9196
Sample 4	30.41668	20.28036	0.269481	0.870522	2.393032	0.14146	30.42625	10.71112
Sample 5	33.86723	26.82	0.238444	0.885972	2.328563	0.1471	33.85967	11.3134
Sample 6	33.87376	26.6994	0.251326	0.880994	2.383493	0.138626	33.88884	11.30407
Sample 7	32.78934	21.19621	0.272167	0.871221	2.321415	0.153528	32.80151	11.17338
Sample 8	30.74709	19.4232	0.280553	0.865168	2.396013	0.142025	30.76144	10.79179
Sample 9	36.54287	47.53872	0.213905	0.895772	2.273179	0.142738	36.47158	11.67279
Sample 10	36.48478	45.97453	0.215153	0.895357	2.254932	0.142317	36.4108	11.68399
Sample 11	34.72277	38.20819	0.232217	0.886996	2.22391	0.152014	34.66159	11.43804
Sample 12	37.81411	50.52082	0.219661	0.892874	2.305562	0.138366	37.74528	11.85984
Sample 13	35.57837	46.71439	0.21929	0.893537	2.246572	0.144089	35.4968	11.53408
Sample 14	30.41668	20.28036	0.269481	0.870522	2.393032	0.14146	30.42625	10.71112
Sample 15	36.40866	51.84204	0.215861	0.895164	2.272802	0.139923	36.31602	11.63677
Sample 16	37.46736	22.46999	0.237995	0.884126	2.443635	0.12142	36.10526	11.62156
Sample 17	38.13699	26.77491	0.227451	0.890406	2.41711	0.128212	38.12547	11.90411
Sample 18	38.10484	25.70152	0.224359	0.891439	2.413012	0.126246	38.07371	11.91899
Sample 19	38.49865	28.8033	0.226147	0.891222	2.416291	0.128914	38.47504	11.95852
Sample 20	35.0261	29.3778	0.236781	0.885221	2.331858	0.1431	34.99969	11.48775

Table 4 Measurement of decision fusion with hybrid technique

Decision rule	GAR (%)	FAR (%)	FFR (%)
AND rule	96	3	1
OR rule	98	1	5
Weighted majority voting	97	2	3

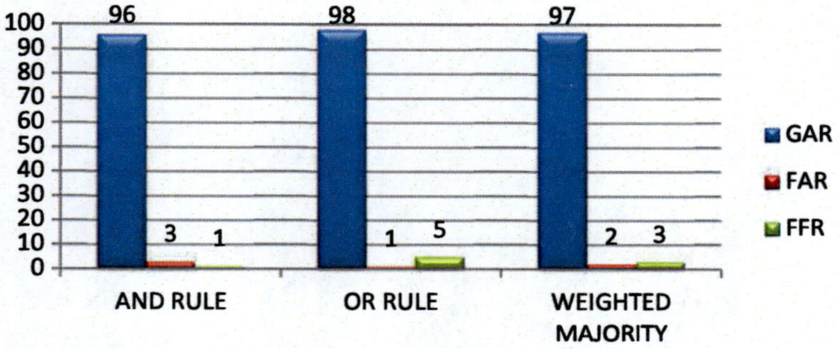

Fig. 13 Fusion rule illustration

Table 5 Performance comparison of neural network classifier with different authentication methods

Parameters/authentication method	Fingerprint authentication	Iris authentication	Face authentication	Fingerprint, iris & face authentication combined (proposed)
Accuracy (%)	92.15	92.06	90.52	94.08
Sensitivity (%)	92.38	92.14	90.12	94.24
Specificity (%)	91.38	91.23	90.24	94.38

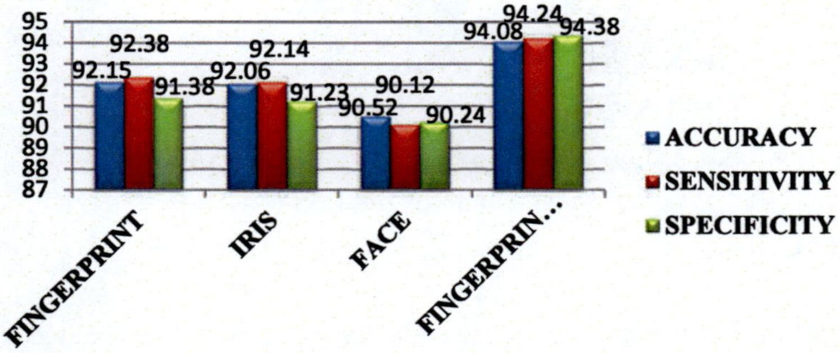

Fig. 14 Neural network classifier performance compared with different authentication methods

Table 6 Performance comparison of K-NN classifier with different authentication methods

Parameters/authentication method	Fingerprint authentication alone	Iris authentication alone	face authentication alone	Fingerprint, iris & face authentication combined (proposed)
Accuracy (%)	94.56	93.23	91.34	96.45
Sensitivity (%)	94.28	93.56	91.54	96.34
Specificity (%)	94.78	93.12	91.19	96.78

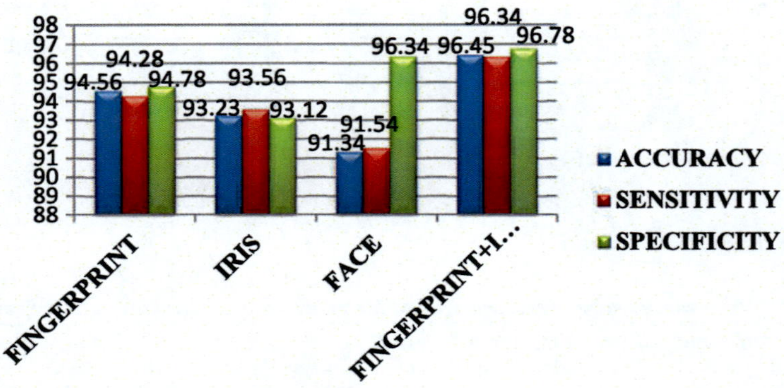

Fig. 15 K-NN classifier performance compared with different authentication methods

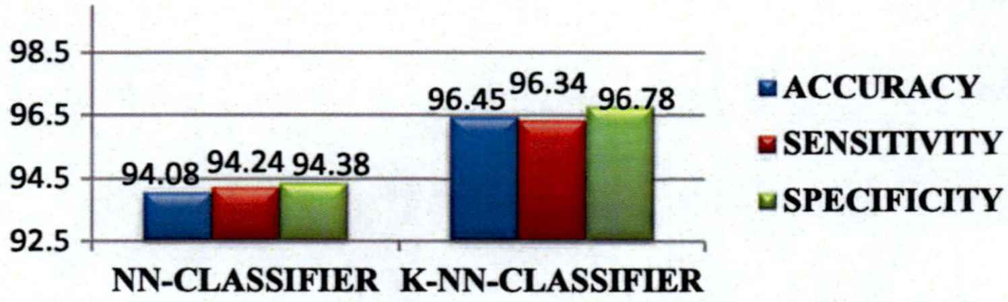

Fig. 16 Performance comparison of neural network classifier and K-NN classifier

5 Conclusion

A single level of authentication may not be sufficient to identify any individual even from a known database, but in multimodal authentication, duplication can be avoided through parallel processing of the biometrics data. The main advantage of the multimodal is decision making with help of fusing of biometrics data. This research paper fused three separate biometrics images such as a face image, an iris image, and fingerprint images together to decide or identify any individual from the known database.

Here, all three types of the image were pre-processed and features were extracted with help of the DWT method. The DWT method uses three level of decomposition steps for each biometrics image and two classifiers (k-NN classifier and neural network classifier) were deployed to identify the individual. Finally, both classifier performances were compared in terms of accuracy, sensitivity, and specificity. The k-NN classifier's accuracy increased by 2.37%, sensitivity increased by 2.1%, and specificity increased by 2.4% when compared to the neural network classifier.

References

1. K. Veeramachaneni et al., An adaptive multimodal biometric management algorithm. IEEE Trans. Syst. Man Cybern.—Part C: Appl. Rev. **35**(3), 344–356 (2005)
2. L. Hong, A. Jain, Integrating faces and fingerprints for personal identification. IEEE Trans. Pattern Anal. Mach. Intell. **20**(12), 1295–1307 (1998)
3. R.W. Frischholz, U. Deickmann, A multimodal biometric identification system. IEEE Comput. **33**(2) (2000)
4. L. Hong, A.K. Jain, S. Panikanti, Can multibiometrics improve performance?, in *Proceedings of IEEE on AutoID*, Summit, NJ, 1999, vol. 10, pp. 59–64
5. M. Vatsa, R. Singh, P. Gupta, Comparison of iris recognition algorithms, in *International Conference on Intelligent Sensing and Information Processing, 2004. Proceedings of* (IEEE 2004), pp. 354–358
6. Y. Yin, L. Liu, X. Sun, et al., SDUMLA-HMT: a multimodal biometric database, in *CCBR 2011* (Springer, Berlin, Heidelberg, 2011), pp. 260–268
7. H. Benaliouche, M. Touahria, Comparative study of multimodal biometric recognition by fusion of Iris and fingerprint. Sci. World J. **2014** (6) (2014). https://doi.org/10.1155/2014/829369
8. C. Dalila, H. Imane, N.A. Amine, Multimodal score-level fusion using hybrid GA-PSO for multibiometric system. Cherifi Dalila and Hafnaoui Imane, Informatica **39**, 209–216 (2015)
9. B.M. Shruthi, M. Pooja, Mallinath et al., Multimodal biometric authentication combining finger vein and finger print. Int. J. Eng. Res. Dev. **7**(10), 43–54 (2013). e-ISSN: 2278-067X/p-ISSN: 2278–800X. www.ijerd.com
10. E. Sujatha Nil, A. Chilambuchelvan, Multimodal biometric authentication algorithm at score level fusion using hybrid optimization. Wirel. Commun. Technol. https://doi.org/10.18063/wct.v2i1.415
11. X. Xu et al., The study of feature level fusion algorithm for multimodal recognition. IEEE Trans. Inf. Forens. Secur. **7**(1), 255–268 (2012)
12. B. Subramaniam et al., Multiple features and classifiers for vein based biometrics recognition. Biomed. Res. (2017). www.Biomedres.info
13. S. Ramkumar et al., Detection of osteoporosis and osteopenia using bone densitometer—simulation study. Mater. Today: Proc. **5**, 1024–1036 (2018)

A Deep Learning-Based Residual Network Model for Traffic Sign Detection and Classification

S. Kiruthika Devi and C. N. Subalalitha

Abstract Traffic sign board recognition is a very significant work for the upcoming driver assistance intelligent vehicle systems. The ability to detect such traffic signs from the real road scenes intensifies the safety of the intelligent vehicle systems. However, automatic detection and classification of traffic signs by such intelligent vehicle systems is a challenging task due to various factors such as variation in light illumination, different viewpoints, colour faded traffic sign, motion blurring, etc. The deep learning models have proved to provide solutions to overcome these factors. This paper proposed deep learning-based residual network for traffic sign detection and classification (DLRN-TSDC) model for effective Indian Traffic Sign Board Recognition. The DLRN-TSDC model makes use of Colour space threshold segmentation technique for the effective identification of sign boards. Simultaneously, pre-processing of the detected traffic sign takes place in three distinct ways such as clipping of edges, image enhancement and size normalization. In addition, the ResNet-50 model is used as a feature extractor and a classifier to determine the final class label of the traffic sign board. Extensive experimental analysis was carried out to validate the effective performance of the DLRN-TSDC model and for the precision, recall, Intersection over Union (IoU) and accuracy scores are 98.76%, 98.92%, 89.56% and 98.84%, respectively.

1 Introduction

Developing an automated traffic sign detection and recognition model is an important need in the field of artificial intelligence-based vehicle systems [1]. In recent days, intelligent vehicle systems are widely gaining importance to assist vehicle drivers. Traffic sign boards convey many important information to drivers such as

S. Kiruthika Devi (✉) · C. N. Subalalitha
Department of Computer Science and Engineering, SRM Institute of Science and Technology, Kattankulathur, Chennai, India
e-mail: kiruthis2@srmist.edu.in

C. N. Subalalitha
e-mail: subalaln@srmist.edu.in

P. Karuppusamy et al. (eds.), *Ubiquitous Intelligent Systems*, Smart Innovation, Systems and Technologies 243, https://doi.org/10.1007/978-981-16-3675-2_5

road conditions, speed limit, maximum height of the vehicle allowed, rules and restrictions, prohibitions, warnings and other useful details for route direction, etc. Hence, automatic detection of traffic signs is very important as it has got the great potential impact on developing intelligent vehicles with driver assistance system, a self-driving car and robot navigation systems. This paper proposes a deep learning model, deep learning-based residual network for traffic sign detection and classification (DLRN-TSDC) model that can automatically detect and classify the Indian Traffic Board Signs.

The automatic traffic sign detection and recognition (TSDR) from complex road scene is quite a difficult task due to various factors. The factors involved in such a scenario are of two categories, namely internal and external. The internal factors are those that are integral part of the traffic signs such as, faded, damaged and mispositioned traffic signs due to the prolonged exposure to environment, whereas the rest of the factors involved in detecting the signs are external factors such as, varying light illumination, shadows falling on signboard bad weather condition, obstacles infront of signboard like trees, vehicle, pedestrians, capturing blurred images and view geometry [2]. Apart from this, building a deep learning-based automatic TSDR system demands for a large data set for training the model. The unavailability of ample Indian traffic dataset makes it a more challenging task.

A deep learning-based automatic TSDR system needs to be robust and should identify the traffic signs at good speed with low computational cost, high accuracy. As deep learning models have proved to be efficient in extracting features and learning the parameters by itself, which leads to effectively detect and classify traffic signs. In this paper, the proposed DLRN-TSDC model aims at improving the accuracy achieved by the state-of-the-art deep learning approaches such as multilayer perceptron (MLP) [3], Iterative Nearest Neighbours-based Linear Project with Iterative-Neighbour Classifier (INNLP + INNC) [4], Gaussian filter, histogram equalization, histogram of oriented gradients with principal component analysis (GF + HE + HOG + PCA) [5], YOLO v3 [6]. For detecting the traffic signs, Colour space threshold segmentation technique is used to fetch the region of interest (RoI) which is pre-processed and fed into ResNet-50 a deep learning (DL) model for feature extraction and classification. Due to the unavailability of sufficient Indian traffic sign dataset, the ResNet-50 model is trained on German Traffic Sign Recognition Benchmark (GTSRB) dataset which is very similar to Indian traffic signs [7, 8].

In the attempt of traffic sign detection and recognition, most of systems are using colour information to segment the traffic sign images from background. The detection of traffic sign using colour information results in low performance owing to the disturbances like poor climate, variations in lighting state and fading of signage. Occluded images result in accuracy of traffic sign prediction and bad environmental ability. Although various machine learning techniques were used in TSDR, feature extraction is a time-consuming process. The advancement of automatic feature extraction in DL models, the most of the existing systems use a very basic DL model for TSDR. The recent work on identification of Indian traffic signs by various DL models includes datasets having very limited number of Indian traffic sign images for training makes the model ineffective.

Hence, developing an efficient deep learning model for recognition of Indian Traffic signs working in real time with high accuracy and minimal computational cost is mandatory. This paper introduces deep learning-based residual network for traffic sign detection and classification model that suits real-time traffic sign detection and recognition. The presented model encompasses diverse subprocesses, namely traffic sign recognition, pre-processing and classification. Primarily, the DLRN-TSDC model makes use of colour space threshold segmentation technique for the effective identification of sign boards. Simultaneously, pre-processing of the detected traffic sign takes place in three distinct ways such as clipping of edges, image enhancement and size normalization. In addition, the ResNet-50 model is used as a feature extractor and classifier for determining the final class label of the traffic sign board. An elaborate experimental analysis has been carried out to validate the effective performance of the DLRN-TSDC model.

The rest of the paper is structured as follows: In Sect. 2 contains of the foundation details of ResNet-50 model. In Sect. 3, the state-of-the-art approaches for automatic TSDR have been discussed, the implementation details, performance evaluation of the proposed DLRN-TSDC model and the comparison with other models are described in Sect. 4. Finally, the conclusion of this experimental analysis and future enhancement is discussed in Sect. 5.

2 Background

The ResNet-50 architecture is shown in Fig. 1 and consists of stacked convolution layers for feature extraction, max-pooling layer, average pooling layer followed by fully connected layer. ResNet-50 is CNN-based DL model that is 50 layers deep and as the layers go deeper, and the parameter learning accuracy of the model will tend to increase. In the Deep Convolution Neural Network (DCNN), beyond certain limit, if we keep on increasing the layer depth, the performance of the model will start to decrease which is termed as vanishing gradient problem [9]. The vanishing gradient issue arises during the training of the DCNN model. The accuracy of the model becomes saturated and starts to degrade as gradient norm of the previous layer

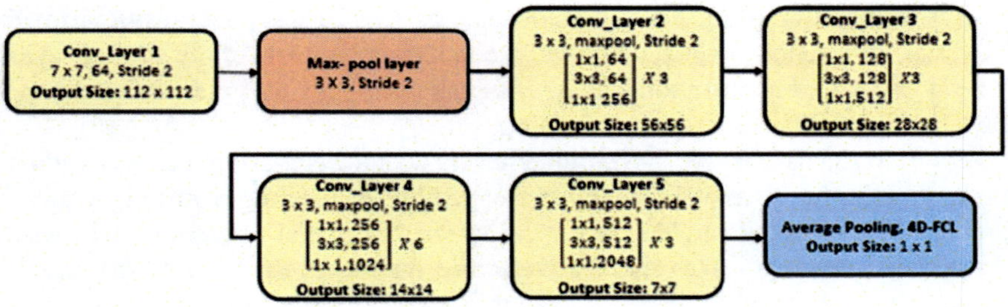

Fig. 1 Architecture of ResNet-50

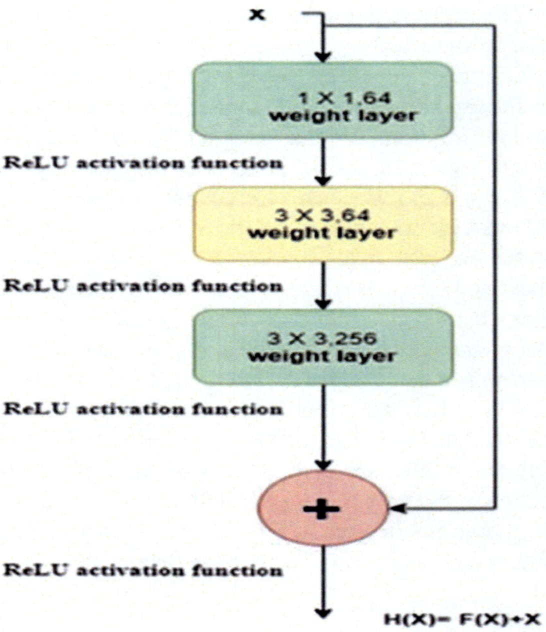

Fig. 2 Residual block of ResNet-50

is reduced to 0. ResNet learning concept attempts to resolve this issue. In ResNet, residual block with skipped connection is used to rectify the problem vanishing gradient as shown in Fig. 2.

The residual block consists of stacked convolution layer, as 1×1 convolution for reducing the dimensions, 3×3 convolution for feature extraction and 1×1 convolution layer for increasing the feature dimensions. Here, the outcome of every residual layer undergoes convolution with the input of the subsequent layer. Consider $H(x)$ be the residual mapping for building the residual block. The residual block determines

$$H(x) = F(x) + x \quad (1)$$

The formulation of $F(x) + x$ is predictable by feed forward neural systems with "shortcut connection", which combines the inputs and outputs of the stacked layers via identity mapping operation with no extra parameter. So, the gradients can simply flow back, which leads to quicker training.

Even thousands of layers can be trained easily with ResNet-50 architecture without major training error as it is having the capability of tackling vanishing gradient problem. Thus, ResNet architecture will increase the neural network performance. As having these qualities of ResNet architecture, it's variant ResNet-50 that was used in our proposed model for better efficiency.

3 Literature Survey

The literature survey has been done on state-of-the-art automatic TSRD systems that use different machine learning and deep learning (DL) techniques built using different benchmark datasets and data set collected in real time. This section also focusses on existing works on image pre-processing techniques and also works that focus on Indian traffic signal detection.

Alturki, A. S. focussed on developing TSDR using Fuzzy Neural Network for traffic sign recognition and Adaptive Thresholding that uses artificial neural network (ANN) and support vector machine (SVM) as classifiers trained on German Traffic Sign Recognition Benchmark (GTSRB) dataset [10]. Author Satılmı.s et al. developed Convolution Neural Network model for TSDR used for mini autonomous vehicle by training the model to identify a required region of interest (RoI) under various perspectives such as different backgrounds, lighting and occlusion. The model was trained on their own created dataset [11]. The detection of traffic sign in outdoor environment includes lighting, occluded, oriented traffic signs which need to be tackled for Advanced Driver Assistance Systems (ADAS). ADAS is established for providing essential data to the drivers by using genetic algorithm for recognition and CNN for classification of traffic symbols [12].

Like traffic sign detection the road lane detection is also an important scenario that needs to be addressed in Intelligence-based vehicle system. Even the road lane detection is more complicated than traffic sign detection due to its internal and external factors like road quality, heavy traffic road, weather condition and falling trees, vehicle shadows on road. Author Toan Minh Hoang et al. proposed a Fuzzy system with line segment predictor algorithms for marking road lanes [13]. Detection of small traffic symbols with good accuracy using multi-scale region-based CNN on Tsinghua-Tencent dataset which consists of 100k small traffic sign images [14, 15].

The traffic sign detection with the elimination for false detected region in Region Proposal Window using Histograms of Oriented Gradients-Boolean CNN. This method is evaluated on real-time environment [16]. The author Guan [17] established a framework for examining traffic signs from mobile Light Detection and Ranging (LiDAR) point clouds as well as digital images. The traffic signs are predicted as mobile LiDAR point clouds on the basis of valid road data as well as size of traffic sign and segmentation using digital image presentation, and the provided images are categorized automatically when completing the normalization.

Along with the traffic sign, the text in English, Chinese character in sign board is trained under the application of regional depth CNN. The Chinese traffic sign dataset and real-time traffic sign captured images are used for examination purpose that has accomplished with maximum accurate recognition rate [18]. The author of the paper [19] developed a traffic sign prediction approach on the basis of capsule network to handle pose and scale-oriented images. This capsule network models yielded better efficiency over traditional CNN on GTSRB dataset. The automatic sign board detection using RGB colour segmentation shape matching was used for prediction followed by SVM classifier for sign classification on Malaysian traffic sign dataset [20].

The Indian traffic sign detection using CLAHE, Haar feature methods focuses on Indian speed limit sign on Laboratory for Intelligent and Safe Automobiles-Traffic Sign an Indian traffic sign dataset [21]. The CNN-based Keras like conv2D, max-pooling layer used for feature extraction followed by fully connected layer as classifier are used for Indian traffic sign detection. For training the CNN model, GTSRB dataset has been used [7]. For India traffic sign detection, speed up robust feature (SURF) is used to extract the traffic sign feature. Nearest neighbour matching-based recognition technique is used for measuring the similarity of extracted feature with Indian Traffic sign Database to classify the class [22].

The next section gives a detailed description of the proposed deep learning-based residual network for traffic sign detection and classification and the experimental analysis carried out.

4 Proposed DLRN-TSDC Model

The proposed DLRN-TSDC model works in two folds, namely detection and classification of traffic signs captured in the real traffic scene, as shown in Fig. 3. The traffic sign in the scene is detected using colour space threshold segmentation technique and pre-processed using three strategies, namely clipping of edges, normalizing the size of the image and improvement of image quality. Finally, ResNet-50 classifies traffic signs. The performance of the DLRN-TSDC model has been experimented with using the German Traffic Sign Recognition Benchmark (GTSRB) dataset [23] containing 51,839 images. A set of 39,209 and 12,630 images are placed in the training and testing dataset as 75 and 25% of the entire dataset.

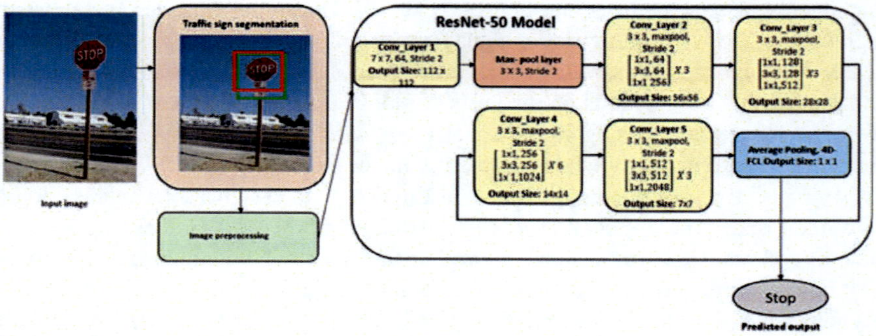

Fig. 3 Working of proposed DLRN-TSDC model

4.1 Traffic Sign Detection

The identification of traffic signs targets at extracting the concerned traffic sign areas from the given test road traffic images. The quality of the test image is usually will not be clear due to the internal and external factors described in Sect. 1. In spite of quality of the captured image, the segmentation needs to be accurately done for better classification. The effective way to segment the traffic sign from the whole image is to consider the shape and colour of features extracted from the traffic signs. This is due to the fact that the traffic signboards are mostly classified into three categories, namely regulatory sign, warning sign and information sign. The shape of the traffic boards will usually be circle, triangle, inverted triangle and rectangle and will mostly be in red, white, blue and yellow colours as shown in Table 1.

Colour is a major characteristic of any traffic sign, and it can be easily determined by the process of colour segmentation. On comparing the RGB and HSI colour spaces, the HSV is found to be beneficial in terms of detection speed. It defines the points in the R, G and B colour space using an inverted cone. Firstly, H represents the variation of colour of the image. The location of spectra colour is indicated by the angles, and diverse colour values signify distinct angles. The angle of R is 0°, G is 120° and G is 240°. Here, S defines the portion of the present colour clarity to the highest clarity with the higher and lower values of 1 and 0, respectively. Besides, V indicates the variation in the brightness of the image. The higher value of 1 defines the white

Table 1 Traffic sign information based on shape and colour

Sign type	Shape	Colour	Samples traffic signs
Mandatory/regulatory signs	Circular	Red, blue	
	Inverted triangle	Red	
	Octagon	Red	
Cautionary/warning signs	Triangle	Red	
Informatory signs	Rectangle	Blue	

colour, and the lower value of 0 indicates the black colour. In the applied HSV colour space, V is a predefined value, whereas H and S are distinct, the HSV colour space has effective brightness ability compared to the variation in brightness condition, and it has low computation complexity. The usual traffic sign colours are red, white, blue and yellow. For satisfying the intended needs of segmentation, it is essential to find the respective ranges of threshold values. The HSV colour segmentation threshold values [24, 25] for the colour red H should be greater than 0.90, S should be greater than 0.40, V should be greater than 0.35, for yellow colour, the H value ranges from 0.50–0.70, S should be greater than 0.40, V should be greater than 0.40, and for colour Blue 0 H value has to range from 0.09–0.18, S should be greater than 0.35, V should be greater than 0.40.

The colour of the traffic signs is mostly identical which makes the segmentation difficult which can be overcome by using binary image with threshold coarse segmentation technique. So, filtering the interferences is required for obtaining proficient identification of RoI [24]. Contour filtering is carried out by investigating the contour examination of the connected regions. The circumference of contours in the connected region is estimated and compared with natural circular marks. Hence, contour that satisfies the requirements are considered, and the remaining is eliminated. This helps in revealing the shape of the test traffic sign board image despite its colour.

4.2 Image Pre-processing

The RoI in traffic sign does not appear exactly in the middle of the image and few background details also exist surrounding the traffic sign. Due to the variation in illumination, the unwanted interference regions increase the complexity and reduces the detection rate. So, pre-processing is needed and is performed at three levels, namely edge clipping, image improvement and normalization. Edge clipping is a significant task in which the irrelevant edge background is removed by bound boxing the RoI. The image quality is improved by eliminating the noise using direct grey scale mechanism, and finally, the image is normalized with the dimension 32×32.

4.3 Sign Classification

Finally, the ResNet-50 model is trained for the classification of traffic signs. The proposed DLRN-TSDC can detect the different general traffic signs and Indian traffic sign images. For the feature extraction and classification of traffic signs, the ResNet-50 model has been trained on German Traffic Sign Recognition Benchmark dataset. Due to the unavailability of sufficient Indian traffic signs images in dataset and also Indian Traffic signal are very similar to the Traffic Signal of United Kingdom, the German Traffic Sign Recognition Benchmark dataset has been chosen. Yet, the

GTSRB dataset is too small and very imbalanced even though, it is a good benchmark dataset for traffic sign detection and recognition for computer vision algorithms. Additionally, the GTSRB dataset augmentation is done in terms of zoom in, zoom out, change in lighting, rotation of images in few degrees to increase the size of dataset, in the way the model can be generalized better. German Traffic Sign Recognition Benchmark (GTSRB) dataset [23] containing 51,839 images in that 39,209 and 12,630 images is placed in the training and testing, respectively. The last layer of pre-trained ResNet-50 model was removed, and SoftMax layer was added on the top of the classifier. All the images used for training are resized into 32×32 dimension. In spite of GPU memory allows larger batch size, the batch size of 256 is found to give optimized result after several experimental analysis and the learning rate of 0.01 was fixed.

4.4 Performance Evaluation

The performance of the DLRN-TSDC model has been experimented using German Traffic Sign Recognition Benchmark (GTSRB) dataset [23] containing 51,839 images. A set of 39,209 and 12,630 images are placed in the training and testing were used. Each image data represents a single traffic sign. The image sizes might be uneven and the traffic signs do not consistently appear at the fixed point of the image. Few sample images from the GTSRB dataset are illustrated in Fig. 4.

Figure 5 shows the sample images with ground truth and detected Indian traffic signs [21]. The green box depicts the ground truth values and the red box detects the traffic signs by the presented model.

The proposed model is evaluated on GTSRB dataset images and Indian traffic sign images using Intersection over Union (IoU), precision, recall and accuracy metrics. IoU targets at finding the correctly detected traffic signs, whereas the precision, recall

Fig. 4 German data set sample images

Fig. 5 Sample images of Indian traffic signs

and accuracy are used to evaluate the classification of the proposed model. The IoU, precision, recall and accuracy are calculated using Eqs. (2–5) given below.

$$\text{IoU} = \frac{\text{Area of overlap}}{\text{Area of union}} \tag{2}$$

$$\text{Precision} = \frac{\text{True positive}}{\text{True positive} + \text{False positive}} \tag{3}$$

$$\text{Recall} = \frac{\text{True positive}}{\text{True positive} + \text{False Negative}} \tag{4}$$

$$\text{Accuracy} = \frac{\text{True positive} + \text{True Negative}}{\text{True Positive} + \text{True Negative} + \text{False Positive} + \text{False Negative}} \tag{5}$$

In Eq. (1), Area of overlap indicates the overlapped area between the ground truth bounding box with predicted bounding box and the area of union indicates area of both ground truth and predicted bounding box.

Table 2 and Figs. 6 and 7 show the comparative results analysis of the proposed model with that of the state-of-the-art models, namely MLP [3], INNLP + INNC [4], GF + HE + HOG + PCA [5] and YOLO V3 [6] models. While observing

Table 2 Comparison of existing with proposed model

Methods	IoU	Precision	Recall	Accuracy
DLRN-TSDC	89.56	98.76	98.92	98.84
MLP	81.73	92.68	93.74	95.90
INNLP + INNC	86.63	98.23	98.41	98.53
GF + HE + HOG + PCA	87.61	98.25	98.43	98.54
YOLO v3	83.32	92.20	91.90	92.10

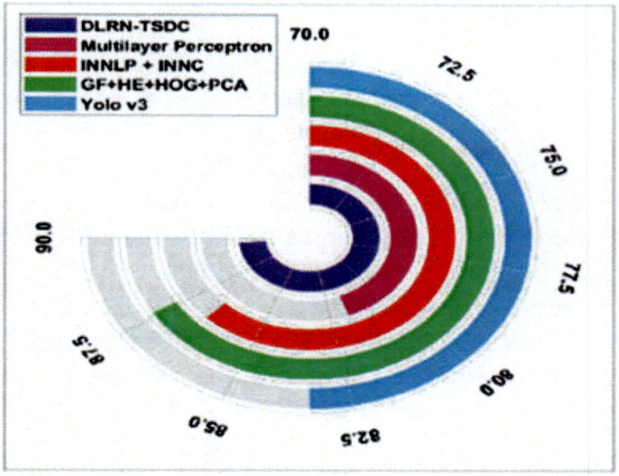

Fig. 6 IoU comparative analysis

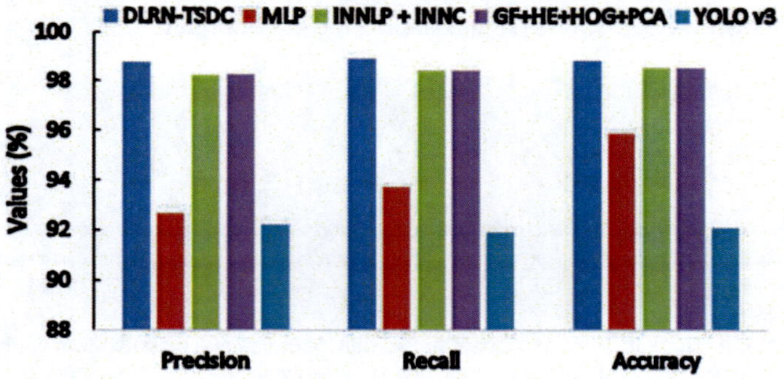

Fig. 7 Comparative analysis accuracy

the IoU and accuracy values, the MLP model has yielded the least IoU value of 81.73% and Accuracy value of 95.9%. As MLP has limited number of layers due to that the better learning may not be possible and therefore it may lead to false or missed classifications. Net to that the YOLO v3 model has accomplished a moderate outcome with the IoU of 83.32% and accuracy of 92.1%. YOLO v3 model was able to detect the sign at good speed but accuracy lacks due to single stage detection.

Though the GF + HE + HOG + PCA model has shown competitive IoU of 87.61% and better accuracy of 98.54%, the detection speed is very low, the presented DLRN-TSDC model performs better than all other models. The presented DLRN-TSDC technique has accomplished a maximum IoU of 89.56% and accuracy of 98.84% because the model was trained with deep number of layers with residual block to avoid vanishing gradient problem. This nature makes the model to learn the parameter efficiently, thereby making the model to perform better. Hence, the DLRN-TSDC model with ResNet-50 architecture outperformed than other models.

5 Conclusion

This paper has presented an DLRN-TSDC deep learning model for TSDR. Firstly, the traffic sign in the image captured is detected using colour space threshold segmentation technique. Secondly, the detected traffic signs are pre-processed in distinct, namely clipping of edges, image quality improvement and normalizing the size of the image ways to improve quality. Finally, ResNet-50 model is employed for the classification of traffic signs and determine the final class label of the traffic sign board. An elaborate experimental analysis was carried out to validate the effective performance of the DLRN-TSDC model using GTSRB dataset images and Indian traffic sign images. The obtained experimental values reveal that the proposed model with the maximum precision, recall, IoU and accuracy of 98.76%, 98.92%, 89.56% and 98.84% due to the fact that DLRN-TSDC model was well trained and tested on GTSRB dataset with very deep ResNet-50. In future, the proposed DLRN-TSDC model can be applied in smart vehicle system, driver assistance system, road safety system, navigation guidance system in real-time environment.

References

1. B. Cyganek, Intelligent system for traffic signs recognition in moving vehicles, in *Lecture Notes in Computer Science* (including subseries *Lecture Notes in Artificial Intelligence and Lecture Notes in Bioinformatics*), 5027 LNAI (2008), pp. 139–148. http://doi.org/10.1007/978-3-540-69052-8_15
2. Y. Saadna, A. Behloul, An overview of traffic sign detection and classification methods. Int. J. Multimedia Inf. Retrival **6**(3), 193–210 (2017). https://doi.org/10.1007/s13735-017-0129-8
3. S.E. Gonzalez-Reyna, J.G. Avina-Cervantes, S.E. Ledesma-Orozco, I. Cruz-Aceves, Eigen-gradients for traffic sign recognition. Math. Probl. Eng. **2013**, 364305 (2013)
4. M. Mathias, R. Timofte, R. Benenson, L. Van Gool, Traffic sign recognition—How far are we from the solution?, in *Proceedings of the 2013 International Joint Conference on Neural Networks (IJCNN)*, Dallas, TX, USA (2013), pp. 1–8
5. S. Vashisth, S. Saurav, Histogram of oriented gradients based reduced feature for traffic sign recognition, in *Proceedings of the 2018 International Conference on Advances in Computing, Communications and Informatics (ICACCI)*, Bangalore, India (2018), pp. 2206–2212
6. A. Shahzad, M. Azeem, M.S. Nazir, X.V. Vo, N.T.M. Linh, N.M.Z. Pastor, S. Dhodary, S. Dakua, S. Umeair, F. Luo, J. Liu, M. Faisal, H. Ullah, G. Sudarmika, I. Sudirman, N. Juliantika, M. Dewi, L. Insiroh, I. Bhawa, et al., No 主観的健康感を中心とした在宅高齢者における 健康関連指標に関する共分散構造分析. E-Jurnal Manajemen Universitas Udayana **4**(3), 1–21 (2019)
7. S.R. Godbole, H.N. Janjal, D. Pawar, S.A. Kanade, A. Ghule, Performance of Keras on Indian traffic signs classification and recognition (2020), pp. 1323–1327
8. J. Cao, C. Song, S. Peng, F. Xiao, S. Song, Improved traffic sign detection and recognition algorithm for intelligent vehicles. Sensors (Switzerland), **19**(18) (2019). http://doi.org/10.3390/s19184021
9. P. Wang, W. Hao, Z. Sun, S. Wang, E. Tan, L. Li, Y. Jin, Regional detection of traffic congestion using in a large-scale surveillance system via deep residual traffic net. IEEE Access **6**, 68910–68919 (2018). https://doi.org/10.1109/ACCESS.2018.2879809

10. A.S. Alturki, Traffic sign detection and recognition using adaptive threshold segmentation with fuzzy neural network classification, in *Proceedings of the 2018 International Symposium on Networks, Computers and Communications (ISNCC)*, Rome, Italy (2018), pp. 1–7
11. Y. Satılmı¸s, F. Tufan, M. Sara, M. Karslı, S. Eken, A. Sayar, CNN based traffic sign recognition for mini autonomous vehicles, in *Proceedings of the International Conference on Information Systems Architecture and Technology*, Nysa, Poland (2018), pp. 85–94
12. A. De la Escalera, J.M. Armingol, M. Mata, Traffic sign recognition and analysis for intelligent vehicles. Image Vis. Comput. **21**, 247–258 (2003)
13. T.M. Hoang, N.R. Baek, S.W. Cho, K.W. Kim, K.R. Park, Road lane detection robust to shadows based on a fuzzy system using a visible light camera sensor. Sensors **17**, 2475 (2017)
14. Z. Liu, J. Du, F. Tian, J. Wen, MR-CNN: a multi-scale region-based convolutional neural network for small traffic sign recognition. IEEE Access **7**, 57120–57128 (2019). http://doi.org/10.1109/ACCESS.2019.2913882
15. P. Saranya, S. Prabakaran, Automatic detection of non-proliferative diabetic retinopathy in retinal fundus images using convolution neural network. J. Ambient Intell. Hum. Comput. (2020). https://doi.org/10.1007/s12652-020-02518-6
16. Z.T. Xiao, Z.J. Yang, L. Geng, F. Zhang, Traffic sign detection based on histograms of oriented gradients and Boolean convolutional neural networks, in *Proceedings of the 2017 International Conference on Machine Vision and Information Technology (CMVIT)*, Singapore (2017), pp. 111–115
17. H.Y. Guan, W.Q. Yan, Y.T. Yu, L. Zhong, D.L. Li, Robust traffic-sign detection and classification using mobile LiDAR data with digital Images. IEEE J. Sel. Top. Appl. Earth Obs. Remote Sens. **11**, 1715–1724 (2018)
18. R.Q. Qian, B.L. Zhang, Y. Yue, Z. Wang, F. Coenen, Robust Chinese traffic sign detection and recognition with deep convolutional neural network, in *Proceedings of the 2015 11th International Conference on Natural Computation (ICNC)*, Zhangjiajie, China (2015), pp. 791–796
19. A.D. Kumar, K. Karthika, L. Parameswaran, Novel deep learning model for traffic sign detection using capsule networks. arXiv 2018, arXiv:1805.04424
20. S.B. Wali, M.A. Hannan, A. Hussain, S.A. Samad, An automatic traffic sign detection and recognition system based on colour segmentation, shape matching, and SVM. Math. Probl. Eng. (2015). https://doi.org/10.1155/2015/250461
21. M. Indumathi, Detection of Indian traffic sign. **2**(10), 184–189 (2016)
22. A. Alam, Z.A. Jaffery, Indian traffic sign detection and recognition. Int. J. Intell. Transp. Syst. Res. **18**(1), 98–112 (2020). https://doi.org/10.1007/s13177-019-00178-1
23. https://www.kaggle.com/meowmeowmeowmeowmeow/gtsrb-german-traffic-sign
24. J. Cao, C. Song, S. Peng, F. Xiao, S. Song, Improved traffic sign detection and recognition algorithm for intelligent vehicles. Sensors **19**(18), 4021 (2019)
25. P. Saranya, S. Prabakaran, R. Kumar et al., Blood vessel segmentation in retinal fundus images for proliferative diabetic retinopathy screening using deep learning. Vis. Comput. (2021). https://doi.org/10.1007/s00371-021-02062-0

AI-Based Automated Fruits and Vegetables Quality Inspection for Smart Cities

Syed Sumera Ali and Sayyad Ajij Dildar

Abstract The adoption of a technology in the food industry is currently occurring with artificial intelligence [AI]. This COVID-19-induced crisis has caused some disruption in selling the food to customers. It is increasingly apparent that, food system was "anti-fragile." Home cooking, meal-kit movement, home delivery, met shops, canteens, etc., all these get shutdown in this pandemic. With automated, the food supply digital technologies like robots, AR, VR, printers, sensors, machine vision, drones, blockchain, IoT, and artificial intelligence are used. Artificial intelligence (AI) refers to the collection of data from sensors and its conversion to comprehensible information. AI can interpret information reducing their need to be involved. AI can also be self-learning and progress beyond human abilities. The use of AI to advance food production is accelerating as the world progresses in post-COVID and expectations of speed, efficiency as well as sustainability are ever-increasing alongside the rapidly growing population. Factors influenced the food sector, where AI has increased their development or even modify the way they work are discussed in this study.

1 Introduction

The evaluation of food quality for inspection process is done by incorporating the imaging methods. The main aim is to obtain an image or several images by using single camera or multiple cameras. The creation of the sensor with high quality gives an enhanced image resolution and quality, and increment in the power computed is necessitated to produce a lot of novel best-suited algorithms, where it employs the image tracking models. The food quality is ensured, when the safety estimation motivates the methods to replace conventional methods with the advanced methods as they are less inconsistent and ineffective. The conventional methods examine the

S. S. Ali (✉) · S. Ajij Dildar
Department of E & TC, Chh Shahu College of Engineering, Aurangabad, Maharashtra, India

S. Ajij Dildar
Department of E & TC, MIT College of Engineering, Aurangabad, Maharashtra, India

P. Karuppusamy et al. (eds.), *Ubiquitous Intelligent Systems*, Smart Innovation, Systems and Technologies 243, https://doi.org/10.1007/978-981-16-3675-2_6

food after a long research. By processing the food quality along with protection, the advanced approaches are required to evaluate the resources used for the food and components at every stage. The enhanced agriculture and food protective issues have been appeared recently with the excess addition of preservatives and toxic residues and creates harmful chemicals through procedures. The approaches have been created for the quality estimation and measurement to conquer the drawbacks of the conventional approaches for spectroscopic and imaging methods. The food structure evaluation employs spectroscopic approaches such as Raman spectroscopy with Fourier transform nuclear magnetic resonance and near-infrared spectroscopy. Furtherly, these approaches are employed in the evaluation of food possessions such as salt, protein, fats, moistures, by providing more accuracy and present linear allocation of quality attributes that is more important for the inspection of food protocols. The technique usually works to achieve linear details, and another method created here is called computer vision. The linear details are obtained from the food's digital image with texture, shape, size, and color. **Automation** increases the expectation in food quality and safety standard. This is being assessed through visual inspection by human inspector, which was tedious and time consuming and affects the evaluation process, also poor quality control, high incidence of industrial accidents. The need to automate industrial processes is driven by several key requirements for competitive success such as improve productivity, products quality and profitability. **Quality** of the food that obtained for the inspection is by using the imaging method. Recently, the quality assessment of the food products includes utilizing some physical techniques. The inspection of food quality can be classified into two, namely non-destructive and destructive. The non-destructive quality checking for the food factory is enhancing the need for the agri-food factory products which is associated with the customer's physical condition as well as community encouragement. The quality of the food is ensured when the safety estimation motivates the method to replace the conventional methods. In **computer vision system**, an image or a video is taken as input, and the goal is to understand the image and its contents. CV uses image processing algorithms to solve some of its tasks. Computer vision and image processing is really a growing research area, which is significant in analyzing the techniques. Computer vision system is the suitable for conventional analysis and quality assurance.

2 Food Industry

2.1 Preprocessing

The processing of food is a labor-intensive business, but one where AI can maximize output and reduce waste, by replacing people on the line, whose only job to distinguish and identify items unsuitable for processing. Decision making of this type at speed requires the sense of sight, smell, and their adaptability to adapt to the changing circumstances. AI brings even more to the table through augmented vision, analyzing data streams either unavailable through human senses, or quantities of data that are overwhelming. Organizations such as TOMRA is already begun to incorporate AI technology into their production processes by including innovative sensor-based sorting machines, detecting, and removing any types of foreign materials from their lines of production, reacting to changes in moisture levels, colors, smells, and tastes of foods.

2.2 *Food Safety*

Reducing the presence of pathogens and detection toxins in food production is a key avenue for AI. The Luminous Group, a Newcastle-based software firm, is developing AI to help prevent outbreaks of pathogens in food manufacturing plants, limiting consumer illness or recalls. Additionally, AI offers the opportunity to increase traceability and consequently, consumer confidence, for example, a KanKan subsidiary consisting of AI-enabled cameras in Shanghai's municipal health agency checks that workers are complying with the safety regulations [1]. This algorithm-based machine learning technology includes facial and object recognition, and "sets the foundation (…) to potentially triple [their] business with the city of Shanghai" [2]. More recently the company added improved facial recognition abilities to account for the mandatory use of a mask, and new body temperature detection, in line with effects of COVID, as detecting increased body temperatures could help in the early detection of a COVID case [3]. This ever-changing project shows an ability to constantly grow and develop a flexibility required today in the world of technology.

2.3 *Supply Chain Efficiencies*

The wave of popularity of food delivery is now incorporating AI to make "recommendations for restaurants and menu items, optimize deliveries," as well as looking into the use of drones. They use Michelangelo, a machine learning platform, for various different tasks. COVID-19 has accelerated the applications of technology to replace human labor, and while smart device food apps, drone and robot delivery, and driverless vehicles all provide new ways to get information and food to the consumer, all of them depend upon AI. Using AI in food supply chain increases productivity and improves the accuracy of information for better decisions [4]. Innovative uses of AI are crucial in moving toward reducing the quantity of food wasted in order to feed the growing world population as efficiently as possible, as well as falling in line with increasingly specific consumer demands and expectations.

2.4 Predicting Consumer Trends and Patterns

AI allows companies to stay competitive within the market, by adapting based on different popular waves of various trends, making predictions about the market. Their data collected includes "up-to-the-minute industry insights, predictions, and emerging food trends based on analysis of billions of social media posts and photos, US restaurant menus, reviews, and recipes."

2.5 Restaurants

The future of restaurants is in peril following this year's COVID-19 outbreak. The explosion of online-based food delivery systems has decreased the focus on the physical experience of restaurants, for example, chat boxes can allow communication with your favorite restaurant without leaving the comfort of your home, all powered by AI. Voice search is another tool useful to allow people to place restaurant orders simply by talking to their screen. AI analytical solutions such as these lead to better consumer experiences and likely to increase sales for restaurants due to the ease with which food orders can be placed. AI is used to increase efficiency and lower costs during the process of food delivery, encouraging restaurants to partner with these companies to ensure the delivery of their food. Automated customer service and segmentation will likely lead to increased accuracy in "creating reports, placing orders, dispatching crews, and formulating new tasks [5]" in a restaurant.

2.6 Designing Better Foods

Food is health has been a mantra for many for a long time, but now with a greater understanding of both human, plant and animal genomes, and it is becoming a reality. Changes in consumer preferences are creating opportunities for AI in food; an example is the growing demand for plant-based alternatives to meat protein, as the world moves toward precision nutrition. Challenges such as achieving consumer acceptable taste and texture qualities have led to creative AI applications. NotCo is a plant-based startup company located in Chile which has been developing its own software company "Giuseppe" a tool used to "predict how to make plant-based materials taste like animal-based products" [6].

Additionally, AI requires skilled IT professionals, which are high in demand and difficult to recruit [7]. Clearly there are costs to retraining programs to adapt to the change in skills required. Finally, the cost of implementing and maintaining AI is very high, which may limit the opportunities for smaller or startup business to compete with already established larger ones.

3 Visual Inspection for Food Products

The purpose of performing a visual inspection for identifying food are …

Determining food or equipment is clean. Changes in packaging has occurred between production runs or not checking raw materials have been stored correctly. Visual inspection is the oldest and most basic method of inspection. It requires no equipment but the naked eye of a trained inspector. It is independent food inspection to examine food in production as a form of quality control.

There are various inspection methods.

Penetrant inspection, hardness inspection, eddy current inspection.

X-ray inspection, computed tomography, visual inspection, ultrasonic testing, magnetic particle inspection. Different ways of sorting …

– Manual Sorting – Conveyer Assisted Sorting – Automated Sorting M/C

4 Visual Food Quality Inspection Using Computer Vision

Use visual inspection of food using computer vision which is smooth and unlike leafy vegetables, which is easier to scan. For solving food products problem, use sorting machine method. All batch of vegetable is collected and put over the warehouse and passed through machine. Onward, the machine performs many steps as follows.. capturing images—scanning images—grading—whereas scanning and grading comes under specific image-based recognition techniques. Defect pattern recognition on vegetable or fruits is possible using machine learning.

5 Artificial Intelligence Food Processing

Artificial intelligence (AI) is being used in the food processing and handling (FP & H) field of the economy. AI has a direct and indirect effect on the FP & H industry.

6 AI in Food Processing and Handling

Food processing is a business that entails sorting farm-fresh food and raw materials, as well as maintaining machinery and different types of equipment. When the final product is ready to ship at the destination end point, the consistency of the product is manually checked, and the decision is made whether or not it is ready to ship.

7 Food Processing

The digital image processing is a promising technology in the agricultural and the food sector, which employed for online quality computation of different food items like fruits, vegetables, fishes, meats, grains, rice, canned food, etc. In recent years, more researches were done on food products….

– Fruit processing methods
– Vegetable processing methods
– Grains quality processing approaches
– Other food processing approaches

FRUITS AND VEGETABLES

Processed fruits and vegetables play vital role in food industry, few are the food challenges.

- Food sufficiency concerns with the agricultural land and its availability
- Food quality concerns with the safety and hygiene food that may be nutrition
- Environmental concerns with the sustainable food production for smart city
- Holistic in food systems on an end-to-end basis
- Focus on local actions to change country international background.

Four major factors play a role in the growth of the food processing sector:

- Domestic demand
- Supply-side advantages
- Export opportunities
- Proactive government policy and support.

Key opportunities in Food Processing

- Technology transfers to reduce wastage
- Aqua-horticulture
- Fruits and vegetable processing
- Processed fruit-based ingredients
- Export potential of processed fruits and vegetables
- Canning, dehydration, pickling, provisional preservation, and bottling.

8 Smart Cities

For the developing any country basic pillar is city must smart to provide the aspirations and needs of the citizens, urban planners. The ecosystem represents by the four pillars for development-institutional, physical, social, and economic infrastructure. This may be long-term goal to develop any cities increasing infrastructure which adds "smartness."

9 Significance and Necessity of Research

In the past few years, many researchers have proposed various methods to solve different problems of food quality inspection and classification in terms of food, quality, automation, computer vision, machine learning, and AI.

For better improvement in terms of qualitative and safe food, we need to take care about food texture, color, size, etc. By adopting new technology and method, good quality food products with quality inspection is possible, which improves efficiency through image processing, recognition, analysis in food industry.

10 Research Motivation

In recent days, the quality of the food is based on the processing concepts and the need for the development of definite quality statements to the nature of the food as well as agricultural food products. Also, it is necessary to take care of the food products in prolonged quality which is aimed at the increasing population. The motivation of this research study is not only to improve the quality of the food but also to classify the quality food and defected food.

11 Identification of Problem

It is observed that in image processing several methods have been described but there are some problems observed in manual/traditional computer vision system like consumption of time is more, computational complexity and the cost of the system are high, real-world problems cannot be solved, online inspection of quality attributes is to be achieved, wavelength is to be reduced without performance loss, robustness has to be increased, and counting off on tree fruits is difficult in the agricultural field. Hence, performance of food products quality inspection system degrades. These are the problem which are identified from previous research. Recently, there is an enormous improvement in the food quality inspection on behalf of the technology. It is important to provide the quality food to the customers by inspecting the quality of the food and also to classify the quality food and the defected food. For this image processing, several methods have been described. There are some problems by using those techniques. The major problems are listed below:

- The noises are perfectly removed in the preprocessing by using the proper approach
- Perfectly analyze the segmentation methods to detect the images
- To develop a light vision method to enhance the scheme using the deep learning and the image processing technique.

- To recognize the real-time quality detection and the ranking of the food on the arranging lines
- To obtain large dataset may be difficult to inspect the quality of the food image.

12 Scope of Research

The aim of research is to design a fruits and vegetable food products recognition and quality analysis system with the help of digitize images which classify the quality food and defected food. For implementation, we have considered the five different types of fruit and vegetable, i.e., tomato, potato, orange, banana, etc. In this way, an "Efficient and Optimized Fruits and vegetable food Product quality" is obtained and inspected in multilayer perception neural network classifier by implementing WOA algorithm using computer vision system. The precise, quick, and intention quality purpose of food products are important to expand. Usually, computer vision is a computerized, non-destructive, and expenditure in procedure. A computer vision is a device used in industrial and agricultural development for enhancing production, expenditure, accessibility, and algorithmic. Therefore, such disadvantages provoked me to accomplish the study work and scope in this field.

13 Research Objectives

The objective of the research is to achieve following requirement for automated system, image processing techniques in the food industry, computer vision system, various segmentation, and image features. The goal of the investigation work is to investigate and develop new techniques and methods for quality inspection of food products.

The objective of this research carried out is summarized as follows:

– Efficient quality inspection – optimizing the error – accuracy increasing
– noise removing – robustness increasing – grading andsorting

- To develop a food quality inspection using ANN classifier used to separate the quality food from defected food and used for ranking the food products.
- To inspect food quality using MLP-WOA classifier used for the classification and to optimize the error.

14 Hypothesis

The DIP is an emerging technology in the agricultural as well as the food sector, which employed for online quality computation of different food items like fruits, vegetables, fishes, meats, grains, rice, canned food, etc. In recent years, more researches were done on food products.

To design an efficient food quality inspection, to optimize the error, and to separate the quality food from defected food in real-time application are the purposes for which we capture the fruits and vegetable images from the digital camera, store them into database, read an image from database, preprocess the image, segment the image and extract the color image and store the extracted features for training. Build neural network the multilayer perception (MLP) for training and recognizing the food products and its quality. Finally, test the system by giving different types of fruits and vegetable as input.

Assumption and testing measurements are on the basis of shape, size, color, sensitivity, specificity, accuracy and error, etc....

15 Research Methodology

Following are the proposed two phase methods explained. One is ANN classifier with BPA algorithm whereas second phase is MLP classifier with WOA algorithm (Fig. 1).

The steps involved in the research methodology of proposed ANN-MLP method are of given as follows.

Preprocessing Histogram equalization.

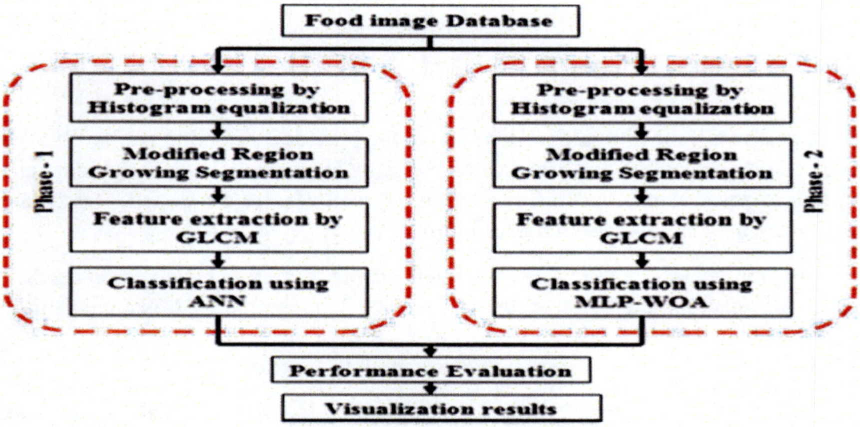

Fig. 1 Comparison of ANN and MLP classification

Segmentation	Modified growing, Enhanced growing segmentation.
Feature extraction	Histogram features, GLCM features.
Classification	ANN-BPA, MLP-WOA classification.

(A) **Implementation of Proposed System Food Quality Classification-Based MLP classifier**

Multilayer perceptron neural network is the most commonly used FFNNs. The multilayer perception consists of three layers: input layer, hidden layer, and the output layer. Here, the input layer in the MLP architecture regarding neural networks includes correlation, contrast, energy, and homogeneity as depicted in Fig. 3. There are several hidden layers in the MLP neural network and thus quality fruits, and the damaged fruits, were obtained from the output layer as shown. The neurons are interconnected, and that connected are characterized as weights that are located in the range [−1,11]. Every layer in this network is represented as,

(B) **Implemented Algorithm: Recognition and Classification of Food Products; Whale Optimization Algorithm (WOA) Technique**

Generally, the swarm intelligent optimization idea is the main element of the whale optimization method, and Mirjalili proposes it in the year of 2016. The metaheuristic whale optimization algorithm is the humpback whales characteristics. The flowchart regarding the WOA-MLP method is described in Fig. 4. Thus, the whale produces bubbles to grasp the smaller fishes. The prey exploitation and explorations are main stages of the whale optimization algorithm (Figs. 2 and 3).

Step 1 Prey encircling:
Step 2 Exploitation stage (attacking bubble net process):
Step 3 Exploration (prey searching process.

16 Design and Procedures Used

To cope up with the difficulties identified in the previously existing methods for the food quality inspection, the effective food quality inspection has to be demonstrated in the proposed work. For this, the following methods are developed. Design and procedure used in this research is as follows…

- The preprocessing of the database images using histogram equalization used.
- Enhanced modified region growing is proposed to segment the broken division of food products.
- GLCM parameters in feature extraction module are used.

The proposed ANN and next MLP classifier is used for ranking the food products….

Fig. 2 Flowchart for proposed WOA-MLP

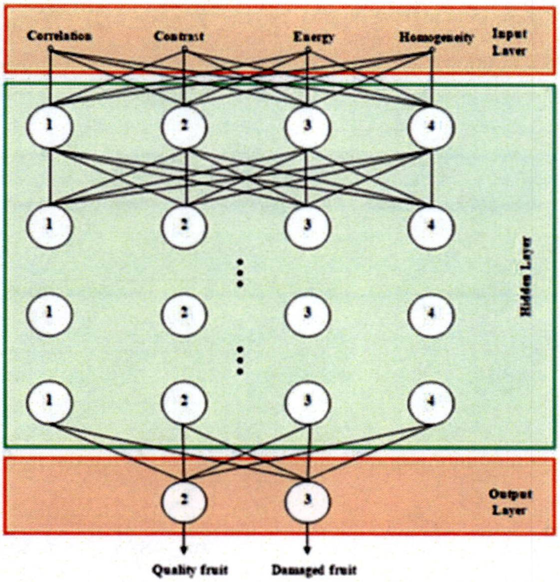

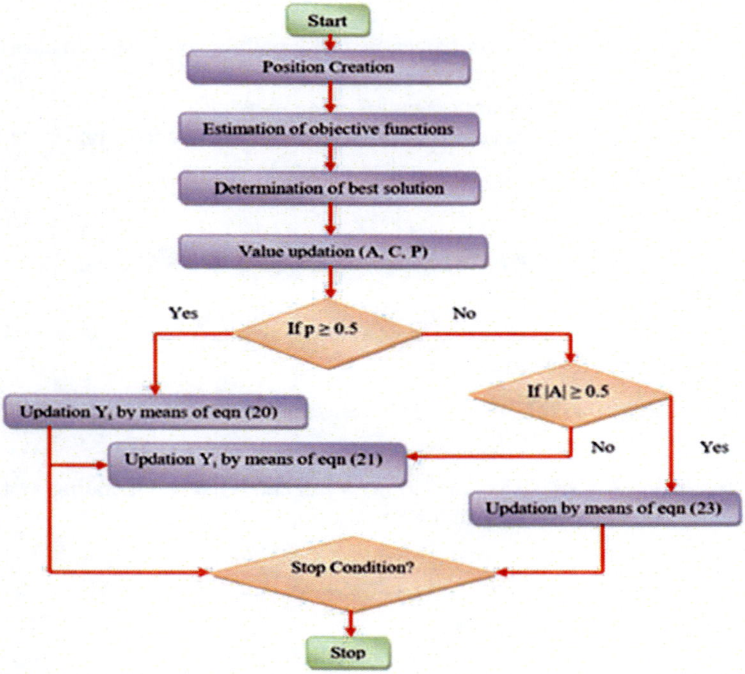

Fig. 3 Method architecture of MLP food quality

Fig. 4 Dataset food images

Food Name	Food Image	Defected Food Image
Banana		
Potato		
Pear		
Melon		
Peach		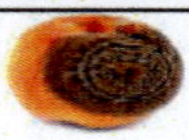
Orange		
Strawberry		

- The proposed whale optimized algorithm with multilayer perception classifier is used.
- The MATLAB is used for implementing research work.
- Many performance metrics used sensitivity, specificity, accuracy and error which were calculated.

17 System Requirements: Software and Hardware Tools

The proposed work is implemented in MATLAB, and the experiment is performed employing a system requirement.

Hardware: Camera with 8MP or higher-Pentium 4 machine or higher of 4 GB of RAM-Intel i-3 Processor, 2.10 GHz.

Software: MATLAB (R2013a and 2017a) or higher-Window Operating system as implemented in MATLAB-Window XP platform.

18 Result and Discussion

The quality of the food products has linked with their attributes such as color, shape, and texture. If the quality of the food is excellent, subsequently it assists in the preservation of food products. There are several color grading systems that help to find out the quality of the food product, and this will help to identify the quality food product from the defected food. While using this grading system, the method helps to give quality food to the customers, thereby customer satisfaction will increase. Thus, this helps in the reduction of the wastage of the food product, and thus, the computational time reduces. For the quality food inspection, the defected food is separated from healthy food by the image processing methods. Here, the seven classes of food diseases are taken for the processing of results. The image processing has four main steps: preprocessing, segmentation, feature extraction, and classification. There are two stages for the quality inspection of food in this research....

In the first stage, effective quality inspection of food processed through image processing concepts. The preprocessing done through the histogram equalization method and segmentation uses the modified region segmentation. The classifier section uses the artificial neural network classifier to classify the defected food from healthy food. Here, the parameters like specificity, sensitivity, accuracy are more compared to other methods. **In the second stage**, only the classification stage is different; instead of the ANN classifier, and MLP-WOA classifier has been used for the classification of defected food from healthy food. Here, the error value is minimized when compared to all other methods. Both approaches are used to separate the quality food from the defected food (Figs. 4, 5 and 6; Table 1).

Table 2 illustrates the performance metrics comparison with various segmentation approaches. The performance metrics such as segmented sensitivity, specificity, accuracy, FPR, FNR are used for the evaluation purpose. It gives high accuracy with low FPR, FNR when compared with other approaches.

To show that the MLP-WOA approach is the best classifier to inspect the food quality, defected foods are separated from healthy food. To make the comparison, the performance metrics are compared with the existing methods (Table 3).

The comparison of performance metrics of various classification methods with the proposed method is shown below in Fig. 3. The proposed approach provides the best result when compared with all other approaches (Fig. 7).

The next is the comparison of the performance metrics such as recall, precision, F-measure, and accuracy. FPR is compared with the various latest results compared with the proposed method. The comparison results shows high F1 and low false-positive rate, etc. Figure 2 shows the performance metrics compared with various existing results and it is plotted (Fig. 8; Table 4).

19 Result Output Compared

The results obtained by the proposed MLP-WOA method are presented. For the image database, four evolutions were performed (Figs. 9 and 10).

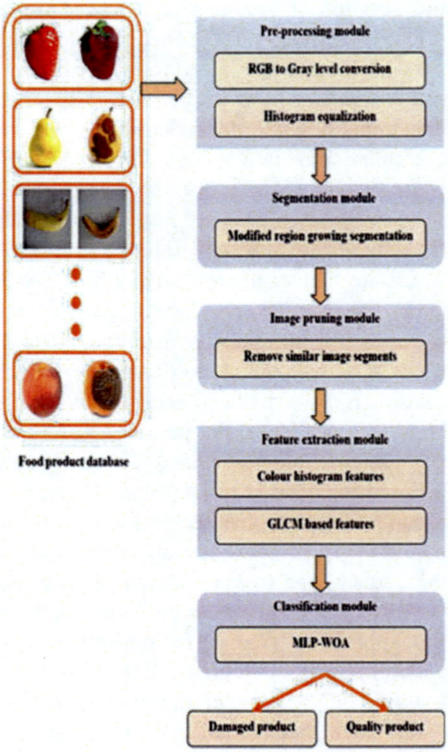

Fig. 5 Proposed MLP food quality ranking method

20 Simulation and Results

MATLAB 2017a : – Training – Testing – GUI

The food detection approach is to detect the defected food from healthy food, and it is useful for agricultural growth. The quality detection in the image processing starts from the preprocessing; the performance metrics of the various preprocessing methods are compared with the proposed histogram equalization. The second stage is the segmentation approach; here, the proposed approach is the modified region growing segmentation and this gives high segmentation accuracy when compared with all other approaches. The feature extraction depends on the color attributes and the method is the GLCM feature extraction approach. The classification techniques such as KNN, SVM, and CNN are compared with the proposed ANN and MLP-WOA approaches. The MLP-WOA approach gives the best accuracy when compared with other methods and other existing results (Figs. 11, 12, 13, 14 and 15).

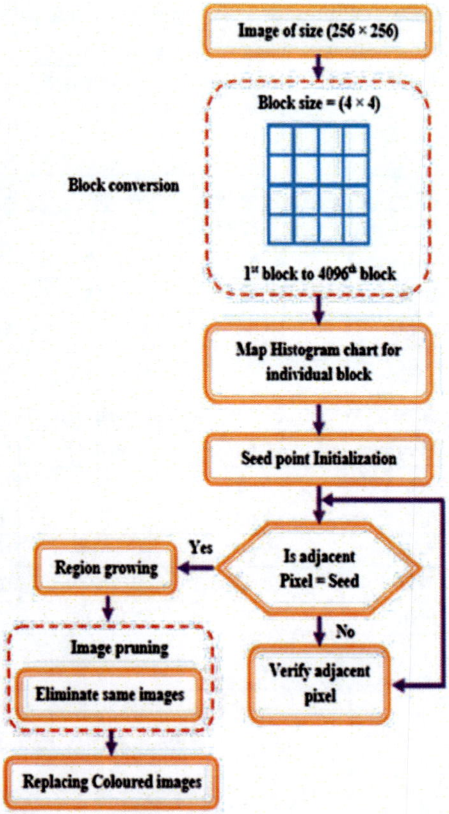

Fig. 6 Flowchart of the enhanced region growing process

Table 1 Performance metrics comparison of various image preprocessing methods

Image name	Performance indexes			
	PSNR	SSIM	Entropy	Contrast ratio
Contrast stretching	39.45	0.9384	3.6478	0.005
Global thresholding	40.98	0.9458	5.4591	0.007
Log transformation	41.56	0.9521	6.3847	0.045
Power law transformation	42.84	0.9584	4.8215	0.052
Histogram equalization (proposed)	43.82	0.9785	2.8475	0.007

21 Conclusion

This conclusion of this research work is to recognize and to address good and bad food products. The quality of the food depends upon color, texture, shape, and size of the food product. The purpose of this research is to detect defected food from

Table 2 Segmentation performance for different approaches

Segmentation performance	Segmentation method				
	Threshold method	Edge-based method	Clustering-based method	Region growing method	Modified region growing method
Sensitivity	0.8756	0.8742	0.8896	0.9877	0.9975
Specificity	0.8523	0.8412	0.8754	0.9689	0.9875
Accuracy	0.8745	0.8968	0.8985	0.9768	0.9985
FPR	0.0432	0.0345	0.0546	0.0235	0.0345
FNR	0.0456	0.0546	0.0254	0.0243	0.0267

Table 3 Performance metrics comparison of various classification schemes

Metrics	KNN	SVM	CNN	ANN	MLP-WOA
Sensitivity	0.7584	0.8219	0.8946	0.9318	0.9614
Specificity	0.8356	0.8586	0.8840	0.9543	0.9682
Accuracy	0.8934	0.8864	0.90	0.96	0.9848
FPR	0.0254	0.0289	0.0364	0.0438	0.0566
FNR	0.0261	0.0298	0.0318	0.0526	0.0587

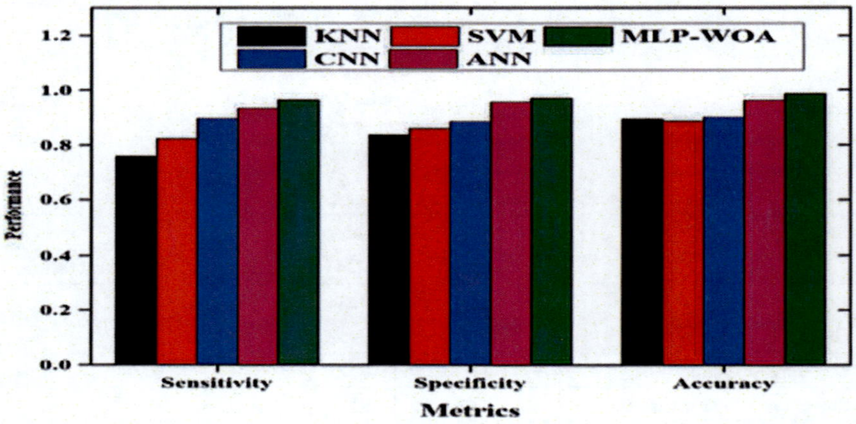

Fig. 7 Performance metrics of classification with various approaches

healthy food. Several algorithms were applied to detect defected food and also to inspect the food quality. Four steps process the food quality inspection used. These four stages are used to classify the quality food and defected food by using different image processing techniques. There are two contributions considered to detect food quality.

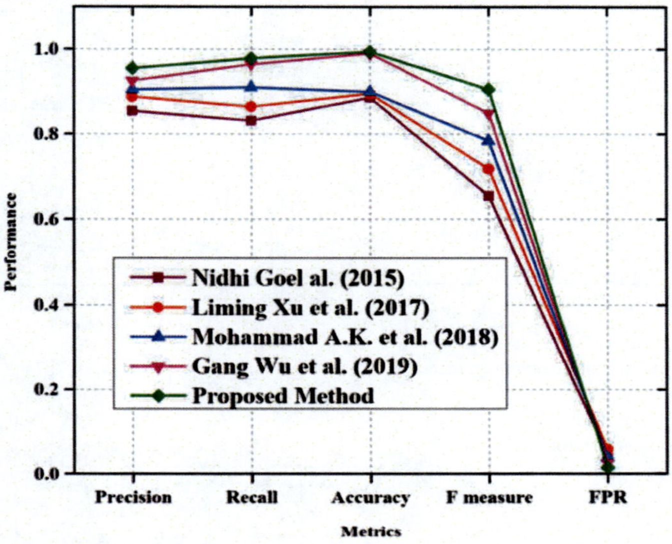

Fig. 8 Performance metrics of classification with various results

Table 4 Comparison of metrics with various existing results

References	Goel et al. (2015)	Xu et al. (2017)	Mohammad et al. (2018)	Gang Wu et al. (2019)	Proposed method
Precision	0.8542	0.8874	0.9048	0.9254	0.9548
Recall	0.8312	0.8643	0.9102	0.9645	0.9785
Accuracy	0.8856	0.8945	0.8996	0.9896	0.9942
F measure	0.6548	0.7185	0.7846	0.8487	0.9054
FPR	0.0459	0.0584	0.0385	0.0258	0.0153

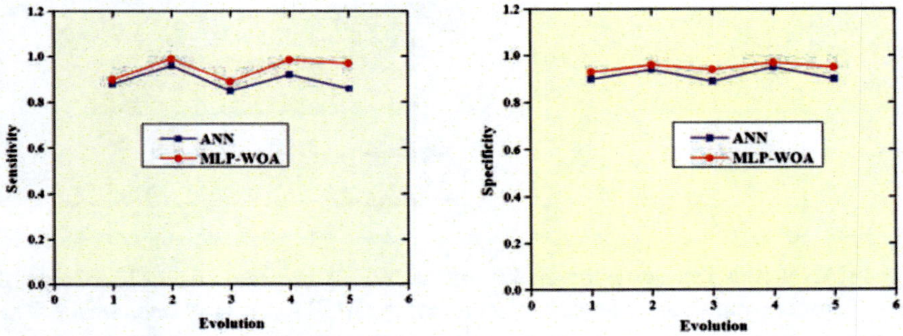

Fig. 9 Comparison of sensitivity, specificity of ANN–MLP

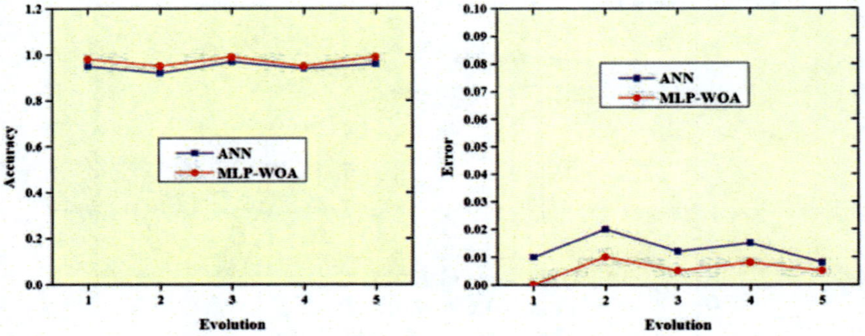

Fig. 10 Comparison of accuracy, error b/n ANN–MLP

Fig. 11 Neural network training

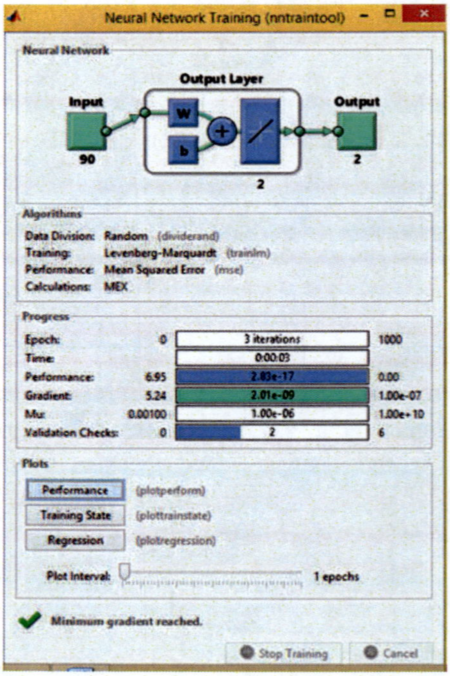

In ANN-BPA (phase I), the quality of the food is inspected using four steps of image processing. The preprocessing converts the RGB to the gray-level image. The modified region growing segmentation is the technique used for the segmentation, whereas it divides the defected food image part into several images. The GLCM is used as the feature extraction process, in which GLCM separates the two attributes such as color histogram for the entire segmentation process, whereas texture features.

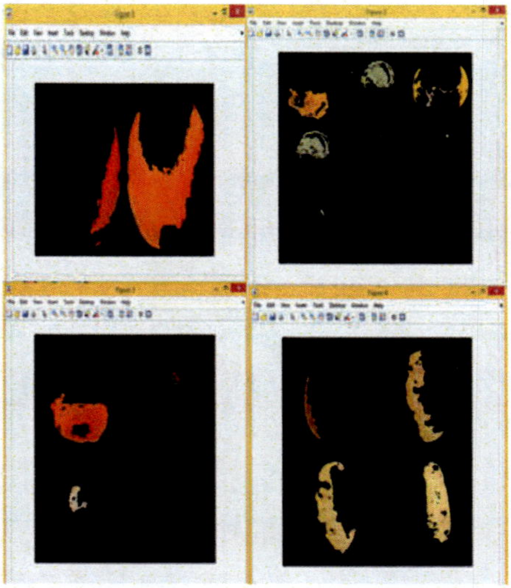

Fig. 12 Segmentation of testing data (Orange)

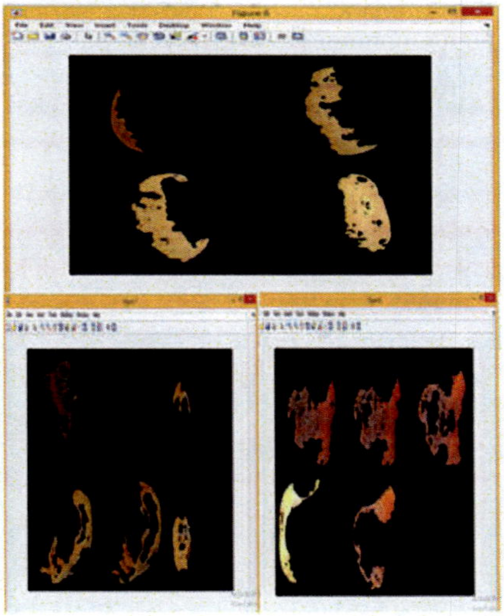

Fig. 13 Segmentation of testing data (Potato)

Finally, the classification stage, here ANN, is used to classify the defected food from healthy food. After the preparation progression is completed, then the system is capable of detecting the food images that are derived from its features. The quality of the food depends upon the grading process that gives ranks to the food based on the quality of the images that are not defective.

In MLP-WOA (phase II), the food quality detection is carried out as same as in the previous phase. The preprocessing uses the same histogram equalization method

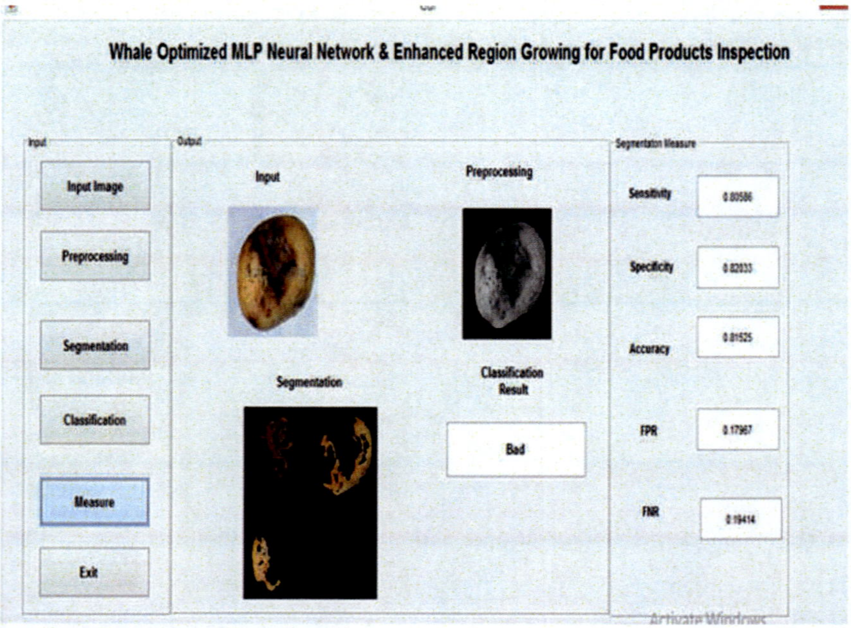

Fig. 14 GUI test data (Bad)

for the removal of unwanted noise and also to reduce computational difficulties. The contrast of the image is improved by utilizing the histogram equalization approach and the strength of the pixel is varied. Then the modified region growing segmentation divides the images into pixels for the feature extraction process. The color and texture are the two attributes that exploit the GLCM process. At last, the food product is classified into quality food and defected food by using an MLP-WOA classifier. The classifier gives better accuracy, sensitivity, specificity, and minimum mean square error when compared with all other methods.

22 Future Scope

The following are the future research directions for the proposed food quality recognition approach.

- Focus on certain faults positioning and restoration alternative to focus on discovering a universal restoring method
- Could weight every label of defects and train a system to establish the order of the restoration method.

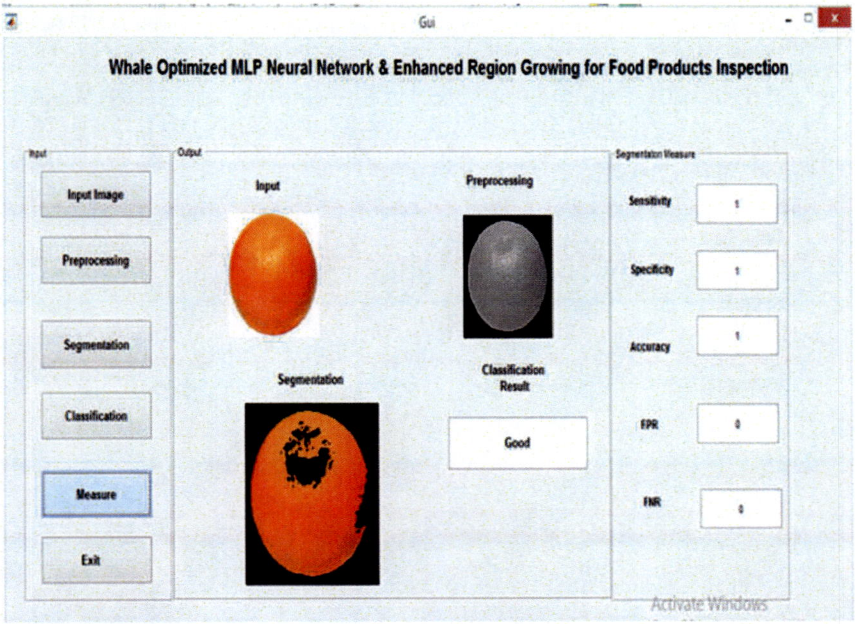

Fig. 15 GUI test data (Good)

- To consider the features such as ripening, acidity, sanitary status, and peroxides, Io estimates the quality.
- Create the best handling approaches and weather management recommendations that contain rapid cooling for sustaining post-harvesting microbial quality and the appearance of fresh fruits in minimizing costs.

References

1. https://www.prnewswire.com/news-releases/remark-holdings-announces-seven-figure-artificial-intelligence-contract-for-facial-and-object-recognition-technology-to-ensure-food-safety-in-shanghai-china-300526557.html.
2. https://www.fastcasual.com/news/restaurant-safety-check-new-ai-platform-watches-reports-violators/
3. https://www.prnewswire.com/news-releases/kankan-ai-upgrades-its-product-technology-to-provide-for-touch-free-temperature-measurement-for-mass-screening-of-high-traffic-areas-301015377.html
4. https://progressivegrocer.com/grocers-embrace-ai-optimize-supply-chain
5. https://spd.group/machine-learning/machine-learning-and-ai-in-food-industry/#Cleaning_equipment_that_does_not_need_disassembling_CIP

6. https://golden.com/wiki/NotCo, https://www.foodmanufacturing.com/home/article/13245042/artificial-intelligence-is-redefining-food-beverage-manufacturing. Aidan Connolly **Follow** Chief Executive Officer at Cainthus, President of AgriTech Capital
7. S. Ali, D. Sayyad Ajij, Whale optimized MLP neural network and enhanced region growing for food product inspection. Int. J. Adv. Sci. Technol. (IJAST) **29** (3), 11155–11174 (18 Mar 2020). http://sersc.org/jourls/inex.php/IJAST/article/view/28011/15461, http://sersc.org/journals/index.php/IJAST/article/view/28011

A Survey on Energy-Efficient Approaches in Wireless Sensor Networks

Ayan Bhuyan and Bobby Sharma

Abstract Wireless sensor networks (WSNs) have been gaining attention from both researchers and user community due to its multitudinous uses, prospects, and possibilities. The areas of application of a WSN may vary from a small scale health-monitoring system (containing a few sensors) to a large scale soil-monitoring system (consisting of thousands of sensors). The deployment of the nodes of a WSN is generally done in hazardous, hostile or hard to reach environment, which makes replacement of power units infeasible. So energy efficiency becomes a major concern in these type of networks. A wide range of literature is available proposing various schemes and protocols pertaining to energy efficiency and network longevity. This paper is intended to provide the reader with a holistic view on some major energy-efficient schemes with classifications based on the network layer affected.

1 Introduction

With the advancement in technology, both the size and cost of electronic circuits have been reduced. Technologies like micro-electromechanical systems (MEMS) have made it possible the construction of small, low-cost, multifunctional, moving sensors called nodes. Each such node is capable of sensing its environment and collecting data like humidity, temperature, pressure, sound, etc. [1, 2]. A number of nodes can be deployed over a region of interest and wirelessly interconnected to form a network of sensors termed as wireless sensor network (WSN). The data collected by these sensors is wirelessly transferred to a base station (BS) which, in turn, is processed to yield useful information to the end user. For example, soil pressure collected at various points may help in predicting earthquake or air pressure

A. Bhuyan (✉) · B. Sharma
Department of Computer Science and Engineering, Assam Don Bosco University, Guwahati, Assam 781017, India

B. Sharma
e-mail: idbobby.sharma@dbuniversity.ac.in

P. Karuppusamy et al. (eds.), *Ubiquitous Intelligent Systems*, Smart Innovation, Systems and Technologies 243, https://doi.org/10.1007/978-981-16-3675-2_7

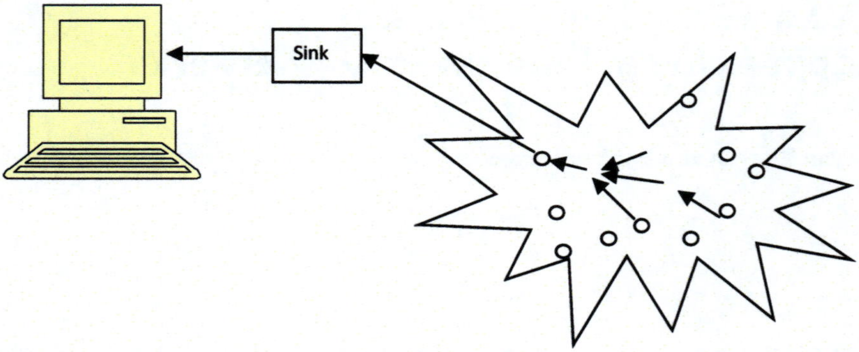

Fig. 1 Wireless sensor network

collected over a region may help in predicting weather. A simple diagram of a WSN is shown in Fig. 1.

Initially, WSNs were designed mainly for military purposes such as intrusion detection and battlefield surveillance. However, over the years, WSN has gained popularity and has found its use over a wide range of applications such as health care monitoring, air pollution monitoring, forest fire monitoring, etc. Since there are no wires involved in data transmission and the deployment of nodes is easy, these types of networks are very suitable over a large domain of applications.

WSNs unlike other traditional wireless communication networks like cellular systems and mobile ad hoc network (MANET)s have unique characteristics such as denser level of node deployment, higher unreliability of sensor nodes, and severe energy, computation, and storage constraints [1]. Also, since these nodes/sensors are usually deployed in hazardous or hard to reach environments, (e.g., battlefield, underwater, etc.), it is sometimes hard or even impossible to recharge or replace the batteries that power them.

Considering these challenges, a plethora of energy-efficient techniques have been proposed over the years. This also leads to a vast scope and need for comparison among the proposed techniques. Although there exists a wide range of literature discussing various energy-efficient techniques, they are confined to a particular layer. This paper provides a more general and rather holistic view of these schemes by categorizing them according to the layers they impact. The paper also introduces the reader to some major causes of energy depletion and techniques used to overcome them. Section 2 discusses the factors affecting energy consumption and network efficiency. In Sect. 3, a taxonomical classification of the widely used energy-efficient schemes is done based on the network layer targeted. Section 4 provides the reader with a summary of the discussed energy saving schemes and their impact on various WSN parameters. Section 5 concludes the literature with some future prospects pertaining to the area of interest.

2 Factors Affecting Network Longevity

As mentioned above, WSNs run on a limited power supply efficiently using which is important to network longevity. In [3], authors have identified two types of energy consumption in a sensor node, viz. useful and wasteful consumption.

(1) **Useful energy consumption** includes processes like **transmission, receiving,** and **sensing** which are useful for proper functioning of the Network.
(2) **Wasteful energy consumption** has been identified to be mainly of four types [4, 5]:

- **Idle listening**: When a node listens to any possible signal that has not arrived, the node is said to be in an idle mode. A node is generally idle for most of the time.
- **Collision**: When two nodes transmit data to a third node at the same time, the data at the receiving end is corrupted due to interference. This is called **collision**. In case there is collision, the entire data needs to be retransmitted which is a major source for energy wastage.
- **Overhearing**: When a node receives data which was meant for another node, the node is said to have overheard.
- **Control packet overhead**: Using too much control packet to prevent collision may also result in wasteful energy consumption. A reasonable trade-off between data and control packets should be maintained while designing a protocol). Considering the above facts, researchers try to make useful energy consumption more efficient while preventing wasteful energy consumption.

It is also noteworthy to mention here, the classic problem of hidden terminal which is a major cause for collision in channel access addressed in Sect. 3.3 (a). In Fig. 2, the node A is transmitting data to node B, but since C is not in range of A, it assumes the transmission channel to be free and also starts a transmission to B.

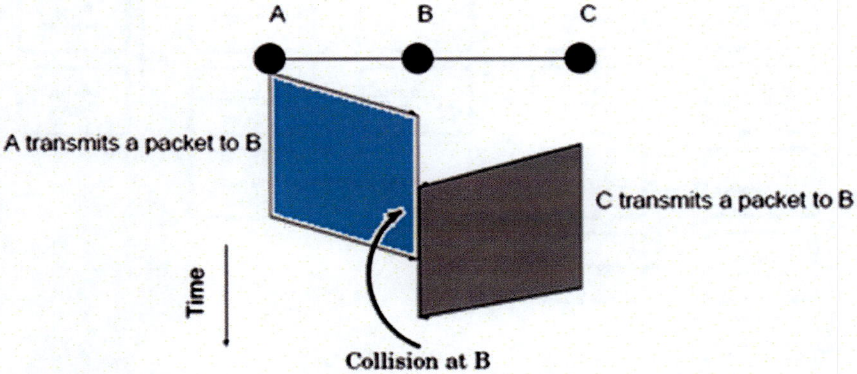

Fig. 2 Hidden terminal problem

This causes a collision at B. It should be noted here that unlike Ethernet, neither of the two sending nodes A or C is aware of this collision, and hence, no preventive measures like retransmission can be taken either by A or C, and hence, there is a loss of data [6].

3 Approaches to Network Longevity

Numerous approaches have been developed over the years by researchers to extend the network lifetime of WSNs. The approaches primarily include minimizing wasteful energy consumption while making useful energy consumption more efficient. Figure 3 shows a classification of the various energy-efficient approaches partially based on the different layers of the OSI model it impacts (although there may be some cross-layer interdependence discussed in Sect. 4) and characteristics with the aid of literature [7–9].

3.1 *Radio Optimization (Physical Layer)*

WSNs are primarily dependent on radio signals for communication, and hence, they are the main source of energy depletion at the physical layer. Researchers have been trying to find the parameters that would result in minimum energy consumption during radio transmission.

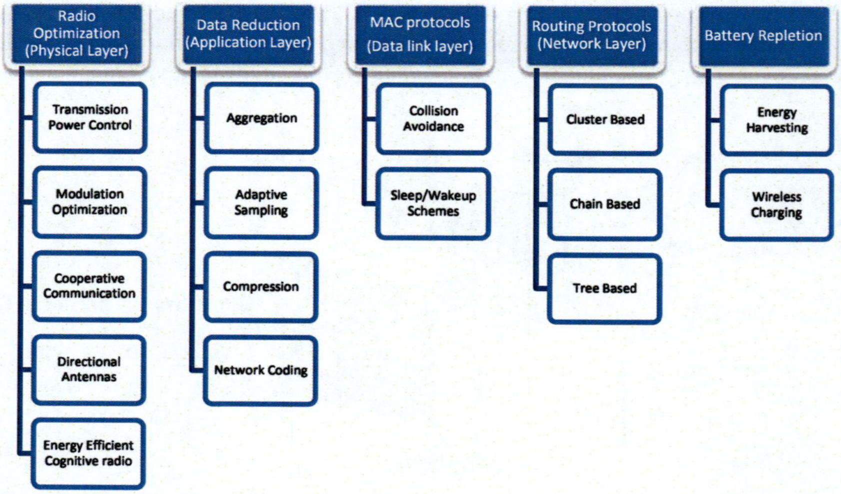

Fig. 3 Classification of energy-efficient schemes

a. **Transmission Power Control (TPC)**: Out of four operating modes of a node, viz. transmission, reception, idle listening, and sleep, measurements show that the highest amount of energy is consumed by transmission [3]. In TPC, the transmission energy is adjusted so as to control energy consumption at the physical layer (employing low power transmission for nearer nodes will help conserve energy). In [10], the authors have developed Cooperative Topology Control with Adaptation (CTCA) which employs game theory to find a Nash equilibrium among parameters like transmission power, number of neighboring nodes, and remaining energy and periodically limits or increases a node's transmission power to decreases the overall power consumption. Figure 4 illustrates this topology control scheme. In Fig. 4a, all the nodes, viz. A, B, and C are transmitting at the default transmission cost proportional to the length of the arrow shown. However, in Fig. 4b, node C increases its transmission power so that it can reach node B. Node A can now reduce its transmission power so as to only reach C which is show in Fig. 4c. Doing so may reduce overall network energy consumption or increase the life time of A. However, limiting transmission power may also affect link quality, time delay, and network connectivity which remains a topic for discussion [7].
b. **Modulation Optimization**: It aims in finding optimal radio modulation parameters that results in efficient transmission and minimum energy consumption. While transmitting data, two main parameters come to play that consumes energy, viz. circuit energy consumption and power consumption by the radio signal. In traditional networks (i.e., long distance communication), the transmission power dominates over circuit energy and hence often ignored. But WSNs are generally dense, and hence, the circuit power consumption also comes into play and cannot be ignored. It becomes essential to identify an optimal trade-off between the two. Also, Cui et al. in [11] showed that for uncoded systems, up to 80% energy can be saved by optimizing the transmission time and the modulation parameters over non-optimized systems. For coded systems, the benefit of coding varies with the transmission distance and the underlying modulation schemes. In [12], authors analyzed three energy consumption with digital modulation schemes namely M-ary QAM (MQAM), M-ary PSK (MPSK), M-ary FSK (MFSK), and MSK and an optimal value of b (no. of bits per symbol was achieved for varying distances between transmitter and receiver). Both the literature [11, 12] confirms that modulation optimization schemes get more efficient when distance increases.
c. **Cooperative Communication**: Overhearing is common in a wireless network. This phenomenon is exploited in cooperative communication, where each node not only transmits its own data but also acts as a relay to another nearby node. In this scheme, a node transmits some of its overheard signal thus increasing the reliability of the network and also not having the nodes to transmit at full power. Though on one hand, it may seem at first glance that cooperative communication may not be energy efficient as a node has to transmit not only for itself but also for its partner, whereas on the other hand, studies show that there is net reduction in transmission power consumption as the baseline transmitting power is reduced

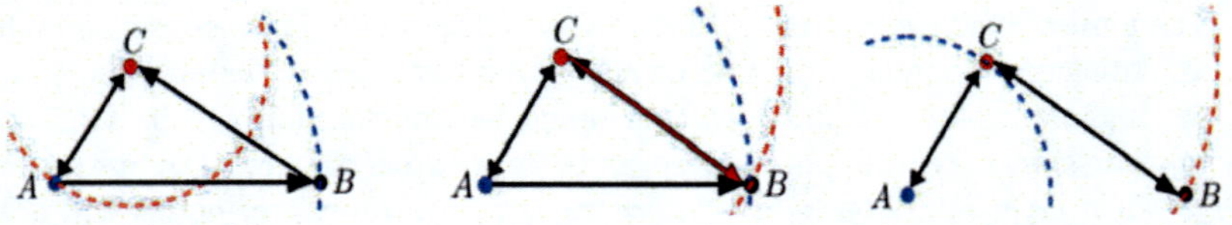

a. Initial transmission range of A and C

b. C increases its transmission power

c. A decreases it's transmission power

Fig. 4 Illustration of cooperative topology

for both nodes [13]. The trade-offs between code rate and transmit power are interesting for this scheme.

d. **Directional Antennas**: In free-space model, a radio wave losses energy proportional to the distance squared. Omni-directional communication is not cost efficient if the network does not require to be fully connected. In this scheme, the radio signal is concentrated toward a particular direction which increases transmission range and throughput. Communication is possible in that direction at a time. Though it requires localization technique for long distance transmission, omni-directional communication can occur in close proximity. In contrast to omni-directional antennas, directional antennas remove overhearing to a great extent and require less power for the same range. Kranakis et al. in [14] derived sufficient conditions for the width of the radio beam to increase signal strength while maintaining the desired connectivity. Though directional antennas increase network longevity, they suffer from time delay and impacts network connectivity.

e. **Energy-Efficient Cognitive Radio**: A key aspect of this scheme is cognition, which is acquired by a series of scanning processes and selecting an unused or better channel within the wireless spectrum. For example, in a greedy scanning process, any channel whose contention is lesser than a predefined threshold is chosen over the currently used channel. The underlying process is expected to increase the spectrum efficiency by using free channels and thereby increasing the energy efficiently. But however, cognitive radio (CR) consumes considerable amount of energy due to its functionalities such as spectrum sensing and underlying adaptable radio technologies such as software-defined radio (SDR) [15]. So, here too, there is a considerable trade-off between the functionalities of CR and its energy efficiency. In [16], the author showed that increasing the predefined contention threshold also increases the energy efficiency

3.2 *Data Reduction (Application Layer)*

Data reduction tries to limit the amount of data to be delivered to the base/sink node. This is generally done in one of the two ways—firstly by limiting the amount of data acquired by the sensors because sensing needs energy and secondly, by discarding redundant or unneeded samples before transmission to reduce the number of bits to be transmitted. Sometimes, both the techniques are used simultaneously to further reduce energy consumption.

a. **Aggregation**: In a **WSN**, there is a high probability of collecting redundant data by the sensor nodes since they sense similar attributes within a specific range and location. In data aggregation, the data from multiple sensor nodes is collected at intermediate nodes and redundant data (if any) is removed. This data, after fusion, is then transmitted to the sink or base station thus preserving transmission energy. Data aggregation can be accomplished is various ways depending on the network organization. In a flat network, the sensor nodes share similar functionalities whereas in a hierarchical network, some nodes are bestowed with special functionalities, which take the burden of data fusion [17]. Examples of hierarchical network may be LEACH, PEGASIS, HEED, etc. which have been discussed in Sect. 2.4. Finding an optimal data aggregation path is a NP-hard problem so in [18], the authors present some suboptimal data aggregation tree generation heuristics and showed the existence of special polynomial time cases.
b. **Adaptive sampling**: While most of the data reduction technique aims in reducing the amount of data to be transmitted, the task of sensing is energy consuming too and may generate unneeded samples that can affect the cost of communication as well as processing. In adaptive sampling, the sampling rate at each sensor is decreased by a certain amount while keeping in mind, the application needs are met in terms of reliability and precision. It is usually applied in cases where the cost of sampling is not negligible in terms of energy consumption, for example, a camera may consume more power than a light sensor in which case the power hungry cameras can be turned on only when the light sensors detect any change. Also, "spatial correlation can be used to decrease the sampling rate in regions where the variations in the data sensed are low. In human activity recognition applications, Yan et al. proposed to adjust the acquisition frequency to the user activity because it may not be necessary to sample at the same rate when the user is sitting or running." [7, 19]
c. **Data Compression**: Data compression is the process of reducing the number of bits required to represent a data or information, which originally required more bits. It is obvious that data compression is advantageous in wireless communication because it requires the transceivers to transmit or receive the same amount of data/information with fewer bits. However, since the sensor nodes consist of limited resources, specialized compression algorithms have to be devised for them to be able to compress the data. Kimura et al. [20] have surveyed compression algorithms specifically designed for WSNs.
d. **Network coding**: In network coding, a node sends linear combination of multiple packets instead of separately sending each packet thus saving energy in transmission. It improves a network's throughput, efficiency, and scalability, as well as resilience to attacks and eavesdropping. Shuo-Yen Robert Li et al. in

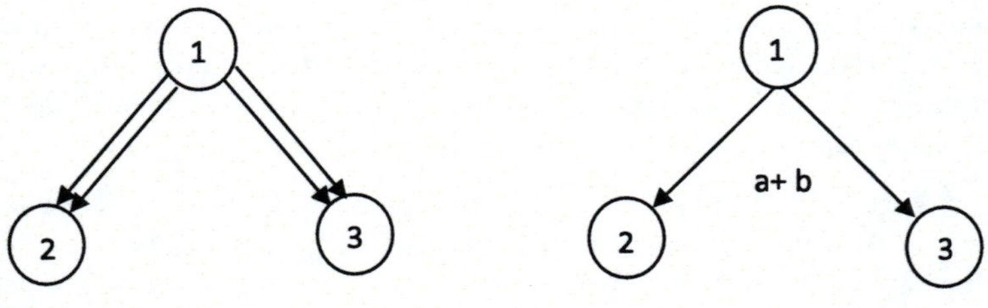

Fig. 5 Network coding

[21] proved that linear coding suffices to achieve the optimum, i.e., the max-flow from the source to each receiving node. To illustrate how network coding works, consider the example given below (Fig. 5):

3.3 Data Link Layer (MAC Protocols)

Since wireless communications are generally contention based, a medium access control (MAC) protocol is necessary for synchronized communication among the nodes. MAC can reduce the energy consumption of a wireless network by a substantial amount without the need to make any extensive changes in the hardware.

a. **Collision Avoidance**: As discussed in Sect. 1, collision remains a major source of energy waste in wireless communication. To address collision and also the hidden terminal problem, protocols like medium access with collision avoidance MACA and power aware multi-access signaling for ad hoc networks PAMAS were developed.

MACA was devised by Phil Karn in the year 1990 and was one of the earliest protocols designed with a motive to solve the hidden terminal problem mentioned in Sect. 2 by introducing a sense of handshaking between the transmitting and receiving nodes. In this protocol, whenever a node has to transmit a packet, it sends an request to send (RTS) signal first and the receiving node, if free, responds with a clear to send (CTS) signal. Upon receiving the CTS signal, the transmitting node starts the transmission. In case, the CTS signal it not received, the sender node goes to a binary exponential backoff state (BEB) and resends the RTS signal after a certain amount of time [22].

Though the MACA protocol seems legitimate and could remove collision to a certain extent, it does not completely remove the hidden terminal problem. A case of collision is show in Fig. 6.

Here, node A sends RTS to B and B responds with a CTS. At the time, B was sending CTS to A, D was sending RTS to C, and there was a collision at C. But the collision did not end here. A started its transmission to B but as there was a collision at C, it did not respond to D's RTS message, and so, D resent the RTS

Fig. 6 Collision in MACA

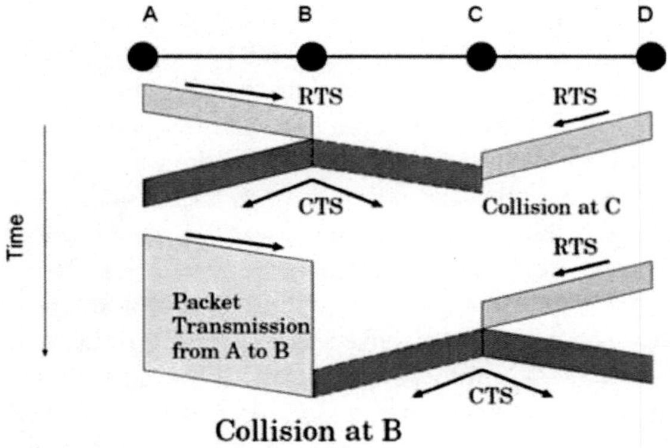

signal again. This time C responded with CTS but this CTS signal caused collision at B, and as the result, A has to retransmit. It was seen in MACA that a control signal may also cause collision and hence PAMAS was developed by Suresh Singh and C. S. Raghavan in 1998 which uses two separate channels for both data and control packets. Though PAMAS removed collision between data and control packets, using two radios in different frequency band set in each sensor node leads to the increase in the sensors cost, size, and design complexity. Also, excessive switching between sleep and wakeup states caused a significant power consumption [5].

b. **Sleep/Wakeup Based**: As mentioned earlier, idle listening is one of the major sources of energy consumption. In cases when the data flow rate is considerably low, sleep/wakeup-based protocols intend to increase energy efficiency by exploiting idle listening and periodically sending a node into sleep mode during which, the power hungry radios remain turned off. For example, turning off the radios of a node for 50% of the time should achieve energy savings up to 50%. As shown in Fig. 7, when a node is not in sleep mode, it is in the listen mode, during which the radios are turned on and communication among nodes takes place

The sleep/wakeup schemes can be categorized into **on-demand**, **asynchronous**, and **scheduled rendezvous** [7]. As the names suggest, in on-demand scheme, a node wakes up only when any other node wants to communicate with it. This is achieved by using two separate radios viz.—a low power radio for waking the node up and a power hungry radio for data transmission. This scheme ensures maximum sleep time, but using two radios increases the network cost. In asynchronous scheme, each node

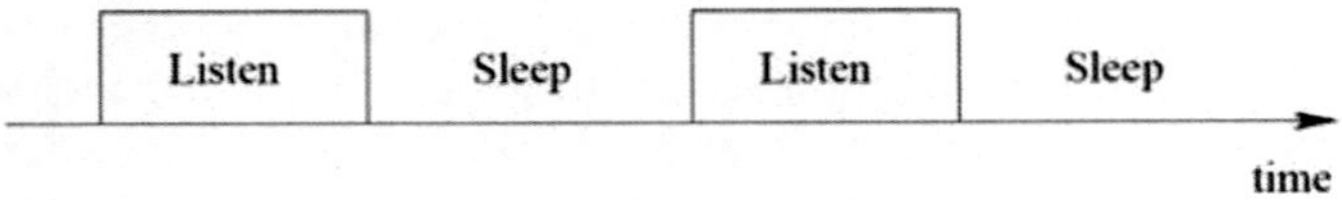

Fig. 7 Periodic listen/sleep

wakes up independently but more frequently so that the listening period between two neighboring nodes may overlap. In case of scheduled rendezvous, neighboring nodes wakeup at the same time to ensure maximum use of the wakeup period and schedules the next wakeup time before going to sleep. But this scheme may suffer collision as all the node wakes up at the same time after a long sleep period.

In [23], the authors developed sensor-MAC (SMAC) protocol based on the *sleep/wakeup-based* scheme. It was designed based on the fact that contrary to traditional wireless communications networks (e.g., voice, data) where each user needed equal time and opportunity, the nodes of a WSN on the other hand work collectively and some nodes may have remarkably more data to transmit than others.

In SMAC protocol, a node which has more data to transmit gets relatively more time to access the medium. In [24], another *sleep/wakeup*-based protocol Threshold sensitive Energy-Efficient sensor Network protocol (TEEN) was developed where the transmitters are turned on only when the change in sensing attribute crossed a certain threshold value defined by the user. But in such a scheme, there may be times when the threshold is never crossed and the user receives no data at all.

3.4 Routing Protocols (Network Layer)

Since WSNs may have several nodes, efficient routing can not only increase energy efficiency but also network reliability and quality. This section discusses a few routing techniques based on network topology.

a. **Cluster Based**: Clustering technique has gained much popularity due to its scalability and its suitability for all types of network [25]. In cluster-based networks, the nodes are arranged in a hierarchical fashion with some nodes being cluster heads (CH) and others being the cluster members as shown in Fig. 8. The CHs are responsible for collecting data from their cluster members and forwarding it the base station (BS) or sink node. The idea is based on decreasing the number of long distance transmission to increase energy efficiency. Data compression and aggregation techniques (as discussed in Sect. 3.2) can also be employed at the CHs to further reduce energy consumption. One of the earliest examples of clustering protocol is low energy adaptive clustering hierarchy (LEACH) [2]. It can be seen that the CHs in LEACH have the additional burden of transmitting data on behalf of its cluster members and thus depletes energy faster. To address this issue, the operation in LEACH is broken down into rounds where the CHs are rotated in each round so as to evenly distribute the energy consumption among the nodes. Each round is further divided in to set up phase, when the cluster formation takes place, and the steady-state phase, when the data is transmitted to the BS. LEACH showed a reduction by a factor of up to 8 compared to its conventional counterparts. However, the CH selection in LEACH was based on a stochastic function and therefore was not reliable enough and suffered from poor cluster formations. Therefore, many successors of LEACH have been proposed

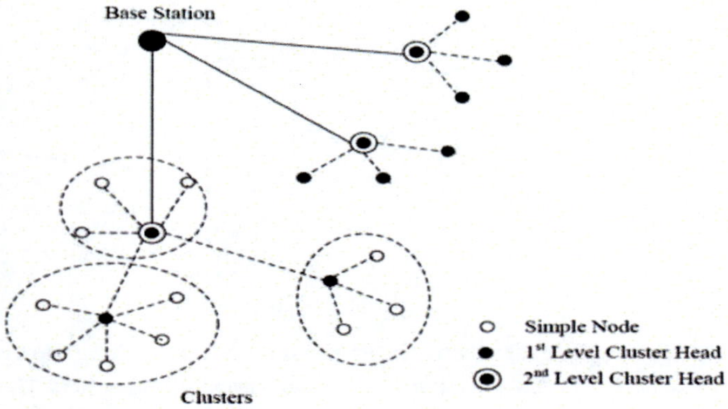

Fig. 8 Clustering topology

by various researchers with modifications in either CH selection or cluster formation. The authors of LEACH themselves proposed LEACH-Centralized (LEACH-C) [8] which employed a centralized control for better CH selection. Another variation HEED [26] takes into account the residual energy of a node while selecting a CH. A node with high residual energy gets more preference than that with a less residual energy. In addition to energy efficiency, clustering algorithms may also improve the network scalability.

In [27], a centralized clustering viz. base station-controlled dynamic clustering protocol (BCDCP) was introduced, which employed dynamic clustering to distribute the energy dissipation evenly among nodes. Extending lifetime of cluster head (ELCH) routing protocol in [28] has self-configuration and hierarchal routing properties. It reforms the existing routing protocols in several aspects and constructs clusters on the basis of radio radius and the number of cluster members. Also, a voting scheme is employed during the CH selection process where each nodes vote their neighbor depending on the ratio between the residual energy and distance from itself. Employing this algorithm makes sure that nodes with high degree of connectivity are chosen as CHs. Equalized Cluster Head Election Routing Protocol (ECHERP), a centralized clustering protocol [29], uses Gaussian elimination algorithm to select a combination of CHs that would extend overall network lifetime.

Since the head of a clustering protocol must bear the extra load of its members, efficient cluster head rotation becomes critical [30]. It uses multilayer clustering to choose between intra and inter cluster communication, rotation of the cluster head, and forwarding node [31]. It uses a combination of grey wolf optimization (GWO) and particle swarm optimization (PSO) to balance the load among the nodes of a WSN.

b. **Chain Based**: Chained-based network topologies are a further improvement of clustering. The flow of data in this protocol is analogous to a chain and hence the name. Unlike clustering protocols with multiple CHs, a chain-based protocol

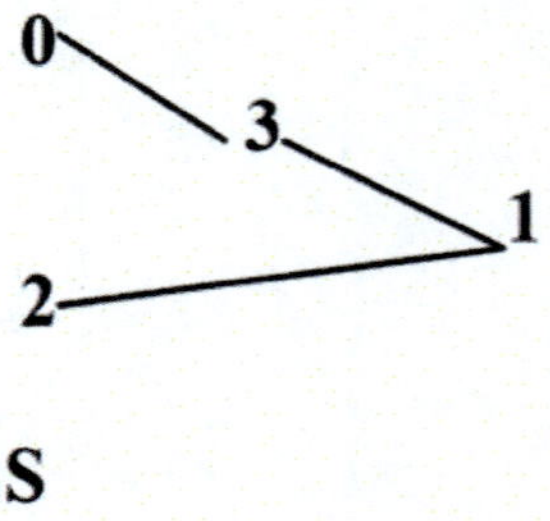

Fig. 9 Routing in PEGASIS

consists of a single leader which is responsible for transmitting the aggregated data of the network. A good example of chain-based protocol is PEGASIS [32] where each node communicates with its closest node available and takes turn being the leader.

Usually, the construction of the chain starts from the node farthest from the BS and the process continues by employing a greedy algorithm until each node is included in the chain. As shown in Fig. 9, node 0 connects to its nearest node 3, node 3 connects to its nearest node other than 0, i.e., 1, 1–2, and so on successively increasing the distance between nodes. In case of node deaths, the dead node is bypassed. For construction of the chain, it is assumed that either the BS or the nodes have global knowledge of the network. After the chain has been formed, the actual data transmission starts. The leader initiates the transmission by passing a token along the chain. As there is only one leader in PEGASIS, it greatly outperforms LEACH in terms of communication overhead [9] but however, they suffer from latency and are not suitable for time critical applications.

c. **Tree Based**: In a tree-based scheme, it can be considered as a number of chains connected together. The structure of the tree may change at each pass/round in a way that would maximize network lifetime. The problem is analogous to finding the classical minimum degree spanning tree which is known to be NP-hard [12]. However, the goal is to find a near optimal solution to keep the resulting tree diameter as small as possible for energy-efficient routing. In [33], the authors introduced shortest hop routing tree (SHORT) protocol. In each round, a node with maximum residual energy and closer to the BS is chosen as a leader. After the leader has been selected, the formation of the tree starts from farthest node from the leader by choosing to transmit to its nearest neighbor. The process of the tree formation is controlled by the BS with the prior knowledge of the position of each node in the network as explained in Fig. 9. Here, the node **h** has been selected the leader. The node **g** is farthest from h and its closest neighbor is found to be b. So, the nodes b and g form a pair (g,b) where g transmits to b. Similarly, the pairs (k,d) and (e,h) are also formed in slot 1, i.e., S1 by respectively, with the decrease in distance from the leader h. Similarly, in S2, the pairs (b,d) and (a,h), and in S3, the pair (d,h) is formed, respectively.

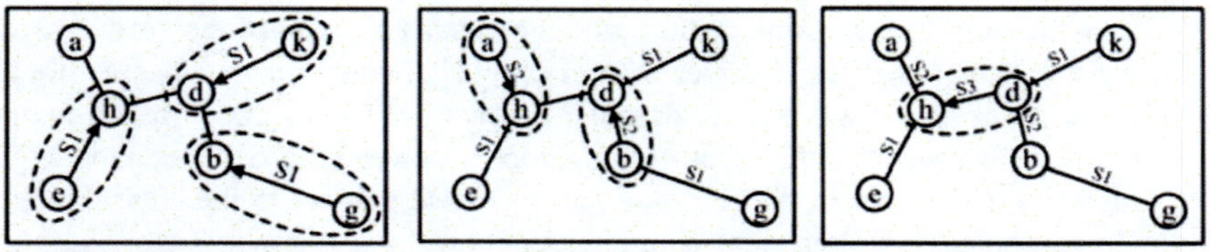

(a) Communication pairs in the first slot (S1), node-h is the leader.
(b) Communication pairs in the second slot (S2), node-h is the leader.
(c) Communication pairs in the third slot (S3), node-h is the leader.

Fig. 10 Process of generation of communication pairs in SHORT

A variation of tree-based routing can be found in [12] where a distributed version of Kruskal's Minimum Spanning Tree (MST) search algorithm is used, which limits the maximum degree of a node is used to find balanced routing spanning trees (Fig. 10).

3.5 Battery Repletion

Energy-efficient methods are prone to degradation of network connectivity and scalability. So, several recent studies focus on battery repletion having a theoretically unlimited supply of power.

a. **Energy Harvesting**: In this scheme, the nodes harvest energy from the surrounding environment which is then either used directly or stored for later use. The process of energy harvesting can be done from various sources such as solar energy, wind energy, heat energy, etc. Compared to conventional networks, energy harvesting techniques yield better network longevity by continuous power supply to the nodes, theoretically for an unlimited amount of time. It is not however completely free from energy constraints. This is because ambient energy may not always be available or may not be enough to suffice the network requirements due to which the nodes often need energy prediction schemes to adjust their behavior dynamically. The nodes can use one or more energy-efficient techniques discussed above between two recharge cycles. For example, a node having solar panels may enter energy conservation mode at night during which it can restrict its sampling rate or increase its sleep time. Emphasis may also be given on the amount of residual energy for example all the nodes may not obtain the same intensity of solar power. A node having low residual energy may restrict long distance transmission or adjust its duty cycle [7]. Also energy harvesting requires additional hardware which may have impact on the cost and node mobility.
b. **Wireless Recharging**: Wireless power transmission is the transmission of electrical energy without the use of wires. Wireless charging in WSNs can be achieved in two ways: magnetic resonant coupling and electromagnetic (EM) radiation. Magnetic resonant coupling is generally used for short distance power

transmission. In electromagnetic radiation, usually a beam of electromagnetic wave is targeted at the receiver. Xie et al. [34] showed that omni-directional power transfer is applicable only to a WSN with ultra-low power requirement because EM waves suffer from rapid drop in power with distance, and high intensity EM waves in the environment may pose a threat to the environment. However, magnetic resonant coupling seems to be a promising technique in addressing the energy issues of WSN. Though this technique was initially used only for short distances, researchers have been able to increase its efficiency and range to several meters. Advancement in wireless energy transmission has also paved a way for energy cooperation [35] where the nodes can share energy; for example, a node having high residual energy may transfer some of its energy to a node having low residual energy. Future WSNs are envisioned to comprise of nodes harvesting energy from the environment and transferring energy to other nodes thus creating a self-sustaining network.

4 Discussions

It should be noted that although a wide range of energy-efficient protocols are available, they are generally targeted on different layers of the network protocol stack. Also network parameters like network delay, throughput, connectivity, reliability, scalability, and even security which are in determining the validity of an energy-efficient scheme are affected differently by different schemes. Although a plenty of survey literature exists on comparison among various protocols, they are usually confined within a particular layer and cross-layer comparisons are rare. Table 1 provides with pros and cons of each technique (except battery repletion) and identifies affected parameters which are crucial in deciding the suitability of a particular scheme.

While all other schemes affect network parameters in some way, battery repletion is not directly related to the any network layer and hence discussed in Table 2.

5 Conclusion

The literature gives a holistic view of the energy-efficient approaches by giving a taxonomical classification of the same based on the network layer affected. Tables 1 and 2 give an overview of the crucial factors that are affected by various schemes thereby helping to decide the credibility of a particular scheme. Since applications areas of WSNs vary widely and factors like network parameters like network delay, throughput, connectivity, reliability, scalability, and even security are highly application specific, the trade-offs needs to be studied carefully before implementing any scheme. For example, time delay may not be of concern in agriculture but might be

Table 1 Energy-efficient schemes—pros and cons

Category	Technique used	Targeted layer	Impacted network parameters	
			Pros	Cons
1 Radio optimization	Transmission power Control	Physical	Collision can be sufficiently reduced by reducing transmission power	May negatively impact network coverage and node connectivity if not used wisely
	Modulation Optimization	Physical	Scalability can be improved if the modulation parameters are set right	May not be effective for dense networks
	Cooperative Communication	Physical	Can reap the benefits of MIMO without multiple antennas	May suffer time delays as data has to travel via multiple hops
	Energy efficient Cognitive radio	Physical	Radio signal quality can be improved by effective channel selection	Very sophisticated and costly Also, choosing a channel from the spectrum requires considerable energy
	Directional antennas	Physical	Improves transmission range by concentrating the angle of view	May impact node connectivity and cause deafness for some nodes
2 Data reduction	Aggregation	Application	Very effective in case redundant data exists	Aggregation and compression consumes time, resource and CPU power
	Compression	Application		
	Network coding	Application	Reduces traffic and number of packets by joining them and broadcasting to multiple nodes	Transmission has to be done at more power even though some of the node may be neighbors
	Adaptive sampling	Physical	Saves a good deal of energy during low sampling rate	Sampling rate has to be chosen wisely for reliability
3 MAC protocols	Collision avoidance	MAC layer	Minimizes collision and retransmission	Collision avoidance using multiple channels is costly
	Sleep/wakeup based	MAC layer	Reduces idle listening	The node may be in sleep mode when data is needed

(continued)

Table 1 (continued)

Category	Technique used	Targeted layer	Impacted network parameters	
			Pros	Cons
4 Routing protocols	Chain based	Link layer	Sufficiently reduces energy consumption	The data follows a longer path and hence may suffer from time delay and also suffers in case of node failure
	Cluster based	Link layer	Reduces the number of long distance transmission and improves scalability	CH selection is critical. Also in some cases, clustering may be less efficient than direct transmission
	Tree based	Link layer	Independent of node failure as there exists multiple data paths	Depending on factors like circuit energy consumption and transmitting amplifier, sometimes multipath routing can be more costly

Table 2 Battery repletion—pros and cons

Category	Technique	Pros	Cons
1. Battery repletion	Energy harvesting	1. Can theoretically supply energy for an unlimited time under ideal conditions 2. Modern nodes are very efficient and hence can be powered even with very low energy source like body heat	1. Dependent on environmental factors and hence all nodes may not harvest the same amount of energy 2. Requires extra hardware for energy harvesting
	Wireless recharging	1. Overhearing can be used as a means of energy to recharge batteries 2. Can be coupled with energy harvesting and exchange energy within nodes	1. Suffer from signal attenuation 2. Obstruction of vision may hamper energy transmission 3. Need extra hardware for receiving and transmitting energy

very crucial in health monitoring. As such chain-based protocols may not be suitable for health-monitoring applications. The trade-off among network parameters remain as future scope for research.

References

1. L. Guo, W. Wang, J. Cui, L. Gao, A cluster-based algorithm for energy-efficient routing in wireless sensor networks, in *Proceedings—2010 International Forum on Information Technology and Applications, IFITA 2010*, vol. 2, issue 02 (2010), pp. 101–103. https://doi.org/10.1109/IFITA.2010.137
2. W. R. Heinzelman, A. Chandrakasan, H. Balakrishnan, Energy-efficient communication protocol for wireless microsensor networks, in *Proceedings of the 33rd Annual Hawaii International Conference on System Sciences*, vol. 1(c) (2000), p. 10. https://doi.org/10.1109/HICSS.2000.926982
3. Z. Rezaei, Energy saving in wireless sensor networks. Int. J. Comput. Sci. Eng. Surv. **3**(1), 23–37 (2012). https://doi.org/10.5121/ijcses.2012.3103
4. A. More, V. Raisinghani, A survey on energy efficient coverage protocols in wireless sensor networks. J. King Saud Univ. Comput. Inf. Sci. **29**(4), 428–448 (2017). https://doi.org/10.1016/j.jksuci.2016.08.001
5. T. Braun, M. Anwander, P. Hurni, M. Wälchli, MAC protocols for wireless sensor networks. Next Gener. Mobile Netw. Ubiquit. Comput. **4**(3), 165–174 (2010). https://doi.org/10.4018/978-1-60566-250-3.ch016
6. S. Singh, C.S. Raghavendra, PAMAS—Power aware multi-access protocol with signalling for ad-hoc networks. ACM SIGCOMM Comput. Commun. Rev. **28**(3), 5–26 (1998)
7. T. Rault, A. Bouabdallah, Y. Challal, T. Rault, A. Bouabdallah, Y. Challal, E. Efficiency, T. Rault, A. Bouabdallah, Y. Challal, Energy efficiency in wireless sensor networks: A top-down survey (2014)
8. A.H. Sodhro, G. Fortino, S. Pirbhulal, M.M. Lodro, M.A. Shah, Energy efficiency in wireless body sensor networks. Netw. Future (December), 339–354 (2018). https://doi.org/10.1201/9781315155517-16
9. L.D.P. Mendes, J.J.P.C. Rodrigues, A survey on cross-layer solutions for wireless sensor networks. J. Netw. Comput. Appl. **34**(2), 523–534 (2011). https://doi.org/10.1016/j.jnca.2010.11.009
10. X. Chu, H. Sethu, Cooperative topology control with adaptation for improved lifetime in wireless sensor networks. Ad Hoc Netw. **30**, 99–114 (2015). https://doi.org/10.1016/j.adhoc.2015.03.007
11. S. Cui, A.J. Goldsmith, A. Bahai, Energy-constrained modulation optimization. IEEE Trans. Wireless Commun. **4**(5), 2349–2360 (2005). https://doi.org/10.1109/TWC.2005.853882
12. R. Anane, K. Raoof, M.B. Zid, R. Bouallegue, Optimal modulation scheme for energy **6613**, 500–506 (2014). https://doi.org/10.13140/2.1.4503.6324
13. A. Nosratinia, T.E. Hunter, A. Hedayat, Cooperative communication in wireless networks. IEEE Commun. Mag. **42**(10), 74–80 (2004). https://doi.org/10.1109/MCOM.2004.1341264
14. E. Kranakis, D. Krizanc, E. Williams, *Directional Versus Omnidirectional Antennas for Energy Consumption and k-Connectivity of Networks of Sensors* (2005), pp. 357–368. https://doi.org/10.1007/11516798_26
15. M. Masonta, Y. Haddad, L. De Nardis, A. Kliks, O. Holland, *Energy Efficiency in Future Wireless Networks: Cognitive Radio Standardization Requirements* (2012), pp. 31–35

16. V. Namboodiri, Are cognitive radios energy efficient? A study of the wireless LAN scenario, in *2009 IEEE 28th International Performance Computing and Communications Conference (IPCCC)* (2009), pp. 437–442. https://doi.org/10.1109/PCCC.2009.5403857
17. R. Rajagopalan, P.K. Varshney, Data-aggregation techniques in sensor networks: A survey. IEEE Commun. Surv. Tutorials **8**(4), 48–63 (2006). https://doi.org/10.1109/COMST.2006.283821
18. B. Krishnamachari, D. Estrin, S. Wicker, The impact of data aggregation in wireless sensor networks. Comput. Syst. 0–3 (2002)
19. G. Anastasi, M. Conti, M. Di Francesco, A. Passarella, Energy conservation in wireless sensor networks: A survey. Ad Hoc Netw. **7**(3), 537–568 (2009). https://doi.org/10.1016/j.adhoc.2008.06.003
20. N. Kimura, S. Latifi, A survey on data compression in wireless sensor networks, in *International Conference on Information Technology: Coding and Computing*, Las Vegas, NV (2005), pp. 8–13
21. S.Y.R. Li, R.W. Yeung, N. Cai, Linear network coding. IEEE Trans. Inf. Theor. **49**(2), 371–381 (2003). https://doi.org/10.1109/Tit.2002.807285
22. V. Bharghavan, A. Demers, S. Shenker, L. Zhang, *Macaw. Proceedings of the Conference on Communications Architectures, Protocols and Applications—SIGCOMM '94*, pp. 212–225. https://doi.org/10.1145/190314.190334
23. W. Ye, J. Heidemann, D. Estrin, An energy-efficient MAC protocol for wireless sensor networks. Proc. Twenty-First Ann. Joint Conf. IEEE Comput. Commun. Soc. **3**, 1567–1576 (2002). https://doi.org/10.1109/INFCOM.2002.1019408
24. A. Manjeshwar, D.P. Agrawal, TEEN: A routing protocol for enhanced efficiency in wireless sensor networks, in *Proceedings—15th International Parallel and Distributed Processing Symposium, IPDPS 2001*, vol. 00, issue (C) (2001), pp. 2009–2015. https://doi.org/10.1109/IPDPS.2001.925197
25. S. Bachchav, Energy efficient technique to improve the sensor network lifetime. **2**(3), 1–7 (n.d.).
26. C.H. Lin, M.J. Tsai, A comment on "HEED: a hybrid, energy-efficient, distributed clustering approach for ad hoc sensor networks." IEEE Trans. Mob. Comput. **5**(10), 1471–1472 (2006). https://doi.org/10.1109/TMC.2006.141
27. S.D. Muruganathan, D.C.F. Ma, R.I. Bhasin, A.O. Fapojuwo, A centralized energy-efficient routing protocol for wireless sensor networks. Commun. Mag. IEEE **43**(3), S8-13 (2005). https://doi.org/10.1109/MCOM.2005.1404592
28. J.J. Lotf, M.N. Bonab, S. Khorsandi, A novel cluster-based routing protocol with extending lifetime for wireless sensor networks, in *5th IEEE and IFIP International Conference on Wireless and Optical Communications Networks, WOCN 2008* (2008). https://doi.org/10.1109/WOCN.2008.4542499
29. S.A. Nikolidakis, D. Kandris, D.D. Vergados, C. Douligeris, Energy efficient routing in wireless sensor networks through balanced clustering. Algorithms **6**(1), 29–42 (2013). https://doi.org/10.3390/a6010029
30. S.R. Mugunthan, Novel cluster rotating and routing strategy for software defined wireless sensor networks. J. ISMAC **2**(02), 140–146 (2020)
31. J.S. Raj, Machine learning based resourceful clustering with load optimization for wireless sensor networks. J. Ubiquit. Comput. Commun. Technol. (UCCT) **2**(01), 29–38 (2020)
32. S. Lindsey, C.S. Raghavendra, PEGASIS: Power-efficient gathering in sensor information systems. IEEE Aerosp. Conf. Proc. **3**, 1125–1130 (2002). https://doi.org/10.1109/AERO.2002.1035242
33. Y. Yang, H.H. Wu, H.H. Chen, SHORT: Shortest hop routing tree for wireless sensor networks. IEEE Int. Conf. Commun. **8**(c), 3450–3454. https://doi.org/10.1109/ICC.2006.255606

34. L. Xie, Y. Shi, Y.T. Hou, W. Lou, H.D. Sherali, S.F. Midkiff, On renewable sensor networks with wireless energy transfer: The multi-node case. Annual IEEE Commun. Soc. Conf. Sensor, Mesh Ad Hoc Commun. Netw. Workshops **1**, 10–18 (2012). https://doi.org/10.1109/SECON.2012.6275766
35. B. Gurakan, O. Ozel, J. Yang, S. Ulukus, Energy cooperation in energy harvesting communications. IEEE Trans. Commun. **61**(12), 4884–4898 (2013). https://doi.org/10.1109/TCOMM.2013.110113.130184

Lean-SE: Framework Combining Lean Thinking with the SDLC Process

Mona Deshmukh and Amit Jain

Abstract Software development process has evolved with the objective to develop methodologies that would adapt to the changing nature of software. Even software has evolved from a scientific equipment to a ubiquitous device. This ubiquitous nature demands it to be more user centric. Developing user-centric products requires strong and continuous customer collaboration and ensuring that no product is developed without a purpose to the user. This paper presents a framework combining lean thinking activities within the SDLC processes models with an aim to strengthen the analysis phase. This framework can be integrated within the waterfall and agile model during the analysis phase to better understand the requirements and transform them into features for the development team. Integration may make the analysis and design phase a bit stretched out, but the resulting requirement will be specific and clear and aid to mitigate changes, improve the end product and user satisfaction.

1 Introduction

Methodologies such as agile have become popular as they are adaptive to change and work in collaboration with customers. Major objective of SDLC and lean is to develop quality product resulting in customer satisfaction. However, there are few concerns that need to be resolved and addressed, and it is usually assumed that the customer knows what needs to be developed and can explain it well but according to Kurchten [1] eliciting customer requirement is a major challenge. Sometimes the user is unable to specify his requirements upfront and is not clear about his needs, and these circumstances lead to insufficient requirement gathering making future changes inevitable. Griffith [2] presents the most relevant reason for software product failure as building or developing a product which nobody wants. This is because the teams are more focused toward the technical aspect of their product and do not empathize with their customers which results in ambiguous requirement gathering. Lean concept is widely practiced in manufacturing industries but is not

M. Deshmukh (✉) · A. Jain
Department of Computer Engineering, SPSU, Udaipur, India
e-mail: mona.deshmukh@ves.ac.in

P. Karuppusamy et al. (eds.), *Ubiquitous Intelligent Systems*, Smart Innovation, Systems and Technologies 243, https://doi.org/10.1007/978-981-16-3675-2_8

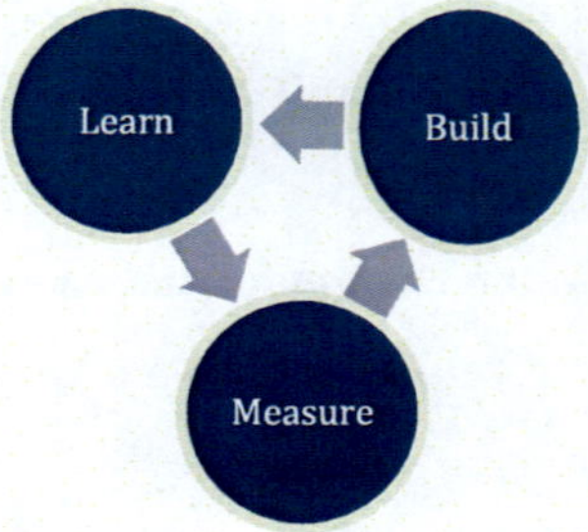

Fig. 1 Lean thinking model

Fig. 2 Waterfall model

popular in the software sector. We aim to integrate the lean thinking concept into the waterfall model and the scrum framework, thereby providing an extended model for the two (Deshmukh et al.). To produce quality work, a process of the human and the working principles must be defined properly (Fig. 1).

The build–measure–learn loop is the major component of the lean model. Its objective is to transform the unknown requirements, assumptions/hypothesis into known ones, thereby guiding the team toward unambiguous requirement gathering. The build–measure–learn loop consists of three phases. Build phase: Goal of this phase is to ideate the customer needs into prototype or a minimum viable product for the customer to test against the hypothesis and assumptions created. Phase Two: MEASURE: This is the second phase which measures the experiments undertaken during the build phase. Phase three: Learn: This phase is a validated learning phase where decisions based on the results obtained from phase two are taken. Based on the results obtained, the team either perishes or preserves the assumptions.

2 Waterfall Model

See Fig. 2.

3 Extended Waterfall Model

Using waterfall model as the baseline, we propose an extended model as shown in Fig. 3. According to Pressman [3], the initial phase that is the communication phase inculcates activities of requirement elicitation, requirement gathering, negotiation,

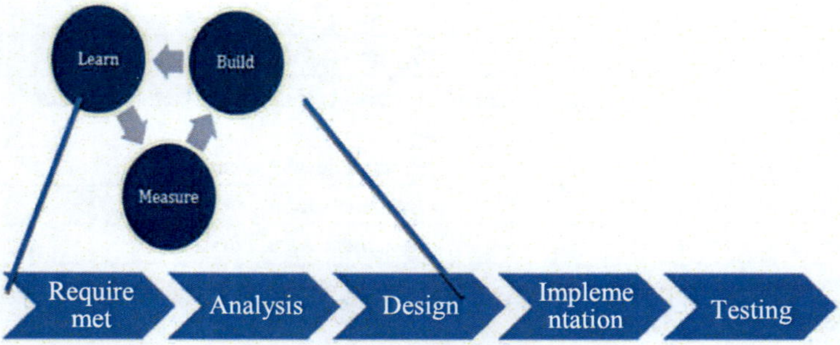

Fig. 3 Extended waterfall model

specification and validation and is all about transforming the unknown to known. Hence all features, functionalities and constraints of a software are finalized and validated here. Since this phase of requirement gathering is the most important phase, it has to be done in a systematic and a specific way. The lean thinking model is all about validated learning, i.e., it ensures that we are building the right thing by keeping the customer in loop.

The proposed extended waterfall model tends to strengthen the initial phases of requirement gathering and analysis by integrating the lean model of learn–measure–build to avoid ambiguous requirement gathering, thereby resulting in features and products which the customer would actually like to use.

4 Scrum Framework

Scrum is an iterative and incremental agile software development framework for managing software projects and product or application development. The key concern of scrum is to avoid developing products that the user will not like to use. Figure 4 shows the Scrum framework, and it follows a plan–build–measure–learn cycle which is similar to the lean thinking model to some extent. Both the models work in close

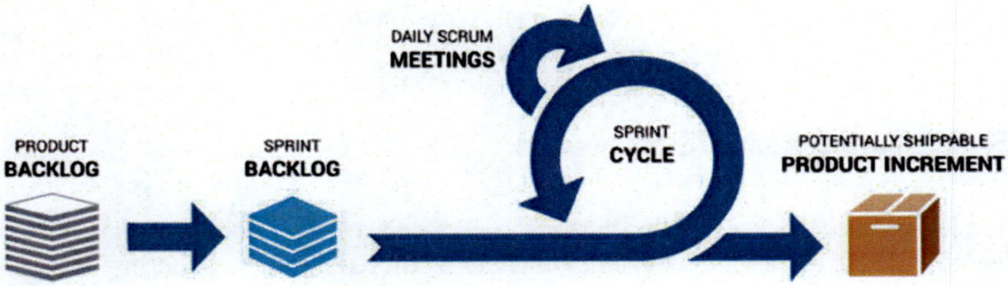

Fig. 4 Scrum framework

Table 1 Lean versus scrum framework

Lean thinking	Scrum
Focuses on customer requirements and needs through validated learning	Focuses on fast delivery of the product
Product discovery	Product development
Generates and develops ideas	Executes ideas
Incremental and iterative	Incremental and iterative
Short development cycles	Short development cycles
Flexible to change	Flexible to change
Follows the build–measure–learn cycle	Follows the plan–build–measure–learn cycle
Empathizes with the customer resulting in better requirements gathering	Requirements are not done upfront and hence may result in rework
Helps learn faster and build right	

collaboration with the customer. Scrum framework consists of a product backlog, sprint backlog, sprint meeting and product increment. Sprint cycle follows a time-boxed iteration during which a potentially releasable product increment is created. Design, build and test activities are performed within the sprint. A sprint begins with a sprint planning meeting and ends with a sprint review and retrospective meetings. A short organizational meeting is held each day in form of a daily sprint. Wherein, each team member has to answer the following three questions: What did you do yesterday? What will you do today? And Are there any impediments in your way? A meeting with project stakeholders to demonstrate the completed solution capabilities from that sprint is called as a sprint review. A sprint retrospective meeting with the project team is conducted to reflect on the experiences of the sprint. Lean thinking methodologies such as agile have become popular as they are adaptive to change and work in collaboration with customers. However, there are few concerns that need to be resolved and addressed. Griffith [2] presents the most relevant reason for software product failure as building or developing a product which nobody wants. This is because the teams are more focused toward the technical aspect of their product and do not empathize with their customers which result in ambiguous requirement gathering. Key concern of Scrum is to avoid products that do not work, whereas key concern of lean is to avoid creating products that people do not like. Table 1 presents a comparison between the Scrum and lean framework.

5 Extended Scrum Framework

The lean model focuses more on customer needs and collaboration through validated learning which makes it the most obvious model for requirement gathering. Integration of the Scrum and lean model will not only strengthen the requirement gathering phase but also mitigate the future changes in requirements from the customer. Figure 5

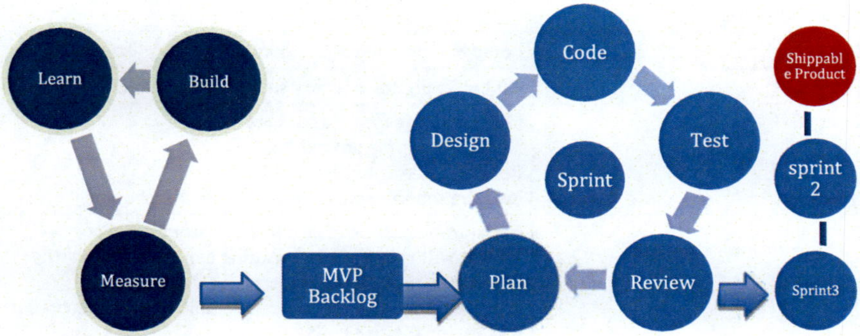

Fig. 5 Extended scrum framework

shows the extended Scrum framework.

Lean thinking is about exploring the problem and testing the possible solutions. Lean thinking helps the team empathize with customers' needs and requirements. Possible solutions are thought upon as assumptions or hypothesis and then passed through the build–measure–learn cycle transforming assumptions into real-world solutions. These validated requirements then enter the Scrum process for execution and delivery. The proposed extended Scrum framework follows the given process: A start-up comes up with a business case. The start-up comes up with a business model, presented in a short business plan. (Build). They start collaborating with customers and ask about features they are expecting in the app. (Measure). They will acquire feedback from customers. (LEARN). From the customer feedback, step 2 will be repeated, and the business plan may be revised until they get it right. Once all the customer requirements are frozen, they can proceed to implement the prototype or an minimum viable product (MVP) for testing. (Build) Prototype is then tested with the customers. (Measure). Customer feedback will be gathered and learned. The learning phase is repeated, making improvements on the prototype until they get the app right.

6 Experiment and Results

The above framework has been used by XYZ Company to digitize the books of an ABC school resulting in a concise and unambiguous requirement gathering. The steps followed to implement the aforesaid framework are mentioned below (Table 2).

Table 2 Findings

Case study	Improve reputation of a school
Proposal	All books digitized (every student gets a tablet)
Benefits	Much better reading experience No more carrying around books Better for taking notes Increase school reputation as tech savvy school
Assumptions	It is a problem for students to carry around books Tablets will provide better reading experience Students will pay for tablet cost Prospective student will have favorable rating for school if we provide digitize books
Classify the assumptions/hypothesis	• Assumption 1: it is a problem for students to carry around books *Low probability that it will be wrong, low impact on solution* • Assumption 2: tablets will provide better reading experience *High probability that it will be wrong, high impact on solution* • Assumption 3: Students will be willing to pay for tablet cost. *High probability that it will be wrong, high impact on solution* • Assumption 4: prospective student will have favorable rating for school if we provide digitize books *High probability that it will be wrong, low impact on solution* *Here, select the riskiest assumption*
Validating assumptions	• Tablets will provide better reading experience *Test: observe and interview. (high cost, high quality)* *campus survey (low cost, low quality)* • Students will pay for tablet cost. *campus survey (low cost, low quality)* *A video and signup link on student portal, (low cost, high quality)* • Prospective student will have favorable rating for school if we provide digitize books *A/B testing on college admission form. (high cost, high quality)* *Survey during the open house for perspective*
Learning from tests	Analyze and think Consolidating/classifying learning The tested and validated assumptions are passed on to the scrum team for development

7 Conclusion

SDLC is all about developing the right product, whereas lean thinking is about building the product right. Agile is adaptive, incremental and flexible to change but the issues with this framework is that requirements are not done upfront. Whereas lean thinking focuses on validated learning which results in fast learning. Integrating the lean model along with the Scrum framework will not only improve the requirement gathering phase, and at same time, it will validate the requirements in collaboration with the customer which may result in understanding the customer needs upfront and improving the user experience resulting in customer satisfaction. Implementation of the proposed extended frameworks may extend the requirement gathering phase but will ensure that the data collected will be of high quality. Adaption and evaluation of proposed frameworks in software organizations can be done as a future work to access their success.

References

1. P. Kurchten, *The Rational Unified Process: An Introduction*, 3rd ed. (Pearson Education, 2004)
2. E. Griffith, Why Startups Fail, According to Their Founders (2014). http://fortune.com/2014/09/25/why-startups-fail-according-to-their-founders/. 26 Sept 2014
3. R. Pressman, *Software Engineering: A Practitioner's Approach* (McGraw-Hill, New York, 1987)
4. P. Middleton, D. Joyce, Lean software management: BBC worldwide case study. IEEE Trans. Eng. Manag. **59**(1), 20–32 (2012)
5. P. Middleton, Lean software development: two case studies. Softw. Qual. J. **4**, 241–252 (2001)
6. M. Poppendieck, T. Poppendieck, *Lean Software Toolkit* (Addison Wesley, 2003)
7. M. Poppendieck, C. Michael, Lean software development: a tutorial. IEEE Softw. **29**(5), 26–32 (2012)

A Comparative Study on Augmented Analytics Using Deep Learning Techniques

M. Anusha and P. Kiruthika

Abstract Image augmentation is the most recognized type of data augmentation and intrinsic development for transforming image diversities in the training dataset that belongs to a similar class as the novel image. In the area of image augmentation handling, a collection of operations is shifting, flipping, zooming, cropping, rotation, and transformation in color space. A wide range of applications frequently used the aspects of deep learning are industry, science, and government domain, namely adaptive testing, image classification, computer vision, object detection, and face recognition and has achieved substantial development and accomplishment of deep learning. This study concentrates on the most important challenges present in the image estimation level that have a significant effect on dimension reduction, pooling, and edge detection. The deep learning methods involved here are convolution neural network (CNN), generative adversarial network (GAN), and deep convolution neural network (DCNN). Finally, a comparative study has performed a massive literature survey on various deep learning models.

1 Introduction

Image data are a pictorial representation of data, which includes two categories of data, one is balanced data, and another one is imbalanced data. The number of positive and negative values will be approximately same in balanced data, and the number of positive and negative values will be highly different in imbalanced data. Depending on the size of the data, it is not very important to train the model and derive useful training data features depending on the size and consistency of the data. The model needs a lot of input data to provide a better solution. To overcome this challenge, image augmentation technique is used [1]. Image augmentation is considered as one of the types of data augmentation techniques. The data augmentation technique helps generate a novel data from the remaining data expand artificially,

M. Anusha · P. Kiruthika (✉)
PG & Research Department of Computer Science, National College (Autonomous), Bharathidasan University, Tiruchirappalli, Tamilnadu, India
e-mail: kiruthikashyne@gmail.com

P. Karuppusamy et al. (eds.), *Ubiquitous Intelligent Systems*, Smart Innovation, Systems and Technologies 243, https://doi.org/10.1007/978-981-16-3675-2_9

where the data augmentation methods are helping to train and show a strong generalization ability of deep learning models [2]. The basic image augmentation techniques are geometric transformation (flipping, rotation, translation, cropping, and scaling), color space transformation (color casting, varying brightness, and noise injection), Kernel filters, images mixing, erase randomly, augmentation in feature space, adversarial training, and augmentation based on GAN, transferring an image style, and self-regulated learning. The image augmentation technique discussed in this study is geometric transformation. Geometric transformation provides a better solution for training the data, such as shifting, flipping, zooming, cropping, rotation, and color space in transformations. A major cause for geometric transformation is horizontal flipping, which is far more common than vertical flipping. The color channel space is one of the techniques used in augmentation, which is quite realistic to implement a relatively modest color augmentation, which includes dividing an atomic color channel and contrast between randomly cropping. Further, the image data translation method has been used by cropping to reduce the dimensions of the input data, and then, translations maintain spatial dimensions of the image. The rotation augmentation is accomplished by rotating the input [3]. The rapid development of deep learning covers machine learning, which is created on neural networks with the knowledge representation. Deep learning helps to solve complex issues for machine learning. The neural network comprises several layers used to describe abstraction data to construct a computational model for deep learning [4]. The knowledge representation is supervised, semi-supervised, and unsupervised. Supervised learning techniques are used for labeled input data to predict the desired output data. Semi-supervised learning is another knowledge representation technique used for labeled data in a small amount and the unlabeled data is large amount during the training process. Finally, unsupervised technique is used for unlabeled data to extract the generative features [5]. Deep learning architectures are deep neural network, convolutional neural network, recurrent neural network, long short-term memory, gated recurrent units, deep belief network, generative adversarial network, and autoencoder is included. This survey concentrates on the generative adversarial network, convolutional neural network, and the deep convolutional neural network that have been applied in the domain of picture classification, computer visualization, object detection, and face recognition [6]. A generative adversarial network is an influential tool for performing unsupervised learning representation to generate new image data from the existing data to fool the discrimination. GAN is used to overcome these problems such as picture to picture translation, image to text synthesis, and picture resolution. The convolutional neural network is most commonly used for picture wise classification, and CNN concentrates on many parameters that reach the enlarged size of networks as well as lacking training data with generalization abilities [7]. A deep convolutional neural network is an effective approach for computer visualization, pedestrian detection, and image segmentation. DCNN retrieves the knowledgeable feature extraction of the training model [8]. This study concentrates on the image level estimation that has a significant effect on dimension reduction, pooling, and edge detection. The rest of the paper is organized as follows Sect. 2 briefly introduces the related work, Sect. 3

shows comparative study, and Sect. 4 is extended with the result and the discussion and ends with a conclusion.

2 Literature Survey

Feng et al. [9] suggested a method for data augmentation by using decoding weights. Due to the collection of training data, the autoencoder is trained, the decoding weights are enabled, and the weights are compared with the samples obtained to produce the augmented samples. The different layers used for this proposed method and their accuracy of VGG-16 50.9%, ResNet-50 52.1%, DenseNet-121 used for two datasets and their outcome is 50.2 and 75.8% of image weight. Chen et al. [10] introduced a novel algorithm for individual re-identification based on self-monitored data increase. Our approach is informed by the recently enacted part-based algorithms, which have accomplished impressive outcomes on the re-identification of the individual. It is important to realize that the intention is not to use split in terms of learning more biased functionality. The algorithm produced an identification rate and average precision of 93.88 and 84.45%, 87.52 and 75.68%, 71.27 and 65.91% in three different types. Liu et al. [11] designed a lightweight transfer learning approach based on ResNet architecture, and it is used for a number of layers that are fused their feature and then take a benefit of the confined layer in various attributes fused them for Softmax regression detection. The FTOTLM method was produced with augmentation techniques for six different datasets accuracy of 99.87, 99.45, 97.45, 97.38, 94.05, and 85.21%. Umer et al. [12] To fix this issue, an innovative approach of data increase technique for extracting features and tasks performed in recognition owed to retina, and region around the retina is merged, then to the enhanced performance of the indicator depending on various conditions that picture comes with data redundancy to generate the data irregularly or the recognition process is difficult. The proposed system was performed in the iris accuracy of (CASIA-dist—99.64%, UBIRIS.v2—98.76%) and particular accuracy of (CASIA-dist—99.64%, UBIRIS.v2—98.76%) for image recognition. Fu et al. [13] built a novel adversarial generative network called fine-grained conditional GAN, to solve the issue of fine-grained depending upon classes in image generation. Fine-grained conditional GAN produces first class-dependent low-resolution images. The consequence is that high-resolution class-dependent images are developed by the generator. Every generator is accompanied by a discriminator. To gain valuable fine-grained data, we create two finer resolution in the fine-grained conditional GAN. The proposed work used for one or more augmentation techniques, and the classification accuracy of dataset 1 is 65.95% and 71.15%, dataset 2 is 75.69% and 79.65%, dataset 3 is 73.39% and 76.16% for high-resolution images. Kaur et al. [14] handled two approaches for these problems, one of the approaches has suggested a paradigm focused on the transition of learning based on the pre-trained AlexNet architecture. Another viewpoint that reflects on the new data augmentation technique is based on the generative adversarial network. Even after the two approaches performed, the Parkinson classification method has increased accuracy. The algorithm was classified with an average accuracy of 89.23% for disease classification. Saini et al. [15] The class imbalanced distribution results in a deterioration in the efficiency of the classification models owing to a bias in the category toward this dominant class to tackle this issue. The author has suggested a different learning technique that includes a deep transfer network in conjunction with the deep convolution generative adversarial network. The algorithm was performed in four different magnification factors, accuracy of 96.5%, 94%, 95.5%, and 93%. Moon et al. [16] Note that an accurate and quick computer-aided detection (CADe) method located on a three-dimensional convolutional neural network (CNN) introduced as another reader for physicians to minimize the duration of the examination and frequency of misdetection. The proposed algorithm has produced a sensitivity of 95.3% for misdetection. Karakanis et al. [17] proposed an innovative method for coronavirus detection in a chest X-ray image dataset used for conditional generative adversarial networks. The author creates a large amount of data using augmentation techniques to overcome the limited dataset. It suggested two models in deep learning by using the availability of the dataset. The proposed binary model has achieved an accuracy of 98.7%, sensitivity of 100%, and specificity of 98.3%, and the three-class model achieved an accuracy of 98.3%, a sensitivity of 99.3%, and specificity of 98.1% for detecting COVID-19. Rai et al. [18] presented the novel deep neural network model with minimum layers and fewer complex to build on U-Net for detecting tumors. This work includes categorizing the brain MR images in regular or irregular classes of 253 high pixel images. The proposed LU-net model performed recall 1.00%, precision 97%, F-score 98%, specificity 95%, and the accuracy of 98% for detecting tumors. Abdelhalim et al. [19] proposed the self-attention mechanism to detect skin melanoma images used for convolutional neural networks based on dimensionality reduction and to achieve better results. The proposed self-attention model performed macro recall 64.7%, AUC 79.3%, macro precision 50.1%, macro F-score 53.4%, average training time in minutes is 16.3, and the accuracy of this model is 66.1% for dimensionality reduction. Hosny et al. [20] A novel deep convolutional neural network approach is exhausting the skin melanoma images to be classified. To overcome the insufficient dataset using augmented techniques, a large dataset is developed. The proposed method to improved the classification accuracy of MED-NODE, DermIS & DermQuest, and ISIC 2017 for 99.29%, 99.15%, and 98.14%. Mzoughi et al. [21] A new classification approach was used for brain tumor MRI images established on a deep convolutional neural network model aimed at low-grade glioma and high-grade glioma. To combine both gliomas information then to reduce weight based on three-dimensional convolutional neural network, the classification approach has produced an accuracy of 96.49% of this dataset. Pasyar et al. [22] The author aims to develop a new hybrid classifier, then verify the liver level based on the liver image dataset handling of a deep convolutional neural network. Clarify the weighted possibility of every class, including its majority voting process. The proposed ResNet50 model has produced an accuracy of 86.4%, first group classified the sensitivity and specificity of 90.9% and 86.4%, and the last group classified the sensitivity of 90.9% and specificity of

81.8%. Loey et al. [23] proposed different deep convolutional neural network models to identify the infected coronavirus patients using chest CT scan images. The author collects possible CT scan images used for traditional augmentation techniques to create a large image dataset. The outcome is classified as COVID or non-COVID. The proposed model was performed in traditional augmentation techniques, and their accuracy of testing is 82.91%, sensitivity of 77.66%, and specificity of 87.62% to found covid-19. Alzubaidi et al. [24] proposed a new deep convolutional neural network model built on a new 754-ft image dataset to detect automatically whether it is healthy or not. The proposed model was performed to an F1-score of 94.5%. Gifani et al. [25] Another problem-solving tool for detecting COVID-19 in chest X-ray images is deep learning techniques. The deep learning techniques enlarged the clarity of image detection. The proposed method produced an average accuracy of 98.93%, a sensitivity of 98.93%, a specificity of 98.66%, a precision of 96.39%, and an F1-score of 98.15% for COVID-19 detection. Nayak et al. [26] diagnosed the brain irregularities in brain MRI datasets utilizing a deep convolutional neural network that focused on the automatic approach, reduced the dimension reduction, and achieved better classification. The proposed approach was performed with a classification accuracy of 100.00 and 97.50% for dimensionality reduction. Abrishami et al. [27] improved the generalization ability based on the augmentation techniques used for deep convolutional neural networks to reduce the computation process. The pre-trained network model for base and target network is based on dataset CIFAR-100 and CIFAR-10 (71.47%, 74.48%, 78.20%) and (64.44%, 78.87%, 83.98%) for computation process. Wang et al. [28] improved the human identification process by consuming a deep convolutional neural network constructed on augmented CPAF images. The proposed CPAFNet model was producing an accuracy of 92.63% of the image identification process.

3 Comparative Study

In this section, comparative study of deep learning models about datasets, merits, and demerits is discussed below for the image augmentation process (Tables 1, 2, and 3).

4 Result and Discussion

Image augmentation is the kind of image handling research that systematically determines the successful states of subjective knowledge at transformations by extracting quantities on image detection. In this comparative survey, it is explored about image augmentation using deep learning techniques. A vast literature survey related to a large dataset, overfitting, number of classifiers, edge detection, pooling, and dimensionality reduction was reviewed and compared with some viewpoints, such as

Table 1 Comparative study with convolutional neural network model

Author/year	Model	Dataset	Merit	Demerit
Mzoughi et al. [21]/(2020)	CNN	Clinical dataset	Kernel size reducing the image weight	Capsule networks for image enhancement
Moon et al. [16]/(2020)	CNN	3-D ABUS	Decrease training time and fault rate	Insufficient 3D images for accurate detection
Rai et al. [18]/(2020)	CNN	MRI	Minimum layer and less complexity	Minimum input layer for classification
Alzubaidi et al. [24]/(2019)	CNN	754-ft images	Minimum computational cost	Algorithm failed for detection
Jain et al. [25]/(2020)	CNN	COVID19 X-ray images	Less requirements for computational process	Sensitivity for detection

Table 2 Comparative study with deep convolutional neural network model

Author/year	Model	Dataset	Merit	Demerit
Hosny et al. [20]/(2020)	DeepCNN	BraTS-2018	Inception module loss 3-classifier	Ensemble model for image identification
Pasyar et al. [22]/(2020)	Deep CNN	ILSVRC	Weighted probability	Algorithm failed to work in three-class classifier
Abrishami et al. [27]/(2020)	DeepCNN	CIFAR100 CIFAR10	Cost reduction while transfer learning	Embedding space for more complexity
Nayak et al. [26]/(2020)	DeepCNN	MD-1, MD-2	Promote automatic feature learning—sequence of hidden layers	Inadequate 3D images for detection
Wang et al. [28]/(2020)	DeepCNN	CPAF	Training time	Kernel size for classification

datasets, methodology, merits, and demerits of the existing models. This paper identifies the core deep learning models and related techniques that have been applied to the image data that is unstructured. Based on the comparative study, a research gap on feature extraction is noted for the image augmentation techniques.

5 Conclusion

At present, extracting knowledge from complex medical image data is more crucial. It is important to use sophisticated analytics techniques in this data-rich age to generate useful knowledge and information about massive, complex datasets. In this survey, a massive image augmented analytics using deep learning techniques, research papers

Table 3 Comparative study with generative adversarial network model

Author/year	Model	Dataset	Merit	Demerit
Abdelhalim et al. [19]/(2020)	GAN	HAM10000	Attention mechanism for image classification	Convincing on 600 * 450 resolution of the image
Loey et al. [23]/(2020)	CGAN	COVID-19 CT scan	Two class classification	Neutrosophic approach
Karakanis et al. [17]/(2020)	CGAN	COVID-19 CXR	Biased model	Pre-trained weights
Kaur et al. [14]/(2020)	GAN	PPMI dataset	Bilateral filter for noise reduction	Overfitting
Saini et al. [15]/(2020)	DCGAN	BreakHis dataset	Global average pooling	Failed in sub-optimal performance

are reviewed for identifying the research gap. The research merits and demerits are thoroughly identified. This survey does not concentrate on image segmentation and image processing. Hence, with the help of this study, it is motivated to proceed with the research work is dimensionality reduction, pooling, noise reduction, large dataset, object detection and overfitting using deep learning models are GAN, CNN, and DCNN for image augmentation.

References

1. J. Ding, X. Li, X. Kang, V.N. Gudivada, A case study of the augmentation and evaluation of training data for deep learning. J. Data Info. Qual. **11**, 1–22 (2019)
2. H.E. Zadeh, K. Koutini, P. Primus, V. Haunschmid, M. Lewandowski, W. Zellinger, B.A. Moser, G. Widmer, On Data Augmentation and Adversarial Risk: An Empirical Analysis. arXiv:2007.02650v1 [cs. LG] (2020)
3. C. Shorten, T.M. Khoshgoftaar, A survey on image data augmentation for deep learning. J. Big Data **6**, 60 (2019)
4. A.R. Pathak, M. Pandey, S. Rautaray, Application of deep learning for object detection. Procedia Comput. Sci. **132**, 1706–1717 (2018)
5. M.Z. Alom, T.M. Taha, C. Yakopcic, S. Westberg, P. Sidike, M.S. Nasrin, M. Hasan, B.C.V. Essen, A.A.S. Awwal, V.K. Asari, A state-of-the-art survey on deep learning theory and architectures. Electronics **8**, 292 (2019)
6. S. Pouyanfar, S. Sadiq, Y. Yan, H. Tian, Y. Tao, M.P. Reyas, M. Shyu, S.C. Chen, S.S. Iyengar, A survey on deep learning: algorithms, techniques, and applications. ACM Commun. Surv. **51**, 1–36 (2018)
7. A. Mikolajczyk, M. Grochowski, Data augmentation for improving deep learning in image classification problem. IIPhDW 1–6 (2018)
8. A. Qayyuma, S.M. Anwar, M. Awais, M. Majida, Medical image retrieval using deep convolutional neural network. Neural Comput. **266**, 8–20 (2017)
9. X. Feng, Q.M.J. Wu, Y. Yang, L. Cao, An auto encoder-based data augmentation strategy for generalization improvement of DCNNs. Neural Comput. **402**, 283–297 (2020)

10. F. Chen, N. Wang, J. Tang, D. Liang, H. Feng, Self-supervised data augmentation for person re-identification. Neural Comput. **415**, 48–59 (2020)
11. S. Liu, G. Tian, Y. Xu, A novel scene classification model combining ResNet based transfer learning and data augmentation with a filter. Neural Comput. **338**, 191–206 (2019)
12. S. Umer, A. Sardar, B.C. Dhara, R.K. Raout, H.M. Pandey, Person identification using fusion of iris and periocular deep features. Neural Netw. **122**, 407–419 (2020)
13. Y. Fua, X. Li, Y. Yea, A multi-task learning model with adversarial data augmentation for classification of fine-grained images. Neural Comput. **337**, 122–129 (2020)
14. S. Kaur, H. Aggarwal, R. Rani, Diagnosis of Parkinson's disease using deep CNN with transfer learning and data augmentation. Multimed. Tools Appl. 1–27 (2020)
15. M. Saini, S. Susan, Deep transfer with minority data augmentation for imbalanced breast cancer dataset. Appl. Soft. Comput. **97**, 1–44 (2020)
16. W.K. Moon, Y.S. Huang, C.H. Hsu, T.Y.C. Chein, J.M. Chang, S.H. Lee, C.S. Huang, R.F. Chang, Computer-aided tumor detection in automated breast ultrasound using a 3-D convolutional neural network. Comput. Methods Programs Biomed. **190**, 1–9 (2020)
17. S. Karakanis, G. Leontidis, Lightweight deep learning models for detecting COVID-19 from chest X-ray images. Comput. Biol. Med. **130**, 1–9 (2021)
18. H.M. Rai, K. Chatterjee, Detection of brain abnormality by a novel Lu-Net deep neural CNN model from MR images. Mach. Learn. Appl. **2**, 1–10 (2020)
19. I.S.A. Abdelhalim, M.F. Mohamed, Y.B. Mahdy, Data augmentation for skin lesion using self-attention based progressive generative adversarial network. Expert Syst. Appl. **165**, 1–13 (2021)
20. K.M. Hosny, M.A. Kassem, M.M. Foaud, Skin melanoma classification using ROI and data augmentation with deep convolutional neural networks. Multimed. Tools Appl. 24029–24055 (2020)
21. H. Mzoughi, I. Njeh, A. Wali, M.B. Slima, A.B. Hamida, C. Mhiri, K.B. Mahfoudhe, Deep multi-scale 3D convolutional neural network (CNN) for MRI gliomas brain tumor classification. J. Dig. Imaging 903–915 (2020)
22. P. Pasyar, T. Mahmoudi, S.Z.M. Kouzehkanan, A. Ahmadian, H. Arabalibeik, N. Soltanian, A.R. Radmard, Hybrid classification of diffuse liver diseases in ultrasound images using deep convolutional neural networks. Inform. Med. Unlock. **22**, 1–27 (2020)
23. M. Loey, G. Manogaran, N.E.M. Khalifa, A deep transfer learning model with classical data augmentation and CGAN to detect COVID-19 from chest CT radiography digital images. Neural Comput. Appl. 1–13 (2020)
24. L. Alzubaidi, M.A. Fadhel, S.R. Oleiwi, O.A. Shamma, J. Zhang, DFU_QUTNet: diabetic foot ulcer classification using novel deep convolutional neural network. Multimed. Tools Appl. **79**, 15655–215677 (2019)
25. P. Gifani, A. Shalbaf, M. Vafaeezadeh, Automated detection of COVID-19 using ensemble of transfer learning with deep convolutional neural network based on CT scans. J. Comput. Assist. Radiol. Surg. **16**, 115–123 (2020)
26. D.R. Nayak, R. Dashb, B. Majhi, Automated diagnosis of multi-class brain abnormalities using MRI images: a deep convolutional neural network based method. Pattern Recogn. Lett. **138**, 385–391 (2020)
27. M.S. Abrishami, A.E. Eshratifar, D. Eigen, Y. Wang, S. Nazarian, Efficient Training of Deep Convolutional Neural Networks by Augmentation in Embedding Space. arXiv:2002.04776v1 [cs.CV] (2020)
28. J. Wang, Y. Li, H. Feng, L. Ren, X. Du, J. Wu, Common pests image recognition based on deep convolutional neural network. Comput. Electron. Agric. **179**, 1–9 (2020)

A Comparative Analysis of Pneumonia Detection Using Various Models of Transfer Learning

Bharat Narayanan, V. A. Ashwin Kuriakose, and K. Sreekumar

Abstract Human lungs consist of compact sacs called alveoli and when a healthy person breathes, the same is filled with air. The same alveoli are filled with pus and fluid for a person with pneumonia, which causes breathing problems and limits the intake of oxygen. This serious disease can affect children very severely. Bacteria and viruses are the main cause of this life-threatening disease, and there are other risk factors also. Mostly affected are the children under the age of two and old-aged people. There are different types of diagnoses done to detect pneumonia. The most commonly used diagnosis is chest X-ray which is used to view inflammation in the lungs. In this article, we use five different types namely DenseNet201, InceptionV3, MobileNet, MobileNetV2, and MobileNetV3Large of lightweight deep convolutional neural networks to find the best transfer learning model. The various models are trained using the chest X-ray image dataset which consists of 1583 normal and 4273 chest X-rays images affected by pneumonia. Based on the results obtained by testing all the five models, it is concluded that the MobileNetV3Large model gives more accuracy to the dataset.

1 Introduction

The transfer learning models have brought a drastic change in image classification, and now, it is also implemented in medical imaging [1] that made a good development in the medical field, where we use the image as data and various researches are going on. It is used for segmentation of tumor [1] and the early detection of many cancers like breast cancer [2], leukemia [3], and malaria [4]; not only in the medical field but is also used for detecting diseases in plants [5], marine animal classification [6] and various other classifications. The various transfer learning models which

B. Narayanan (✉) · V. A. Ashwin Kuriakose · K. Sreekumar
Department of Computer Science and IT, Amrita School of Arts and Sciences, Amrita Vishwa Vidyapeetham, Kochi, India

K. Sreekumar
e-mail: sreekumar@asas.kh.amrita.edu

P. Karuppusamy et al. (eds.), *Ubiquitous Intelligent Systems*, Smart Innovation, Systems and Technologies 243, https://doi.org/10.1007/978-981-16-3675-2_10

include ResNet50 [7], VGG16 [8], etc., are used to train the ImageNet dataset for creating productive models for classifying the images in the dataset [1]. There are many vital parts in the human body and chest and which have many tissues that help provide data on the human body which are helpful for the diagnosis of different lung diseases, fracture in ribs, and different injuries and which can be easily determined by looking through X-ray images of the chest. The models of transfer learning can easily detect pneumonia without human assistance [1]. Modern science is far developed by implementing different features of image processing for the analysis of different diseases and diagnosing them easily. In the present medical science, the precious source of data is images, and analysis of this type of data is a difficult one. For easy analysis of these types of data, we use compute vision software that uses modern deep learning algorithms to analyze images easily and we can easily detect the diseases efficiently. Currently, we all are dealing with a pandemic like COVID-19 which causes many difficulties to us and this disease may also lead to Pneumonia and easy screening [9], and rapid detection is needed the doctors need to confirm that this Pneumonia is caused by COVID-19 or another infection. By using transfer learning models, it is easy to understand the type of pneumonia [10]. The pulmonary infection in the lungs causes pneumonia, and it is also determined by doctors using X-ray images of the chest. Based on the studies of WHO, pneumonia is mostly affected in children [11] and may lead to death. In India, 10 million cases are reported per year and yearly detection will help the person and recovery will be easy; if not treated early, death can also occur. The death rate of pneumonia from 1990 to 2017 is shown in Fig. 1.

This dataset consists of 5856 images. As mentioned above, there are two categories as normal–1583 images and pneumonia–4273 images. By doing a comparative study on different transfer learning models on this dataset, we aim to find the best model.

2 Related Works

Doctors always rely on the vital information provided by diagnostic medical imaging about the patient. The radiologist can help diagnose illnesses such as appendicitis, pneumonia, and the effects of trauma by reading medical images of the body, and this has given good results. In this section, a brief outline of living composition is disclosed.

At the end of 2019, the world was affected by a pandemic COVID-19, which was caused by SARS-CoV-2. The virus generally affected the respiratory system which was similar to that of pneumonia. This created tension for physicians to identify COVID-19 as well as pneumonia patients. To ease this tension, the help of technology was used. And chest X-ray was the most accurate to dissolve this problem. Chest X-ray was fast and effective in identifying COVID-19, pneumonic patients as well as healthy despite the age. Adding on, it provided accurate results in 12 min which is quite promising. Compared to other baseline models, a chest X-ray dataset showed better in the following aspects such as macro-average precision, recall, F1-Score,

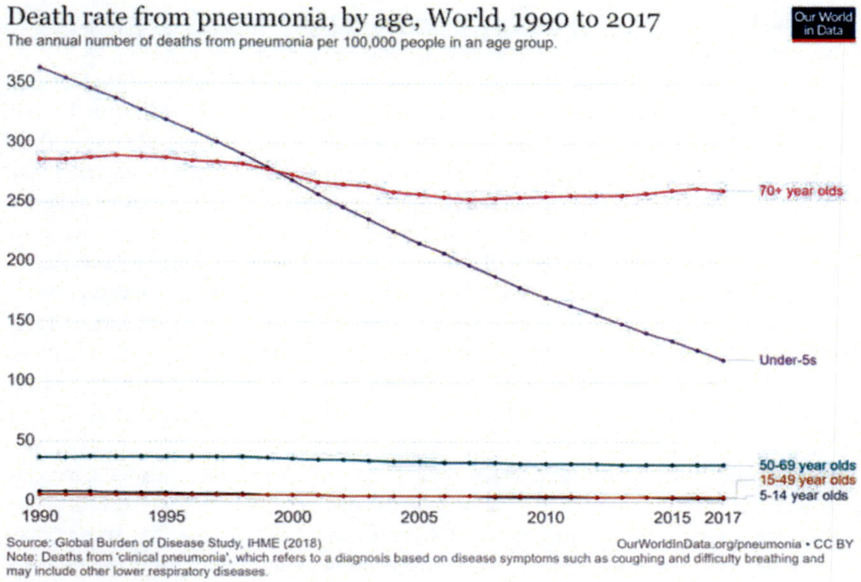

Fig. 1 The death rate of pneumonia from 1990 to 2017 [12]

and AUC. Also, it was able to identify infected with both COVID-19 and pneumonic patients. This result gives a clear image that chest X-rays can be a lot helpful in identifying and visualizing SARS-CoV-2 and certain pneumonia [10].

The chest X-rays are the best method for pneumonia diagnosis, even though there are other methods for pneumonia diagnosis such as CT of the lungs, ultrasound of the chest, needle biopsy of the lung, and MRI of the chest. Already there are several image detection techniques proposed by different authors. Various work has already been done for detecting different diseases with the help of deep learning techniques as stated by Shen [13]. Mehra [2] and Lenin [14] proposed a model that uses transfer learning for the classification of breast cancer. Mukti [5] also proposed a methodology for the detection of plant diseases. It is also an urban myth that is believed to be the most commonly used method for radiological investigation. CXR is used in wide applications such as suspected metastasis, suspected pulmonary embolism, pneumothorax, chronic dyspnea, exclude radiopaque foreign bodies, and so on.

A deep convolutional neural network (DCNN) architecture is used to do a binary classification of pneumonia images with the help of fine-tuned versions of (VGG16, VGG19, DenseNet201, Inception_ResNet_V2, Inception_V3, Resnet50, MobileNet_V2, and Xception). The work was carried using a chest X-ray and CT dataset which holds 5856 images, where 4273 were pneumonic and 1583 were healthy. The fine-tuned versions of Resnet50, MobileNet_V2, and Inception_Resnet_V2 displayed acceptable performance along with an increase in training and validation accuracy of more than 90 percent. On the other hand, the remaining tuned versions exhibited low accuracy of around 84% [15].

Furthermore, the convolutional neural network is an artificial intelligence learning method incited by the fathomless structure of the mammal brain. Multiple hidden layers in deep structure allow abstraction of multiple levels of the features. A deep network is instructed layer by layer, making it more effective. Convolution and subsampling are introduced in this technique to uproot top to bottom of features of the captured data. When it comes to computer vision, biological computation, fingerprint enhancement DCNN is off the charts based on its performance scale.

Another work uses the mathematical investigation of cough sounds to detect pneumonia. In this model, cough sounds are captured using bedside microphones from 91 patients having illnesses like pneumonia, asthma, and bronchitis. The wavelet features are then mined from all the sounds and trained a logistic classifier to discrete pneumonia from the other respiratory illnesses. This model got a sensitivity of 94% and a specificity of 64% [16].

Nowadays, the evaluation of chest X-rays is one of the prime methods for screening and detecting respiratory-related illnesses. Medical experts prefer to use a chest X-ray when it comes to detecting pneumonia disease. Even so, chest X-ray has some errors when it comes to imaging, i.e., blurred images and overlapping of organ boundaries, which can affect the detection of disease. Furthermore, these are rectified using the novel hybrid system by combining ACNN-RF. This combination has increased the accuracy to 97%. Results claim that it is far better than the primitive method being used [17].

3 Methodology

Deep learning methods have brought an outstanding change in the arena of computer vision and digital image processing. In medical imaging, the implementation of deep learning has led to the growth of medical science and early detection of disease, and new modern facilities help the human to increase his lifespan. Deep learning is now a part of every human being. Day by day, science has to grow and new technology is being implemented in medical imaging for making immense achievements in image segmentation, classification, and detection. There are different methods implemented in the detection of various cancers for rapid identification and to the necessary needs. The images from the dataset are taken and data preprocessing is done and various transfer learning models are applied and then classified into normal and pneumonia, and we compare various transfer learning models like InceptionV3, DenseNet201, MobileNet, MobileNetV2, and MobileNetV3Large on the Chest X-Ray dataset (Fig. 2).

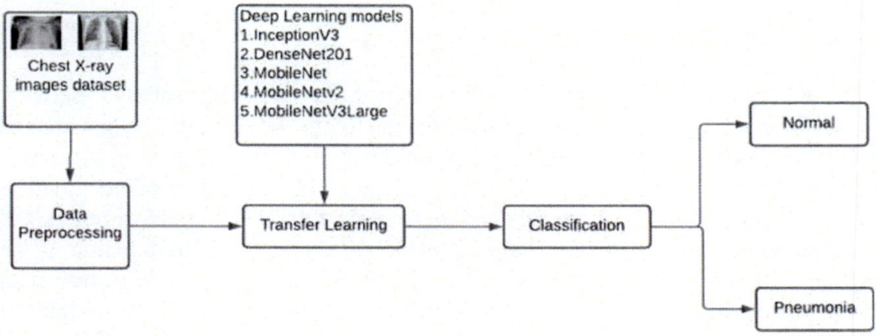

Fig. 2 Conceptual Diagram

3.1 InceptionV3

It is a category of convolutional neural network that differs from CNN in its architecture. In 2014, the inception came into existence and now there are four versions of it, and from version to version, it improves its power. The main feature of it is error rate is low compared to other models. Its architecture is 42-layers deep. The number of the network layer and neurons was increased to get high performance, but there were many disadvantages, like overfitting. To overcome this issue, GoogLeNet introduced an inception model. The main feature of it was the large size convolution kernel was replaced with a small size convolution kernel. Another feature is a new layer by the name batch normalization layer was introduced to normalize the output in each layer. The image input is taken as 299 * 299 * 3 and we get the output as 8 * 8 * 2048. Figure 3 represents the architecture of the InceptionV3 model. We can load many pre-trained models on above 10 lakhs of images from the database known as ImageNet. This network has high feature representation for a broad range of images.

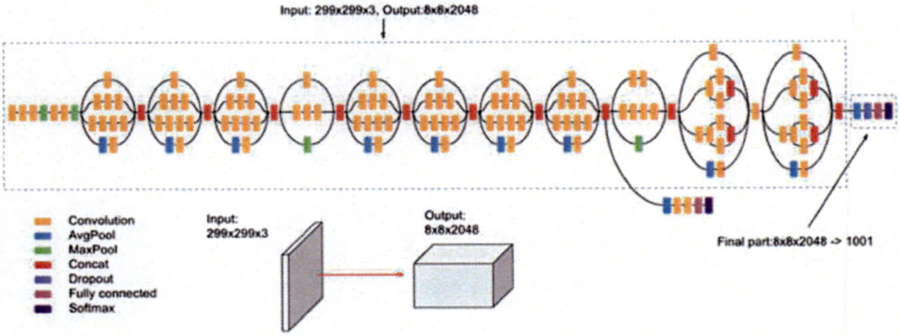

Fig. 3 The architecture of InceptionV3 [18]

3.2 DenseNet201

The main feature of DenseNet201 is we can stack a pre-trained variant of the network trained one more than 1 million images from the ImageNet dataset. This pre-trained network can convert images into 1000 object categories. This trained network will be capable of representing a wide range of images. The input size of the image is 224 * 224 and the network has 201 layers. DenseNet201 was formed from ResNet, Changes were made to ResNet and DenseNet is made, and the main feature is the layers in the network are connected. But in old models, if there are *n* number of layers there will be *n* number of connections. In some networks, we can see that the information vanishes before it reaches its correct destination, because of a distant path between the input and output layers.

As shown in Fig. 4, it is the architecture of DenseNet201, and the output of the previous layer is directed as the input to the next layer. This is done through composite operation, and this operation consists of different layers. The technical word for this type of connection is a feed-forward manner. Each layer gets the other input from all the preceding layers, and each layer transfers its feature maps to the next layer and this makes the model dense. By this architecture, the channels used for the connection will be less because of these features.

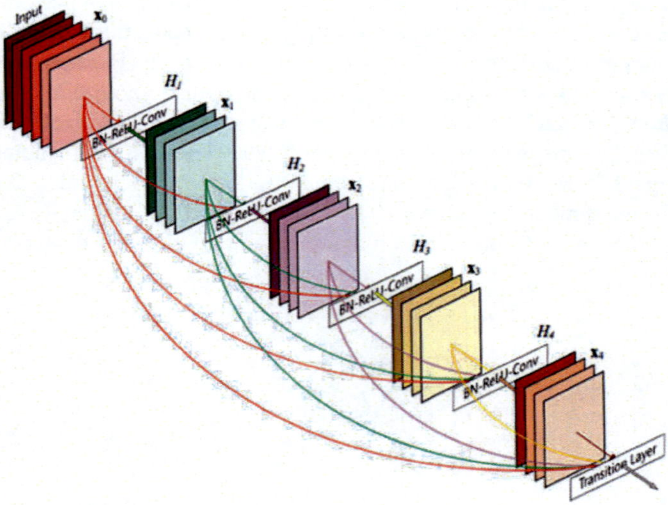

Fig. 4 The architecture of DenseNet201 [19]

3.3 MobileNet

MobileNet is a sleeked design that utilizes depth-wise distinct convolution to build a lightweight profound convolution neural network that gives a productive model for embedded vision applications. It reduces the number of parameters when we relate it with other nets. It uses separable convolution which will reduce the model size and complexity. Single convolution is performed on each input channel. In MobileNet, we can see that after each convolution batch normalization and ReLU are applied.

3.4 MobileNetV2

It is the upgrade version on Mobilenet which has the input size of the image is 224 * 224. Compared to other nets, here, a combination of two 1D convolutions with two kernels is done here, which implies less memory and parameters that are needed for training this model which gives a small and capable model. Same as the first version, we use separable convolution as an effective building block. The new features like linear bottlenecks among the layers and the shortcut connection among the bottlenecks area are added in V2.

3.5 MobileNetV3Large

It is the third version of MobileNet introduced in 2019 in ICCV in Korea. When compared to V2, V3 large is faster and more accurate in object detection. V3 large focuses on high resource use. The models were developed by implementing platform-aware NAS and NetAdapt. MobileNetV3 does not use any advanced blocks. These models are more efficient on GPU than CPU. There is one more model for V3 which is MobileN3 small, which uses low resource.

4 Experimental Analysis and Result

4.1 Dataset and Implementation

In this study, we compare the various models of transfer learning on the Kaggle dataset of chest X-ray images which consists of 1583 normal and 4273 pneumonia images. The dataset is separated into training and testing sets. For training, 4685 images are taken, and for validation, 1171 images are selected. The dataset consists of X-ray images of the chest that are screened by the specialist which are readable images with high quality. In Fig. 5, it shows (a) normal chest X-ray and (b) shows

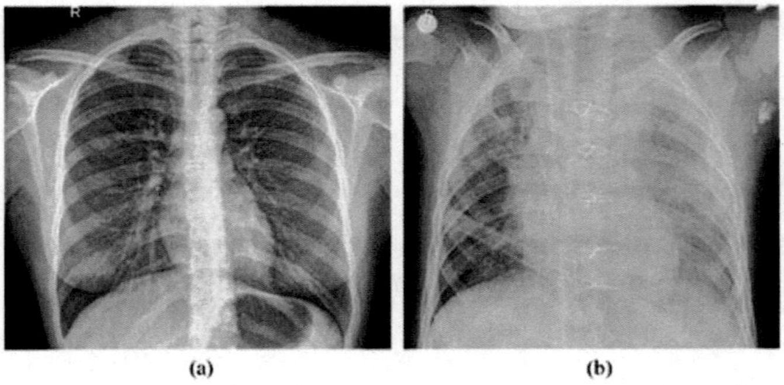

Fig. 5 **a** Chest X-ray image normal person, **b** chest X-ray image pneumonia-affected person

the chest X-ray which is pneumonia effected.

We have taken five transfer learning models for the process, and the models which are pre-trained from that we want to find which one is better. The process of feature extraction and classification is done here. The data was divided for training set and validation set as 80% and 20%, respectively, from the validation set test batches are created. The various factors like accuracy, loss, and validation loss is taken for the comparison of the five models.

4.2 Results

We have selected various transfer learning models for our comparative study on the dataset to detect pneumonia. After the implementation, the results of the five models are taken and MobileNetV3Large achieved 96.43% accuracy, MobileNetV2 achieved 93.75% of accuracy, DenseNet201 achieved 92.86% accuracy, MobileNet acquires 91.07%, and inceptionV3 achieved 91.07%. Table 1 shows the result obtained on

Table 1 Result of five transfer learning models on chest X-ray images

Model	Accuracy (%)	Loss (%)	Validation loss (%)
MobileNetV3Large	96.43	8.21	10.45
MobileNetV2	93.75	12.73	16.70
MobileNet	91.07	17	19.03
DenseNet201	92.86	18.36	20.11
InceptionV3	91.07	23.04	18.47

various transfer learning models with accuracy, loss, and validation loss.

InceptionV3

It is a transfer learning model from Google that has been created for easy image analysis, the input image size of this model is 299 * 299 with a color depth of 3 and achieved test accuracy of 91.07%. Figure 6a shows the train and validation accuracy of this model. Figure 6b shows the training and validation loss.

DenseNet201

The transfer learning model has 201 layers and each layer has the information of the layer next to it and each layer is connected. The input size of the image in the model is 224 * 224 and the color depth is 3 and we obtained a test accuracy of 92.86%. Figure 7a depicts the training and validation accuracy of this model. Figure 7b shows the training and validation loss.

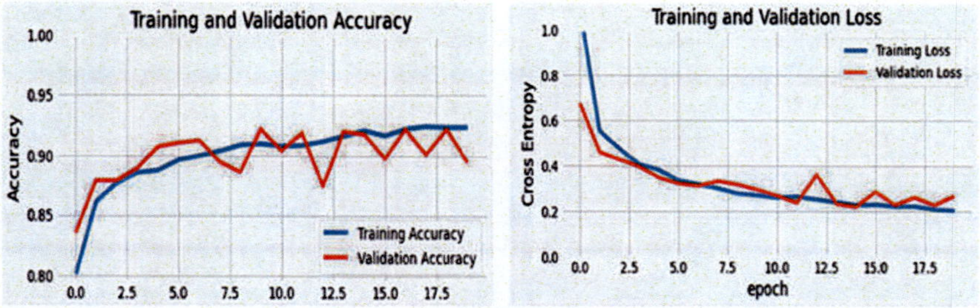

(a) (b)

Fig. 6 **a** Plot of InceptionV3 with training and validation accuracy in each epoch, **b** plot of InceptionV3 with training and validation loss in each epoch

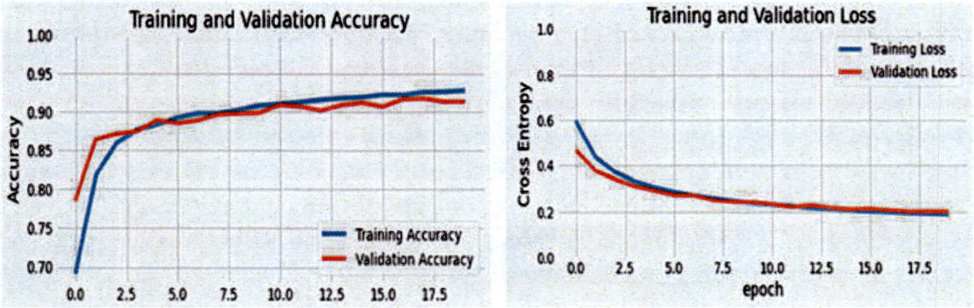

(a) (b)

Fig. 7 **a** Plot of DenseNet201 with training and validation accuracy in each epoch, **b** plot of DenseNet201 with training and validation loss in each epoch

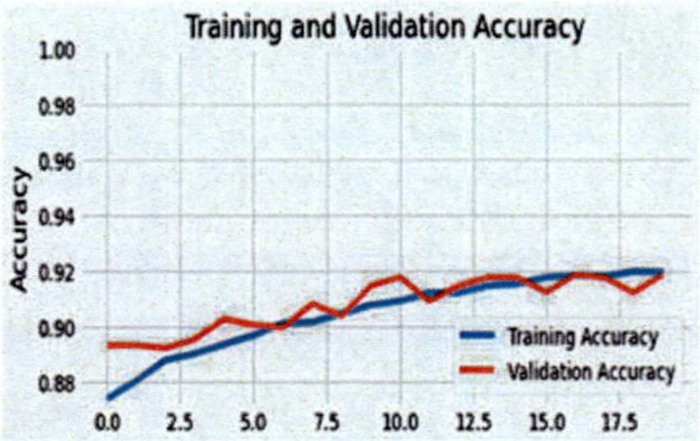

(a)

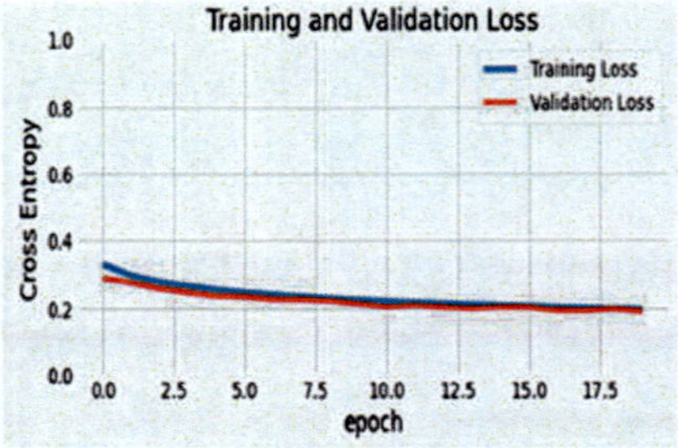

(b)

Fig. 8 **a** Plot of MobileNet with train and validation accuracy in each epoch, **b** plot of MobileNet with train and validation loss in each epoch

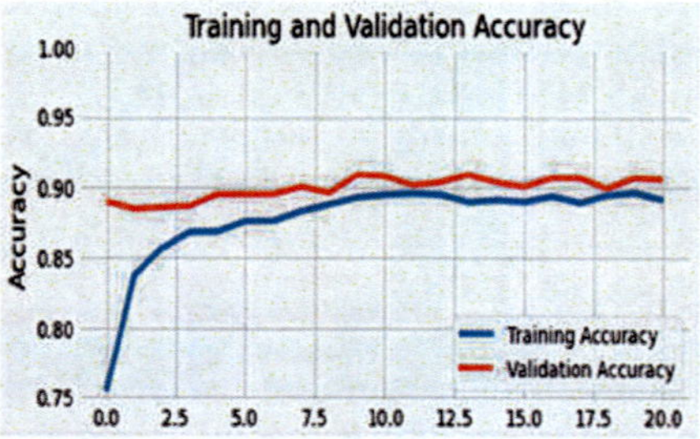

(a)

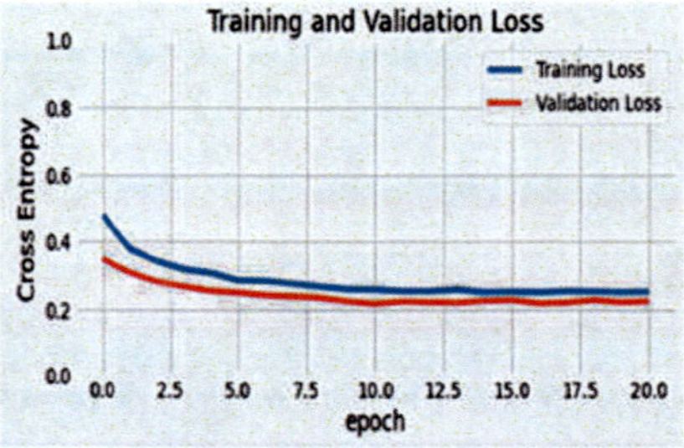

(b)

Fig. 9 **a** Plot of MobileNetV2 with train and validation accuracy in each epoch, **b** plot of MobileNetV2 with train and validation loss in each epoch

MobileNet

This model is 30 layers deep with an input image of size 224 * 224 and color depth is 3 we obtained an accuracy of 91.07%. Figure 8a shows the training and validation accuracy of this model and Fig. 8b shows the training and validation loss.

MobileNetV2

Depth-wise separable convolution is used here which means 1D convolution is applied on two kernels that use less memory and fewer parameters and we get a test accuracy of 93.75%. Figure 9a shows the training and validation accuracy and Fig. 9b shows the training and validation loss.

5 MobileNetV3Large

It is the new version of mobilenet and this model is with image input size as 224 * 224 and color depth as 3. Figure 10a shows the training and validation accuracy, and

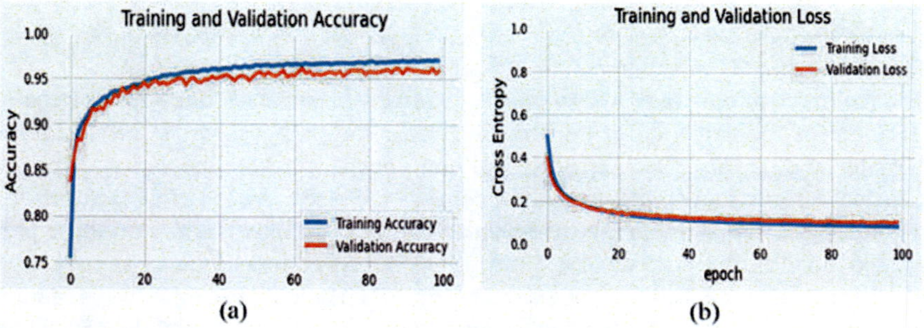

Fig. 10 **a** Plot of MobileNetV3Large with training and validation accuracy in each epoch, **b** plot of MobileNetV3Large with train and validation loss in each epoch

```
Predictions:
 [1 0 1 1 1 1 1 0 1 1 1 1 1 0 1 0 1 1 1 1 1 0 1 1 1 1 0 0 1 0 1 1]
Labels:
 [1 0 1 1 1 1 1 0 1 1 1 1 1 0 1 0 1 1 1 1 1 0 1 1 1 1 0 0 1 0 1 1]
```

Fig. 11 MobileNetV3Large model predicted against test-batch

Fig. 10b shows the training and validation loss of this model. We got a test accuracy of 96.43%.

After the comparison of various transfer learning models, MobileNetV3Large gets 96.49% of test accuracy which is better than other models. This model is fast when compared to other models. Figure 11 shows the prediction of MobileNetV3Large test sets.

6 Discussion

Now the backbone of medical imaging is deep learning, and with its support, medical field has received many benefits like early detection of disease that may cause the death of humans. Pneumonia is one type of life-threatening disease, and it should be early detected, so it will be helpful for the doctors for the treatment. In every field of medical science, computer-aided diagnosing system has to be implemented to make diagnosing and treatment easier. The main complication in this field is the unavailability of the dataset for different diseases, there should be proper datasets for getting better results. The limitation of our proposed work is it can only identify whether it is pneumonic or normal, but it cannot identify whether the pneumonia is viral or bacterial. The models we took for the comparison take less time for training and give better accuracy. The validation loss for the models is less and it improves the model. We have done feature extraction techniques in our work as a future work feature selection can be done instead of extraction.

7 Conclusion

The main aim of our study was to conduct a review of various models using transfer learning for the detection of pneumonia from X-ray images. MobileNetV3Large, MobileNetV2, MobileNet, DenseNet201, and InceptionV3 are the various types of transfer learning models we used in the comparison study. We took these models for our study because the training time needed for these models is low, and these pre-trained models give more accuracy. InceptionV3 showed less accuracy as compared to all the other models, i.e., 91.07% accuracy, 23.04% loss, and 18.47% validation loss. DenseNet201 and MobileNetV2 got almost the same accuracy, that is, 92.86 and 93.75%. MobileNetV3Large shows the best performance, 96.43%, as compared to all the other models and it is the best model on this dataset. Additionally, the obtained results showed that MobileNetV3Large gave the best performance with an accuracy of 96.43% in comparison with the rest of the models used in this analysis which is less than 94%. In the current scenario, we know that people accept things with better accuracy and less time for the operation. When we compare the models, we can understand that the MobileNetV3Large took less time for the training on this dataset than other models, and the training loss is less for the same model when compared with others. When we look at all the fields for the comparison like loss, validation loss, accuracy, and training time, MobileNetV3Large gives better results than other models, and we concluded that MobileNetV3large is the better model on chest X-ray images dataset.

References

1. G. Labhane, et al. Detection of pediatric pneumonia from chest x-ray images using cnn and transfer learning, in *2020 3rd International Conference on Emerging Technologies in Computer Engineering: Machine Learning and Internet of Things (ICETCE)* (IEEE, 2020)
2. R. Mehra, Breast cancer histology images classification: Training from scratch or transfer learning? ICT Express **4**(4), 247–254 (2018)
3. Y. Li, et al., Accuracy of deep learning for automated detection of pneumonia using chest X-Ray images: a systematic review and meta-analysis. Computers Biol. Med. 103898 (2020)
4. A. Reddy, S. Bharadwaj, D. Sujitha Juliet, Transfer learning with ResNet-50 for malaria cell-image classification, in 2019 International Conference on Communication and Signal Processing (ICCSP) (IEEE, 2019)
5. I.Z. Mukti, D. Biswas, Transfer learning based plant diseases detection using ResNet50, in *2019 4th International Conference on Electrical Information and Communication Technology (EICT)* (IEEE, 2019)
6. X. Liu, et al., Real-time marine animal images classification by embedded system based on mobilenet and transfer learning, in *OCEANS 2019-Marseille* (IEEE, 2019)
7. S.M.H. Hossain, S.M. Raju, A.R. Ismail, Predicting pneumonia and region detection from X-Ray images using deep neural network (2021). arXiv preprint arXiv:2101.07717
8. S.S. Yadav, S.M. Jadhav, Deep convolutional neural network based medical image classification for disease diagnosis. J. Big Data **6**(1), 1–18 (2019)
9. E. Verenich, et al., Improving explainability of image classification in scenarios with class overlap: application to COVID-19 and pneumonia (2020). arXiv preprint arXiv:2008.02866

10. J.E. Luján-García, et al. Fast COVID-19 and pneumonia classification using chest X-ray images. Mathematics **8**(9), 1423 (2020)
11. A. Saraiva, A. Andrade, et al., Classification of images of childhood pneumonia using convolutional neural networks. BIOIMAGING (2019)
12. https://ourworldindata.org/pneumonia
13. D. Shen, Wu. Guorong, H.-I. Suk, Deep learning in medical image analysis. Annu. Rev. Biomed. Eng. **19**, 221–248 (2017)
14. L.G. Falconí, M. Pérez, W.G. Aguilar, Transfer learning in breast mammogram abnormalities classification with mobilenet and nasnet, in *2019 International Conference on Systems, Signals and Image Processing (IWSSIP)* (IEEE, 2019)
15. K.E. Asnaoui, Y. Chawki, A. Idri, Automated methods for detection and classification pneumonia based on X-ray images using deep learning (2020). arXiv preprint arXiv:2003.14363
16. K. Kosasih, et al., Wavelet augmented cough analysis for rapid childhood pneumonia diagnosis. IEEE Trans. Biomed. Eng. **62**(4), 1185–1194 (2014)
17. H. Wu, et al., Predict pneumonia with chest X-ray images based on convolutional deep neural learning networks. J. Intell. Fuzzy Syst. Preprint, 1–15 (2020)
18. https://cloud.google.com/tpu/docs/images/inceptionv3onc--oview.png
19. https://pytorch.org/assets/images/densenet1.png

Performance Enhancement of Suspension System of an Electric Vehicle Using Nature Inspired Meta-Heuristic Optimization Algorithm

Megha Khatri, Pankaj Dahiya, and Akshat Chaturvedi

Abstract To achieve the stability for in-wheel suspension system of electric vehicle, the feedback controller gains of proportional-integral, proportional-integral-derivative, and cascaded proportional-integral-proportional-derivative controller are tuned using a meta-heuristic based flower pollination algorithm to obtain the vehicle handling stability and control. Simulations using proposed algorithm are compared with other methods to showcase the effectiveness of the selected algorithm. The results are indicating better stabilization of suspension systems in regard to displacement and acceleration of vehicle body and wheel.

1 Introduction

The demand of electrical vehicles has been accelerated in recent years primarily due to the increase in consumable fuel prices and consumption. The air quality index of the metro cities is dropping with the air pollution caused by conventional vehicles. The gasoline vehicles convert approximately 17–21% of the fuel to power the wheels whereas about 59–62% of electrical energy obtained from the grids by electric vehicles can be converted into useful mechanical energy as per the U.S. Environment Protection Agency [1]. Therefore, the usage of electrical vehicle is future because it reduces air pollution, helps to improve climate changes and global warming. However, the batteries used in these vehicles have disposal issues because of harmful chemicals used in it [2, 3].

The mechanical propulsion system of electric vehicles may have amalgamated motor driven and in-wheel motor driven arrangement. These in-wheel electric motors are installed to obtain measurable high torque. Apart from this, the configuration has

M. Khatri (✉) · A. Chaturvedi
School of Electronics and Electrical Engineering, Lovely Professional University, Phagwara, Punjab, India
e-mail: megha.25035@lpu.co.in

P. Dahiya
Department of Electronics and Communication Engineering, Delhi Technoogical University, New Delhi, India

P. Karuppusamy et al. (eds.), *Ubiquitous Intelligent Systems*, Smart Innovation, Systems and Technologies 243, https://doi.org/10.1007/978-981-16-3675-2_11

simple design, fast response to the controllers and ability to generate forward and reverse torques without affecting the driveshaft [4]. It also provides the flexibility to control each wheel independently [5]. However, it has some drawbacks such as air gap eccentricity of the motor, unsprung weight, which reduces the comfort, road holding and performance.

In past decades, the suspension system of conventional vehicles is improved, but still suffers with the suspension nonlinearities, external disturbances and uncertainty. An active suspension control and optimization for electric vehicles are two crucial and effective methods to deal with these issues [6–9]. The electro-hydraulic, electromagnetic and regenerative active suspension systems stabilize the vehicle motion and provides comfortable ride with safety [10]. The non-linearity disturbances and uncertainty with suspension systems can be regulated by utilizing various control methods like sliding mode control, robust H∞ control, preview control, optimal control used for the controllers [11, 12].

In this article, the meta-heuristic approach is implemented to obtain the optimum control of the active suspension of electrical vehicles with in-wheel drives. The comparison between proportional-integral linear quadratic regulator (PI-LQR) [13–15], proportional-integral linear matrix inequality (PI-LMI) [16], proportional-integral-derivative flower pollination algorithm (PID-FPA) and proportional-integral-proportional-derivative flower pollination algorithm (PIPD-FPA) based controllers for active suspension damper has been carried out [17–21]. The supremacy of the proposed control algorithm in terms of vehicle body displacement, vehicle body acceleration, wheel displacement, wheel acceleration has been realized under step and sinusoidal excitation. Thus, enhance the vehicle handling stability and control.

The article is structured as follows: in Sect. 2 the operating model of an active suspension system. Section 3 is about the control algorithm for the suspension system, followed by Sect. 4 with discussion on the experimental validation of the proffered controller structure.

2 Electric Vehicle Suspension System

An active suspension system with spring and damper is shown in Fig. 1. The ride comfort as specified in ISO 2631 can be quantified by vehicle dynamic response [22]. The parameters discussed in the introduction must be optimized to suppress vehicle vibration for road building stability and mechanical structural damage. The two degree of freedom suspension model with vertical motion of sprung (m_1) and unsprung mass (m_2); vertical displacements x_1, x_3; road disturbance $w1$; damping coefficients (c_1 and c_2); suspensions and wheel stiffness k_1 and k_2, respectively [23–27]. The dynamic model is expressed in state space Eq. 1.

$$\dot{x}(t) = Ax(t) + Bu(t) + Fw(t) \tag{1}$$

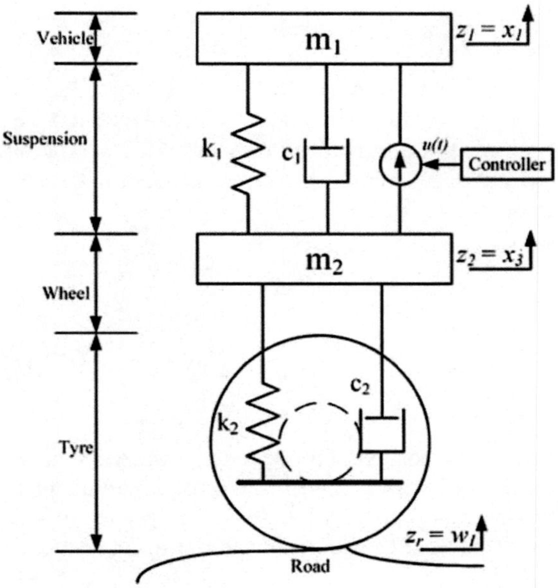

Fig. 1 Suspension system

where the vehicles state vector $x(t) = \left[z_1\ \dot{z}_1\ z_2\ \dot{z}_2\right]^T$.

$$A = \begin{bmatrix} 0 & 1 & 0 & 0 \\ -\frac{k_1}{m_1} & -\frac{c_1}{m_1} & \frac{k_1}{m_1} & \frac{c_1}{m_1} \\ 0 & 0 & 0 & 1 \\ \frac{k_1}{m_2} & \frac{c_1}{m_2} & -\frac{k_1+k_2}{m_2} & -\frac{c_1+c_2}{m_2} \end{bmatrix}, B = \left[0\ \frac{1}{m_1}\ 0\ -\frac{1}{m_2}\right]^T,$$

$$F = \begin{bmatrix} 0\,0\,0\,\frac{k_2}{m_2} \\ 0\,0\,0\,\frac{c_2}{m_2} \end{bmatrix}^T$$

In this model, an idle control input, i.e., step signal and sinusoidal input are applied to the suspension system with a state feedback controller that is equivalent to the idle control.

3 Controller Design

An estimated state feedback controller with controllable and observable LTI state-space under observation is

$$\dot{X} = AX + BU + \Gamma P_d \tag{2}$$

$$Y = CX \tag{3}$$

where X = state vector, U = control vector, P_d = disturbance vector, Y = output vectors, A = system matrix, B = control matrix, C = output matrix, and Γ = disturbance matrix of suitable dimensions [28–32]. The full state vector feedback control law in Eq. (4), and the performance index J is computing the error minimization is presented in Eq. (5)

$$U^* = -K^* X, \tag{4}$$

$$J = \int_0^\infty [X^T Q X + U^T R U] \frac{1}{2} \mathrm{d}t \tag{5}$$

where Q = positive semi-definite symmetric state, R = positive definite symmetric control cost weight matrix and must assure the determinacy as: $Q \geq 0$ and $R > 0$. However, post disturbance steady state values are obtained by replacing the terms Γ and P_d in Eq. (2), with redefined states and controls as in Eq. (6)

$$\dot{X} = AX + BU + \Gamma P_d X(0) = X_0 \tag{6}$$

The application of Pontryagin's minimum principle for finite time problems offers continuous time algebraic matrix Riccati equation [30]:

$$PA + A^T P - PBR^{-1} B^T P + Q = 0. \tag{7}$$

From Eq. (7) P = positive definite symmetric matrix is obtained, whereas the feedback gain matrix K^* computed using MATLAB software in Eq. (8), which reduces error referring Eq. (5) is computed using the solution of Eq. (7)

$$K^* = R^{-1} B^T P \tag{8}$$

Tuning of controller using flower pollination algorithm.

The objective function decided based on the system parameter setting to ensure stability and to boost the parameters of the controller selected is integral time absolute error (ITAE) [28–30]. The concept of flower pollination algorithm is described with the help of following four important rules [31, 32]

1. Global pollination includes biotic fertilization where pollinators follow Levy's flight movement given in Eq. (9).
2. Local pollination including abiotic fertilization is presented in Eq. (11).
3. The consistency of the outcome depends on the reproduction probability and similarity of involved parameters that means connection of the pollinators such as birds, insects with the variety of flowers.

4. The probability function superintend the switching from local to global pollination

The mathematical expression of the global pollination using Lévy flight behavior is

$$X_i^{t+1} = X_i(t) + \gamma L(\rho)\big(P_* - X_i^t(t)\big) \tag{9}$$

where X_i^t = pollen i at iteration t, P_* = present best solution found in the present population, X_i^{t+1} = prospective solution for repetition at $t + 1$, γ = scaling factor decides the step size, $L(\rho)$ = step size taken from Lévy distribution presented as [33].

$$L \sim \frac{\rho\tau(\rho)\sin(\pi\rho/2)}{\pi}\frac{1}{s^{1+\rho}} \tag{10}$$

where $\tau(\rho)$ = standard gamma function and the dispersal is applicable for large steps $S > 0$. Case the arbitrary number is insignificant in comparison with the switching probability (p), then local pollination occurs that is expressed below

$$X_i^{t+1} = X_i(t) + \epsilon\big(X_a^t(t) - X_b^t(t)\big) \tag{11}$$

where X_a^t and X_b^t = flower consistency in the event of local pollination. The step size can be drawn using following equation.

$$S = \frac{A}{B^{\frac{1}{p}}},\ A \sim n\big(0, \sigma^2\big)\ \&\ B \sim n(0, 1) \tag{12}$$

Following the Gaussian distribution having random numbers A and B with variance of σ^2 and zero mean. The variance is calculated as

$$\sigma^2 = \left\{\frac{\tau(1+\rho)}{\rho\tau\left[\frac{1+\rho}{2}\right]}\cdot\frac{\sin(\pi\rho/2)}{2^{(\rho-1)/2}}\right\}^{1/\rho} \tag{13}$$

Considering the selected problem executed for 100 iterations, population size of 30 with modification probability 0.06 and interbreeding probability 0.80 to ensure the best response from the chosen algorithm with respect to linear quadratic regulator and linear matrix inequality methods. For FPA, the initial value of population size ($N = 10$) and switching probability ($p = 0.5$) and the algorithm is presented in Fig. 2 [24]. After multiple runs, the function values are evaluated by keeping one parameter fixed. The optimal solution is decided based on the value of parameter J which is compared with different algorithms.

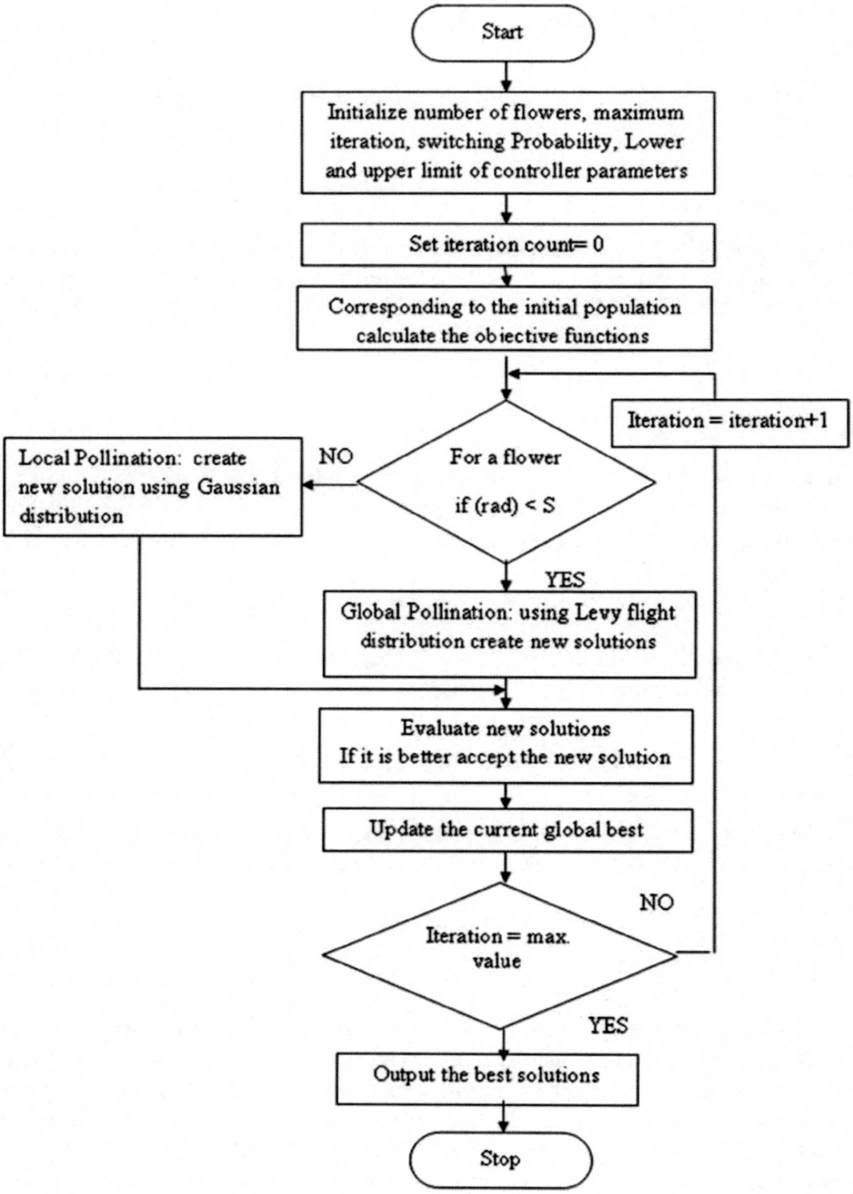

Fig. 2 FPA for control optimization

4 Results and Discussions

To validate the accomplishments of proffered control algorithm for active suspension; the step and sinusoidal road excitation are employed at 4 Hz to judge the ride performance. The system parameters are system parameters taken are $m_1 = 2.45$ kg, $m_2 = 1$ kg, $k_1 = 900$ N/m, $k_2 = 2500$ N/m, $c_1 = 7.5$Ns/m, $c_2 = 5$Ns/m for designing the optimal controller R is taken as 1 and Q is taken as 100 * I where I is identity matrix of size 4 × 4, whereas controller gains found by solving the associated Ricatti equation is $k = [0.05555, 6.7935, -27.03015, -2.657478]$.

However, the controller gains found to be k = [896.31, 3.37, − 1311.49, 8.082] with LMI and by applying FPA algorithm to the controller, gains are found to be $K_p = 10.7090$, $K_i = 1.4085$ for PI controller, $K_p = 12.5754$, $K_i = 2.1696$, $K_d = -0.0173$, for PID controller and $K_{p1} = 9.1466$, $K_i = 6.0110$, $K_d = 0.1327$, $K_{p2} = 7.4444$ for PIPD controller.

The dynamic response for step disturbance is shown in Fig. 3 with step disturbance $z_r = 0.01$ (m) and the vehicle displacement of suspension system is compared with PI, PID and PIPD controllers. It has been observed that at the response of PIPD controller tuned with FPA reach to the steady state in the very first transient cycle followed by PID-FPA, PI-FPA, PI-LQR and PI-LMI. Similarly, the vehicle body acceleration under step excitation; the response of proposed algorithm with PIPD, PID, PI is superior to the conventional PI controller with LQR and LMI as shown Fig. 4.

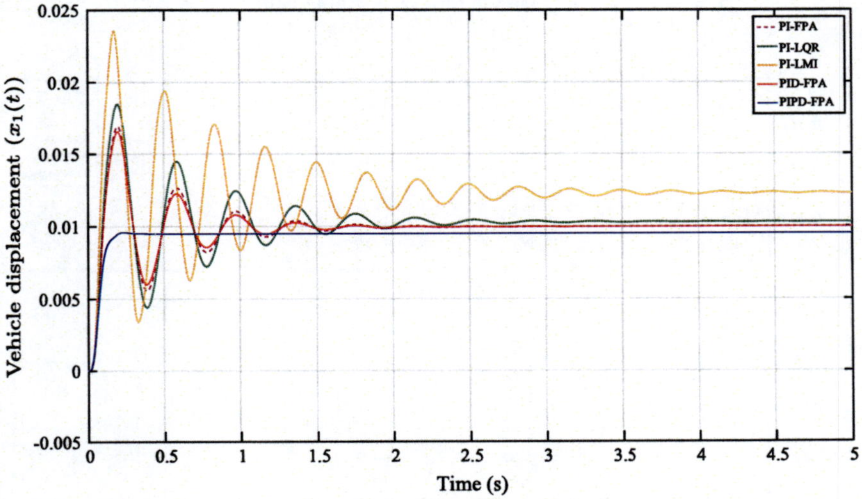

Fig. 3 Vehicle body displacement with different controller structures applicable to suspension system

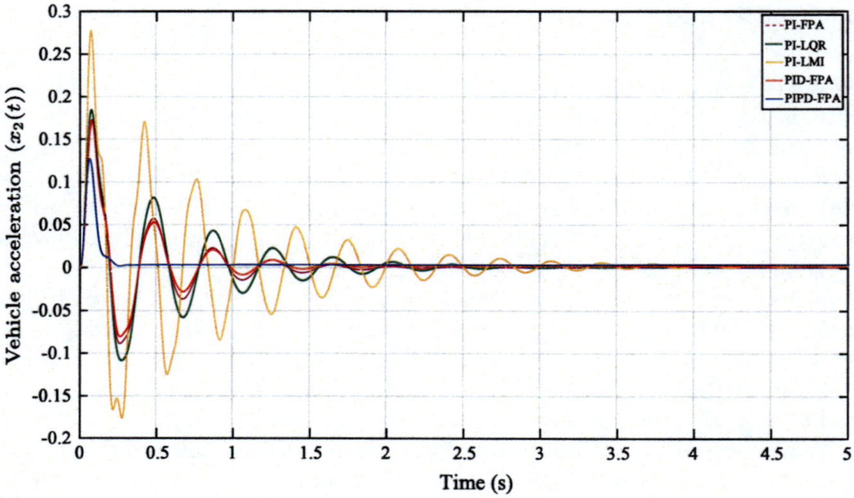

Fig. 4 Vehicle body acceleration with different controller structures applicable to suspension system

There is always external disturbance in the active control system due to displacement sensor noise which generates unstable controller output. The wheel displacement under bumpy road step excitation the PIPD-FPA controllers settle with parameters in less than and 0.5 s and provide the comfortable ride as shown in Fig. 5 and in

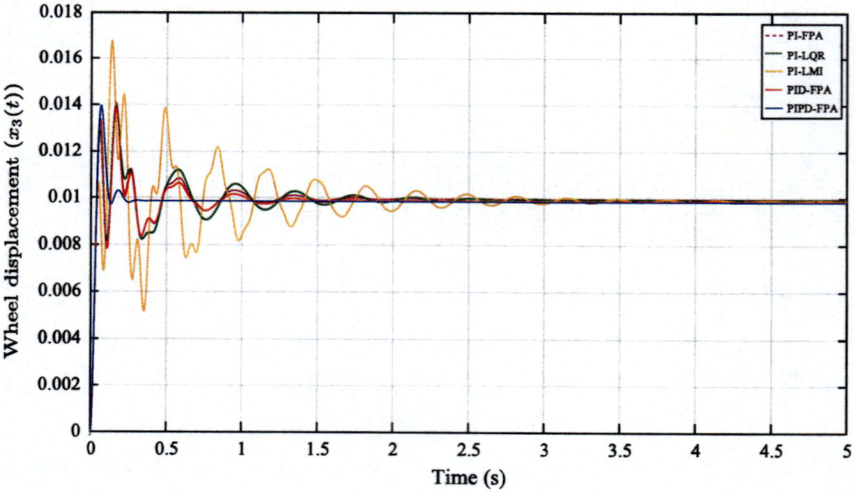

Fig. 5 Wheel displacement of the vehicle with different controller structures applicable to suspension system undergoing step excitation

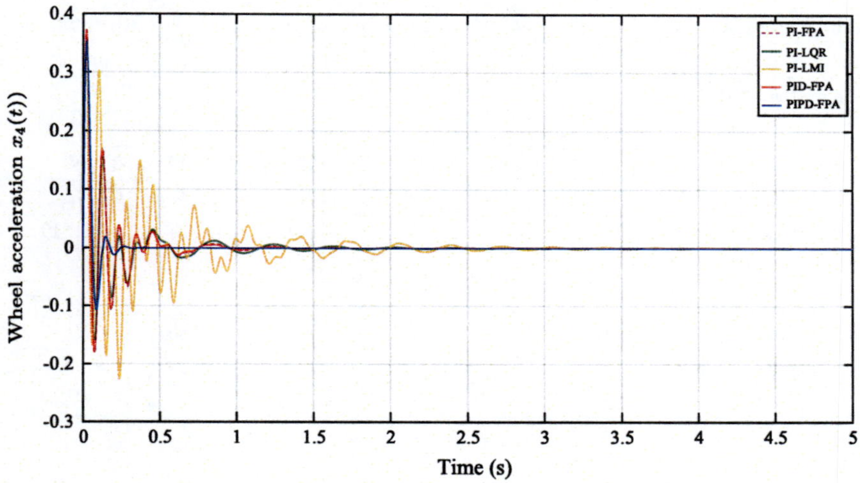

Fig. 6 Wheel acceleration of the vehicle with different controller structures applicable to suspension system

Fig. 6 the response of PIPD-FPA in vertical wheel acceleration is also appreciable. The controller parameters such as peak overshoot, peak undershoots, settling time and integral time absolute error have been elaborated in Table 1. The ITAE in case of PIPD-FPA is minimum, followed by PID-FPA, PI-FPA, PI-LQR and PI-LMI. The peak overshoot, i.e., 0.0096 is also quite less. Thus, the system performance is high with FPA optimized PIPD controller under expected instabilities.

For the sinusoidal disturbance $z_r(t) = 0.002\sin(6\pi t)$, the vertical displacement of vehicle body decreases in the active suspension system with PIPD-FPA controller as shown in Fig. 7 and in Fig. 8 the acceleration of the suspension system at input frequency of 4 Hz. The wheel displacement and acceleration under sinusoidal excitation are presented in Figs. 9 and 10. In all the discussed active suspension parameters, the response of PIPD controller with flower pollination algorithm is superior in acquiring the overall stability.

5 Conclusions

The dynamic model of suspension system of electric vehicle is presented, and the vertical motion control has been analyzed. The feedback PI, PID, PIPD controller is tuned using flower pollination algorithm. The proposed optimization technique is effectively reducing the vertical fluctuations and vibration acceleration which interns decrease the wear and tear of the whole system, extend the life of bearings and improves the comfort level of electric vehicles. With comparison to the

Table 1 Comparative performance analysis of different controllers for the applied step disturbance

Parameter		$x_1(t)$	$x_2(t)$	$x_3(t)$	$x_4(t)$
Controller (PI-FPA)					
ITAE	0.02034				
Settling time (s)		1.6026	1.6622	1.1649	0.7130
Peak overshoot (ms)		0.0169	0.1722	0.0141	0.3722
Peak uvershoot (ms)		0	− 0.0885	0	− 0.1819
Controller (PI-LQR)					
ITAE	0.04088				
Settling time (s)		2.3409	2.2519	1.5758	1.0726
Peak overshoot (ms)		0.0185	0.1829	0.0139	0.3575
Peak uvershoot (ms)		0	− 0.1092	0	− 0.1444
Controller(PI-LMI)					
ITAE	0.1059				
Settling time (s)		3.1899	3.0802	2.6602	2.0896
Peak overshoot (ms)		0.0236	0.2759	0.0168	0.3502
Peak uvershoot (ms)		0	− 0.1779	0	− 0.2259
Controller(PID-FPA)					
ITAE	0.01636				
Settling time (s)		1.4115	1.4761	0.9908	0.6593
Peak overshoot (ms)		0.0166	0.1698	0.0140	0.3722
Peak uvershoot (ms)		0	− 0.0829	0	− 0.1796
Controller(PIPD-FPA)					
ITAE	0.0007497				
Settling time (s)		0.1709	0.2137	0.2072	0.2272
Peak overshoot (ms)		0.0096	0.1238	0.0141	0.3527
Peak uvershoot (ms)		0	− 0.0015	0	− 0.1073

LQR and LMI the proposed method has the best dynamic characteristics under the same optimization conditions.

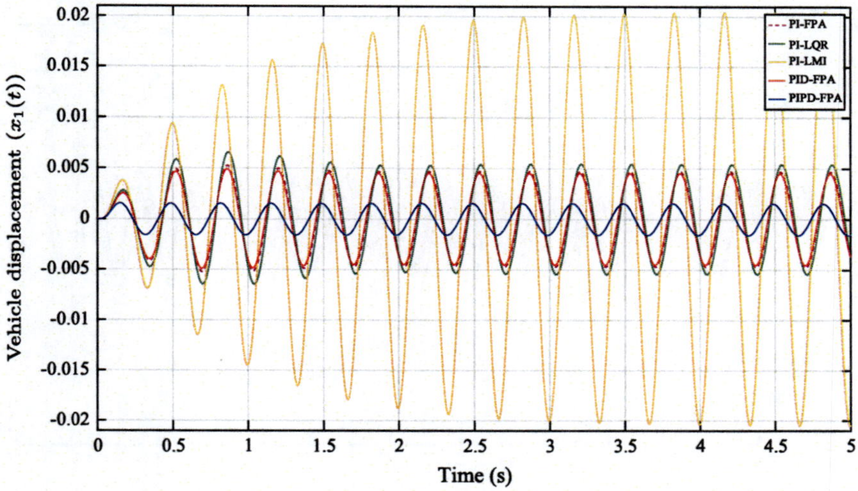

Fig. 7 Vehicle body displacement with different controller structures applicable to suspension system undergoing sinusoidal wave excitation

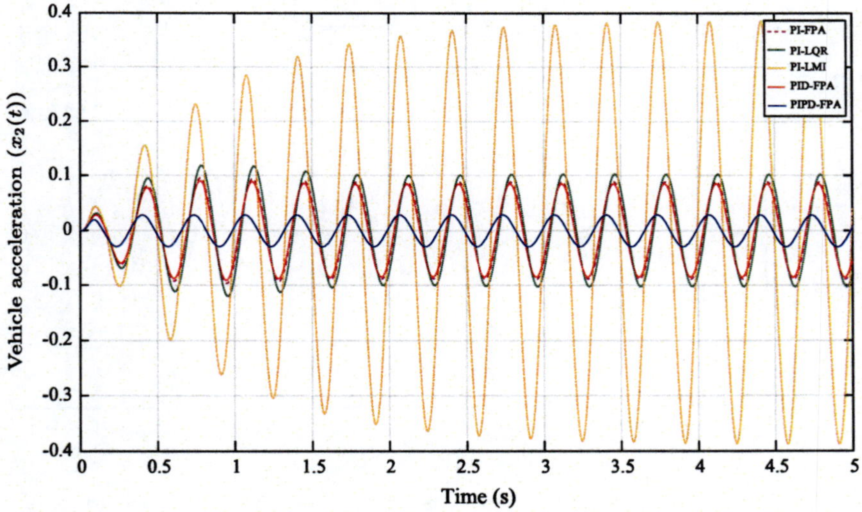

Fig. 8 Vehicle body acceleration with different controller structures applicable to suspension system undergoing sinusoidal wave excitation

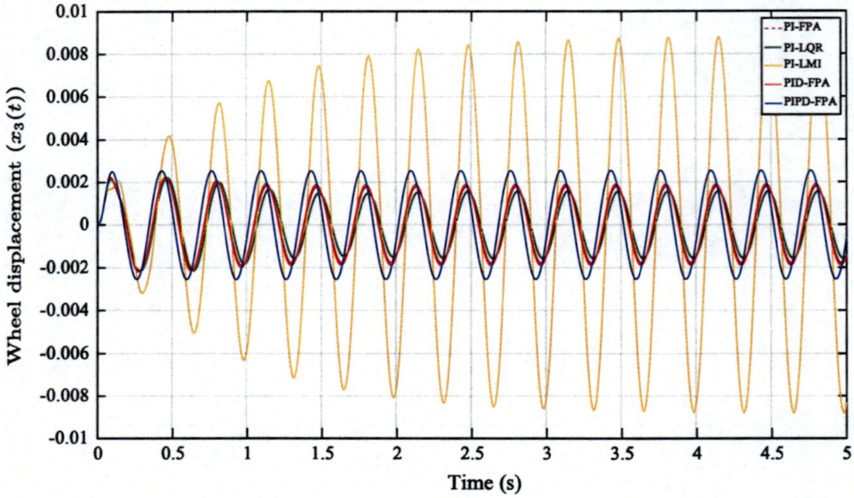

Fig. 9 Wheel displacement of the vehicle with different controller structures applicable to suspension system undergoing sinusoidal wave excitation

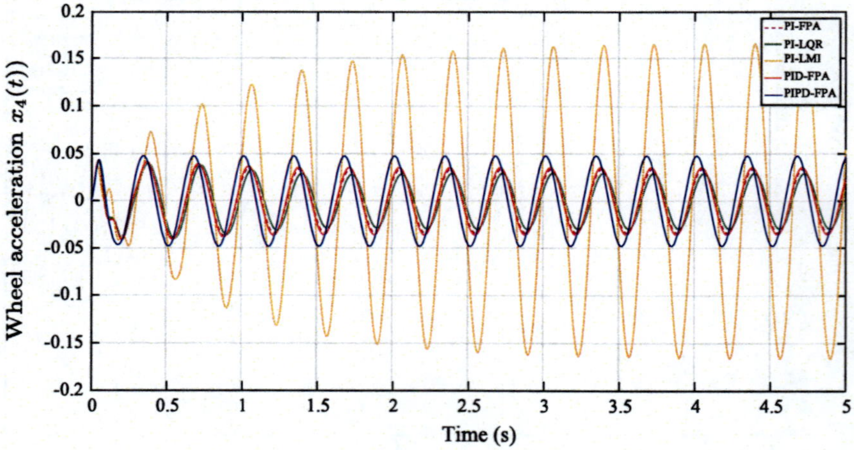

Fig. 10 Wheel acceleration of the vehicle with different controller structures applicable to suspension system undergoing sinusoidal wave excitation

References

1. W. Sun, Y. Li, J. Huang, N. Zhang, Vibration effect and control of in-wheel switched reluctance motor for electric vehicle. J. Sound Vib. **338**, 105–120 (2015)
2. Y. Wang, Y. Li, W. Sun, L. Zheng, Effect of the unbalanced vertical force of a switched reluctance motor on the stability and the comfort of an in-wheel motor electric vehicle. Proc. Inst. Mech. Eng. Part D J. Automob. Eng. **229**, 1569–1584 (2015)
3. X.D. Xue, K.W.E. Cheng, J.K. Lin, Z. Zhang, K.F. Luk, T.W. Ng, N.C. Cheung, Optimal control method of motoring operation for SRM drives in electric vehicles. IEEE Trans. Veh. Technol. **59**, 1191–1204 (2010)
4. A. Kulkarni, S.A. Ranjha, A. Kapoor, A quarter-car suspension model for dynamic evaluations of an in-wheel electric vehicle. Proc. Inst. Mech. Eng. Part D J. Automob. Eng. **232**, 1139–1148 (2018)
5. R. Vos, I.J.M. Besselink, H. Nijmeijer, Influence of in-wheel motors on the ride comfort of electric vehicles, ın *Proceedings of the 10th International Symposium on Advanced Vehicle Control (AVEC10)*, 22–26 Aug 2010, Loughborough, United Kingdom. pp. 835–840 (2010)
6. Y. Wang, Y. Li, W. Sun, C. Yang, G. Xu, FxLMS method for suppressing in-wheel switched reluctance motor vertical force based on vehicle active suspension system. J. Control Sci. Eng. (2014)
7. B. Li, H. Du, W. Li, Fault-tolerant control of electric vehicles with in-wheel motors using actuator-grouping sliding mode controllers. Mech. Syst. Signal Process. **72**, 462–485 (2016)
8. S. Ayari, M. Besbes, M. Lecrivain, M. Gabsi, Effects of the airgap eccentricity on the SRM vibrations, ın *IEEE International Electric Machines and Drives Conference. IEMDC'99. Proceedings (Cat. No. 99EX272)* (IEEE, 1999), pp. 138–140
9. Y. Wang, P. Li, G. Ren, Electric vehicles with in-wheel switched reluctance motors: coupling effects between road excitation and the unbalanced radial force. J. Sound Vib. **372**, 69–81 (2016)
10. A. Tanabe, K. Akatsu, Vibration reduction method in SRM with a smoothing voltage commutation by PWM, ın *2015 9th International Conference on Power Electronics and ECCE Asia (ICPE-ECCE Asia)* (IEEE, 2015), pp. 600–604
11. N. Nakao, K. Akatsu, Controlled voltage source vector control for switched reluctance motors using PWM method. Electr. Eng. Japan. **198**, 27–38 (2017)
12. D. Tan, H. Wang, Q. Wang, Study on the rollover characteristic of in-wheel-motor-driven electric vehicles considering road and electromagnetic excitation. Shock Vib. (2016)
13. J. Wu, A simultaneous mixed LQR/H∞ control approach to the design of reliable active suspension controllers. Asian J. Control. **19**, 415–427 (2017)
14. M.M. ElMadany, Z.S. Abduljabbar, Linear quadratic Gaussian control of a quarter-car suspension. Veh. Syst. Dyn. **32**, 479–497 (1999)
15. K.-Y. Lian, C.-H. Chiang, H.-W. Tu, LMI-based sensorless control of permanent-magnet synchronous motors. IEEE Trans. Ind. Electron. **54**, 2769–2778 (2007)
16. A. Draa, On the performances of the flower pollination algorithm–qualitative and quantitative analyses. Appl. Soft Comput. **34**, 349–371 (2015)
17. R.O. Abdel, B.M. Abdel, I. El Henawy, *A New Hybrid Flower Pollination Algorithm for Solving Constrained Global Optimization Problems* (2014)
18. E. Burzo, *PID Control: New İdentification and Design Methods* (Springer, 2010)
19. Y. Li, F. Chai, Z. Song, Z. Li, Analysis of vibrations in interior permanent magnet synchronous motors considering air-gap deformation. Energies **10**, 1259 (2017)
20. J.-W. Jung, V.Q. Leu, T.D. Do, E.-K. Kim, H.H. Choi, Adaptive PID speed control design for permanent magnet synchronous motor drives. IEEE Trans. Power Electron. **30**, 900–908 (2014)
21. H. Jing, R. Wang, C. Li, J. Wang, N. Chen, Fault-tolerant control of active suspensions in in-wheel motor driven electric vehicles. Int. J. Veh. Des. **68**, 22–36 (2015)

22. R. Wang, H. Jing, F. Yan, H.R. Karimi, N. Chen, Optimization and finite-frequency H∞ control of active suspensions in in-wheel motor driven electric ground vehicles. J. Franklin Inst. **352**, 468–484 (2015)
23. X. Shao, F. Naghdy, H. Du, Enhanced ride performance of electric vehicle suspension system based on genetic algorithm optimization, in *2017 20th International Conference on Electrical Machines and Systems (ICEMS)* (IEEE, 2017), pp. 1–6
24. M. Liu, F. Gu, Y. Zhang, Ride comfort optimization of in-wheel-motor electric vehicles with in-wheel vibration absorbers. Energies **10**, 1647 (2017)
25. F. Tahami, S. Farhangi, R. Kazemi, A fuzzy logic direct yaw-moment control system for all-wheel-drive electric vehicles. Veh. Syst. Dyn. **41**, 203–221 (2004)
26. K. Hartani, A. Draou, A. Allali, Sensorless fuzzy direct torque control for high performance electric vehicle with four in-wheel motors. J. Electr. Eng. Technol. **8**, 530–543 (2013)
27. H. Zhao, B. Gao, B. Ren, H. Chen, Integrated control of in-wheel motor electric vehicles using a triple-step nonlinear method. J. Franklin Inst. **352**, 519–540 (2015)
28. Z. Shuai, H. Zhang, J. Wang, J. Li, M. Ouyang, Lateral motion control for four-wheel-independent-drive electric vehicles using optimal torque allocation and dynamic message priority scheduling. Control Eng. Pract. **24**, 55–66 (2014)
29. Z. Shuai, H. Zhang, J. Wang, J. Li, M. Ouyang, Combined AFS and DYC control of four-wheel-independent-drive electric vehicles over CAN network with time-varying delays. IEEE Trans. Veh. Technol. **63**, 591–602 (2013)
30. P. Dash, L.C. Saikia, N. Sinha, Flower pollination algorithm optimized PI-PD cascade controller in automatic generation control of a multi-area power system. Int. J. Electr. Power Energy Syst. **82**, 19–28 (2016)
31. X.-S. Yang, Flower pollination algorithm for global optimization, in *International Conference on Unconventional Computing and Natural Computation* (Springer, 2012), pp. 240–249
32. X.-S. Yang, M. Karamanoglu, X. He, Flower pollination algorithm: a novel approach for multiobjective optimization. Eng. Optim. **46**, 1222–1237 (2014)
33. R. Storn, K. Price, Differential evolution–a simple and efficient heuristic for global optimization over continuous spaces. J. Glob. Optim. **11**, 341–359 (1997)

Advancements in Healthcare: Multi-Agent Based Intelligent Sensor Approach

Kushagra Singh Bisen

Abstract The paper studies the applications of multi-agent systems and sensors in healthcare. It goes through the current advancements in the field by introducing the literature and discussing applications using the technology in healthcare. The paper proposes a multi-agent system in a healthcare environment. It then conclude by discussing the merits and limitations of the technologies involved in the domain and their future scope.

1 Introduction

Healthcare is a rapidly changing environment with constraints being lower cost metric and resource metric and improved, dynamic feature requirements from the users as well as the hospitals. The healthcare system in nations is put to devise procedures to combat the increasing population of the elderly. The number of elderly, according to a study by the United Nations in 2015 is increasing drastically. The population of the elderly (aged over 80) is supposed to increase from 125 million in 2015 to 202 million in 2030 and to 434 million in 2050 [1]. The metric is of vital importance as many developing countries still do not have their life expectancy above 80 [2]. In conclusion, nations need major advancements in healthcare to assist the elderly to pursue an independent living. A variety of strategies could be deployed to counter the imminent growth in demand aforementioned in the report [1]. The government could decide to sanction major budgets to healthcare to address the elderly although total revenue of a country does not change over a year, cost-cutting from other necessary sectors could bring unpleasant outcomes through the course of time. The paper [3] describes a case study addressing the huge spending in the USA over healthcare. The reasons for overspending were the unnecessary health-insurance expenditure

K. S. Bisen (✉)
Université Jean Monnet, 10, Rue Tréfilerie, 42023 Saint-Étienne, France

Ecole des Mines Saint Etienne, 158 Cours Fauriel, 42023 Saint-Étienne, France
e-mail: kushagra.bisen@etu.univ-st-etienne.fr

P. Karuppusamy et al. (eds.), *Ubiquitous Intelligent Systems*, Smart Innovation, Systems and Technologies 243, https://doi.org/10.1007/978-981-16-3675-2_12

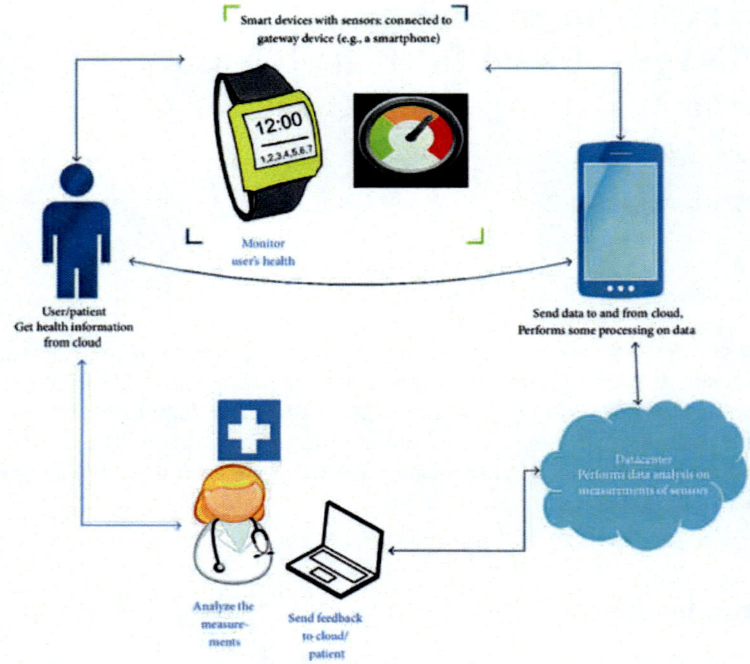

Fig. 1 The block diagram of a smart healthcare system

and poorly executed response to medical emergencies which encourage inefficient and low-effort services [3].

Multi agent-based intelligent sensors can be deployed in a healthcare environment to increase productivity as well as the quality of the services. Novel advancements in technology will be tailor-made for applying in the current infrastructure resulting. Investing in technological advancements will promote well being of the patient and healthcare worker, increase the accuracy of measurements and simultaneously will not cripple the current funds and the infrastructure [4]. The current research direction is the implementation and analysis of intelligent sensors through various methods to assist and generate data for further studies and research [5]. Technologies normally deployed in a minimal healthcare environment is seen in automating labour intensive and inefficient processes such as maintaining records. Advancements in healthcare could be seen in applications as remote monitoring systems made possible by the use of intelligent and distributed agents. The paper focuses on the present advancements made to ensure the endurance of the system to support the growing population of the elderly (Fig. 1).

1.1 Intelligent Agents

Intelligent agents are entities simulating and performing the duties which were meant to be performed by a user. They are allowed to make their own decisions by being autonomous [6] They were visualized as entities with a written set of rule to act accordingly, the visualization was effective to an instant but had its drawbacks. The particular visualization was the single state and some use cases needed the agent to change its state in applications where an external trigger event or another agent was presently interacting with the same environment. The paper [7] shows the use of feeding the agent model by collecting parameters (knowledge) which will ensure the abundance of information about the agent. As the general research direction shifted toward Machine Learning or applying Machine Learning to use, an optimum way to address this problem was presented [8]. The solution is by enabling the agent to learn the activities and patterns in the training period. The patterns will be stored in a knowledge base to be recreated when needed [6]. A feedback loop is set up with the knowledge base to test the working of the agent which if found inappropriate is fed back to the knowledge base. The agent continuously learns and trains itself while working and interacting in the environment. The feedback mechanism will ensure the perfect performance of the agent after a certain number of successful iterations.

An agent can thus be defined as an autonomous entity able to perceive the environment directly or indirectly and act accordingly to achieve the goal by following a certain set of rules. A dynamic state feedback controller for an input affine non-linear system which stabilizes the point in output space was presented by the paper [9] to yield a decentralized controller for a multi-agent system shows a mechanism for a feedback loop. Another approach for the development of an multi-agent system using feedback loop as a concept to identify organizations, complex systems are modeled with ease by using a feedback loop by enabling the cause-effect loop in between the micro and the macro of the system [10]. The approach proposed in the paper [10] consists of defining a loop pattern to provide activities and guidelines to help identify the candidates during the analysis phase of the design. In healthcare, cognitive agents are used which maintain an explicit model of an environment and have multiple goals to be accomplished and are able to change their plans and implement behaviour according to the environment [11]. They have distinctive properties such as (1) they contain heterogeneous components and are loosely coupled together (2) they require security reliability and privacy in their applications. (3) Patient's records are used by many to assist other entities or agents. The rules and artifacts are used used to assist other entities. Integration of agents together is a feature. Daily care systems are constituted of members of various other sub-departments to formulate a multidisciplinary team to plan and execute different objectives. The organization sector in the planning of any healthcare system is vital, as any error could lead to a life-threatening accident (Fig. 2).

Intelligent Sensors Healthcare as any other IoT application needs to have multiple sensors to collect information. Application of healthcare systems requires *context-awareness* to adapt to the environment. Intelligent sensors applied have low cost,

Fig. 2 Flow diagram for an intelligent agent

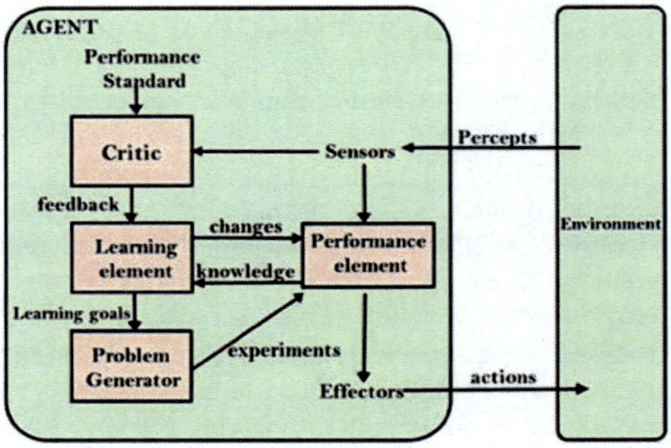

are low in size. In a cooperation network, the collection of organized nodes is called a sensor network. They are visualized as an agent as they are capable of collaborating with other agents to detect the environment and also detecting, processing the information they have received from the environment. Wireless sensor networks have two properties separating them from the others, (1) Agents are homogenous, i.e., the nodes, agents and sensors are the same. There is no hierarchy or priority between the agents. (2) Agents are abundant. In the real-world application of agents, a huge number of agents will aid measuring data patterns with accuracy, making the network system trustworthy [12] (Fig. 3).

There are numerous limitations to body area sensor networks described by [13], (1) Agents are heterogenous, the tools and the sensors are put directly on the body resulting in the installation of two or more same sensors to measure different metrics. Moreover installing the sensor in or onto the target patient's body increases the incentive of making the sensors small as possible yet retaining high accuracy. Installation of two sensors on the different parts of the body can be integrated as it seems like a positive option, but the cost and increase in size will be an unavoidable demerit. (2) The agents are made small in number, fulfilling the requirement of the sensor to be small and rigid to be placed with ease. Increase in the number of agents would require an increase in battery consumption and decreased redundancy also

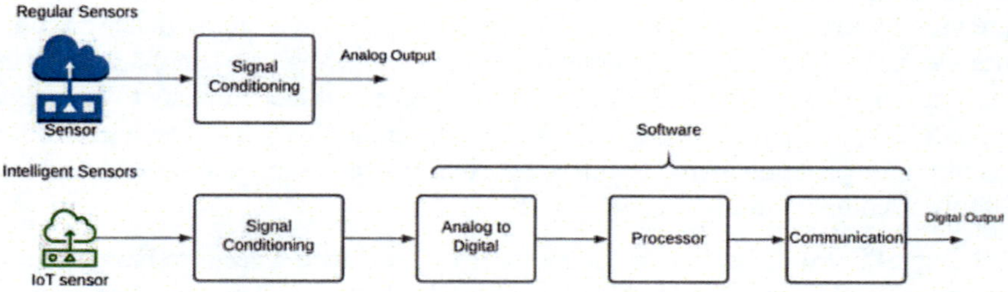

Fig. 3 Novelty of intelligent sensors

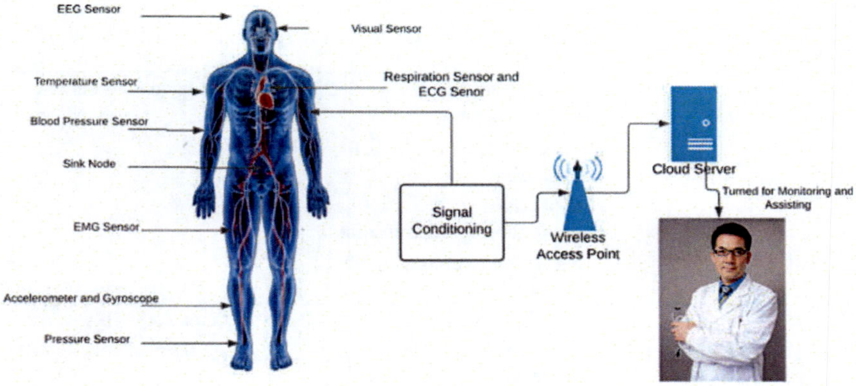

Fig. 4 Body sensor area network

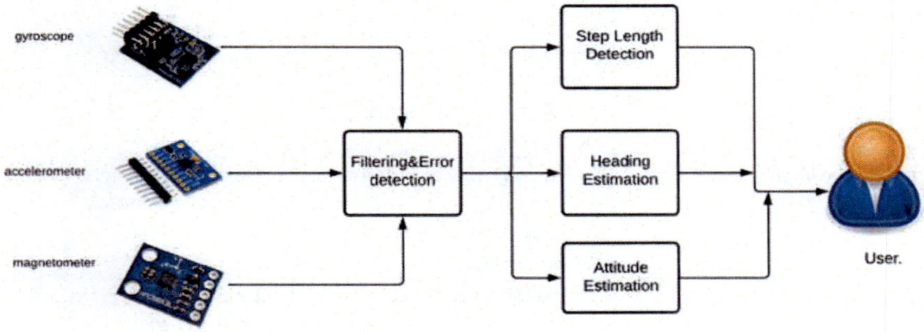

Fig. 5 Sensor Fusion Technique

increase inaccuracy in measurement by planning and communication. (3) Intelligent sensor's communication with other agents is facilitated by signal strength which is challenging due to the small size constraint in the body area sensor networks even more when we realize that human body is a mobile subject (Fig. 4 and 5).

Sensor Fusion Techniques Sensors used in body area sensor networks are deployed for Human Activity Recognition are accelerometers, gyroscope and magnetometers. Healthcare applications face huge obstacles being, (1) complexity and variety of daily activities, (2) inter-subject and intra-subject variance for the same activity, (3) performance and Privacy constraint, (4) gathering data is tough, (5) computational efficiency in embedded systems and portable devices [14]. Sensor fusion techniques bring out accuracy in measuring the data. It is evident that one single sensor is not able to measure the activity which can be deviated due to an external trigger event. Sensor fusion techniques solve this issue by merging the input from various sources. Merging various sources combined with data fusion and mining techniques provide advantages, as (1) reduction in noise of the sample, (2) reduced uncertainty in data, (3)

Table 1 Comparison of various sensor fusion techniques

Technique	Pros	Cons
Fuzzy logic	(1) Flexible (2) Adaptable (3) Deals with Simple Input Sensors (4) Used in Monitoring/Classification (5) Can be used with other techniques	(1)Can't handle dependency
Bayesian	(1) It is simple (2) Better in multi-sensor inputs (3) Dependecy between sensor is allowed (4) Used in fall detection applications	(1) limited scalability (2) requires medical knowledge
Markov process	((1) built on binary sensor outputs (2) Used in predicting daily activities	(1) High complexity (2) Requires other fusion methods
Dempster-Shafer based	(1) Dealing with uncertainities (2) Works well with limited sensor inputs	(1) Complexity proportional to the number of sensors
Threshold technique	(1) Produces good results (2) Applicable in various scenarios	(1) Limited literature in healthcare domain

increase in robustness, (4) integration of prior knowledge to input signals [15]. The fusion of input gets tough as the number of sensors increase. Fusion uses techniques such as Bayesian Estimation, Kalman Filters and Particle Filtering Techniques [16]. According to [17], sensor fusion can be categorized into data-level, feature-level and decision level. If the raw data is combined directly it is data-level, if features are extracted and then fused together then it is feature-level and decision level fusion deals with including machine learning and data mining techniques. Comparison of various sensor fusion techniques has been done in Table 1.

1.2 Multi Agent Systems

Multi-agent systems are a popular standard for the formulation, programming and simulation of distributed and complex systems where involved entities are autonomous agents. These agents involve many agents who are coupled together to solve goals, way beyond their capacity. Agents are bounded with a specific set of rules and characteristics. They have the following characteristics, (1) Data shared over the system is decentralized. (2) There is no hierarchy in the system between the agents. (3) The computations taking place are asynchronous. (4) The agents can be visualized into anything. (5) The autonomous entity is not aware of it's surroundings and agents in entirety. The resulting system involves agents to work together to reach a goal. Multi-agent systems can be applied to a vast area of applications in healthcare, which is a database monitoring system, distributed sensor computers and entities to do data mining and knowledge base. In a multi-agent system paradigm, there are two classes of agents, based on their autonomous capabilities, (1) Cognitive Agents with capabilities of taking decisions by themselves and using the decisions by referring the knowledge base. (2) Reactive agents, responding to an trigger event and knowledge base, are < *conditions, agents* > based [18] . The intelligent agents allow, (1) repetition of monotonous tasks. (2) making recommendations to the recommender system installed in the healthcare device. (3) using real-time data to extract high detailed information. The agents' coordination in the healthcare system is made by exchanging data and reaching goals by collaboration. The coordination is done by exchanging data, providing partial plans and analysing the constraints between agents for work. Intelligent Sensor Networks are used to realize multi-agent systems to solve healthcare systems (Fig. 6).

Multi-agent systems are developed using various programming frameworks. Programming multi-agent systems require a collective agent-oriented programming, organization oriented programming and environment-oriented programming, these are brought in existence together into a concrete programming framework named *JaCaMo* [19]. The framework was built over three existing platforms, (1) Jason for programming autonomous agents. (2) Moise for programming agent organizations and (3) CArtAgo for programming shared environments. *JaCa* is designed to be

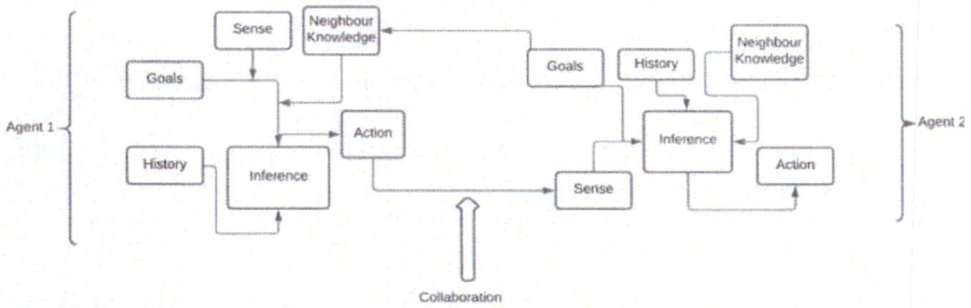

Fig. 6 Interaction between agents

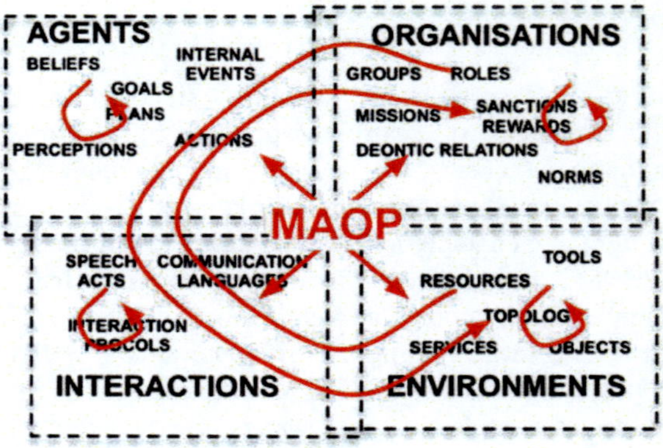

Fig. 7 *JaCaMo* programming framework

programmed as a set of agents working cooperatively under a shared environment. Programming of agents as well as simultaneously encapsulating the logic of control of tasks that are meant to be executed, as an abstraction which provides the actions and functionalities to do their tasks. *JaCa* is realized to execute and implement the agents and CArtAgo is present to program and execute the environments [20]. The different programming synergies are represented together by conceptual mappings that [19] identified in the definition of the integrated approach. The interaction between workspace and environment is defined based upon actions and percepts [21]. On the perception side, the artifact properties and events are mapped into agent percepts [19]. BDI agents' observable state is mapped through percepts, into the belief base of agents who are observing the artifact [19]. Jason rules allow for connections between artifact observable events and make it easy to program agents in response to a state change (Fig. 7).

2 Multi-agent Systems in Healthcare

Problems in Healthcare Numerous problems in healthcare share the same characteristics. Studying these characteristics helps providing solutions to solve these problems.

- Methods and knowledge required to tackle a problem is distributed at many locations.
- To solve a problem, a collaboration between various departments with varying skills and functions is required.
- Healthcare is complex, we can not find a one-step software solution for it.
- Accessing true related medical information is important to build a solution.

Why Multi-agent Systems? Multi-Agent systems provide a distributed, robust method to design solutions where each worker can be visualized as an agent.

- Multi-Agent Systems are distributed. The components are in different locations with varying knowledge and rules to solve tasks. They thus offer an inherent way to attack problems.
- Agents can communicate and collaborate with one another.
- Medical problems are complex. Multi-agent Systems provide methods to divide it into sub-problems to solve.
- Agents can be used to provide information to actors in the system by retrieving information from various sources, e.g., *Internet Agents.*
- Agents are able to do tasks which can be useful in providing a future-proof solution.
- Agents are autonomous, being able to take decisions by themselves exactly like in the healthcare environment.

2.1 Agent Architecture in Healthcare

Healthcare system can be visualized as distributed sets of departments working together. Department is depicted with a conjunction < *resource/interaction* >. Resources to consider are, 1) Human(healthcare workers, accountants, etc.), 2) Material (chair, medicines and equipment), 3) Information Database (records, measurements, personal information) (Fig. 8).

The agents work in the background and provide help to users whose login data is stored in cloud environment. The architecture can be implemented with the *JaCaMo* [19] framework (Table 2).

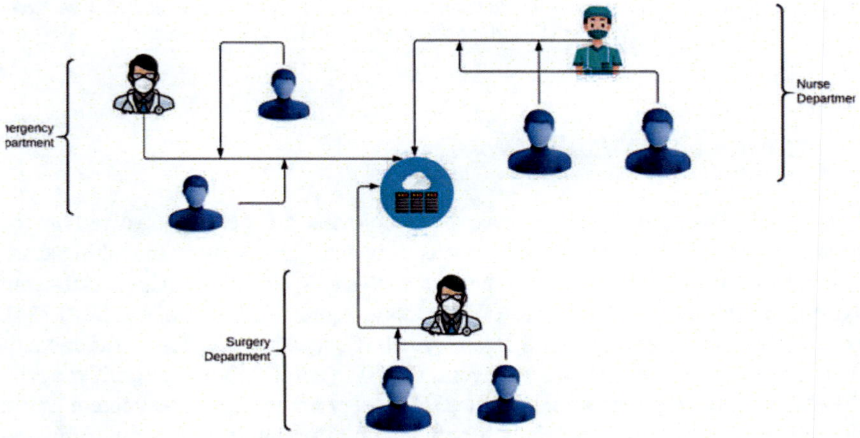

Fig. 8 Multi-Agent based Interaction between Departments

Table 2 Agent functions in healthcare

Agents	Tasks
Manager agent	(1) Managing and assigning resources in facility
Patient agent	(1) Maintaining and Upliftment of health status (2)Choosing hospital for treatment. (3) Managing health status and reporting to nurse agent
Doctor agent	(1) Consulting expert agents when can not make a decision (2) Conducting diagnosis and analyzing data from sensors (3) Returning result of the test to the patient agent (4) Doctor decides final treatment and updates department and patient about the treatment
Nurse agent	(1) Locating patient and room via RFID and assisting (2) Follow doctor's directive for treatment
Discharge agent	(1) Enable efficient discharge of patient (2) Help to find another healthcare service
Service agent	(1) Delivering services and medicines to the patient (2) Collaborate over human resources.
Access agent	(1) Providing methods to access services around the hospital. (2) Informing patient over services and doctors available. (3) Managing the resources and schedule
Administrator agent	(1) Collaborate with a doctor to share images and data Search for collaborations and manage remote workers
Emergency medical agent	(1) Informing the emergency ward about the incoming patients
Lab result agent	(1) Notifying laboratory, a doctor about developments in results
Calendar agent	(1) Providing the schedule of the doctors and other healthcare workers in a particular department and sharing for collaboration

2.2 *Current MAS Based Projects*

Diagnosing Diseases Computer-based technologies are heavily involved in the healthcare process in diagnosis, from measuring data using sensors to making sense out of data using machine learning and data analysis. Applications using multi-agent systems in this domain are, IHKA [22], HealthAgents [22, 23] and ODHS [24]. IHKA [22] was based on five different typed of agents with different functionalities, (1) query knowledge retrieval agent, (2) UI agent, (3) Query optimizer agent, (4) Query knowledge adaption agent and (5) Query knowledge procurement agent, the broken case if everything fails it will search different sources for information autonomously. OHDS [24] uses existing data to predict and assist doctors for future disease outbreaks. It uses hierarchy with feature-based ontology development for organizing research in the medical field. Ontology can be visualized as a CPU to share effective information as soon as possible. The knowledge that is available on the internet is described in an ontology standard by extracting it with fuzzy techniques and algorithms. HealthAgents [22, 23] deals with classifying brain tumor patterns by incorporating multi-agents over a distributed network of databases. They develop recognition methods for RNA/DNA and examine the quality of a new dataset of sample values and give it a score. They develop a distributed global repository for accessing data related to brain tumours using multi-agent systems.

Assistance to Elderly Providing assistance to the elderly by automating homes and tasks is essential. Such applications often require multiple applications for assistance, making it a tedious task. Interesting application in assistance are CASIS [25], TeleCARE [26] and K4CARE [27]. CASIS [25] is a multi-agent framework with a goal to deliver context-aware assistance making it service-oriented. It enables remote healthcare workers to monitor the health of the elderly. It interacts with numerous applications present, providing context-aware result by detecting the state of the elderly through the sensors being equipped with safety. K4CARE [27] was a project to integrate computer science into healthcare by building a knowledge graph. TeleCARE [26] was funded by the European Union to implement a framework through websites to provide supervision. Users were elderly and healthcare professionals and framework was incorporated for the use of both. Project discarded TCP/IP protocols and used multi-agent paradigm due to, (1)by using multi-agents, the framework could distribute the rules where it was actually required, providing privacy and autonomous capabilities to the independent agent. (2) greater flexibility is ensured by a distributive framework. The project was a prototype requiring huge testing before it's rollout. The accuracy of the systems can be improved by recommender systems [28] and providing a feedback loop after training it with a deep learning model.

Hospital Applications These applications tend to ease to work for healthcare workers. This is fulfilled by making access to information by the doctors and nurses easy. Automation in accessing user's information in a context-free manner. It will help the doctor as the department to get familiar with incoming emergency patient's vitals. The projects which use multi-agent systems in automation of hospital are, ERMA [29], Akogrimo [30] and CASCOM [31]. Akogrimo [30] is a project made to integrate intelligent sensor networks in hospitals to make it smart. These sensor networks are flexible and scalable to adapt to any bandwidth and environment. People who can benefit from this technology are, (1) patients with chronic diseases, requiring a mobile monitoring system through a smartphone application. (2) supplier agents who supply medicines and equipment to departments and patients. The application can recognize heart attacks early with parameters and offer treatment with the supervision of a medical professional. It can be deployed for emergency detection and providing subsequent rescue decisions. ERMA [29] assists in diagnosis by providing suggestions to the healthcare professional in a catastrophic health complication like a heart attack. It has a knowledge base and provides suggestions by integrating data through fuzzy logic, trend analysis and qualitative logic. CASCOM [31] is a similar project incorporating web and multi-agents in the context-aware healthcare environment.

Limitations Multi-agent systems have numerous technical and developmental issues associated such as user's acceptance of the application. Decentralization is ineffectual in scenarios where we may need to shut down the whole application in retaliation to unpleasant behavior. Security and privacy risks are there when agents share classified information with one another in a decentralized environment. These limitations are the reasons why the integration of multi-agent systems in healthcare is quite slow. The multi-agent-based intelligent sensor technology is widely accepted in computer science but social, economical and ethical issues are preventing it's widespread in healthcare. There is also a huge difference between literature and real applications based on multi agent-based sensors.

Future Scope Multi Agent-based Intelligent Sensors will improve the efficiency of healthcare systems. As computers become more complex with an increase in computational power, we will see advancements[32]. Integration of sensors within the human body as proposed by the paper [33] will pave way for the development of human prototypes with embedded sensors to monitor, assist and improve quality of life. As multi-agent systems are decentralized, future applications can include them in some departments where they are most useful. Implantable sensors are important as they measure data continuously with the constraint being size and if the quality of signals emitted are good for human's health. Improvements in battery technologies or wireless electricity transfer mechanisms can recharge the embedded sensor without taking it out of the body. This will lead to a better prediction of disease with deep learning models due to increased sample size.

3 Conclusion

Multi-agent systems coupled up with Intelligent sensors build up an interesting research domain with inter-disciplinary applications. The rising population of elderly and an increasing strain on the healthcare system requires assistance applications which are provided by multi agent-based intelligent sensors. However, unavoidable limitations such as not enough emphasis on the privacy and security of delicate data being interchanged between agents in a decentralized environment are halting the integration as a first choice to build a healthcare system. There is a lack of methodology to economically evaluate a health service built. Lack of data and methodology is an obstacle to obtaining funding for the projects and is preventing researchers as well as industries to move forward in this direction.

References

1. United Nations, World Population Ageing [Report] (2015). WPA2015 Report. Retrieved 02 May 2021, from https://www.un.org/en/development/desa/population/theme/ageing/WPA2015.asp
2. Statista. July 2020. Survey Period : 2018. Average life expectancy in industrial and developing countries for those born in 2020 [by gender]. Retrieved 11/02/2021 from https://www.statista.com/statistics/274507/life-expectancy-in-industrial-and-developing-countries/
3. T.G. Bentley, R.M. Effros, K. Palar, E.B. Keeler, Waste in the U.S. Health Care System: a conceptual framework. Milbank Q **86**(4), 629–659 (2008). https://doi.org/10.1111/j.1468-0009.2008.00537.x
4. V. Simpkin, E. Namubiru-Mwaura, L. Clarke, et al., Investing in health R&D: where we are, what limits us, and how to make progress in Africa. BMJ Global Health **4**, e001047 (2019)
5. Y. Yin, Y. Zeng, X. Chen, Y. Fan, The internet of things in healthcare: an overview. J. Ind. Inform. Integr. **1**, 3–13 (2016). ISSN 2452-414X. https://doi.org/10.1016/j.jii.2016.03.004, https://www.sciencedirect.com/science/article/pii/S2452414X16000066
6. C. Chang, Y. Chen, Autonomous intelligent agent and its potential applications. Computers Ind. Eng. **31**(1–2), 409–412 (1996). ISSN 0360-8352, https://doi.org/10.1016/0360-8352(96)00163-5, https://www.sciencedirect.com/science/article/pii/0360835296001635
7. H. Skov-Petersen, Feeding the agents—collecting parameters for agent-based models (2005)
8. T. Panayiotopoulos, N.Z. Zacharis, Machine learning and intelligent agents, in *Machine Learning and Its Applications (ACAI 1999)*. Lecture Notes in Computer Science, vol. 2049, ed. by G. Paliouras, V. Karkaletsis, C.D. Spyropoulos (Springer, Berlin, 1999). https://doi.org/10.1007/3-540-44673-716
9. F.D. Brunner, H. Dürr, C. Ebenbauer, Feedback design for multi-agent systems: a saddle point approach, in *2012 IEEE 51st IEEE Conference on Decision and Control (CDC)*, Maui, HI (2012), pp. 3783–3789. https://doi.org/10.1109/CDC.2012.6426476
10. G. Basso, M. Cossentino, V. Hilaire, F. Lauri, S. Rodriguez, V. Seidita, Engineering multi-agent systems using feedback loops and holarchies. Eng. Appl. Artif. Intell. **55**, 14–25 (2016). ISSN 0952-1976, https://doi.org/10.1016/j.engappai.2016.05.009. https://www.sciencedirect.com/science/article/pii/S0952197616300999
11. M. Wooldridge, N.R. Jennings, Intelligent agents: theory and practice. Knowl. Eng. Rev. **10**(2), 115–152 (1995). https://doi.org/10.1017/S0269888900008122
12. J.A. Stankovic, Wireless sensor networks. Computer **41**(10), 92–95 (2008). https://doi.org/10.1109/MC.2008.441
13. M. Hernandez, L. Mucchi, Survey and coexistence study of IEEE 802.15.6^{TM}-2012 Body Area Networks, UWB PHY, in *Body Area Networks Using IEEE 802.15.6*, ed. by M. Hernandez, L. Mucchi (Academic, New York, 2014), pp. 1–44. ISBN 9780123965202, https://doi.org/10.1016/B978-0-12-396520-2.00001-7. https://www.sciencedirect.com/science/article/pii/B9780123965202000017
14. O.D. Lara, M.A. Labrador, A survey on human activity recognition using wearable sensors. IEEE Commun. Surv. Tutor. **15**(3), 1192–1209 (2013). https://doi.org/10.1109/SURV.2012.110112.00192
15. F. Demrozi, G. Pravadelli, A. Bihorac, P. Rashidi, Human activity recognition using inertial, physiological and environmental sensors: a comprehensive survey. IEEE Access **8**, 210816–210836 (2020). https://doi.org/10.1109/ACCESS.2020.3037715
16. H.F. Nweke, Y.W. Teh, U.R. Alo, G. Mujtaba, Analysis of multi-sensor fusion for mobile and wearable sensor based human activity recognition, in *Proceedings of the International Conference on Data Processing and Applications (ICDPA 2018)* (Association for Computing Machinery, New York, NY, USA, 2018), pp. 22–26. https://doi.org/10.1145/3224207.3224212
17. H. Medjahed, D. Istrate, J. Boudy, J.-L. Baldinger, B. Dorizzi, A pervasive multi-sensor data fusion for smart home healthcare monitoring, in *2011 IEEE International Conference on Fuzzy Systems (FUZZ)* (IEEE, 2011), pp. 1466–1473

18. J. Lee, M. Barley, eds., *Intelligent Agents and Multi-Agent Systems*, 1st ed. (Springer, Berlin). https://www.springer.com/gp/book/9783540204602
19. O. Boissier, R.H. Bordini, J.F. Hübner, A. Ricci, A. Santi, Multi-agent oriented programming with JaCaMo. Sci. Computer Program. **78**(6), 747–761 (2013). ISSN 0167-6423. https://doi.org/10.1016/j.scico.2011.10.004. https://www.sciencedirect.com/science/article/pii/S016764231100181X
20. The JaCaMo Project Homepage, http://jacamo.sourceforge.net. Last accessed 11 Feb 2020
21. A. Ricci, M. Piunti, M. Viroli, Environment programming in multi-agent systems: an artifact-based perspective. Autonom. Agent Multi-Agent Syst. **23**, 158–192 (2011). https://doi.org/10.1007/s10458-010-9140-7
22. Z.I. Hashmi, S. Sibte, R. Abidi, Y. Cheah, An intelligent agent-based knowledge broker for enterprisewide healthcare knowledge procurement, in *Proceedings of 15th IEEE Symposium on Computer Based Medical Systems (CBMS'2002)*, Maribor (Slovenia) (2002)
23. M. Croitoru, B. Hu, S. Dasmahapatra, P. Lewis, D. Dupplaw, A. Gibb, M. Julia-Sape, J. Vicente, C. Saez, J.M. GarciaGomez, R. Roset, F. Estanyol, X. Rafael, M. Mier, Conceptual graphs based information retrieval in HealthAgents. Computer-Based Med. Syst. **7**(20–22), 618–623 (2007)
24. M. Hadzic, E. Chang, M. Ulieru, Soft computing agents for e-health applied to the research and control of unknown diseases. Inform. Sci. **176**, 1190–1214 (2006)
25. W. Jih, J.Y. Hsu, T. Tsai, Context-aware service integration for elderly care in a smart environment, in *2006 AAAI Workshop on Modeling and Retrieval of Context Retrieval of Context*, ed. by D.B. Leake, T.R. Roth-Berghofer, S. Schulz (AAAI Press, Menlo Park, CA, 2006), pp. 44–48
26. UNINOVA - INSTITUTO DE DESENVOLVIMENTO DE NOVAS TECNOLOGIAS, A Multi-agent Tele-supervision System for Elderly Care (2001). Retrieved 02 June 2021, from https://cordis.europa.eu/project/id/IST-2000-27607
27. K4CARE, K4CARE project Web site (2007). Retrieved 14 Feb 2021. http://www.k4care.net
28. S. Zhang, L. Yao, A. Sun, Y. Tay, Deep learning based recommender system: a survey and new perspectives. ACM Comput. Surv. **52**(1), 38, Article 5 (2019). https://doi.org/10.1145/3285029
29. S.L. Mabry, C.R. Hug, R.C. Roundy, Clinical decision support with IM-agents and ERMA multi-agents, in *Proceedings of the 17th IEEE Symposium on Computer-Based Medical Systems (CBMS 2004)*, Bethesda, MD (2004), pp. 242–247
30. Akogrimo, Akogrimo project Web site (2007). Retrieved 14 Feb 2021, from http://www.mobilegrids.org
31. M. Schumacher, H. Helin, CASCOM: intelligent service coordination in the semantic web. Birkhauser Boston (2008)
32. E. Mollick, Establishing Moore's Law. IEEE Ann. History Comput. **28**(3), 62–75 (2006). https://doi.org/10.1109/MAHC.2006.45
33. E. Musk, Neuralink: an integrated brain-machine interface platform with thousands of channels. J. Med. Internet Res. **21**(10), e16194 (2019). PMID: 31642810, PMCID: 6914248, https://www.jmir.org/2019/10/e16194, https://doi.org/10.2196/16194

Comprehensive Analysis on Security Threats Prevalent in IoT-Based Smart Farming Systems

G. Jeba Rosline, Pushpa Rani, and D. Gnana Rajesh

Abstract Smart farming is a vital notion for the development of agriculture and food processing industries globally. Industrial revolution in computing and digital network turned agriculture industry to a digitalized and automated technology. Manual and mechanical tools are replaced by tools that are controlled by mobile phones, drones and Web-based applications. IoT is the major applicant in smart farming that controls sensors in devices and data analysis by remote servers. Security is deployed as built-in mechanisms in the devices or as software tools implemented in the mobile devices, sensor systems and machines that are remotely controlled. Protocol-based security is provided to data that are collected from the fields and to the data that are transmitted to remote servers for processing. However, in recent years vulnerabilities in smart farming applications are demoralized which ensued smart farming systems being victimized to cyber-attacks. This research work provides insight analysis on several security threats that are being subjugating smart farming devices and processes. This paper will provide intrigue on vulnerabilities existing in smart farming systems and the threats that exploit them.

1 Introduction

Smart farming encompasses automation in agriculture and food processing systems. As information technology has stimulated to the radical revolution in industrial development, there is a tremendous advancement in the agriculture, manufacturing of tools used in forms, food production and preservation industries. The swift in increasing population, unpredicted climatic conditions, decline in availability of natural resources and restraints in pest control are the major hurdles in the modern

G. J. Rosline
Mother Theresa Women's University, Kodaikanal, India

P. Rani
Department of Computer Science, Mother Theresa Women's University, Kodaikanal, India

D. Gnana Rajesh (✉)
University of Technology and Applied Sciences, Al Mussanah, Sultanate of Oman

P. Karuppusamy et al. (eds.), *Ubiquitous Intelligent Systems*, Smart Innovation, Systems and Technologies 243, https://doi.org/10.1007/978-981-16-3675-2_13

agriculture. Modern techniques and tools helped food production to a revolutionary lead. However, there are challenges in developing farming systems that are sustainable, eco-friendly, secured and more productive [1]. Digital revolution has led to the innovation in information and communication technology [2]. Internet of things (IoT) is one such area where technology connects man, machines and methods [3]. It has fetched applications and tools for data analysis on various subprocess in agriculture such as climate prediction, irrigation control, soil testing, pesticide application, weed patching, disease detection and crop growth rate to predict quality and quantity of the agricultural products. These tools collect data from several devices in the fields and send data to the cloud for data analysis [4].

Farmers can access the analytical reports on agricultural entities such as soil characteristics, rain, climatic conditions, water requirements, soil enrichment, chemical compositions in soil, quality of farm produce and preserving timeline of farm products and so on. Data analysis can be done based on the present and historical data which will give analytical report that helps in critical decision making. Monitoring of sensors, collection of data and distribution of diagnostic reports to various units of the form are done via wireless networks or mobile phones [5].

Although security is provided to an optimum level in various levels of IoT implementation, potential threat exists beyond this security which is provided to the data and devices in smart farming. As the smart farming devices and processes are simple enough for the farmers to get trained themselves, there is a lack of security awareness and training to the farmers. There are encounters to be handled in data processing as the automated farming equipment is not completely centralized. As an illustration, we can perceive that there is a lack in security while the readings of the moisture sensors are collected individually and transferred to the cloud for further processing and analysis [6].

It is mandatory to analyze the security risks involved in smart farming and mitigate as farming and food production involve a huge capital investment that gives key indication to the development of a nation.

2 Automation in Agriculture with Internet of Things

Internet of things empowers the automation of agriculture-related manual tasks which in turn impacts the supply chain management, marketing of farm produce, improved quality and enormous quantity. Due to various factors such as deforestation, destruction of cultivation lands by industries, real estates and other business elements, the requirement for producing farm products with high quality became crucial. IoT constituent includes devices with sensors, tools for data acquisition and analysis, memory for data storage and wireless sensor network. IoT is trending as a global network revolution that involves millions of devices, sensors, interfaces and computers. IoT can help the farmers in reducing their wastage, increase production and improve the quality of the product.

With the assistance of IoT, smart farming processes and subprocess can ensemble various technology such as wireless sensing, precision computing, big data analytics, communication protocols, Web application services, positioning systems and Internet search engines. It helps farmers to control and monitor water irrigation, fertility of the soil, moisture level, applications of pesticides, prediction of growth and variation in climatic conditions such as rain and heat [7].

Agriculture turned into smart-agriculture with the help of IoT. The production of agriculture products and supply management are completely taken care by smart farming systems using cloud computing and big data analytics. Critical decisions are made on farm events such as soil enhancement, irrigation control, pest control prediction of crop growth, quality enhancement and harvesting with the help of decision support systems incorporated in smart farming systems. For instance, a hydrometer placed in the farm continuously monitors the humidity and sends the data through a wireless sensor network to the data processing units where the application checks the threshold. If it reaches the threshold, the signal is sent through the wireless sensor network to induct the irrigation process [8].

3 Security Challenges in Smart Farming Systems

In smart farming systems, security is one of the imperative requirements as the farmers are very sensitive about revealing the data related to their predicted yield, growth rate, soil conditions, water availability and so on. Security challenges exist on the data storage, processing, access control, authentication and TCP/IP network transmission. Security incidents in smart farming might be unintentional or intentional [1]. Automated devices that are used for monitoring and controlling various agricultural processes such as irrigation, pesticide application and moisture control need to be secured from unauthorized access. Smart farming is an immense area to implement IoT as there are enormous requirements for devices that monitor and control farming and marketing of food produce. There are bounteous challenges existing in interaction between the farming equipments and security devices. For example, a sensor that is used to determine the level of water needs to be monitored as the compromise in security of this device will mislead the amount of the irrigation schedule which in turn will harm the growth of plants [9]. The security challenges in smart farming systems embrace access control for farmers from different platforms in order to manage and control their own data, interoperability between various devices and applications, internet connectivity issues and data security [7].

Most of the IoT devices are not capable of handling security problems. This may lead to various cyber-attacks that might harm the data and lead to availability problems. Availability is very crucial in farming systems as there are processes such as sowing of seeds, irrigation schedules, soil enhancements, pest control services, cattle health monitoring and harvesting processes are time-critical. Decision support systems that provide analytical results in smart farming applications might be affected with the tarnished availability in the system.

Security challenges are the foremost apprehension as there are more potential security threats present in smart farming systems. As IoT-based smart farming systems are becoming more diverse in implementing technologies, the varied data increases the prospects of security attacks on the data. Greater diversity in data increases the vulnerability; thus, the system has to incur more security threats [10].

These threats might exploit the vulnerabilities in the mobile communications, wireless networks, sensing devices, medium of transmission, data acquisition devices and the communication protocols [11]. These security threats might pose on physical security, authentication, confidentiality, integrity and availability. The following sections present an insightful analysis on the impact of cyber-attacks to smart farming systems and the security threats that are commonly prevailing in smart farming systems which are exploited by the vulnerabilities in the IoT devices, network communications, data analysis tools and other applications in smart farming systems [12].

4 Impact of Cyber-Attacks on Smart Systems

A cyber-attack exploits vulnerability in computer devices, software applications, internet services and network communication to disrupt the operations of any computerized system. Rise in the application of internet of things for various automation processes is directly proportional to the raise in the number of cyber-attacks exploiting smart IoT systems [8, 13]. A cyber-attack in smart farming is usually performed to create a denial of service by spreading on malicious programs or data in the data processing systems, intercepting wireless sensor network communication, breach security settings in IoT devices, false authentication in remote sensing devices and malfunction of IoT devices in the fields. The motive of these attacks from individuals or a cybercrime activist groups might influence the growth and development of smart farming systems.

Even though security is implemented in different constituents of smart farming systems, incidences to the cyber-attacks are extensive. The following study by Crowe [8] on 60 cybersecurity establishments infers that even with high percentage of security implementation and widespread security awareness training to the users, the probability of the cyber-attack success rate is higher. Small- and medium-sized smart farms are more vulnerable to cyber-attacks than the large farms. [13]. Table 1

Table 1 Security breaches

Security mechanism breached	Percentage of bypassing security (%)
Antivirus software system	100
Firewall	95
Email filters	77
Anti-malware protection	52

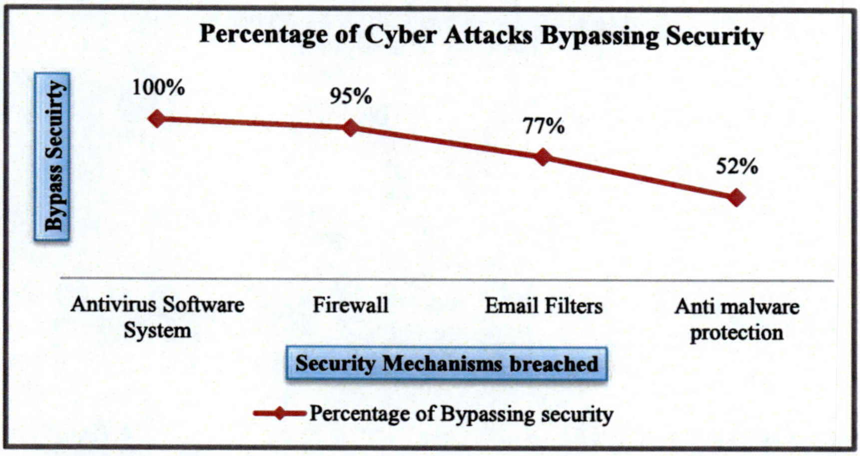

Fig. 1 Percentage of cyber-attacks bypassing security

Table 2 Cyber security attacks

Number of cyber security companies studied (victim of ransom ware attack)	Attack succeed	Attack failed
60	20	40

illustrates the bypassing of security mechanism by cyber-attacks (Fig. 1; Table 2).

5 Security Threats to Smart Farming Systems

Figure 2 shows the smart farming system and the entities that are involved in accomplishing smart farming processes using IoT. Security provided to smart farming system inclines the security provided to each one of these entities. The threats that exploit security in each of these rudiments need to be addressed. The following sections in this paper will exemplify the security threats that are abused in cyber-attacks most commonly in smart farming systems.

5.1 Physical Security Threats

In smart farming system, physical security to devices is limited as the fields are vast in space and natural factors like heat, rain, cold, snow and wind are not predictable precisely. This may lead to any anonymous access to the devices due to the damage in

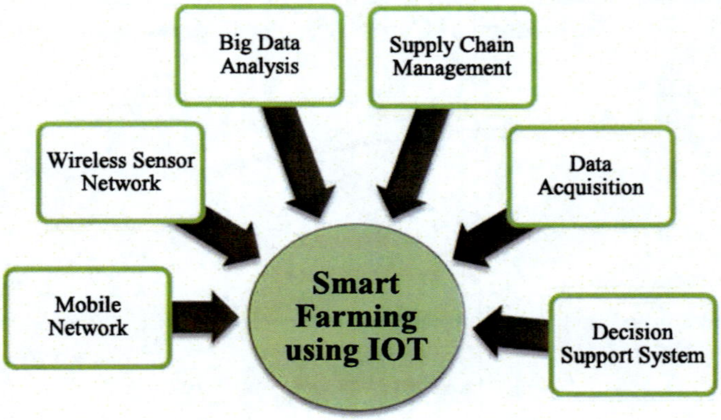

Fig. 2 Smart farming using IoT

fences, malfunction of the sensing devices, delay in data collection and unavailability of connections to devices in the field such as sensor devices, cameras, radars, drones and transmission towers. The above might delay the decision making process in the data centers or lead to inadequate decisions by the DSS applications on the workstations, mobiles or tablets used by the farmers.

5.2 Data Tampering

Data tampering refers to the unauthorized modifications of data. Data collected from the field by various devices and sensors are transmitted to the cloud for further processing. There are potential threats for these data to be tinkered which might lead to fault decision making such as increasing water level, changes in harvest schedule, wrong prediction in climatic changes, erroneous health report of the cattle and so on. Furthermore, this might affect the integrity of the systems that in turn leads to lack of reliability. This will affect the further process of smart farming systems such as food supply management, estimation of cost of the crop to be harvested and order processing of the farm's produce [12, 14]. Furthermore, the farmers are very sensitive to the facts that affect their goodwill. Possibilities include tampering of the data on the duration of crop growth before harvest, level of pesticide given, the amount of chemicals in the soil or health report of the cattle probably affects the selling price of the farm goods to be supplied to the market.

5.3 Eavesdropping

Mobile networks and wireless networks are used in most of the smart farming systems for communication among devices and farmers. The inherent weakness in the protocols used in these communications might allow a third party to intercept the data that are transmitted to the cloud. Unless there is a secured protocol standard such as transport layer security (TLS), it would be hard to escape from sniffing, message interception or modification by the eavesdroppers [15].

5.4 Spoofing

Spoofed addresses are used by some hackers in order to gain access to the network devices in the farm. By this, they are getting authentication to communicate with the devices in the field, wireless sensors network and data center, as the spoofed source address is a trusted site or host.

5.5 Phishing

Farmers are not greatly skilled in information security. Most of the security procedures and tools are self-learned by the farmers. Their concern to data security is less as they lack awareness on information security. Social engineering mistunes human mind to support unauthorized access knowingly or unknowingly. Phishing is a human factor that a psychologically mistuned user in the system or anyone close to the network can gather sensitive information from the farm. For instance, an unsolicited email that asks for the details of crop growth rate providing an attractive offer for their new farm products. Smart farming users require enthusiastic and periodic security awareness training.

5.6 Denial of Service

Many of the security issues in IoT components lead to failure in accessing the field devices that involve data acquisition and sensor devices. Likewise, they disrupt the access to data from the cloud, prevent internet access to the mobile devices, PCs and tablets. Farming system might be unable to run their DSS applications in mobile devices and tablets. Most of the threats exploited by attackers are aimed at the disruption of farming services that creates denial of services.

5.7 Inherent Threats in Network Devices

Many of the network devices like switches, routers, access points, transmission towers and unmanned automated devices such as drones have vulnerabilities in their functions such as weak authentication, signal distortions and reachability problems. These vulnerabilities create impacts on the integrity and availability of the smart farming.

5.8 Password Threats

Smart farms are multiuser systems. Multiple devices and multiple users require a vast password management. As there are no systematic applications for user and password management among smart farming users, more probability for password attacks such as dictionary attacks, brute force, password guessing, weak passwords and sharing of passwords. Lack of encryption of passwords opens vulnerability during the data transmission by the protocols in the network layer [15]. User awareness and security alertness messages will support the farming system to maintain a secured password management.

6 Conclusion and Future Scope

The first line of security to the smart farming systems should be given based on the potential threats available to the applications in smart farming systems. In this paper, we illustrated the significance to analyze the security threats to the IoT-based smart farming systems. This research work provides a deep understanding on the security threats underlying the various layers of IoT-based smart farming systems. Many advanced information security solutions are available that can mitigate any attack caused by the above threats. However, as cost is also a vivacious factor in implementation of security mechanisms, this research work helps to design a security model that can be scalable, cost effective and sustainable. Based on this analysis, a new model can be developed to plan the prevention methods for attacks that involves these threats. Implement the following procedures as a basic line of defense against attacks and prevent exploitation of the vulnerabilities in the hardware and software used. Security principles such as obscurity, limitation, layering and diversity should be followed in the network management. Likewise, practice the following procedures in the security administration [16].

- Accounting of authorized and unauthorized hardware and software.
- Secured installation and configuration of software applications and IoT devices.
- Client authentication in servers.
- Closing down of unused ports.

- Access control on privileges by the administrator.
- Periodic vulnerability assessment.
- Audit Log maintenance.
- Immediate recovery procedures.
- Systematic backup procedures.
- Spam filters and content filtering for emails
- Secured protocols and Web browsers.
- Perimeter defense using firewall filters.
- Encryption for data security.

References

1. A.R. de Araujo Zanella, E. da Silva, L.C.P. Albini, Security challenges to smart agriculture: current state, key issues, and future directions. Array, 100048 (2020)
2. S. Jaiganesh, K. Gunaseelan, V. Ellappan, IOT agriculture to improve food and farming technology, in *2017 Conference on Emerging Devices and Smart Systems (ICEDSS)* (IEEE, 2017), pp. 260–266
3. W.A. Devanand, R.D. Raghunath, A.S. Baliram, K. Kazi, Smart agriculture system using IoT. Int. J. Innov. Res. Technol. **5**(10) (2019)
4. R. Kumar, M. Kajjidoni, M. Kumar, Smart agriculture system using IoT, in *Third International Conference on Current Trends in Engineering Science and Technology (ICCTEST-2017)* (2017)
5. N. Kaewmard, S. Saiyod. Sensor data collection and irrigation control on vegetable crop using smart phone and wireless sensor networks for smart farm, in *2014 IEEE Conference on Wireless Sensors (ICWiSE)* (IEEE, 2014), pp. 106–112
6. S.K. Choudhary, R.S. Jadoun, H.L. Mandoriya, Role of cloud computing technology in agriculture fields. Computing **7**(3) (2016)
7. A. Nayyar, E.V. Puri, Smart farming: IoT based smart sensors agriculture stick for live temprature and moisture monitoring using Arduino cloud computing & solar technology, in *Conference: The International Conference on Communication and Computing Systems (ICCCS-2016)* (2016)
8. J. Crowe, Survey: ransomware vs. traditional security. Barkly Stats and Trends (2016). Retrieved from https://blog.barkly.com/ ransomware-attacks-bypassing-antivirus
9. T. Baranwal, P.K. Pateriya, Development of IoT based smart security and monitoring devices for agriculture, in *2016 6th International Conference-Cloud System and Big Data Engineering (Confluence)*. (IEEE, 2016), pp. 597–602
10. V. Hassija, V. Chamola, V. Saxena, D. Jain, P. Goyal, B. Sikdar, A survey on IoT security: application areas, security threats, and solution architectures. IEEE Access **7**, 82721–82743 (2019)
11. C. Brewster, I. Roussaki, N. Kalatzis, K. Doolin, K. Ellis, IoT in agriculture: designing a Europe-wide large-scale pilot. IEEE Commun. Mag. **55**(9), 26–33 (2017)
12. J.H. Ziegeldorf, O.G. Morchon, K. Wehrle, Privacy in the internet of things: threats and challenges. Secur. Commun. Networks **7**(12), 2728–2742 (2014)
13. J. West, A prediction model framework for cyber-attacks to precision agriculture technologies. J. Agric. Food Inform. **19**(4), 307–330 (2018)
14. A.B. Pawar, S. Ghumbre, S. (2016, December). A survey on IoT applications, security challenges and counter measures, in *2016 International Conference on Computing, Analytics and Security Trends (CAST)* (IEEE, 2016), pp. 294–299
15. D. Glaroudis, A. Iossifides, P. Chatzimisios, Survey, comparison and research challenges of IoT application protocols for smart farming. Computer Networks **168**, 107037 (2020)

16. Public-Private Analytic Exchange Program (2018). Threats to Precision Agriculture. Retrieved from https://www.dhs.gov/sites/default/files/publications/2018%20AEP_Threats_to_Precision_Agriculture

Detection of Brain Tumors—A Comparative Analysis of Various Transfer Learning Methods

N. K. Rahul, Sandeep Suresh, and K. Sreekumar

Abstract Brain tumors are among the most aggressive of common diseases and can lead to drastic reduction of the lifespan of those affected—effective diagnosis and treatment planning thus become highly important. Broadly, the methods used to diagnose tumors in the brain are computed tomography scan, magnetic resonance imaging scan and ultrasound imaging. Brain tumor detection is a crucial and difficult task in the medical image processing field, and it requires handling large amount of data. Manual classification generally results in false prediction and diagnosis. Magnetic resonance imaging is the imaging technique used to diagnose the brain tumor. In this paper, various transfer learning models such as MobileNet, InceptionV3, ResNet50 and VGG19 are applied to train the model to detect brain tumors from magnetic resonance images and compare these methods. The above models are trained on the BraTS 2015 dataset and observed accuracy rates of 90.54%, 85.96%, 95.42% and 91.69%, respectively.

1 Introduction

A mass of abnormal cells that grow within the brain is called a brain tumor. The brain is protected by a rigid skull. So, any growth within its constraints gives rise to many severe clinical conditions. Brain tumors could be either cancerous in nature (malignant) or without cancerous nature (benign) and are categorized into primary and secondary. Primary brain tumors, usually benign, arises from within the brain whereas secondary—also known as metastatic—brain tumors, occur when cancer cells that form elsewhere—breast, lungs etc. spread into the brain. According to a study, the incidence of tumors in the central nervous system in India ranges from

N. K. Rahul (✉) · S. Suresh · K. Sreekumar
Department of Computer Science & IT, Amrita School of Arts and Sciences, Amrita Vishwa Vidyapeetham, Kochi, India

K. Sreekumar
e-mail: sreekumar@asas.kh.amrita.edu

P. Karuppusamy et al. (eds.), *Ubiquitous Intelligent Systems*, Smart Innovation, Systems and Technologies 243, https://doi.org/10.1007/978-981-16-3675-2_14

five to ten cases per 100,000 population [1]. In 2018, brain tumors were ranked as the tenth most common kind of tumor among Indians.

Computer vision techniques are bringing a huge result in many areas of the medical domain like surgery and therapy of different diseases. Medical imaging is basically different techniques that can be employed in a non-invasive manner to examine the body. It is the process in which the interior part of the body is imaged, and it is used for clinical analysis and medical intervention. It includes the processes of formation, enhancement, visualization, analysis and management. Medical image processing is one of the most challenging tasks, in that we are specializing in brain tumor detection. Segmentation of the tumor from magnetic resonance imaging brain images is an exigent task. All over the world, researchers are engaged in research to find more efficient algorithms and methodologies. Neural network-based segmentation is one of the structured ways which gives remarkable outcomes as well. Since this area of medical image processing is a wide one, future innovations can be implemented in this area; a lot of researches can be performed (Fig. 1).

Previous works on brain tumor detection are based on detecting with the help of a single method or architecture. So this work aims at improving the efficiency of detection of brain tumor and to find out which transfer learning model exhibits the best performance. Transfer learning models refers to the process in which a model is trained for a specific problem, and it is used on another problem in some other way. These are pre-trained models which are already trained on bigger datasets such as ImageNet. These models can be embedded with our model for improved

Fig. 1 Brain and other nervous system cancer stats in SEER 13 (1992–2018) [2]

performance. The transfer learning models MobileNet, InceptionV3, ResNet50 and VGG19 are evaluated on the BraTS 2015 dataset based on the accuracy and loss metrics.

2 Related Works

There are different methods for the detection of brain tumors. Madhesh et al. [3] has suggested an efficient method for the automatic identification and segmentation of brain parts from magnetic resonance image slices using the histogram of oriented gradients features and support vector machine classifier. This helps in identifying the region of interest which makes it easier in diagnosing the diseases.

Das et al. [4] proposed a computer-aided system to detect tumors and segment them. At the preprocessing stage, noise removal and gamma law transformation are performed. Then during the segmentation phase, the prepossessed image is converted to a binary image, and finally, the image is split and segmented.

Patil et al. [5] compare various works which proposed different techniques to detect and segment brain tumors from magnetic resonance images that include SOM clustering. K-mean clustering, fuzzy C-mean technique and curvelet transform.

Hossain et al. [6] applied various traditional classifiers such as support vector machine, K-nearest neighbor, multilayer perceptron, logistic regression, naive Bayes and random forest which were implemented in scikit-learn. Later on, it was switched on to convolutional neural network implemented using Keras and tensorflow as it yields a better performance than the above-mentioned classifiers. In their work, the convolutional neural network gained an accuracy of 97.87%.

Swati et al. [7] made use of a pretrained deep convolutional neural network model and proposed a block-wise fine-tuning strategy which is based on transfer learning. This is a generic method that requires minimal preprocessing which achieved an accuracy of 94.82%. The interchangeability of learning from normal images to medical brain magnetic resonance images is shown in this model.

Ronneberger et al. [8] applied a U-net architecture which splits it into three different segmentation methods which includes segmentation of neuronal structures. The paper yields an average IOU of 92% for precision segmentation by building a fully convolutional network architecture which works with a very few training images.

Ullah Khan et al. [9] applied various ResNet models on two different datasets to evaluate the performance of various ResNet models. ResNet18, ResNet50, ResNet101, ResNet152 were compared for the best accuracy and the results were 152 layered. ResNet152 achieved the high accuracy compared to other models.

Wang et al. [10] introduced Dense-MobileNet using dense blocks for image classification. By using dense blocks, two models are proposed to improve the basic structure of MobileNet. It also shows that Dense2-MobileNet gives more efficient results. The Dense1-MobileNet model has less accuracy compared with the MobileNet model, but it reduces the number of parameters and calculation time by nearly half.

Rehman et al. [11] performed a study on brain tumor classification using transfer learning and convolutional neural network architectures such as AlexNet, GoogleNet and VGGNet. Features and patterns were extracted from the magnetic resonance image slices and attained the best accuracy using the VGG16 network.

3 Proposed Methodology

Our approach makes use of transfer learning with the aid of different intense learning models such as MobileNet, InceptionV3, ResNet50 and VGG19. First, the images from the dataset were augmented to increase the number of images and the features were collected. Then, the images are classified as tumors or non-tumorous. The models are then assessed by the help of accuracy and loss metrics. The following section gives the details of the dataset used, data augmentation techniques used and also the feature extraction techniques (Fig. 2).

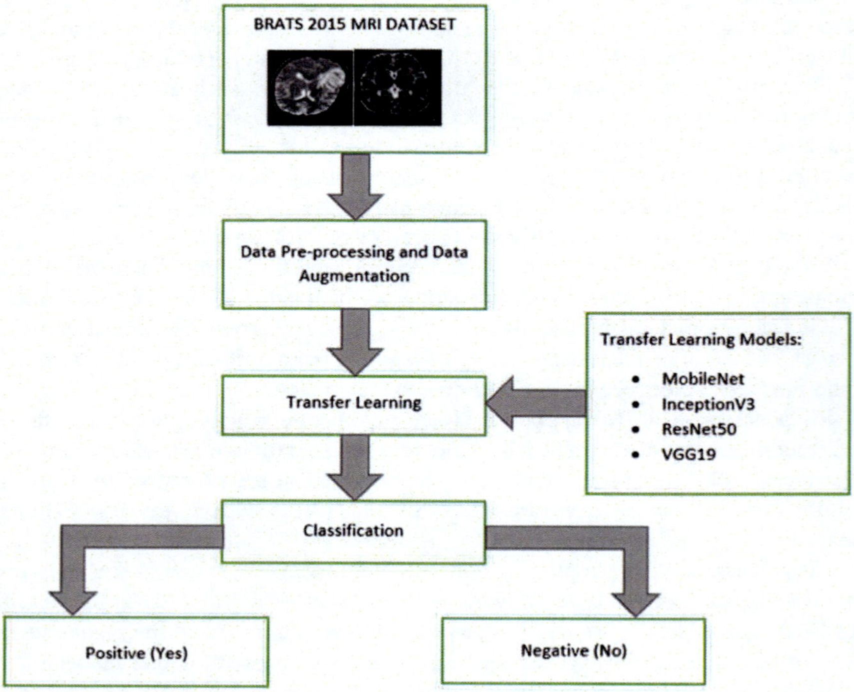

Fig. 2 Conceptual diagram

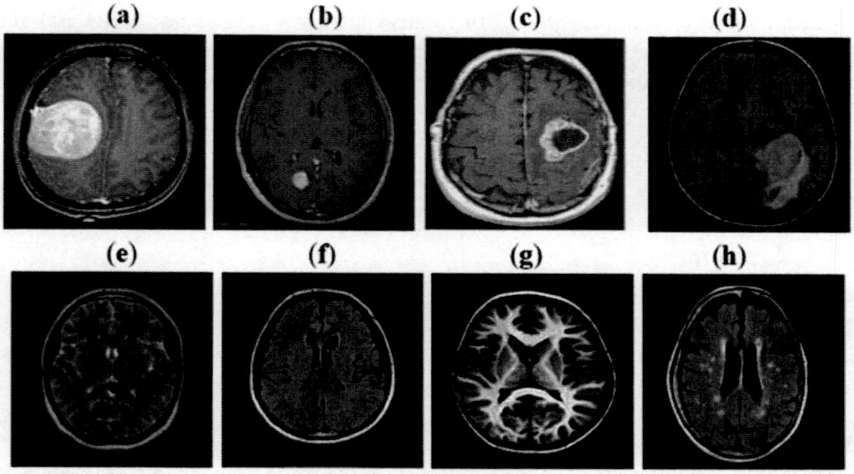

Fig. 3 Sample images from BraTS 2015 dataset. Tumorous magnetic resonance images (**a–d**) and non-tumorous magnetic resonance images (**e–h**)

3.1 Dataset

Magnetic resonance images (MRI) of brain tumors used in the model were acquired from BraTS 2015 dataset, a public dataset, which is designed for image classification. Different types of MRI images are generated by introducing changes in the arrangement of the radiofrequency pulses. Based on the factors such as repetition time and time to echo, the magnetic resonance images can be of four types of weighted images. They are:

- T1—these are produced by the help of shorter time to echo and repetition times.
- T1c
- T2—these are produced by the help of longer time to echo and repetition times.
- FLAIR (Fluid Attenuated Inversion Recovery)—these are similar to T2 weighted images with a difference that the time is very longer than T2.

The dataset used here consists of 253 magnetic resonance images of the brain in which 155 images with positive results and 98 images with negative results. All these images are in a resolution of 240X240 pixels (Fig. 3).

3.2 Data Augmentation

The dataset is a smaller one which consists of only 253 images which is not enough to train the network. This leads to achieving a result with low precision and overfitting. As the dataset is limited to a small range, data augmentation techniques such as rotation, width shifting, height shifting, shear intensity, brightness, horizontal and vertical flip and nearest fill are used. Data augmentation also helps to tackle the data imbalance issue in the data. After augmentation, the total number of images was increased to 2065.

3.3 *Feature Extraction and Classification Using Transfer Learning*

Transfer learning methods are commonly used with a limited dataset. It makes use of a pretrained network with a large dataset, and it is then used to train a new network. The different transfer learning models that we analyzed to detect brain tumor are:

MobileNet

MobileNets are classes of efficient models for mobile and embedded vision applications which is developed on a streamline architecture. It makes use of depth wise separable convolutions in order to construct lightweight deep neural networks. Extracting features from the input image is the job done by MobileNet by converting the pixels from the image to features and then passed to other layers (Fig. 4).

InceptionV3

InceptionV3 is a Convolutional Neural Network which is most commonly useful in image and object detection. It is a 48-layer deep convolutional neural network. It is one of the Inception model with a lot improvements such as Label smoothing, factorized 7 * 7 convolutions and the use of auxiliary classifiers to propagate label information lower down the network (Fig. 5).

ResNet50

ResNet50 is a 50-layer deep Convolutional Neural Network. This model consists of 5 stages which have both convolution and identity block. Both the blocks have 3 convolution layers each. It is also known as residual learning (Fig. 6).

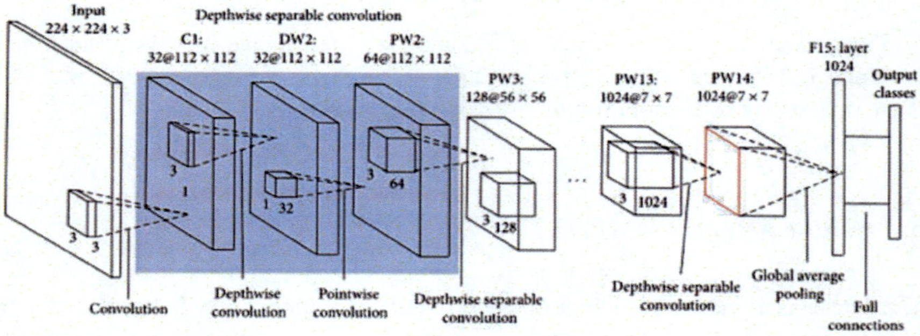

Fig. 4 MobileNet architecture [11]

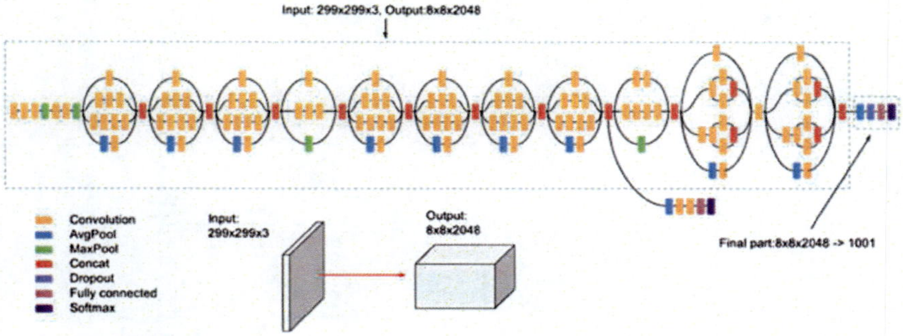

Fig. 5 InceptionV3 architecture [12]

VGG19

VGG19 is a deep Convolutional Neural Network which consists of 19 layers. Here the network is built using only 3 × 3 convolutional layers which are placed on top of each other as a stack on the basis of increasing depth. The 19 layers include 16 convolutional layers, 3 fully connected layers, 5 MaxPool layers and a SoftMax layer (Fig. 7).

4 Experiment and Result Analysis

Through this work, we have done a comparative study of various transfer learning models to detect brain tumor. From those various results, the best result to detect brain tumor is identified. For the experiment, the dataset used consisted of a total of 2065 images that was produced after augmentation from a total of 253 from the BraTS 2015 dataset in which 1085 were tumorous images and 980 were non-tumorous images. The dataset was divided into two sets—training which had 80% and validation which had 20%. Also, test batches were produced from the validation set. For this analysis, we took four transfer learning models to identify which provided the best result for the detection of brain tumor based on the accuracy and loss metrics. A batch size of 52 with 100 epochs were used to train the model. The learning rate for compiling was set to 0.0001 with the Adam optimizer. Here, in this work after the instantiation of the model, an input image of size 240 * 240 is passed to the model and the convolutional base is frozen to prevent weight updation. Then, GlobalAveragePooling layer is added to convert the features into a single 1024-element vector per image. Dropout and early stopping were done to forestall overfitting. From above training and compiling, MobileNet achieved an accuracy of 90.54%, InceptionV3 achieved an accuracy of 85.96%, ResNet50 achieved an accuracy of 95.42% and VGG19 achieved an accuracy of 91.69%. From the above results, it is found that ResNet50 attains the highest result than others in terms of accuracy and loss. Figure 8 displays the performance metrics for the training models.

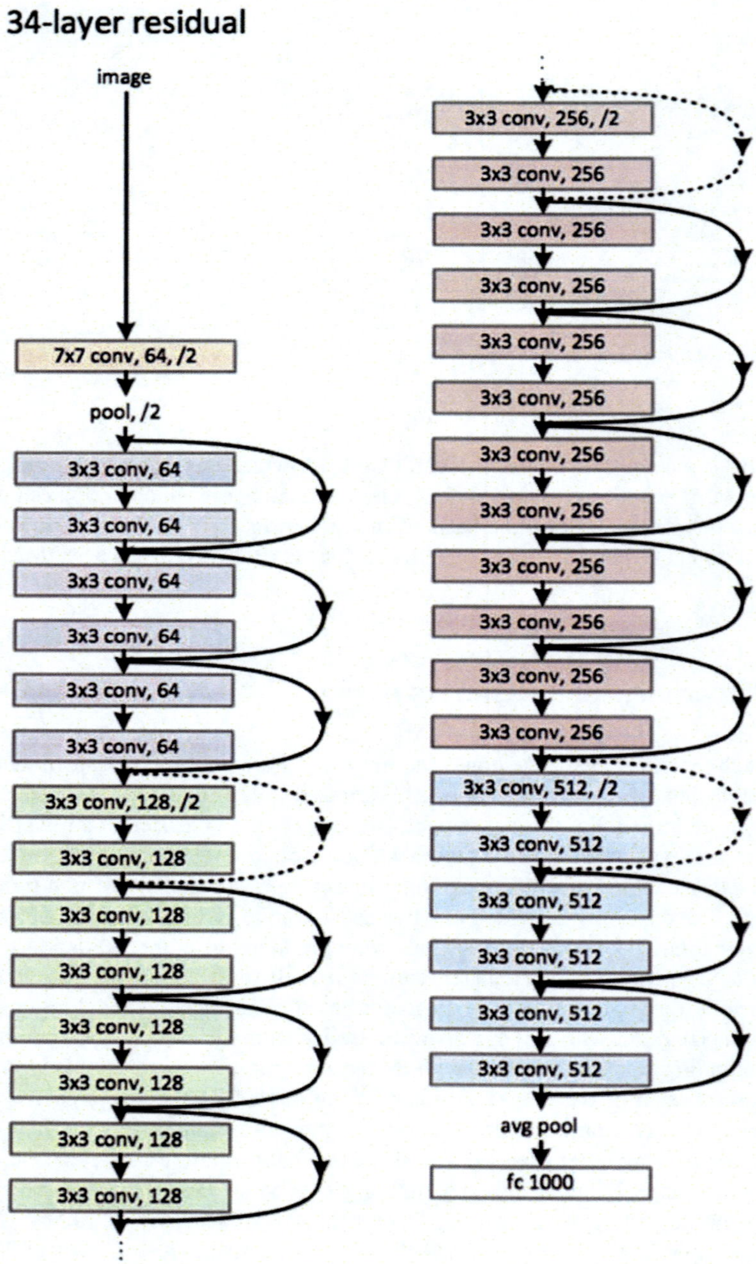

Fig. 6 ResNet50 architecture [13]

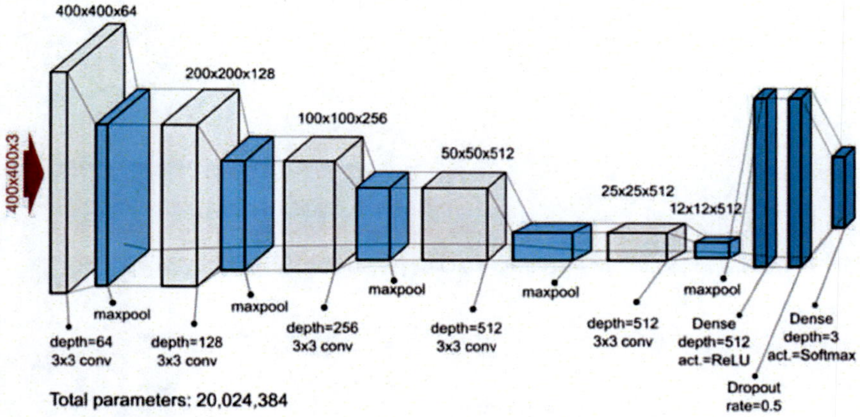

Fig. 7 VGG19 architecture [1]

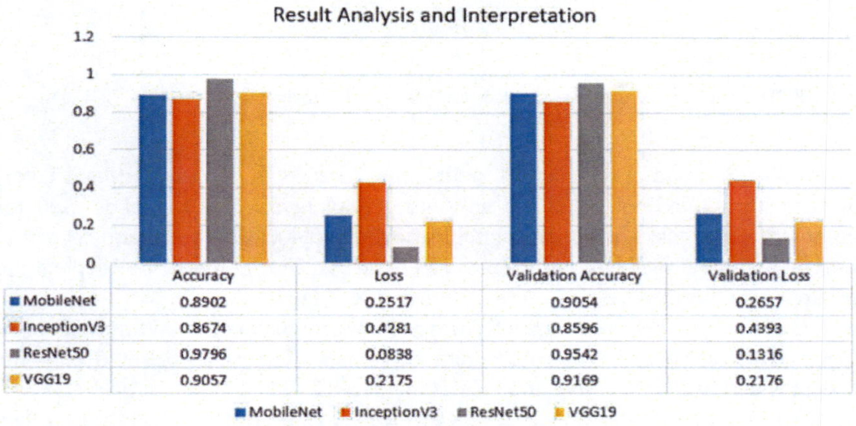

	Accuracy	Loss	Validation Accuracy	Validation Loss
MobileNet	0.8902	0.2517	0.9054	0.2657
InceptionV3	0.8674	0.4281	0.8596	0.4393
ResNet50	0.9796	0.0838	0.9542	0.1316
VGG19	0.9057	0.2175	0.9169	0.2176

Fig. 8 Result of various transfer learning models in BraTS 2015 dataset

MobileNet is a simplified architecture that makes use of depth-wise separable convolutions in order to create lightweight deep convolutional neural networks and provides an efficient model for mobile and compact vision applications. It is a 30-layer deep transfer learning model. This model achieved an accuracy of 90.54%. The graph below depicts the accuracy and loss for training and validation (Fig. 9).

InceptionV3 is a commonly used image recognition model which is composed of various symmetric and asymmetric building blocks such as convolution, average pooling, max pooling, concats, dropouts and fully connected layers. This model achieved an accuracy of 85.96%. The graph below depicts the accuracy and loss for training and validation (Fig. 10).

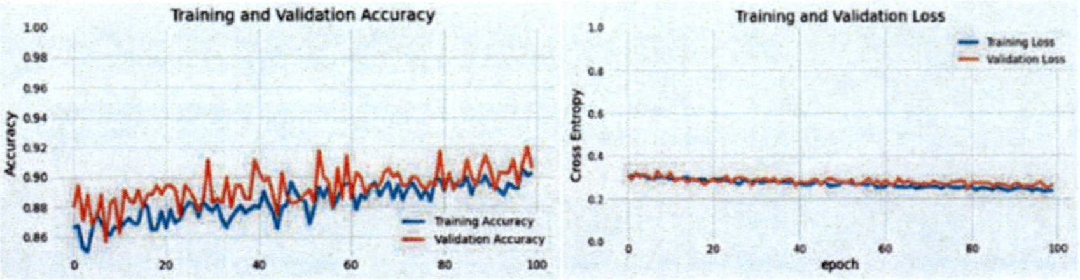

Fig. 9 Training and validation—accuracy and loss graphs (MobileNet)

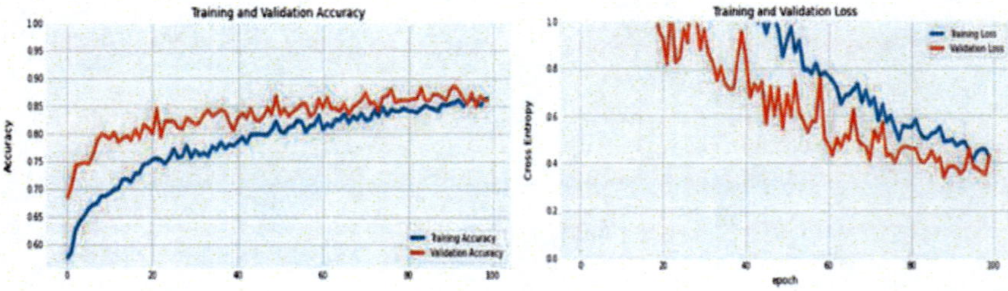

Fig. 10 Training and validation—accuracy and loss graphs (InceptionV3)

ResNet50 is one of the ResNet models that consists of 48 convolutional layers along with a MaxPool layer and an AveragePool layer. This model is used for computer vision tasks like image classification, object localization and object detection. This model achieved an accuracy of 95.42%. The graph below depicts the accuracy and loss for training and validation (Fig. 11).

VGG19 uses a 224 * 224 sized RGB input image which is passed through the 19 layers. This model achieved an accuracy of 91.69%. The graph below depicts the accuracy and loss for training and validation (Fig. 12).

After each model was trained, it was verified for its performance on the new data using a test set. From the above results, it can be clearly identified that ResNet50 is the best model that gave the highest accuracy with the lowest loss. The prediction for the best model among the four which is the ResNet50 is given below (Fig. 13).

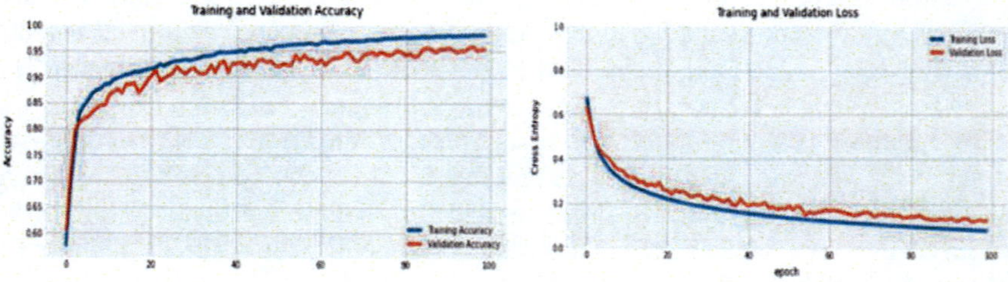

Fig. 11 Training and validation—accuracy and loss graphs (ResNet50)

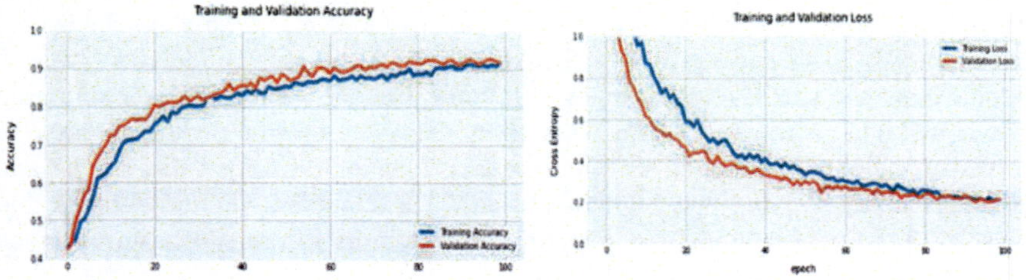

Fig. 12 Training and validation—accuracy and loss graphs (VGG19)

```
Predictions:
 [1 0 1 0 1 0 0 0 1 0 1 1 1 0 0 1 0 0 0 0 0 0 0 0 1 1 0 1 1 0 0 0]
Labels:
 [1 0 1 0 1 0 0 0 1 0 1 1 1 0 0 1 0 0 1 0 0 0 0 0 1 1 0 1 1 0 1 0]
```

Fig. 13 ResNet50 prediction for test set

5 Discussion

We find that computer-based systems can help in diagnosing the tumor in the early stage, and that makes a way for better treatment. This study highlights the fact that deep learning has a lot of significant applications and the ability to handle and interpret very large amounts of data can enhance the efficiency of humans, especially in the analysis of medical images. The objective of this study is to compare different models of transfer learning for the detection of brain tumors and identify the best. Through this work, a comparison of different transfer learning models such as MobileNet, InceptionV3, ResNet50 and VGG19 which are applied to the convolutional neural network model for the detection of brain tumors is performed and the best one among the four is found out. As this work is a miniature of the real-world activity, it may not work well with bigger and complex situations. Also, the limitation of dataset is a major shortcoming which is encountered in the field of medical image processing. In the future, an upcoming research could be performed with a large amount of data that can be trained for using it in real life situations to detect brain tumor which will be a great achievement in the field of medical science.

6 Conclusion

The aim of this study was to analyze and compare various transfer learning models for the detection of brain tumor from magnetic resonance images. The transfer learning models used for the comparative study are MobileNet, InceptionV3, ResNet50 and VGG19. The above models were trained on the BraTS 2015 dataset and achieved an accuracy and loss of 90.54 and 26.57% for MobileNet, 85.96 and 43.93% for InceptionV3, 95.42 and 13.16% for ResNet50 and 91.69 and 21.76% for VGG19. MobileNets are small, low-power models. InceptionV3 requires more computing power and a large amount of data. VGG19 models are slow to train and are large in terms of weights. Compared to other models, ResNet50 avoids false results and provides faster training and higher accuracy. The comparative study shows that ResNet50 is the best model for detecting brain tumors in terms of both accuracy and loss.

References

1. J. Jaworek-Korjakowska, P. Kleczek, M. Gorgon, Melanoma thickness prediction based on convolutional neural network with VGG-19 model transfer learning, in *Proceedings of the IEEE/CVF Conference on Computer Vision and Pattern Recognition Workshops* (2019)
2. T. Carvalho, et al., Exposing computer generated images by eye's region classification via transfer learning of VGG19 CNN, in *2017 16th IEEE International Conference on Machine Learning and Applications (ICMLA)* (IEEE, 2017)
3. K. Vikram, H.P. Menon, D.M. Dhanalakshmy, Segmentation of brain parts from MRI image slices using genetic algorithm. *Computational Vision and Bio Inspired Computing* (Springer, Cham, 2018), pp. 457–465
4. P. Das, et al., Computer aided system for brain tumor detection and segmentation. Int. J. Eng. Manage. Res. (IJEMR) **5**(5), 392–395 (2015)
5. Ms. Patil, et al., A review paper on brain tumor segmentation and detection. IJIREEICE **5**, 12–15 (2017)
6. Shah, F. Muhammad, in *Brain tumor detection using convolutional neural network*. Dissertation, Ahsanullah University of Science and Technology (2019)
7. Z.N.K Swati, et al., Brain tumor classification for MR images using transfer learning and fine-tuning. Comput. Med. Imag. Graph. **75**, 34–46 (2019)
8. O. Ronneberger, P. Fischer, T. Brox, U-net: convolutional networks for biomedical image segmentation, in International Conference on Medical Image Computing and Computer-Assisted Intervention (Springer, Cham, 2015)
9. I.D. Mustafa, M.A. Hassan, A. Mawia, A comparison between different segmentation techniques used in medical imaging. Am. J. Biomed. Eng. **6**(2), 59–69 (2016)
10. R.U. Khan, et al., Evaluating the performance of resnet model based on image recognition, in Proceedings of the 2018 International Conference on Computing and Artificial Intelligence (2018)
11. W. Wang, et al., A novel image classification approach via dense-MobileNet models, in *Mobile Information Systems 2020* (2020)
12. https://datascience.stackexchange.com/questions/33022/how-to-interpert-resnet50-layer-types/47489
13. https://cloud.google.com/tpu/docs/inception-v3-advanced
14. Chandra, J. Naveen, V. Bhavana, H.K. Krishnappa, Brain tumor detection using threshold and watershed segmentation techniques with isotropic and anisotropic filters, in *2018 International Conference on Communication and Signal Processing (ICCSP)* (IEEE, 2018)
15. M. Havaei, et al., Brain tumor segmentation with deep neural networks. Med. Image Anal. **35**, 18–31 (2017)
16. B. Devkota et al., Image segmentation for early stage brain tumor detection using mathematical morphological reconstruction. Procedia Computer Sci. **125**, 115–123 (2018)
17. D.S. Prabha, J. Satheesh Kumar, Performance evaluation of image segmentation using objective methods. Indian J. Sci. Technol. **9**(8), 1–8 (2016)

18. C. Wang, et al., Pulmonary image classification based on inception-v3 transfer learning model. IEEE Access **7**, 146533–146541 (2019)
19. A. Rehman, et al., A deep learning-based framework for automatic brain tumors classification using transfer learning. Circ. Syst. Signal Process. **39**(2), 757–775 (2020)
20. https://seer.cancer.gov/statfacts/html/brain.html
21. https://www.nhp.gov.in/world-brain-tumour-day2019_pg
22. https://paperswithcode.com/method/inception-v3
23. https://www.kaggle.com/andrewmvd/brain-tumor-segmentation-in-mri-brats-2015
24. https://www.hindawi.com/journals/misy/2020/7602384/

Synthesis and Research of Orthonormal Functions Based on Chebyshev–Legendre Polynomials

for Simulation

Vadim L. Petrov

Abstract The special characteristics of dynamic systems put forward new requirements for the creation of stable algorithms for describing the dynamic characteristics of these systems. Spectral methods for dynamic characteristics analyzing make it possible to create models that can be used to solve problems of identification and diagnostics of technical systems, for example, data channels, power lines, electric or hydraulic drives. Impulse response functions, correlation and autocorrelation functions are used as dynamic characteristics of systems. Spectral models are determined on the basis of the well-known Fourier integral in the basis of functions, the justification of which is also very important. The phased implementation of the transformation procedures, the normalization of the Chebyshev–Legendre polynomials made it possible to synthesize the transformed generalized orthonormal Chebyshev–Legendre functions that retain their properties on the argument interval $[0, \infty]$. These functions can be used to approximate the impulse response functions of dynamic systems. The research of the properties of the synthesized orthonormal functions made it possible to establish their recurrence formulas, which form the basis of computational procedures in spectral mathematical models. The obtained results allow ensuring the uniqueness of mathematical models, their connection with other operator models (for example, Laplace), stability in determining the parameters of models, the implementation of computational procedures and create universal algorithms for identification and diagnostics.

1 Introduction

The special characteristics of dynamic systems put forward new requirements for the creation of stable algorithms for describing the dynamic characteristics of these systems [1–4]. Spectral methods for analyzing dynamic characteristics make it possible to create models that can be used to solve problems of identification and diagnostics of technical systems, for example, data channels, power lines, electric

V. L. Petrov (✉)
National Research Technological University "MISiS", Leninsky Prospect 4, 119049 Moscow, Russia

P. Karuppusamy et al. (eds.), *Ubiquitous Intelligent Systems*, Smart Innovation, Systems and Technologies 243, https://doi.org/10.1007/978-981-16-3675-2_15

or hydraulic drives [5–8]. Power systems in special conditions, such as mining, are particularly in need of control methods based on spectral simulating the condition of electrical equipment [9, 10]. Such researches can be seen also in the economic field [11, 12]. The presence of effective mathematical support for stable algorithms for the mathematical description of the dynamic characteristics of systems for spectral models is the main condition for the success of the creation and functioning of spectral models [2, 3].

Impulse response functions, correlation and autocorrelation functions, the spectral representation of which determines the mathematical models, are used as the dynamic characteristics of the systems. Spectral models are determined on the basis of the Fourier integral, the justification of which is also very important [13, 14]. The use of orthonormalized functions of Chebyshev functions as such a functional basis makes it possible to ensure the uniqueness of the models, their connection with other operator models (for example, Laplace), stability in determining the parameters of the models, the implementation of computational procedures, etc. Therefore, the synthesis and research of orthonormal functions for the purposes of mathematical modeling of dynamic characteristics are an urgent task [14, 15].

The purpose of the research is to synthesize new systems of orthonormal functions that can be used for simulations of dynamical systems.

If we use the impulse response functions of a dynamical system as a dynamic characteristic, then its mathematical model will be:

$$h_\delta(\tau) = \sum_{j=0}^{\infty} \mu_j \Phi_j(\tau),$$

where

$h_\delta(\tau)$	impulse response functions;
$\Phi_j(\tau)$	a system of functions, such as orthogonal or orthonormal functions;
μ_j	coefficients of the Fourier expansion of the $h_\delta(\tau)$ in the basis of functions $\Phi_j(\tau)$;
$\Phi_j(\tau)$	has the properties of an orthonormal or orthogonal function on the interval of the argument τ [0, ∞ [.

The model scheme is shown in Fig. 1.

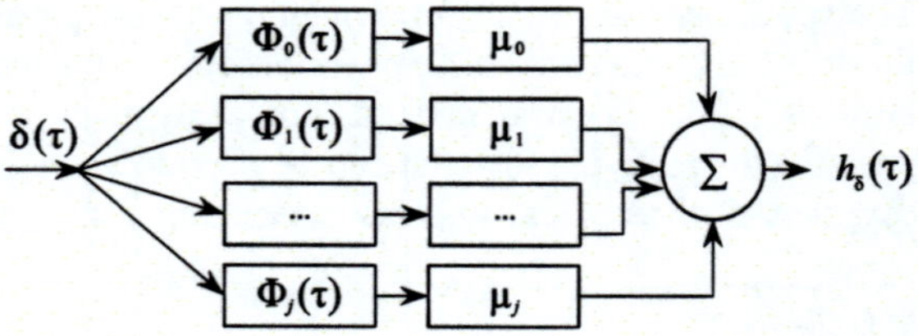

Fig. 1 Model scheme of impulse response functions, where $\delta(\tau)$—the Dirac function

2 Synthesis of Orthonormal Chebyshev–Legendre Functions

The formula for determining the standardized Chebyshev–Legendre polynomials has the form [12–14]:

$$P_n(x) = \frac{1}{n!2^n}\frac{d^n}{dx^n}\left[\left(x^2-1\right)^n\right]. \tag{1}$$

These polynomials are orthogonal on the interval $[-1, 1]$.

Through n-fold differentiation, we obtain an algebraic formula for determining the Chebyshev–Legendre polynomials:

$$P_n(x) = \frac{1}{2^n}\sum_{k=0}^{\left[\frac{n}{2}\right]}\frac{(-1)^k\Gamma(2n-2k+1)}{\Gamma(k+1)\Gamma(n-k+1)\Gamma(n-2k+1)}x^{n-2k}, \tag{2}$$

where

$\left[\frac{n}{2}\right]$ is the operation of allocating the integer part of the number $\frac{n}{2}$;
$\Gamma(k)$ gamma function.

For orthogonal Chebyshev–Legendre polynomials, one can define another algebraic formula. To do this, we carry out the differentiation operation using the Leibniz formula:

$$\frac{\mathrm{d}^n}{\mathrm{d}x^n}\left[(x-1)^n(x+1)^n\right] = \sum_{k=0}^{n} C_n^k\frac{\mathrm{d}^k}{\mathrm{d}x^k}\left[(x-1)^n\right]\frac{\mathrm{d}^{n-k}}{\mathrm{d}x^{n-k}}\left[(x+1)^n\right]. \tag{3}$$

Considering that

$$\frac{\mathrm{d}^k}{\mathrm{d}x^k}\left[(x-1)^n\right] = \frac{\Gamma(n+1)}{\Gamma(n-k+1)}(x-1)^{n-k} \quad \text{and}$$
$$\frac{\mathrm{d}^{n-k}}{\mathrm{d}x^{n-k}}\left[(x+1)^n\right] = \frac{\Gamma(n+1)}{\Gamma(k+1)}(x+1)^k,$$

we obtain the following expression from formula (2):

$$P_n(x) = \frac{\Gamma(n+1)}{2^n}\sum_{k=0}^{n}\frac{C_n^k}{\Gamma(k+1)\Gamma(n-k+1)}(x-1)^{n-k}(x+1)^k. \tag{4}$$

The resulting formula does not include the determining of the integer part of the number; therefore, it is more convenient to use.

At the next stage, we define the norm (1) for finding the orthonormal Chebyshev–Legendre polynomial.

The highest coefficient of the polynomial $P_n(x)$ is determined from formula (1) and it is equal to

$$\frac{2n(2n-1)\cdots(n+1)}{n!\cdot 2^n}=\frac{(2n)!}{(n!)^2\cdot 2^n}.$$

The norm of a polynomial is determined by solving the following integral [14]:

$$\|P_n\|^2=\int_{-1}^{1}P_n^2(x)\mathrm{d}x=\frac{\Gamma(2n+1)}{[\Gamma(n+1)]^2\cdot 2^n}\int_{-1}^{1}x^n P_n(x)\mathrm{d}x=\frac{2}{2n+1}$$

The formula for the orthonormal Chebyshev–Legendre polynomial is determined after the implementation of the normalization procedure from formula (4):

$$\hat{P}_n(x)=\sqrt{\frac{2n+1}{2}}\,P_n(x)=\sqrt{\frac{2n+1}{2}}\frac{\Gamma(n+1)}{2^n}\times\sum_{k=0}^{n}\frac{C_n^k}{\Gamma(k+1)\Gamma(n-k+1)}(x-1)^{n-k}(x+1)^k. \quad (5)$$

It is necessary to carry out the substitution of the $x=1-2y$ for polynomial $P_n(x)$.

Such a substitution makes it possible to form a new polynomial with orthogonal properties on the interval [0; 1], and fulfill the following conditions:

$$\int_{0}^{1}(-2)P_m(1-2y)P_n(1-2y)\mathrm{d}y=\begin{cases}\frac{2}{2n+1} & \text{for } m=n;\\ 0 & \text{for } m\neq n.\end{cases}$$

The resulting polynomial (5) in practical applications is called the shifted Chebyshev–Legendre polynomial. Using formulas (1) and (3), we obtain expressions for determining the shifted Chebyshev–Legendre polynomials:

$$\dot{P}_n(y)=\frac{1}{2^n}\sum_{k=0}^{[\frac{n}{2}]}\frac{(-1)^k\Gamma(2n-2k+1)}{\Gamma(k+1)\Gamma(n-k+1)\Gamma(n-2k+1)}(1-2y)^{n-2k}\quad \text{or}$$

$$\dot{P}_n(y)=[\Gamma(n+1)]^2\sum_{k=0}^{n}(-1)^{n-k}\frac{y^{n-k}(1-y)^k}{[\Gamma(k+1)]^2[\Gamma(n-k+1)]^2}. \quad (6)$$

To determine norm (6), it is necessary to use the binomial representation formula, according to which

$$(1-y)^k = \sum_{j=0}^{k} \frac{(-1)^j \Gamma(k+1) y^j}{\Gamma(j+1)\Gamma(k-j+1)}.$$

The following formula can be obtained by using (6)

$$\dot{P}_n(y) = [\Gamma(n+1)]^2 \sum_{k=0}^{n} \frac{(-1)^{n-k}}{[\Gamma(k+1)]^2[\Gamma(n-k+1)]^2} \times \sum_{j=0}^{k} (-1)^j \frac{\Gamma(k+1)}{\Gamma(j+1)\Gamma(k-j+1)} y^{n-k+j}.$$

The integral for determining the norm of the shifted Chebyshev–Legendre polynomials is determined taking into account that the leading term of formula (6) is equal to $\frac{(-1)^n\Gamma(2n+1)}{\Gamma(n+1)\,!}$

$$\|\dot{P}_n(y)\| = \frac{(-1)^n\Gamma(2n+1)}{(\Gamma(n+1))!} \int_0^1 y^n \cdot \dot{P}_n(y)\mathrm{d}y = \frac{1}{2n+1}.$$

Established norm allows us to represent orthonormal shifted Chebyshev–Legendre:

$$\hat{\dot{P}}_n(y) = \frac{\sqrt{2n+1}}{2^n} \sum_{k=0}^{[\frac{n}{2}]} \frac{(-1)^k\Gamma(2n-2k+1)}{\Gamma(k+1)\Gamma(n-k+1)\Gamma(n-2k+1)} (1-2y)^{n-2k};$$

$$\hat{\dot{P}}_n(y) = \sqrt{2n+1}[\Gamma(n+1)]^2 \sum_{k=0}^{n} (-1)^{n-k} \frac{y^{n-k}(1-y)^k}{[\Gamma(k+1)]^2[\Gamma(n-k+1)]^2}. \quad (7)$$

The subsequent transformations of the considered polynomials are carried out by implementing the substitution $y = e^{-u \cdot t}$ in (6), (7), where u is the scale parameter.

The following formula is obtained after transformations:

$$P_{sn}(u,t,n) = [\Gamma(n+1)]^2 e^{-utn} \sum_{k=0}^{n} \frac{(e^{ut}-1)^k}{[\Gamma(k+1)]^2[\Gamma(n-k+1)]^2}. \quad (8)$$

A necessary condition for the classical orthogonal polynomials with a unit weight is defined as follows:

$$\int_a^b P_n(x)P_m(x)\mathrm{d}x = 0,$$

where $n \neq m$, a AND b—orthogonality interval.

This condition for the transformed Chebyshev–Legendre functions (8) has the form:

$$\int_{0}^{\infty} P_{sn}(u,t,n)P_{sm}(u,t,n)H_p(u,t)\mathrm{d}t$$
$$= \int_{0}^{\infty} P_{sn}(u,t,n)P_{sm}(u,t,n)\left(ue^{-ut}\right)\mathrm{d}t = 0. \tag{9}$$

Thus, by artificially splitting the function $H_p(u,t)$, one can obtain a system of orthogonal Chebyshev–Legendre functions:

$$P_{n\infty}(u,t) = (-1)^n\sqrt{u}[\Gamma(n+1)]^2 e^{-ut(n+\frac{1}{2})}\sum_{k=0}^{n}\frac{(1-e^{ut})^k}{[\Gamma(k+1)]^2\Gamma[(n-k+1)]^2}. \tag{10}$$

These functions, as well as the Chebyshev–Legendre polynomials, have unit weight. The orthogonality of the obtained functions is proved by solving (9).

Orthonormal Chebyshev–Legendre functions are determined taking into account the calculated norm of the shifted polynomials:

$$\hat{P}_{n\infty}(u,t) = (-1)^n\sqrt{u(2n+1)}[\Gamma(n+1)]^2 \times e^{-ut(n+\frac{1}{2})}\sum_{k=0}^{n}\frac{(1-e^{ut})^k}{[\Gamma(k+1)]^2\Gamma[(n-k+1)]^2}. \tag{11}$$

Binomial representation $(1-e^{ut})^k = \sum_{j=0}^{k}\frac{(-1)^j\Gamma(k+1)e^{utj}}{\Gamma(j+1)\Gamma(k-j+1)}$ allows obtaining another formula for the orthonormal Chebyshev–Legendre functions (11)

$$\hat{P}_{n\infty}(u,t) = (-1)^n\sqrt{u(2n+1)}[\Gamma(n+1)]^2 e^{-ut(n+\frac{1}{2})} \times \sum_{k=0}^{n}\left\{\frac{1}{\Gamma(k+1)[\Gamma(n-k+1)]^2}\sum_{j=0}^{k}\frac{(-1)^j e^{utj}}{\Gamma(j+1)\Gamma(k-j+1)}\right\}. \tag{12}$$

3 Research of Orthonormal Functions Chebyshev–Legendre

Theorem 1 *The recurrent formula of transformed generalized orthonormal Chebyshev–Legendre functions:*

$$2\cdot(n+1)^2\sqrt{\frac{1}{(2n+3)(2n+1)(2n+1)^2}}\hat{P}_{n+1}(u,t)$$
$$=\left(1-2e^{-ut}\right)\hat{P}_n(u,t)-n\cdot\sqrt{\frac{1}{(2n-1)(2n+1)}}\hat{P}_{n-1}(u,t). \qquad (13)$$

To prove this theorem, we use the recurrence formula for orthonormal polynomials, according to which the condition [12–14]:

$$\lambda_n\hat{P}_{n+1}(x)=(x-\eta_n)\hat{P}_n(x)-\lambda_{n-1}\hat{P}_{n-1}(x),$$

where $\lambda_n=\frac{\mu_n}{\mu_{n+1}}$; μ_n—the highest coefficient of an orthonormal polynomial $\hat{P}_n(x)$; η_n—coefficient determined by the type of polynomial.

Coefficients of the form $\hat{P}_n(x)=\mu_n x^n+\nu_n x^{n-1}\ldots$ can be determined using the following expressions for orthonormal Chebyshev–Legendre polynomials

$$\mu_n=\frac{1}{\Gamma(n+1)^2}\frac{\Gamma(2n+1)}{\sqrt{2^{2n+1}(2n+1)}},$$
$$\nu_n=\frac{\Gamma(2n)\sqrt{n(2n+1)}}{\sqrt{2^{2n+1}\Gamma(n+1)^3\Gamma(n)}}.$$

Parameter λ_n in a three-term recurrence formula is defined as:

$$\lambda_n=\frac{\mu_n}{\mu_{n+1}}=2\frac{(n+1)^2}{(2n+2)}\sqrt{\frac{1}{(2n+3)(2n+1)}}.$$

The equations, which are compiled under the condition of equality of the coefficients for the same powers of the argument, are solved for η_n:

$$\lambda_n\nu_{n+1}=\nu_n-\eta_n\mu_n.$$

The solution to this equation gives: $\eta_n=0$.

Taking into account the previously performed substitutions when carrying out transformations over the Chebyshev–Legendre polynomials, we obtain the final expression from the three-term recursive formula.

Recurrent formulas (13) for functions (12) are important for the implementation of computational procedures in mathematical models.

4 Simulation Using Functions Chebyshev–Legendre

The coefficients of the expansion of the impulse response function in the basis of the synthesized orthonormal Chebyshev–Legendre functions are determined in accordance with the following expression:

$$\chi_i = \int_0^\infty h_\delta(\tau)\hat{P}_{i\infty}(u,\tau)\mathrm{d}\tau,$$

where $h_\delta(\tau)$—impulse response function of the identified dynamic system.

In this case, the impulse response function is represented by the following series [11, 14]:

$$h_\delta(\tau) = \sum_{j=0}^{\infty} \chi_j \hat{P}_{j\infty}(u,\tau). \tag{14}$$

The use of (10) allows obtaining a general formula for the orthogonal spectral model of the impulse response function:

$$h_\delta(\tau) = e^{-\frac{u\tau}{2}} \sum_{j=0}^{\infty} (-1)^j \chi_j \sqrt{u(2j+1)}[\Gamma(j+1)]^2$$
$$\times e^{-u\tau j} \sum_{k=0}^{j} \frac{(1-e^{u\tau})^k}{[\Gamma(k+1)]^2[\Gamma(j-k+1)]^2}.$$

Thus, the applicability of the synthesized orthogonal Chebyshev–Legendre functions in the problems of nonparametric identification of dynamical systems was demonstrated.

5 Assessment of the Effectiveness of Modeling

For evaluating the effectiveness of model synthesis, the researcher can use the standard deviation of the restored impulse response function

$$F(u) = \int_0^\infty \left[h_\delta(\tau) - \sum_{j=0}^{\infty} \chi_j \hat{P}_{j\infty}(u,\tau) \right]^2 \mathrm{d}\tau.$$

The square of the normalized dispersion can be also used as:

$$\sigma_n^2(u) = \frac{\int_0^\infty \left[h_\delta(\tau) - \sum_{j=0}^{\infty} \chi_j \hat{P}_{j\infty}(u,\tau)\right]^2 \mathrm{d}\tau}{\int_0^\infty h_\delta^2(\tau)\mathrm{d}\tau}. \tag{15}$$

The unique properties of the orthonormal Chebyshev–Legendre functions allow determining a simpler expression for the normalized dispersion

$$\sigma_n^2(u) = 1 - \frac{\sum_{i=0}^{n} \chi_i^2}{N_h^2},$$

where N_h^2—square of the impulse response function norm.

The highest confidence of the spectral model (14) is achieved by selecting the optimal values of the parameter u. The minimum condition (15) is traditionally used as a criterion for selecting optimal parameters.

6 Conclusion

The proposed synthesis algorithm for Chebyshev–Legendre orthonormal functions allows one to create a model functional basis for constructing spectral models of dynamic systems.

Recurrence relations for orthonormal Chebyshev–Legendre functions form the basis for the synthesis of computational procedures in mathematical models.

The obtained results make possible to ensure the uniqueness of mathematical models, their relationship with other operator models (for example, Laplace), stability in determining the parameters of models, the implementation of computational procedures and create universal algorithms for identification and diagnostics.

The obtained research results are applicable for modeling stable and physically realizable linear dynamic systems. The presented research methods can be used for the procedures of linearization of dynamical systems.

The research methods used in the issue have been successfully tested in the construction of mathematical models of electric drive systems for machines and equipment.

The main direction of further research will be the development of mathematical support for the implementation of procedures for parametric and nonparametric identification of dynamic systems. The methodological support was successfully used in the educational programs for the training of mining engineers of electrical engineering specialization [17, 18].

References

1. L. Ljung, *System Identification: Theory for the User*, 2nd edn. (Prentice-Hall, Englewood Cliffs, NJ, 1999)
2. T. Chen, L. Ljung, Regularized system identification using orthonormal basis functions. European Control Conference, ECC **2015**, 1291–1296 (2015). https://doi.org/10.1109/ECC.2015.7330716
3. P. Heuberger, P. Van Den Hof, B. Wahlberg, Modelling and identification with rational orthogonal basis functions, in *Modelling and Identification with Rational Orthogonal Basis Functions* (2005), pp. 1–397. https://doi.org/10.1007/1-84628-178-4
4. G. Pillonetto, G. De Nicolao, A new kernel-based approach for linear system identification. Automatica **46**(1), 81–93 (2010). https://doi.org/10.1016/j.automatica.2009.10.031
5. J.S. Bendat, A.G. Piersol, *Engineering Applications of Correlation and Spectral Analysis* (Wiley-Interscience, New York, 1980)
6. G. Rogers, *Power System Oscillations* (Kluwer, 2000)
7. P. Korba, Real-time monitoring of electromechanical oscillations in power systems: First findings). IET Gener. Transmission Distrib. **1**(1), 80–88 (2007). https://doi.org/10.1049/iet-gtd:20050243
8. D. Łuczak, K. Nowopolski, Identification of multi-mass mechanical systems in electrical drives, in *Proceedings of the 16th International Conference on Mechatronics*, Mechatronika (2014), pp. 275–282. https://doi.org/10.1109/MECHATRONIKA.2014.7018271
9. A.B. Sadridinov, Analysis of energy performance of heading sets of equipment at a coal mine. Gornye nauki i tekhnologii = Mining Sci. Technol. (Russia) **5**(4), 367–375 (2020) https://doi.org/10.17073/2500-0632-2020-4-367-375
10. F.P. Shkrabets, Electric supply of underground consumers of deep energy-intensive mines. Gornye nauki i tekhnologii = Mining Sci. Technol. (Russia) **3**, 25–46 (2017). https://doi.org/10.17073/2500-0632-2017-3-25-42
11. H. Lütkepohl, *Impulse Response Function* (The New Palgrave Dictionary of Economics, 2008)
12. A. Hatemi-J, Asymmetric generalized impulse responses with an application in finance. Econ. Model. **36**, 18–22 (2014). https://doi.org/10.1016/j.econmod.2013.09.014
13. P. Eykhoff, *System Identification: Parameter and State Estimation* (Wiley-Interscience, London, 1974)
14. P.K. Suetin, *Classical Orthogonal Polynomials* (Nauka, Moscow, 1979). ((in Russian))
15. G. Szegö, *Orthogonal Polynomials*, 4th ed. (American Mathematical Society Colloquium Publication, American Mathematical Society, Providence, RI, 1975), p. 23
16. V.L. Petrov, Identification of models of electromechanical systems using analysis methods in bases of continuous orthonormal functions. Mekhatronika, avtomatizatsiia, upravlenie **10**, 29–36 (2003). ((in Russian))
17. V.L. Petrov, Federal training and guideline association on applied geology, mining, oil and gas production and geodesy—a new stage of government, academic community and industry cooperation. Gornyi Zhurnal **9**, 115–119 (2016). https://doi.org/10.17580/gzh.2016.09.23
18. V.L. Petrov, Training of mineral dressing engineers at Russian Universities. Tsvetnye Metally **7**, 14–19 (2017). https://doi.org/10.17580/tsm.2017.07.02

Driver's Drowsiness Detection System Using Dlib HOG

Athira Babu, Shruti Nair, and K. Sreekumar

Abstract For human beings, sleep is a key requirement. The secret of humankind's physical well-being is sleep. In a study on sleep, researchers have proved that adults from the age of eighteen and above must get seven to nine hours of sleep a day. Drowsiness is the root cause of the hazardous road accidents. If drivers are notified as drowsy at the correct instant of time, we can prevent the majority of road accidents that took place in the world. New strategies are introduced by the researchers to detect the drowsiness of the driver and each technology has its own merit and demerit. This paper uses Python and Dlib models to build a drowsiness identification model. We aim to integrate both face detection and head pose detection which makes this an ideal detection method. In the proposed system, a laptop is used, using which real-time video is recorded. Head-pose detection along with face detection helps to increase accuracy. For dataset video input, the proposed system gives a maximum accuracy rate of 94.51%.

1 Introduction

In India, the majority of road accidents takes place due to driver fatigue. Driver's drowsiness is the main cause of increasing death rates. Various surveys have shown that about 20 percent of all road accidents are fatigue-related [1]. The numbers of passing and wounds rate keep on emerging annually, and most of this happens due to drowsy state of the driver while driving. Many people lose their life due to drowsiness. Fatigue decreases the driver's ability to control the vehicle in decision making. In the early afternoon, after lunch and at midnight, the exhaustion and sleepiness of the driver are much greater than at other times. Drinking alcohol, addiction to opioids, and the use of hypnotic drugs can all lead to loss of consciousness. Introduction

A. Babu (✉) · S. Nair · K. Sreekumar
Department of Computer Science & IT, Amrita School of Arts and Sciences, Amrita Vishwa Vidyapeetham, Kochi, India

K. Sreekumar
e-mail: sreekumar@asas.kh.amrita.edu

P. Karuppusamy et al. (eds.), *Ubiquitous Intelligent Systems*, Smart Innovation, Systems and Technologies 243, https://doi.org/10.1007/978-981-16-3675-2_16

of a good drowsiness detection system can make the driver realize his drowsiness while they are driving, and it can save their life. Signs which are helpful for fatigue detection can be split into three main categories in driver face monitoring systems:

(1) Eye region-related symptoms.
(2) Symptoms related to the area of the mouth.
(3) Head-related symptoms.

Usually, the strategies for detecting drowsy drivers are classified into three forms: vehicle-based measurements, behavioral-based, and the third one is based on physiological measures.

Vehicle-based measurements: A variety of parameters are continuously tracked, including lane direction deviations, steering wheel movement, acceleration pedal pressure, etc., and any change in these factors crosses a defined threshold clearly shows a markedly increased likelihood of drowsiness of the driver [1].

Behavioral-based measures: A camera is deployed to monitor the yawning, eye closure, eye blinking, head pose, etc. of the driver. A quick alarm is generated if any of the mentioned symptoms occur [1].

Physiological-based measures: It is a meticulous method to detect drowsiness of the driver as it is closely related to physiological signals. Electrocardiogram (ECG) and Electrooculogram (EOG) help to examine physiological and cognitive states in humans. ECG is used to assess the driver's health status and level of drowsiness. ECG records the heart electrically. The ECG sensor is used for the detection of drowsiness to extract heart rate variability data [2]. Electrooculography signals can be measured using different systems, which when transmitted to electronic devices such as smartphones, an alarm will be raised if the signal goes beyond a particular value [3]. By examining the pulse rate, heartbeat and brain information, we can identify if the driver is drowsy [4].

The paper aims to develop a way to alert drowsy drivers while driving. Here, the Dlib model is trained using 68 shape landmarks predictor. The drowsiness features of a driver are extracted which alerts the driver on time if the driver is detected with the symptoms of fatigue. The proposed system is efficient for real-time driver drowsiness detection.

2 Related Work

Many instances were recorded mainly for developing the system which can identify factors based on various elements like heart rate, grip quality and movement [5]. A paper described an algorithm that mainly used to examine a driver's drowsiness level by examining his changes in eye state [6]. In the paper on improved fatigue detection system, upon having the image of the face, the image is sent to support vector machine support (SVM), a classifier that identifies whether the facial image is fatigued or not. If the result given by the classifier is stated as fatigue, the alert unit informs the driver that he is drowsy. They focused only on the eye and mouth regions and ignored the rest. By concentrating more on the eye and mouth, they reduced unwanted characteristics in the feature set [7]. In paper [8], a model for facial feature recognition was proposed. Very serious conditions are divided into two modes: online and offline modes. In online mode, mobile devices with the help of computer vision libraries like Opencv and Dlib are used in real-time to detect the state of the driver. The hazardous conditions while driving must be noted in real-time. The mobile apps are focused extensively on calculating facial features in accident-prone situations and make decisions based on the identified visual behavior state [8]. Offline mode is based on a statistical analysis done before in hand and also based on previously collected data. The mobile apps are focused on extracting and evaluating facial features in hazardous situations and make decisions based on the identified visual behavior state [8].

A paper described a new algorithm and representation which can contribute to the wider application of digital image processing and computer vision. They used integral images to compute a fine set of image features. To accomplish real scale invariance, all face detection systems must function on multiple image frames [9]. There is no need to calculate a multi-scale image pyramid. So, it has proved that the time taken by integral image for face detection and the time taken by the image pyramid for computation are almost similar. For effective face detection, a classifier called AdaBoost is used for feature selection. Advantage of using the AdaBoost classifier is that we can give extremely large and complex features as input for the learning process. It also mentioned a cascade of classifiers that can increase the accuracy rate of detection. The cascade of classifiers helps to reduce the computation time [9].

In a recent paper, a more advanced technology called multitask ConNN model was used [10]. Driver's level of drowsiness is predicted by evaluating driver's eye closure time or percentage of eye closure (PERCLOS) and the frequency yawning of the driver or frequency of mouth (FOM). The driver's eye and mouth details were fetched more accurately by using the Dlib algorithm. For the estimation of fatigue parameters, the system was trained using multitask ConNN models. The number of frames to be used and frequency range to be used are kept fixed. The fatigue is then indicated as "intense, less intense and not intense" based on the fatigue parameter. This model is stated as a powerful model as it creates one and only one single ConNN model instead of creating a separate ConNN model that would otherwise constitute two different architectures [10]. In another paper, the application was installed on an Android phone. Using Dlib, they found the facial landmark. On calculating the ear aspect ratio, they found the distance between the eyelids and they determined whether the driver is drowsy or tired [11]. Another paper [12] first detects the face using Viola Jones algorithm and then the image is generated. Extended Sobel operator was used to localize and filter the eyes of the driver. Sobel operator was used to find the curvature of the eyelids. Concavity is used to note whether the eyes are open or close. A concave upward curve indicates that the eyes are closed, and a concave downward curve indicates that the eyes are open [12].

3 Proposed Methodology

In this paper, we proposed a drowsiness detection system using Python, OpenCv and Dlib library.

Brief introduction of OpenCv and Dlib library is as follows.

3.1 Open Computer Vision (OpenCV)

OpenCV [13] is the immense open-source library for computer vision, machine learning and image processing, and it now plays an important role in real-time activity in today's systems. The aim is to be able to process relevant data stored in an image or a video. OpenCV is used in various real-time applications such as motion detection, automated inspection and surveillance, interactive art installations and medical image analysis, which can be performed by using computer vision and image processing algorithms [13].

3.2 Dlib Library

Dlib is a toolkit for C++, which mainly comprises of machine learning algorithms and some are very good tools for building real-time applications. It has a pretrained facial landmark detector which computes the region of 68 *x*–*y* coordinates that trace facial landmarks in the face region [14]. The identification of facial landmarks is an important topic in terms of estimation of facial zone forms. The Dlib library was used in this research to detect and map the faces of the drivers in real-time videos. Relevant facial structures, thus, were detected using shape estimating techniques on the face region. The Dlib library uses its pretrained facial landmark detector to detect and localize facial landmarks. It mainly includes two form predictor models [15] which is prepared by the i-Bug 300-W dataset that each localize 68 and 5 landmark points lies within a face zone; 68 facial landmarks have been used in this approach (as shown in Fig. 1).

Dlib uses the histogram of oriented gradients (HOG)-based face detector. HOG is particularly convenient for object detection because object shape is characterized using the local intensity gradient distribution and edge direction.

Steps on how HOG works are followed:

Step 1. HOG divides the face into a number of connected cells (Fig. 2).

Step 2. For each cell, it creates a histogram (Fig. 3).

Step 3. Then, it merges all the cells to form one histogram which is unique for each individual face (Fig. 4).

Fig. 1 Processed 68 facial landmarks on a detected face traced [16]

Fig. 2 HOG face division

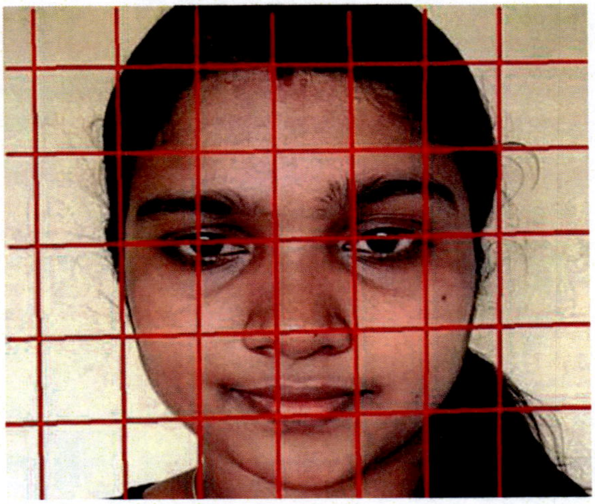

It can marvelously describe the edge characteristics of any object. It performs some operations on localized cells that allows the movement of the subject to be ignored. In our proposed model, the HOG-based face identifier will find the location of face from the real-time video. It is effectively convenient for face detection because it can extraordinarily describe contour and edge characteristics in various objects.

Both the eyes and mouth coordinates will be used in computing the aspect ratio which is based on the Euclidean distance for both the eyes and the mouth region (Fig. 5).

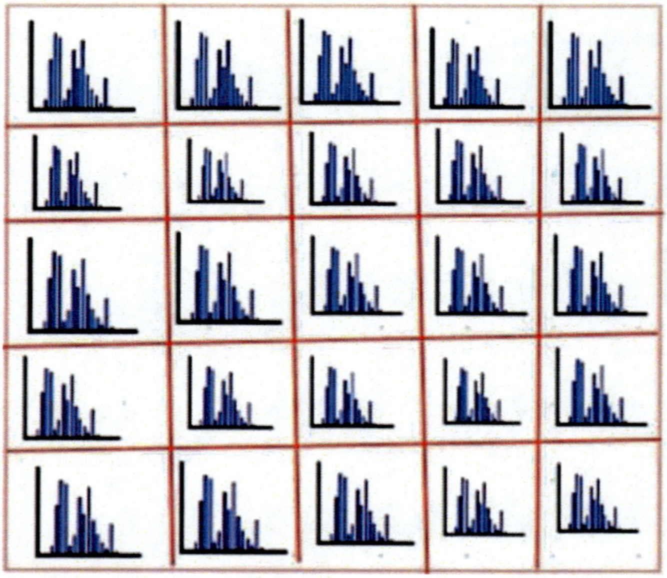

Fig. 3 Histogram for each cell

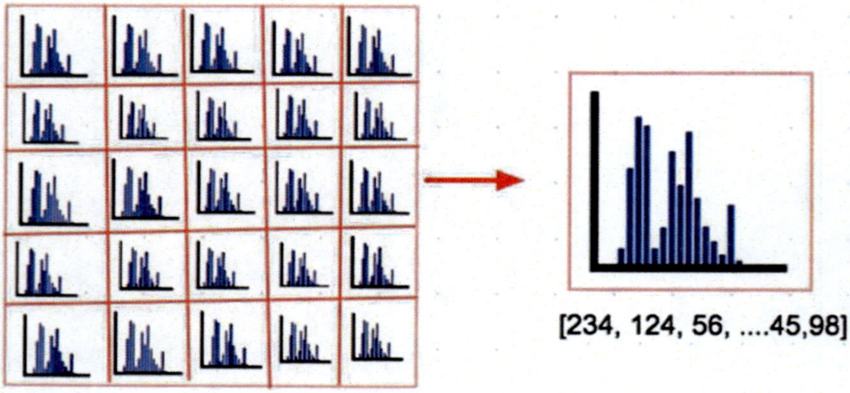

Fig. 4 Merged histogram

$$\mathrm{EAR} = \frac{|\mathrm{CD}| + |\mathrm{EF}|}{2 * |\mathrm{AB}|} \tag{1}$$

Eye aspect ratio (EAR) is computed for both the eyes from the formula above to detect eye movement (1). And also, mouth aspect ratio (MAR) is computed using the mouth coordinates to detect the yawing of the driver and the formula below (2):

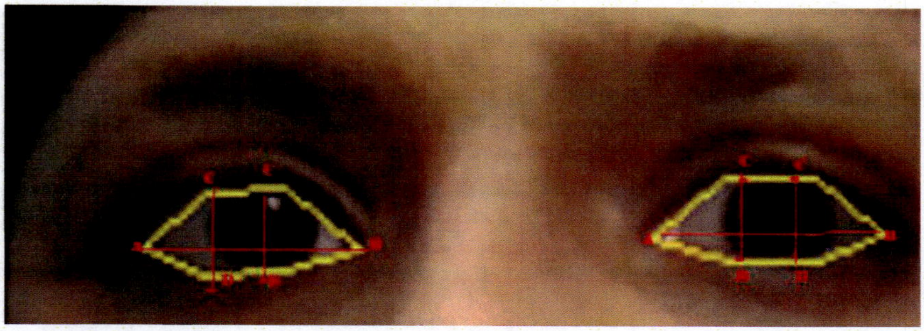

Fig. 5 Eye coordinates

$$\text{MAR} = \frac{|\text{CD}| + |\text{EF}| + |\text{GH}|}{3 * |\text{AB}|} \tag{2}$$

Here, we find the relative position of human's head, with respect to a camera. The reference frame here is the field of the camera. Head pose estimation helps to predict the pose of a human head. Head pose estimation is often referred to as a perspective-n-point problem or PnP problem in computer vision [16] (Figs. 6 and 7).

Firstly, an image will be taken from the video. The left side of the equation $s\ [u\ v\ t]^t$ denotes the 2D image taken from the video taken webcam. The right side of the

Fig. 6 Mouth coordinates

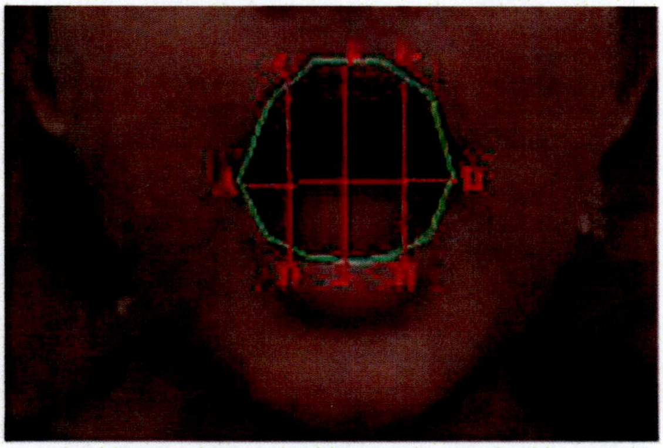

$$s\begin{bmatrix} u \\ v \\ 1 \end{bmatrix} = \begin{bmatrix} f_x & \gamma & u_0 \\ 0 & f_y & v_0 \\ 0 & 0 & 1 \end{bmatrix} \begin{bmatrix} r_{11} & r_{12} & r_{13} & t_1 \\ r_{21} & r_{22} & r_{23} & t_2 \\ r_{31} & r_{32} & r_{33} & t_3 \end{bmatrix} \begin{bmatrix} x \\ y \\ z \\ 1 \end{bmatrix}$$

Fig. 7 PnP problem statement

equation, the first portion is the camera matrix where f (x, y) is the focal length γ is the skew parameter which is given 1 in the code. (u_0, v_0) are the center of our image. The middle portion, r and t, represents rotation and translation, and the final portion denotes the 3D model of the face [16]. Generally, OpenCV provides two APIs to solve PnP, (1) solvePnP and (2) solvePnPRansac.

In this paper, solvePnP is used. It mainly uses four input parameters which are objectPoints, imagePoints, cameraMatrix and disCoefs. By resolving this PnP, the API returns a rotation matrix, translation matrix and a success message. OpenCv contains an API named RQDecomp3 × 3. This helps to calculate RQ decomposition with the help of given rotations. This function is used to decompose the left 3 × 3 submatrix of a projection matrix into a camera and a rotation matrix. It mostly gives back 3 rotation matrices, one for each and every axis, and the 3 Euler angles in degrees which can be used in OpenGL. Usually, more than one sequence of rotation regarding these three principal axes which gives results in the same orientation of any subject. Returned tree rotation matrix and 3 Euler angles are only one of the possible solutions. In short, this RQDecomp3 × 3 is basically used to extract the Euler angles. Euler angles have three parameters: roll, pitch and yaw. These parameters describe the motion of an object within 3D space.

4 Experimental Results

The final output of the drowsiness identification system shows the video input feed (from the real-time video). On screen, it shows the calculated aspect ratio based on the computed values and it will show the alert message. If the result of the EAR is smaller than the specified threshold value, then on the screen the warning "Wake up!" flashes along with an alarm sound (as given in Fig. 8).

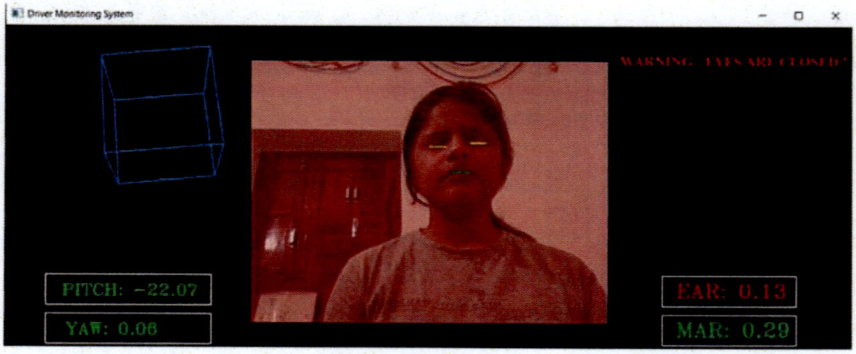

Fig. 8 Eye detection

Same as the eye detection, the mouth aspect ratio is calculated then. If the mouth aspect ratio is smaller than the specified threshold value, then "Don't Yawn" warning message is shown on the screen along with an alarm sound (as given in Fig. 9).

If the head bends more than the prescribed threshold of Euler angles, then an alert message "Please Look Straight" will be shown on the screen along with the alarm sound as shown in Fig. 10.

And, when the head of the driver bends in some position and the eyes are closed, then an alert message" Wake up!" is shown on the screen along with an alarm sound (Fig. 11).

The accuracy rate of this paper is 94.51%. As the output is taken using a real-time video, which helps in increasing the accuracy rate.

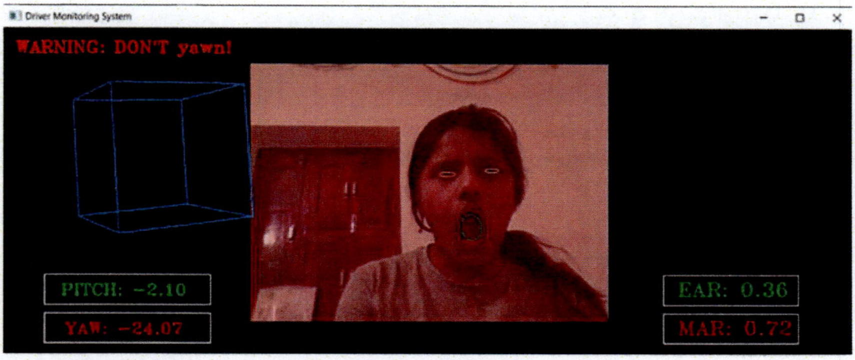

Fig. 9 Yawning detection

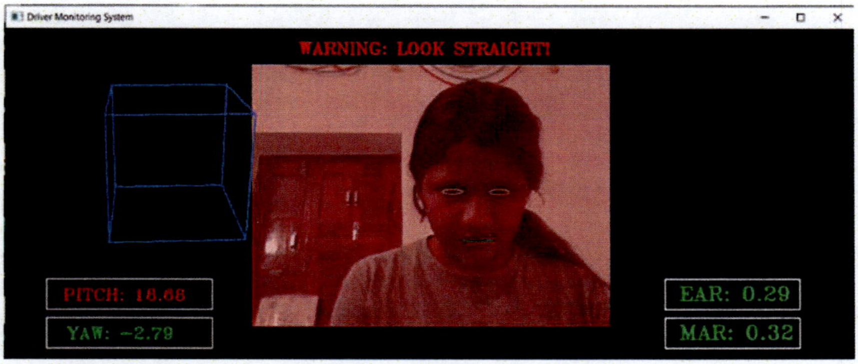

Fig. 10 Head pose detection

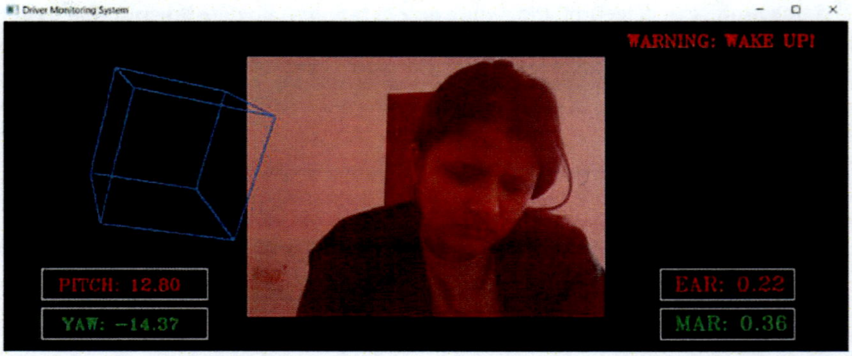

Fig. 11 Drowsiness detection

5 Conclusion

We have analyzed one of the key causes of the road accidents: drowsiness. The suggested solution monitors eyes, mouth and head position of the driver and then notifies him when his eyes are closed or the number of yawning exceeds the limit and/or his head bends downwards or toward the left or right sides, in order to prevent him of losing control of the vehicle. In the proposed system, facial landmarks are detected using Dlib. The facial landmarks include eyes, mouth, head pose, etc. Dlib provides better "frontal face detection." Dlib is based on the principle of histogram of oriented gradients (HOG) and support vector machine (SVM). The Euclidean distance method is used to find the distance between eyes. Also, the same method is used for finding the mouth distance. By calculating the mouth distance, we can find the frequency of yawn count. Then, the head position of the driver is estimated using Euler angles which is extracted by the API RQDecomp 3×3. If the eyes closure exceeds the threshold value or if the person exceeds the specific yawn count or if the head is bent downwards, upwards or toward left or right side for particular time, then the driver will be notified that he is drowsy and should stop driving the vehicle. The speediest method on CPU is Dlib HOG. The proposed system takes less time to detect that the driver is drowsy. The proposed system works slightly for non-frontal faces and shows better performance for frontal face detection. Dlib HOG detects faces which are bigger in size and fails to detect faces that are small in size. Since we cannot predict the size of the drivers face beforehand, it is a demerit of this system. The accuracy rate of the system seems to be 94.51%. Future work should be based on a system that can detect faces at odd angles. For real-time video CNN based face detection can be used for better results.

References

1. V. Saini, R. Saini, Driver drowsiness detection system and techniques: a review. Int. J. Computer Sci. Inform. Technol. **5**(3), 4245–4249 (2014)
2. M. Gromer, D. Salb, T. Walzer, N.M. Madrid, R. Seepold, ECG sensor for detection of driver's drowsiness. Procedia Computer Sci. **159**, 1938–1946 (2019)
3. Z. Ma, B.C. Li, Z. Yan, Wearable driver drowsiness detection using electrooculography signal, in *2016 IEEE Topical Conference on Wireless Sensors and Sensor Networks (WiSNet)* (IEEE, 2016), pp. 41–43
4. M. Awais, N. Badruddin, M. Drieberg, A hybrid approach to detect driver drowsiness utilizing physiological signals to improve system performance and wearability. Sensors **17**(9), 1991 (2017)
5. J. Ahmed, J.-P. Li, S. Ahmed Kran, R. Ahmed Shaikr, *Eye Behavior Based Drowsiness Detection System* (School of Computer Science & Engineering, VESTC, Chengdu 611731, China)
6. Y. Du, P. Ma, X. Su, Y. Zhang, Driver fatigue detection based on eye state analysis, in *11th Joint International Conference on Information Sciences* (Atlantis Press, 2008)
7. R. Gupta, K. Aman, N. Shiva, Y. Singh, An improved fatigue de- tection system based on behavioral characteristics of the driver, in *2017 2nd IEEE Interna- tional Conference on Intelligent Transportation Engineering (ICITE)* (IEEE, 2017), pp. 227–230
8. I. Lashkov, A. Kashevnik, N. Shilov, V. Parfenov, A. Shabaev, Driver dangerous state detection based on OpenCV & dlib libraries using mobile video processing, in *2019 IEEE International Conference on Computational Science and Engineering (CSE) and IEEE International Conference on Embedded and Ubiquitous Computing (EUC)* (IEEE, 2019), pp. 74–79
9. P. Viola, M.J. Jones, Robust real-time face detection. Int. J. Comput. Vision **57**(2), 137–154 (2004)
10. B.K. Savaş, Y. Becerikli, Real time driver fatigue detection based on svm algorithm, in *2018 6th International Conference on Control Engineering & Information Technology (CEIT)* (IEEE, 2018), pp. 1–4
11. S. Mehta, S. Dadhich, S. Gumber, A. Jadhav Bhatt, Real-time driver drowsiness detection system using eye aspect ratio and eye closure ratio, in *Proceedings of International Conference on Sustainable Computing in Science, Technology and Management (SUSCOM)*, Amity University Rajasthan, Jaipur, India (2019)
12. I. Teyeb, O. Jemai, M. Zaied, C.B. Amar, A drowsy driver detec- tion system based on a new method of head posture estimation, in *International Conference on Intelligent Data Engineering and Automated Learning* (Springer, Cham, 2014), pp. 362–369
13. G. Bradski, The OpenCv library. Dr Dobb's J. Software Tools **25**, 120–125 (2000)
14. http://dlib.net/
15. https://github.com/davisking/dlib-models
16. https://medium.com/datadriveninvestor/training-alternative-dlibshape-predictor-modelsusing-python-d1d8f8bd9f5c

Sentiment Analysis of Covid Vaccine Tweets Using Different Text Classification Models

R. Rahul, C. S. Aravind, and T. Remya Nair

Abstract Internet has turned into an online learning network and a platform to exchange ideas and reviews. Social networking sites are quickly gaining popularity as it allows users to have a discussion, share, and express their views on subjects across the globe. It is generally known that social media and social networks are the best tools to collect knowledge about the viewpoint and the thoughts of people on entirely different subjects as they pay a considerable amount of time on social media networks to express opinions and interests. The proposed model detects the sentiments of tweets and using Twitter sentiment, and this research work has also attempted to perform classification on different models using TF-IDF vectorization. Python3 and its libraries are used for implementation.

1 Introduction

Sentimental mining is a significant research area, where it is considered as an analytical method for categorizing the texts as positive, negative, or neutral. It is already known that there are large amount of posts on social networks, so collecting people's views and opinions is considered as a heuristic activity. Sentiment analysis has many applications in various fields, such as collecting input from consumers and reviews via social media networks to enhance the product quality and it can also be used to review social media based on trending topics. In a recent survey, authors emphasize various situations on how incorrect information is shared in social networking websites. From here, analysis of different tweets can be considered as a valuable tool for decision makers and healthcare providers to assess and address the needs of communities.

R. Rahul (✉) · C. S. Aravind · T. R. Nair
Department of Computer Science and IT, Amrita School of Arts and Sciences, Amrita Vishwa Vidyapeetham, Kochi, India

T. R. Nair
e-mail: remyat@asas.kh.amrita.edu

P. Karuppusamy et al. (eds.), *Ubiquitous Intelligent Systems*, Smart Innovation, Systems and Technologies 243, https://doi.org/10.1007/978-981-16-3675-2_17

As the world is currently facing a pandemic situation, people are eager to know about the progress of COVID-19 vaccine production, and people share their thoughts and updates on social media vaccines that are very insightful. Now, across the globe, various groups, institutions, and government organizations are using technology to interact with each other on a number of issues related to the COVID vaccine. It is also desirable to understand public sentiment in this epidemic situation that is why the proposed research work has chosen sentiment analysis on COVID vaccine tweets.

Twitter is an American platform for microblogging and social networking, where users post tweets known as messages and connect with others. Twitter also provides an API for retrieving the tweets. The Twitter datasets are accessible from Kaggle, which are available for the public.

2 Literature Review

It is possible to describe analysis of feelings as a process that automates the mining of text through natural language processing with behaviors, emotions, views, and feelings from text, tweets, and database sources (NLP). Sentiment analysis requires text views to be grouped into categories such as "positive," "negative," and "neutral." It is also defined as the study of subjectivity, point of view, mining, and extraction of appraisals. Figure 1 shows tweets with its respective sentiment found through analysis.

Bagheri et al. [2] application of sentimental analysis and how to relate to Twitter and run sentimental analysis queries are demonstrated in this paper. This provides some interesting results. The key conclusion of this paper is that the neutral feeling

Sentiment	Tweets
Negative	@united is the worst. Nonrefundable First class tickets? Oh because when you select Global/FC their system auto selects economy w/upgrade.
	@united I will not be flying you again
Neutral	@VirginAmerica my drivers license is expired by a little over a month. Can I fly Friday morning using my expired license?
	@VirginAmerica any plans to start flying direct from DAL to LAS?
Positive	@VirginAmerica done! Thank you for the quick response, apparently faster than sitting on hold ;)
	@united I appreciate your efforts getting me home!

Fig. 1 Example of tweets based on service of an airplane company with its corresponding sentiment [1]

for tweets is substantially strong, clearly illustrating the shortcomings of current works.

Kharde et al. [3] have used lexicon-based approaches and machine learning, along with other methods and some assessment metrics. They came to a conclusion that both Naive Bayes and SVM have better accuracy compared to others and can be considered as simple learning methods.

Sarlan et al. [4] Twitter sentiment analysis is structured to examine the perceptions of customers about the marketplace's vital performance. In this, a program uses a more precise machine-based learning approach to analyze a sentiment; it has been used along with natural language processing techniques.

Alsaeedi et al. [5] have used dictionary-based approaches, ensemble and machine learning approaches were explored for Twitter sentiment analysis methods. Furthermore, Twitter sentiment analysis of the hybrid and ensemble method analysis were discussed.

Nemes et al. [6] applied recurrent neural network (RNN) which is used to classify tweets based on different approaches. They have also classified the tweets into positive and negative emotions by applying some recurrent emotional prediction neural networks, where they analyzed the correlation between words. Different texts were grouped into a much more articulated emotional intensity scope instead of positive and negative extremes.

Kamaran et al. [7] in this paper, there are two keywords chosen for to assess polarity and subjectivity, scan for #coronavirus and #COVID-19 tweets. TextBlob python library sentiment analysis techniques applied to 530,232 tweets collected. The findings showed that there was a substantially high neutral toll of over 50 and 19% for both coronavirus and COVID-19 keywords for polarity.

Chandra et al. [8] suggested a novel meta-heuristic technique based on cuckoo scan and K-means. The method proposed was used to classify the best cluster heads from the Twitter dataset's sentiment material. On various Twitter datasets, the efficacy of the proposed method was checked and different cuckoo search models and two n-gram approaches compared to particle swarm optimization. Result analysis verifies the current approaches that are outperformed by the proposed process.

Wagh et al. [9] The machine can also compute the frequency of each word in the tweets. Also, the use of a supervised approach to machine learning helps to produce outcomes. Twitter is a broad data source that makes it more attractive for sentiment analysis to be carried out. Stanford University's publicly available data collection, which includes a total of 4 million tweets, is analyzed. Analyze the findings, understand the trends, and give a review of people's views.

3 Proposed System

The structured methodology of the proposed system in this research is shown in Fig. 2. Mainly, the proposed system has been divided into four stages, which include

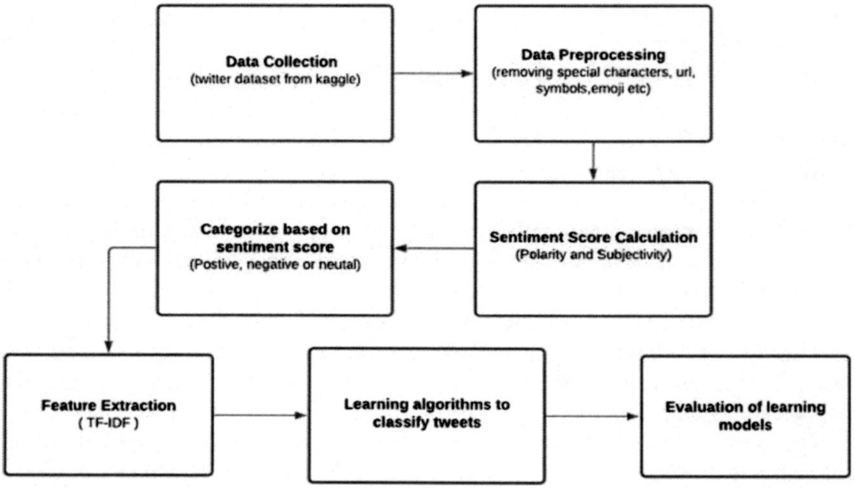

Fig. 2 Workflow diagram of approach

data collection, data preprocessing, data analysis, and text classification. A python-based platform known as natural language processing (NLP) has been widely used in this paper. Data analysis process is divided into two stages, wherein the first section of the research includes sentiment analysis of tweets. Polarity and subjectivity scores of each tweet are found using different python libraries and methods. Second section mainly includes TF-IDF approach for feature extraction, and different classification models are used to find out accuracy.

3.1 Data Collection

Data extraction is the process of extracting or retrieving different types of data from a number of sources, many of which may be poorly organized or completely unstructured. Data extraction allows information to be consolidated, processed, and optimized so that it can be processed in a single place for transformation. There are many ways to extract data. We can use Twitter API using tweepy python library. In this paper, we have used the COVID_vaccine-based Twitter dataset from Kaggle. Kaggle is an online community of data scientists and machine learning practitioners [10]. Dataset used here is a comma-separated value (CSV) file which contains about 38,459 rows and 13 columns including user_name, user_location, user_description, …., text, hashtags, source and is retweet. We only need tweets for analysis, so the text column is selected. CSV file is read in python as a data frame with the help of libraries (pandas).

3.2 Data Preprocessing

Preprocessing is the next step after data extraction. In the collected Twitter dataset, there will be a significant amount of noise that needs to be filtered. Preprocessing is an important step in natural language processing (NLP).

Data extracted from Kaggle usually contains large numbers of noise such as emojis, URLs, hashtags, stop words, which are not needed for analysis. Figure 3a shows collected tweets without performing preprocessing steps.

Following are the steps used:

- Removed URLs (e.g.: https://www.gmail.com) Hashtags (e.g.: #topic) and Username
- Changed letter casing
- Performed tokenization, stemming, normalization.
- Removed punctuations, symbols, numbers, and unwanted whitespaces
- Stop word removal.
- Removed different Emojis from the tweets.

Letter casing: Converting all uppercase to lowercase.

Tokenizing: The words that are separated by spaces are created as tokens.

Noise removal: Unwanted characters are eliminated.

Normalization: All the texts are converted into similar grades using normalization through a series of tasks which helps to refine the text match.

Stop words: Stop words are those words which don't add much meaning to a sentence, e.g.: "a," "and," "but," "after," "had," "happen," etc.

Remove Emoji: Emojis are read as characters and this may cause noise in data, for example:

Grinning Face Emoji is read as "U0001F600."

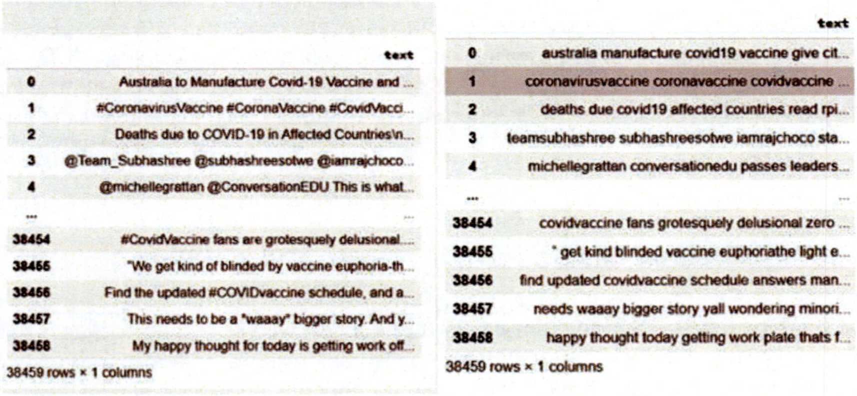

	text
0	Australia to Manufacture Covid-19 Vaccine and ...
1	#CoronavirusVaccine #CoronaVaccine #CovidVacci...
2	Deaths due to COVID-19 in Affected Countries\n...
3	@Team_Subhashree @subhashreesotwe @iamrajchoco...
4	@michellegrattan @ConversationEDU This is what...
...	...
38454	#CovidVaccine fans are grotesquely delusional....
38455	"We get kind of blinded by vaccine euphoria-th...
38456	Find the updated #COVIDvaccine schedule, and a...
38457	This needs to be a "waaay" bigger story. And y...
38458	My happy thought for today is getting work off...

38459 rows × 1 columns

(a)

	text
0	australia manufacture covid19 vaccine give cit...
1	coronavirusvaccine coronavaccine covidvaccine ...
2	deaths due covid19 affected countries read rpi...
3	teamsubhashree subhashreesotwe iamrajchoco sta...
4	michellegrattan conversationedu passes leaders...
...	...
38454	covidvaccine fans grotesquely delusional zero ...
38455	" get kind blinded vaccine euphoriathe light e...
38456	find updated covidvaccine schedule answers man...
38457	needs waaay bigger story yall wondering minori...
38458	happy thought today getting work plate thats f...

38459 rows × 1 columns

(b)

Fig. 3 Twitter dataset with noise (**a**) and Twitter dataset after preprocessing (**b**)

	text
0	australia manufacture covid19 vaccine give cit...
1	coronavirusvaccine coronavaccine covidvaccine ...
2	msnbc well let ' qualify would anyone party ge...
3	dna zooms charts 1st week hear vaccines episod...
4	' covid19millionares covid19 corona covidvaccine
...	...
19581	tomfolanmd lets turn 2020 shitshow around cov...
19582	got moderna covidvaccine today ready end pande...
19583	covidvaccine fans grotesquely delusional zero ...
19584	find updated covidvaccine schedule answers man...
19585	happy thought today getting work plate thats f...

19586 rows × 1 columns

Fig. 4 Filtered tweets

Stemming: Affixes from words are terminated to acquire root words. Commonly used and most reliable technique is Porter stemmer.

Figure 3(b) shows tweets after performing preprocessing steps mentioned above. There will be plenty of tweets available even after preprocessing data that we don't need for analysis. So, we're filtering out tweets containing the keyword "COVID vaccine" into a new CSV file. Out of 38,458 tweets, 19,586 tweets containing the keyword "COVID vaccine" were collected. After filtering, Fig. 3 highlights the data frame (Fig. 4).

3.3 Sentimental Analysis

After the noise is removed from the Twitter dataset, we use python packages and functions for analysis of its sentiment. We're also using a python library called TextBlob. TextBlob is a library that is used for complex operations and analysis of textual data. It uses the Natural Language Toolkit (NLTK) to perform its functions. NLTK is a library that provides access to lexical resources and allows users to work

	text	Polarity	Subjectivity
0	australia manufacture covid19 vaccine give cit...	0.40	0.800000
1	coronavirusvaccine coronavaccine covidvaccine ...	0.70	0.600000
2	msnbc well let ' qualify would anyone party ge...	-0.10	0.600000
3	dna zooms charts 1st week hear vaccines episod...	0.00	0.000000
4	' covid19millionares covid19 corona covidvaccine	0.00	0.000000
...	...	...	...
19581	tomfolanmd lets turn 2020 shitshow around cov...	0.00	0.000000
19582	got moderna covidvaccine today ready end pande...	0.20	0.500000
19583	covidvaccine fans grotesquely delusional zero ...	-0.55	1.000000
19584	find updated covidvaccine schedule answers man...	0.25	0.283333
19585	happy thought today getting work plate thats f...	0.80	1.000000

19586 rows × 3 columns

Fig. 5 Polarity and Subjectivity representation of tweets

with classification, categorization, etc. Tweets are then analyzed with the help of the TextBlob package to generate polarity and subjectivity values.

Polarity is between 10 [−1,1] where 1 is positive, 0 is neutral, and −1 is negative. Subjective sentences refer to personal opinion, sentiment, or judgment. Subjectivity is a float between 0 and 1 (Fig. 5).

After getting polarity values tweets are passed through another function to determine whether they are positive, neutral, or negative tweets (i.e., if the polarity is less than 0, it is said to be negative, greater than 0 is positive, and equal to 0 is negative) (Fig. 6).

3.4 TF-IDF Approach on Different Classification Models

Text classification is a technique that classifies text-based data into predefined categories, such as tweets, reviews, posts, and blogs. Sentiment analysis is a form of text classification, where textual data predicts user attitudes or feelings about any product.

We are designing a sentiment analysis model that will use the method of generating TF-IDF features and will be able to predict the accuracy of user sentiment.

TF-IDF can be defined as a product of TF & IDF (Fig. 7).

With numeric data, mathematical methods such as machine learning and deep learning work well. However, natural language is made up of words and phrases.

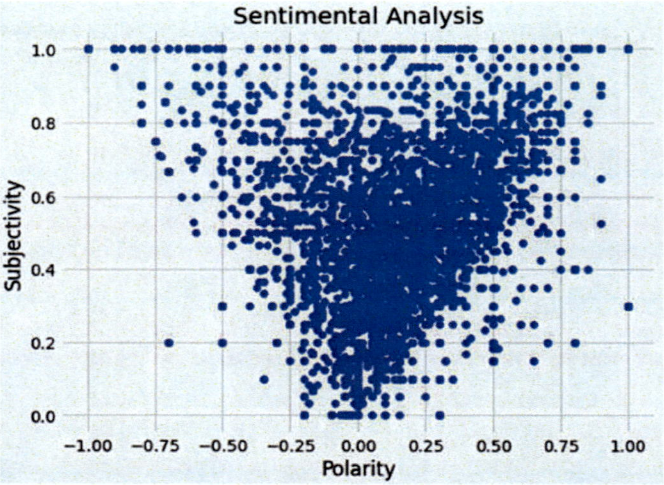

Fig. 6 Subjectivity and polarity scatter graph

$$\textit{Term Frequency} = \textit{Frequency of a word in the document} \div \textit{Total words in the document}$$

$$\textit{Inverse Document Frequency} = \log(\textit{Total number of docs}) \div \textit{Number of docs containing the word}$$

Fig. 7 TF-IDF formula

We, therefore, need to translate text to numbers before we can construct a model of sentiment analysis. For converting text to numbers, different methods have been developed. The Bag of Words, Word2Vec, and N-grams are some of them.

Here, we have used a N-grams approach with TF-IDF for converting text to numbers.

For example, consider two documents:

$D1 =$ "When US COVID case is at high"
$D2 =$ "High COVID case in US"

By applying a nGram_Range (1,2), it will create a set such as.

$V =$ ["'When US', 'US COVID', 'COVID case', 'case are', 'are at', 'at high', 'High COVID', 'COVID case', 'case in', 'in US'"].

Term Frequency values:

$D1 = [1\ 0\ 1\ 1\ 0\ 1\ 1\ 0\ 1]$
$D2 = [0\ 1\ 0\ 1\ 1\ 0\ 0\ 1\ 0]$

TF-IDF values:

$D1 = [1.40546511, 0, 1.40546511, 1, 0, 1.40546511, 1.40546511, 0, 1.40546511]$

$$D2 = [0, 1.40546511, 0, 1, 1.40546511, 0, 0, 1.40546511, 0]$$

After converting text to numbers using N-gram approach, we have used a TfidfVectorizer class which is a part of the sklearn module that is used to generate feature vectors that contain TF-IDF values. Attribute called max_features is used that specifies the most occurring number of words on which we create a feature vector. Most occurring words play an important role in classification. Fit transfer method is used in TfidfVectorizer class and is passed to a preprocessed dataset to convert our dataset to TF-IDF function vector.

Before creating the actual model, we need to divide our dataset to testing and training sets. We have used 20% of the dataset as training set and 80% as a testing set. After creating testing and training sets, we need to evaluate the model for which we have used a different text classification model to train the dataset.

K Neighbors

One of the forms of classification of neighbors is the K-Neighbors classification. In this methodology, a general internal model is not constructed. It is only determined by the neighbors who are classes to each other by simple majority vote. The point's nearest neighbors are allocated to the question point.

The best option for the K value is highly data based. Generally, it is to be said that K demonstrates the effect of noise but makes the classification limits less distinct.

Naive Bayes

This classifier works by grouping the various Bayes Theorem classification algorithms. It is a group of different algorithms that share a common concept. On the basis of strong assumptions, it works. The major benefit of this algorithm is that to find the parameters, and it only requires very tiny training data.

Random Forest

It is one of the best classification algorithms which is able to classify large amounts of data with accuracy. It is a group learning method for classification and regression that constructs a number of decision trees at training time and delivers a class that is the class output mode of each tree. In this RF classifier, the number of decision trees in RF is set to 100 and the tree depth has been set as none [11].

Extra Tree

It is a type of ensemble learning technique that combines the results of many decorrelated decision trees collected to produce the results of their classification. It is very similar in concept to the random forest classifier and varies only in the way decision trees are built in the forest. Each decision tree is drawn from the original training sample in the extra trees forest. A random sample of k characteristics from the set of features is then given at each test node, from which each decision tree must choose the best feature to divide the data on the basis of some mathematical criteria [12].

Fig. 8 Representation of Confusion Matrix

	Predicted Value			
Actual Value	N	0	1	2
	0	TP	FN	FN
	1	FP	TN	TN
	2	FP	TN	TN

Linear SVC

The purpose of the linear SVC is to match the data that you provide by returning the best fit hyperplane that divides or categorizes your data. From there, after you get the hyperplane, you can feed a few features to your classifier to see what the "expected" class is like[13].

4 Experimental Results and Analysis

The sentimental analysis produced a graphical result that shows the positive, negative, and neutral tweets.

On the basis of sentimental analysis on our 19,586 tweets dataset, we found that about 39.7% tweets were positive and 11.4% were negative and 48.9% were neutral.

After converting all this text into numbers, we perform linear support vector classification, naive Bayes, random forest, KNN, extra tree classification over this data which generates confusion matrix and accuracy of the predicted sentiment (Fig. 8; Table 1).

5 Conclusion

We performed different preprocessing techniques for cleansing noise from the dataset and found sentiments of tweets based on polarity values. Later on, we used the TF-IDF approach to convert the dataset into numeric feature vectors and with the help of different classifiers, we trained our model. The above models were trained on python and achieved accuracy of 81% in naïve Bayes, extra tree 94%, random forest 91%, K-Neighbors 62%, linear SVC 93%. Table 1 shows the results obtained on each models. The above values show that extra tree and linear SVC provide a better accuracy for Twitter sentiment classification. Figure 9 shows the confusion matrix

Table 1 Generated accuracy of each classifier

Methods	–	Precision	Recall	$F1$ score	Support	Accuracy
Naive Bayes	Positive	0.82	0.85	0.84	789	0.81
	Negative	0.98	0.18	0.31	225	
	Neutral	0.79	0.92	0.85	945	
Extra Tree	Positive	0.93	0.94	0.94	779	0.94
	Negative	0.72	0.89	0.80	183	
	Neutral	0.98	0.93	0.95	997	
Random Forest	Positive	0.93	0.92	0.93	1573	0.91
	Negative	0.90	0.60	0.72	426	
	Neutral	0.90	0.97	0.94	1919	
K-Neighbors	Positive	0.32	0.91	0.47	276	0.62
	Negative	0.17	0.76	0.28	51	
	Neutral	0.98	0.56	0.72	1632	
Linear SVC	Positive	0.95	0.92	0.93	1573	0.93
	Negative	0.92	0.66	0.77	426	
	Neutral	0.90	0.99	0.94	1919	

```
array([[ 41, 115,  69],
       [  0, 865,  80],
       [  1, 115, 673]], dtype=int64)
```

(a)

```
array([[ 39, 181,   5],
       [  4, 922,  19],
       [  8, 529, 252]], dtype=int64)
```

(b)

```
array([[163,  31,  31],
       [  7, 926,  12],
       [ 13,  40, 736]], dtype=int64)
```

(c)

```
array([[ 283,   93,   50],
       [   6, 1891,   22],
       [  18,  115, 1440]])
```

(d)

```
array([[ 254,  103,   69],
       [  15, 1871,   33],
       [  14,  104, 1455]])
```

(e)

Fig. 9 Confusion matrices for **(a)** Naïve Bayes, **(b)** KNN, **(c)** Extra Tree, **(d)** Linear SVC, and **(e)** Random Forest obtained on each model. Different machine learning, deep learning concepts, and methods can be used to get better results.

References

1. https://ipullrank.com/step-step-twitter-sentiment-analysis-visualizing-united-airlines-pr-crisis
2. H. Bagheri, M. Johirul Islam, Sentiment analysis of twitter data. arXiv preprint (2017). arXiv:1711.10377
3. V. Kharde, Sonawane, Sentiment analysis of twitter data: a survey of techniques. arXiv preprint (2016). arXiv:1601.06971
4. A. Sarlan, C. Nadam, S. Basri, Twitter sentiment analysis, in *Proceedings of the 6th International conference on Information Technology and Multimedia* (IEEE, 2014)
5. A. Alsaeedi, M. Zubair Khan, A study on sentiment analysis techniques of Twitter data. Int. J. Adv. Comput. Sci. Appl. **10**(2), 361–374 (2019)
6. L. Nemes, A. Kiss, Social media sentiment analysis based on COVID-19. J, Inform. Telecommun. 1–15 (2020)
7. K.H. Manguri, R.N. Ramadhan, P.R. Mohammed Amin, Twitter sentiment analysis on worldwide COVID-19 outbreaks. Kurdistan J. Appl. Res. 54–65 (2020)
8. A.C. Pandey, D.S. Rajpoot, M. Saraswat, Hybrid step size based cuckoo search, in 2017 Tenth International Conference on Contemporary Computing (IC3) (IEEE, 2017)
9. B.J. Wagh, J.V. Shinde, P.A. Kale, A Twitter sentiment analysis using NLTK and machine learning techniques. Int. J. Emerg. Res. Manage. Technol. **6**(12), 37–44 (2018)
10. Z. Luo, M. Osborne, T. Wang, An effective approach to tweets opinion retrieval. World Wide Web **18**(3), 545–566 (2015)
11. A. Mitra, Sentiment analysis using machine learning approaches (lexicon based on movie review dataset). J. Ubiquitous Comput. Commun. Technol. (UCCT) **2**(03), 145–152 (2020)
12. H. Wang, Emotional analysis of bogus statistics in social media. J. Ubiquitous Comput. Commun. Technol. (UCCT) **2**(03), 178–186 (2020)
13. https://www.geeksforgeeks.org/ml-extra-tree-classifier-for-feature-selection
14. R. Xia, C. Zong, S. Li, Ensemble of feature sets and classification algorithms for sentiment classification. Inf. Sci. **181**(6), 1138–1152 (2011)
15. S.A. Grover, Twitter data-based prediction model for influenza epidemic, in *Proceedings of IEEE 2nd International Conference on Computing for Sustainable Global Development, India* (2015), pp. 873–879
16. K. Sailunaz, R. Alhajj, Emotion and sentiment analysis from twitter text. J. Comput. Sci. **36**(101003), 1–18 (2018). J. Samuel, G. Ali, M. Rahman, E. Esawi, Y. Samuel, COVID-19 public sentiment insights and machine learning for tweets classification. Information **22**(4), 1–21 (2020)
17. S. Naz, A. Sharan, N. Malik, Sentiment classification on twitter data using a support vector machine, in *2018 IEEE/WIC/ACM International Conference on Web Intelligence (WI)* (IEEE, 2018)
18. T. Pranckevičius, V. Marcinkevičius, Comparison of naive bayes, random forest, decision tree, support vector machines, and logistic regression classifiers for text reviews classification. Baltic J. Mod. Comput. **5**(2), 221 (2017)
19. Y. Al Amrani, M. Lazaar, K.E. El Kadiri, Random forest and support vector machine-based hybrid approach to sentiment analysis. Procedia Comput. Sci. **127**, 511–520 (2018)
20. https://pythonprogramming.net/linear-svc-example-scikit-learn-svm-python
21. Step-By-Step Twitter Sentiment Analysis: Visualizing Multiple Airlines' PR Crises [Updated for 2020] | iPullRank
22. https://en.wikipedia.org/wiki/Kaggle

An Empirical Analysis to Explore the Best Algorithm for Covid-19 Dispersion

Athira Jayan, T. S. Sethulakshmi, and Prasanna Kumar

Abstract The scourge of novel-coronavirus 2019 disease (COVID-19) has been created a devastating situation throughout the world. On the grounds to the endorsed Data of COVID-19, the proposed learn analyses the dispersion rate of coronavirus in Kerala for the coming year. This study applies several algorithmic programs approaches of data mining includes naïve Bayes, J48 tree, random tree forest to find the best classifier for predicting spreading rate of COVID-19. In this proposed approach, this paper used WEKA tool for analyzing and comparing results. This study rather to realistically divulge the outpouring of epidemic novel coronavirus across the state. The computational outcome using integrated algorithmic tools affords a better clarity of the contagion situation and transports a suggestive and comparative method to subside the out bust.

1 Introduction

Data mining (DM) is an approach to discover useful information from hefty quantity of data. Data mining emphasizes an assortment of algorithmic techniques that set of heuristics that create a model from data. Models are created from discovering and extracting patterns from stored data. This paper focuses on the prediction of coronavirus disease. The COVID-19 is a highly infectious disease originated from the SARS-CoV-2 virus. The primary case was indication in Wuhan, the capital city of Hubei prefecture in China. Within few weeks, virus has contiguously broadened to different regions of the world. On January 30, 2020, the primary case of covid-19 in India was declared. The covid-19 pandemic was confirmed in Thrissur district of Kerala, which was also the primary altogether of India.

A. Jayan (✉) · T. S. Sethulakshmi · P. Kumar
Department of Computer Science and IT, Amrita School of Arts and Sciences, Amrita Vishwa Vidyapeetham, Kochi, India

P. Kumar
e-mail: prasannakumarcv@asas.kh.amrita.edu

P. Karuppusamy et al. (eds.), *Ubiquitous Intelligent Systems*, Smart Innovation, Systems and Technologies 243, https://doi.org/10.1007/978-981-16-3675-2_18

For this paper, chiefly studied [4] 'outbreak trends of coronavirus disease 2019 in India: a prediction.' This paper uses the information of China to predict the end result in India within the next twenty-two days a prophetic model is built using WEKA to predict the per day reckon of inveterate cases, healthier cases, fatality cases from April 4, 2020. COVID-19 dataset has been collected from Kaggle. Period progression predicting is showcased on the in sequence collected, and a replica is ready. It over that the quantity of accumulative deep-rooted findings in India is probably going to extend at a hurried change after April 6, 2020. According to the foretell representation, India might have virtually millions of dyed-in-the-wool cases by the last of May 2020. This may be subsided if the distinctive condition and Government of India policies become approving to manage the virus. The quantity of death cases from COVID-19 is foretold to extend around April 5, 2020. Scrutiny is performed on definite cases. The popular metrics for estimation are root mean square error (RMSE) and mean absolute error (MAE).

The main intention of this swot up is to predict the spreading rate of coronavirus in the coming year if not proper vaccine is used. For the treatment of COVID-19, no particular vaccine or antibiotic is discovered in the world. Virus causes from human to human one from infected via globule generated once a contaminated person coughs, sneezes, speaks or breathes. And also infected by touching a surface where outburst of microorganisms occur and then with your unsterilized hands are being automatically used to poke your eyes, nose or mouth will lead to dangerous situations.

The goal of this paper is to find the best classifier algorithm for the COVID dispersal rate prediction using the data mining tool called WEKA. WEKA is a data mining tool that has different sets of classification algorithms. Through WEKA required data is extracted from the COVID dataset for accurately building the predictive model, with the help of classification algorithms. For the dataset, data collected from Dashboard of Government of Kerala COVID-19 battle. Firstly, classified the collected COVID-19 dataset and the classified result are compared through the crossing point of WEKA. The objective of this paper is to identify a best classifier algorithm accurately from the refined dataset.

2 Related Work

In order to analyze and evaluate data, this proposal used the data mining tool called WEKA. Explorer and experimenter are the interface available in WEKA that has used in this study. Dataset includes 100 and above confirmed coronavirus cases per day. District wise data is collected along with the date of COVID confirmation. After the lockdown, the spreading rate is increased day by day. Before the lockdown, the rate of confirmed cases in Kerala was much lower than in India. Close contact of people is the main reason for rapidly increasing the COVID cases. For reducing, the dataset this paper decided a threshold of 100 and above confirmed cases. The dataset started collecting from June 5, 2020 0.7 months (June to December) COVID dataset is collected from the website of government of Kerala. The data collected in excel

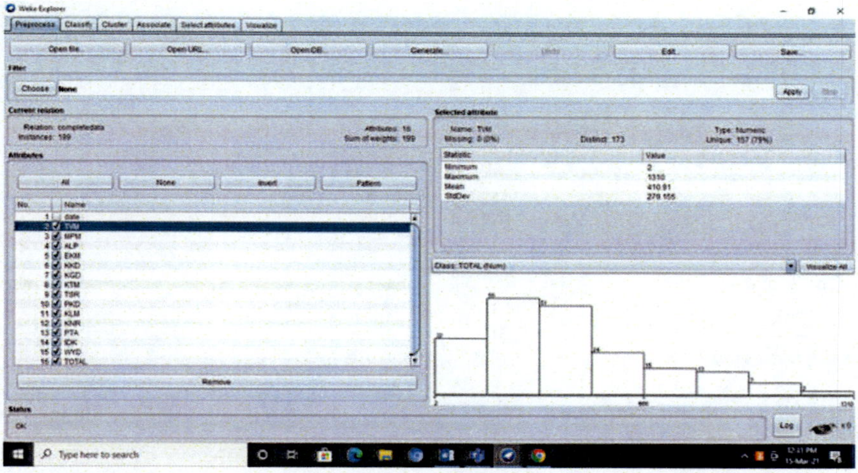

Fig. 1 Screenshot of CSV data for pre-processing

format is converted into CSV format. CSV file is uploaded in WEKA through experimenter. The collected data may contain unwanted things that lead to wrong analysis. So, data pre-processing technique is used to remove unwanted and noisy data, which is done through the various steps of data pre-processing. Classification is performed on the pre-processed data. This study tends to utilize a broadly used classification tool which is a data mining modus operandi that implies objects in an assemblage to goal classes. Classification important aspiration is to precisely envisage the target class for each case in the data. Three main algorithms are extensively used for the progression of classification which are naïve Bayes, J48 tree, and random forest algorithm. Figure 1 shows the screenshot view of CSV data opened in explorer interface for data pre-processing.

3 Proposed Work

Software dataset file format	Datasets purpose	Weka data mining technique	Classification algorithm	Operating system
WEKA	COVID-19	Explorer	Naïve Bayes	Windows 10
CSV	Classification	Experimenter	J48 tree Random forest	

- *The precession is righteously classified, and it shows the percentage of test accuracy fruitfully.*

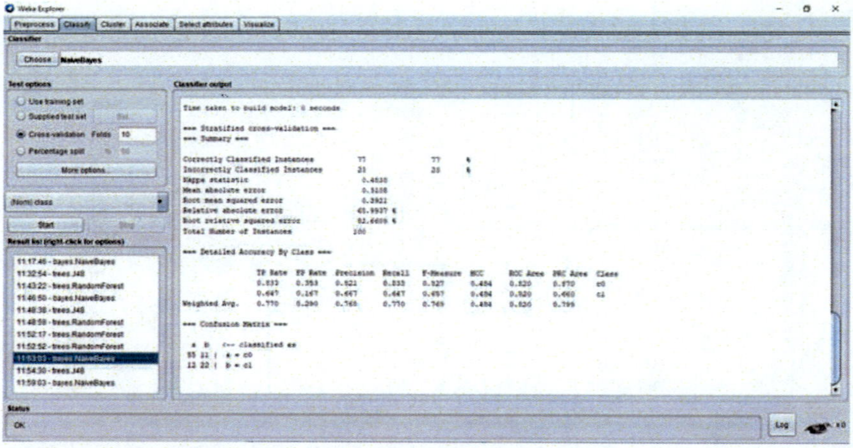

Fig. 2 Output of naïve Bayes classification

- ***Improper precessions calculations that it shows the percentage of test correctness not properly.***
- ***While in the case of absolute error detection using mean-shows multiple errors to categorize the classification accuracy.***

3.1 Naïve Bayes

It is a cataloging technique on the basis of Bayes' Theorem with autonomy among predictors. In other terms, a naive Bayes classifier presumes that the existence of a unambiguous feature in a class is unrelated to the presence of any other feature.

After applying naïve Bayes algorithm, this paper attained accuracy of 77% for 77 correctly classified instances. The study produces 0.3108 as mean absolute. The whole instance for engendering the model is 0 s and obtained ROC area is 0.820 (Fig. 2).

3.2 J48 Tree

The algorithmic precession model called J48 is one in the midst of the most effectual machine learning algorithms to inspect the data emphatically and incessantly. When it is used for case in point purpose, inefficient ways of memory usage are experienced and use up the performance and exactness in classifying medical data. A lot more cases as J48 algorithms consumed, and all these cases, this paper got accuracy of 77%

for 77 properly classified instances. Hence, this retains a MAE as 0.2641, elapsed duration is taken to build this model is 0.03 s, and ROC area is 0.753 (Figs. 3 and 4).

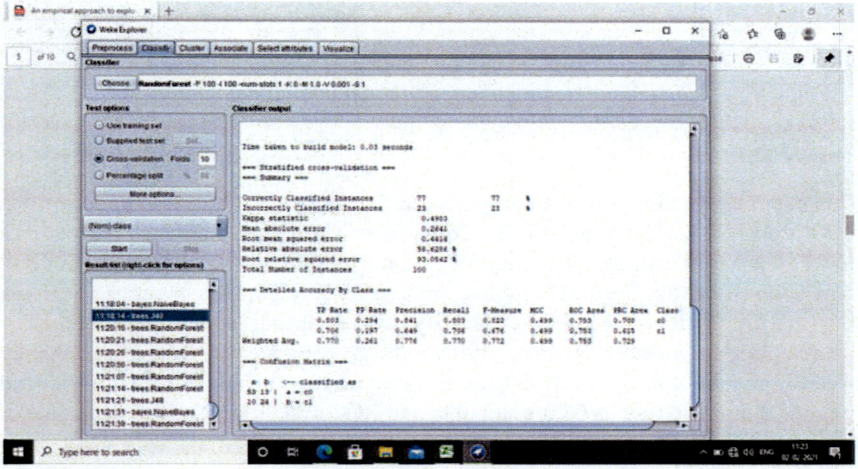

Fig. 3 Output of J48 tree classification

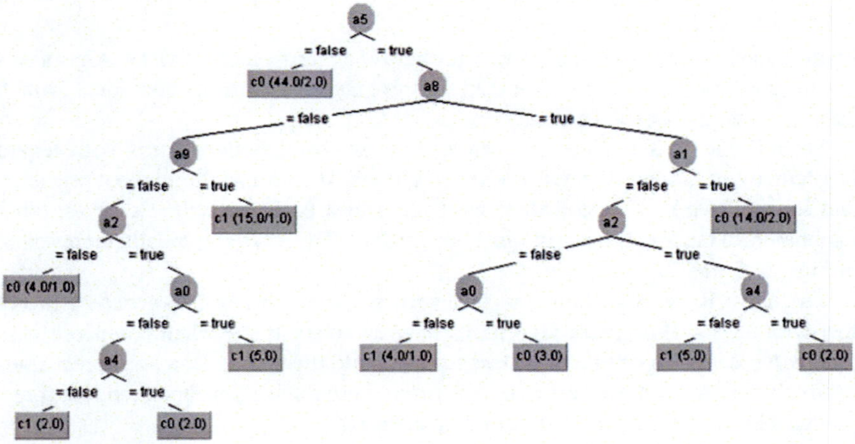

Fig. 4 Decision tree obtained from J48 tree

Table 1 Detailed test result

Classifier	Accuracy	Precision	Recall	RMSE	*F*-measure
Random Forest	83	0.827	0.830	0.3582	0.826
J48 Tree	77	0.776	0.770	0.4414	0.772
Naïve Bayes	77	0.768	0.770	0.3921	0.769

3.3 Random Forest

This paper uses random forest at every node for selecting k attribute to permit the class probabilities for judgment. Random forest is an algorithmic program comes under supervised learning; it is generally used for both classification problems and regression. By using the data samples, random forest algorithm generates decision trees and the result gets analyzed and finally chooses the best solution via voting. For the 83 correctly classified instances, random forest generates classification accuracy of 83%. As predominantly proceed, in the output mean absolute error is 0.2785 and the model erection time is 0.03 s and obtained ROC area is 0.897 these are stated in output.

4 Results

In this paper, data analysis is done with the help of experimenter interface. Algorithms that are used for experimenting the data are naïve Byes, J48 tree, random forest, which uses test sets to classify the data.

From Table 1, it is clear that random forest shows better results with regard to accuracy in comparison with other classifiers. It also has the highest precision and lowest RMSE. Lower RMSE worth indicates higher match. However, other parameters of random forest have got higher values. These are graphically represented in Figs. 5, 6, and 7.

The naïve Bayes algorithm has 77% accuracy in 0 s time. 77% accuracy in J48 tree with 0.03 s. Both these algorithms have same accuracy, when compared with time naïve Bayes algorithm takes less time to build the model. Random forest algorithm gives 83% accuracy within 0.03 s. From these three algorithms, random forest provides highest accuracy with minimum error rate.

5 Conclusion

The goal of this paper is to find the best classifier algorithm for the dispersal rate of COVID-19 virus in Kerala state. For this approach, 7 months of 100 and above confirmed cases per day data is collected. This study applies naïve Bayes, J48 tree,

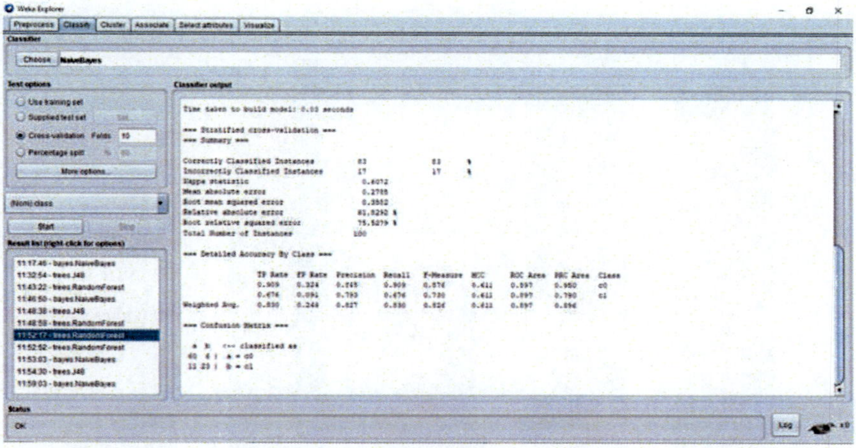

Fig. 5 Output of random forest classification

Fig. 6 Classifiers accuracy values

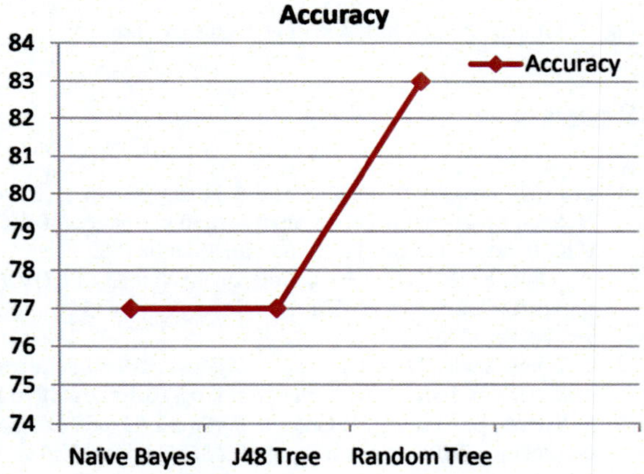

random forest for prediction. The dataset is evaluated using these three algorithms, and the different accuracies are compared. Best classifier is identified based on the time taken to build the model, correctly classified instances, ROC area and mean absolute error. From this analysis, the paper found that random forest algorithm has highest accuracy of 83% so this paper explores that random forest is the best classifier for finding Covid-19 dispersion. The fastest algorithm is the naïve Bayes with 0 s.

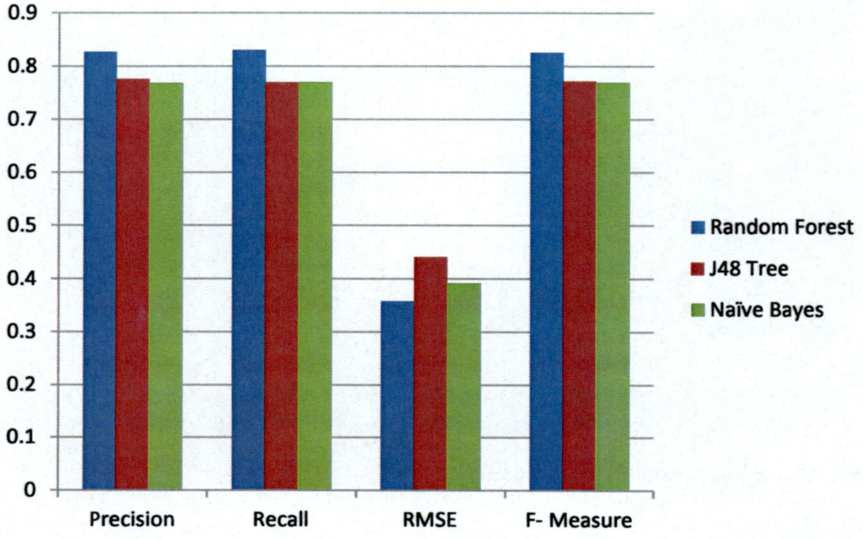

Fig. 7 Graphical representation of accuracy by class

References

1. L. Li, Z. Yang, Z. Dang, C. Meng, J. Huang, H. Meng, D. Wang, G. Chen, J. Zhang, H. Peng, Y. Shao (2020) Propagation analysis and prediction of the COVID-19. Infect. Dis. Model. **5** (2020). https://doi.org/10.1016/j.idm.2020.03.002
2. A. Thomar, N. Gupta, Prediction for the spread of COVID-19 in India and effectiveness of preventive measures (2020). Sci. Total Environ. **728** (2020). https://doi.org/10.1016/j.scitotenv.2020.138762
3. R. Singh Yadav, Data analysis of COVID-2019 epidemic using machine learning methods: a case study of India (2020). https://doi.org/10.1007/s41870-020-00484-y
4. S. Tiwari, S. Kumar , K. Guleria, Outbreak Trends of Coronavirus Disease-2019 in India: a prediction (2020). https://doi.org/10.1017/dmp.2020.115
5. S.S. Jayesh, Analysing the Covid-19 Cases in Kerala: a visual exploratory data analysis approach (2020). https://doi.org/10.1007/s42399-020-00451-5
6. K.A. Shakil, S. Anis, M. Alam, Dengue disease prediction using weka data mining tool (2015)
7. A. Gola, R.K. Arya, Animesh, R. Dugh, Review of forecasting models for coronavirus (COVID-19) pandemic in India during country-wise lockdowns, 11 Aug 2020. https://doi.org/10.1101/2020.08.03.20167254
8. S. Kumar, Monitoring novel corona virus (COVID-19) infections in India by cluster analysis, 19 May 2020
9. I. Al-Turaiki, T. Mohammed Almutairi, Building predictive models for MERS-CoV infections using data mining techniques (2016). https://doi.org/10.1016/j.jiph.2016.09.007
10. D. Mahesh Matta, M.K. Saraf, Prediction of COVID-19 using machine learning techniques, May 2020
11. P. Radanliev, D. De Roure, R. Walton, Data mining and analysis of scientific research data records on Covid-19 mortality, immunity, and vaccine development—in the first wave of the Covid-19 pandemic. Daiabetes Metab. Syndrome Clin. Res. Rev. **14**(5), 1121–1132 (2020)
12. Kerala COVID-19 battle, Government of Kerala Dashboard. https://dashboard.kerala.gov.in/

13. A. Jamwal, S. Bhatnagar, P. Sharma, Coronavirus disease 2019 (COVID-19): current literature and status in India (2020). https://doi.org/10.20944/preprints202004.0189.v1
14. Z. Ceylan, Estimation of COVID-19 prevalence in Italy, Spain, and France. Sci. Total Environ. **729** (2020). https://doi.org/10.1016/j.scitotenv.2020.138817
15. I. Ali, O.M.L. Alharbi, COVID-19: disease, management, treatment, and social impact. Sci. Total Environ. **728** (2020). https://doi.org/10.1016/j.scitotenv.2020.138861

A Deep Learning Approach to Predict Academic Result and Recommend Study Plan for Improving Student's Academic Performance

Ayon Roy, Md. Raqibur Rahman, Muhammad Nazrul Islam, Nafiz Imtiaz Saimon, MAqib Alfaz, and Abdullah-Al-Sheak Jaber

Abstract Predicting the academic results and preparing the study plan are crucial concerns for students to improve their academic performance. The existing literature mainly focused to predict the academic results and how best the teacher can design a specific course for improving the student's academic performance. However, the process of recommending study plan for a specific student based on his/her predicted results is not well investigated. Therefore, the objective of this study is to propose artificial intelligence (AI)-based models to predict the academic results and recommend study plan accordingly to improve the student's performance. As outcomes, this study proposed two models based on sophisticated deep learning algorithms and artificial neural networks namely, result prediction and recommending study planner. The proposed result prediction and study-planner models showed the accuracy of 97.02% and 99.8%, respectively, on training datasets, and also 92.94% and 87.65%, respectively, on test datasets. A Web-based system for predicting results and recommending study plan is also developed based on the proposed models.

1 Introduction

A good study plan is a crucial requirement for students to improve their academic performance. Good planning for a day-to-day study schedule requires a lot of effort and time. Students always encounter hard time to formulate an appropriate solution regarding their study plan. Even if they can come up with a plan, the plan may not be effective enough for improving their academic results. Sometimes they fail to give equal emphasis on all subjects. Sometimes the plan fails due to its improper execution. In such cases, an automated study planner could be of great use for students. A good study plan will give them the motive to be responsible for their study and eventually will help them to improve their academic performance. In addition to that, if students are able to know their prospective cumulative grade point average (CGPA), then it

A. Roy (✉) · Md. R. Rahman · M. N. Islam · N. I. Saimon · M. Alfaz · A.-A.-S. Jaber
Department of Computer Science and Engineering, Military Institute of Science and Technology, Mirpur Cantonment, Dhaka 1216, Bangladesh

P. Karuppusamy et al. (eds.), *Ubiquitous Intelligent Systems*, Smart Innovation, Systems and Technologies 243, https://doi.org/10.1007/978-981-16-3675-2_19

will help them to emphasize more on their studies by motivating them to identify their shortcomings and to make necessary preparations.

The application of machine learning (ML) and deep learning (DL) technologies for predicting experimental results in different contexts is quite popular. A lot of studies have already adopted ML and DL means for predicting academic performance of students. For example, Pushpa et al. [1] proposed a machine learning approach to predict a student's class result, while Aliponga [2] tried to find out the factors having greater impacts on a student's academic success. Jing [3] applied support vector machine (SVM) algorithm to predict the result of Chinese College English Test Band 4 (CET-4). In another study, Cheon et al. [4] proposed an automated lesson planner for teachers to plan their lessons. However, most of these studies did not focus on student's perspective to improve their academic performance on their own.

Therefore, the objective of this study is to predict prospective CGPA of a student and to recommend useful study plan for improving students' academic performance. To attain these objectives, two artificial neural network (ANN) models, one for CGPA prediction and another for recommending automated study plan, were developed and trained accordingly. The study-planner model is capable of figuring out the appropriate time for studying a specific subject automatically. This model takes four factors as input regarding a particular subject, i.e., required creativity level, memorization level, computation level and analysis level. Each of these 4 parameters has a range between 1 and 5. Based on these inputs, the system recommends an effective time interval for that particular subject. The CGPA-prediction model takes all the previous semester results of a student as input. It then predicts the prospective CGPA the student may secure in the upcoming semester. A Web application was also developed where these two prediction models were integrated [5].

The rest of the paper is organized as follows: Sect. 2 discusses the related works; Sect. 3 briefly presents the methodology that includes data acquisition and preprocessing, model creation and training, associated algorithms and the overview of the developed system; Sect. 4 provides the concluding remarks with the limitations and future work.

2 Literature Review

A number of studies have been conducted focusing on improving student's academic performance that includes various aspects of academic activities. This section briefly introduces the related works.

2.1 Predicting Study Completion Time

A limited number of studies are focused to predict the time of study completion. For example, Putri et al. [6] proposed a prediction model obtained from a data classification process using a decision tree. The model was developed with the C4.5 algorithm which gave an accuracy rate of 82.24% in predicting whether a student will be able to graduate on time or not. They found that the most influencing factor in predicting a student's graduation time is GPA in the second year. The prediction accuracy was not that much satisfactory. Wibowo et al. [7] focused almost on a similar concept and adopted C4.5 algorithm. The only difference they proposed was to make a decision support system dashboard that notifies corresponding faculties whether their students will graduate in time. Cahaya et al. [8] proposed an unsupervised learning approach using K-Medoids algorithm to predict the length of a study time of university students and showed an average prediction accuracy of 99.58%. However, these studies did not explicitly focus on improving academic performance of students.

2.2 *Academic Result Prediction*

Some studies predicted student's academic results. For example, Putpuek et al. [9] proposed a comparison between prediction models developed for predicting the final grade point average (GPA) of students. They evaluated decision tree algorithms, i.e., C4.5, ID3, naïve Bayes and K-nearest neighbor data mining techniques to analyze the data according to the CRISP-DM process. Zollanvari et al. [10] utilized machine learning techniques to develop a GPA prediction model based on a set of self-regulatory learning behaviors instead of analyzing the performance of students in previous semesters. Again, Nasiri et al. [11] proposed an educational data mining (EDM) case study based on the data they collected from the learning management system (LMS) to develop a model for predicting academic dismissals and GPA of their students.

2.3 *Academic Performance Improvement*

A number of other studies have been conducted focusing on improving student's academic performance. Sa et al. [12] proposed a predictive system to predict the student's performance of a specific course named "TMC1013 System Analysis and Design," that assists the lecturers to identify students who are anticipated to have bad performance in that particular course. The proposed system offers student performance prediction by means of data mining technique. Sokkhey and Okazaki [13] compared different prediction models developed for evaluating students' performance in Mathematics. Tripathi et al. [14] differentiated between SVM classifier and naïve Bayes algorithm in terms of accuracy and execution time while implementing prediction models for evaluating students' performance. They found that naïve Bayes shows more accuracy than SVM, and SVM takes less execution time than naïve Bayes. Widyahastuti and Tjhin [15] also tried to pitch a comparative discussion between linear regression and multilayer perceptron in terms of accuracy, performance and error rate in similar context. Their work showed that multilayer perceptron is better than linear regression. Amra and Maghari [16] worked pretty much on a similar concept to find out which algorithm is better between KNN and naïve Bayes. The experimental results showed that naïve Bayes is better than KNN by receiving the highest accuracy value of 93.6%. Hasan et al. [17] proposed a machine learning model to predict student's performance and tested the model using K-nearest neighbors, decision tree classifier, SVM, random forest classifier, gradient boosting classifier and linear discriminant analysis algorithms. The study found that K-nearest neighbors and decision tree classifier models showed the highest accuracy of 89.74% and 94.44%, respectively.

Grounding on this literature survey, a few important concerns were observed. Firstly, existing studies focused on predicting the final GPA of students, whereas this study aimed to propose a prediction model that is able to predict a student's CGPA for the upcoming semesters. For example, if a student has just passed his 3rd semester, he/she will be able to know his/her CGPA in the 4th semester. Secondly, most of the research that worked for the improvement of academic performance mainly focused on a particular subject. On the other hand, the proposed system puts equal emphasis while anticipating study plans. Thirdly, most of the researches were carried out considering the teachers' point of view, while the system proposed in this study considered students' perspective so that they can improve their academic performance on their own. Finally, unlike the existing studies, this study not only proposed the prediction models but also developed an application named MIST.AI [5].

3 Methodology

3.1 Data Acquisition and Preprocessing

Two datasets were used for training and testing the 'Study-Planner' model and the 'CGPA-Prediction' model. To train and test the study-planner model, the 'Study-Preferences' dataset was created by taking students' general preferences for studying a particular subject or course. Four parameters: *Creativity, Memorization, Computation and Analysis (CMCA)* were introduced to generalize any course. Each CMCA tuple denotes a particular subject. Alternatively, any subject or course can be represented by a unique CMCA tuple. Each of the four parameters of the CMCA tuple has the range 1–5, where 1 being the least difficult and 5 being highly difficult. For example, a subject like mathematics can have the CMCA value of (4, 1, 5, 4) that indicates, mathematics requires more effort for 'Computation,' while it requires less effort for 'Memorization.'

The key idea here is that by getting different TimeSlots for a few CMCA values for a particular student, we can predict the preferred time for any CMCA for that particular student. With this, we can predict study preferences for any subjects or courses for any student. Students were asked about what the best time is they feel to be most focused to study a particular subject, given the CMCA values. For this dataset, a total of 258 time preferences were collected for different CMCA values (i.e., for different courses).

The final 'Study-Preferences' dataset consists of Creativity, Memorization, Computation, Analysis and TimeSlot data columns. The first four columns (CMCA) are used for the features and the TimeSlot is used for the label. For the TimeSlots, instead of directly feeding time as a string or as a general time format, continuous ranges of time were converted to discrete TimeSlots so that we can feed the data to the model with ease. For example, time from 12:00 P.M. to 1:59 P.M. (2 h,) was considered as TimeSlot 1, 2:00 P.M. to 3:59 P.M. was considered as TimeSlot 2, 4:00 P.M. to 5:59 P.M. was considered as TimeSlot 3 and so on. And with this, a total of 12 TimeSlots (from 1 to 12) were used each having 2 h duration.

To train and test the CGPA-prediction model, the 'CGPA-dataset' was used, which was provided by the authors' institute. The dataset comprised term-wise (semester-wise) results of previously graduated students. The dataset was curated and only consisted of term-wise results without any personal information for privacy reasons. The final dataset consists of the following fields: [one_one, one_two, two_one, two_two, three_one, three_two, four_one, four_two, FinalCGPA] where the first eight fields denote the result (CGPA on the scale of 4.00) on respective terms and the last field, 'FinalCGPA' denotes the final result before graduating.

3.2 Building the Models

In this section, the process of building two deep learning models to predict CGPA for a term ('CGPA-Prediction' model) and to predict the best time for studying a particular subject or course ('Study-Planner' model) is presented.

3.2.1 CGPA-Prediction Model

The 'CGPA-Prediction' model was developed for predicting the upcoming term's result (CGPA on the scale of 4.00), given the previous all terms' results. Since the output (CGPA) is a continuous number, a regression deep learning model was used to predict the next term's result. Here, the model for predicting the result for the final term ('4–2') is discussed, given results of all previous terms ('1–1','1–2','2–1','2–2','3–1','3–2','4–1') as inputs to the model. The model has a total of two dense hidden layers which have 64 and 64 neurons for each hidden layer. Both the hidden layers used rectified linear unit (ReLU) as an activation function. The activation function of a node defines the output of that node given an input or set of inputs. ReLU is chosen since it is computationally less expensive because it involves simpler mathematical operations [18]. The output of the hidden layer is then forwarded to the output layer of one neuron (Result). The output from the neuron of the output

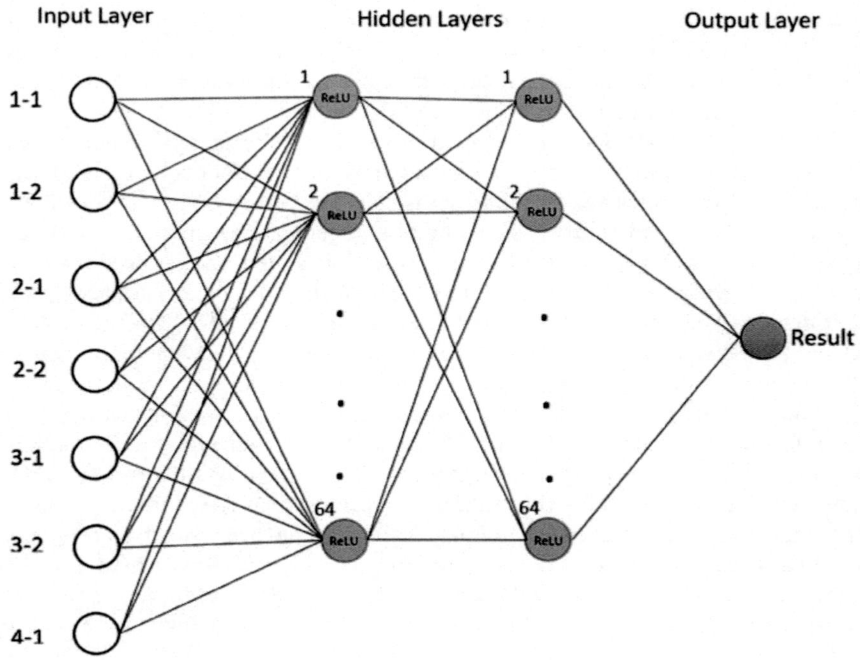

Fig. 1 CGPA-prediction model

layer gives the predicted result for the final term ('4–2'). The model was designed with mean square error (MSE) as loss function and used RMSprop for the optimizer and used mean absolute error (MAE) and mean squared error (MSE) for the metrics. The CGPA-prediction model is shown in Fig. 1.

The NN-based CGPA-prediction model was trained and tested with the 'CGPA-dataset.' The dataset was first split in a 4:1 ratio, where 80% of the main dataset was considered as a training dataset and the remaining 20% was a test dataset. The model was then trained for 1000 epochs with a validation split of 20% (0.2) with early-stopping. The mean absolute error with respect to epochs for both training and validation can be shown in Fig. 2 and the mean squared error with respect to epochs for both training and validation can be shown in Fig. 3.

The observed accuracy scores of 'CGPA-Prediction' model for both training and test dataset is shown in Table 1.

3.2.2 Study-Planner Model

The 'Study-Planner' model predicts the best time for studying a subject, given the CMCA values of that subject. Since the output is a TimeSlot out of 12 TimeSlots, a classification deep learning model was used to predict the best time for studying a

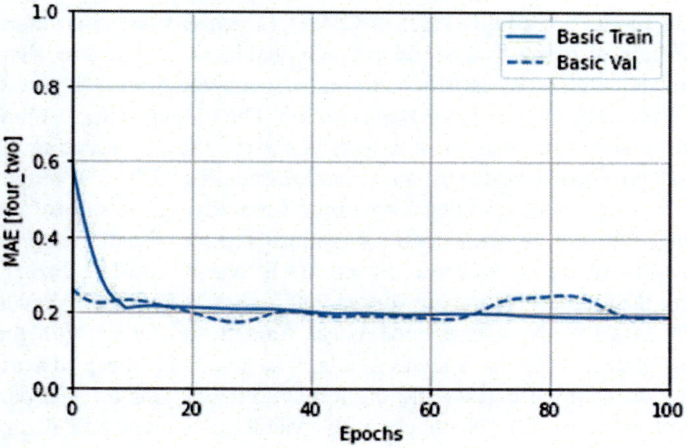

Fig. 2 Mean absolute error versus epochs

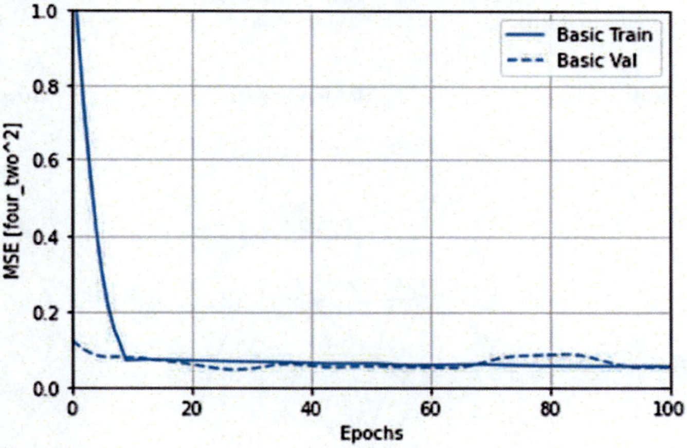

Fig. 3 Mean squared error versus Epochs

Table 1 CGPA-prediction model accuracy metrics

Dataset	Loss	Mean absolute error (MAE)	Mean squared error (MSE)
Training	0.029770	0.137765	0.029770
Validation	0.057799	0.187378	0.057799
Test	0.0706	0.2231	0.0706

subject. The model takes four values of CMCA (*Creativity, Memorization, Computation, Analysis*) as inputs to the model. The model has a total of three dense hidden layers which have 250, 175 and 150 neurons, respectively. Hidden layers used rectified linear unit (ReLU) as an activation function. The output of the last hidden layer was then forwarded to the output layer with 12 classes each representing a TimeSlot. The output layer had SoftMax as an activation function [19]. The output class of the output layer gives the predicted best time for studying the subject. The model was designed with sparse categorical cross entropy as loss function and used Adam Optimizer [20] for the optimizer and used MAE (Mean Absolute Error) and used accuracy for the metrics. Neural network model for Study-Planner is shown in Fig. 4.

The NN-based study-planner model was trained and tested with the 'Study-Preferences' dataset. The dataset was first split as 88–12%, where 88% of the main dataset was the training dataset and the remaining 12% was the test dataset. The model was then trained for 500 epochs with a validation split of 10% (0.1). Then the model was evaluated against the test dataset, and the accuracy of the model is shown in Table 2.

The observed accuracy for both the CGPA-prediction model and the study-planner model are shown in Table 2.

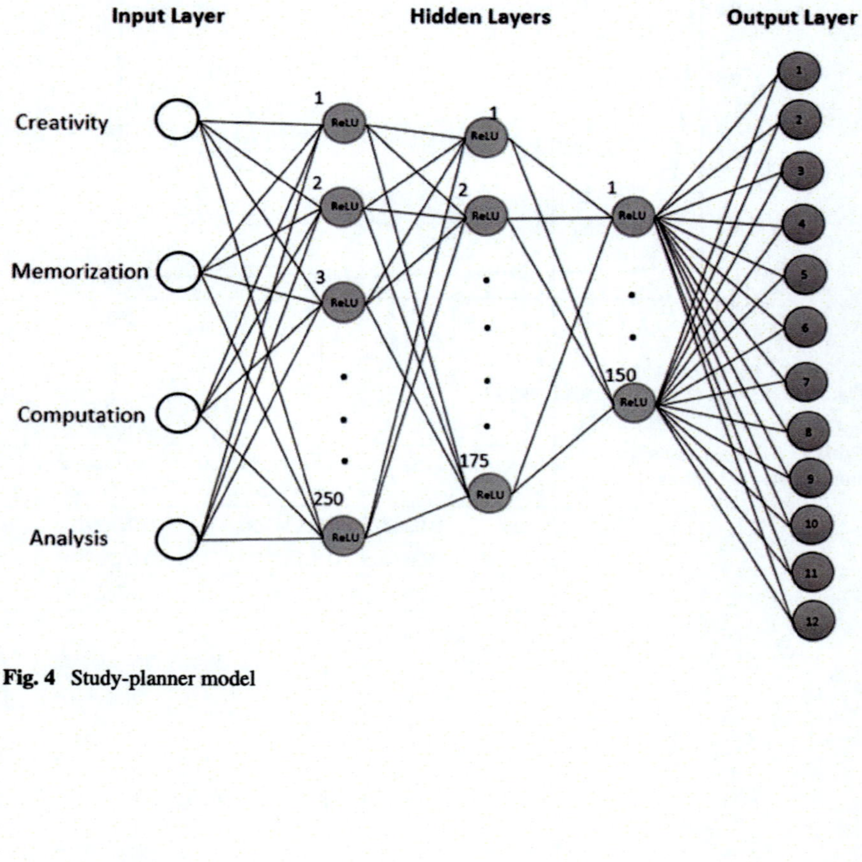

Fig. 4 Study-planner model

Table 2 Accuracy for both models

Model	Dataset	Accuracy (%)
CGPA-prediction model	Model on training dataset	97.02%
	Model on test dataset	92.94
Study-planner model	Model on training dataset	99.8
	Model on test dataset	87.65

3.3 Developing a Web Application

A Web application system [5] was developed integrating the proposed models which also includes some additional features (voice assistance, AI motivation, weather update, reminders, etc.).

For the CGPA-prediction model, the user interface (UI) takes results of previous terms as inputs and feeds them to the CGPA-prediction model, considering the number of previous results is given. The output from the model is the predicted result for the next term. The system calculates CGPA for each term with the help of GPA and credit for a particular term. A few UI of the application for predicting CGPA is shown in the Fig. 5.

For the study-planner model, when a course teacher registers a new course for a term, he/she needs to enter CMCA (*Creativity, Memorization, Computation, Analysis*) values for that course on the range of 1 to 5, that indicates how creative that

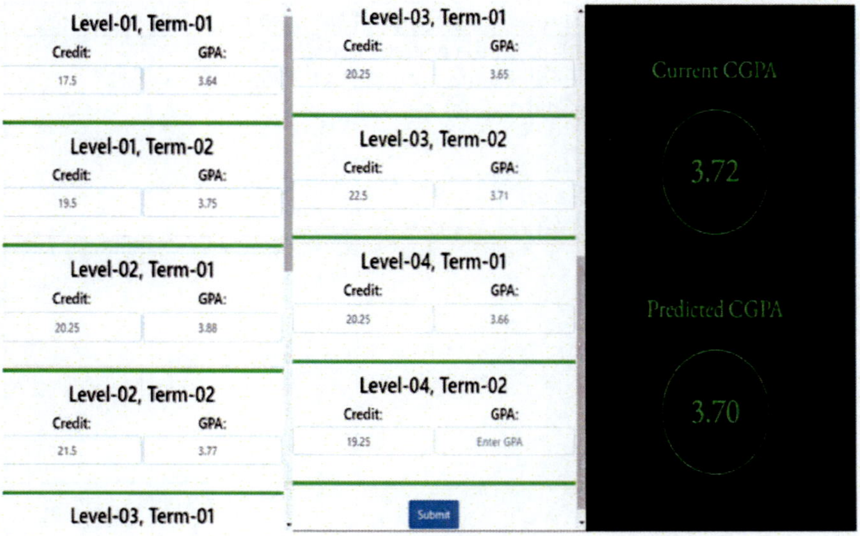

Fig. 5 UI of Web application system for CGPA-prediction

course is, how much memorization is needed for that course, how much computation skill is required and how much analysis skill is needed for that course. These values are then fed to the study-planner model and the output of the model is a class representing a particular TimeSlot for the time range. This timeslot represents the best time to study this subject or course. Then, this timeslot is stored on the database. When a student registers this course for the term, the automated study planner uses this timeslot and makes an effective study plan for the week (with one day break) for that student. To avoid multiple courses having the same TimeSlot, two algorithms were used (See Algorithm 1 and Algorithm 2).

Algorithm 1 helps to distribute each course to each day of the week (except one day for break). This helps students to distribute studies throughout the week. Also, this algorithm helps to reduce conflicts of having two or more subjects having the same TimeSlots by distributing each subject to each day of the week. But still, this process cannot entirely avoid all the conflicts when the number of courses are greater than the number of weekdays (e.g., courses. Length > 7). For this reason, *Algorithm 2* is used after applying *Algorithm 1* so that conflicts do not occur at all for a greater number of courses.

Algorithm 2 makes the nearest best TimeSlot for a particular subject, if another subject already occupies that TimeSlot for that day of the week. It calculates the minimum time distance from TimeSlot.MIN(1) to TimeSlot.MAX(12) for finding the best TimeSlot for a particular subject when the time predicted by the model is already occupied by other similar subjects. As such, *Algorithm 2* tries to pick the nearest free time for the respective subject without having any conflicts with other subjects.

Algorithm 1 Distribute courses to the days of the week:

```
Input: CoursesLen: length of all the courses registered
Output: day: An Array of distributed courses to the days of the week
function dayDistributor(CoursesLen)
    initialize day = []
    initialize di = 1
    for i = 0 to CoursesLen do
        day[i] = di
        if di = 6 then
            di = 1
        else
            di = di + 1
        endif
    endfor
    return day
endfunction
```

Algorithm 2 Create weekly study planner:

```
Input: courses: Array of tuples each tuple having course-name and time-slots
of predictions from the model for that course
Output: weekDays: A 2D Array with time slots for 6 weekdays(1 day for
break)
function createWeekStudyPlan(courses)
   initialize weekDays = A 2D Array for keeping time slots for 6 weekdays
   day = dayDistributor(courses.Length)
   for course = 0 to courses.Length do
      initialize timeSlot = 0
      if not courses[course].timeSlot exists in weekDays[day[course]] then
         timeSlot = courses[course].timeSlot
      else
         initialize minDistOrg = 20
         for dayTime = TimeSlot.MIN to TimeSlot.MAX do
            if not dayTime exists in weekDays[day[course]] then
               if minDistOrg > | dayTime- courses[course].timeSlot | then
                  minDistOrg = | dayTime - courses[course].timeSlot |
                  timeSlot = dayTime
               endif
            endif
         endfor
      endif
      weekDays[day[course]].push(timeSlot)
   endfor
endfunction
```

Using these two algorithms with the study-planner model, weekly study plans are created for every student. The UI for the study planner is shown in the Fig. 6.

4 Discussion and Conclusion

The main goal of this study was to develop a system for students that helps them understand their progress and recommends study plan accordingly. In order to achieve that goal, a cumulative grade point average (CGPA) prediction model and a study-planner model were proposed. The study planner and CGPA predictor are basically two deep neural network (DNN) models. Both the CGPA-prediction model and study-planner model are used for the improvement of studies and productivity for students. The CGPA-prediction model can be used to predict the next term's result based on previous terms results. This will motivate a student to give more attention and concentration on their studies. The study-planner model can be used to create study plans for a student by appointing subjects at a specific time that a student feels the most focused to study that particular subject. A Web application was also developed incorporating

Fig. 6 UI of Web application system for study-planner model

these two models along with some other features, e.g., voice command support—to interact with the app using voice command; attendance tracker—to keep track of attendance status (whether they become non-collegiate/dis-collegiate); reminder—to keep the user updated about tasks; weather status update—to get to know about the current weather in the current location; motivational speech generator—to get motivated at times when a student feels low.

This study is a bit different from existing works. Putri et al. [6] proposed a prediction model to predict whether their students will graduate in time or not, and it gave 82.24% accuracy. They tried to find the influencing factors that affect a student's graduation time. Similarly, Wibowo et al. [7] proposed to make a decision support system dashboard that will notify the corresponding faculties whether their students will graduate in time or not. In this study, we did not focus on the graduation time. Rather, we focused on predicting student's CGPA so that they get to know whether they should maintain their current pace, or they should speed up a bit in their study. Again, some existing works focused on predicting a student's GPA. For example, Putpuek et al. [9] proposed a comparative discussion between prediction models that predict the final grade point average (GPA) of students, while Nasiri et al. [11] proposed an educational data mining (EDM) case study to develop a model that is able to predict academic dismissal and GPA of their students. These models can predict a student's GPA for the final semester. Our system is more effective than these models as students can know about their prospective CGPA in the next semester. Later on, we studied some other research that focused on academic performance. Sa et al. [12] proposed a predictive system that is able to predict the student's performance in a specific course so that the instructors get to identify students who are anticipated to have bad performance. Similarly, Sokkhey and Okazaki [13] discussed different methods that were used in developing prediction models with a view to evaluate their student's performance in mathematics. These studies focused on a particular subject. But our proposed system treated all subjects equally while planning the study schedule. Also, above works were carried out focusing on the teacher's point of view so that they can take care of their students properly. But our system lessens the load from the teacher's head. Using our system, students will get the motive to improve on their own. It will not only help them to perform well at studies but also will make them more responsible.

Currently, this work uses CMCA parameters mentioned before for predicting the best time for studying any subject. Other inherent parameters like students' likeness or priority of a subject, subject's importance, etc., were not considered. In this work, CGPA results are predicted based on the result patterns from previously graduated students. Also, a student's sudden motivation to improve CGPA, sudden seriousness in studies and careers, etc., may appear to be important factors in affecting a student's result.

In the future, we aim to add more parameters and factors to these models to get more accurate results. We plan to include individual students' own preferences to study a particular subject so that the automation of study planner can be more effective. We also aim to keep adding results of students as they graduate and retrain the CGPA-prediction model so that the model can predict more accurately.

References

1. S.K. Pushpa, T.N. Manjunath, T.V. Mrunal, A. Singh, C. Suhas, "Class result prediction using machine learning, in *2017 International Conference On Smart Technologies For Smart Nation (SmartTechCon)*, Bangalore, pp. 1208–1212 (2017). https://doi.org/10.1109/SmartTechCon.2017.8358559
2. J. Aliponga, Key predictors of student academic success: the case of 2011 and 2013 students, in *2016 5th IIAI International Congress on Advanced Applied Informatics (IIAI-AAI)*, Kumamoto, Japan, pp. 501–504 (2016). https://doi.org/10.1109/IIAI-AAI.2016.14
3. Z. Jing, The study on the result prediction and comparison of College English Test Band 4 in China based on Support Vector Machine, in *2011 3rd International Conference on Computer Research and Development*, Shanghai, China, pp. 239–243 (2011). https://doi.org/10.1109/ICCRD.2011.5763904
4. J.-P. Cheon, J.-M. Paek, S.-G. Han, C.-H. Lee, Automated lesson planner system for ICT education, in *International Conference on Computers in Education, 2002. Proceedings*, Auckland,

New Zealand, vol.1, pp. 485–489 (2002). https://doi.org/10.1109/CIE.2002.1185985
5. MIST.AI web application. https://mist-ai.herokuapp.com/
6. D.Y. Putri, R. Andreswari, M.A. Hasibuan, Analysis of students graduation target based on academic data record using C4.5 algorithm case study: ınformation systems students of Telkom University, in *2018 6th International Conference on Cyber and IT Service Management (CITSM)*, Parapat, Indonesia, pp. 1–6 (2018). https://doi.org/10.1109/CITSM.2018.8674366
7. S. Wibowo, R. Andreswari, M.A. Hasibuan, Analysis and design of decision support system dashboard for predicting student graduation time, in *2018 5th International Conference on Electrical Engineering, Computer Science and Informatics (EECSI)*, Malang, Indonesia, pp. 684–689 (2018). https://doi.org/10.1109/EECSI.2018.8752876
8. L. Cahaya, L. Hiryanto, T. Handhayani, Student graduation time prediction using intelligent K-Medoids Algorithm, in *2017 3rd International Conference on Science in Information Technology (ICSITech)*, Bandung, pp. 263–266 (2017). https://doi.org/10.1109/ICSITech.2017.8257122
9. N. Putpuek, N. Rojanaprasert, K. Atchariyachanvanich, T. Thamrongthanyawong, Comparative study of prediction models for final GPA score: a case study of Rajabhat Rajanagarindra University, in *2018 IEEE/ACIS 17th International Conference on Computer and Information Science (ICIS)*, Singapore, pp. 92–97 (2018). https://doi.org/10.1109/ICIS.2018.8466475
10. A. Zollanvari, R.C. Kizilirmak, Y.H. Kho, D. HernáNdez-Torrano, Predicting students' GPA and developing intervention strategies based on self-regulatory learning behaviors. IEEE Access **5**, 23792–23802 (2017). https://doi.org/10.1109/ACCESS.2017.2740980
11. M. Nasiri, B. Minaei, F. Vafaei, Predicting GPA and academic dismissal in LMS using educational data mining: a case mining, in *6th National and 3rd International Conference of E-Learning and E-Teaching*, Tehran, Iran, pp. 53–58 (2012). https://doi.org/10.1109/ICELET.2012.6333365
12. C. Li Sa, D.H.b. Abang Ibrahim, E. Dahliana Hossain, M. bin Hossin, Student performance analysis system (SPAS), in *The 5th International Conference on Information and Communication Technology for The Muslim World (ICT4M)*, Kuching, Malaysia, pp. 1–6 (2014). https://doi.org/10.1109/ICT4M.2014.7020662
13. P. Sokkhey, T. Okazaki, Comparative study of prediction models on high school student performance in mathematics, in *2019 34th International Technical Conference on Circuits/Systems, Computers and Communications (ITC-CSCC)*, JeJu, Korea (South), pp. 1–4 (2019). https://doi.org/10.1109/ITC-CSCC.2019.8793331
14. A. Tripathi, S. Yadav, R. Rajan, Naive Bayes classification model for the student performance prediction, in *2019 2nd International Conference on Intelligent Computing, Instrumentation and Control Technologies (ICICICT)*, Kannur, India, pp. 1548–1553 (2019). https://doi.org/10.1109/ICICICT46008.2019.8993237
15. F. Widyahastuti, V.U. Tjhin, Predicting students performance in final examination using linear regression and multilayer perceptron, in *2017 10th International Conference on Human System Interactions (HSI)*, Ulsan, pp. 188–192 (2017). https://doi.org/10.1109/HSI.2017.8005026
16. I.A. Abu Amra, A.Y.A. Maghari, Students performance prediction using KNN and Naïve Bayesian, in *2017 8th International Conference on Information Technology (ICIT)*, Amman, pp. 909–913 (2017). https://doi.org/10.1109/ICITECH.2017.8079967
17. H.M.R. Hasan, A.S.A. Rabby, M.T. Islam, S.A. Hossain, Machine learning algorithm for student's performance prediction, in *2019 10th International Conference on Computing, Communication and Networking Technologies (ICCCNT)*, Kanpur, India, pp. 1–7 (2019). https://doi.org/10.1109/ICCCNT45670.2019.8944629
18. A.F. Agarap, Deep learning using rectified linear units (relu).arXiv preprint arXiv:1803.08375 (2018)
19. X. Liang, X. Wang, Z. Lei, S. Liao, S.Z. Li, Soft-margin softmax for deep classification, in *The International Conference on Neural Information Processing*, pp. 413–421. Springer (2017)
20. G.x. Cuı, D.-k. Lı, Research on handwritten digit recognition based on adam optimizer self-encoding. J. Jiamusi Univ. (Nat. Sci. Ed.) **154**(03), 11 (2018)

Deep Learning-Based Legal System Architecture for Africa: An Architectural Study

L. Rajesh, V. Lakshmi Narasimhan, and Moemedi Lefoane

Abstract Legal information processing attracts attention from a number of organizations globally which includes research institutions; specific areas of interest ranges from representation of legal data, such as prior court cases, for countries which adopt common law system. These legal datasets typically include legislative acts or statutes, which vary from one country to another. Mining this legal data in order to extract useful information poses formidable challenges, which include coming up with ways for storing datasets, which are continuously generated and growing exponentially every year. Additional challenges include keeping up with amendments of the statutes and invalidating old statutes. This paper details a system architecture containing several key subsystems toward design and development of Legal Humanities for Africa, which is digital, query-able and easy to use and navigate by both lay user and experienced professionals. The Legal Humanities of Africa architecture has three modules, namely, knowledge base, knowledge engine and HCI module. The knowledge base handles the legal data dictionary, glossary and metadata in a domain-specific manner, while the knowledge engine handles processing data from the legal cases or statutes. Various approaches to computational linguistics, such as natural language processing or information extraction, have been used for natural language processing, including finding part of speech in text that contains most informative terms that are useful in computing similarity between prior cases to user queries. Submodules in knowledge base are also employed as needed to help optimize the process of extracting useful information to users usually in terms of relevant cases relating to specific legal matters at hand. Other techniques employed include several aspects of machine learning, such as unsupervised learning approaches to identify and cluster prior cases so that similar cases are clustered together, thus, making the process of matching user queries to prior cases easy. The HCI module provides user-specific (lay vs. experts), domain-specific and other anchor desks as required in a typical large application that can be commonly used by many types of users. A parametric evaluation of the performance of the Cloud-based Legal Architecture

L. Rajesh (✉)
Sri Sankara Arts and Science College, Tamil Nadu Kanchipuram, India

V. Lakshmi Narasimhan · M. Lefoane
University of Botswana, Gaborone, Botswana

P. Karuppusamy et al. (eds.), *Ubiquitous Intelligent Systems*, Smart Innovation, Systems and Technologies 243, https://doi.org/10.1007/978-981-16-3675-2_20

indicates that the system can enhance its use by both professionals and commoners; details are provided in this paper. It is hoped that the Legal Humanities of Africa architecture will become the benchmark architecture for Africa at large.

1 Introduction

Legal information processing has attracted interests from a number of organizations that include research institutions, private organizations and governments and special interest groups, such as researchers through dedicated conferences like Jurix [1]. The Jurisin workshops [2] are continuously working on various problems in legal information processing and data and information extraction. Several other forums, such as the Forum for Information Retrieval Evaluation [3], also provide research challenges that attempt to address different aspects of legal information extraction.

This paper details a system architecture for legal information processing called the Legal Humanities of Africa architecture, which has several advanced subsystems contained in it. These systems provide several capabilities and functions that enhance the usability of the system for both a lay user and experienced professionals. The rest of the paper is organized as follows: Sect. 2 provides an overview of the issues in relation to legal information systems—called Legal Humanities for Africa, while the in-depth technical details of the proposed legal system architecture for digital humanities of Africa is provided in Sect. 3. The Legal Architecture per se is described in Sect. 4, followed by parametric evaluation of the architecture in Sect. 5. The conclusion summarizes the paper and offers scope for future research in this arena.

2 Overview of the Issues and Related Work

The legal domain is fraught with several issues which include, but not limited to, handling archival cases, legal precedence, ever-changing statutes, nation-specificity and technology, to name a few. Even the glossary of terms and their underlying metadata have been changing since the days of the *Magna Carta*. Technology in particular is currently able to assist the legal domain, but a few countries have been trying to adopt them. Unfortunately, most African countries are still trying to fathom the use of IT to their respective legal domains, although a few small exceptions may be there. The questions that arise include the following:

1. Can an African (or South African Development Community—SADC)-specific glossary, dictionary and metadata be developed for their legal domain?
2. What kind of IT architecture and technologies could be usefully employed in their IT systems?
3. What kind of subsystems are required in the African legal IT architecture?

4. How does one update all systems and subsystems with the ever-changing legal ecosystem?
5. With growing challenges, how can state-of-the-art approaches such as deep learning be employed in this domain.

This paper addresses some of these questions, besides providing a wide variety of ideas for the African Legal Architecture, through the following sections.

3 Technical Details: Legal Humanities of Africa

The proposed Legal Humanities of Africa has three major components, namely:

- Knowledge Base
- Knowledge Engine, and
- HCI Module.

These major components are explained in the following subsections.

3.1 Knowledge Base

The knowledge base module deals primarily with the processing and storage of legal precedence. This is where the previous court cases, statutes and legal dictionaries are processed. The processing includes indexing of the previous court cases and converting to an index that can be used to match previous cases to user queries. While processing and generating index, the following submodules optimize the process of indexing with the aim of improving retrieval effectiveness (see Fig. 1).

3.1.1 Legal Cases

Legal cases are prior court case judgments that form the law in countries adopting the common law system such a Botswana and other African countries. For example, in Botswana, legal cases are available on the government portal for precedents [4]. With advancements in digital technologies, this means that year after year these cases are generated by the Judiciaries and are captured in digital formats as well as archived. For example, Table 1 shows an extract showing the structure and format of a legal case from eLaws website.

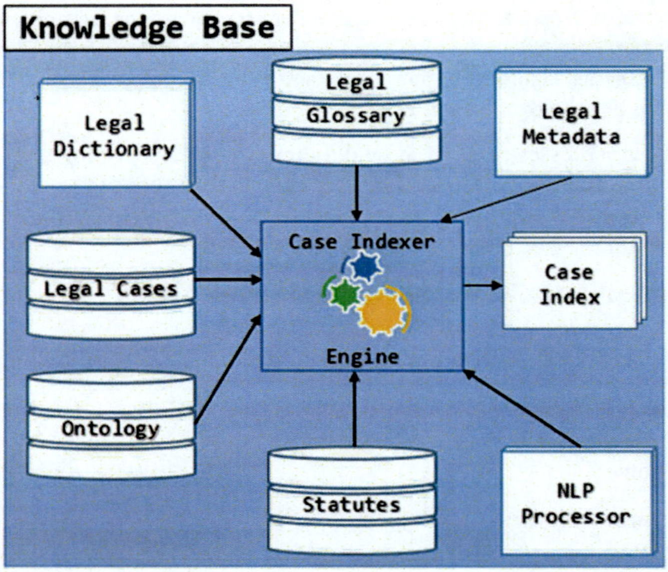

Fig. 1 Architecture of the knowledge base

Table 1 Parameters for evaluating legal humanities information systems architecture

S. No.	Explanation	Symbol	Average value	Max value
1	Size of PDF Legal file	*a*	0.001 GB	1 GB
2	Number of Retrievals per case per day	*b*	3	5
3	Number of cases per day	*c*	20	35
4	Number of follow-up visits per client/case	*d*	3	5
5	Number of Lawyer-initiated retrievals per day	*e*	5	7
6	Number of Judge-initiated retrievals per day	*f*	5	7
7	Number of sub-specialties in a given Legal system	*g*	4	6
8	Number of special issues per case	*i*	5	7
9	Number of reports to be generated per day	*n*	40	60
10	Number of compliance requirements per day	*t*	1	3

S. No.	Explanation	Symbol	Relative cost units
1	Data storage cost per GB per month	$C1$	25
2	Data access cost per GB	$C2$	2
3	Encryption cost per file (1 MB)	$C8$	5
4	Decryption cost per file (1 MB)	$C9$	5
5	Compliance management costs per compliance requirement	$C12$	500
6	Average report generation cost per report	$C13$	10

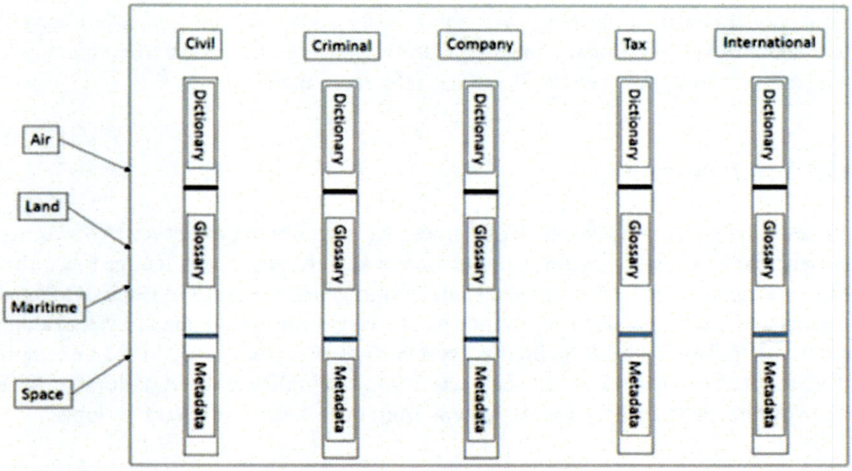

Fig. 2 Architecture of legal dictionary, glossary and metadata

3.1.2 Statutes

Statutes are typically passed by legislative authorities and also form part of the law. They capture the essence of policies in specific countries and proscribe what is prohibited and command the help guide on how to approach specific matters should they arise. This module captures all statutes along with their time frames of applicability.

3.1.3 Legal Dictionary, Glossary & Metadata

Customized legal dictionary and corresponding legal glossary of terms and related metadata sets are vitally important in this architecture for the purpose of searching, sorting, analyzing, collating and visualizing relevant documents. These three entities will be held in a domain-specific manner as shown in Fig. 2. Automatic generation of each of the dictionary terms, glossary terms and metadata words are additional tasks. Currently, many attempts are being made to standardize legal dictionary, glossary and related metadata, but as the legal field is nation-specific, deriving a set of common issues is non-trivial.

3.1.4 Legal Ontology

Ontology refers to the study of classification schemes, while legal ontology is specific to the classification requirements for the legal profession—vide [5, 6], while [7] illustrates how to choose the right type of ontology for a given architecture. Typical words contained in an ontology are orthogonal words, and further, an ontology is also domain- and subdomain-specific. Currently, many attempts are being made to standardize ontologies (see [5–7] and the references therein).

3.1.5 NLP Processor

A natural language processing (NLP) processor is vitally important for two reasons, namely, (i) to generate automatic linguistic translation from one language to another (this is a must in the African context) and (ii) to generate common English to legal linguistic bidirectional meta-translation so that laypeople can use the IT system along with experts. Several NLP facilitation systems are available (see [9] for an example of such NLP frameworks), but most are for pure English only. A tailorable NLP processor needs to be developed, perhaps from open-source software systems.

3.1.6 Lega XML & Indexing of Documents

Legal XML [10] is for representing legal documents in a standardized way, while Metalex Standard is, "*meant as an interchange format for legal documents. It differs from other existing metadata schemes in two respects: It is language and jurisdiction independent and it aims to accommodate uses of XML beyond search and presentation services.* [11]" The xmLegesEditor: an open-source visual XML editor for supporting legal National Standards [12], which can be customized to a particular country.

More informative terms are identified and indexed, which can then be used to compute similarities between user queries and previous court case judgments. In this module, preprocessing activities, such as removal of *stopwords*, *stemming* or lemmatization of terms, are performed. Services from NLP processor are also requested as necessary in this module. Natural language processing services might include (automatically) tagging paragraphs in legal court cases with parts of speech (PoS) tag in order to improve retrieval effectiveness. Lefoane et al. [13, 14] address this aspect of indexing of prior court case judgments using K-nearest neighbor search to find cases similar to a current case. The score of this unsupervised learning approach is then used to rank cases according to how closer they are to the current case. Such approaches need to be investigated further so as to find out how they can be used in the legal domain. For studies involving information retrieval from legal documents approaches, an open-source platform—Terrier can be used to facilitate indexing [15].

3.2 Knowledge Engine

The knowledge engine of the African Legal Architecture provides the brain for the IT system (see Fig. 3).

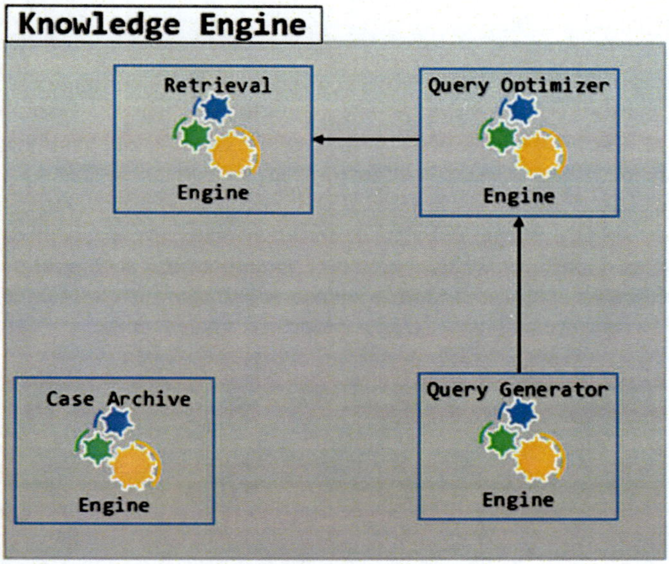

Fig. 3 Architecture of the knowledge engine

3.2.1 Query Generator Engine

In this system, user queries need not be written in SQL or any computing language but in simple plain English. The query engine is capable of using the domain-specific metadata and glossary, and it can also distribute the query to multiple databases as selected by a user. A user can also force a query to go across domains, in which case, appropriate glossaries and dictionaries must be used to translate key terms and phrases—this process is aided by the user-interface design wherein a user can intercede using several drop-down menu of terms, i.e., words and phrases.

3.2.2 Query Optimizer Engine

The query optimizer engine is able to perform query splicing, query pipelining and enhancing query performance using such techniques as time, CPU and memory optimizations [16]. Query distribution over multiple databases can also be optimized using a variety of techniques [17]. Using multilingual datasets, one could also optimize queries over multiple languages.

3.2.3 Retrieval Engine

Lefoane et al. [13, 14] address this aspect by investigating how unsupervised learning approaches such as K-nearest neighbor search as well as topic modeling (latent Dirichlet allocation) affect retrieval effectiveness. Open-source platforms such as Scikit learn are available to facilitate research involving machine learning [18]. The effect of these techniques was compared with information retrieval term weighting models such as BM25 and TFIDF. The results of these are inconclusive and call for further work on these techniques and other approaches, such as argumentation mining [19], other approaches include keyphrase extraction from legal text [20] and may play a critical role in identifying the most informative part of a legal text.

3.2.4 Legal Case Archive Engine

The Legal Case Archive Engine handles archiving of cases into the systems and then triggers Case Indexer Engine to regenerate the index including the just archived legal cases.

3.2.5 HCI Module & Anchor Desk Design

The HCI module deals with two tasks: archiving of prior court case judgments into the system [21], and the second task involves the interface that is used to capture user queries. The queries are processed accordingly and ultimately the system returns matching prior cases as results to the user. The HCI and anchor design module has several subcomponents as described below:

- **Query Composer**: permits query to be composed using several small set of queries and/or statements.
- **Word Recommender (for Query Suggestion words and query autocomplete):** provides appropriate legal words so that laypeople can come to terms with legal vocabulary.
- **Metadata Recommender**: provides several metadata (and also alternative words) so that both lay user and experts can channelize their queries properly.
- **Drop-Down Menu Manager**: provides drop-down menu of various entities, such as glossary, dictionary and meta-data.
- **Results Display Module**: displays the final results with appropriate articulation (or highlighting) of words and phrases as per the user query.
- **Display Composer**: provides a mechanism for user to organize the final results the way they want.
- (pack of) **Automatic Linguistic Translators**: provide mechanism/s for translation of both queries and results in their language of user-choice.

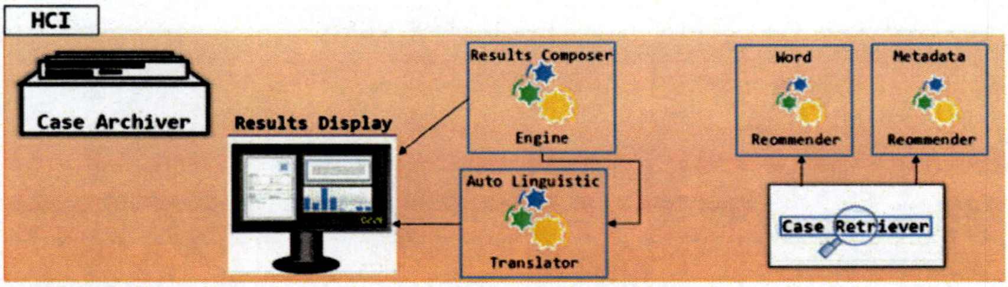

Fig. 4 HCI architecture of the legal system

- **Common English to Legal Linguistic Bidirectional Meta-Translator**: provides mechanism/s to translate both queries and results from legal English to common English (or language of user choice) in a bidirectional manner.

Anchor desk design is aimed at generating HCI that is specific to a given user profile or role. Even the HCI architecture can be composed dynamically, along with the user-preferred color settings for each module (Fig. 4).

4 Proposed Architecture

Figure 5 shows the proposed architecture. This architecture borrows ideas from information retrieval platform architecture (Terrier) [15]. This includes approach to information retrieval that involves indexing as well as retrieval. The proposed build onto these ideas by proposing implementation of domain-specific modules depicted by the figure.

5 Parametric Evaluation of the Cloud-Based Legal Architecture

A parametric model-based evaluation of the Legal Architecture using a Cloud-based system has been carried out. Table 1 provides typical parameters used for the evaluation of the Legal Architecture, which have been obtained after discussions with several experts. Table 2 provides a list of performance indicators and their values, and it is hoped that these indicators will provide the way forward for the advancement of such systems in various countries.

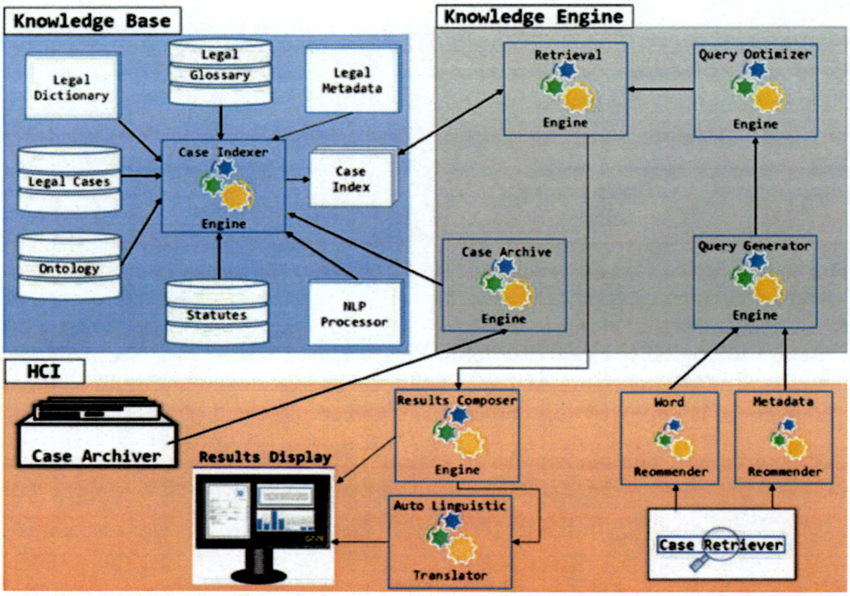

Fig. 5 Overall architecture of legal humanities of Africa

Table 2 Performance indicators for the Cloud-based legal humanities information systems architecture

S. No.	Metric name	Symbol	Formula	Typical average value	Max. value
1	Average bandwidth used per day	PI-2	$a * b * c * i$	0.30	1225
2	Average cost of security per client	PI-4	$(C8 + C9) * b * e * d$	450.00	1750
3	Average report generation cost per day	PI-6	$C13 * n$	400.00	600
4	Average compliance requirement cost per day	PI-7	$C12 * t$	500.00	1500
5	Average network usage cost = Average execution time client cost + visit cost + specialty-related cost + InfoSec cost + data access and storage cost	PI-8	$(a * b * c) + (d * f) + (g * i) + (C8 + C9) * i + (C1 + C2) * b * c * i$	8185.06	33,397

6 Conclusions

This paper proposes a system for addressing issues in legal information processing as well as representation and its parametric performance evaluation. A parametric evaluation of the performance of the Cloud-based Legal Architecture indicates that the system can enhance its use by both professionals and commoners. This domain attracts several dedicated research workshops such as Jurisin workshops that work on addressing many of the problems in legal information processing. While there are a number of conferences dedicated to this arena, Africa does not seem to be represented well or does not seem to be advancing in research in this domain. As such, there is a need to mobilize and promote research in legal information processing for Africa. It is hoped that this paper will provide a starting point toward developing a common Africa-wide or SADC-wide IT infrastructure for Legal Humanities of Africa.

References

1. JURIX—The Foundation for Legal Knowledge Based Systems. http://jurix.nl/conferences. 26 March 2019
2. International Workshop On Juris-Informatics, http://www.iaail.org/?q=article/jurisin-2018-12th-international-workshop-juris-informatics. 26 March 2019
3. Forum for Information Retrieval Evaluation—Information Retrieval from legal documents. https://sites.google.com/view/fire2017irled26 March 2019
4. http://elaws.gov.bw/. 26 March 2019
5. C. Cardellino, M. Teruel, L.A. Alemany, S. Villata, Legal NERC with ontologies, Wikipedia and curriculum learning. http://www.aclweb.org/anthology/E17-2041. 18 Feb 2019
6. Ontologies for Legal Domain. https://core.ac.uk/download/pdf/15604384.pdf. 18 Feb 2019
7. V. Leone, L. Di Caro, S. Villata, Legal Ontologies and How to Choose Them: the InvestigatiOnt Tool. http://ceur-ws.org/Vol-2180/paper-36.pdf. 18 Feb 2019
9. E. Loper, S. Bird, NLTK: the natural language toolkit, in *Proceedings of the ACL-02 Workshop on Effective Tools and Methodologies for Teaching Natural Language Processing and Computational Linguistics - Volume 1 (ETMTNLP '02)*, vol. 1 (Association for Computational Linguistics, Stroudsburg, PA, USA), pp. 63–70 (2002). https://doi.org/10.3115/1118108.1118117
10. Legal XML, http://www2.law.columbia.edu/johnson/lda/readings/SMULRLegalXMLAndStandards.pdf. 18 Feb 2019
11. Metalex: An XML standard for legal documents: https://www.researchgate.net/publication/228828366_Metalex_An_XML_standard_for_legal_documents. 18 Feb 2019
12. xmLegesEditor: an OpenSource Visual XML Editor for supporting Legal National Standards. http://www.xmleges.org/ita/images/articoli/art17.pdf. 18 Feb 2019
13. M. Lefoane, T. Koboyatshwene, L. Narasimhan, KNearest neighbor search approach to legal precedence retrieval, in *Twelfth International Workshop on Juris-Informatics* (JURISIN 2018)
14. M. Lefoane, T. Koboyatshwene, L. Narasimhan, L. Dirichlet, Allocation field based retrieval of prior case judgments, in *3rd International Conference On Internet, Cyber Security And Information Systems*, pp. 61–64 (2018)
15. I. Ounis, G. Amati, V. Plachouras, B. He, C. Macdonald, C. Lioma, Terrier: a high performance and scalable information retrieval platform, in *Proceedings of ACM SIGIR'06 Workshop on Open Source Information Retrieval (OSIR 2006)*, 10 Aug 2006. Seattle, Washington, USA

16. S. Wang, E. Rundensteiner, S. Ganguly, S. Bhatnagar, State-Slice: New Paradigm of Multi-query Optimization of Window-based Stream Queries. http://davis.wpi.edu/dsrg/PROJECTS/CAPE/publication/vldb06-slicejoin.pdf. 18 Feb 2019
17. Distributed Query Processing, https://link.springer.com/referenceworkentry/10.1007%2F978-0-387-39940-9_704. 18 Feb 2019
18. F. Pedregosa, G. Varoquaux, A. Gramfort, V. Michel, B. Thirion, O. Grisel, M. Blondel, P. Prettenhofer, R. Weiss, V. Dubourg, J. Vanderplas, A. Passos, D. Cournapeau, M. Brucher, M. Perrot, É. Duchesnay, Scikit-learn: machine learning in Python. J. Mach. Learn. Res. **12**(November 2011), 2825–2830 (2011)
19. R.M. Palau, M.-F. Moens, Argumentation mining. Artif. Intell. Law **19**(1), 1–22 (2011)
20. T. Koboyatshwene, M. Lefoane, L. Narasimhan, Machine learning approaches for catchphrase extraction in legal documents, Working notes of FIRE 2017—Forum for Information Retrieval Evaluation, Bangalore, India, December 8–10, pp. 95–98 (2017)
21. A Guide to User Interface Design, https://devforum.roblox.com/t/a-guide-to-user-interface-design/47526. 18 Feb 2019

SoloDB for Social Media's Big Data Using Deep Natural Language with AI Applications and Industry 5.0

B. Sita Devi and M. Muthu Selvam

Abstract Deep natural language processing is an algorithmic approach that enables computers to understand language using patterns, purpose, adequate experience, and a natural human data extraction context. It goes beyond a strategy for syntax and depends on a semantic strategy. Industry 5.0 democratizes the co-production of information from Big Data, building on the current symmetrical innovation concept. The Industrial Revolution 5.0 is transforming companies into working through human and computer cooperation with the massive amount of data. Industry 5.0 develops human expertise and accuracy of the computer and will be creative and satisfy consumer needs with the final product. Big data produces usable data and analyzes the best data suited for the good of the industry. The industry has no benefit from NLP converted knowledge without Big Data. Currently, users can get information from several Web sites and do not have enough time to scan all Web sites. Data is distributed with various forms of data such as education, cinema, and politics. Numerous Web sites and social media (WhatsApp, Twitter, etc.) data are distributed in the world. These different data are collected via social media and stored in one SoloDB database that allows the user to access it quickly and easily. With the authorization of the administrative process, the database information can be accessed. Deep natural language, Big Data, and artificial intelligence will be discussed, and the results will be evaluated using the Industry 5.0 private database. The combination of computer and person would make it easy to access information from the database in a customized manner.

1 Introduction

Big Data is a combination of all structured, unstructured, and semi-structured data with a vast and extensive collection of data from different sources, including social media. Deep learning is a synthetic brain (AI) function that imitates the functioning of human talent in the processing of facts and producing decision-making patterns. Deep learning methods are translating text from one language to another language, these

B. S. Devi (✉) · M. Muthu Selvam
Department of Information Technology, Vels Institute of Science, Technology and Advanced Studies (VISTAS), Chennai, India

P. Karuppusamy et al. (eds.), *Ubiquitous Intelligent Systems*, Smart Innovation, Systems and Technologies 243, https://doi.org/10.1007/978-981-16-3675-2_21

methods are real methods and results with the combination of artificial intelligence, and in recent days, deep natural language processing is used to translate speech recognition. Deep natural language promises better performance, new approaches to model, improvement in technology, and speed up. Not only can deep learning help to pick and extract features, but also to create new ones. The idea of deep learning algorithms is based on the notion of today's artificial neural networks and the preparation of such algorithms. The availability of abundant data and data has made the world simpler power for computing.

Deep studying constructions and algorithms take by now accomplished splendid success in arenas such as computer vision and prescient and additionally the sample recognition. Recent NLP studies, following this trend, are now an increasing number of targets on the use of new deep learning methods. Machine studying strategies aimed at NLP issues have centered on shallow models skilled for many years on very excessive dimensional and sparse characteristics (e.g., SVM and logistic regression). In the ultimate few years, neural networks based on dense vector representations have produced the most appropriate effects for a range of NLP tasks. This sample is sparking the growth of phrase embedding and deep learning techniques. Deep gaining knowledge of permits for multi-level automatic characteristic illustration learning [1].

Machine learning is used by the interdisciplinary area of computer science and linguistics, natural language processing (NLP), to accomplish the ultimate objective of artificial intelligence. To put it simply, it requires computers, language or text, to comprehend human language. To construct a parse tree that distinguishes parts of speech within a sentence, computers must first be trained in the grammatical rules of the language. When computers can understand the very basics of language conventions, it is possible to analyze simple questions and commands at a high rate of success. The process of natural language processing is NLU and NLG. Two main methods used in the development of natural languages are syntax and semantic analysis. The syntax is the arrangement of words in a sentence to make grammatical sense. Based on grammatical legislation, NLP uses syntax to determine the meaning of a language. The use and importance of words are part of semantics. NLP uses algorithms to understand the context and form of sentences.

Scientists used algorithms in the previous year to translate the text, now that deep learning performs the same role. Look at the use of systems to process human languages to perform the valuable role of natural language processing. Processing natural language is an interdisciplinary method that uses artificial intelligence, cognitive processes, etc. to create software to promote the interaction between computer and human language (Fig. 1).

Natural language processing, in its simplest form, is the ability of a computer/system to completely understand and interpret human language in the same way as an individual does. The processing of natural languages is also an effective approach to creating an efficient framework for handling linguistic feedback through different words, phrases, and texts in the natural environment. Various grammatical principles and linguistic techniques such as derivations, infections, grammatical

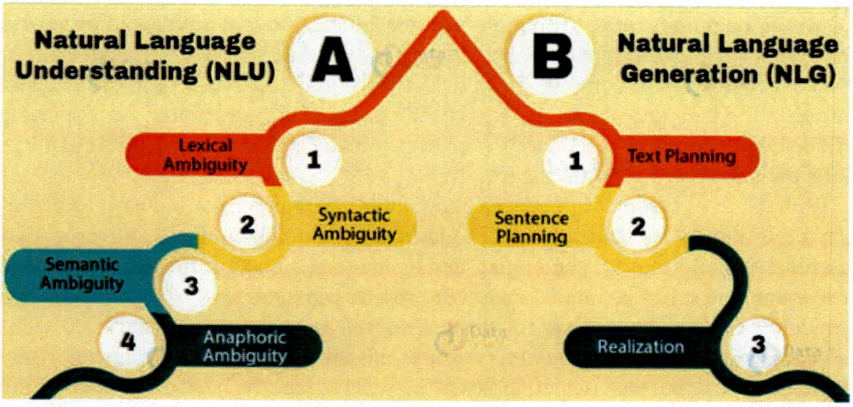

Fig. 1 Components of natural language processing (*source* https://data-flair.training/blogs/ai-natural-language-processing/)

tenses, a semantic system, lexicon, corpus, morphemes, tenses, etc., are also used in the production of the natural language [2].

The view on the role of computers in the management of enterprises is also shifting. From sensors and technical process automation to data integration and visualization and intellectual help, the perspective on the methods and means of industrial automation is evolving. Even though Industry 4.0 will be able to show its results, and achievements no earlier than in 2020–2025, researchers are already starting to talk about Industry 5.0, which will be based on self-learning machines, copying the human's actions or other robots, and continuous optimization of production algorithms.

Industry 5.0 acknowledges that in dealing with increasing customization in an integrated robotized production process, humans and machines must be interconnected to meet the technological complexity of the future. Industry 5.0 is expected to influence the social, ecological, and economic worlds. Industrial robots are a major aspect of the Fifth Industrial Revolution, involving the possibility of a human being to customize and mass-scale develop a product with the help of advanced robotic capabilities. In a shared workspace, a collaborative robot is a type of robot intended to physically communicate with humans. Additive technologies are beginning to shape Industry 5.0 centered on the Industry 4.0 sector and the fifth generation network.

5G wireless technology is expected to provide more consumers with higher peak data rates of multi-Gbps, ultra-low latency, more reliability, huge network capability, improved connectivity, and a more uniform user experience. Higher productivity and enhanced efficiency enable new user experiences and put new industries together.

Industry 5.0 will sort each development process smarter and further flexible; this purpose will be accomplished by the fifth generation. Across the 5G network, 5G's advantages in pace, real-time output, and data reliability are growing quickly and versatile technical messages. As a mobile communication, the transmission would transform mobile communication as a whole. The use of the 5G network in Industry

5.0 would be omnipresent. Although Industry 4.0 has brought automated processes and systems to the forefront of manufacturing and the commercial Internet of Things, the 5.0 industry will be compliant with industrial robotics applications, more efficient among individuals and machines. Via the realization of the potential of these systems will be highly received by humanity precise automated processes with vital cognitive skills the human brain's thinking [3].

With the integration of humans and computers, the full aim of Industrial Revolution 5.0 is to meet consumer needs. The human brain collaboration and the speed of machines with the latest technologies, and techniques such as deep natural language processing and artificial intelligence will provide consumers with the best creative product. A personalized view for customers will be given by Industry 5.0 technology. Digital transformation and technical advances provide industries with new challenges. With human contact, Industry 5.0 will reshape the landscape of emerging technology.

The need for personalization as well as mass customization of goods for consumers would be addressed by Industry 5.0. In a method known as cognitive computing, it will facilitate and then incorporate human intelligence and thought processes in computers. Smart factory cobots will also be intelligent enough to understand the needs of the human operator, decide whether they want assistance, and assist them accordingly. Also, in two separate ways, it will be good for the workforce: (a) rate was significantly higher and (b) providing in-production value-added assignments.

Nowadays, Social networking sites are famous and used by most people in the world. Social networking sites are moving on seconds to seconds, minutes to minutes, and daily so everyone likes to use it and gather information through it for many purposes. Social media is used to update new and latest things every day. So, every individual automatically likes and using it in day-to-day life. Social media (WhatsApp, Facebook (social networking), Twitter (social networking and microblogging), to name just a few, have changed the world and are among the most popular social media websites. Two major different steps to analyze through social data are: The first step is to gather the data generated on networking sites by users and then analyze that data. Businesses using this form of data analysis need to take many factors into account, including how to differentiate between social data and emotions and time relevance.

The Fifth Industrial Revolution will go hand in hand with to better use of human brainpower and imagination to increase process e-science, humans, and machines by integrating intelligent systems with workflows. As in Industry 4.0, the primary concern is Industry 5.0, a synergy between humans and autonomous machines, would be about automation. The autonomous workforce will be responsive to human purpose and desire and will be aware [4].

Humans are good at some things, and machines are good at others. A constant back and forth between humans and machines are improving the ability of all to learn. Applications for text analytics and science techniques derive from a decade of research on measurements that speed up machine learning.

2 Literature Review

Khaled proposed the natural language learning process and its consequences in educational settings were discussed. The author researched how NLP can be used to develop the process of education with scientific computer programs. The results showed the effectiveness of linguistic methods such as grammar, syntax, and textual patterns that are reasonably productive in the educational context for learning and evaluation [2].

Bryndin stated the international scientific innovation group, which started the formation and technical implementation of industry control automation systems 5.0 based on the cognitive virtual mind, proposed the study of artificial intelligence, additive technology, and 5G networks [3].

ElFar et al. checked the production and processing of algae in Industry 4.0 from the point of view of industry and processing of algae, as well as the paradigmatic shift from the point of view of Industry 4.0 which was well established in Industry 5.0. Industry 5.0's effects on the new facets of market opportunities and the environment, as well as the possibility of achieving SDGs, have also been substantially studied [5].

Nahavandi described a range of main features and concerns about the Industry 5.0 that each manufacturer might have. Moreover, it introduces many advances that researchers have accomplished for use in Industry 5.0 the apps and environments. Finally, the influence of Industry 5.0 on the automotive and manufacturing sectors from an economic and growth point of view, the overall economy is discussed, arguing that Industry 5.0 is going to produce more jobs than it can take away [4].

Young et al. examined important deep learning models and techniques that have been used for different NLP duties and provide a walk-through of their development [1].

Demir et al. studied human–robot collaboration for low-level tasks with an emphasis on robot creation, concentrating on human–robot co-working organizational problems. In this report, from the organizational and human employee viewpoint, we address the potential problems related to human–robot co-working. We believe that many upcoming organizational robotics research studies will be the focus of the problems identified in this study [6].

Martynov et al. developed technology that helped the world to step into Industry 4.0. The promising technologies essential for the organization of the digital industry in businesses and the collection of technologies needed to ensure the transformation from the current state of the industry to Industry 4.0 and then to Industry 5.0. A formal overview of Industry 4.0 and Industry 5.0 is also given, allowing the problem to be presented as a mathematical problem [7].

Revathy and Madhavu proposed new structure that provides the NLP-based method that shows the search for the role of relevance and the Harmony Comm Generation author. The author population generation shows the groups of writers who use the framework to search for similar documents needed by the user [16].

Matsuda et al. described an Industry 4.0 Cell Output Model Comparison and a new Autonomous Distributed Agent Society 5.0 Mechanism [8].

Skobelev and Borovik proposed technology developed in organizations where authors work, from IoT to emerging intelligence. The integration of these innovations will ensure the transition from Industry 4.0 to Industry 5.0 [9].

Özdemir and Hekim proposed Big Data with artificial intelligence and cobots. Industry 4.0 is a high-tech manufacturing automation technique that uses IoT to build the Smart Factory. Extreme automation until there are Internet viruses that can fully penetrate interconnected networks until everything is linked to everything else. New social and political power systems are created by intense connectivity. They could lead to authoritarian governance if left unchecked [10].

Hasan et al. developed natural language processing (NLP) based preprocessed data framework to evaluate sentiment, and integrated the model definition of Bag of Words (BoW) and Term Frequency-Inverse Text Frequency (TF-IDF) [7].

3 Framework Construction for SoloDB

The projected structure is to develop the idea which is very helpful to Internet users to get information quickly in **SoloDB** in one place. The detailed view of the device proposed is shown in Fig. 2 which is a comprehensive summary of the overall scheme.

The proposed framework module is divided into two modules: the user module and the admin module. The admin module gives access to and maintains the database information securely and the user module can access data by getting permission from the admin with restricted access. The methodology conveys an idea of SoloDB data which is collecting information and places in one place for future Internet users to get information easily and faster pace. To develop this system, we need three major processes: first one is collecting information, second is cleaning data, and the third one is the NLP concept, and the last one is SoloDB which is the developed system.

Social networking is the source of supplying users with a lot of knowledge anywhere in the world, but it is a little difficult for users to use various Web sites for another reason. A new system must build and position all data in one place and access it to solve this issue. So, here is the need to build SoloDB to provide Internet lovers with Twitter Data, WhatsApp data, and Facebook Data in one location for easy access.

There is no database like this given by the current framework. In one database, a new recommendation framework generates social media info. The goal input is handled by the method of NLP. For grouping Twitter, Facebook, and WhatsApp info, K-means clustering is used. The records are groups that use a clustering algorithm. The data is clustered and recorded and created in various clusters.

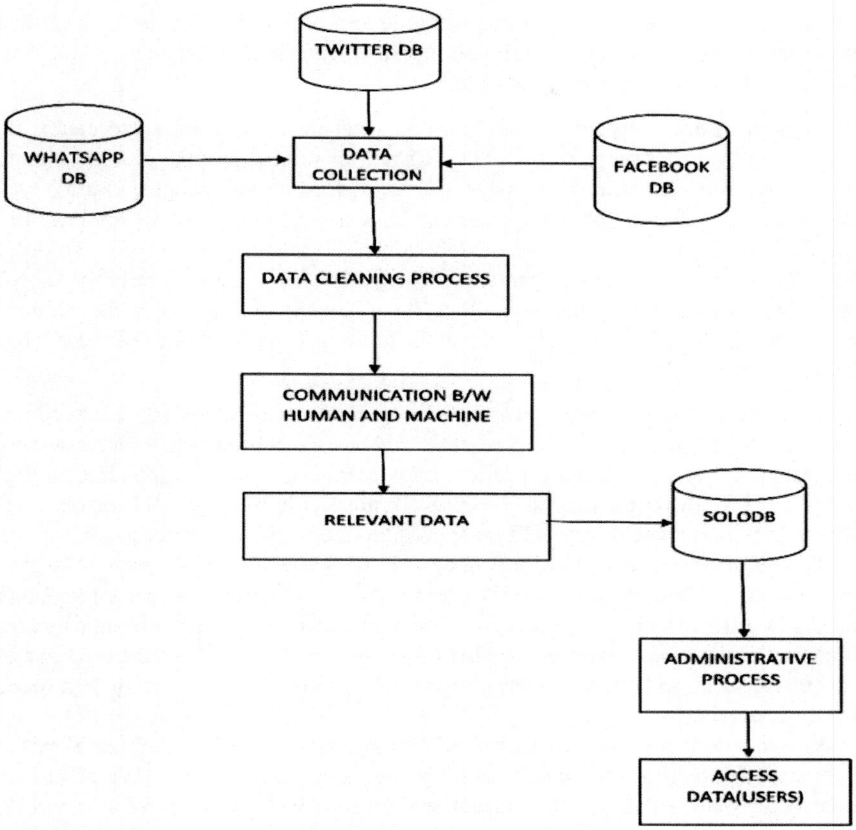

Fig. 2 Architecture of SoloDB

3.1 Gathering Data

Collecting information is a systematic method that can gather data from different sources and that data is used to analyze the required purpose. Programming languages, such as Python and R, and science tools, such as SAS, provide API interaction packages and have libraries to communicate with most major digital platforms. The strength of using software development tools such as Python allows you to gather loads of data quickly. The various datasets can be collected in various methods, some of which are described here to understand how social media information can be collected.

Online Data: Collecting data from the Web is cheap, self-administered, with a very low risk of data errors.

Data collection method: Different methods and techniques are used to collect different social media data for different purposes, and some of them are WhatsApp Data, Twitter Data, and Facebook Data.

WhatsApp data: To access WhatsApp data, first we have to organize data from WhatsApp into four sections or 'tasks'. Check for related groups, join the phone, archive backup, and extract messages (Download messages, images, videos, etc.) The WhatsApp data collection process can be done with four ways to get data. The first is manually used for all group discussions. The second is the Web WhatsApp, scraping/automation, to get information. Third, WhatsApp stores locally the rooted phone with a database containing all the data on the phone. Rooting by the phone is not recommended. It is not indicated that the last one is the Jailbroken because it has a stand-alone machine.

Twitter Data: Twitter offers several different methods of accessing Twitter info, including the Search API and the Streaming API, are the most effective for recovering tweets. Since version 1.1 of the Twitter API, all requests must be signed into Twitter using the OAuth protocol. Both types of Twitter APIs return JSON-format data. Every 15 minutes, the Search API presents a limited number of requests. Streaming APIs provide developers with low latency access to the worldwide stream of Twitter, but restricted access to all tweets. Twitter provides various endpoints for streaming tailored for the form of use: media, consumer, and platform. There are some restrictions in terms of the maximum number of tweets per hour for both the Search and Streaming APIs, and either of them does not ensure that all tweets can be purchased for the analysis [11].

Visualization must allow data manipulation and filtering interactively on a Twitter data analytics platform to quickly identify anomalies and outliers. Also, at various time resolutions, it must be able to visualize streaming data (per years, months, weeks, days, hourly, in real-time). Processed data must also allow predictive analytics, such as regression, forecasts, clustering, machine learning, to help decision-making processes. The ability to extract processed data from the platform must be available, so that it can be processed later independently to determine how it functions in the study, prediction, or early warning of events based on social media data [11].

Twitter data can be used to train the different algorithms [12]. Twitter user information contains the following information: user name, random tweet, account profile, image, and location information in a Twitter dataset consisting of 20,000 rows. There are four key ways to access Twitter information: Retrieve the public API from Twitter, find an existing dataset for Twitter, Twitter Buy, and access or buy from the provider of a Twitter service.

Facebook data: Researchers distinguish between data collection within the Facebook platform and beyond it [8]. The data inside Facebook will be interactions (likes, comments, and scrolling, and/or clicking), content uploaded from the Web site created, pages visited, actions, and behavior. A digital footprint is the mobile devices-PC ID, location, contacts, SMS content, etc., and desktop or laptop, operating system, browser form, etc. This section deals with information gathered by

Facebook outside of its website. The four main ways that this happens are cookies, mobile phones, other Facebook businesses, and Facebook partners.

It is a very significant and important task for data collection to continue some of the concepts or processes to be designed and implemented for further use. Different social media data is collected easily using Python tools and can be used for research purposes.

3.2 *Data Cleaning*

Data cleaning is the method by which corrupt or inaccurate information from a recordset, table, or database is detected and corrected (or removed) and refers to the identification of missing, wrong, inaccurate, or irrelevant parts of the data and then the substitution, alteration, or deletion of dirty or coarse data [8]. The categories of data cleaning are duplicate data, abnormal data, and incomplete data.

Big Data provides grid workers with a range of data fusion and data cleaning solutions, which are the basis for grid data mining and analysis. A standardized data file storage format is proposed based on the features of grid data, and a multi-source file formatting and file identification solution is offered. With emerging technology advances, generating, collecting, and storing large datasets is becoming easier. Although these massive datasets are useful for the use of Big Data analytics to obtain valuable insights, dirty data is also a major challenge. All of the knowledge generated is not useful. Human input, sensor data, weblog data, and other data sources are used by various software and applications. Such incorrect or inaccurate decisions will cost businesses enormous losses. And not just companies, in all data-oriented industries, dirty data will lead to incorrect decisions and faulty analysis: banking, healthcare, smart city, hazard management, education, governance, satellite, etc. [13].

The first step is to analyze the data after collecting the data from different sources, to find that the data provided is noisy or unknown information. This phase is necessary because the findings would be inaccurate and complicated by performing research on noisy or uncleaned data, leading to incorrect conclusions. There is some dirt in the social info, too. The data is not standardized as it is obtained from various sources, containing null values, incomplete data, as well as some missing values and an inconsistent sample date format. Thus, by using Python and R, data analysis was performed to validate the dirt in the data and then by using some data cleaning techniques and algorithms, and procedures for obtaining well-structured data that will be used for the study or visualization. Various processes, techniques, and tools for data cleaning can be used to render it standardized. This makes the dataset more accurate, right, and valuable [13] (Fig. 3).

Duplicate data: Duplicate instances of data are mainly extracted from repeated records produced by the detection system.

Abnormal data: It is valid or non-compliant data. Invalid data is the null values, and non-compliant data is the data that violate the rule. An example is outside the scope value.

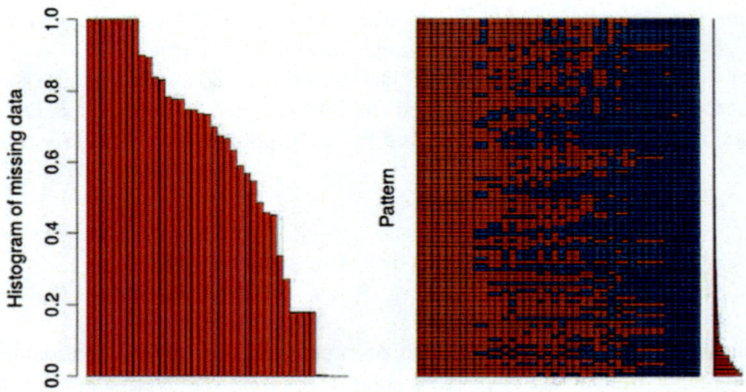

Fig. 3 Missing data histogram (*source* https://stats.stackexchange.com)

Incomplete data: It will deal with missing data. Again, it has a different method of detecting the missing data. The first is the missing data heatmap: the missing data can be pictured through a heatmap if there are a smaller number of functions. The second technique is the missing data percentage list: You can create a percentage list of missing data for each feature if there are many features in the dataset. The third is a missing data histogram: When there are many features, a missing histogram of data is also a method.

Some data that can be handled with care will not be available in the dataset. When there is no value applied to the data, unnecessary data is uninformative/repetitive, and irrelevant duplicates will be the distinctive type of unnecessary type. Two main types of duplicate data exist based on all functions and based on key features. To find out the inconsistent details, the information must be explored in various ways. It relies on observations and experience most of the time. To run and patch them all, there is no set code. Four inconsistent data types are capitalization, formats, and categorical values: There is a restricted number of values for a categorical function. Often for reasons such as typos, there might be other beliefs and addresses: the address feature might be a headache, since individuals who enter the information in the database frequently do not follow a standard format.

3.3 Natural Language Processing

Natural language processing (NLP) is a type of artificial intelligence that, by simulating human language skills, helps machines read the text. NLP methods use several methods to extract entities, interactions, and understand the meaning, including linguistics, semantics, statistics, and machine learning, to allow an understanding of what is being said or written in the context. NLP lets computers understand sentences as they are spoken or written by a person, instead of interpreting single words or combinations of them. It uses several methodologies to decipher linguistic ambiguity, including automatic summarization, part-of-talk, ambiguity, entity extraction, and relationship extraction, as well as ambiguity and natural language comprehension.

In reality, a typical human–machine interaction using natural language processing will be (a) a person speaking to a computer, (b) audio is captured by the computer, (c) conversion of audio to text occurs, (d) production of the data from the document, and (e) conversion of data to audio takes place.

Together with deep learning, recent advances in machine learning (ML) have allowed computers to do quite a lot of useful stuff with NLP. Besides, it has helped to write programs to perform things like the translation of language, semantic comprehension, and emotion recognition. Although there is a drawback, computers do not yet have the same intuitive understanding of natural language that humans do. "Reading between the lines" read between the lines. That is why it is justifiable to doubt that they will not be able to do a better job than humans.

3.4 Relevant Data

Relevant data is the data that is very popular among users from different countries and the different Web sites in the whole world. Relevant data is the popular data that will be stored in the database for future use. The client can get the information from this database only after the registration process is completed.

3.5 SoloDB

Social data is nowadays available publicly for a different purpose on various Web sites. This information is the information that users publicly share, their images, videos, some personal information, and location, etc. User shared public information will be used as information to analyze the customer behavior with the help of various tools and techniques. Social data analysis is a real-time method and another critical challenge is to decide how it can be accessed anywhere. The proposed SoloDB model would collect information that is more common and accessible on various Web sites among users worldwide. This well-known data from Twitter, WhatsApp, and Facebook is collected from well-known Web sites, and SoloDB is located in one location. Data from a single DB can be accessed faster than the different databases. The most wanted or popular data only provided by SoloDB. So, it is a database with a unique concept.

Collected data will be incomplete or missing, so the cleaning process is carried to make it valuable information. Different techniques and tools are available in the market to clean the data, and this process is very important for a further useful purpose. Natural language processing is a vital concept to understand the interaction between machine and human. Natural language processing with artificial intelligence is going to hit the world in Industry 5.0. The Fifth Industrial Revolution will go hand in hand with to better use of human brainpower and imagination to increase process e-science, humans, and machines by integrating intelligent systems with workflows.

Relevant data is the data that is very popular among users from different countries and the different Websites in the whole world. So, taking the relevant data from a different Web site is the major task of the database. After the relevant data is collected, then that data will be placed in one database which is a major concept here because all the popular data is going to place in one place for users' access. Relevant data is now ready to keep in one database that is the SoloDB.

SoloDB is the database where all the popular data is processed for easy access in one single location for users who do not have much time to spend on different Web sites or different uses for browsing purposes. It is the database where all data for faster access for users is available in one room. The database gathers the information that is known to users from various social networks and holds it in one SoloDB database and gives access to users who with administrative approval can access that specific data.

NLP-based scanning is used to make searching easier and more data available. More data is provided by the NLP-based search, and the author group generation is also generated, to encourage the authors to connect.

The administrative purpose here is to give access only to requested users not publicly. It is a private database where only paid users can access this information with administrative permission. SoloDB is a private database, so it is very fast to access and get the information in one database.

Users can access the information only after getting permission from the admin. Only those who are paid are going to get access to this database and can get information quickly and easily at a faster pace. Nowadays, faster access is very much important for the user who is accessing social data. Keep in mind this concept, SoloDB was developed to place the popular data in one place. SoloDB will save the time of the user and gives the information at a faster pace.

4 AI and Its Applications in Industry 5.0

With industry, AI offers the promise of accelerated processing of large data volumes and deep machine learning. In growth, the applications are infinite. AI and voice-controlled assistance, both at work at home and in the car, are an important part of almost any aspect of the future.

Facebook uses advanced machine learning to do everything from serving your content to identifying your face in pictures to targeting users with ads. AI is a core component of the popular social networks you use every day. AI is used by Instagram (owned by Facebook) to classify visuals. There is a range of AI-powered instruments to deliver insights from the social media accounts and audience of your company. This also involves using AI's power to evaluate social media posts on a scale, to understand

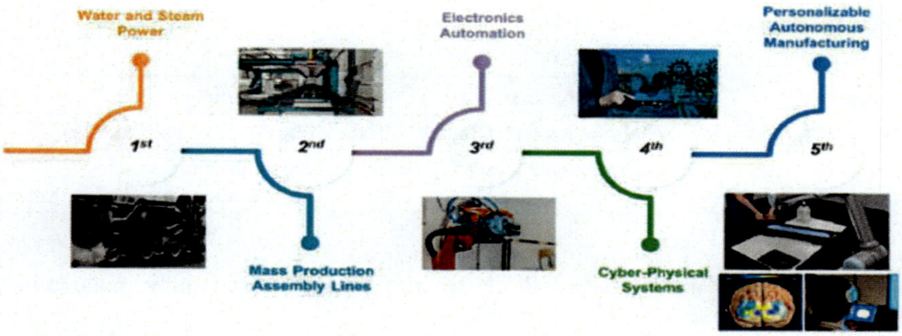

Fig. 4 Industrial Revolutions (*source* pixabay)

what they mean, and then to gain insights based on that data. Like other modes of automation, the human–machine interface (HMI) and the close and mutually beneficial interaction between the two will be the most critical aspect of applying robotics to Industry 5.0. Robots can learn from people and share their abilities to perform tasks that operators do not or do not need to do.

Figure 4 shows the Industrial Revolution from first to fifth. Collaborative robots will work together with technicians conducting routine, intensive activities in the future of Industry 5.0. Carrying out a water spider's duties, refilling parts at each stop, and conducting regular production equipment maintenance. While Industry 4.0 is still the biggest innovation in the minds of most manufacturers, it is still vital to keep an eye on the future.

Technology is continually evolving, and to stay competitive, production must advance with it. Manufacturers will probably benefit from what Industry 5.0 has to offer with the increase in demand for quality custom-made hands-on products, and maybe it will reduce the inherent fear that automation has to replace most manufacturing employees. New skills are required, but in the long run, the collaborative workforce would be advantageous for everyone. Keep an open mind for all the changes.

Industry 5.0 is being marketed as a step forward and an improvement in human–machine cooperation. The superfast precision of automated technology, combined with an individual's critical thinking ability and imagination, would lead to greater collaboration between the two. The theory is that Industry 5.0 creates even higher-value jobs than Industry 4.0, because people have taken responsibility for preparing back, or the job requiring innovative thinking. Artificial intelligence simulates human intelligence, which is handled by machines and, in particular, computer systems. The technology is mostly used to handle more conventional, monotonous tasks with machines that can make suggestions that people can trust, although that is changing (Table 1).

For industry, two visions are currently emerging. "Human–robot co-working" is the first one. In this vision, wherever and wherever possible, robots and humans

Table 1 Comparision of I 4.0 and I 5.0

	Industry 4.0	Industry 5.0
Inspiration	Mass production	Smart society
Involved technologies	AI, Robotics, IoT, Cloud, Big Data	Human robot collaboration
Research areas	Organizational research process	Smart environments, Organizational research process

would work together. Humans will concentrate on tasks that require imagination, and the rest will be performed by robots [6].

5 Discussion

In this section, the everyday use of social media is carried out daily. Data is collected from the Kaggle platform after preprocessing and analysis.

Figure 5 shows the chart with social media Big Data using Industry 5.0 with NLP

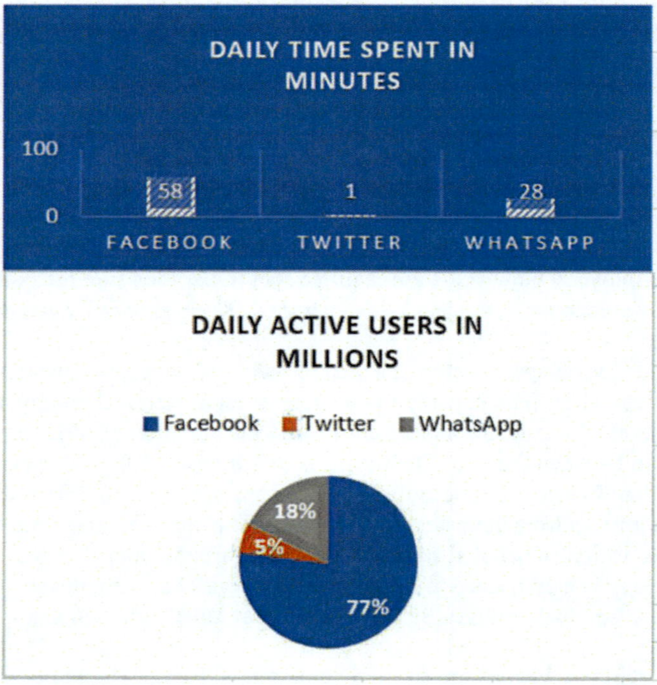

Fig. 5 Social media daily data usage comparison

which can be used to decrease the time spent on social media using the proposed methodology. K-means clustering algorithm is used to group data in one database with more accuracy level. The social media data used regularly for Facebook in the current method is on average 58 min per day, Twitter is on average 1 min per day, and WhatsApp is on average 28 min per day, so the overall time spent on various social media is on average 87 min per day. This is one use at a time.

In recent times, a huge and huge amount of data is available on social media. Near collaboration between workers and machines, the introduction of Industry 5.0 technology, and the development of artificial intelligence in order not to replace humans, but to accelerate their performance.

Increasing performance and decreasing the time spent on social media is the goal here. SoloDB is a private database, so that the data can be accessed easily and at one location without using multiple login page applications and with the trending data at preferential. The objective of the experiment was to develop a SoloDB that would decrease the total time spent on social media by 50%.

6 Conclusion

Digital transformation and technical advances are giving industries new challenges. With the emerging technology and human contact, Industry 5.0 is going to reshape the globe. Deep natural language and artificial intelligence with Big Data using a private database in Industry 5.0 is a major concept. The combination of machine and human will make a personalized way of access the information from the database with ease. For all of the other purposes, Industry 5.0 would reach the world with human and machine interaction, for instance, if it is manufacturing, then the machine will operate with the instruction provided by the machine if live interaction is still possible to access if appropriate. In this chapter, the innovative idea is recommended such as all the popular data is collected from social Web site and stored in one database that is the SoloDB database. This is a private database, so the access speed is high, and only popular data is there. This database is the paid database where only registered users can access the data with administrative permission. With the increasing prominence of digital information and an uncountable number of applications in practice, a strong SoloDB Database framework has been developed for online users. In the future, this database will be utilized for different purposes with public access permission.

References

1. T. Young, D. Hazarika, S. Poria, E. Cambria, Recent trends in deep learning based natural language processing [Review Article]. IEEE Comput. Intell. Mag. **13**(3), 55–75 (2018). https://doi.org/10.1109/MCI.2018.2840738

2. D. Khaled, Natural language processing and its use in education. Int. J. Adv. Comput. Sci. Appl. **5**(12), 72–76 (2014). https://doi.org/10.14569/ijacsa.2014.051210
3. E. Bryndin, Formation and management of Industry 5.0 by systems with artificial intelligence and technological singularity. Am. J. Mech. Ind. Eng **5**(2), 24–30 (2020). https://doi.org/10.11648/j.ajmie.20200502.12
4. S. Nahavandi, Industry 5.0-a human-centric solution. Sustainability **11**(16) (2019). https://doi.org/10.3390/su11164371
5. O.A. ElFar, C.K. Chang, H.Y. Leong, A.P. Peter, K.W. Chew, P.L. Show, Prospects of Industry 5.0 in algae: customization of production and new advance technology for clean bioenergy generation. Energy Convers. Manag. **X**(April), 100048 (2020). https://doi.org/10.1016/j.ecmx.2020.100048
6. K.A. Demir, G. Döven, B. Sezen, Industry 5.0 and human-robot co-working. Procedia Comput. Sci. **158**, 688–695 (2019). https://doi.org/10.1016/j.procs.2019.09.104
7. V. Kumar, C. Khosla, Data cleaning–a thorough analysis and survey on unstructured data, in *Proc. 8th Int. Conf. Conflu. 2018 Cloud Comput. Data Sci. Eng. Conflu.* 2018, pp. 305–309 (2018). https://doi.org/10.1109/CONFLUENCE.2018.8442950.
8. Data Cleaning in Python: https://towardsdatascience.com/data-cleaning-in-python-the-ultimate-guide-2020-c63b88bf0a0d/ Lianne & Justin @ Just into Data/ Data Cleaning in Python
9. K. Matsuda, S. Uesugi, K. Naruse, M. Morita, Technologies of production with society 5.0, in *2019 6th International Conference on Behavioral, Economic and Socio-Cultural Computing (BESC)* (2019). https://doi.org/10.1109/BESC48373.2019.8963541
10. Skobelev, Borovik, On the way from Industry 4.0 to Industry 5.0: from digital manufacturing to digital society. Int. Sci. J. "Industry 4.0" **II**(6), 307–311 (2017)
11. V. Özdemir, N. Hekim, Birth of Industry 5.0: making sense of big data with artificial intelligence, 'the internet of things' and next-generation technology policy. Omi. A J. Integr. Biol. **22**(1), 65–76 (2018). https://doi.org/10.1089/omi.2017.0194
12. Twitter data collection tutorial using Python: https://towardsdatascience.com/twitter-data-collection-tutorial-using-python-3267d7cfa93e
13. D. Cenni, P. Nesi, G. Pantaleo, I. Zaza, Twitter vigilance: a multi-user platform for cross-domain Twitter data analytics, NLP and sentiment analysis, in *2017 IEEE SmartWorld, Ubiquitous Intelligence & Computing, Advanced & Trusted Computed, Scalable Computing & Communications, Cloud & Big Data Computing, Internet of People and Smart City Innovation (SmartWorld/SCALCOM/UIC/ATC/CBDCom/IOP/SCI)*, pp. 1–8 (2018). https://doi.org/10.1109/UIC-ATC.2017.8397589

Comparative Analysis of Local Binary Descriptors for Plant Discrimination

Rose Mary Titus, Rona Stephen, and E. R. Vimina

Abstract Weed management is one of the prime obstacles faced by most of farmers nowadays. Efficient weed detection methods will cut back the price of weed management. Feature extractors have an important role in the domain of computer vision. The feature extracting algorithm takes the image as its input, and then it gives back the feature descriptors of the image that can be used to discriminate one feature from another. In software systems, there are various binary descriptors that are widely used for face recognition, plant discrimination, fingerprint detection, etc. This paper shows the performance comparison of different binary descriptors like local directional relation pattern (LDRP), local directional order pattern (LDOP), and local binary pattern (LBP) with support vector machine (SVM) for the image set classification. The results indicate that the sequence of LBP and SVM together produce a better accuracy of 84.51% in "bccr-segset" plant leaf database when compared to LDOP which produced an accuracy of 75% and LDRP with an accuracy of 75.56%.

1 Introduction

The growth of weeds among the crops has always been a huge headache for the farmers which reduces the yield and the productivity of the crops drastically. Weeds, which are an equivalent of pests, use the equivalent nutrients that crop plants use, usually in a proportion that is similar to what the actual crops use. They additionally use resources like water, nutrients and which are intended for the crops. The more similar their requirements are, the more aggressively they will compete for the resources. Identification and classification of the plant leaves supported their options do not seem to be that abundant straightforward within the agricultural field. Perhaps 2 or a lot of plant leaves will have an identical texture. So, distinctive and classifying

R. M. Titus (✉) · R. Stephen · E. R. Vimina
Department of Computer Science and IT, Amrita School of Arts and Sciences, Amrita Vishwa Vidyapeetham, Kochi, India
e-mail: KHSCI5MCA16108@asas.kh.amrita.edu

R. Stephen
e-mail: KHSCI5MCA16107@asas.kh.amrita.edu

P. Karuppusamy et al. (eds.), *Ubiquitous Intelligent Systems*, Smart Innovation, Systems and Technologies 243, https://doi.org/10.1007/978-981-16-3675-2_22

the plant leaf supported their texture mistreatment, our eye might not perpetually get a success. However, today several techniques are accessible for distinctive and classifying the leaves supporting their extracted feature. During this paper, we tend to aim to use completely different descriptors for feature extraction and create a comparative study supporting those descriptors. For that, we tend to employ a dataset known as "bccr-segset" [1, 2] that contains four subclasses namely (1) background, (2) corn, (3) canola, and (4) radish at four growth stages with 24,000 pictures used for the purpose of training the classification model and the rest of the 6000 images is used for its validation. "Bccr-segset" is a divided dataset in which segmentation is done by using ExG-ExR (Excess inexperienced minus Excess Red Indices) methodology. Three local binary descriptors like local directional relation pattern (LDRP), Local binary pattern (LBP), and local directional order pattern (LDOP) that are used for the extraction of features from the images, and we use SVM for the classifying the images into the respective categories. Then, the performance of the four descriptors is evaluated to support their accuracy.

2 Literature Review

Precise weed separation has a great role in the removal of weeds from crops in agriculture. In [2], a computationally feasible and robust weed discrimination program is created and tested against three crops which are radish, corn, and canola. The devised program predicts the combination of local binary feature descriptors which is used for the extracting important information from the images and SVM method for classifying the plants into multiple classes. We use a similar segmented "bccr-segset" dataset for our analysis.

Binary descriptors are powerful feature descriptors, they are used for many computer vision applications such as face recognition, plant discrimination, and fingerprint detection.

In [3] the authors use various binary descriptors for facial emotion classification.

Some of the present native descriptors think about solely a few immediate native neighbors for feature extraction. So, they do not seem to be able to utilize wider native data. In [4], the authors have projected a binary descriptor LDRP to utilize the broader native data.

In [5], they introduce multiple approaches based on LBP for discrimination of weeds from crops using the features extracted from the dataset. Instead of converting a color image into a grayscale image, they proposed some methods which use the red and green color channel of the images.

In [6], features of plant leaves are extracted by using the combination of local binary pattern (LBP) descriptor with histogram of oriented gradients (HOG) descriptor, and it uses SVM for classifying images into multiple classes. By using the Flavia leaf dataset, they found that the accuracy obtained by the group of LBP and HOG with SVM classification is higher than both HOG with SVM and LBP with SVM methods. The native descriptors have gained a whole lot of attention because of their increased discrimination skills. It has been verified that the inclusion of multiscale native neighbors advances the efficiency of the feature extractor.

In [7], the author has projected a way to formulate a neighborhood extractor with the help of multiscale neighborhood by extracting local directional order pattern (LDOP) from the intensity values at various measures in an exceedingly explicit direction.

In [8], the performance of seven classification models like k-nearest neighbor, decision tree, Naïve Bayesian, logistic regression, C4.5, support vector machine, and linear classifier are compared and they have found that DT, k-NN, C4.5, and SVM altogether performed greater than LogR, Naïve Bayesian.

3 Materials and Methods

The proposed system uses various local binary descriptors such as LDRP, LBP, and LDOP for the extraction of features from "bccr-segset" [1, 2] (which is the dataset we use) and uses SVM for their classification. Figure 1 shows the work flow of the algorithm.

3.1 Dataset

For this experiment, we use "bccr-segset" [1, 2] as the dataset used for our approach. Dataset consists of 24,000 images which will be used for the purpose of training and the rest of the images are used for validation. These images are divided into four different classes as canola, corn, radish, and background as given in Table 1, and each image in each class is captured at four different stages of growth as shown in Fig. 2. Each of the analyzed images is classified into these corresponding groups. Each class has got a different spec. The images are discriminated against based on their size, color, and texture.

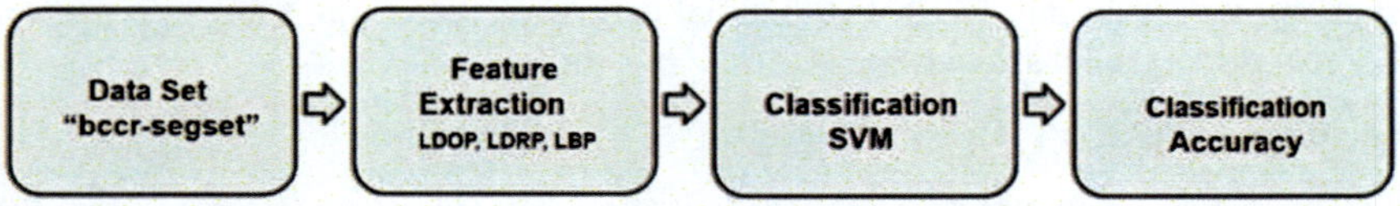

Fig. 1 Steps to measure the performance of feature descriptors

Table 1 Training set (24,000 plant images), Validation Set (6000 plant images)

Training dataset					Validation dataset			
	Stage1	Stage2	Stage3	Stage4	Stage1	Stage2	Stage3	Stage4
Canola	963	800	3088	1149	90	100	1021	289
Corn	663	1901	2272	1164	221	322	520	437
Radish	1363	1359	1056	2222	201	200	450	649
Background	6000 images				1500 images			

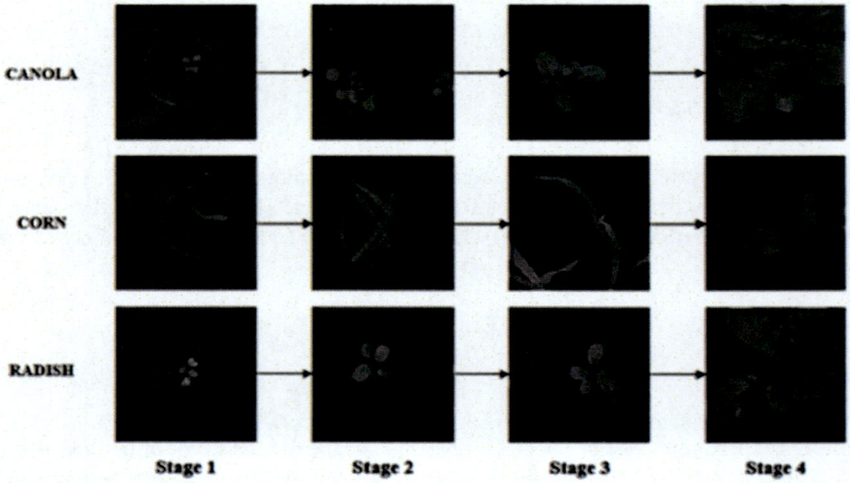

Fig. 2 Image of canola, corn, and radish at four different stages of growth

3.2 *Feature Extraction*

This method is to narrate the important information enclosed in a pattern in order to ease the task of classifying the pattern. The fundamental goal of these descriptors is to get the most relevant info from the initial knowledge and exhibit that information in a low-dimensional space (Fig. 3).

Local binary pattern (LBP)

LBP is treated as an effective tool for the extraction of features from images. By using LBP, we can extract robust features of a plant-based on their texture, and classify them according to those features. LBP algorithms are constructed in order to distinguish the objects in an image. The feature vector calculation is mainly done by taking the local neighborhood of a pixel and by segmenting its local structure and considers them as binary numbers (Fig. 4).

Fig. 3 Segmented image of canola, corn, and radish

Fig. 4 LBP image of canola, corn, and radish

LBP is calculated by finding the intensity of the center pixel and its eight neighbors; then, they are compared with each other as shown in Fig. 5. If the intensity of the neighboring pixel is less than that of the central pixel, then we denote it by "0," otherwise as "1." Then, a chain of binary code is formed from the resultant matrix.

The histogram is built by using this obtained binary number, which is used for showing the obtained texture of the image, but it is only capable enough to cover a small area, which is one of the main issues of LBP operators in this case. LBP operator is not capable to take the important features from an image for a small 3 × 3 neighborhood. Therefore, by increasing the pixel count and the size of circular neighborhood, we can increase the performance of the descriptor. Using textures of various scales, we easily improve the efficiency of LBP operators.

Mathematical expression of LBP is given as:

$$\mathrm{LBP}_{P,R} = \sum_{P\;0}^{P\;1} S(g_p - g_c)2^P \text{ where } s(x) = \begin{cases} 1, & x \geq 0 \\ 0, & x < 0 \end{cases}$$

6	0	1
3	3	8
2	4	5

1	0	0
1		1
0	1	1

10011101 157

Fig. 5 LBP representation of an image

g_{cc} It represents the value of the center pixel.
g_{pp} It represents a circularly symmetric neighborhood gray value.
P It is the count of pixels surrounding the circular neighborhood.
$s(x)$ It stands for a step function for brink.

Local directional order pattern (LDOP)

LDOP is calculated by finding the link between a pixel and its neighbors. The center pixel value has to be converted into a range of neighboring orders. LDOP feature extractor is constructed by computing the histogram of LDOP values.

For the calculation of LDOP, the first step is to extract all the local neighborhoods of every pixel, and the pixel which resides on the right side of the central pixel will be taken as the first neighbor, and the remaining other neighbors are calculated based on that first neighbor in a counterclockwise direction with respect to the center pixel. In step 2, all the local neighborhood pixel relation with the center pixel is encoded in a particular direction, i.e.; it encodes all the neighborhood pixels' distance with the center pixel and the obtained distinct value between the neighborhood pixels and center pixel in directional order can be used to compute the index value, in order to represent the order in a single value. LDOP is an essential descriptor for uniform robust illumination. LDOP descriptor is designed based on the directional information obtained from local directional order. In step 3, due to the mismatch of ranges, it has become difficult to compare the relationship between the center pixel and the neighborhood pixels. In order to overcome this issue, a center pixel transformation scheme is introduced to transform the value ranges. The capability of descriptor tolerance toward noise will increase by center pixel transformation. In step 4, the LDOP feature vector is constructed. LDOP will capture all the needed information using narrow neighborhood pixels for lower values and also captures needed information using wider neighborhood pixels for the higher values.

$$\mathrm{LDOP}_{x,y,R} = \sum_{K=1}^{N} wkx\delta k_{x,y,R}$$

LDOP is computed using histogram as follows:

$$\mathrm{LDOP}_R(\zeta) = \frac{1}{d_x^R x d_y^R} \sum_{y=R+1}^{d_y-R} \zeta\left(\mathrm{LDOPm}_{x,y,R}\zeta\right)$$

Local directional relation pattern (LDRP)

Most of the descriptors consider their neighboring pixel which leads to the reduction of discriminative ability. But some feature extractors increase their dimension for computation by using a wide range of neighbors. In order to increase the discriminative power, directional information is needed. Some descriptors become very much complicated when it uses filters to use directional gradient images. This issue is solved by using an LDRP.

In order to calculate LDRP, the first step is to extract all the local neighborhoods of every pixel. The pixel which is on the right side of the central pixel is taken as the first neighbor, and all the other neighbors are calculated based on that first neighbor in a counterclockwise direction. In the second step, the relation between local neighbors at different sizes are used to calculate the directional knowledge for increasing the discriminative ability. The third step is to figure out what relationship exists between the pixels at the center and the native directional code. In step four, the feature vector for LDRP is constructed by discovering how many times the LDRP values occurred throughout the image. And the last step is multiscale LDRP in which multiscale neighborhood features are used for making the LDRP descriptor more discriminant.

The LDRP calculation for the pixel (i, j) is as follows,

$$\mathrm{LDRP}_{i,j}^{N,M} = \sum_{K=1}^{N} \left(P_{i,j}^{N,M}(k) \right) \times \xi(k)$$

where

ξ is a weight function calculated by using the equation;

$$\xi(\eta) = 2_{\eta} - 1$$

N is the no. of directions.
M is the number of neighbors in each direction.

3.3 *Classifier*

After extraction, the final step is to classify the images into different classes. Various machine learning models are used for classification such as SVM, Bayesian, and K-nearest neighbor method... In accordance with the comparison between these models, SVM performs more accurately than other classifying methods.

Support Vector Machine (SVM)

SVM is a supervised machine learning algorithm that helps in classification or regression problems. SVM will eliminate over-fitting, and robust noise. The performance of SVM classifiers is more accurate than the other algorithms. It aims to seek out the excellent boundaries between the attainable outputs. SVM tries to maximize the separation boundaries between your information points depending on the labels or categories you have outlined. SVM can compare the testing set with the coaching set and provides a correct classification of objects. In order to do the classification using SVM, we need to draw the ideal line which separates the dataset into multiple classes. We can draw an infinite number of lines from which we need to select the best suitable one. The hyperplane for which the margin is maximum is considered as the optimal hyperplane. For multiclass classification, an equivalent principle is used when breaking down the multiclassification problem into multiple binary classification issues. The concept is to map information points to high-dimensional space to realize mutual linear separation between each two categories. This can be referred to as a one-to-one approach that breaks down the multiclass problem into multiple binary classification issues. Another approach one will use is one-to-rest. In that approach, the breakdown is set to a binary category per each class. The high performance of SVM in classification is used for several applications like face recognition, identification of weeds, and distinguishing diseases poignant leaves in crops.

3.4 Performance Metrics for Plant Classification

Classification algorithm can be accessed by calculating its accuracy, and it is defined as follows:

$$\text{Classification Accuracy}(\%) = \frac{\text{Number of correct classification}}{\text{Total number of samples}} \times 100\%$$

In order to estimate the performance of each class after classification, a confusion matrix is calculated, from which accuracy, precision, recall, and F1 score can be computed. In order to calculate all these performances, we figure out the TP, TN, FP, and FN from the confusion matrix. Each value in the confusion matrix represents the number of predictions where it is classified. When the classifier predicts the positive class as positive, we denote it by TP (True Positive), negative class as negative is denoted by TN (True Negative), negative class as positive is denoted by FP (False Positive), and positive class as negative denoted by FN (False Negative). A normal confusion matrix has four entries as shown in Fig. 6, but here in this paper, we use the confusion matrix of a multiclass model for calculating the performance.

$$\text{Precision} = \frac{\text{True Positive}}{\text{True Positive} + \text{False Positive}}$$

$$F1\text{Score} = \frac{2 \times \text{True Positive}}{(2 \times \text{True Positive}) + \text{False Negative} + \text{False Positive}}$$

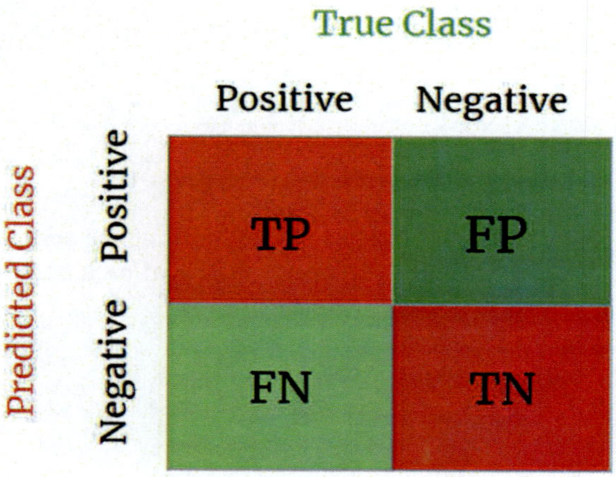

Fig. 6 Confusion matrix for binary classification

4 Experimental Results and Analysis

We implemented our algorithm in MATLAB using the "bccr-segset" database. 80% of the image set is used for the training and the rest of the image set is used for the validation. The confusion matrix obtained after the classification of the images using LBP, LDOP, and LDRP descriptors are shown in Fig. 7, respectively.

	A	B	C	D
A	1499	0	1	0
B	11	812	496	181
C	0	1	1397	102
D	138	0	0	1362

LBP

	A	B	C	D
A	472	0	542	486
B	0	1500	0	0
C	162	5	1311	22
D	0	161	122	1217

LDOP

	A	B	C	D
A	1047	2	414	37
B	496	623	63	318
C	132	0	1363	5
D	0	0	0	1500

LDRP

Fig. 7 Confusion matrix of various classes; A: Canola, B: Corn, C: Radish, D: Background

Table 2 Classification accuracy

Descriptor	Precision	$F1$ score	Accuracy
LBP	85.5	83.5	84.51
LDOP	75	74	72.44
LDRP	75.56	75.54	73.82

Accuracy is calculated after calculating the performance of each descriptor after SVM classification. The performance of each descriptor on the SVM classifier is shown in Table 2.

5 Conclusion

In the agricultural field, weed management is the major issue faced by all farmers. With the help of efficient weed discriminating methods, the cost of weed management can be reduced drastically and thereby increasing the better growth of crops. By using computational descriptors, it is more advantageous to identify the weed-affected crop. Here, in this paper, we present a performance analysis of three various local binary descriptors such as LBP, LDRP, and LDOP. The performance of each of these descriptors is analyzed and evaluated with an SVM classifier. These descriptors are applied on our dataset "bccr-segset," the results indicate that LBP + SVM gives 84.51% of accuracy more than other descriptors, and at last, the performance of the system is evaluated.

References

1. https://data.pawsey.org.au/download/Weedvision/public/LBP-SVM-analysis/bccr-set/bccr-segset%20dataset.rar
2. V.N.T. Le, B. Apopei, K. Alameh, Effective plant discrimination based on the combination of local binary pattern operators and multiclass support vector machine methods. (IPA), vol. 6, issue 1 (2019)
3. R. Arya, E.R. Vimina, An evaluation of local binary descriptors for facial emotion classification. (ICICSE), pp. 193–204 (2020)
4. Shiv Ram Dubey1 "Local directional relation pattern for unconstrained and robust face retrieval" (MTA), no.78, (2019): 28063–28088 / arXiv:1709.09518v1
5. Muammer Turkoglu, Davut Hanbay "Leaf-based plant species recognition based on improved local binary pattern and extreme learning machine" Journal of Physica A: Statistical Mechanics and its Applications Vol. 527, (2019), 121297
6. M.A. Islama, Md.S.I. Yousufb, M.M. Billahc, Automatic plant detection using HOG and LBP features with SVM. J. Int. J. Comput. (ISSN), 2307–4523 (2019)
7. S.R. Dubey, Local directional relation pattern for unconstrained and robust face retrieval, MTA **79**, 6363–6382 (2020)
8. R. Entezari-Maleki, A. Rezaei, B. Minaei-Bidgoli, Comparison of classification methods based on the type of attributes and sample size. J. Convergence Inf. Technol. **4**(3), 94–102

9. S. Sivasakthi, Plant leaf disease identification using image processing and SVM, ANN classifier methods. J. Anal. Comput. (2020). ISSN 0973–2861
10. V. Vishnoi, K. Kumar, B. Kumar, Plant disease detection using computational intelligence and image processing. J. Plant Dis. Prot. **127** (2020). https://doi.org/10.1007/s41348-020-00368-0
11. J. da Rocha Miranda, M. de Carvalho Alves, E. Ampelio Pozza, H.S. Neto, Detection of coffee berry necrosis by digital image processing of landsat 8 oli satellite imagery. J. Appl. Earth Observ. Geoinf. **85**, 101983 (2020). ISSN 0303–2434
12. P. Sharma, Y.P.S. Berwal, W. Ghai, Performance analysis of deep learning CNN models for disease detection in plants using image segmentation. Inf. Process. Agric. **7**(4), 566–574 (2020). ISSN 2214–3173
13. S. Giraddi, S. Desai, A. Deshpande, Deep learning for agricultural plant disease detection, in *ICDSMLA 2019*, vol. 601 (2020). ISBN: 978-981-15-1419-7
14. L.C. Ngugi, M. Abelwahab, M. Abo-Zahhad, Recent advances in image processing techniques for automated leaf pest and disease recognition -a review. Inf. Process. Agric. (2020). ISSN 2214-3173
15. M.E. Pothen, M.L. Pai, Detection of rice leaf diseases using image processing, in *2020 Fourth International Conference on Computing Methodologies and Communication (ICCMC)*, Erode, India, pp. 424–430 (2020). https://doi.org/10.1109/ICCMC48092.2020.ICCMC-00080

Ensuring Security in IoT Applications by Detecting Sybil Attack

Gayathri M. Menon, N. V. Nivedya, and Nima S. Nair

Abstract Internet of Things (IoT) is growing every day and has become a major part in the development and advancement of technology. Any IoT device is a transformation of almost all the substantial things which are connected to the Internet for proper transmission of information. IoT is used in various sectors, including agriculture, healthcare, making Web sites, security, etc. However, it is susceptible to Sybil attacks where the attacker generates false peer identities in order to compromise the system's disproportionate share. In this paper, we model the Sybil attack in the Cooja simulator and evaluate the behavior of the nodes and reach a conclusion on the active masquerade by using the trust value scheme. Further, we discuss the AODV routing protocol which helps in deciding the correct route for sending packets and enhances the process of detecting Sybil nodes in the network.

1 Introduction

The Internet of Things is basically a mesh of tangible devices which when connected to the Internet helps in collecting needed information and to share them whenever necessary. Mainly, it consists of all sorts of devices like sensors, processors, and other hardware devices used for communication. All this involves an easy and better lifestyle and offers a sustainable way of living as well. As IoT networks also hold confidential information, they are prone to lose the strength of security by some unauthorized information. Since IoT networks also hold sensitive data, they are vulnerable to malicious entities launching security attacks. Through forging several identities that may be used to breach the network, a malicious actor may target the network.

Now the main focus is to make homes and offices IoT friendly for more progress, better, and easier living. There are several domains in which IOT is being applied, like street lights, traffic, wearables, children's toys, and also in items as extreme

G. M. Menon (✉) · N. V. Nivedya · N. S. Nair
Department of Computer Sciences and IT, Amrita School of Arts and Sciences, Amrita Vishwa Vidyapeetham, Kochi, India

P. Karuppusamy et al. (eds.), *Ubiquitous Intelligent Systems*, Smart Innovation, Systems and Technologies 243, https://doi.org/10.1007/978-981-16-3675-2_23

as a driverless truck. These systems can be loaded with sensors and other physical devices, which have mainly been used to collect information and check for its efficient working. Wireless sensor networks (WSNs) are nodes of sensors that are interconnected and interacted wirelessly to gather information about the environment. The usage of WSN helps in corroborating the security of the IoT devices [1]. Mainly, WSNs are used in IoT applications for efficient communication. The concept of IoT can be generalized to network of connected things. These can be recognized in various sectors of industries. It can used in small daily use purposes like wearables, trackers, voice assistants which are very useful in every family, for every consumer. It can also be embedded on large equipment like robots, airplanes, other large-scale machines. Hence, the use of IoT nowadays is widespread and is growing day by day. Due to the increase in the consumption of IoT, it has become vulnerable to various malpractices.

Security is one of IoT's biggest problems. In certain cases, IoT devices happen to collect information that can be very delicate and should not be accessed by unauthorized individuals. Hence, the security of such data is rather crucial, but sometimes the safety of IoT devices get exploited. Although it is said that IoT straddles the line between the two worlds, which means that there can be destructive consequences of shattering the security of data. The IoT bridges the void between the digital world and the physical world, which ensures that there can be harmful real-world implications of breaking into computers. Hacking the secured data can be very easy in any domain of application. Different attacks can occur, which can lead to great exploitation of data and identity protection. One such attack is the Sybil attack. The attacker fakes his identity in a Sybil attack and uses it to influence the network in order to capture sensitive data or to perform some malicious act on the network. The Sybil node disguises itself as a legit node and starts communication with other nodes. So, there is an ardent need to identify such nodes, so that they do not do any harm to the application. For instance, in a monitoring IoT, device used in a factory can be exploited. The use of such device is to monitor the temperature differences which is very important for making the factory products. Any mistake in inference would lead to total failure of the production. Therefore, proper working of such devices is vital. However, in the devices, there can be a Sybil attack wherein a fraudster can fake the identity of the motes in the simulation and can pass wrong messages which can eventually lead to huge damage. Such attacks in the network should be identified and removed.

In this paper, we are going to discuss about the Sybil attack that are very dangerous to every IOT application and how can they be detected with the help of a sink node which collects all the send from the senders present in the particular wireless network.

We use the IPv6 protocol and use the sink node to collect the data in the simulation. We evaluate the functioning of nodes in Cooja simulator [2] and analyze its impact on network performance in terms of throughput and delay. In a general outline:

- Simulating a Sybil attack and analyzing its impact on network performance.
- Gathering needed data and then evaluating the simulation and differentiating.
- Further implying how the AODV protocol can be used in detecting Sybil attacks.

2 Related Works

Since the issue of Sybil attack is very important, essential measures need to be taken which are considered by many researchers. Sathish Kumar et al. [3] has cited a paper which presents the Internet of Things (IoTs), which provides capacities for the identification and connection of global physical objects into a unified scheme. Severe concerns are a part of IoTs raised over access to device-related personal information and individual confidentiality. The paper gives the detailed architecture of IoT and the issues raised in each layer. Further, it discusses about how the IoT can be effectively used in various areas and how it is helpful. This survey summarizes the safety of IoT threats and concerns about the privacy. To build the IoT platform across the Internet, Čolakovića and Mesud Hadžialić [4] has shown the various visions. Furthermore, in their paper, they have discussed enhancing technologies and overviewing the challenges faced in future research. This paper discusses about the various domains where IoT is used. In some domains, even cloud computing provides services to these domains. Deployment of various computing like fog computing and mobile edge computing in these domains can help in the security.

The security privacy in IoT highlighted by Deep et al. [5] identifies the issues in each layer of IoT and specifies the crucial security requirements, which needs proper use of authentication and gives a critical understanding of recent security solutions. The paper elaborates the challenges faced in keeping the IoT devices secured, such as the complexity, bandwidth, and power consumption, and also about the solutions for security in each layer. For example, in the network layer, the usage of mutual authentication method is for safe transmission of packets. They have suggested methods and protocols like SVELTE [6] for the sink-hole, selective forwarding attacks. However, Evangelista et al. [7] presented a study illustrating Sybil's strengths and weaknesses. Approaches used for detecting the attack is used for the propagation of IoT content in this paper. In order to assess its effectiveness and efficiency in an IoT network, an analysis of the LSD solution was conducted. Based on its conduct, Murali et al. [8] explained the Sybil attack in terms of energy consumption, PDR, and traffic control. They also examined the performance of the algorithm regarding its understanding and thoroughness. The Ad-Hoc network environment is a concept used in IoT which enhances its capabilities. This feature is done with routing protocols which are discussed in detail by Xin et.al. [9] In their paper, they analyze the behavior of such protocols like AODV. They evaluate their performance which is used in IoT devices for searching the proper routes.

3 Sybil Attack

In a Sybil attack, an opponent generates false or stolen identities to act as a few separate nodes in the peer-to-peer network. The existence of malicious conduct in the network will affect the integrity of data, the usage of resources, and the overall performance of the network. By overcoming group-based voting strategies and fault-tolerant schemes, Sybil attack will dramatically reduce network efficiency. Thus, they can have a serious effect on the daily operation of wireless and communication systems. By impersonating as an honest node, the Sybil node attempts to connect with neighboring nodes and these nodes make illegal activities in the network region.

Some of the honest nodes, as shown in Fig. 1, get influenced by the Sybil nodes and perform an attack on the honest nodes. While in other cases, the honest nodes are directly attacked by the Sybil nodes.

Therefore, these Sybil nodes can be categorized depending on the behaviors of attack.

Direct Sybil Attack. In this, Sybil nodes share data directly with legitimate nodes, which allow a Sybil node to influence the other node to get the communicated message.

Indirect Sybil Attack. In this, Sybil nodes communicate indirectly with legitimate nodes. In this, there is an intermediate node which is the one which is actually under the Sybil influence and communicates the legitimate node.

These can also be categorized on the basis of their behavior or social nature with other nodes as shown in Fig. 2.

SA-1. This attack usually exists between the social and sensing domain. It focuses on manipulation of options and popularity.

SA-2. Its main aim is to attack volatile users' privacy; it can build a social connection with Sybil identities and normal users.

SA-3. It is the same as SA2, but the impact of this is within a small area or period of time. Table 1 illustrates the description for the same.

The Sybil attack as discussed above can come in different behaviors, and each attack can be depended on its nature. This attack of forging the identity can be a real threat to not only the small sectors but can affect every industry using IoT. IoT which has various sensors using network is where these attacks cause the trouble. The nodes in these networks are said to be vulnerable to such attacks. The Sybil node entering

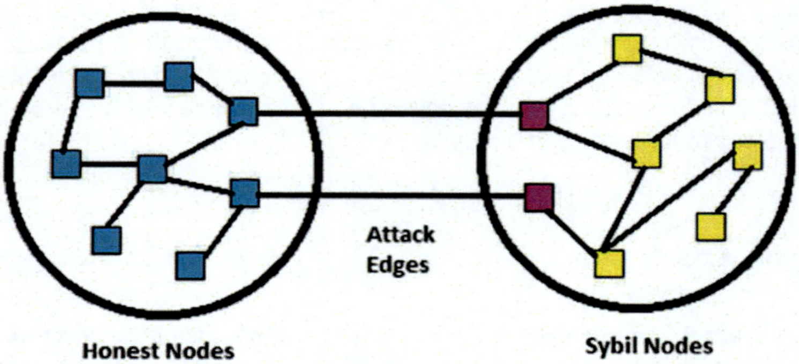

Fig. 1 Sybil attack

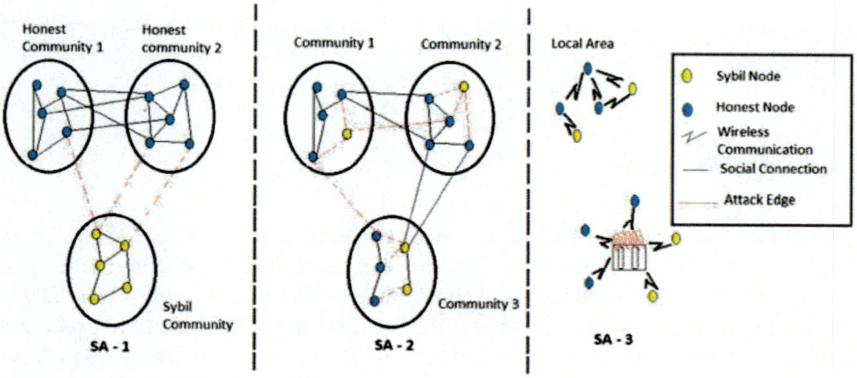

Fig. 2 Sybil attack in social graph [10]

Table 1 Sybil attack types description [10]

Various Sybil attack	Social graph structure	Attack aim	Behavior judgment
SA-1	Exists in same region or community and limited attack edges	Biased report or comment is uploaded maliciously	Normal user and frequently specific behaviour is repeated
SA-2	Connect tightly with normal users and more attack edges	User privacy dissemination spam malware attack	High frequency behaviour purposely repeated
SA-3	Normal users may be connected with Sybil	Local popularity manipulation and spam in mobile environment	Specific behaviour frequently

the network influences these nodes and make an attack in whichever way as per the nature of the attack. The effect of such attack is very dangerous. As the use of IoT is widespread, the Sybil attack can happen in any area of industry, including Web sites using IoT and small gadgets. Recently, there have been multiple fake reviews in various social networking sites due to Sybil attacks which affected the sites with their exposure. It also affects in massive level in each sector, even effecting the external affairs of each countries across the globe.

For example, in agriculture sector, there is an ardent need of using IoT devices. Embedding agriculture and the IoT is often referred to as smart farming. The smart farming is necessarily used to check the weather conditions, soil fertility which helps in the management of the production of crops and help in good yield. However, if there is any anomaly in the devices, it can lead to crop failure and loss in business. The sensors used in these devices can be exploited to any Sybil attack, and the Sybil nodes can forge and change the working of these devices. Like for the automatic working of any process, such as irrigation or harvesting by machines can be stopped or done inefficiently. This may cause the damage in effective amount. Therefore, Sybil node is to be detected in such IoT devices for the smooth and profitable functioning of every system.

3.1 Implementation

This system model is created with the Cooja tool under Contiki OS which is done using a virtual machine [11]. We have created a model with about 23 nodes. The sensor nodes behaving under AODV and one base station for all nodes are constructed. The features and benefits of the Contiki Cooja Simulator have prompted us to choose it over the other popular simulators. Cooja is a compatible simulator that allows nodes not only to be software but also to be hardware at different levels. Cross-level simulation helps the simulation to take place at various sensor points. It is an erudite tool to give the result accurately.

Firstly, we initialize the window through VirtualBox. After this, we have to add the motes. Motes are sensor nodes or any wireless device that are used for communication. The sky mote is selected for the transmission of data in the network. In our work, in order to disperse the nodes of the network in the selected region, we carried out the simulation with a single sink and a random network topology. The sink node used here is mainly obtaining information from other nodes in a wireless network. All others will be sender nodes. We have created an environment with following simulation setup as given in table 2.

As the sink node collects the data, the other UDP motes start the communication with each other. Each mote will send a HELLO message to each other as per the REQ sent. The purple node indicates the Sybil node. A great benefit of Cooja is that the motes used in a simulation use the same firmware as real physical devices. The network window displays the layout of the network motes as well as the network traffic, as shown in Fig. 3.

The collected sensor data are then forwarded to sink nodes. By using multi hop forwarding, the sink node which was used for data collection in the network can be helpful in the transfer of data. In Fig. 4. The Sensor map displays that all the data is accumulated by the sink node and how the sensor nodes communicate with each other.

Table 2 Simulation setup specifications

Total motes	26
Topology	Random
Simulation time	15 min
Simulation area	1000 m
Network	IPv6
Transport	UDP

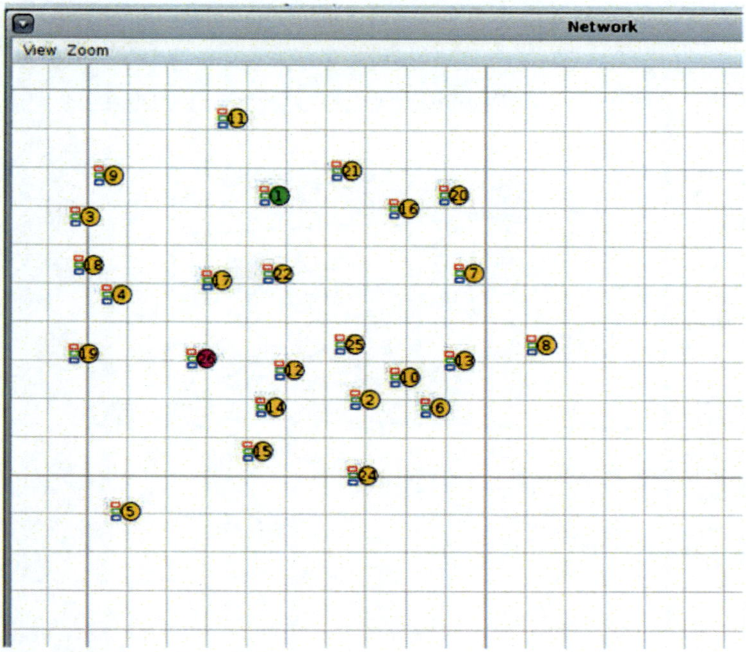

Fig. 3 Network graph illustrating the motes

3.2 *Evaluation*

The sink node which gets the data stores the values in a node table (Fig. 5.) which stores all the information about the nodes and gives the values of the parameters needed for the detection.

When it comes to performing detailed analysis of a malicious attack, we need to collect their data to understand not only the nature of the attack, but the meaning of what happened at the time of the attack. Which users, software, and segments of the network were active? During this process, accurate, historical playback also becomes critical.

The proposed scheme is to make sure that the Sybil node is detected and prevented. It also enhances the detection process by evaluating the performance with the calculation of average delay in the transmission of data packets, the throughput, and other necessary factors to assess the standard of service of the routing protocol. With the help of these factors, from Fig. 5, we get to understand the nature of the attack and how it has it affected. We calculate trust values for all nodes based on the data which was calculated by evaluating their network performance.

The following parameters are considered:

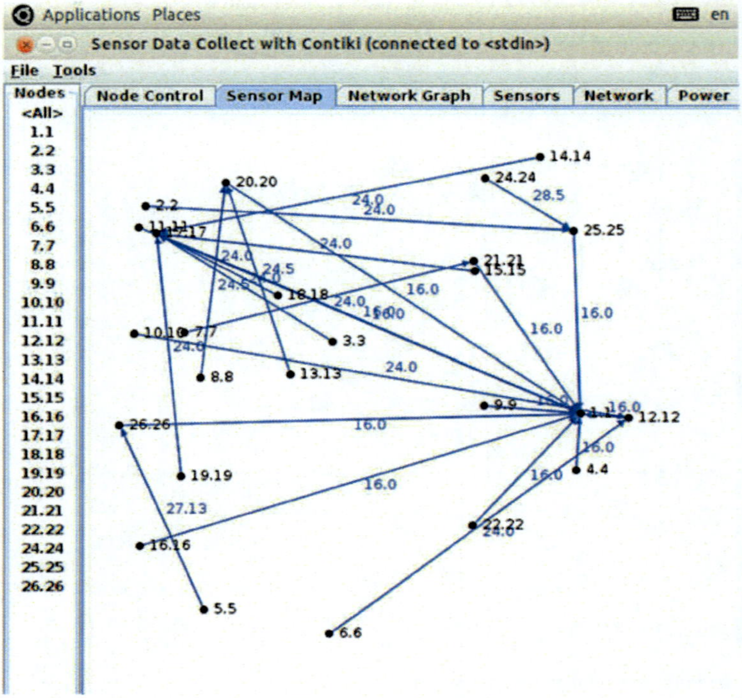

Fig. 4 Sensor Map

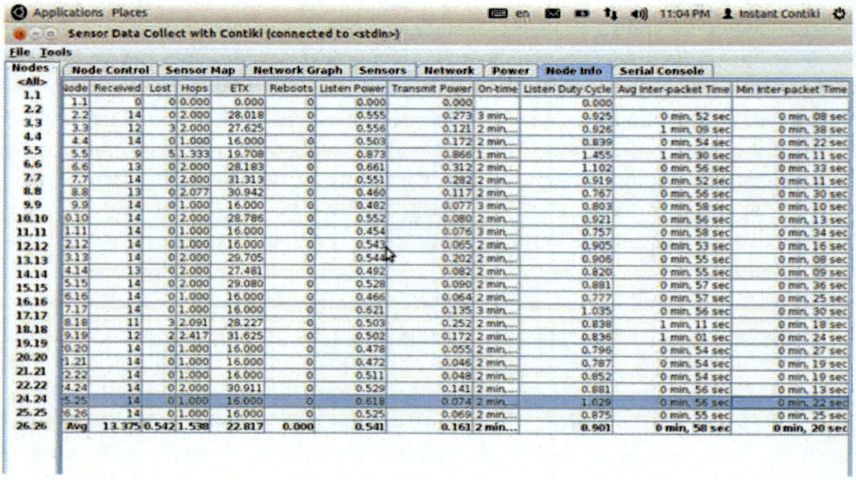

Node	Received	Lost	Hops	ETX	Reboots	Listen Power	Transmit Power	On-time	Listen Duty Cycle	Avg Inter-packet Time	Min Inter-packet Time
1.1	0	0	0.000	0.000	0	0.000	0.000		0.000		
2.2	14	0	2.000	28.018	0	0.555	0.273	3 min,...	0.925	0 min, 52 sec	0 min, 08 sec
3.3	12	3	2.000	27.625	0	0.556	0.121	2 min,...	0.926	1 min, 09 sec	0 min, 38 sec
4.4	14	0	1.000	16.000	0	0.503	0.172	2 min,...	0.839	0 min, 54 sec	0 min, 22 sec
5.5	9	5	1.333	19.708	0	0.873	0.866	1 min,...	1.455	1 min, 30 sec	0 min, 11 sec
6.6	13	0	2.000	28.183	0	0.661	0.312	2 min,...	1.102	0 min, 56 sec	0 min, 33 sec
7.7	14	0	2.000	31.313	0	0.551	0.282	2 min,...	0.919	0 min, 52 sec	0 min, 11 sec
8.8	13	0	2.077	30.942	0	0.460	0.117	2 min,...	0.767	0 min, 56 sec	0 min, 30 sec
9.9	14	0	1.000	16.000	0	0.482	0.077	3 min,...	0.803	0 min, 58 sec	0 min, 10 sec
0.10	14	0	2.000	28.786	0	0.552	0.080	2 min,...	0.921	0 min, 56 sec	0 min, 13 sec
1.11	14	0	1.000	16.000	0	0.454	0.076	3 min,...	0.757	0 min, 58 sec	0 min, 34 sec
2.12	14	0	1.000	16.000	0	0.543	0.065	2 min,...	0.905	0 min, 53 sec	0 min, 16 sec
3.13	14	0	2.000	29.705	0	0.544	0.202	2 min,...	0.906	0 min, 55 sec	0 min, 08 sec
4.14	13	0	2.000	27.481	0	0.492	0.082	2 min,...	0.820	0 min, 55 sec	0 min, 09 sec
5.15	14	0	2.000	29.080	0	0.528	0.090	2 min,...	0.881	0 min, 57 sec	0 min, 36 sec
6.16	14	0	1.000	16.000	0	0.466	0.064	2 min,...	0.777	0 min, 57 sec	0 min, 25 sec
7.17	14	0	1.000	16.000	0	0.621	0.135	3 min,...	1.035	0 min, 56 sec	0 min, 30 sec
8.18	11	3	2.091	28.227	0	0.503	0.252	2 min,...	0.838	1 min, 11 sec	0 min, 18 sec
9.19	12	2	2.417	31.625	0	0.502	0.172	2 min,...	0.836	1 min, 01 sec	0 min, 24 sec
0.20	14	0	1.000	16.000	0	0.478	0.055	2 min,...	0.796	0 min, 54 sec	0 min, 27 sec
1.21	14	0	1.000	16.000	0	0.472	0.046	2 min,...	0.787	0 min, 54 sec	0 min, 19 sec
2.22	14	0	1.000	16.000	0	0.511	0.048	2 min,...	0.852	0 min, 54 sec	0 min, 19 sec
4.24	14	0	2.000	30.911	0	0.529	0.141	2 min,...	0.881	0 min, 56 sec	0 min, 13 sec
5.25	14	0	1.000	16.000	0	0.618	0.074	2 min,...	1.029	0 min, 56 sec	0 min, 22 sec
6.26	14	0	1.000	16.000	0	0.525	0.069	2 min,...	0.875	0 min, 55 sec	0 min, 25 sec
Avg	13.375	0.542	1.538	22.817	0.000	0.541	0.161	2 min...	0.901	0 min, 58 sec	0 min, 20 sec

Fig. 5 Node table

Packet Loss Ratio (PLR)

Packet delivery ratio of a network is the ratio between the total number of packets received by a node and the total number of packets sent to that node. [12]

$$\text{PLR} = (\text{Total pkts received /Total packets sent}) * 100 \tag{1}$$

Throughput (T)

We can describe throughput abstractly as the product of the number of packets, the size of the packets, and the integer 8 (for conversion), divided by the total simulation time in seconds.

$$\text{Throughput} = (\text{No. Of delivered packets} * \text{size} * 8) / \text{Total duration of simulation} \tag{2}$$

Delay (D)

It is possible to measure the latency as a distinction between the time the packet was sent from the source and the time it was received at the destination. Thus, by summing the total received time and the total sent time separately and then finding the difference, the total latency can be determined for all packets. The average packet delay can be derived from this.

$$\text{Packet Delay} = \text{Total latency/ Total packet received} \tag{3}$$

These values are then calculated as the trust values for the network. Trust depends upon the ratings of consecutive nodes in the WSN [13]. The trust value of adjacent sensor nodes is calculated in this scheme, and one trust threshold value is fixed. If the trust value of the sensor node is less than the threshold, then the node is considered a Sybil node otherwise normal node [14]. The throughput usually decreases if there is any Sybil attack in the network. It also further affects the packet transmission in the network, as shown in Fig. 6, which means the network delay increases. Therefore, calculating the throughput and delay helps in this scheme.

4 Detecting Sybil Attack Using AODV

AODV (Ad-hoc On-demand Distance Vector) is a routing protocol for ad-hoc networks as well as mobile ad-hoc networks [15]. It is used in an environment where it goes through all types of network behaviors like traffic, link failures, communications, etc. It is an on-demand routing protocol, which means the route is only established when the nodes want to communicate with each other or wants to send requested packets. Although it is said that destination sequenced distance vector) has

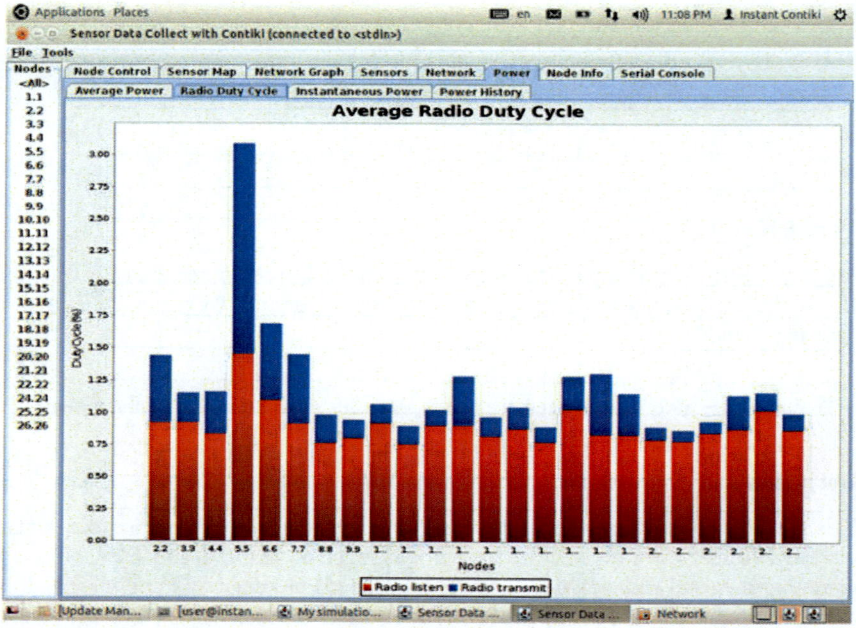

Fig. 6 Radio message transit graph indicating the delay in sending messages

low packet delivery ratio, AODV is normally used for the network communications and the physical altercations does not affect its speed or throughput [16].

These factors are used as vector value for the algorithm. These trust values along with the hop count can be further used in detecting in the route to a destination. We are here integrating the trust values and the AODV algorithm in order to find the path with lower hop count and to detect the Sybil node [17]. If there is any variance or any anomalous outcome in the trust values, then the likelihood of the node being Sybil increases topology. AODV utilizes routing tables to store routing information, as shown in Fig. 7. The route table stores values as vector: < trust values, destination, next, hope count > . The RREP (Route Reply) message is sent only when the reply is to be sent.

In this protocol whenever a node is created, it is likely to keep a list of lists of the nodes which it has to send the packet such nodes are the beginning nodes of the route which later helps in the continuation of the network. These nodes keep a track of the sequence number which is set to be increasing whenever there is a change in the network environment which indicates that there might have been an attack. The nodes then decide upon the route for the correct destination by checking its routing table. The packet is forwarded if an appropriate route is found from the table. If it fails to find one, it initiates a Sybil node locating process.

The neighboring nodes help in identifying the Sybil nodes. For example, let us consider that a Sybil node fakes its identity and masquerade as one of the honest

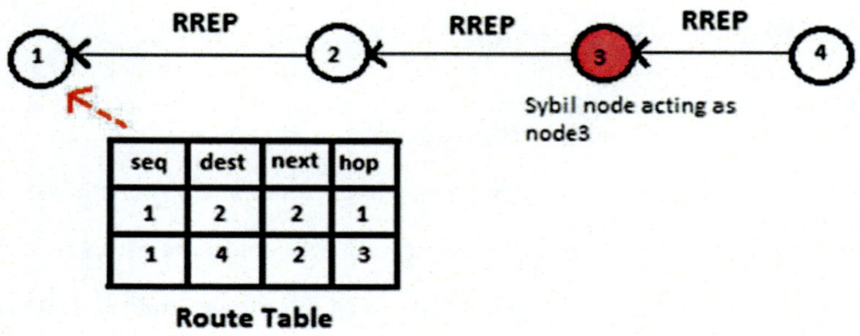

Fig. 7 AODV implementation

nodes which is supposed to get a packet from another node [17]. In AODV, the path established checks for the trust value as well as hop count, according to which it selects a path with no Sybil node. So, by checking the behavior or finding any dubious act in the network, the neighboring node also can help in detecting Sybil attacks. AODV helps the nodes in the network about any potential connection break in the route discovered. Further by removing the Sybil node, the chance of having a delay in the network using the AODV algorithm reduces. Once the RREP message is received to the sender node, the process of routing is terminated [18]. This mechanism using the trust values helps in the secure routing. These measures keep changing and hence has to be updated in the table. The relation between each node must be strong, so that they can trust each other whenever the communication is done and that they could easily identify the Sybil node or any alterations in the network environment.

5 Conclusion and Future Work

We discuss the Sybil attack that can occur on any IoT device in this paper and would be harmful to the application. The primary job is to detect the Sybil node in any network of sensors. We implemented the simulation in a Cooja simulator where the sensor network is created with a number of honest and Sybil nodes. The usage of sink nodes is for implementing the collection of data in WSN. With all the values collected, we can identify the Sybil nodes with the trust values and their behavioral profiling. In addition, the AODV protocol is discussed to detect and remove the Sybil attack with the help of the trust values obtained and the hop count which will be used for the routing path in the network. We observe that our proposed work helps for detecting the Sybil attack without affecting the network throughput or delay. However, the effect of sensor node mobility has not been taken into account in this work, so this work will be extended for few mobile sensor nodes, so that we get to know how in GSM and other mobile networks this scheme can be implemented by adding other features into it. Further, the IoT and cloud can be integrated together for security purposes using blockchain [19].

References

1. U.S. Raj, K. Dhamodharan, R. Vayanaperumal, Detecting and preventing Sybil attacks in wireless sensor networks using message authentication and passing method. Sci. World J. (2015)
2. Cooja Simulator Manual Version 1.0, https://www.napier.ac.uk/~/media/worktribe/output-299955/cooja-simulator-manual.pdf
3. J. Sathish Kumar, D. Patel, A survey on internet of things: security and privacy issues. Int. J. Comput. Appl. (2014)
4. A. Čolaković, M. Hadžialić, Internet of Things (IoT): A Review of Enabling Technologies, Challenges, and Open Research Issues
5. S. Deep, X. Zheng, A. Jolfaei, D. Yu, P. Ostovari, A.K. Bashir, A survey of security and privacy issues in the Internet of Things from the layered context
6. S. Raza, L. Wallgren, Thiemo Voigt: SVELTE: Real-time intrusion detection in the Internet of Things. Raza, S., Wallgren, L., & Voigt, T. , SVELTE: Real-time intrusion detection in the Internet of Things. Ad Hoc Netw. **11**(8), 2661–2674 (2013). https://doi.org/10.1016/j.adhoc.2013.04.014
7. D. Evangelista, F. Mezghani, M. Nogueira, A. Santos, Evaluation of Sybil Attack Detection Approaches in the Internet of Things Content Dissemination. IEEE Xplore/Wireless Days (WD) (2016)
8. S. Murali, A. Jamalipour, A lightweight intrusion detection for Sybil attack under mobile RPL in the Internet of Things. IEEE Internet Things J. **7**(1) (2020)
9. H.-M. Xin, K. Yang, *Routing Protocols Analysis for Internet of Things*. IEEE
10. P. Singhal, P. Sharma, S. Rizvi, Thwarting Sybil Attack by CAM Method in WSN Using Cooja Simulator Framework. Science Publishing Corporation (2018)
11. M. Adithya, P.G. Scholar, B. Shanthini, Security analysis and preserving block-level data DE-duplication in cloud storage services. J. Trends Comput. Sci. Smart Technol. (TCSST) **2**(02), 120–126 (2020)
12. A.S. Joseph Charles, K. Palanisamy, QoS measurement of RPL using Cooja simulator and Wireshark Network Analyzer. Int. J. Comput. Sci. Eng. (2018)
13. R. Singh, J. Singh, R. Singh, TBSD: s defend against Sybil attack in wireless sensor networks. Int. J. Comput. Sci. Network Secur. (2016)
14. D. Kumaria, K. Singha, M. Manjul, Performance evaluation of Sybil attack in cyber physical system, in *International Conference on Computational Intelligence and Data Science (ICCIDS 2019)*
15. P.K. Maurya, G. Sharma, V. Sahu, A. Roberts, M. Srivastava, An overview of AODV routing protocol. Int. J. Mod. Eng. Res. (IJMER)
16. V.P. Patil, Performance evaluation of on demand and table-driven protocol for wireless ad hoc network. Int. J. Comput. Eng. Sci. (IJCES)
17. A. Rajan, J. Jithish, S. Sankaran, Sybil attack in IoT: modelling and defenses, in *IEEE Xplore/ 2017 International Conference on Advances in Computing, Communications and Informatics (ICACCI)*
18. H. Simaremare, A. Abouaissa, R.F. Sari, P. Lorenz, Secure AODV routing protocol based on trust mechanism. Wireless Networks and Security (2013)
19. B. Vivekanadam, Analysis of recent trend and applications in block chain technology. J. ISMAC **2**(04), 200–206 (2020)

Borda Count Versus Majority Voting for Credit Card Fraud Detection

M. Aswathi, Aiswarya Ghosh, and Leena Vishnu Namboothiri

Abstract As financial fraud is increasing day by day, cardholders have been affected seriously by a lot of economic losses. To detect fraud, mostly machine learning algorithms are used. The research proposed in this paper utilizes the European bank transaction dataset, and it is highly imbalanced. Henceforth, three class imbalance techniques—SMOTE, SMOTE + TOMEK, SMOTE + ENN—were used for removing imbalance in the dataset and five machine learning algorithms such as logistic regression, support vector machine, random forest, decision tree, and K-nearest neighbors are applied. Random forest provides better results when compared with other classifiers. Later, by applying two voting methods, viz. Borda Count and majority voting, on each class imbalance technique, the performance is evaluated and compared based on different parameters such as precision, accuracy, F1 score, as well as Matthews correlation. This paper explains the significance of majority voting over Borda Count in detecting fraudulent transactions.

1 Introduction

For years, credit card utilization has increased rapidly, so credit card fraud is rising at a rapid pace. The reason for such illegal transactions might be to get items without giving money. Fraud detection is an application of anomaly detection, which is characterized by a large imbalance between the classes. Also, the transaction patterns often change their statistical properties over time. Machine learning is considered as one of the most successful techniques used for creating a fraud detection algorithm for fraud identification. The researchers need more concentration to decrease financial loss and increase accuracy.

M. Aswathi (✉) · A. Ghosh · L. V. Namboothiri
Department of Computer Science and IT, Amrita School of Arts and Sciences, Amrita Vishwa Vidyapeetham, Kochi, India

L. V. Namboothiri
e-mail: leena@asas.kh.amrita.edu

P. Karuppusamy et al. (eds.), *Ubiquitous Intelligent Systems*, Smart Innovation, Systems and Technologies 243, https://doi.org/10.1007/978-981-16-3675-2_24

The credit card can be used for physical usage or virtual or online usage. Physical usage requires an individual using the credit card to directly pay for purchases in any store. While virtual or online usage is where the owner of the card uses the credit card to pay online over the internet for purchased goods by simply entering the credit card information required. Card account numbers are generally the primary account number written on the card which includes: expiry date, card type, number of the card, verification code, and cardholder's name. Fraud can be committed for various reasons, such as the intent of entertainment, the manipulation of a company or organization, revenge, financial loss, and identity damage. There are also different kinds of fraud, such as bankruptcy fraud and identity theft. The two kinds of credit card fraud are both online and offline fraud. Offline credit card frauds are those where the credit card of a person is lost or stolen. If the data is compromised by an attacker or hacker and used to perform illegal acts, it is referred to as online fraud. With the rapid growth of technology, the use of the internet is growing significantly. This substantially contributes to too many fraudulent credit card transactions.

Machine learning algorithms perform much better when the count of instances of each class is approximately equivalent, i.e., whenever the count of instances of one class vastly outnumbers the other, challenges emerge. In our paper, the dataset we used contains valid cases that are much high compared to fraud cases.

Figure 1 depicts that out of 284,807 transactions, only 492 fraud (approximately 0.2 percent) transactions are there.

This paper aims to evaluate the performance of voting methods [1]—Borda Count as well as majority voting in machine learning algorithms, using various class imbalance techniques [2] to assess which one is most appropriate in detecting fraud.

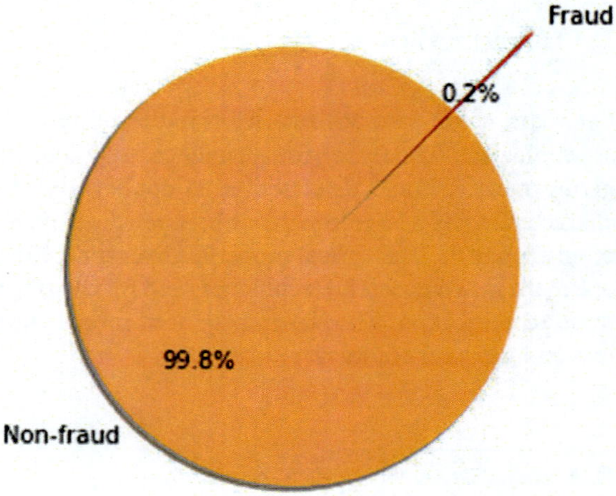

Fig. 1 Illustration of class imbalance

2 Literature Review

Nileena Thomas et al. in their studies have shown that logistic regression and K-nearest neighbors have an accuracy of 0.98 and 0.94, respectively. The results obtained indicate that during the detection of fraud, accurate accuracy with 0.999 is obtained by using the random forest classifier. By modifying the voting process, while combining random forest with Borda Count, the random forest classifier provides much better exact outcomes. [3]. Van Erp et al. explain the Borda Count. It explains that the strategy which works very great on the small multilayer perceptron mix is the Borda Count. It is less difficult. The Borda Count performs great on bigger ensemble sizes, so it consequently turns into a supplement to the product rule and sum rule. The majority voting affecting is low conversely with several voting techniques. In this way, it is desirable overtraining the product rule, sum rule, or the Borda Count rather than plurality voting [4]. Varmedja et al. in their studies for the identification of fraud cases examined several machine learning algorithms and indicates that the random forest classifier performed better by evaluating recall, precision, and accuracy [5]. Suresh K Shirgave et al. reviewed various machine learning algorithms to detect fraud in credit card transactions. They have selected the supervised learning technique, random forest, to classify the alert as fraudulent or authorized. This classifier will be trained using feedback and delayed supervised samples. Next, it will aggregate each probability to detect alerts, and they proposed a learning rank approach, where the alert will be ranked based on priority [6]. Vaishnavi Nath Dornadulaa et al. established a pioneering approach for the detection of credit fraud. Depending on transactions and extracting behavioral trends, these customers are clustered to create an account for each cardholder, and afterward, classification algorithms are introduced to three categories and subsequent score ratings are established for each classifier category. For handling the imbalance, they tried SMOTE and the classifiers used by them are random forest, logistic regression, and decision tree and found that the Matthews correlation coefficient is the best one [7].

Sain et al. have used class imbalance methods such as SMOTE, Tomek links, and combine sampling where the classifier used is the support vector machine. With the help of F-measure and roc, they concluded that combine sampling is accurate compared to the other two [8]. Maniraj et al. explain how to apply machine learning to achieve better fraud detection results along with the algorithm, pseudo code, description, and observations from experiments. Based on the machine learning algorithms, performance is increased even when more data is put in it over time. This large percentage of accuracy is predicted by the vast difference between successful and real transactions [9]. Lakshmi S V S et al. in their paper used machine learning algorithms such as decision tree, random forest, and logistic regression to identify fraud and non-fraud transactions. Using accuracy, specificity, sensitivity, and error rate, the efficiency of the model is validated, and random forest performed well with 95% [10]. Randhawa et al. in their study provides a technique in Machine Learning for Credit Card Fraud Detection. Standard models were first used and later hybrid classics appeared, using AdaBoost and majority voting methods. The publicly accessible datasets were used to survey the viability of the model and included datasets used by the financial sector. Numerous voting techniques achieved a decent score of 0.942 including noise at 30% [11].

Ali et al. in their study provide an outline about class imbalance and the indispensable consequences that arise and also explain the key factors that impede the efficiency of the classifier while handling dataset imbalance [12]. Raj et al. in their study show that applying support vector machine optimization to recurrent neural networks will overcome the nonlinear regression estimation problem. With the help of three dynamic optimization algorithms such as the differential evolution algorithm, gravitational search algorithm, and artificial bee colony algorithm, a hybrid prediction model is constructed. And the model will enhance the accuracy and pace in determining the optimal values of support vector machine parameters [13]. Mohammed et al. in their study, two techniques are introduced to fix the issue of class imbalance using oversampling and undersampling techniques, and performance metrics are evaluated by applying them to machine learning models. The result shows that oversampling obtained better scores with respect to undersampling for various classifiers [14]. Zorkeflee et al. used a hybrid technique of undersampling (FDUS) and oversampling(SMOTE) to deal with imbalanced datasets, where FDUS is a fuzzy logic applied undersampling technique, and they concluded that FDUS + SMOTE performs better than standalone techniques with the help of performance metrics like G-mean and F-measure [15]. Awoyemi et al.: it is a comparative study of imbalanced credit card fraud data by using three machine learning classifiers such as K-nearest neighbor, logistic regression, and naïve Bayes. The extremely imbalanced dataset are sampled using a hybrid approach and the performance is evaluated [16]. Chandy et al. explain that while the need for the workload is linked to the allocation of resources, the suggested method forecasts the workload by utilizing a random forest algorithm. And a genetic algorithm is assigned to allocate the resources. The findings obtained show that the resource used in the proposed approach are accuracy observed while prediction and system-level features [17].

3 Proposed System/Material and Methods

Figure 2 illustrates the proposed system that uses class imbalance techniques with distinct machine learning algorithms like decision tree, K-nearest neighbor, random forest, logistic regression, and support vector machine. The first step is the collection of data. After the data has been collected, the preprocessing stage is done. Here, the data contains columns starting with 'V' that is obtained after principal component analysis. The column 'Time' is the time elapsed between each transaction, and this field is not significant to determine whether a transaction is genuine or not. Hence, dropping the 'Time' column. The 'Amount' column needs to be standardized since all other columns are obtained after principal component analysis. After preprocessing, the resulted data is split into train and test data where train data contains 70% of the

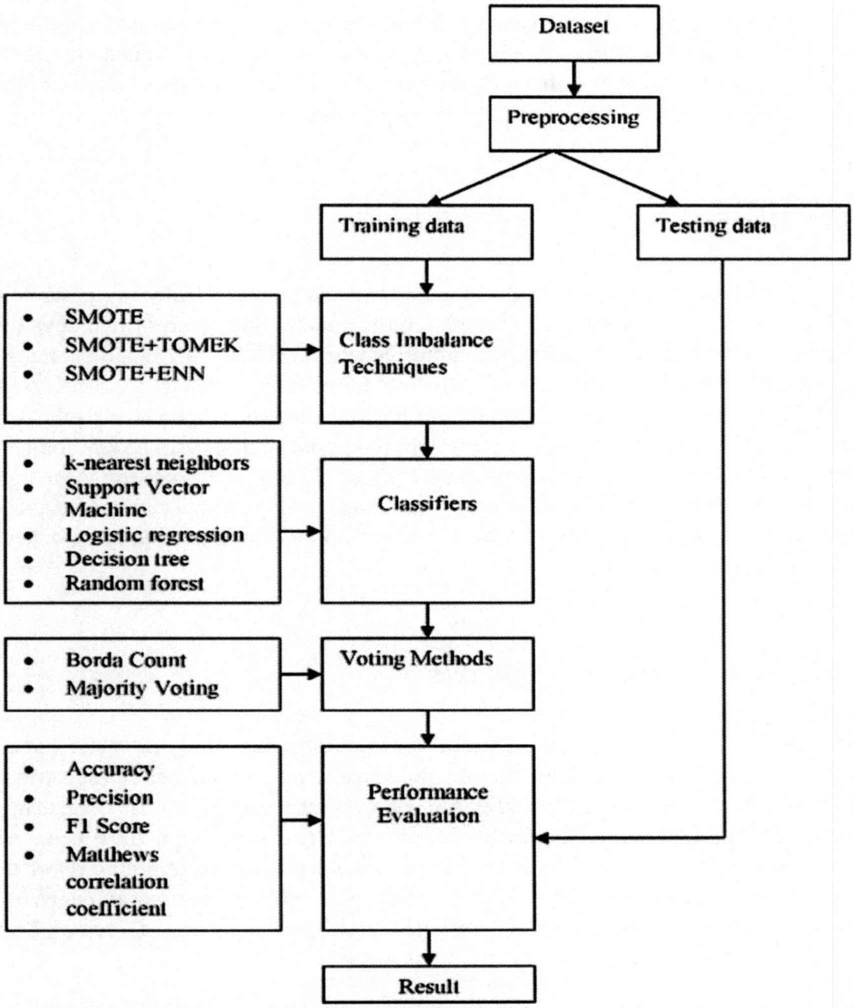

Fig. 2 Illustration of proposed work

original dataset. If the dataset is imbalanced, the model will be considered as either overfitting or underfitting. It is often faced when we do classification. To make a model ideal, we have to have a balanced dataset to obtain higher accuracy. Here, the dataset is highly imbalanced, and to handle this, class imbalance techniques are used. The different class imbalance techniques like SMOTE [18] and hybrid techniques such as SMOTE + TOMEK [19], SMOTE + ENN are applied to the train set, and then they are classified with the help of machine learning algorithms such as decision tree, support vector machine, K-nearest neighbor, logistic regression, and random forest.

Later, voting methods—Borda Count and majority voting—are applied to determine whether the credit card is defaulted or not. At last, a comparison is made between Borda count and majority voting with the help of certain performance measures like accuracy, precision, F1 Score, and Matthews correlation.

3.1 Dataset

The dataset[20] used here can be downloaded from Kaggle, and it covers transaction information in September 2013 through credit cards that occurred in two days by European cardholders containing 31 features, in which we have 492 fraudulent transactions out of 284,807 transactions. The dataset is extremely unbalanced, accounting for 0.172 percent of all transactions with the positive class (fraud). Input variables are numeric due to the transformation of principal component analysis. The key components derived with principal component analysis are V1 to V28, 'Time' and 'Amount' are the only attributes that have not been transformed with principal component analysis. The feature 'Class' takes only binary values: value 1 indicates fraud transaction and 0 otherwise.

3.2 Class Imbalance Techniques

An issue with machine learning is that the total instances of a data class (positive) are much less than that of the total instances of another data class (negative). The main approach for imbalanced classification is resampling which is done using oversampling, undersampling, and hybrid of both. Undersampling is the process of eliminating some of the instances of majority class whereas oversampling refers to copying some of the instances of the minority class. Firstly, we used an oversampling (Smote) technique, along with two hybrid techniques that are—Oversampling + Undersampling (SMOTE + Tomek, Smote + ENN).

SMOTE. One of the main problems with imbalanced classification is that the number of samples in the minority class is very less to make the judgment effectively. To address this issue, minority class examples can be subjected to oversampling. The oversampling method is performed by copying the entries in the minority class of the training set before fitting any model. It only makes the imbalanced data into the balanced form and no other extra knowledge is provided. This method is to create a novel minority class.

SMOTE + Tomek. SMOTE is a type of oversampling that instantiates possible new instances. A procedure for defining sets of immediate neighbors is referred to as Tomek Links. SMOTE and Tomek Link is a merging of both oversampling and undersampling process, where SMOTE method will be first used to over-sample the minority classes toward a stable allocation, so it defines and extracts examples in Tomek Links from the majority classes.

SMOTE + ENN. SMOTE–ENN is a preprocessing algorithm that, by resampling the data space, re-balances class distribution. This approach incorporates the under-sampling of majority classes with an oversampling of the minority classes to resolve the shortcomings associated with implementing each of them separately.

3.3 Voting Methods

Every classification model must make a smooth judgment that means a vote that is added and evaluated to turn up at a tough choice on input for each voting algorithms. The key benefit of a voting system is that it allows the combination of a broad range of classifiers without much intellect about underlying methodologies of classification model processing. Here, we have used two voting methods that are Borda Count and majority voting.

Borda Count. To all of the alternative solutions sorted based on the preferences, the voters have to determine a ranking position, and based on this, every alternative gets credit for every vote. It covers the total ranking of each voter to assess the result.

Majority Voting. The candidate, who earned the majority which is greater than 50% of the votes, and a candidate has a majority of first-choice votes, then that candidate wins the election. When no majority candidate is available, no outcome is created, then the majority criterion does not apply. The actual benefits of this voting method are its low mistake count and no difficulty. A majority vote-based classification algorithm is used for improving performance.

4 Experimental Results and Analysis

AUC Graph. A ROC curve is utilized to assess the accuracy of a classification prediction. The bigger the zone underneath the ROC curve, the higher the accuracy is. If it is increasingly centered on the accuracy, we tried some algorithms for taking care of the issue.

Here, in Fig. 3, the more area beneath the ROC curve is for random forest, i.e., AUC—98.48, other classifiers values are K-nearest neighbor—93.21%, support vector machine—97.38%, decision tree–88.53%, and logistic regression—97.70%.

Here in Fig. 4, the more area beneath the ROC curve is for random forest, i.e., AUC—98.13, other classifiers values are support vector machine——7.38%, logistic regression–97.71%, decision tree—88.87%, and K-nearest neighbor—93.21%.

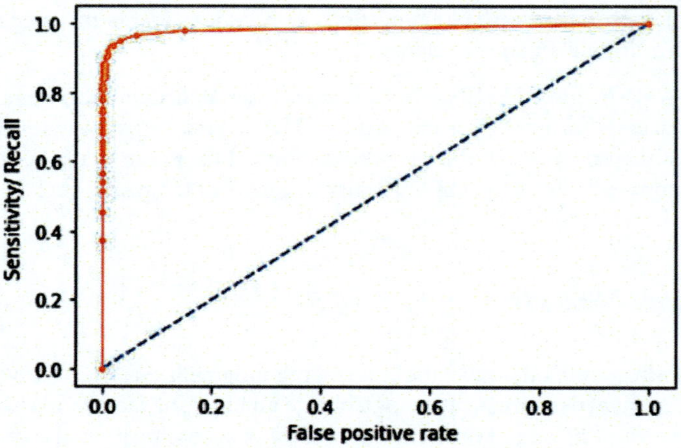

Fig. 3 ROC-Random Forest of SMOTE

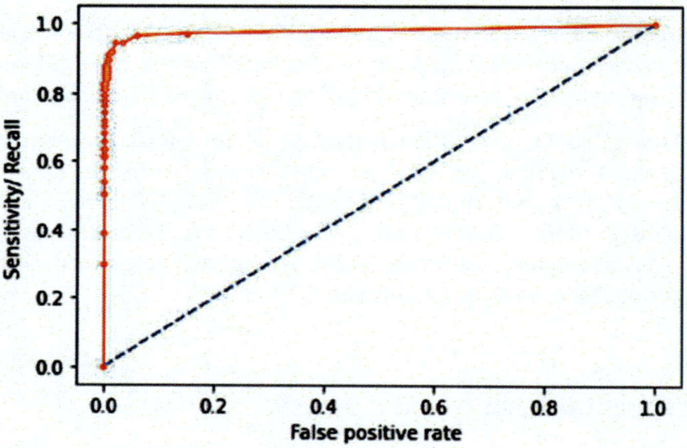

Fig. 4 ROC-Random Forest of SMOTE + TOMEK

Here in Fig. 5, the more area beneath the ROC curve is for random forest, i.e., AUC—97.96%, other classifiers values are K-nearest neighbor–93.54%, logistic regression—97.15%, support vector machine—97.38%, and decision tree—88.17%.

This research work attempts to balance the dataset by using SMOTE, SMOTE + TOMEK, and SMOTE + ENN techniques and found from these experiments that, relative to other classification algorithms, random forest provides better results.

Figures 6 and 7 depict the performance analysis of Borda Count and majority voting. Comparing the results shown above, majority voting produced slightly better results than Borda count by applying SMOTE, SMOTE + TOMEK, and SMOTE

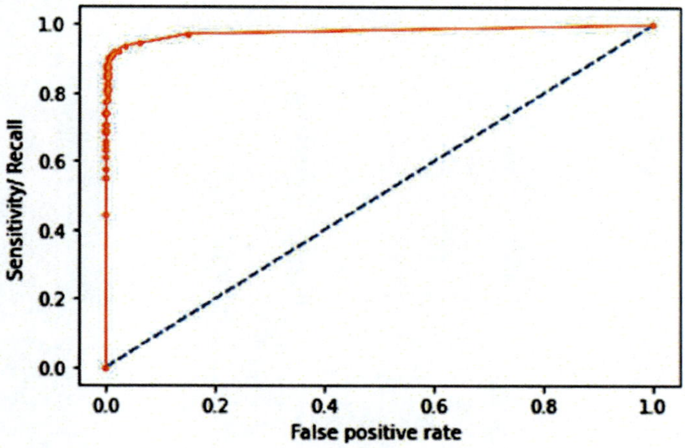

Fig. 5 ROC-Random Forest of SMOTE + ENN

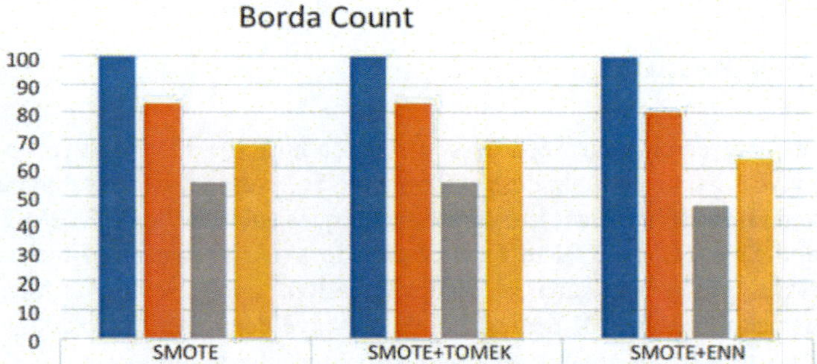

	SMOTE	SMOTE+TOMEK	SMOTE+ENN
Accuracy	99.859555493	99.859555493	99.815081399
F1 Score	83.298148083	83.390228702	80.203661884
Precision	54.794520548	54.751131222	46.718146718
Matthews Correlation	68.226899777	68.484097769	63.242208356

Fig. 6 Performance analysis of Borda Count

+ ENN techniques. As we can see here, SMOTE and SMOTE + TOMEK have produced approximately similar results compared to SMOTE + ENN.

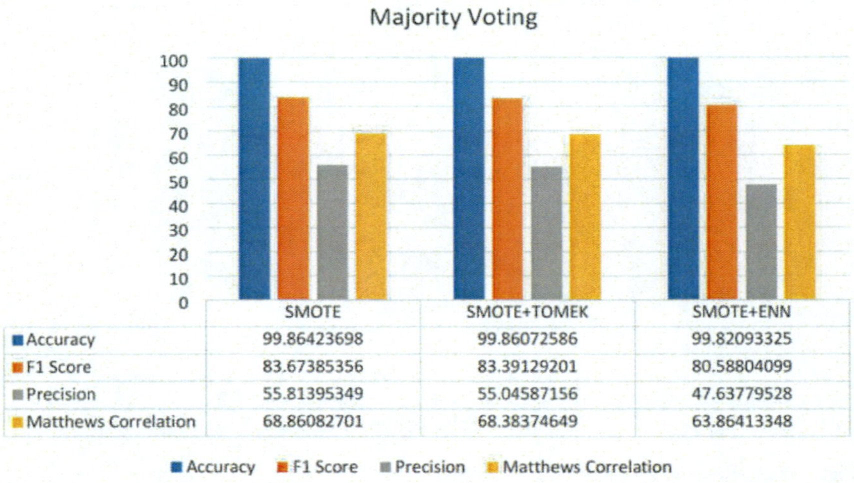

	SMOTE	SMOTE+TOMEK	SMOTE+ENN
Accuracy	99.86423698	99.86072586	99.82093325
F1 Score	83.67385356	83.39129201	80.58804099
Precision	55.81395349	55.04587156	47.63779528
Matthews Correlation	68.86082701	68.38374649	63.86413348

Fig. 7 Performance analysis of Majority Voting

5 Conclusion

In our research, data preprocessing is essential as the distribution proportion of classes plays a vital role in model performance. So, preprocessing is done in primary stage and later applied class imbalance techniques such as SMOTE, SMOTE + TOMEK, and SMOTE + ENN to balance data to avoid skewness of the dataset. Machine learning classifiers such as logistic regression, support vector machine, random forest, decision tree, and K-nearest neighbors have been used for classification purposes. And to achieve greater accuracy and determine which voting method gives better results, voting methodologies like majority voting and Borda Count are used. Borda count can violate both the majority and Condorcet criterion, whereas the downside of majority voting is that if there is no majority candidate, the outcome will be null, and thereby sample is eliminated. Later on, the comparison was made and concluded that majority voting performs well compared to Borda Count. And also found that applying SMOTE and SMOTE + TOMEK techniques produced better results than applying SMOTE + ENN technique for credit card fraud detection. The SMOTE approach has the advantage of avoiding overfitting problems and SMOTE + TOMEK is a remedy to solve the limitations of SMOTE. It is observed that Matthews correlation coefficient has been increased for Borda Count when SMOTE + TOMEK technique is applied.

6 Discussion

A skewed dataset refers to a dataset where the total instances of a data class are much less than that of the total instances of another data class. The drawback of skewness in the dataset is always unpredictable. Hence, it is necessary for future researchers to focus more on the skewness of the dataset and should introduce more class imbalance techniques.

Acknowledgements Sincere gratitude and thanks to Dr. E.R Vimina, HOD of Computer Science and IT, for her leadership and guidance throughout our research. To Leena Vishnu Namboothiri, our guide, for her valuable guidance, advice, and support. Moreover, for all of our friends for their cooperation and moral support.

References

1. F. Leon, S.-A. Floria, C. Bădică, Evaluating the effect of voting methods on ensemble-based classification., in *2017 IEEE International Conference on INnovations in Intelligent SysTems and Applications (INISTA)* (IEEE, 2017)
2. A. Gosain, S. Sardana, Handling class imbalance problem using oversampling techniques: a review, in *2017 international conference on advances in computing, communications and informatics (ICACCI)* (IEEE, 2017)
3. N. Thomas, J. Jayalakshmi, E.S. Sreelakshmi, L.V. Namboothiri, Implementation of Random Forest and proposal of Borda Count in credit card fraud detection. Int. J. Emerg. Technol. **11**(2), 536–540 (2020)
4. M. Van Erp, L. Vuurpijl, L. Schomaker, An overview and comparison of voting methods for pattern recognition, in *Proceedings Eighth International Workshop on Frontiers in Handwriting Recognition* (IEEE, 2002)
5. D. Varmedja, et al., Credit card fraud detection-machine learning methods, in *2019 18th International Symposium INFOTEH-JAHORINA (INFOTEH)* (IEEE, 2019)
6. S. Shirgave et al., A Review On Credit Card Fraud Detection Using Machine Learning. Int. J. Sci. Technol. Res **8**, 1217–1220 (2019)
7. V.N. Dornadula, S. Geetha, Credit card fraud detection using machine learning algorithms. Procedia Comput. Sci. **165**, 631–641 (2019)
8. H. Sain, S.W. Purnami, Combine sampling support vector machine for imbalanced data classification. Procedia Comput. Sci. **72**, 59–66 (2015)
9. S. Maniraj, et al., Credit card fraud detection using machine learning and data science. Int. J. Eng. Res. 8.09 (2019)
10. S.V.S.S. Lakshmi, S.D. Kavilla, Machine learning for credit card fraud detection system. Int. J. Appl. Eng. Res. **13**(24 Pt. 1), 16819–16824 (2018)
11. K. Randhawa, et al., Credit card fraud detection using AdaBoost and majority voting. IEEE Access **6**, 14277–14284 (2018)
12. A. Ali, S.M. Shamsuddin, A.L. Ralescu, Classification with class imbalance problem. Int. J. Adv. Soft Comput. Appl. **5**(3) (2013)
13. J.S. Raj, J. Vijitha Ananthi, Recurrent neural networks and nonlinear prediction in support vector machines. J. Soft Comput. Paradigm (JSCP) **1**(01), 33–40 (2019)
14. R. Mohammed, J. Rawashdeh, M. Abdullah, Machine learning with oversampling and undersampling techniques: overview study and experimental results, in *2020 11th International Conference on Information and Communication Systems (ICICS)* (IEEE, 2020)

15. M. Zorkeflee, A. Mohamed Din, K.R. Ku-Mahamud, Fuzzy and smote resampling technique for imbalanced data sets, pp. 638–643 (2015)
16. J.O. Awoyemi, A.O. Adetunmbi, S.A. Oluwadare, Credit card fraud detection using machine learning techniques: a comparative analysis, in *2017 International Conference on Computing Networking and Informatics (ICCNI)* (IEEE, 2017)
17. A. Chandy, Smart resource usage prediction using cloud computing for massive data processing systems. J Inf Technol **1**(02), 108–118 (2019)
18. N.V. Chawla, et al., SMOTE: synthetic minority over-sampling technique. J. Artif. Intell. Res. **16**, 321–357 (2002)
19. M. Zeng, et al., Effective prediction of three common diseases by combining SMOTE with Tomek links technique for imbalanced medical data, in *2016 IEEE International Conference of Online Analysis and Computing Science (ICOACS)* (IEEE, 2016)
20. https://www.kaggle.com/mlg-ulb/creditcardfraud

Comparative Study of Multiple Feature Descriptors for Detecting the Presence of Alzheimer's Disease

Ben Nicholas, Akhil Jayakumar, Basil Titus, and T. Remya Nair

Abstract Medical image processing has a very important role in medical diagnosis where a doctor can compare the scanned image of his patient with a heap of images and find the result of the image that matches with it. With the help of feature descriptors, we can make the process of image classification much more efficient. By implementing various feature descriptors, we are able to identify Alzheimer's at the very early stages which helps the entire curing process faster. This paper presents the comparison of various binary descriptors such as local binary pattern (LBP), local wavelet pattern (LWP), histogram-oriented gradients (HOG), local bit plane decoded pattern (LBDP) along with K-nearest neighbour (KNN) for its classification. The results indicate that the combination of LBP and KNN together produce a better accuracy of 91.21% in "Alzheimer's Dataset" (Alzheimer's Dataset (4 class of Images) https://www.kaggle.com/tourist55/alzheimers-dataset-4-class-of-images [1]) when compared to other descriptors.

1 Introduction

In the medical field, images play a vital role in the detection of diseases and management of the same. As technology grows, different types of images are generated with the help of different image capturing mechanisms in order to diagnose the disease accurately. Even after all these inventions, the doctors are still struggling to properly diagnose the diseases as the number of medical images are increasing day by day and thus it is necessary to have more precise and effective image retrieving mechanisms. In order to properly classify the images, we use some feature extractors which extract information from each image, and then, these features of that image are compared with the database images. Depending on these extracted features the performance of the CBIR system may vary drastically. In this paper, we use different sets of descriptors for the extraction of features from the images and create a comparative study to show how these descriptors perform in this scenario. Four feature descriptors like

B. Nicholas (✉) · A. Jayakumar · B. Titus · T. Remya Nair
Department of Computer Science and IT, Amrita School of Arts and Sciences, Amrita Vishwa Vidyapeetham, Kochi, India

P. Karuppusamy et al. (eds.), *Ubiquitous Intelligent Systems*, Smart Innovation, Systems and Technologies 243, https://doi.org/10.1007/978-981-16-3675-2_25

local wavelet pattern (LWP), local binary code (LBP), local bit plane decoded pattern (LBDP) and histogram-oriented gradients (HOG) are used for extracting features from the images, and we use K-nearest neighbour (K-NN) for the classification of images. Then, their performance is measured on the basis of some matrices.

2 Literature Review

For the conservation of cognitive scope and for keeping public health in society, Alzheimer's diagnosis is playing an important role. During the later stages of a person's life, he or she comes across damage in the nerve cells, but when this process continues uncontrollably, they will have difficulty in doing basic human intellectual works [2]. A person who is diagnosed with AD will show symptoms such as poor memory and sometimes even forgets the language they speak [3]. Early detection of the disease will help the patients to recover from it and will be able to retrieve their cognitive functions [4]. MRI is one of the most popular machines that is used to recognize signs of the disease, and it is a time-consuming methodology mainly because it is having a manually reviewing mechanism and this slows down the whole process [5]. Malloy et al. reviewed the available methods for the diagnosis of Alzheimer's in their paper [6] and states that with the help of computers many cases were able to detect at an early stage itself. In [7] Rabeh et al. proposed a system that can be used to diagnose the presence of dementia in an early stage. They designed the system using support vector machines for classification and was able to acquire an accuracy of 90.66%. Whereas in [8], A. Khan et al. conducted a study where they analysed the performance of many machine learning algorithms such as AR-Mining, SVM, etc., in which linear SVM outperformed every other method and in [9]. Deepika Bansal et al. conducted the analysis of random forest, naïve Bayes, j48 in which j48 came up top. Patil et al. in their paper [10] demonstrated the application of K means and some customized algorithms for classification. In their study, they found that when using KNN along with Shearlet transformation produced a greater classification accuracy. In [11], Kim et al. have pointed out the importance of feature extraction in classification to produce a more accurate result. He recommended the use of HOG feature extraction in his paper. Nisha et al. in their paper [12] compared how feature extractors like HOG and SURF performed in the detection of Alzheimer's. They found out that better results were achieved when both HOG and SURF were used together. In [13], A. Francis et al. analysed the performance of local descriptors in this scenario and found out that local descriptors such as LBP performed better than most of the global descriptors. In the paper [14] by Shiv Ram Dubey, he recommended an LWP by feature descriptor for medical CT image retrieval. Based on his studies, he came to the conclusion that the feature descriptor he proposed had a better performance compared to the already existing feature descriptor. In the paper [15] by Shiv Ram Dubey, for biomedical image indexing as well as retrieval, an LBDP-based feature description is recommended. He performed three experiments on biomedical image retrieval for investigating the power and productivity of it, and he found out that it outperforms the state-of-the-art feature descriptors, and also the retrieval time is reduced drastically with an enhanced performance.

3 Proposed System

In this section, we examine the working of the four feature descriptors LBP, HOG, LDBP and LWP which we use for the extraction of features from the MRI scan image set, and the use of K-nearest neighbour classification method for classification of the images into the respective classes. For this experiment we use "Alzheimer's Dataset", it consists of 6400 images in total, and in which 5888 images are used for training and the remaining 512 images are used for the purpose of testing. The images belong to four different classes such as non-demented, very mild demented, mild demented and moderate demented as shown in Fig. 1. The number of samples in each set are shown in Table 1. The whole process is summarized in Fig. 2.

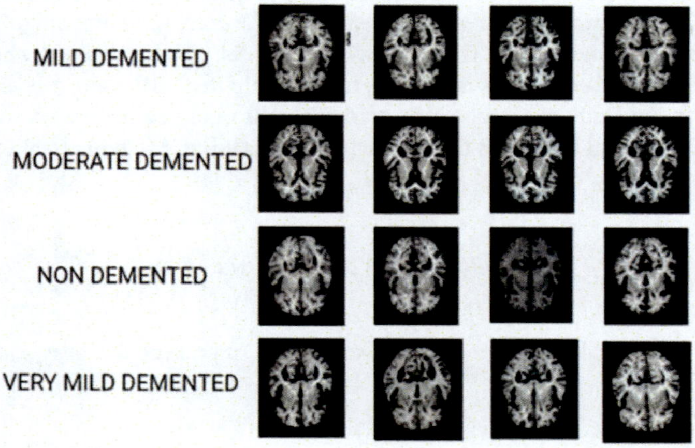

Fig. 1 Image categories

Table 1 Number of samples in each class

	Non-demented	Very mildly demented	Mildly demented	Moderately demented
Training set	2944	2061	824	60
Test set	256	179	72	4

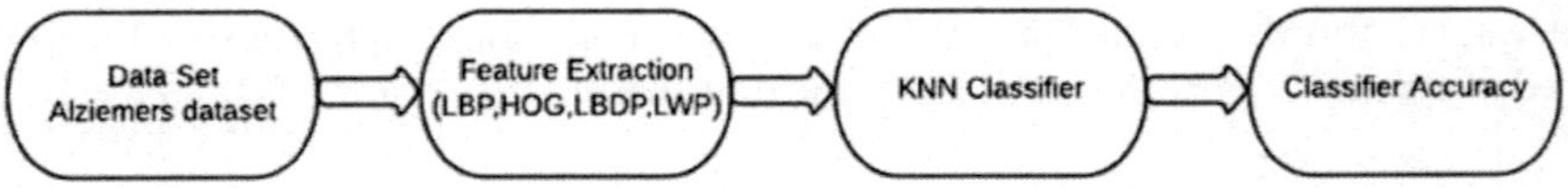

Fig. 2 Process summary

3.1 Feature Descriptors

It encodes some information obtained from the image into a series of numbers (which is also known as feature vectors) and it will act as a unique identity for that corresponding image with which we can distinguish it from other images. Even if we perform some transformation on the image, the feature vector remains the same.

Local binary pattern

For a pixel, the LBP code is calculated in accordance with its neighbours. In order to get the LBP code, we compare the intensities of a pixel's eight neighbours with the intensity of the centre pixel. We denote the value of neighbours as 0 if the intensity of the pixel in the centre has a greater intensity than the corresponding neighbour pixel otherwise we denote it as 1. Then, we will have a set of binary codes, and we will convert this binary code to its decimal form to obtain the LBP code, it is also shown in Fig. 2. In our study, we built a histogram from the obtained feature vector and this resultant histogram is used as the metrics for the training. Mathematical expression of LBP is given as:

$$\mathrm{LBP}_{P,R} = \sum_{P0}^{P1} S(g_P - g_c)2^P \text{ where } s(x) = \begin{cases} 1, x \geq 0 \\ 0, x < 0 \end{cases}$$

where g_c is the value of the centre pixel and g_p represent the grey value of its neighbours.

Histogram Oriented Gradients (HOG)

This descriptor mainly focuses on the shape of the object or its structure. The whole image is broken down into little regions and for every one of them, the gradients and orientation are calculated. In order to calculate HOG, the first step is to preprocess the data. Here we have to convert the size of the image to a ratio of 1:2 (width × height). Then the next step is to find the gradient values of each pixel. To calculate the gradient in y-direction, we have to subtract the pixel value which is on the left side with the pixel on the right side. Likewise, we have to subtract the value of the pixel lies below from the value above the pixel. With these calculated gradients we will now calculate the magnitude and orientation for each pixel. The equation to calculate magnitude and or is given below.

$$\text{Total Gradient Magnitude} = \sqrt{\left[(G_x)^2 + \left(G_y\right)^2\right]}$$

$$\text{Orientation is given as, } \Phi = a\tan\left(G_y \div G_x\right)$$

With the calculated magnitude and orientation now, we are able to calculate the histogram for the given image.

Local wavelet pattern (LWP)

LWP finds its value by finding a relationship between a pixel and its local neighbours as well as finding a relationship between its local neighbours. Local wavelet decomposition process generates a binary pattern, by comparing the values of the local neighbours and the transformed centre pixel. LWP begins by performing the Local Neighbourhood extraction followed by performing the centre pixel transformation and the Local Wavelet Decomposition. The output of the process is a Local wavelet pattern (LWP) along with a LWP feature vector (Fig. 3). It is calculated as:

$$\text{LWP}^{i.j.l}_{R,N} = \left[\text{LWP}^{i.j.l}_{R,N,1}, \text{LWP}^{i.j.l}_{R,N,2}, \ldots, \text{LWP}^{i.j.l}_{R,N,t}\right]$$

where $\text{LWP}^{i.j.l}_{R,N,t} = \sin\left(\Delta^{i,j,l}_{R,N,t}\right)$.

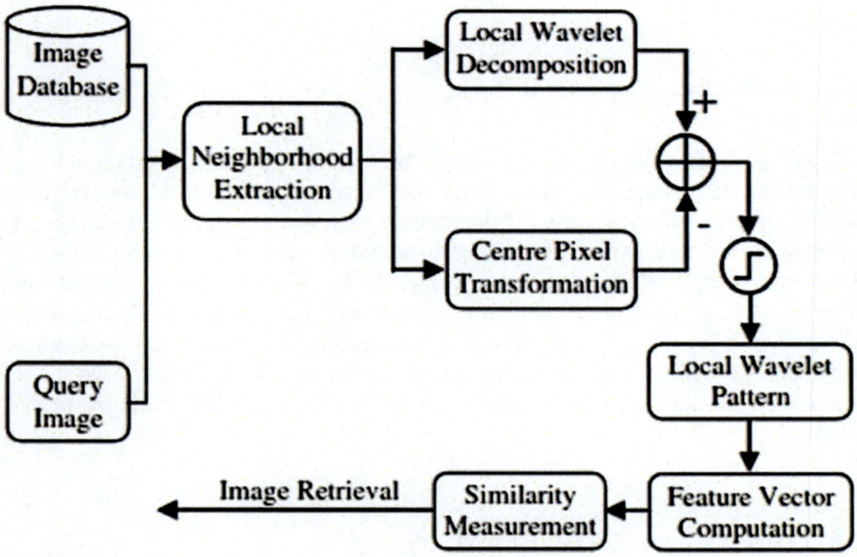

Fig. 3 LWP process summary

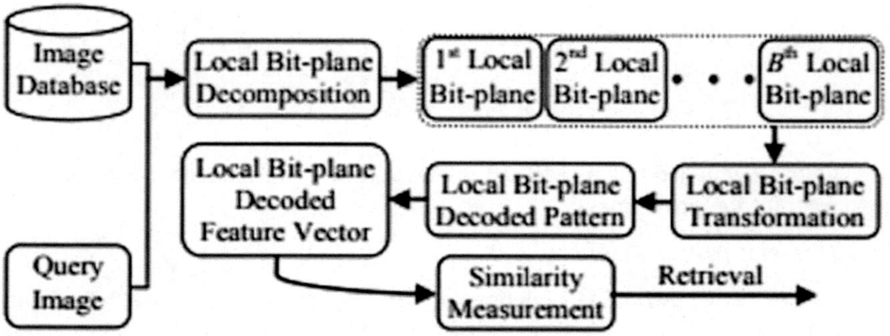

Fig. 4 LBDP process summary

Local Bit-plane Decoded Pattern (LBDP)

LBDP is generated by finding a binary pattern using the difference of the local bit-plane transformed values with the centre pixel's intensity value. For the calculation of LBDP, first, we have to decompose the values of the neighbouring pixels into bit-planes. Then it takes the local information in each bit-plane independently and makes some local bit transformation and with the obtained result it creates the LBDP binary code. Then these binary codes are converted into histogram to produce the feature vector of the input image (Fig. 4).

3.2 *KNN Classifier*

It is an algorithm, which stores all the available classes and classifies an input image into its respective class based on some similarity measure. The algorithm relies heavily on distance measure so if the feature comes in very different scales, it is better to normalize the feature otherwise its output may also vary. In order to classify the images using KNN the first step is to measure the distance between the querying image and the current image from the data set and then we will add the obtained distance with the index of the class to an ordered set. Then, based on the distance and index we will sort it in ascending order. Lastly, we select the first K number of elements from this sorted set and fetch its labels, then return the mode of these K labels.

3.3 *Performance Metrics for the Classifier*

The performance of the classification algorithm is measured based on some metrics such as accuracy, precision, recall and $F1$ Score. In order to calculate all of these, the first step is to generate its confusion matrix and then obtain true positives, true negatives false positives, and false negatives variables. Here, since this is a multi-class classification problem, we have to find the result of all those matrices for each and every class. Then, the average of those individual results is taken as the final result. The equation for the calculation of accuracy, precision, recall and $F1$ score are given below;

$$\text{Classification Accuracy (\%)} = \frac{\text{Number of correct classification}}{\text{Total number of samples}} \times 100\%$$

$$\text{Recall (class)} = \frac{\text{TP(class)}}{\text{TP(class)} - \text{FN(class)}}$$

$$\text{Precision(class)} = \frac{\text{TP(class)}}{\text{TP(class)} + \text{FP(class)}}$$

$$F1\ \text{Score(class)} = \frac{2\text{TP(class)}}{2\text{TP(class)} + \text{FN(class)} + \text{FP(class)}}$$

4 Experimental Analysis and Results

We designed and implemented our algorithms in MATLAB using the "Alzheimer's" [1] database. The performance of our classification algorithm was also assessed with the help of MATLAB. The whole data set was given as input and then 92% of the total data was given for training and the rest 8% is used for validation. Then using multiple feature descriptors like LBP, HOG, LBDP, and LWP features were extracted from both the training and validation set. Then using KNN classification we classified the images in the test set to their corresponding classes. Figure 5 shows the resultant confusion matrix after implementing our algorithm. Table 2 shows how all these descriptors have helped in the performance of KNN Classification.

5 Conclusion

Alzheimer's is a major health concern and is very important to detect them in early stages. Image retrieval plays a great role in the early diagnosis of the disease. In this paper, we compared 4 feature extractors (LBP, HOG, LBDP, LWP) with KNN classifiers and we came to find that the combination of KNN with LBP gives the maximum accuracy of 91.21%, which is greater than the accuracy obtained when compared against other feature descriptors described in this paper.

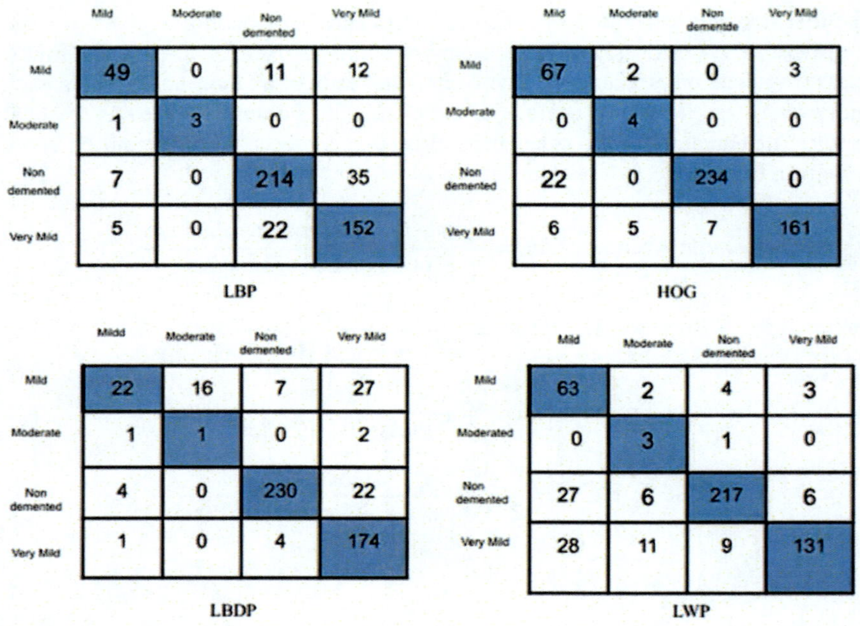

Fig. 5 Confusion matrix obtained after classification

Table 2 Performance comparison

Descriptor	Precision	Recall	$F1$ Score	Accuracy
LBP	74.1	85.32	78.30	91.21
HOG	93.58	75.4	80.27	86.73
LBDP	60.0	63.9	58.02	83.56
LWP	61.02	63.4	65.13	81.00

References

1. Alzheimer's Dataset (4 class of Images) https://www.kaggle.com/tourist55/alzheimers-dataset-4-class-of-images
2. N.A. Mathew, R. Vivek, P. Anuranjan, Early diagnosis of Alzheimer's disease from MRI images using PNN, in *2018 International CET Conference on Control, Communication, and Computing (IC4)*, pp. 161–164 (2018)
3. L. Yue et al., Auto-detection of alzheimer's disease using deep convolutional neural networks, in *2018 14th International Conference on Natural Computation, Fuzzy Systems and Knowledge Discovery (ICNC-FSKD)*, Huangshan, China, pp. 228–234 (2018). https://doi.org/10.1109/FSKD.2018.8687207
4. T. Warnita, N. Inoue, K. Shinoda, Detecting Alzheimer's Disease Using Gated Convolutional Neural Network from Audio Data, pp. 1706–1710. https://doi.org/10.21437/Interspeech.2018-1713

5. R. Varatharajan, G. Manogaran, M. Priyan, R. Sundarasekar, Wearable sensor devices for early detection of Alzheimer disease using dynamic time warping algorithm. Clust. Comput. **21**(1), 681–690 (2017)
6. P. Malloy, S. Correia, G. Stebbins, D.H. Laidlaw, Neuroimaging of white matter in aging and dementia. Clin. Neuropsychol. **21**(1), 73–109 (2007)
7. A.B. Rabeh, F. Benzarti, H. Amiri, Diagnosis of Alzheimer diseases in early step using SVM (support vector machine), in *2016 13th International Conference on Computer Graphics, Imaging and Visualization (CGiV)*, Beni Mellal, pp. 364–367 (2016). https://doi.org/10.1109/CGiV.2016.76
8. A. Khan, M. Usman, Early diagnosis of Alzheimer's disease using machine learning techniques: a review paper, in *2015 7th International Joint Conference on Knowledge Discovery, Knowledge Engineering and Knowledge Management (IC3K)*, Lisbon, pp. 380–387 (2015)
9. D. Bansal, R. Chhikara, K. Khanna, P. Gupta, Comparative analysis of various machine learning algorithms for detecting dementia. Procedia Comput. Sci. **132**, 1497–1502 (2018)
10. C. Patil et al., Using image processing on MRI scans, in *2015 IEEE International Conference on Signal Processing, Informatics, Communication and Energy Systems (SPICES)*, Kozhikode, pp. 1–5 (2015). https://doi.org/10.1109/SPICES.2015.7091517
11. S. Kim, K. Cho, Fast calculation of histogram of oriented gradient feature by removing redundancy in overlapping block. J. Inf. Sci. Eng. **30**, 1719–1731 (2014)
12. S. Nisha, S.A. Nisha, A study on surf and hog descriptors for Alzheimer's disease detection. Int. Res. J. Eng. Technol. **4** (2017)
13. A. Francis, I. Alex Pandian, Review on local feature descriptors for early detection of Alzheimer's disease, in *2018 International Conference on Circuits and Systems in Digital Enterprise Technology (ICCSDET)*, Kottayam, India, pp. 1–5 (2018). https://doi.org/10.1109/ICCSDET.2018.8821115
14. S.R. Dubey, S.K. Singh, R.K. Singh, Local wavelet pattern: a new feature descriptor for image retrieval in medical CT Databases. IEEE Trans. Image Process. **24**(12), 5892–5903 (2015). https://doi.org/10.1109/TIP.2015.2493446
15. S.R. Dubey, S.K. Singh, R.K. Singh, Local bit-plane decoded pattern: a novel feature descriptor for biomedical image retrieval. IEEE J. Biomed. Health Inform. **20**(4), 1139–1147 (2016). https://doi.org/10.1109/JBHI.2015.2437396

IoT-Based Integrated Smart Home Automation System

N. Satheeskanth, S. D. Marasinghe, R. M. L. M. P. Rathnayaka, A. Kunaraj, and J. Joy Mathavan

Abstract Humans being warm-blooded always prefer to adjust surroundings according to their comfort and convenience. The categories of comfortness include thermal comfort, visual comfort and hygienic comfort. The thermal comfort is related to with maintaing optimum surrounding temperature and humidity. Visual comfort relates with luminance intensity and colors. Hygienic comfort is related to the quality of air. The proposed smart home automation system functions to monitor all the parameters within the desired range that is widely accepted. Smart home automation provides assistance to the elderly and physically challenged people. It can control electrical appliances in home such as bulbs, fan, air conditioner and heater. Simultaneously, the proposed system intended to recognize gas leakage which accounts for significant domestic accidents. Proposed smart home automation is designed to function using the Internet of Things (IoT) which enables controlling the parameters from a distance. The proposed system uses NodeMCU-ESP8266 microcontroller board for IoT communication and data storage.

1 Introduction

The process of involving automation to make the existing process convenient while reducing the process speed has evolved over the recent years. Smart home automation is becoming increasingly useful and widely preferred because of its certain features and convenience. Smart home automation system not only monitors the process but also controls it with advancement. Smart home automation proves to provide energy efficiency apart from appliances monitoring by maintaining the specified parameters

N. Satheeskanth (✉) · S. D. Marasinghe · R. M. L. M. P. Rathnayaka · A. Kunaraj · J. Joy Mathavan
Faculty of Technology, University of Jaffna, Jaffna, Sri Lanka

A. Kunaraj
e-mail: kunaraj@tech.jfn.ac.lk

J. Joy Mathavan
e-mail: joy@tech.jfn.ac.lk

P. Karuppusamy et al. (eds.), *Ubiquitous Intelligent Systems*, Smart Innovation, Systems and Technologies 243, https://doi.org/10.1007/978-981-16-3675-2_26

at optimum level. Home automation provides satisfaction and comfortable environment to the user. The paper aims at designing a smart home automation using web server and Wi-Fi technology. The devices can be turned ON/OFF using a personal computer (PC) through Wi-Fi.

The proposed system comprises of a smartphone and PC acting as command center at user end and NodeMCU-ESP8266 microcontroller board, Wi-Fi module and relay circuit. The outline of the proposed home automation system is shown in Fig. 1. Arduino Nano acts as controlling device. The data sent from user end—either PC or Android mobile—will be received by NodeMCU microcontroller connected with the controlling device Arduino Nano board. Then, arduino NANO reads the data, processes it and decides the switching function of relays which are connected with electrical appliances on the other end. The home automation is designed to work on two modes of operations, namely automatic control and manual control. LDR and PIR sensors are used to de-energized the appliances when the user is not around and not used the appliance for a specified period of time. In the designed home automation system, multiuser access is also allowed abiding the security protocols.

When operated in auto mode, output of DHT11 sensor, PIR sensor and LDR control the relay modules. These sensors check the temperature and humidity, human presence and light intensity, respectively. Users can adjust the room temperature based on the preset temperature values. When the actual temperature reaches the predefined temperature, the temperature conditioning appliance, the air cooler starts working. If the gas level increases inside the home, it will be indicated on the panel board and the buzzer alarms. The block diagram of designed home automation system is shown in Fig. 2.

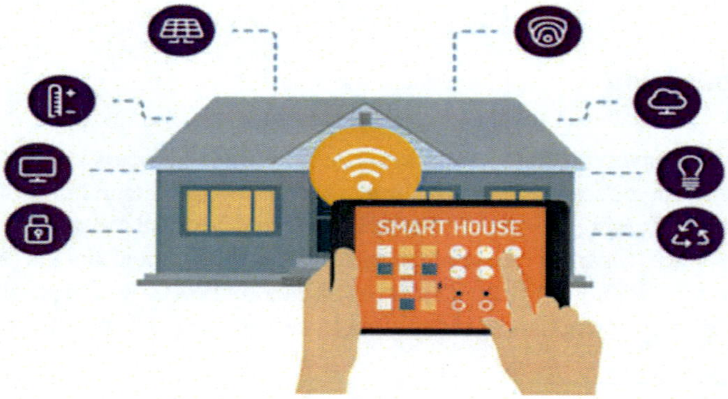

Fig. 1 Smart home automation system. *Source* iotnewsportal.com/homes/securing-the-smart-and-connected-home-with-iot

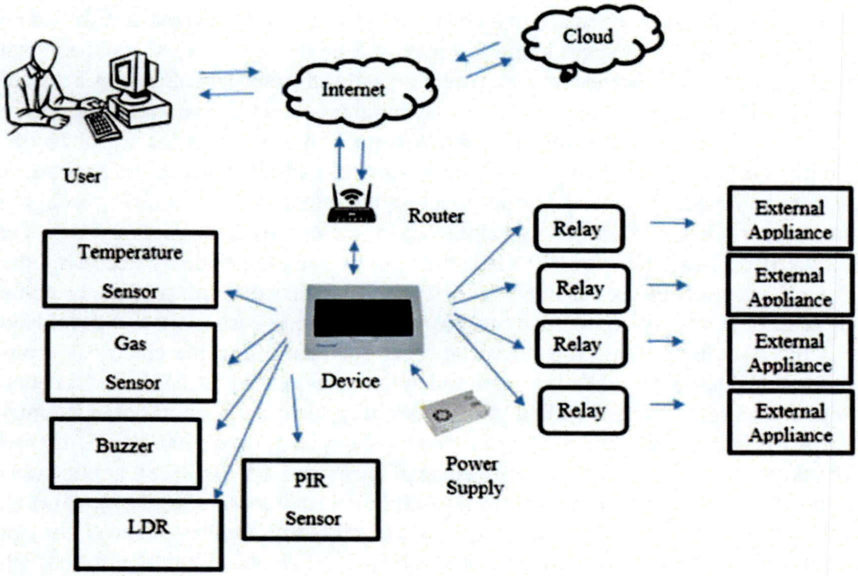

Fig. 2 Block diagram of the designed system

2 Literature Review

A. A. Zaiden et al. surveyed the existing limitations and utilized approaches in the communication protocols used for the smart home automation system [1]. Communication components used for IoT-based smart home automation can be summarized as wireless sensor network (WSN) connectivity, IP technology-based, ZigBee model and state transfer architecture. A. A. Zaiden suggested to select technology and components while reducing the power consumption, ensuring safety and security, achieving accurate and reliable management of devices and improving user experience. J. Chhabra and P. Gupta proposed additional security measures in the home automation system using voice control for authorization of home automation [2]. S. Badabaji and V. S. Nagaraju proposed an IoT-based home automation system using microcontroller LPC2148 for controlling the relay switching and global system for mobile communication GSM for communication with the mobile and PC [3]. E. N. Ganesh in his research used combination of Bluetooth and GSM communication methods to control the home automation systems through web-based applications [4]. In his research, the electrical appliances were controlled by Bluetooth when the user is indoor and electrical appliances were controlled using GSM when the user is outdoor. Since most cell phones and laptops have Bluetooth as an inbuilt application, the system costs can be reduced drastically. The apparatus can be screened and controlled by the clients from far off spots by just sending a SMS through GSM. But, such a framework has limits in two cases. Bluetooth has a restricted reach and less data rate, and GSM is expensive, which is a direct result of SMS costs that should be borne by the client [5]. Smart home automation depended on sensors' input and can consequently control home apparatus utilizing android-based cell phones as a distant regulator. Here, Bluetooth is utilized as the communication protocol and Raspberry Pi is utilized as the microcontroller. Wi-Fi is utilized to connect the Raspberry Pi microcontroller to the cell phone, which is associated with all the electrical appliances used in the house. Raspberry Pi would get all the information of sensors through a local server. However, in this technique, client cannot send the commands to the Raspberry Pi controller directly using the android mobile phone by accessing the server if he is outside the Wi-Fi range [6, 7]. D. Anandhavalli et al. proposed a home automation system along with environmental monitoring system. It was developed utilizing Arduino Mega 2560 microcontroller for controlling the electrical appliances and Bluetooth module for communication purposes [5]. D. M. Konidala et al. proposed a similar system utilizing RF ID-based application in smart home automation for privacy and counter security threats. They suggested that RF ID tagged consumer items, RF ID reader-enabled appliances and RF ID-based applications would interact with each other to create smart home environment [8]. N. David et al. in their research used few sensors and switches to control home appliances through web portal. The web portal controls Arduino by passing data and instructions to it [9]. Since Bluetooth has limited reach and Arduino Mega is costlier than Node MCU, the utilization of this combination is not advisable for smart home applications. R. K. Kodali and S. Soratkal analyzed about Message Queuing Telemetry Transport (MQQT)-based home automation framework utilizing Wi-Fi module ESP8266 [10]. Sensors and actuators associated with MQTT and ESP8266 are utilized for controlling and observing household appliances. Wi-Fi was utilized as the communication protocol between the gadgets and the prototype. The electrical gadgets are governed by MQTT utilizing Wi-Fi module ESP8266. ESP8266 is programmed using Arduino IDE and MQTT, which brought about low data transmission and low power utilization. ESP8266 board was less expensive than Raspberry Pi, Arduino UNO and other similar microcontrollers. Nonetheless, the solitary disadvantage with this framework is that switching, security and safety measures were overlooked, and the created model was not approved. Ravi kishore kodali et al. proposed IoT-based smart security system and smart home automation system. TI-CC3200 launchpad board and PIR sensor act as a smart security system, and TI-CC3200 launchpad board and electrical relay system synchronized with IoT act as smart home system [11]. Daneshwari jotawar et al. proposed a similar smart home automation system using esp32 Node MCU system that integrates the surveillance camera for security automation and sensors for home automation [12]. Yet, IoT-based PC and mobile access that is required for fully automation were not fully implemented and emphasized. Satish palaniappan et al. in their study described different communication methods that can be utilized in a home automation which includes Wi-Fi, GSM, Bluetooth and ZigBee [13]. Comparative analysis of the above study reveals that each communication method has their own merits and demerits starting with cost, speed, number of devices that can be connected and real-time operation. Jennifer S. Raj et al. proposed a series of clustering along with neural and fuzzy algorithm with the shortest path for energy efficient and enhanced performance [14]. The above method was proposed to overcome the limitations in conventional methods such as over energy consumption, frequent failures, less packet delivery ratio and delay. Smys smys proposed intrusion detection mechanism against sink hole attack, eavesdropping and denial of service attack [15]. The proposed method detects attack using highly sensitive hybrid neural network model in IoT applications.

3 System Description

The proposed smart home automation comprises of temperature, humidity, gas, PIR and LDR sensors. The system is integrated with Internet through Wi-Fi module NodeMCU-ESP8266. The flow diagram depicting the working of the proposed model is shown in Fig. 3. Parameters sensed by sensors are read by the system once the connection is established. The threshold levels for the sensor operations are predefined. Sensed parameters are passed to the web server and subsequently stored in the cloud. The parameters thus sensed are also displayed on the LCD screen. Using the obtained data, the situation in the home can be continuously monitored from anywhere and anytime. In the proposed smart home automation system, temperature humidity level and cooking gas leakage in the house can be monitored. If the temperature exceeds the predefined threshold value, the cooler turn ON automatically and it turns OFF when the temperature is back to the predefined value. Similarly, when there is a leakage of gas in the house, alarm is turn ON alerting the user about the leakage.

PIR motion sensor is used to detect the presence of humans, and turn ON the bulbs and turn it OFF once the user leaves the room if did not return within a specified time period. Another added feature in the smart home automation is integration of LDR (light-dependent resistor); incase if user forgot to turn OFF the outdoor lights in the morning, once the day time arises, the bulbs will be turned OFF automatically. Further, dimmer circuit is used to reduce the light intensity of the room according to the user preferred values and requirements. The attractive feature is that all these controls can be performed via mobile or Internet accessed PC. The user after installing the smart home automation can monitor all the electrical appliances through web portal or mobile application. If any of the home lights or electrical appliances are left turned ON without noticing, user can still observe and turn OFF all those appliances from anywhere by accessing the web portal through dedicated IP address. Then the user needs to log in to the system using correct credentials, and then, he can turn ON or OFF the electrical appliances according to his desire (Table 1).

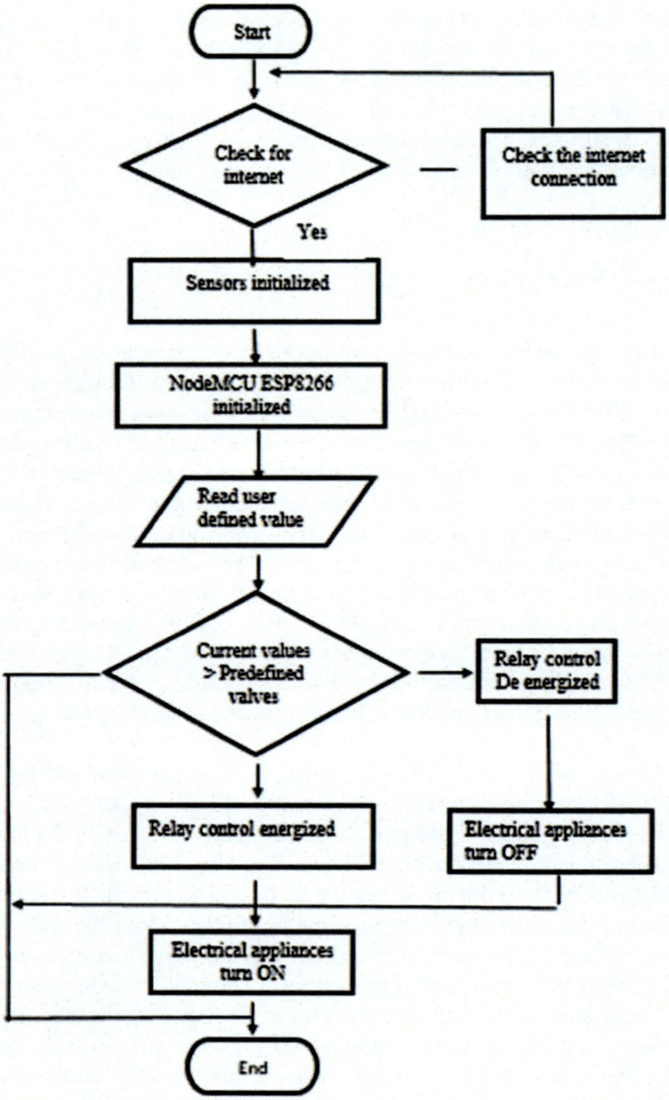

Fig. 3 Flow diagram showing the working of proposed smart home automation

3.1 Motion Detection System

PIR-based motion detection sensors are used here. Human beings in general emit 9–10 μm thermal energy on daily basis [16]. The PIR sensor shown in Fig. 4 functions in a way to capture those thermal energy and generate outputs based on the provided

Table 1 Specification of the hardware components

Hardware components used	Specification
Arduino Nano	Microcontroller: AT-Mega 328 Operating voltage: 5 V Flash memory 32 KB Clock speed 16 MHz Analog I/O pins 8 Digital I/O pins 22
Wi-Fi module node MCU ESP8266	Microcontroller: Tensilica 32 bit RISC CPU Xtensa LX106 Operatting voltage: 3.3 V Input voltage: 7–12 V Flash memory: 4 MB
PIR sensor HC SR501	Working voltage: DC 4.8–20 V Working current: 50 μA (idle) to 65 mA (full active) Detection range: 3–7 m Detection angle: 120°
DHT11-Temperature and humidity sensor	Operating voltage: 3.5–5.5 V Measuring current: 0.3 mA Temperature range: 0–50 °C Humidity range: 20–90% Accuracy: ±1 degree and ±1%
MQ 2 sensor gas sensor	Operating voltage: 5 V Output voltage: 0–5 V (analog) and 0 V or 5 V (digital)
LDR sensor	Light resistance at 10 lx (at 25 °C) 8–20 kΩ Dark resistance at 0 lx 1 MΩ Maximum voltage (at 25 °C) 150 V Ambient temperature range −30 °C to +70 °C

inputs. PIR sensor comprises of two slots which detect the movement of creatures passing by. When the warm body (human or creature) crosses by, the first slot is intersected and a positive differential pulse is generated. Then a negative differential pulse is generated when the second slot is intersected. Generated pluses are then sent to Arduino Nano controller. Arduino Nano depending on the output of PIR sensor decides whether to switch ON or OFF the light. Based on the controlling signals generated by Arduino Nano, relays either energized or de-energized powering ON or OFF the lights.

In case, if the user wanted to control the bulbs using web portal, corresponding button in the web portal is clicked and NodeMCU Wi-Fi module is activated which sends signal to Arduino Nano controller. Arduino then sends signal either to energize or de-energize the relays turning ON and OFF the lights.

DC–DC buck–boost converter is used to step down the 12 V input voltage supplied by AC-to-DC converter to steady 5 V for the operation of Arduino Nano controller

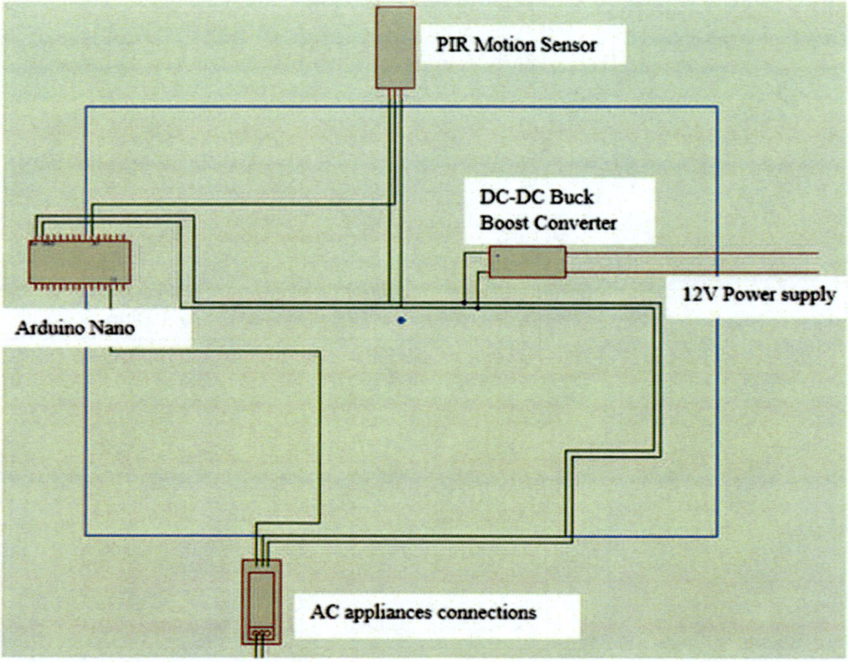

Fig. 4 Motion detection system

and sensors, since 5 V is enough for the working of Arduino Nano. Excess voltage can damage the Arduino Nano microcontroller.

3.2 Temperature and Humidity Monitoring System

DHT-11 temperature and humidity sensor reads the surrounding temperature and humidity level separately and transfers the information to the microcontroller as information [8]. The data pin of temperature and humidity sensor DHT11 is associated with pin A0 of the Arduino Nano board. VCC and ground of DHT-11 temperature and humidity sensor is associated with the VCC and ground of Arduino Nano board. Schematic of the connection diagram is shown in Fig. 5. When temperature reaches predefined or above the predefined value, these values are sensed by temperature and humidity sensors. Output of the sensors is fed to the microcontroller which in turn triggers the relay to switch on the fan and cooling device in an attempt to bring down the temperature.

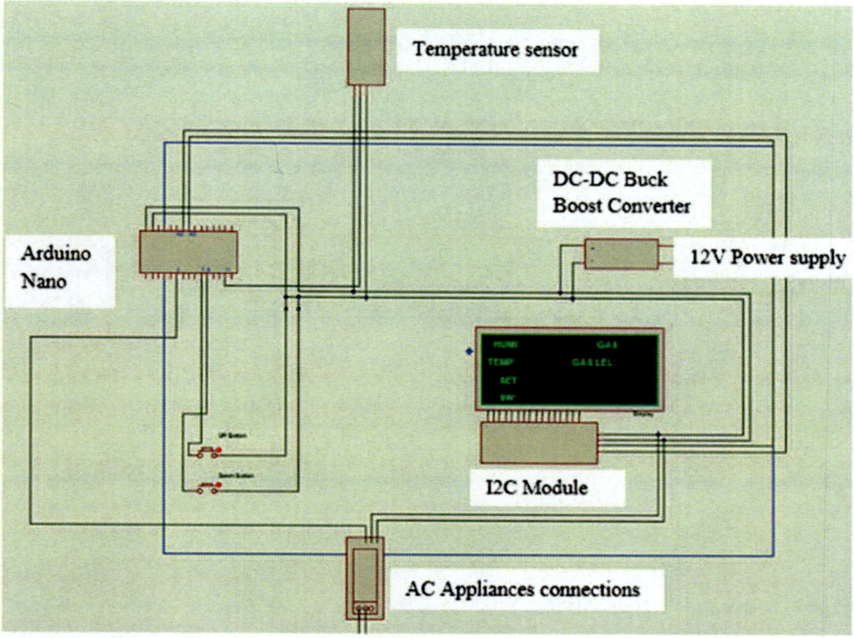

Fig. 5 Temperature and humidity detection system

3.3 Gas Leakage Detection System

Gas sensor can identify carbon monoxide, LPG (liquefied petroleum gas), methane, smoke, liquor, hydrogen and propane in the range of 200 to 10,000 ppm (parts per million) [17]. Pin A0 of MQ2 gas sensor is connected to the Arduino Nano board. VCC and ground of MQ2 gas sensor is connected to the common VCC and ground of Arduino Nano board as shown in Fig. 6. If the LPG gas leakage is detected, M2 gas detection sensor is activated which in turn provide signal to the microcontroller to activate the alarm, thus alerting the user about the gas leakage.

3.4 Automatic Daylight Power Saver System

A photoresistor [photoconductive cell or light-dependent resistor (LDR)] is a light-controlled variable resistor. A high-resistance semiconductor is used for photoresistor. If there is the presence of light, the resistance of the LDR would be very low probably in the range of few Ohms. In case of darkness, the resistance of the LDR rises high to around a few mega ohms. This property of the LDR is used here to turn ON the light in the night time and turn OFF it in the day time. The schematic diagram

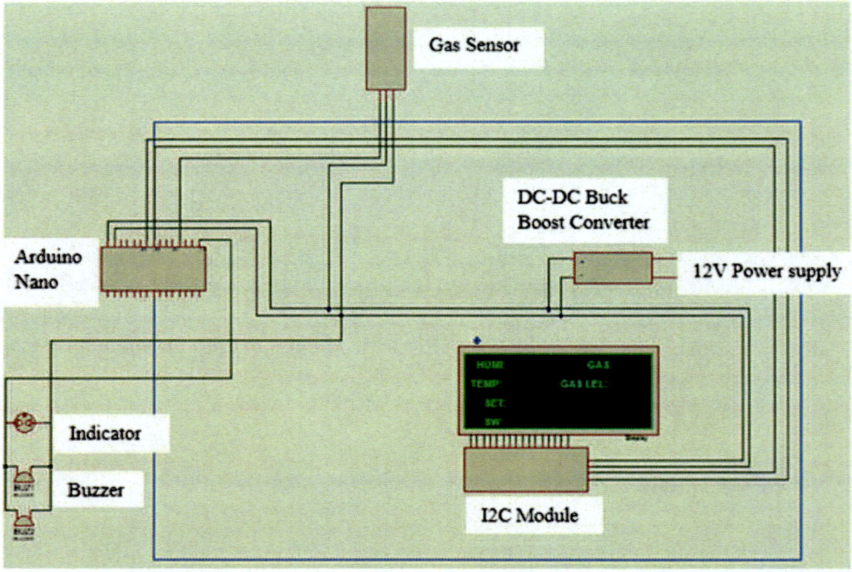

Fig.6 Gas leakage detection system

showing the integration of LDR sensor with the microcontroller and relay system is shown in Fig. 7.

3.5 *Device Control System*

In this project, the relays are used for connection between various electrical equipments for control in accordance with the input signal. It is connected with the fan and bulbs which acts as outputs. Relays are utilized in numerous applications due to their relative straightforwardness and long life and demonstrated high reliability. Relays are used to secure, regulate and control the power supply of the electrical appliances. The relay module is working at around 5 V level.

3.6 *Authentication Interface System*

The central server can be accessed by the user authorized by the "user name" and a "password." Central server gives the client the essential information stockpiled in the database. After accessing the central server, based on the available information, the client would then be able to make queries or send commands. The IoT devices

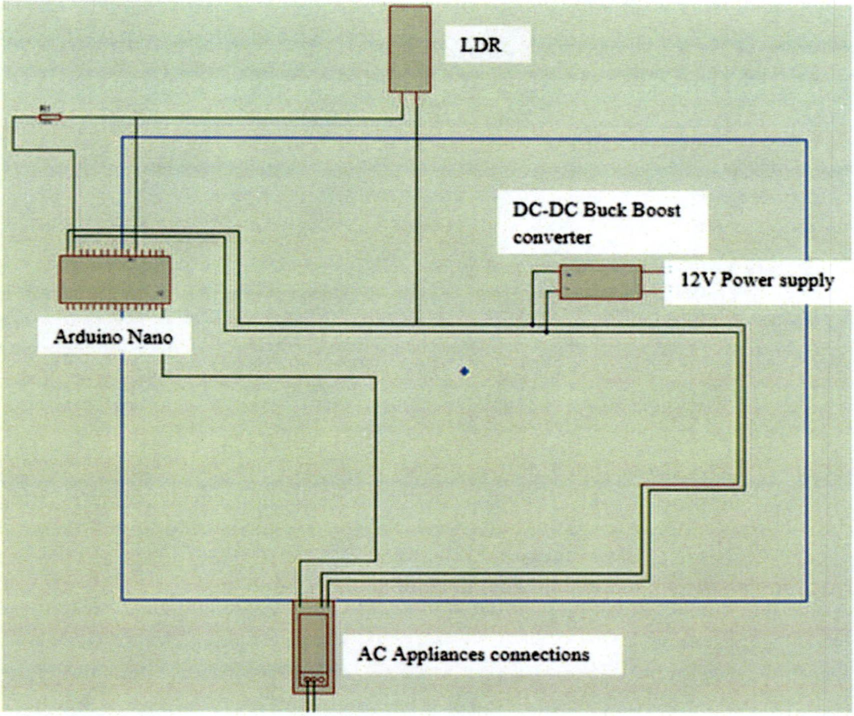

Fig. 7 Automatic daylight power saver system

usually have an authentication method, it can be used for user administration or it can be used to connect the device to a central controller. When the user enters the correct information, it directs to open the web page. On this web page, the appliances in the house can be controlled. The passwords can be tried until the correct password is guessed. The username and password can be changed by the user. The above-mentioned authentication interface is shown in Fig. 8.

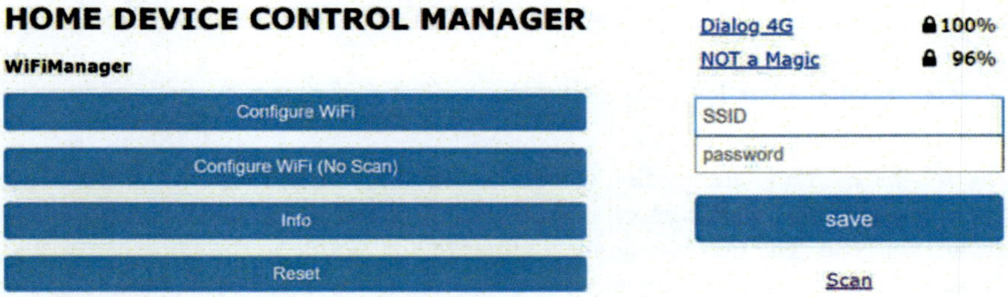

Fig. 8 Authentication interface

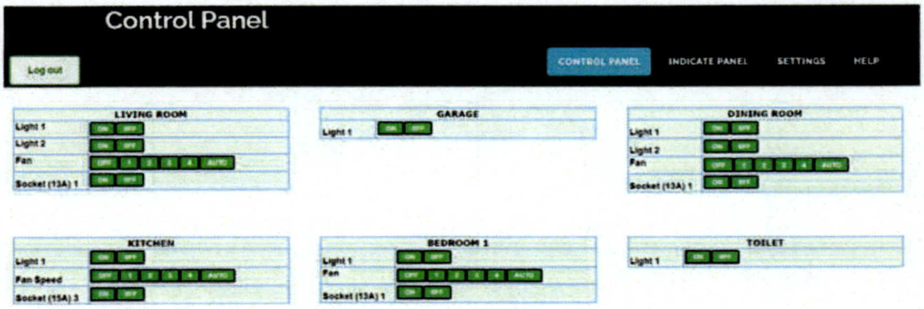

Fig. 9 User interface

3.7 User Interface

The appliances like fan, lights and some other loads are connected to the NodeMCU modules. NodeMCU module is used here for easy portability, since complete wiring of the home cannot be done for automation. Nodes are distributed in the rooms to manage the appliances available in each room in parallel. The control of relays is done through the control panel user interface as shown in Fig. 9.

4 Results and Discussion

The proposed smart home automation system includes temperature and humidity sensing elements, gas detection elements, motion detection elements and light intensity detection elements. The designed prototype function effectively detects the temperature, humidity, gas leakage, human presence, entry to the living space and differentiating the day from night. Output from the sensors is given to the Arduino Nano microcontrollers which is used to control the relay system shown in Fig. 8. Based on the control inputs, relays are either energized or de-energized, thus turning ON or OFF the electrical appliances. Finalized prototype of the proposed smart home automation system is shown in Figs. 10 and 11 (Fig. 12).

5 Conclusion and Future Scope

The IoT-based home automation system was developed and tested. All the home electrical appliances were controlled using the online web portal specifically designed for this purpose. The user can connect either PC or mobile to the same network as the module so exchange of signal takes place frequently while being around home. When user is away from home, the online control of the device can be accessed through PC with a Internet connection by simply logging in using the IP address

Fig. 10 Hardware implementation of relay system

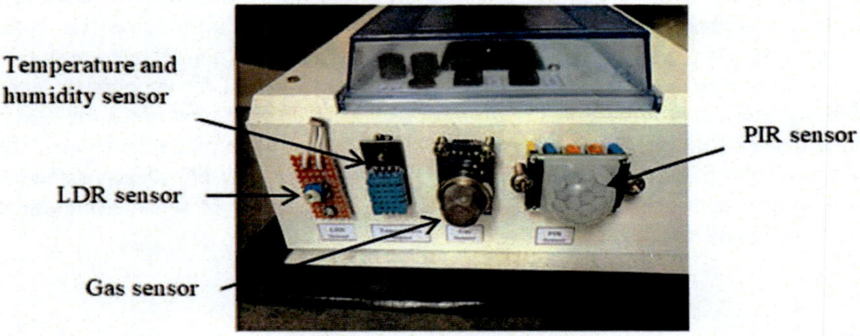

Fig. 11 Hardware implementation

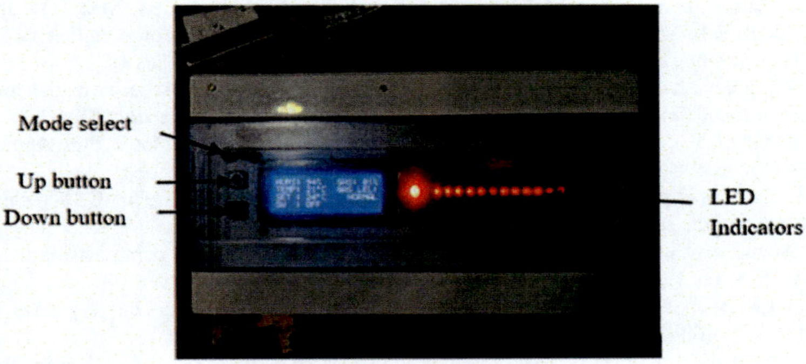

Fig. 12 Front view of the final device

and correct credentials. Proposed smart home automation system proves to function effectively for the elderly and physically challenged people when no one is around to take care of them. This smart home automation also proves to be effective for office workers who leave to office in rush. The status of the electrical appliances can be monitored and controlled from the office easily. Limitations include chances of hacking the system as the proposed system is not designed with high security features. This system is built using Arduino Nano controller which includes limited features and functionalities. For effective control and increased security, Raspberry Pi can be used.

As the future scope, number of features can be added to this system including adding motor to control window drapes, fire sensor to alert and control fire accidents, rain alerting system, etc. Gas leakage system described in this research paper is another area to improve. In this research work, the leakage can be detected and alarmed, but cannot be controlled. In future researches, a method to control this leakage and prevent any hazardous outcomes is to be done. In this research work, air cooling system is discussed. As the future scope, based on detected temperature values, heater system is to be added to heat the house. If the temperature goes down below a certain value (as like winter season of Western countries), the heater should work to bring up the temperature of the house to the predefined state. The proposed plan of the smart home is entirely adaptable and can be effortlessly extended and applied to bigger structures by expanding the number of sensors, measured parameters and controlling devices.

References

1. A.A. Zaidan, B.B. Zaidan, M.Y. Qahtan, O.S. Albahri, A.S. Albahri, M. Alaa, F.M. Jumaah, M. Talal, K.L. Tan, W.L. Shir, C.K. Lim, A survey on communication components for IoT based technologies in smart homes. Telecommun. Syst. **69**(1), 1–25 (2018)
2. P. Gupta, J. Chhabra, IoT based smart home design using power and security management, in *International Conference on Innovation and Challenges in Cyber Security* (2016)
3. S. Badabaji, V.S. Nagaraju, An IoT based smart home service system. Int. J. Pure Appl. Math. **119**(16), 4659–4667 (2018)
4. E.N. Ganesh, Implementation of IOT architecture for SMART HOME using GSM technology. Int. J. Comput. Tech. **4**(1) (2017)
5. D. Anandhavalli, N.S. Mubina, P. Bharathi, Smart home automation control using Bluetooth and GSM. Int. J. Informative Futuristic Res. 2547–2552 (2015)
6. B.D. Labus, A smart home system based on sensor technology. Facta Univ. Electron. Energ. **29**(3), 451–460 (2015)
7. M.L. Sharma, S. Kumar, N. Mehta, Smart home system using IoT. Int. Res. J. Eng. Technol. **4**(11) (2017)
8. D.M. Konidala, Security frame work for RFID-based applications in smart home environment. **7**(1), 111–120 (2011)
9. N. David, A. Chima, A. Ugochukwa, E. Obinna, Design of a home automation system using arduino. Int. J. Sci. Eng. Res. **6**(6) (2015)
10. N. Kodali, S. Soratkal, MQTT based home automation system using ESP8266, in *IEEE Region 10 Humanitarian Technology Conference* (2016)

11. R.K. Kodali, V. Jain, S. Bose, L. Boppana, IoT based smart security and home automation sytem, in *International Conference on Computing, Communication and Automation, ICCCA* (2016)
12. D. Jotawar, K. Karoli, M. Biradar, N. Pyruth, IoT based smart security and home automation. Int. Res. J. Eng. Technol. (IRJET)**07**(08) (2020)
13. S. Palaniappan, N. Hariharan, N.T. Kesh, S. Vidhyalakshimi, S. Angel Deborah, Home automation systems—a study. Int. J. Comput. Appl. **116**(11) (2015)
14. J.S. Raj, A. Basar, QoS optimization of energy efficient routing in IoT wireless sensor networks. J. ISMAC. **01**(01), 122–3 (2019)
15. S. Smys, A. Basar, H. Wang, Hybrid intrusion detection system for ınternet of things (IoT). J. ISMAC. **02**(04), 190–199 (2020)
16. F. Vatansever, M.R. Hamblin, Far infrared radiation (FIR): its biological effects and medical applications. US National Library of Medicine, National İnstitute of Health. https://www.ncbi.nlm.nih.gov/pmc/articles/PMC3699878/
17. K. Keshamoni, S. Hemanth, Smart gas level monitoring, booking & gas leakage detector over IoT, in *International Advance Computing Conference* (IEEE, 2017)

Reinforce NIDS Using GAN to Detect U2R and R2L Attacks

V. Sreerag, S. Aswin, Akash A. Menon, and Leena Vishnu Namboothiri

Abstract Network attacks have been a headache since the days of the network. But with the advancement of technology, computers have proven to be more effective at detecting attacks. Machine learning and deep learning technologies have made it even more efficient. NIDS were very good at detecting attacks but was unable to detect alternating new. Adversarial attacks have become more common and more difficult to detect today. Similarly, not all attacks are known to be detectable using the same ML algorithm. Also, the lack of the number of 'attack' category training of these ML models is not much efficient. In this paper, we look at the U2R and R2L attacks, and an approach using GAN, the machine learning framework to enhance the efficiency of NIDS in detecting these attacks through adversarial training. For that, the KDD dataset is utilised. Since there are other attacks on this dataset, this research work changed it into a useful way through data preprocessing. The proposed research work has shown that by training the GAN model, that is, by using the existing dataset to generate the attacks and tune the existing dataset and retrain the NIDS to enhance its accuracy and detection rate.

1 Introduction

The growth of Internet technology is most useful and, at the same time, misfortune is in the cyber domain. Compared to the past, the value of the Internet technologies is observed in cyber security, but another aspect of the same technology is its rival. And network attacks are the topmost among them. In the past, cyber-attacks were identified by the manually using well-crafted rules, but with the advent of the ML and DL, NIDS also underwent a transformation. ML-based NIDS also provided relief to cyber security. But at the same time, adversarial attacks began to form on, that could

V. Sreerag (✉) · S. Aswin · A. A. Menon · L. V. Namboothiri
Department of Computer Science and IT, Amrita School of Arts and Sciences, Kochi, Amrita Vishwa Vidyapeetham, Kochi, India

L. V. Namboothiri
e-mail: leena@asas.kh.amrita.edu

P. Karuppusamy et al. (eds.), *Ubiquitous Intelligent Systems*, Smart Innovation, Systems and Technologies 243, https://doi.org/10.1007/978-981-16-3675-2_27

not detect using existing NIDS. Therefore, ML-based NIDS do not work as well. Adversarial attacks can be very dangerous sometimes in terms of cyber security. They have the potential to fool the models and mispredict the input. So the attacker can easily fool the model by input some adversarial attacks to the network so it would not identify it as an attack and breaches happen. The existing NIDS model is not capable of categorising the adversarial attacks. Since they work on with the data they provide during the training phase, adversarial attacks always penetrate through these models. Generative adversarial network (GAN) is a ML framework designed by Ian Goodfellow, and his team in 2014 has become a milestone in the cyber world. It is an unsupervised learning method, which makes use of convolution neural network (CNN) to identify and learn the patterns from input data, so that it can generate new data/samples. Since the meat and potato of any ML models is the data. Enormous number of datasets can be obtained from many sources. But the main problem with network attacks dataset is the insufficient amount of the anomalies in it. A large difference is observed in the number of normal to anomaly data in any datasets. Here, the KDD+ dataset has been selected. It is because new type of attacks is evolving day by day and the NIDS are not being able to classify it correctly. Since any ML, DL, AI models work efficiently on the basis of data provided, these models require a balanced dataset.

Studies are ongoing for a defensive mechanism against these adversarial attacks. Rather than doing a defensive mechanism, we are more into improving the efficiency of the models by retraining the same (Fig. 1).

In GAN, it consists of two sub-models which are the generator and the discriminator, the function of the generator is to generate new examples from the given dataset and the function of discriminator is to classify the generated examples as real or fake examples. We know that most of the ML models are designed to work and process on the problem sets in which training and the test data are generated on the same statistics distribution. The example which has been generated artificially using the GAN is the adversarial examples; the specialty of adversarial examples is that it looks similar to the normal example but contains some noise which leads to misclassification on the ML model. So what we are going to do is adversarial training that is to train the same ML model not only with the same statistical data but also

Fig. 1 Proposed architecture

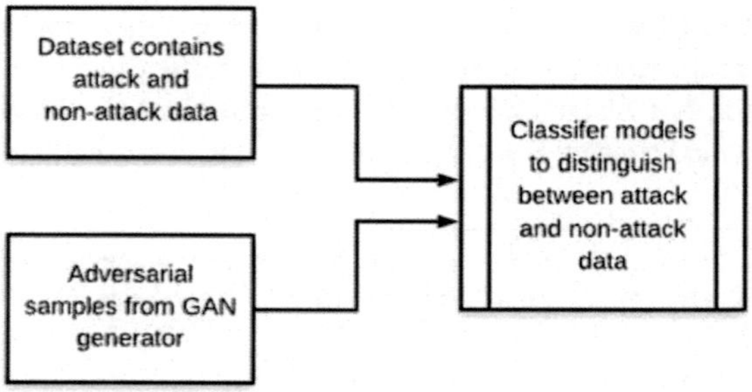

with the adversarial examples (Fig. 1). As a result of the adversarial training, we can make the model more effective.

2 Related Works

Many studies have been undertaken to improve the efficiency of the NIDS in the network security areas. Recent days, GAN has been widely used in generating adversarial attacks and train the models for better classification. Machine learning models have been a boon to the network areas in detecting anomalies. One of the techniques used is the adversarial training. The traditional intrusion detection methods have limitations in detection like accuracy rate and precision rate. Therefore, they proposed a more accurate and efficient intrusion detection method to improve the intrusion detection capability in the wireless network environment. A DBN-SVM combined wireless network intrusion detection model based on deep learning was proposed. Yang et al. [1] They proposed a deep learning-based model with the combination of DBN and SVM where feature extraction is done using DBN and classification is done by SVM; therefore, the precision, accuracy and recall of this method show better results than the other methods.

Qiu et al. [2] In this paper, they proposed a new deep learning (DL)-enabled security authentication scheme by implementing blind feature learning (BFL) and lightweight physical layer authentication (LPLA) to overcome the security issues with the wireless multimedia sensors. Analysis verifies that the proposed system can guarantee the privacy with high accuracy and precision rate of the wireless multimedia sensors and also achieve lightweight authentication.

Vinayakumar et al. [3] In this paper, an effective deep learning approach is proposed by modelling a deep neural network (DNN) by combining NIDS and HIDS to detect cyber-attacks proactively. As the result, this is the only framework which has the capability to collect network and host level activities using DNN to detect attack more accurately.

Yang Xin et al. [4] done survey report that defines major literature surveys on machine learning (ML) and deep learning (DL) methods for network analysis of intrusion detection and gives small tutorial description of both the methods. This paper discusses about the cyber threats that are increasingly evolving with the growth of the Internet, and the cyber security situation is not positive. A network security system consists of a security system for a network and a security system for a computer. Firewalls, anti-virus apps and intrusion prevention mechanisms are part of most of these systems (IDS). Without representative data, the ML and DL approaches do not work, and collecting such a dataset is challenging and time-consuming. There are several issues with the current public dataset, however, such as uneven knowledge, material that is obsolete and the like. The production of research in this field has largely been restricted by these issues.

Buczak et al. [5] This paper introduces the findings for computer security implementations of a literature study of machine learning (ML) and data mining (DM) approaches. In this survey paper, algorithms and related training data are vulnerable to a number of security attacks, triggering a substantial output decline, independent of effective implementations of machine learning algorithms in many contexts, such as facial recognition, malware detection, autonomous driving and intrusion detection. Therefore, a systematic survey with a range of machine learning techniques on security issues is observed.

Li et al. [6] In this paper, it endorses a machine learning framework for identifying and detecting DGA (domain generation algorithm) domains to ease the unknown threats; for that, it uses a two-level model and prediction mode. Using a DNN model, the proposed system is enhanced by, handling the huge data. The deep learning model is proposed to classify the DGA domains and normal domains and compare it with deep learning model with machine learning methods. Finally, it compares DNN classification model with the first-level classification in our machine learning framework and the long short-term memory (LSTM) model.

Liu et al. [7] has done that this survey paper describes a focused literature survey of machine learning (ML) and data mining (DM) methods for cyber analytics in support of intrusion detection. The collection of technology and procedures are cybersecurity built to safeguard computers, networks, programs and knowledge from attack, illegal entry, alteration or damage. It is helpful for an IDS to be able to access network and kernel-level data to be able to perform anomaly detection and misuse detection.

Rasool et al. [8] In this paper, they have proposed cyber-pulse, an add-on module in the application layer of the SDN controller extension for securing the SDN control channel against LFA utilises machine learning and deep learning techniques, to select appropriate traffic features for accurate classification in a large volume of traffic data. Finally, the Cyber-Pulse was able to appropriately describe traffic flows displaying LFA features and effectively mitigates the attack.

Kumar et al. [9] They had proposed a defined anti-jamming protocol for vehicular traffic environments and concentrated on the localisation of vehicles in delimitated jamming environments, in a machine learning. Without any interference, like noise or jamming, the VANET achieves 99.91% accuracy in locating the car. Improved accuracy, high throughput, higher delivery ratio of packets and decreased packet loss ratio, VANET efficiency has been improved.

Al-Qatf et al. [10] In this paper, they proposed an effective deep learning approach, STL-IDS, based on the STL framework, a combination of sparse autoencoder for the effective representation of raw dataset (NSL-KDD), and SVM based on self-taught learning for classification. Their experimental outcomes indicate that the model shows enhanced accuracy of SVM classification and accelerated training and testing times. It also displays strong results in the two-category and five-category grouping.

Raj et al. [11] The paper mainly focuses on the issues related with the nonlinear regression estimation which they have been solved by the successful implementation of the novel neural network technique termed as SVM. They also provide a multi-application prediction model. System reliability also is predicted as the result of applying SVM learning algorithm to RNN. After the analysis, the performance of the proposed system is faster and accurate than the existing system.

Velswamy [12] has proposed a hybrid algorithm for selecting virtual machines for scheduling applications based on GSA and NSGA. Selecting the proper vm is the major task because the users are paying for the resources on the basis of the time used. In this paper, the algorithm calculates the total utilisation and completion time and assigns the vm after normalising the retrieved data. As the result of the algorithm, it provides optimal solution on energy consumption, response time and the cost.

Wang [13] In this paper, an intelligent system is been developed using ANFIS with the help of the sensors available in the various electronic devices, as the result, it allows the various devices to take dynamic decisions while using. On using them over the internet, the close security analysis feature ensures the safety of the devices.

3 Proposed Methodology

We proposed a system in enhancing the current NIDS of detecting the network anomalies especially U2R and R2L using GAN. Our approach makes use of the GAN aided with the IDS. The proposed model consists of several phases through which we obtained a better classification after the adversarial training. Data preprocessing and GAN training are the most important phases of our model. We used the Google Colab as tool implementing our python based model. Classification models such as Decision Tree, Random Forest, KN Neighbour, Logistic Regression, SVM and different Naive Bayes are deployed in the model. Each produces different accuracy values before and after the GAN training. The architecture model consists of two neural network (generator and discriminator) based on which the adversarial samples are generated. In training, the GAN the noise for each model is set to 9. Initially the input dimension is the set to the number of features in the dataset and output dimension is set to 2 so as to increase the stable training.

The proposed system comprises of 6 steps, which are:

1. EDA (Exploratory Data Analysis)
2. Data Preprocessing
3. Train the NIDS
4. Train the GAN
5. Generate the adversarial attacks
6. Adversarial training.

3.1 Exploratory Data Analysis

The purpose of this EDA is to find acumen which will serve us, for data cleaning/preparation/transformation which can be finally used as a machine learning algorithm (Fig. 2).

Through EDA, we can understand what is our actual data beyond the formal format. It will provide us with a crystal clear idea about the data we use since data

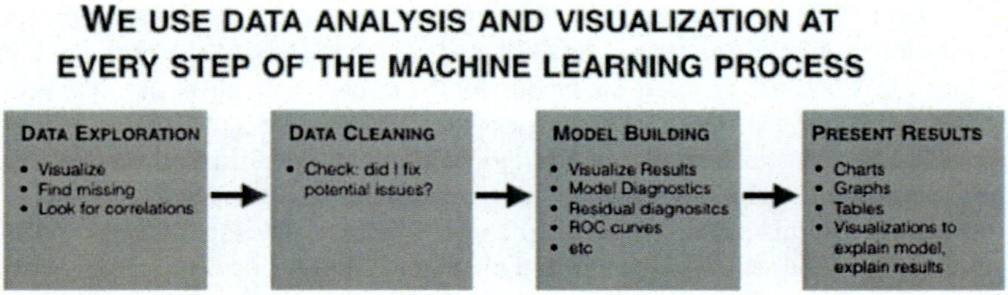

Fig. 2 Exploratory data analysis

are the fuel of any ML model. Using EDA we can identify the relationship between each and features; for the easy understanding we can make over the numerical/tabular data into easily understandable graphs. Further, it aids in determining if the statistical techniques you are considering for data analysis are appropriate.

The correlation matrix gives us the statistical relationship between each features in the dataset (Fig. 2). At the same time, it can be used to analyse the dependency of variables.

This scatter plot Fig. 3 represents the relationship between the continuous values, the state and the label. It depicts how on value affect the other through the datasets. Similarly using various visualisation methods, we can get the variables relation using this EDA.

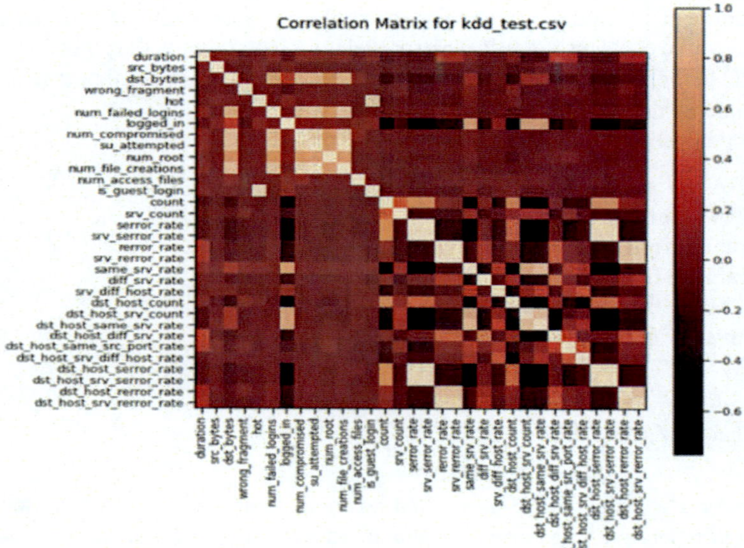

Fig. 3 Correlation matrix of test data

3.2 Data Preprocessing

Data preprocessing is an important step before training every ML model. It is a data mining technique that comprises transforming raw data into an understandable format. Since real-time data are incomplete, lacking in certain features may likely have many errors. Solving such issues, data preprocessing is used.

In this model, both the training and test data have been encoded and normalised. Then, the data are separated based on the category. At first during preprocessing, the non-numeric features are identified and they are encoded; also, the binary features are to object types. For the numeric data, it is categorised and then normalised using min–max method. Now the dataset is saved and ready to use.

3.3 Train NIDS

NIDS is the key component of this thesis; we test our models using these NIDS. Here, we specified the attack category (U2R and R2L) and the ML models used for classifying such as SVM, Decision Tree, KN Neighbour, Logistic Regression, Naive Bayes and Random Forest. During this phase, the data are split (formatted) and used for the model to use. During training and testing of NIDS, we can specify which model to be used. We ran our dataset through all the models and returned the accuracy and detection rate of each model.

The accuracy and detection rate in increasing with the number of epochs.

3.4 Train GAN

The very next step after running the IDS is the training of the GAN model. Here, Generator: The generator of a GAN learns to produce fake data by incorporating feedback from the discriminator. The generator takes in to make the discriminator classing its output as real. Discriminator: The discriminator in a GAN is simply a classifier which distinguishes real data from the data created by the generator. It could use any network architecture appropriate to the type of data its classifying.

Then, we train the GAN using all the 3 models. Then, these models are saved for the future purpose. A 100 epochs is used to train the GAN.

Here, we see that the GAN IDS are inefficient in detecting the adversarial attacks (Fig. 4). Adversarial attacks are always hard to detect by the NIDS (Fig. 5).

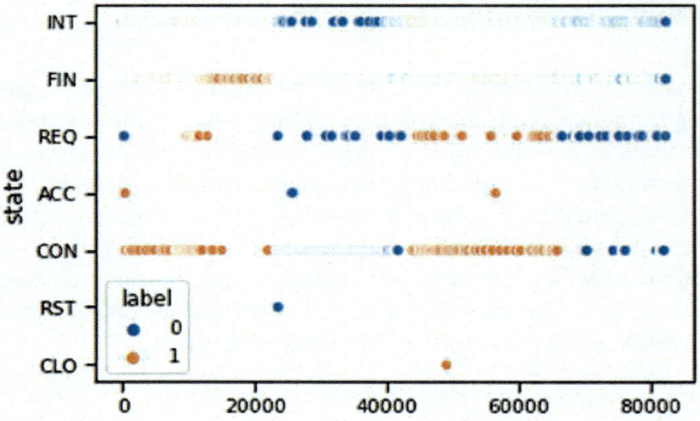

Fig. 4 Scatter plot (state versus lable)

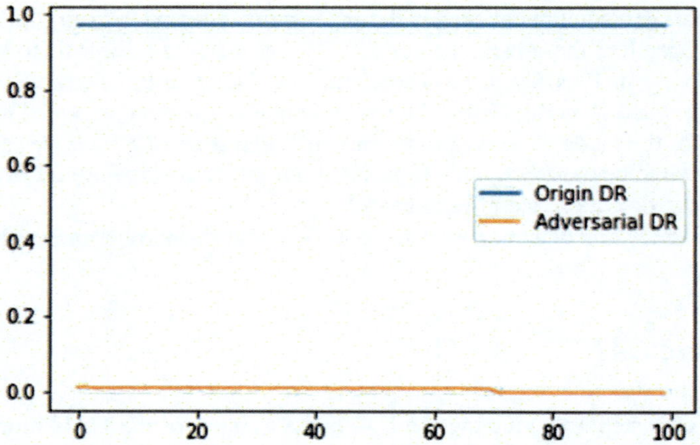

Fig. 5 Comparison of original and adversarial detection rates in GAN model

3.5 Generate Adversarial Attacks

After training the GAN, the next step is to generate these attack samples and save to dataset. After preprocessing and running GAN functions, the attack samples are generated and saved to datasets. Here, we use the above saved models to generate the attack samples. We will generate both the Dos and U2R and R2L attack categories in this phase.

Then, we compare these datasets to get the information.

3.6 *Adversarial Training*

Now the last phase is to retrain the NIDS using the dataset that we generated above and with the original dataset. So by combining these, we get more attacks samples in the new dataset. Here also, the 3 models are used and train the ids. It tries to simulate the distribution of the data when GANs generate adversarial instances, and tries to generate from that distribution. During this step, the ids is now learned to detect the possible adversarial attacks that might occur in the network.

4 Experiment and Result Analysis

The working of GAN-based classifier consists of two neural networks which is a generator and a discriminator. The generator is to generate attack sample in a candidate. And the task performed by the discriminator is to produce the binary output in which '0' represents 'normal' and '1' represents an 'attack categories'. The attack samples generated by the generator are sent to the discriminator hoping that it will misclassify them as the 'normal' class. The discriminator will receive the generated samples along with data samples from the actual dataset as the input and tries to predict the normal as 0 and others as 1. This works as an iterative process at the end of each iteration the samples is been generated at every 10 epochs from the generator 'attack' class. We use loss function to calculate the discriminator loss and the generator loss.

Since we are focusing on U2R and R2L attacks, firstly, run the IDS under different models to generate the Accuracy, Detection Rate (DR) and Runtime of each model for the input dataset. We have used the machine learning's sklearn models such as Decision Tree, Random Forest, KNN, Logistic Regression, SVM and different Naive Bayes models.

Here, we can see that the accuracy, detection rate and the time taken for processing of dataset by the IDS are shown in Table 1. On examining the table data, we can find

Table 1 IDS model results

Model	Accuracy (%)	DR (%)	Runtime (s)
Complement NB	82.89	41.26	0.54
Decision Tree	81.27	21.82	0.68
Bernoulli NB	80.87	19.92	0.55
Random Forest	79.33	11.38	2.88
KN Neighbour	77.51	77.51	22.25
Multinomial NB	76.80	0.71	0.50
Logistic Regression	76.74	0.44	1.11
SVM	76.67	0.20	5.71
Gaussian NB	39.18	78.05	0.55

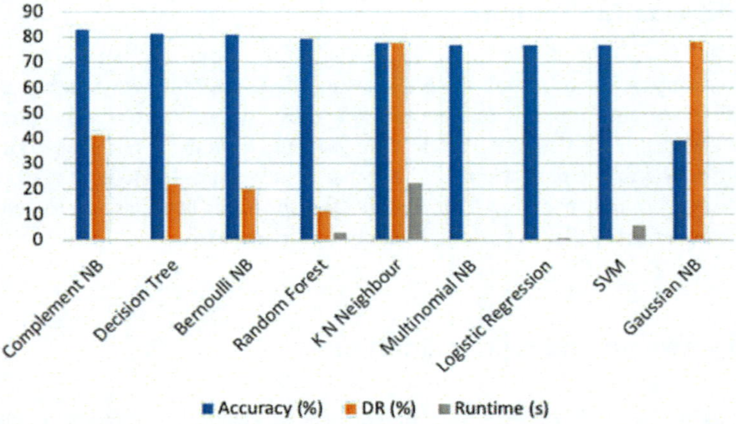

Fig. 6 IDS models results

that Complement NB model has the highest accuracy rate of 82.89% while Gaussian NB model has the highest detection rate 78.05% and the multinominal NB model has the least runtime of 0.50 s. On comparing with other models, the model with best result is Complement NB because it has highest accuracy rate of 82.89% and an average detection rate of 41.26 and runtime of 0.54 s (Fig. 6).

Then using the same saved models, the above generated adversarial samples are then extracted. At each iteration, we generate samples at each 10 epochs the 'attack' class from the generator. In the last part of our experiment, we perform the IDS train using the generated attack samples. Combining the original dataset with the generated adversarial samples, we successfully able to improve the NIDS in classifying the attacks. This indicates that when we train the IDS with more no of attacks samples, the more the accuracy and decision rate (Fig. 7).

After retraining the model with the generated attack samples, the accuracy of each model has been increased (Table 2). This indicates that when the number of data samples increases the accuracy of model to classifying attacks also increases. The DT shows the highest accuracy with a percentage of 93.09% and the lowest is the Gaussian NB with an 74.22%. Even though the Gaussian NB is still the lowest among, the difference on before and after adversarial training has a wide range. In general, the NB models shows with the minimum accuracy rate on comparing with other models (Fig. 8).

Data deficiency can be resolved by generating data artificially. Even though the adversarial attacks change its shape frequently, we can push the IDS in identifying the adversarial attacks by the technique of Adversarial training.

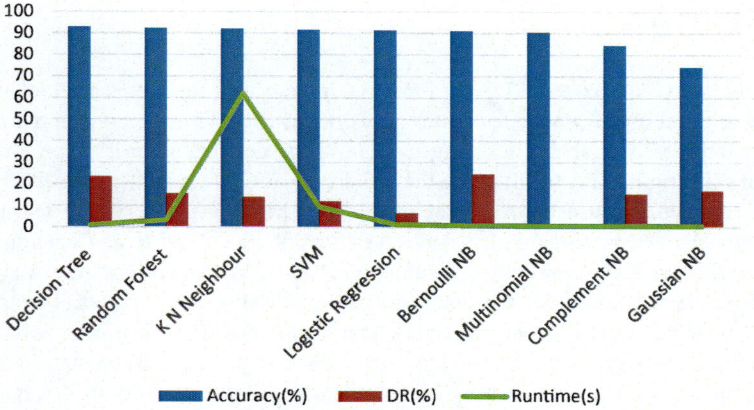

Fig. 7 IDS retrained results

Table 2 IDS retrained results

Model	Accuracy (%)	DR (%)	Runtime (s)
Decision Tree	93.09	23.45	0.90
Random Forest	92.39	15.66	3.31
KN Neighbour	92.17	14.2	62.05
SVM	91.67	12.04	9.59
Logistic Regression	91.33	6.68	1.32
Bernoulli NB	91.16	24.68	0.78
Multinomial NB	90.66	0.72	0.70
Complement NB	84.4	15.33	0.72
Gaussian NB	74.22	16.95	0.70

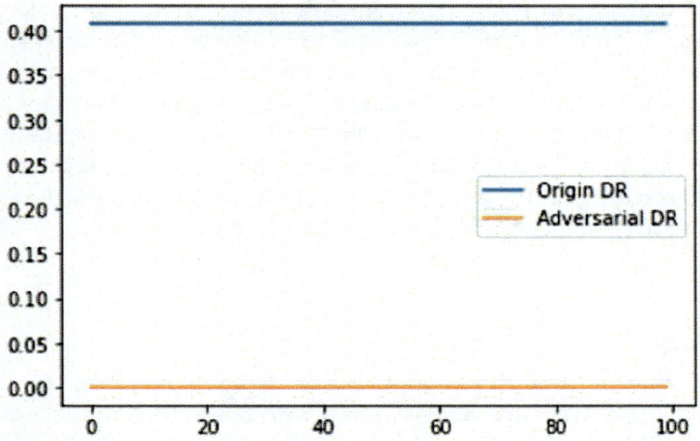

Fig. 8 IDS models results

5 Discussion

We find that GAN-based NIDS are effective in detecting the adversarial attacks in a network that makes a way for better network security. This methods demonstrate the importance of machine learning models in the network security. The ultimate aim of this approach is to illustrate the efficiency of GAN in generating the adversarial samples and use them for the adversarial training for detecting the adversarial attacks. A comparison of various classification models, such as Decision Tree, Random Forest, KN Neighbour, Logistic Regression, SVM and different Naive Bayes has done via this work, for the detecting and classifying the adversarial attacks in U2R and R2L attacks. Since this work is a scaled-down version of a real-world operation, it may not work well in larger or more complicated circumstances. Also, the adversarial attacks are varying in nature and it can take any form, this may not effective in real-time data. In the future, a vast amount of data that can be equipped to use it in real-life scenarios to detect adversarial attacks, which would be a huge accomplishment in cyber security, will be carried out in future research.

6 Conclusion

Adversarial attacks have become a nightmare to the cyber security. The randomness in its nature and faking the identity make it undetectable by the traditional IDS. Since no such IDS exist in identifying the adversarial tats the only way is to make them even more efficient in identifying those attacks. Adversarial training is one of the effective ways against the adversarial attacks. In this paper, we discussed how, using the GAN attack samples, we can improve an ML-based NIDS for a network intrusion detection. By training the GAN using the original data and generating the adversarial samples and then retrains, the ML models have significantly changed the accuracy. Even for network intrusion detection system, adversarial attacks can cause harm. With adversarial examples, classified and trained it, to make it robust. This particular form of adversarial attack, however, is efficient in improving attack detection and Improving the precision of the NIDS. In the future, we like to improve the IDS by increasing its DR. Even though we could improve the accuracy rate, the DR is considerably less after adversarial training. This may be due to the robustness of the adversarial samples to begin with the fake identity.

References

1. H. Yang, G. Qin, L. Ye, Combined wireless network intrusion detection model based on deep learning. IEEE Access **7**, 82624–82632 (2019). https://doi.org/10.1109/ACCESS.2019.2923814

2. X. Qiu, Z. Du, X. Sun, Artificial intelligence-based security authentication: applications in wireless multimedia networks. IEEE Access **7**, 172004–172011 (2019). https://doi.org/10.1109/ACCESS.2019.2956480
3. R. Vinayakumar, M. Alazab, K.P. Soman, P. Poornachandran, A. Al-Nemrat, S. Venkatraman, Deep learning approach for intelligent intrusion detection system. IEEE Access **7**, 41525–41550 (2019). https://doi.org/10.1109/ACCESS.2019.2895334
4. Y. Xin et al., Machine learning and deep learning methods for cybersecurity. IEEE Access **6**, 35365–35381 (2018). https://doi.org/10.1109/ACCESS.2018.2836950
5. A.L. Buczak, E. Guven, A survey of data mining and machine learning methods for cyber security intrusion detection. IEEE Commun. Surv. Tutorials **18**(2), 1153–1176 (2016). http://doi.org/10.1109/COMST.2015.2494502
6. Y. Li, K. Xiong, T. Chin, C. Hu, A machine learning framework for domain generation algorithm-based malware detection. IEEE Access **7**, 32765–32782 (2019). https://doi.org/10.1109/ACCESS.2019.2891588
7. Q. Liu, P. Li, W. Zhao, W. Cai, S. Yu, V.C.M. Leung, A survey on security threats and defensive techniques of machine learning: a data driven view. IEEE Access **6**, 12103–12117 (2018). https://doi.org/10.1109/ACCESS.2018.2805680
8. R.U. Rasool, U. Ashraf, K. Ahmed, H. Wang, W. Rafique, Z. Anwar, Cyberpulse: a machine learning based link flooding attack mitigation system for software defined networks. IEEE Access **7**, 34885–34899 (2019). https://doi.org/10.1109/ACCESS.2019.2904236
9. S. Kumar, K. Singh, S. Kumar, O. Kaiwartya, Y. Cao, H. Zhou, Delimitated anti jammer scheme for internet of vehicle: machine learning based security approach. IEEE Access **7**, 113311–113323 (2019). https://doi.org/10.1109/ACCESS.2019.2934632
10. M. Al-Qatf, Y. Lasheng, M. Al-Habib and K. Al-Sabahi, Deep learning approach combining sparse autoencoder with SVM for network intrusion detection, IEEE Access **6**, 2843–52856 (2018) https://doi.org/10.1109/ACCESS.2018.2869577
11. J.S. Raj, J.V. Ananthi, Recurrent neural networks and nonlinear prediction in support vector machines. J. Soft Comput. Paradigm (JSCP) **1**(01), 33–40 (2019)
12. K. Velswamy, A stochastic development of cloud computing based task scheduling algorithm. J. Soft Comput. Paradigm 41–48 (2019). http://doi.org/10.36548/jscp.2019.1.005
13. H. Wang, Sustainable development and management in consumer electronics using soft computation. J. Soft Comput. Paradigm 49–56 (2019). http://doi.org/10.36548/jscp.2019.1.006

Hand Gesture Recognition Using CNN

S. Preetha Lakshmi, S. Aparna, V. Gokila, and Prithviraj Rajalakshmi

Abstract This research work emphasizes the utilization of machine learning and convolution neural network (CNN) to recognize hand gesture lively, in spite of variations in hand sizes and spatial position in the image by providing our own personalized system inputs as a dataset representing the gestures according to the classes developed and to implement our model that will identify and classify the gesture into one of the defined categories. CNN utilizes three layers, where two are hidden layers and another one is convolution. The proposed model has been designed with three classes containing personalized gestures. The classes considered here are first-aid, food, and water. This model can be used for in-flight comfort facilities by travelers and also where there is a need for the use of these gestures.

1 Introduction

Humans recognize body and hand gesture easily. Hand motions are an essential part of communication. Growth of many automated technologies like computer vision, machine learning, neural networks, and deep learning will help in gesture controlling and recognition. It is interesting that preprogrammed human motion recognition from images captured using camera plays a significant role in the advancement of the artificial intelligent vision systems. For image classification problems, convolution neural networks (CNN) algorithms are typically applied. The usage of CNN architecture is a combination of layers that converts the image to a form that can be filtered quickly without suppressing the required characteristics so as to achieve exact output. An image classifier takes a picture as an input and classifies it into one of the possible categories that was trained to identify. Here, the design is built in a way that will help us identify and classify it.

Our idea is to implement our model in an application that will use a webcam (or an external camera) as an input device and it will then identify and classify the gesture into one of the categories that require to be defined. These gestures may also be used

S. Preetha Lakshmi (✉) · S. Aparna · V. Gokila · P. Rajalakshmi
Department of Computer Science and IT, Amrita School of Arts and Sciences, Kochi, Amrita Vishwa Vidyapeetham, Kochi, India

P. Karuppusamy et al. (eds.), *Ubiquitous Intelligent Systems*, Smart Innovation, Systems and Technologies 243, https://doi.org/10.1007/978-981-16-3675-2_28

to assist individuals, who have trouble in handling operating systems or devices. Classes are first-aid, food, and water. A folder path is given, where these gestures are stored according to their respective classes. Training and testing are performed on this dataset. Images will be read from the subfolder files where datasets are classified and stored. These images are read through gray channel, and in preprocessing state, resizing of image is done by passing them into an interpolation filter. During prediction of the tested image, the image is taken from the loaded model and displays its corresponding label, and during real-time testing, the image is captured and read from the webcam and then follows the same testing process for predicting the input's label (first-aid/food/water). In the comparative study of other papers, they have performed CNN on ASL and like gestures used to control home appliances but our model is unique which we developed by ourselves. This model can be used for in-flight comfort facilities by travelers as it provides the gestures for essential needs while traveling and also can be used where there is a need for the use of these gestures.

2 Literature Review

This paper uses the concept of convolutional neural network (CNN) to actually train the machine for recognizing gestures [1]. This mainly make use of two image bases having 24 gestures each, CNN for classification process and some kind of segmentation techniques. For separating the hand motions from the scene and also to delete the errors, for color segmentation, used neural networks, accompanied with morphological operations and a polygonal approximation. And they also made a logical operation, i.e., AND, which also included segmentation masks. The exact images produced the essential details of palm and the fingers. This paper generated successful results that had low computational costs. This mainly focused on static hand gesture recognition. The paper put forward a technique to present 32 separate letters and figures from American sign language [2]. The grant which improves the lives of the disabled is the sign language recognition technique. Principal component analysis is a method to get dimensionality cutback. The paper implementation is having a sequence of steps. Thus, sign language system is made using PCA and SVM followed by testing. The expected output was achieved with an accuracy of—80% using PCA. From the 2012 ImageNet competition, the network 'AlexNet' is continuously used to different types of computer vision tasks [3]. The paper had been solved with less computation and also came to the conclusion that best quality results can be achieved with low field resolution as 79 * 79. This is very useful to detect small objects. The paper focuses on the idea of CNN effect on large scale image recognition system [4]. Usage of deep convolutional networks for extensive image classification was evaluated. This research paper was all about the method of static hand gesture recognition using CNN [5]. In the dataset, data enhancing such as shearing, scaling, zooming, rotation, width shift, and height was provided. The design with augmented data achieved approximately 4% excess than the design without augmentation which had 92.87%

accuracy. This paper came up with a motion identification technique using convolutional neural networks [6]. The procedures involved for a good feature extraction are the usage of contour generation, morphological filters, polygonal approximation, and segmentation during preprocessing. Training and assessing are performed with many CNN. For validation of the robustness of the suggested technique, during training, all intended convergence graphs and metrics obtained are discussed and analyzed. This paper suggested CNN approach to recognize hand gestures from a camera picture of human task activities [7]. In order to achieve robustness, skin model, hand location and orientation arrangement to obtain CNN training and testing results. They assume a Gaussian mixture model (GMM) is used to train the model of skin to robustly eliminate an image's non-skin colors due to light problems. The hand model focuses on an ordinary pose with the rectification of hand location and orientation. The processed images are then used for CNN training. They suggested a validation technique to identify human movements that show robustness with different hand positions and light conditions. This paper put forward that by using a webcam to automatically route the region of interest (ROI), i.e., the hand region, and identify hand gestures for control of home devices (to build smart homes) or fields of human and computer interaction [8]. They then used framework subtraction to check the ROI and then use of the algorithm of kernelized correlation filters (KCF) to classify the ROI detected. To find multiple hand movements, the resulting image of the ROI is then resized and then placed into the deep convolution neural network (CNN). In this analysis, two profound CNN architectures are made and AlexNet and VGGNet are made of these architectures. Then, the above technique is again replicated to get an instant impact, and until and unless the hand is shifted out from the camera range, method execution resumes. This paper proposed that human hand gestures are noticed and interpreted using the classification methods of CNN [9]. Using mask image, the image's hand area is isolated from the whole image. The adapted histogram equalization method is used to increase the distinctiveness of each pixel in an image. This paper makes use of connected component analysis algorithm for extracting the finger tips from the image of the segmented hand. To distinguish the image into various classes, the resulted segmented finger regions are then provided to the CNN classification algorithm. This paper marks higher performance using state-of-the-art methods. Integrated RNN layers into the FCNs [10]. They developed a network which is end-to-end connected and detected human skin. Here, they selected gestures related American sign language (ASL) and applied deep learning on 24 hand gestures for recognition [11]. They showed that stacked denoising autoencoder and CNN are capable of learning hand gestures classification tasks and produce results with lower error rates and proposed training on skin features based on stacked autoencoders learning algorithm [12]. Experiments show that this algorithm reduced the difficulty in identifying skin pixels present in foreground skin area in their data sets. The proposed feature fusion-based CNN and a system that analyzes on three different benchmark datasets were compared using CNN [13]. Experiments was performed using depth data, gray scale, and binary also with two distinct validation techniques. Their process is based on exemplar's production and how a dataset (standard ASL) collected from five individuals and image include variations in hand posture and lighting [14]. Work is done using moment invariants. They focused on problem of collecting data from signers and on fingerspelling after which translating them into a written format [15]. SLR involves two main challenges that entail gathering of isolated examples of letters and separately translating letter one by one, which was trained on the SVM classifiers by applying vectors that consist of the Euclidean space from fingertips to palm center. Next one involved translating a word shown in sign and transform into a series of letters. It was achieved by a process, i.e., from the leap controller a frame, was extracted and identified the sign in frame and determined when the signs representing a letter were formed.

3 Material and Methods

3.1 Proposed System

What we proposed focuses on implementing the recognition of hand gestures using machine learning and CNN by providing our own personalized framework that can be used for in-flight comfort facilities for travelers and also where there is a need for the use of these gestures. The dataset used in this paper for the purpose to train and test is personalized inputs of dataset gathered through the mobile camera or webcam by clicking on images of our own hands that represent gestures or signs according to the classes we developed. It will then classify the gesture and categorize it into one of the groups we have identified during testing. During prediction of the tested image, the image is taken from the loaded model and display its corresponding label it belongs to, and during real-time testing, the image is captured and read from the webcam and then follows the same testing process for predicting the inputs label. Water, food, and first-aid are the courses we took. Thumb sign is used to portray drinking water, while 'L' shaped (using pointing finger and thumb) is used portray first-aid and food is depicted by sign two (number two).

We use the CNN approach to understand hand gestures. The CNN algorithm is used to classify an image based on various characteristics and make it possible to differentiate it from its respective classes. By passing the input images across different layers, the CNN technique works. In CNN, we used three layers, two are hidden layers and one layer is convolution; in these hidden layers, machine learning process is done. The layers involve operation of convolution, ReLU, pooling, and fully connected layer for appropriate results to be obtained. In the CNN method, a classifier called 'softmax' is already developed that performs by applying probability and determines into which class the tested image must be categorized. The CNN design mechanism is based on the number of alternate convolution operations and pooling layers, the number of neurons in each layer and the activation function selection. (rectified linear unit and softmax) (Fig. 1).

- *Convolution operation layer*

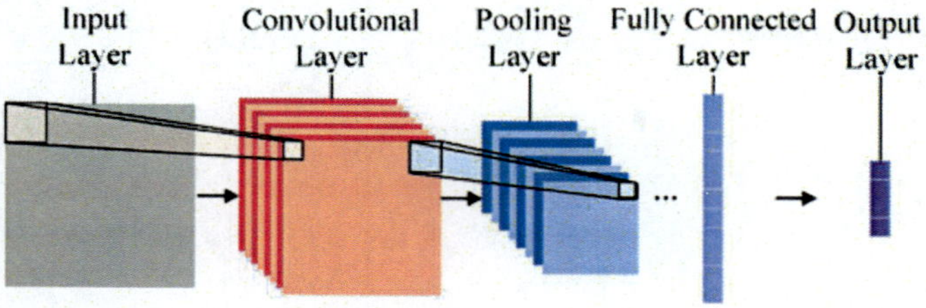

Fig. 1 CNN layers—an illustration

This layer is to obtain authentic features and from the input images and to migrate them to next layer. The convolution layer maintains a spatial pixel relationship that senses an image's features. The convolution layer thus maintains a pixel spatial relationship that senses an image's characteristics. To remove convolved features, the selected filter is then applied to the image. The filter just moves across input image and then sends the output to the convolved map. We carry out distinct filters with multiple convolution inputs, ending in multiple convolved maps. Then these converted maps are combined to form the final execution of the layers of convolution (Fig. 2).

- *Relu*
  This layer performs by passing the acquired output from the convolution layer to turn our input nonlinear with the means of this activation function. Noise is removed from the convolved feature and is substituted it with the value 0. Rectified linear unit (ReLU) is proven to provide the best solutions to the lack of gradient problems. Derivative of relu = $R(z) = \max(0, z)$. If $z <= 0$, then $R(z) = 0$, else if positive $R(z) = z$. Figure 3 shows this activation function.
- *Pooling layer*
  The objective of the layer is to isolate maps of features. It is achieved by summing/averaging values in the convolved feature maps that have been acquired. The goal of the pooling layer is to produce flexible convolved feature production by minimizing spatial dimensionality. The pooling layer functions as the filter slides over the transformed feature with the map size > pooling filter. 'Max pooling' is one of the most commonly used pooling methods. It operates by manipulating each patch's ceiling value taken from the convolved feature. The pooling layer operation is shown in Fig. 4.

Fig. 2 Convolution layer operation on pixels

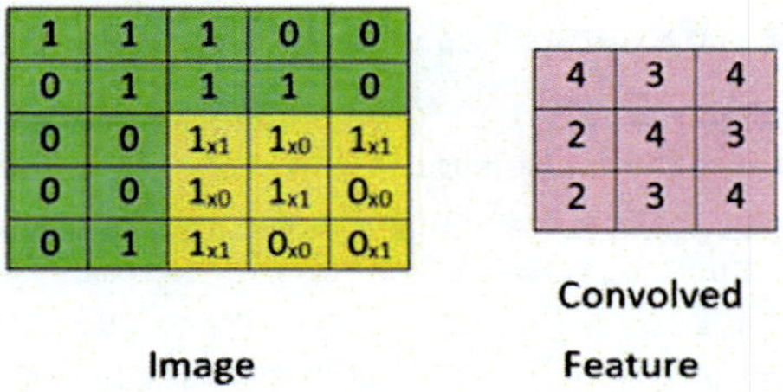

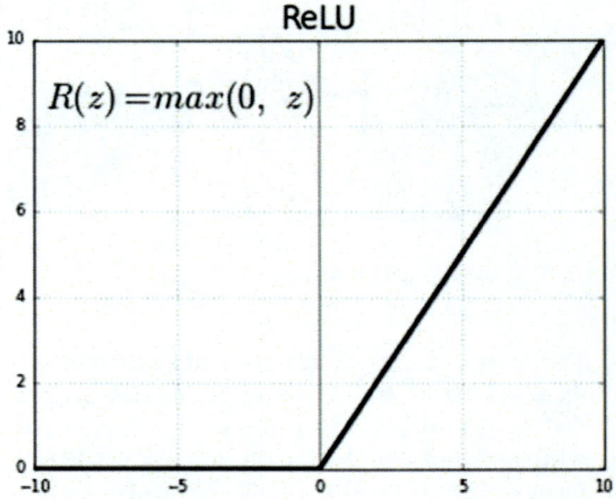

Fig. 3 Activation function ReLU

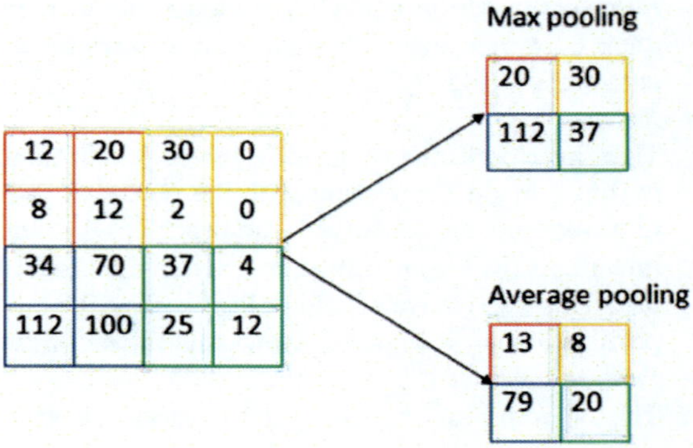

Fig. 4 How pooling layers operates

- *Fully connected layer*
  This layer purpose is to reassign the pooled feature map from 2D structure to 1D vector. The function of this layer totally depends upon the results from the convolution and pooling layer. This is the final layer in which all the feature maps are used and prepared for the classification part (Fig. 5).

4 Experiment and Result Analysis

In the Python version 3.6 library, we formulated and implemented our CNN algorithm and used keras/TensorFlow (CNN training). Imported Python libraries—opencv, numpy, matplotlib. Images will be read from the subfolder files where datasets are classified and stored. These images are read through gray channel, and in preprocessing state, resizing of image is done by passing them into an interpolation filter.

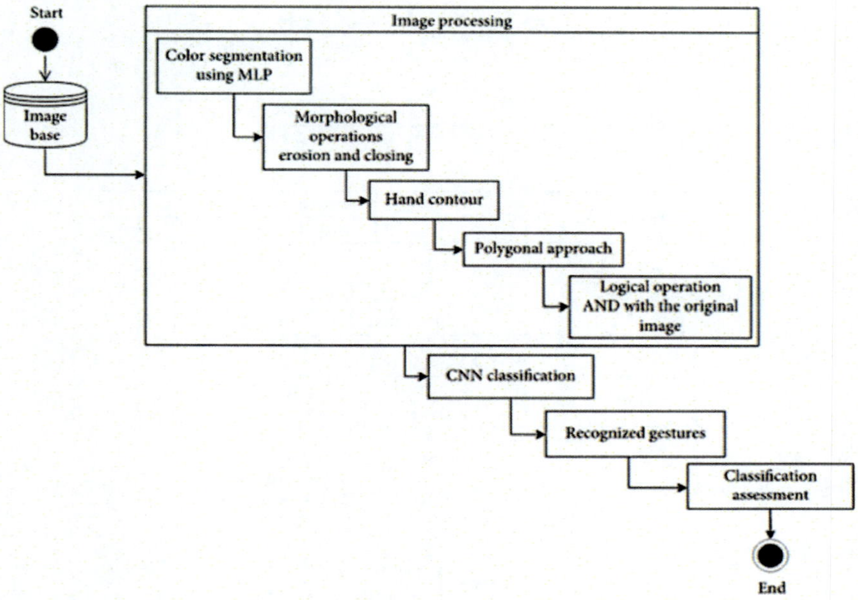

Fig. 5 Sequence of processes in our methodology

The whole dataset was divided into sets for training and testing. For training, 70% of our dataset is used and the remaining 30% is used for testing; this is achieved by importing the scikit-learn Python library. Model will be evaluated on that 30 percentage of dataset. Categorized using one-hot-encoding. Imported sequential model from keras library.

The CNN method already contains a built-in classifier called '**softmax**' which helps to classify the tested images by using probability to which class image has to be assigned (Fig. 6).

Where 'zi' is the input vector and can take any real value. '*K*' is number of classes in multi-class classifier. Where each entries range will be (0–1) and ensures that all the output values of the function will add up to 1, thereby creating a correct probability distribution (Fig. 7).

We used cross-entropy metrics as the inputs are categorized into their respective classes. Inputs will be trained and will show what is its corresponding label (class it belongs to). Imported Adam optimizer as learning parameter to learn how to read this model. Learning rate starts from 0.05 and it modifies the initialized learning rate by optimizing it enabling the gradient descent to converge to global minimum

Fig. 6 Equation of softmax function

$$\sigma(\vec{z})_i = \frac{e^{z_i}}{\sum_{j=1}^{K} e^{z_j}}$$

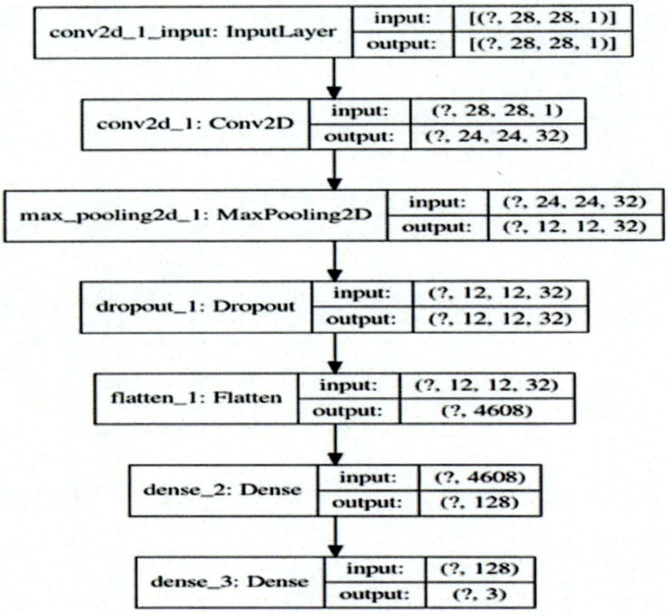

Fig. 7 Our model architecture

successfully. The number of **epochs** is how many times of data passes through the training, our epochs = 10. The **batch size** is how many samples to be passed at a time and processed before the model is updated, our batch size = 64.

Confusion matrix used for the performance of classifier or model of classification on a set of tested images for which its true value of its respective classes is known. Standard equation to find its accuracy for multi-class model is (TP + TN)/(TP + TN + FP + FN). Figure 8 shows an confusion matrix of our model, and Fig. 9 shows classification report of predictions.

4.1 Gestures

See Figs. 10, 11, 12, and 13.

4.2 Formulas

Accuracy evaluation metrics

Fig. 8 Confusion matrix of our multi-class model

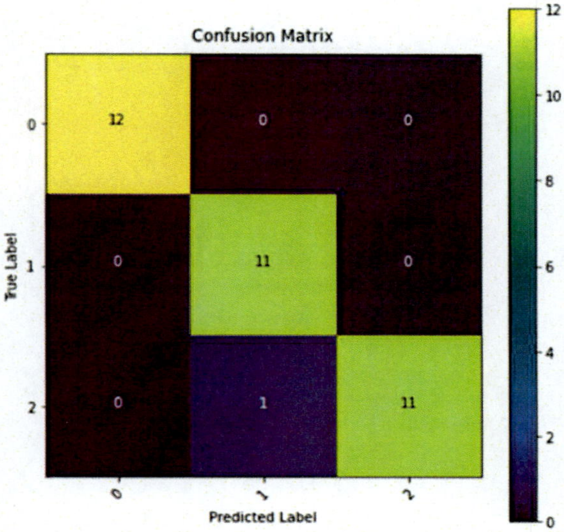

	precision	recall	f1-score	support
0	1.00	1.00	1.00	12
1	0.92	1.00	0.96	11
2	1.00	0.92	0.96	12
accuracy			0.97	35
macro avg	0.97	0.97	0.97	35
weighted avg	0.97	0.97	0.97	35

Fig. 9 Predictions

Fig. 10 Water

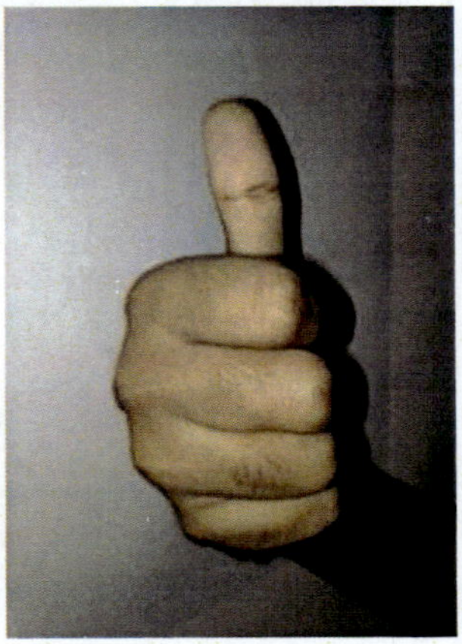

Fig. 11 Food

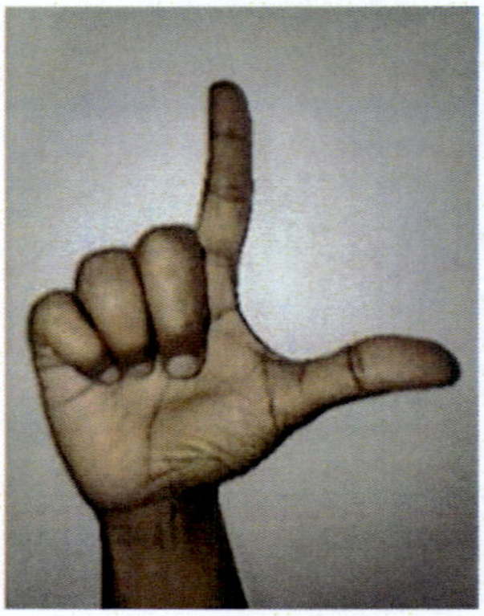

Fig. 12 First-aid

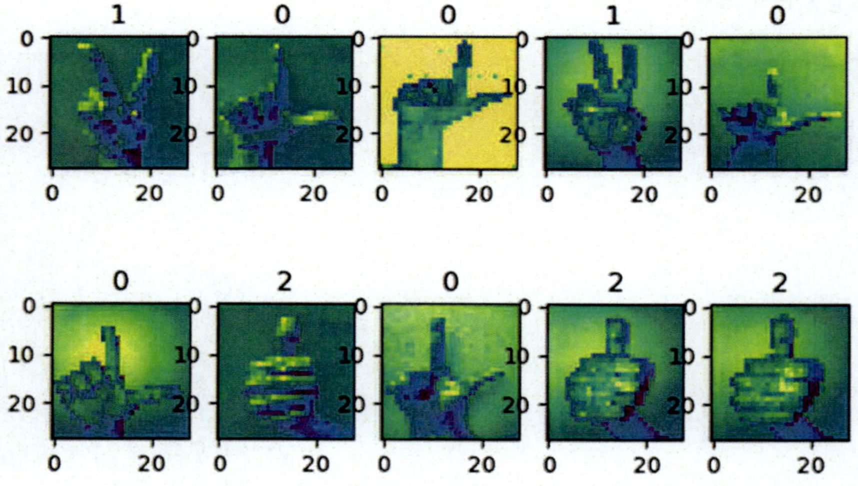

Fig. 13 Training images with its numerical labels (0—First-aid, 1—Eat, 2—Drink)

$$A = (c/t + r) * 100$$

where 'c' represents total number of correct classifications. 't' represents number of correct input and number of incorrect input is represented by 'r.'

Descriptor	No: of images for training	No: of classes	Accuracy
CNN	81	3	97.14%

5 Conclusion

We came toward the conclusion that CNN method is the best way to provide the optimal result needed and within less processing time. We introduced a new model with customized inputs and that seem to have more scope in-flight services. We analyzed that CNN provides an accuracy of 97.14% of recognition rate with error value of 2.86% on training set and accuracy of 97.14% on validation set. The accuracy can be improved more. We were able to collect personalized dataset from limited number of people as it should be an image clicked with similar lighting condition and properly shown hand gestures as per our assigned classes of gestures to get more accuracy. Later, we would be focusing on that and future works include experimenting with various transfer learning model, various activation functions, and better feature engineering.

References

1. R.F. Pinto Jr., C.D.B. Borges, A.M.A. Almeida, I.C. Paula Jr., Universidade Federal do Ceará, Sobral, Ceará: Static hand gesture recognition based on convolutional neural networks 62010-560 (2019)
2. N.A. Ming-Hsuan Yang, M. Tabb, Extraction of 2d motion trajectories and its application to hand gesture recognition. IEEE Trans. Pattern Anal. Mach. Intell. **24**(8), 1061–1074 (2002)
3. C. Szegedy, V. Vanhoucke, S. Ioffe, J. Shlens, Z. Wojna, Rethinking the inception architecture for computer vision, in *Proceedings of the IEEE Conference on Computer Vision and Pattern Recognition*, Las Vegas, NV, USA (2016), pp. 2818–2826
4. K. Simonyan, A. Zisserman, Very deep convolutional networks for large-scale image recognition. http://arxiv.org/abs/1409.1556 (2014)
5. Static hand gesture recognition using convolutional neural network with data augmentation, https://www.researchgate.net/publication/333618617_Static_Hand_Gesture_Recognition_using_Convolutional_Neural_Network_with_Data_Augmentation
6. Static hand gesture recognition based on convolutional neural networks, https://www.researchgate.net/publication/336427446_Static_Hand_Gesture_Recognition_Based_on_Convolutional_Neural_Networks
7. Human hand gesture recognition using a convolution neural network, https://www.researchgate.net/publication/286791813_Human_hand_gesture_recognition_using_a_convolution_neural_network

8. An efficient hand gesture recognition system based on deep CNN, https://ieeexplore.ieee.org/document/8755038
9. An efficient method for human hand gesture detection and recognition using deep learning convolutional neural network, https://doi.org/10.1007/s00500-020-04860-5
10. H. Zuo, H. Fan, E. Blasch, H. Ling, Combining convolutional and recurrent neural networks for human skin detection. IEEE Signal Process. Lett. **24**(3), 286–293 (2017)
11. O.K. Oyedotun, A. Khashman, Deep learning in vision based static hand gesture recognition. Neural Comput. Appl. **28**(12), 3941–3951 (2017)
12. Y. Lei, W. Yuan, H. Wang, Y. Wenhu, W. Bo, A skin segmentation algorithm based on stacked autoencoders. IEEE Trans. Multimedia **19**(4), 740–749 (2017)
13. S.F. Chevtchenko, R.F. Vale, V. Macario, F.R. Cordeiro, A convolutional neural network with feature fusion for real-time hand posture recognition. Appl. Soft Comput. **73**, 748–766 (2018)
14. A.L.C. Barczak, N.H. Reyes, M. Abastillas, A. Piccio, T. Susnjak, A new 2D static hand gesture color image dataset for ASL gestures. Res. Lett. Inf. Math. Sci. **15**(4356), 12–20 (2011)
15. M.W. Cohen, N.B. Zikri, A. Velkovich, Recognition of continuous sign language alphabet using leap motion controller, in *Proceedings of the 2018 11th International Conference on Human System Interaction (HSI)*, Gdansk, Poland (2018), pp. 193–199

An Integrated Three-Port DC–DC Modular Power Converter with Multiple Renewable Energy Sources Suitable for Low and Medium Power Applications

R. Sekar, D. S. Suresh, and H. Naganagouda

Abstract In the recent energy scenario, the electrical sources are operating in the integrated form rather than stand-alone system. Preferably, the renewable energy sources are synchronized with the existing sources and contributed to the supply of power. But the fact is that the renewable energy sources are intermittent in nature. So, we are constructing a power electronic module with the facility of integrating multi-sources in providing the power to the common grid in continuous manner. Besides, the common grid connected with live loads has the fluctuations in supplying load frequently. Electrically, this situation is termed as transient operating condition. With this, the power electronic module must have the additional features such as handling the frequent variations in the load side and to manage the transients in the load/grid. By taking all this consideration, the research on constructing power electronic module is carried out with the results of above-mentioned facility; in addition, the power/energy handling method is also explained. The multi-port DC–DC converter with unique topology is constructed and validated with the experimental results.

1 Introduction

The actual scenario in power sector is they are trying to match the load demand with the available power that are collected from various sources. The energy sector consultants are trying at most to find the alternative energy resources, which are consistency in nature to meet the ever-increasing energy demand with high quality of power. Even this research work is an attempt to resolve the one of the issues come

R. Sekar (✉)
Channabasaveshwara Institute of Technology, Gubbi-Tumkur, VTU-Belagavi, Karnataka, India

D. S. Suresh
Department of ECE, Channabasaveshwara Institute of Technology, Gubbi-Tumkur, VTU-Belagavi, Karnataka, India

H. Naganagouda
National Training Centre for Solar Technology, KPCL, Bangalore, Karnataka, India

P. Karuppusamy et al. (eds.), *Ubiquitous Intelligent Systems*, Smart Innovation, Systems and Technologies 243, https://doi.org/10.1007/978-981-16-3675-2_29

across when the renewable energy resources are tied up with the existing energy resources.

In the present-day energy development, the higher educational institutions and other research and development laboratories operated by the government and non-government organizations encourage the research work in the area of the renewable energy system and its subsystem developments. Since the power demand is increasing day by day, designing the power system for meeting the load is becoming a complex task [1]. The inclusion of renewable energy sources along with the battery and other load buffering devices in the existing power system makes the problem density still worser. Due to the innovations and advancements in power electronics, handling those technical glitches becoming easier, further the cost involved in establishing the entire system is decreased and in turn, it slashes the power tariff to the consumers.

As mentioned earlier, the usage of the battery and other storage equipments in the existing power system is unavoidable; besides electric vehicles also use the storage equipment. Establishing the facility in a remote locality for charging the battery used in electric vehicles is trending now. It is found that there is a wide scope in designing the entire facility with the existing power system by adding multiple renewable energy sources [2]. Multiple energy resources and their synchronization will bring lots of benefits such as improvising system performance, lowering the cost, keeping the sources away from the load fluctuations, and enhancing the dynamics of the system. With these benefits, integrating more than one renewable energy source and managing the power effectively on the load side are a matter of interest [3].

Before integrating the resources electrically, the individual sources involved are analyzed through their lively behavioral characteristics. The various DC–DC converter topologies with their isolation features were investigated, in which the multi-converter topology and multi-port converter topologies are the two major categories [4, 5]. The prior one is meant for connecting the source and load by means of linking with various converters; the later one is using some of the common equipment in its structure such as a high-frequency isolation transformer and filter circuits, through which the cost involved is low and increases its power density. Therefore, the multi-port converter topology for renewable energy systems turned the researcher's attention toward it.

Developing a DC–DC multi-port converter using unique topology with the necessary subsystem equipment results in the full-fledged structure of the converter; provision for constructing the centralized controller is an added advantage in this topology. This research work incorporates the common controller circuit that includes the MPPT technique, ZVS scheme, and an effective switching control technique configuration using PWM techniques, which will add the advantage of using the central controller.

2 Proposed Circuit Description and Working

The power circuit mentioned in Fig. 1 with the unique topology has three ports; individual ports are connected to sources, battery, and load. The ports are integrated using the high-frequency transformer. The Port-2 consists of a supercapacitor and battery for buffering purposes, and it engages the load during the transients and the source fails to supply the load.

The boost converter components, L1, SW-1, and L2, SW-2, are connected with the sources; the diodes are used to ensure the unidirectional current flow from the source. The capacitor C1 in the Port-1 discharges and isolates the entire Port-1 from other ports. Sometimes, the switches are forced to carry the circulating current, so they are selected to the maximum current rating of the sources with the suitable tolerance values. The switch SW-3 is meant for controlling the current flow of the entire Port-1, this switch will come into action whenever the sources generate more power than the required.

As mentioned in the previous paragraphs, the advanced switching scheme using PFM and PWM duty ratio control method is incorporated for the power circuit operation with the reference of renewable energy sources output, load conditions, and the storage port component circumstances. For a better understanding, the power circuit performance is presented through various modes of operation.

MODE-1

The mode-1 operation is well described in Fig. 2. Whenever the sources are not generating the required electrical output, the sources must be isolated and the other sources must engage the load because it is defined that the proposed topology will ensure the reliable operation and the load will be engaged continuously.

Further, the power management facility ensured the sources short-circuit current value and switched ON the SW-1 and SW-2 repeatedly with the defined switching frequency. With this, the renewable sources are not engaging the load, rather the storage components will take care of the load. Though the sources are kept isolated,

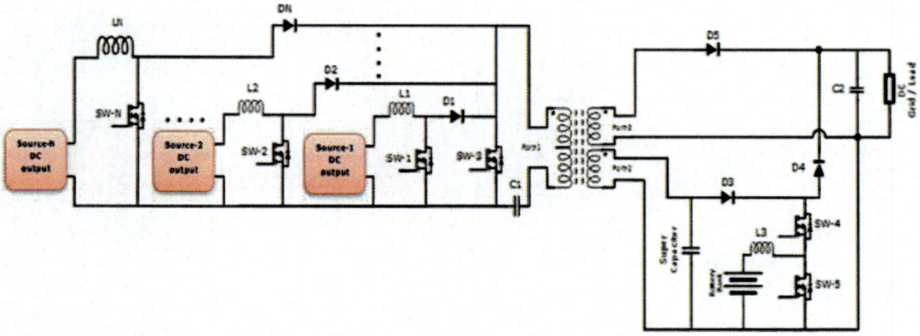

Fig. 1 Proposed power circuit with 'N' resources

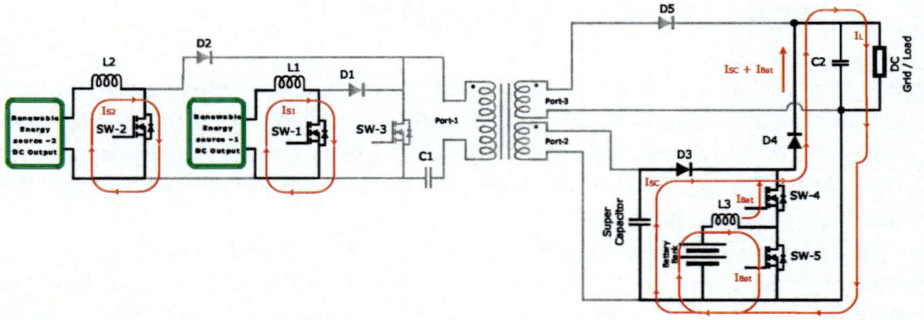

Fig. 2 MODE-1: power circuit operation

the circulating current with the minimum value is used for charging the respective inductor.

MODE-2

During this mode, the sources are generating sufficient amount of electrical output due to the adequate irradiation and wind flow available in the environment, through which the Port-3 will carry the required power to the load without any interruption. At the same time, the Port-2 storage components also switched ON for charging. Further, the supercapacitor is engaging the load continuously; the supercapacitor dischargement is shown in the dotted arrow marks. In this mode of operation, the source-side switches are kept in the OFF condition, so that the switching losses are considerably minimal (Fig. 3).

MODE-3

During the summer and rainy sessions, the solar PV panels and wind generator electrical outputs exceed their maximum values. In that situation, the electrical power circuitry must be operated in a safe mode to avoid damages. But the challenge is

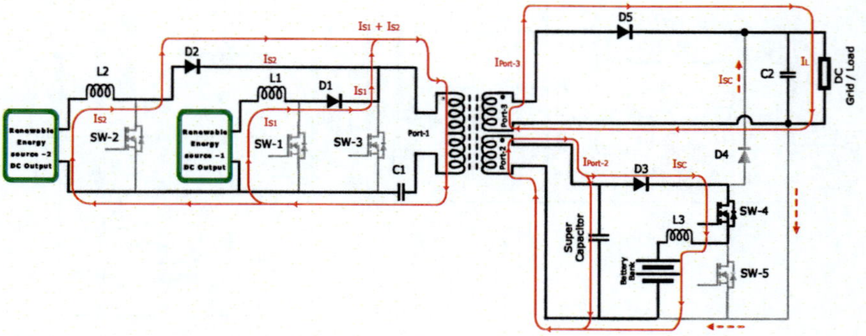

Fig. 3 MODE-2: power circuit operation

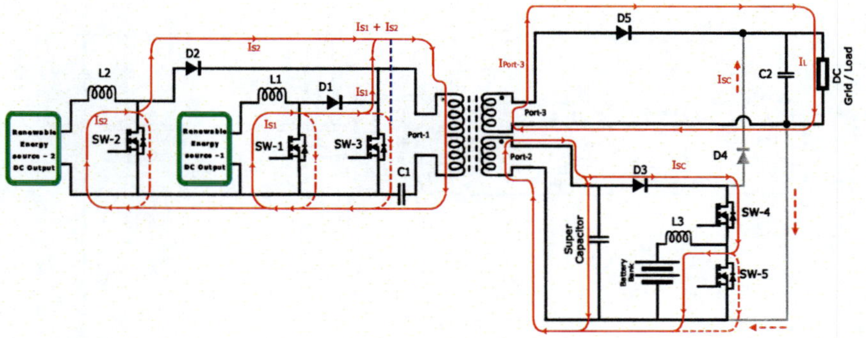

Fig. 4 MODE-3: power circuit operation

continuous power delivery to the load. By considering the reliable operation during these abnormal conditions, the power circuit has the facility to subside the excess power by operating the middle switch SW-3 with the required/calculated duty ratio. Simultaneously, the source-side switches such as SW-1 and SW-2 also operate with the defined safe duty ratio values (Fig. 4).

Also, the storage port equipments are turned ON to charge condition by turning ON their respective switches. In turn, it is operating as a load for the sources, and certainly the excess power generation will be controlled and the entire system is electrically safe operating zone.

MODE-4

This mode of operation is about when the PV and wind sources are not generating the electrical output, which is required to satisfy the load, but at the same time, a considerable/lower amount of power generation is in the sources side. This situation occurs whenever there is a minimum amount of irradiation and wind flow. In general, whenever the sources are not connecting with the load, it must be isolated from the load and other parts of the circuitry. But in this mode, even though the sources are not generating the electrical output, the sources will be utilized to the maximum extent.

In this mode of operation, both the sources' optimization is very high. Besides, the switches are ON and OFF frequently with the appropriate duty ratio, which results in the boosted output from the source side. Port-2 switches are in OFF condition, so the battery is not charging in this mode of operation. But the supercapacitor is engaged with the load which is marked using the arrow marks (Fig. 5).

MODE-5

This mode of operation is the extension of previous mode, in which the Port-2 supercapacitor is in the charging mode. The boosted electrical output from the source port is shared by Port-2 and Port-3. At the same time, the supercapacitor is connected with the load through which the changes in the load side will be taken care by the supercapacitor.

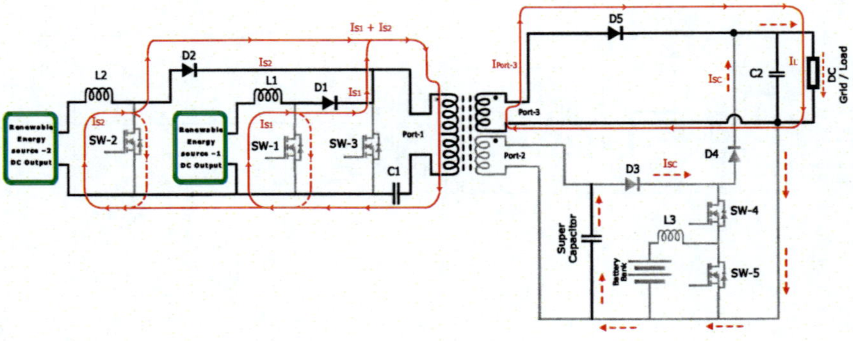

Fig. 5 MODE-4: power circuit operation

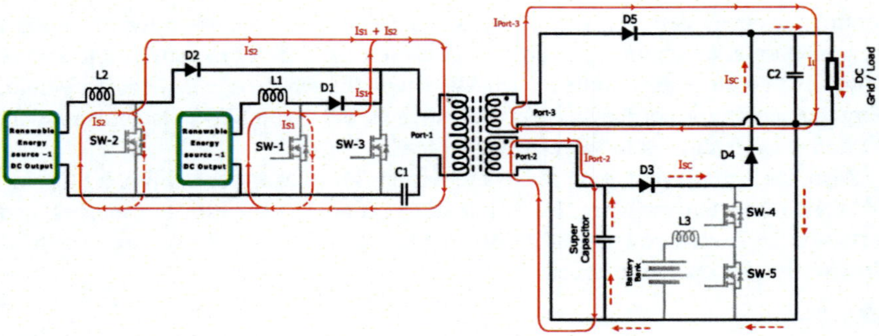

Fig. 6 MODE-5: power circuit operation

In this mode, the supercapacitor charging and discharging is shown. The supercapacitor control is on the switches SW-4 and SW-5, alongside, the entire Port-2 control can be taken by those two switches. For the safety reasons, the SW-4 and SW-5 are operating in non-simultaneous operation (Fig. 6).

As a summary of all the working modes, the tabular column, which is presented below, emphasizes the switching sequences of various modes given above (Table 1).

3 Power Circuit Analysis

To derive the theoretical analysis results, the possible operation modes and their circuits are taken for the discussion. In this proposed power circuit operation, by considering the duty cycle there are two possible conditions such as switching ON and OFF of the boost mode operation. With this, the analysis is taken forward (Fig. 7),

For the analysis, the tabular column No. 2 contents are essential,

Table 1 Switching sequences during all the modes of operation

Switching sequence					
	S1	S2	S3	S4	S5
Mode-1	ON	ON	OFF	ON	ON
Mode-2	OFF	OFF	OFF	ON	OFF
Mode-3	ON	ON	ON	ON	ON
Mode-4	ON	ON	OFF	OFF	OFF
Mode-5	ON	ON	OFF	OFF	OFF

ON indicates—turn ON and OFF, based on the duty ratio. OFF indicates—the switches are in OFF condition during the entire mode of operation. Though the switching sequence is same, functionality-wise the modes-4 and 5 are different

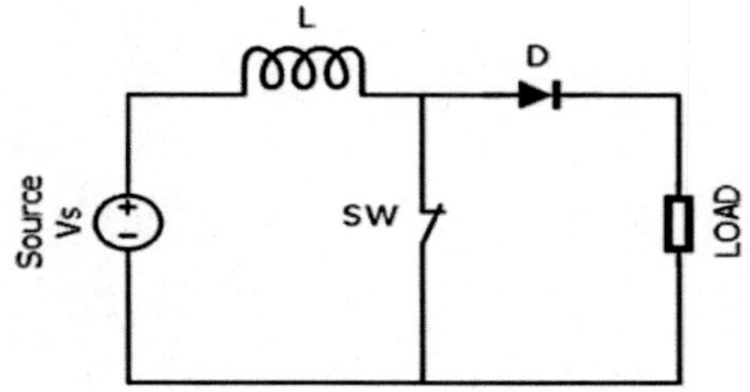

Fig. 7 Circuit operation in Port-1 and Port-2 switches is closed

The total switching period is T, and the switch is closed during T_{ON} and opened during $T_{OFF} = (1 - T_{ON})$

The duty ratio is defined as,

$$(D) = \frac{T_{ON}}{T}$$

When the mode of the switch is closed,
the KVL for the path V_s, L, and SW, is

$$V_S = V_L = L\frac{diL}{dt}$$

$$\frac{diL}{dt} = \frac{V_S}{L}$$

$$\frac{\Delta iL}{\Delta t} = \frac{\Delta iL}{D} = \frac{V_S}{L}$$

$$(\Delta iL)_{closed} = V_S.\frac{D}{L} \tag{1}$$

When the mode of the switch is opened (Fig. 8),

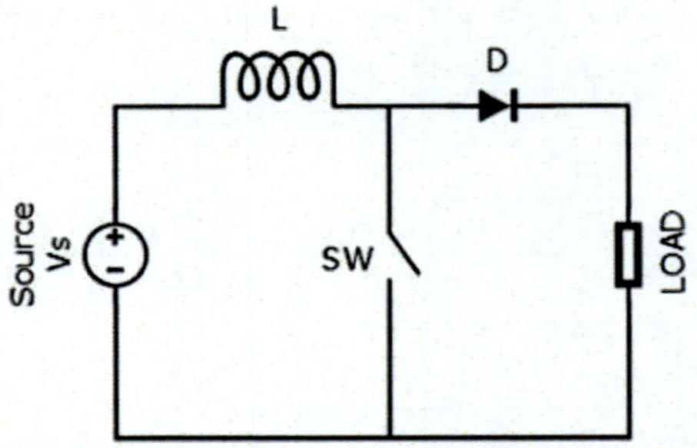

Fig. 8 Circuit operation in Port-1 and Port-2 switches is opened

the inductor voltage can be calculated as,

$$V_L = V_S - V_O = L.\frac{\Delta iL}{\Delta t} \tag{2}$$

$$\frac{\Delta iL}{\Delta t} = \frac{(V_S - V_O)}{L}$$

$$\frac{\Delta iL}{\Delta t} = \frac{\Delta iL}{(1-D)} = \frac{(V_S - V_O)}{L}$$

$$(\Delta iL)_{\text{open}} = \frac{(V_S - V_O)(1-D)}{L}$$

The total current during both switches is closed and opened resulting to zero, so the expression can be written by adding Eqs. (1) and (2)

$$(\Delta iL)_{\text{Open}} + (\Delta iL)_{\text{Closed}} = 0$$

$$V_S.\frac{D}{L} + \frac{(V_S - V_O)(1-D)}{L} = 0$$

$$V_S(D + 1 - D) - V_O(1-D) = 0$$

$(T_{on} + T_{off}) = 1$ the expression can be written as,

$$(V_S) - V_O(1-D) = 0$$

$$(V_s) = V_o(1-D)$$

$$(V_S) = V_O(1-D)$$

$$V_O = \left(\frac{V_S}{1 - D}\right)$$

With this expression, it is concluded that the output of this circuit completely relies on the duty ratio selected. The highest/maximum range of duty ratio selected for the switches will yield the maximum output. To get more clarity on operating this novel topology designed for three-port converter, operation is divided into five different modes of action, which are discussed exhaustively in Sect. 2.

4 Power Flow Management and Control

The block diagram shown below represents the power/energy management and control mechanism along with respective feedback regulators for the proposed research work (Fig. 9).

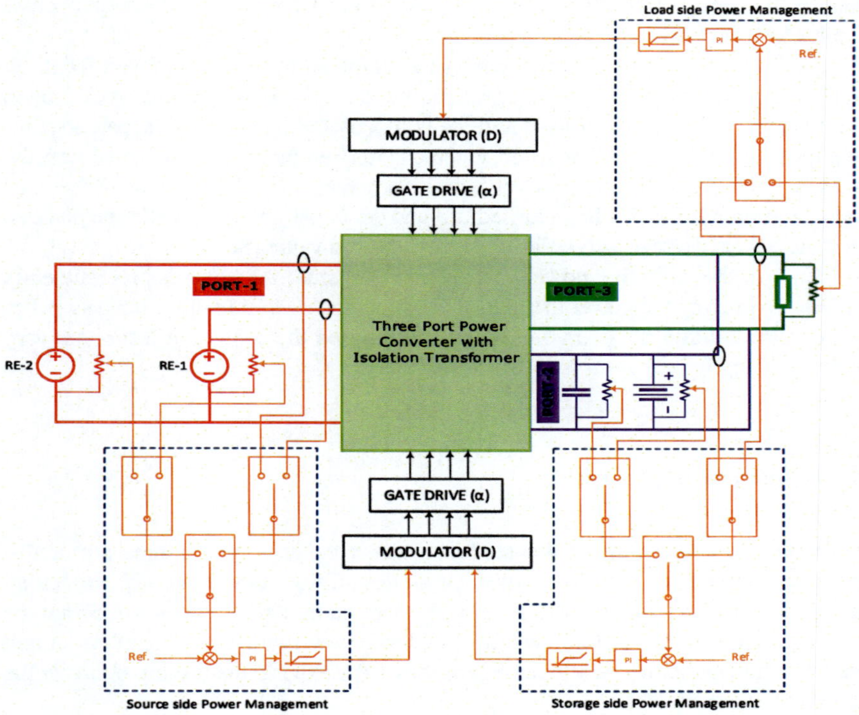

Fig. 9 Block diagram of power flow management and control

To reduce the complexity of the power circuit operation, the decentralized power management segments are deployed such as source-side power management, storage-side power management, and load-side power management. The voltage and current estimations of the individual sources from the specific estimating points are distinguished; at that points the voltage and current estimation gadgets are kept ready for recovering the actual quantities [6]. The actual functionality of energy management and its control structure are explained below.

The research is mainly considering renewable energy as input sources, preferably solar and wind energy. This energy management scheme is incorporated with the MPPT techniques; by varying the duty ratio, the maximum power utilization from the sources is achieved. Besides, we are well known that the sources are intermittent and they cannot handle the load power at all times [4]. Whenever the sources are delivering the power to the load, the respective switches are kept open and the entire power can be managed by the sources themselves. During the high-power generation, the storage equipment will be turned on for charging and then the rest of the controlled/required power will be delivered to the load [7]. Whenever the source outputs are lesser than the required, the respective switches must be turned ON with the calculated duty ratio (*D*). Through this, the Ton will be varied to achieve the desired output across the load.

The SoC of the energy storage port is measured all the time and given feedback to the moderator of the power management scheme. Also, the transients occurring in the load are precisely supervised and the compensated by using the supercapacitor and the sources are in idle condition, and the battery elements are connected with the load by turning ON the respective switch. If the battery is not charged up to the rated value, the switch SW-5 will be turned ON and the boost operation will take place so that the battery voltage will build-up to the required value [8].

Through this effective power management scheme, the sources are efficiently utilized to provide the power to the load [9]. In all the operating modes, the switches are operated within the operating range and ensured the switching losses are very minimum/negligible.

5 Experimental Results

The systematic procedure is followed for constructing the proposed three-port multi-input converter and its subsystems such as designing a new topology, modeling, simulation, and to validate the previously derived/modeled values; its prototype is constructed in the scale-down form and validated with the help of its output waveforms. The below-mentioned figure shows the experimental setup done for the proposed research work.

The converter is proposed to interface with hybrid renewable energy systems operating as a stand-alone DC micro-grid power grid. The specifications of the components involved in constructing the model are shown in Table 2. The control strategies are fed into the Arduino Mega-2560 controller for generating the five

Table 2 Components and their assumptions for the analysis

Component existing in the power circuit	Assumed as
All the renewable energy sources	DC sources (V_s)
Battery	DC source (V_s) during port-2 operation
The primary winding of the transformer	Load during Port-1 operation
The inductor current	Continuous
Overall, all the components used	Ideal components

different isolated pulses for driving the switches. The acceptable duty ratio values are calculated by the controller automatically and drive the switches (Fig. 10).

As discussed in the previous sections, the experimental model is constructed with the necessary components. To energize the control circuit, the step-down transformer is used for reducing the voltage from 230 to 12 V, through which the required inputs for controller and the triggering circuits are taken. The power circuit consist of five MOSFET switches (IRF540) with the designed inductors.

The triggering module is individually isolated with the TLP230 opto-isolator. The filtered output of the power circuit is connected with isolation transformer. The

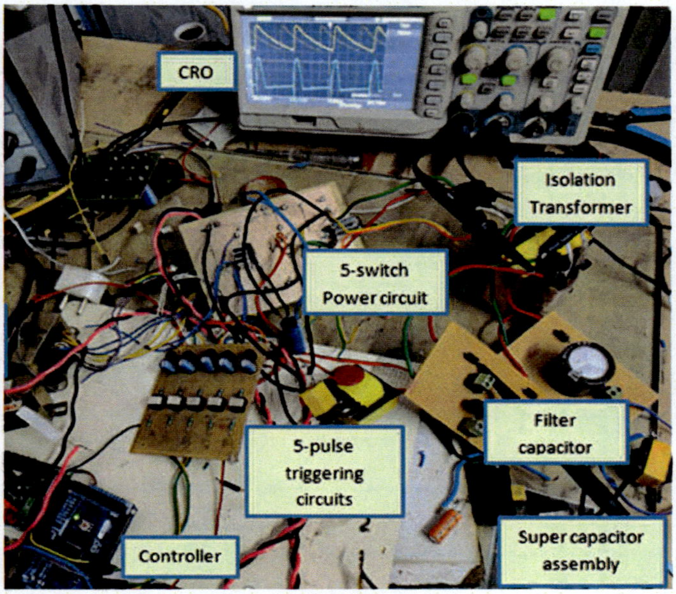

Fig. 10 Experimental setup of the proposed research work

secondary of the transformer consists of storage port and load port with the diode arrangements as shown in Fig. 1.

The DC output from the sources is used to pass through the inductor, and based on their magnitude of the output its respective switches will be triggered which is decided by the central controller with the defined duty ratio. It results in with the required output voltage across the primary of the isolation transformer. The waveform shown in Fig. 11 is the output of the source with the defined switching frequency. The vertical line appears during every switching, which is due to the inductor dischargement in the boost mode of operation.

The current flows through the sources are measured by connecting the ammeter in the source side. The waveform shown in Fig. 12 shown the above is the current waveform of any one of the sources. In this waveform, the lines appearing in every switching are similar to the voltage form shown in Fig. 11. The transient line appearing in every switching can be reduced by selecting/altering the duty ratio of the switches appropriately.

For the verification of the results, the load connected in the secondary of the transformer in the Port-3 is resistive load. The waveform shown in Fig. 13 is the output voltage measured across the load. It shows that the waveform is pure DC. The voltage and current waveforms appearing in Figs. 11 and 12 are controlled by the Port-2 and the filter circuit is connected in the Port-3. It shows that the output of the proposed circuit is delivering the power to the load in all the situations/modes. So, the reliable power supply to the load is ensured.

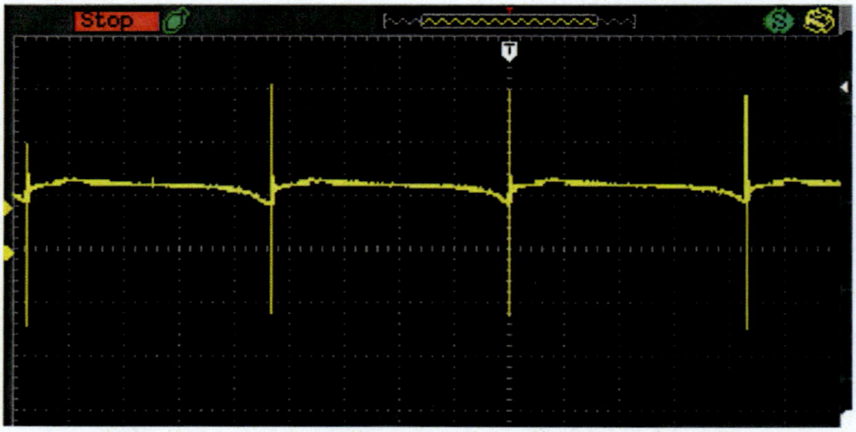

Fig. 11 Output voltage waveform from the sources

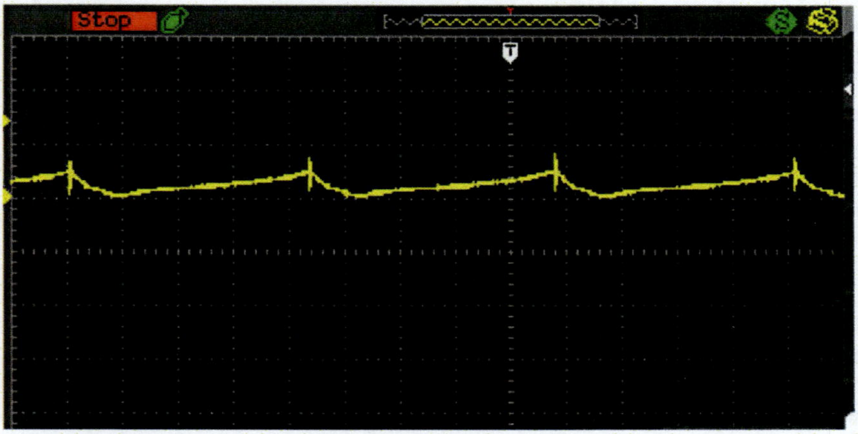

Fig. 12 Current waveforms of the sources

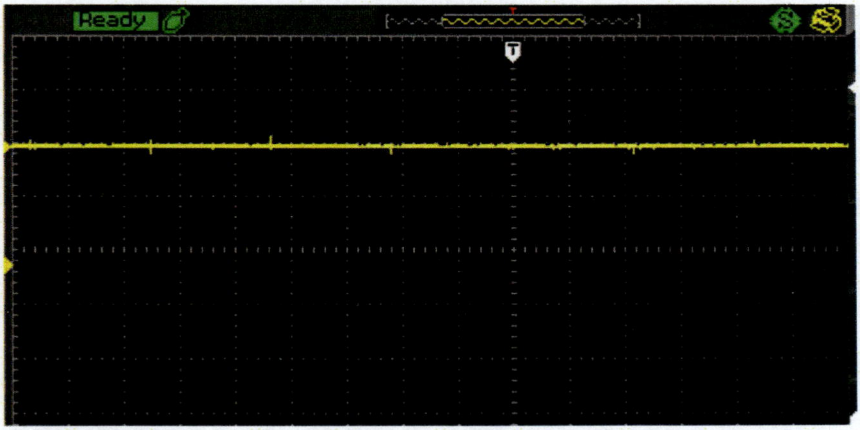

Fig. 13 Voltage across the load (R-load)

6 Conclusion

An isolated topology employed on multi-port converter embedding with multi-input source facility has been proposed in detail. This unique topology was explained through possible modes of operation. In addition, the transient and steady-state variations occurring in the load side are appropriately compensated using the battery and supercapacitor as explained in the modes of operation.

The simple analysis on the power circuit was carried out; using this, the relativity between the input and output was analyzed in terms of duty ratio required for switching. The entire operation of the power circuit was focused on controlling the output voltage.

To illustrate the actual working, the voltage and current waveforms of the proposed converter are shown in Figs. 11, 12 and 13 as references. In all the modes of operation, the output waveform was observed as filtered DC output with the aid of its filter components. The miniaturized prototype model (open loop) was constructed, and the experimental tests were conducted. And it was found that the results derived were in line with the theoretical outputs.

With this, it is evidenced that in all the modes of operation, the output waveforms derived from the various modes of operation are maintained as a constant DC output. Further, during all the modes of operation, the load is engaged, and for testing the transients, the load connected in the Port-3 is switched ON and OFF consequently, during the instants the supercapacitor was engaging the load. And it was ensured with the waveform derived.

To demonstrate more on the uniqueness of the research work carried out, primarily the structure/frameworks of the power circuit are taken into the consideration along with the improvement in overall efficiency of the system, wherein the common structure is embedded with all the source, load, and storage components with compact structure. The number of switches is comparatively lower, so the switching loss claimed in the system is less in turn improving efficiency. Further, this model/structure can be taken up further into MIMO systems.

References

1. A.K. Bhattacharjee, N. Kutkut, I. Batarseh, Review of multiport converters for solar and energy storage integration. IEEE Trans. Power Electron. **34**(2), 1431–1443 (2019)
2. H. Tao, A. Kotsopoulos, J.L. Duarte, M.A.M. Hendrix, Family of multiport bidirectional DC-DC converter. IEE Proc. Electr. Power Appl. **153**(3), 451–458 (2006)
3. H. Wu, K. Sun, S. Ding, Y. Xing, Topology derivation of non-isolated three port DC-DC converters from DIC and DOC. IEEE Trans. Power Electron. **28**(7), 3297–3307 (2013)
4. A. Agrawal, R. Gupta, Power management and operational planning of multiport HPCS for residential applications. IET Gener Trans. Ditrib. **12**(18), 4194–4205 (2018)
5. B. Wang, L. Xian, V. Kanamarlapudi, K.J. Tseng, A. Ukil, H.B. Gooi, A digital method of power-sharing and cross-regulation suppression for single-inductor multiple-input multiple-output DC–DC converter. IEEE Trans. Ind. Electron. **64**(4), 2836–2847 (2017)
6. B. Vidales, M. Madrigal, D. Torres, High stepping DC/DC topology for voltage source converters in low power renewable energy applications, in *IEEE PES Transmission & Distribution Conference and Exposition-Latin America*, Mexico (PES T&D-LA) (2016)
7. M.B.F. Prieto, S.P. Litrán, E.D. Aranda, J.M.E. Gómez, New single input, multiple output converter topologies: combining single-switch non-isolated dc-dc converters for single-input, multiple output applications. IEEE Ind. Electron. Mag. **10**(2), 6–20 (2016)
8. O. Ray, A. Prasad, S. Mishra, A. Joshi, Integrated dual output converter. IEEE Trans. Ind. Electron. **62**(1), 371–382 (2015)
9. Z. Qian, O. Abdel-Rahman, I. Batarseh, An integrated four-port DC/DC Converter for renewable energy applications. IEEE Trans. Power Electron. **26**(7), 1877–1887 (2010)

10. H. Wu, K. Sun, L. Zhu, Y. Xing, An interleaved half-bridge three-port converter with enhanced power transfer capability using three-leg rectifier for renewable energy applications. IEEE J. Emerg. Sel. Top. Power Electron. **4**(2), 606–616 (2016)
11. F. Blaabjerg, K. Ma, Future on power electronics for wind turbine systems. IEEE J. Emerg. Sel. Top. Power Electron. **1**(3), 139–151 (2013)
12. J. Han, S.K. Solanki, J. Solanki, Coordinated predictive control of a wind/battery microgrid system. IEEE J. Emerg. Sel. Top. Power Electron. **1**(4), 296–305 (2013)
13. X. Zhang, T.C. Green, The new family of high step ratio modular multilevel DC-DC converters, in *Applied Power Electronics Conference and Exposition (APEC), 2015 IEEE*, Charlotte, NC (2015), pp. 1743–1750
14. S. Falcones, R. Ayyanar, X. Mao, A DC-DC multiport converter based solid-state transformer integrating distributed generation and storage. IEEE Trans. Power Electron. **28**(5), 2192–2203 (2013)
15. W. Hu, H. Wu, Y. Xing, K. Sun, A full-bridge three-port converter for renewable energy application. IEEE Xplore (2014)

Predictive Modeling for the Classification of Child Behavior from Children Stories

A. G. Hari Narayanan and J. Amar Pratap Singh

Abstract Emotions finding from stories is a wide range area of research with lot of different applications. Through this research work, we are trying to predict the effect of emotion from stories with the help of ensemble classifiers. Stories are an essential part of childhood. It is an effective way for children to understand the environment and things happening around the world. The story telling will help them to develop good manners and intellectual power. It will definitely impact the behavior of children like through the situations described in the story. The basic emotions in kids square measure joy, fear, anger, disgust, surprise, disappointment and neutral. The degree of those emotions depends upon the essential character of kid. Classification is an important data mining technique which is used here to classify the sentences in the stories based on the emotion reflected in the child. Here, we are examining the efficiency of classification algorithms for creating the prediction models from children stories. For that, we are using both single and ensemble classifiers, which help us to make good comparison for story-based emotion experiment because it shows 80% accuracy and takes only less time to build model using both classifiers.

1 Introduction

Every young mind likes to hear and enjoy stories. It is an effective and brilliant way to develop good manners in children as well as an easy way to handle them. The most important point is that there are no side effects for this method not like television cartoons and mobile games that harms the children's eye sight, their concentration

A. G. Hari Narayanan (✉)
Department of Computer Application, Noorul Isalm Centre For Higher Education, Kumaracoil, Thuckalay, Kanyakumari, Tamilnadu 629180, India

Department of Computer Science and IT, Amrita School of Arts and Sciences, Kochi, Amrita Vishwa Vidyapeetham, Kochi, India

J. Amar Pratap Singh
Department of Computer Science and Engineering, Noorul Isalm Centre For Higher Education, Kumaracoil, Thuckalay, Kanyakumari, Tamilnadu 629180, India

P. Karuppusamy et al. (eds.), *Ubiquitous Intelligent Systems*, Smart Innovation, Systems and Technologies 243, https://doi.org/10.1007/978-981-16-3675-2_30

power, etc. The emotions and behavior have a close connection. The basic behaviors, which child expresses, are happiness, violence, laziness, bulling, short tempering, yelling, stealing, disrespect, self-esteem, etc. [1]. These behaviors are the result of basic six emotions of Ekman [2]. These emotions and behaviors like their experimental feeling, expressions, change in motives and goals, actions, etc., definitely influenced by the stories they read and hear. This can be amendment over time because the kid grows and depends on his atmosphere. Kid's behavior and emotions are also a result of his growing environment, social experience, and cultural context.

In this paper, we are examining the efficiency of classification algorithms for creating prediction models from these children stories. For that, we are extracting the sentences and emotions reflected in them from children stories.

Data mining is that the method of extracting helpful information from the big quantity of information. Data mining is the most important part in KDD (Knowledge Data Discovery). The important functions are

- Concept Descriptions.
- Association Rules.
- Classification.
- Prediction.
- Clustering.
- Sequence Discovery.

There are different data mining or analysis techniques can be applied on a training data set to build a model which we are concentrating here are Classification and prediction. Prediction of class label is done by classification and prediction of continues valued functions is done by prediction. Here, we are concentrating on the combination of these two, i.e.; classification model for prediction. Main DM method which used to classify a large population of records to classify according to a predefined model is called Classification. The two types of classifiers we are using here are single and ensemble classifiers.

Single classifiers: It is a type of classifier which will classify the data according to the training set which has class labels. It is more focused on single classification. Experiments and studies employing a single classifier are used in this paper. Data Mining and its predictive tasks are officially addressed and analysed, with the outcomes compared again to determine the most distinguishable techniques of choosing.

Table 1 Single classifiers

Classification category	Algorithm
Trees	Random Tree
	J48
Rules	ZeroR
Bayes	Naïve Bayes
Functions	SMO

Table 2 Ensemble classifiers

Classification category	Algorithm
Trees	Random Forest
Meta	Vote
	Bagging
	AdaBoost
	Stacking

Ensemble classifier: It is an advanced version of single classification. The prediction of this classifier is based on multiple classes which will improve the performance of classification. It is a combination of classification methodologies, and the result will be an integrated value of these. Ensemble learning is a general meta approach to machine learning that combines predictions from different models to improve predictive. In most cases, ensemble approaches provide more accurate results than a single model. In a number of machine learning competitions, the winning solutions used ensemble approaches.

2 Related Work

Bharat Deshmukh, Ajay S. Patil and B. V. Pawar in the paper ‘Comparison of Classification Algorithms using WEKA on Various Data sets’ has done that Data mining classification technique in different data sets to compare the performance of these algorithms. It has been done with ADTree, Bayes Network, Decision Table, J48, Logistic, Naive Bayes, NBTree, PART, RBFNetwork and SMO algorithms. The data sets which used are as follows: bank, car, breast cancer, credit –g and diabetics. This paper concludes that perfection of algorithms depends upon the data set which we use. No algorithm is best suit for all classification [3].

Jasmine Bhaskar A, Sruthi KA, Prema Nedungadi B have done the paper ‘Hybrid Approach for Emotion Classification of Audio Conversation Based on Text and Speech Mining’. It is through with emotion classification of audio supported both speech and text. They have used Natural Language Processing, Support Vector Machines, WordNet Affect and SentiWordNet. They conclude that the precision of emotion classification will be much higher compared to the classification of text and audio alone [4].

Aleksandr Sboev, Tatiana Litvinova, Dmitry Gudovskikh, Roman Rybka, Ivan Moloshnikov in the paper ML-based techniques are used by Author Gender Using Topic-Independent Features’ have done the text classification according to Russian language. They say that the best classification algorithm for their work is ReLU, which is suitable for Russian language [5].

The analysis of emotion has been done with sentences and context information in the stories. It has been mainly done with the HMM model to classify the emotions [6].

Emotion recognition has been done with YouTube comments by exploiting the emotional states of users from the comments and the classification of it is done with Point wise Mutual Information measure. The emotion classification was based on text [7].

Statistical and machine learning methods have become good choices to address many natural language processing problems. Some researchers formulated the emotion identification as a classification problem [8–11]. The classes could be either the seven classes discussed above or two classes: neutral and emotional. The features employed include bag of words, N-Gram, dependencies, punctuations, and position information, etc.

3 Experiment (Proposed Work)

The aim of this work is to examine the performance of the classification models. The classification is based on the child emotions like joy, fear, anger, disgust, surprise, sadness and neutral from the sentences in the stories. The data set which is used for this experiment has sentences and the emotion reflected in it. Surely, the stories have an impact on the child behavior. So, we are creating different models for the prediction of probable emotions reflect in children. For this, we have selected both single and ensemble classifiers. For this experiment, we have chosen a small data set. The experiment is done with the help of Weka tool which is a powerful tool used for data mining. Weka is a set of data mining-related machine learning techniques. The algorithms can be used to directly apply to a dataset or invoked from Java code. Data pre-processing, classification, regression, clustering, association rules, and visualisation are all available in Weka. It's a resource that's open to the public.

STEP 1: Import data set in WEKA.
STEP 2: Text Preprocessing.
STEP 3: Run various ensemble classifiers.
STEP 4: Finally compare the results (Fig. 1).

Steps in proposed Methodology:

- Using the Weka tool to apply various forms of classification techniques, such as single and ensemble classifier algorithms.
- To step on to the next stage of implementation, compare all the experimental outcomes of all these forms of classification methods.
- The numerous single and ensemble classifiers are more precise than the data set of classifiers based on rules.
- Comparative analysis of outcomes using precision parameters, time of execution,
- Evaluation of results produced by classifiers of single and ensemble.
- Find the best classifier to create an enhanced classification model using the story data set with optimal efficiency and accuracy.

Fig. 1 Methodology

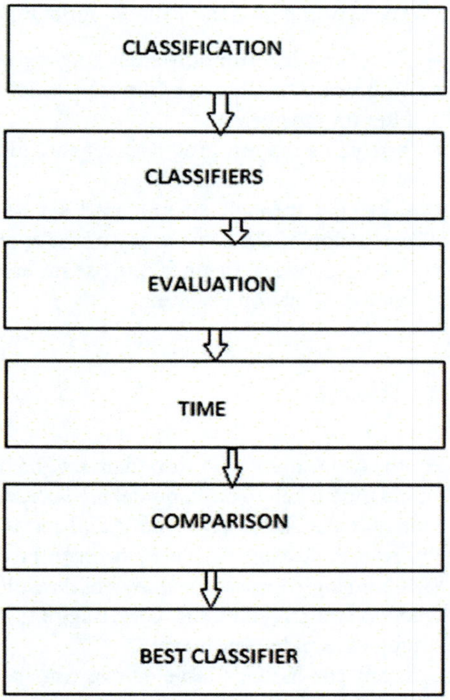

3.1 Data Set and Working

Storyberries.com [12] and childhood101.com [13] both sites are online collection of quality stories, comics, fairy tales and poems for children. Storyberries offers both classic and contemporary stories with lot emotional contents.

Quality stories from Storyberries.com and Childhood101.com were included in the data set. Too much emphasis is placed on these sites for two key reasons. Then, using the seed words for each group, we extracted data from the above sources. The first group consists of emotional content and style for readers, whereas the second group employs the structuring of emotional sentence annotations. The clearly recognizable facial expressions of emotion reflect these categories: anger, joy, sadness, fear, disgust, and guilt. In the sense of a specific emotion, we took words widely used and considered them to be the seed words. Next, we extracted information from the above sources using the seed words for each group.

The type of words used in text classification (Ekman's six emotions) is mainly based on content words and n-grams. Here, we have done using the StringToWord-Vector (STWV) filter in WEKA to analyze the data set. We are trying relate the large set of sentences from the above stories with anger, joy, sadness, fear, disgust and guilt emotions set.

The experiment done with the following steps:

- To collect data from different languages so that a simple data set can be created.
- To prepare the data for learning, which includes using the StringToWordVector filter to transform it.
- Analyzing the resulting data set and, ideally, using attribute selection to enhance it.
- Checking through an unbiased set of samples, which will give us a robust evaluation of the consistency of the approaches to specific examples.
- To learn and use the most precise model as obtained from the preceding stage for our classification program.

4 Result

For the experiment with story data set, the best single classification algorithms based on the time taken to build model are Random Tree, ZeroR, Naïve Bayes. Among the ensemble classifiers, the Vote, AdaBoost and Stacking show the best results.

From this result (Table 3), we can analyze that the Random Tree, ZeroR, and SMO in single classifiers show good result according to the time taken to build the model on training data and Vote, Bagging, AdaBoost, and Stacking show best results among ensemble classifiers.

From this Result (Table 4), we can analyze that the Random Tree, ZeroR, and SMO in single classifiers show good result according to the time taken to build the model on training data and Vote, Bagging, AdaBoost, and Stacking show best results among ensemble classifiers.

Accuracy is the most important variable for building up a model. So, from the above Result (Table 5), it is clear that the Random Tree and SMO show the 100% accuracy in single classifiers and Random Forest shows highest accuracy among ensemble classifiers.

Table 3 Time to check model on training data

Type of classifier	Algorithm	Time (s)
Single	Random Tree	0
	J48	0.01
	ZeroR	0
	Naïve Bayes	0.01
	SMO	0
Ensemble	Random Forest	0.01
	Vote	0
	Bagging	0
	AdaBoost	0
	Stacking	0

Table 4 Time taken to test model on training data

Type of classifier	Algorithm	Accuracy (%)
Single	Random Tree	100
	J48	26.6667
	ZeroR	26.6667
	Naïve Bayes	86.6667
	SMO	100
Ensemble	Random Forest	100
	Vote	26.6667
	Bagging	80
	AdaBoost	33.3333
	Stacking	26.6667

Table 5 Based on accuracy of classification

Type of classifier	Algorithm	Time (s)
Single	Random Tree	0
	J48	0.01
	ZeroR	0
	Naïve Bayes	0
	SMO	0.14
Ensemble	Random Forest	0.04
	Vote	0
	Bagging	0.01
	AdaBoost	0
	Stacking	0

5 Conclusion

From the overall point of view, the Random tree among the single classifier shows the best result to classify and to create a model with our data set. It shows 100% accuracy, zero seconds to build the model as well as zero seconds to test model on training data. Among the ensemble classifiers, the Random Forest is best in the accuracy to create the model but it takes some time to build the model and to test model on training data. Bagging is the second best algorithm suitable for our experiment because it shows 80% accuracy and takes only less time to build model and test model on training data compared to Random Forest. These results may vary according to the data set used and the size of the data set. Thus, we conclude that for every data set there is one or more algorithms show better results. It is our concern to select the best algorithm according to our needs and parameters.

References

1. V.V. Sruthy, A. Saju, A.G. Hari Narayanan , Predictive methodology for child behavior from children stories. J. Eng. Appl. Sci. **13**(5), 4597–4599 (2018)
2. P. Ekman, Universals and cultural differences in facial expressions of emotions, in *Nebraska Symposium on Motivation*, vol 19 (1972), pp. 207–283
3. B. Deshmukh, A.S. Patil, B.V. Pawar, Comparison of classification algorithms using WEKA on various datasets. Int. J. Comput. Sci. Inf. Technol. (IJCSIT)**4**(2), 85–90 (2011)
4. J. Bhaskar, S. Ka, P. Nedungadi, Hybrid approach for emotion classification of audio conversation based on text and speech mining
5. A. Sboev, T. Litvinova, D. Gudovskikh, R. Rybka, I. Moloshnikov, Machine learning models of text categorization by author gender using topic-independent features
6. Z. Zhang, M. Dong, S.S. Ge, Emotion analysis of children's stories with context information
7. D. Yasminaa, M. Hajarb, Al Moatassime Hassanaa, Using YouTube comments for text-based emotion recognition
8. R.A. Calix, S.A. Mallepudi, B. Chen, G.M. Knapp, Emotion recognition in text for 3-d facial expression rendering. IEEE Trans. Multimedia **12**(6), 544–551 (2010)
9. D. Ghazi, D. Inkpen, S. Szpakowicz, Hierarchical versus flat classification of emotions in text, in *Proceedings of the NAACL HLT 2010 Workshop on Computational Approaches to Analysis and Generation of Emotion in Text* (Association for Computational Linguistics, 2010), pp. 140–146
10. https://www.storyberries.com/tag/feelings/
11. https://childhood101.com/books-about-emotions/
12. C. Strapparava, R. Mihalcea, Learning to identify emotions in text, in *Proceedings of the ACM Symposium on Applied Computing*. ACM (2008), pp. 1556–1560
13. C.O. Alm, D. Roth, R. Sproat. Emotions from text: machine learning for text-based emotion prediction, in *Proceedings of the conference on Human Language Technology and Empirical Methods in Natural Language Processing* (Association for Computational Linguistics, 2005), pp. 579–586

Morse Tool—A Digital Communication Aid for Visually Impaired

Manish Tiwari, Gaurav Kumar, Megha Chambyal, and Sheilza Jain

Abstract Morse code is not only an elementary way of communication, but also a very convenient way for visually impaired people to communicate. Morse code alphabets are a combination of dots and dashes or *dits* and *dahs* and are based on international standards. In this paper, we will introduce a Morse Tool aka *Morse pen*, which is composed of five keys—dots, dashes, space, reset, and send, to help the visually impaired. In accordance with International Morse code, users will first make combinations of dots and dashes. The Morse Tool then decodes the input with the help of an on-board microcontroller into ASCII and transmits the same wirelessly via Bluetooth to a paired Smartphone device. "Unwired lite," an android application used to display the text in a Smartphone. Now the text is in the Smartphone, and it can be used by many applications like a standard text editor, a messaging platform like WhatsApp or email.

1 Introduction

The numbers of blind people have been increasing due to a number of reasons including eye diseases and traffic accidents [1]. From an ancient time, representation of information is not found in printing format; it is resembled in aural format. Partially, it facilitates the people to acquire knowledge during an era of printing [2]. One-third of blind population of world lives in India. Among 15 million blind people living in India, 2 million are children, and out of these, only 5% receives education [3]. Visual impairment limits people's ability to interact with the surrounding world. Losing the sense of sight, the blind has to depend on other sensory organs, making it extremely tricky for them to communicate [4, 5]. But to live, communication is necessary. This blindness forces a visually impaired person to build a strong ability to make constructive use of other sense such as, to read information the blind person uses the sense of touch, and the sense of touch can be developed to interpret some divergent patterns like Braille [2]. Several techniques are implemented to assist the blind and

M. Tiwari (✉) · G. Kumar · M. Chambyal · S. Jain
Department of Electronics Engineering, J.C. Bose University of Science and Technology (Formerly known as YMCAUST), Faridabad, Haryana, India

P. Karuppusamy et al. (eds.), *Ubiquitous Intelligent Systems*, Smart Innovation, Systems and Technologies 243, https://doi.org/10.1007/978-981-16-3675-2_31

visually challenged [6]. Some technologies such as Morse code, Perkins Brailler, Pac Mate, and Orbit reader aid the blind in communicating with the surroundings. Augmentative and alternative communication (AAC) devices use signals from a patient and eventually convert them to a specific type of data that can be transmitted, but those tools are very expensive and probably not available to most people [7]. Braille script-based devices require prior knowledge of braille in order to use, and it requires more keys which makes it cumbersome for regular usage.

The simplicity of Morse code is that it requires less resources compared to other codes that depend on keyboards with 100 or more keys, in terms of complexity involved in bringing the contraption work [8]. Morse code consists of dots and dashes. The only thing that is vital for the Morse code is timing, as the code relies on precise intervals of time between dots and dashes, between letters, and between words [9–11]. Morse code is used worldwide so as to provide reliable communication through wires for military, overseas shipping, and even the railway road. What makes the Morse code so widely preferred is that any person can easily tap Morse code using his/her finger [10].

In this paper, we will introduce a "Morse Tool" aka "Morse pen" based on Morse code, which can be used by novice blind for texting.

2 Overview of Proposed Solution

Morse Tool device acts as an interface for the visually impaired people to digitally communicate with other people. The development of Morse Tool device from scratch is discussed in the sections below. It explains the complexity of the defined problem and the working of the proposed solution, "Morse Tool." Morse lends more simplicity and universality to our system when compared to other Braille-based devices. The functioning of our system has been described with the help of Figs. 1 and 2.

The proposed device is a digital communication writing system consisting of a specially designed tool to be used by people who are visually impaired. This tool or pen consists of switches for various purposes while operating the device. Initially on start, the device is in idle state with a lit-up LCD showing the name "Morse Code Generator." To communicate using the device, the user uses the pen to input the message string. The string is entered character by character; each character is entered in the form of dots and dash combination as per the Morse code table shown in Fig. 4. After entering Morse code of one character, a long press of dot and dash

Fig. 1 Block diagram of the system

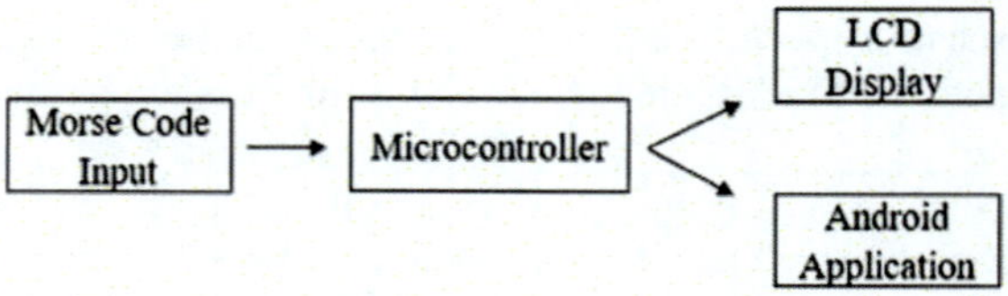

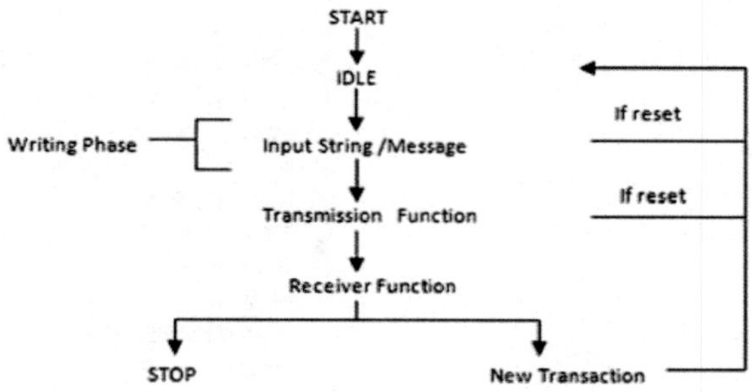

Fig. 2 Functioning of the system

key initiates MCU to decode the Morse code. The MCU reads the entered data by the user, (the dots and dashes combination), and converts back into corresponding characters based on Morse code. These characters after being decoded by MCU are simultaneously updated on the LCD screen. Once the whole message string is completed, a complete message is visible on the LCD display. The user is then required to activate the Bluetooth connectivity of the Smartphone. To enable the Bluetooth connectivity feature, the device has an inbuilt Bluetooth module which connects to the Smartphone. Now, the whole message string can be transmitted from the MCU to the Smartphone via the send button on the device. The message is received on phone via android application, which can be used to read, send and can be utilized for other purposes as well. Above discussed functioning of device is shown in Fig. 3.

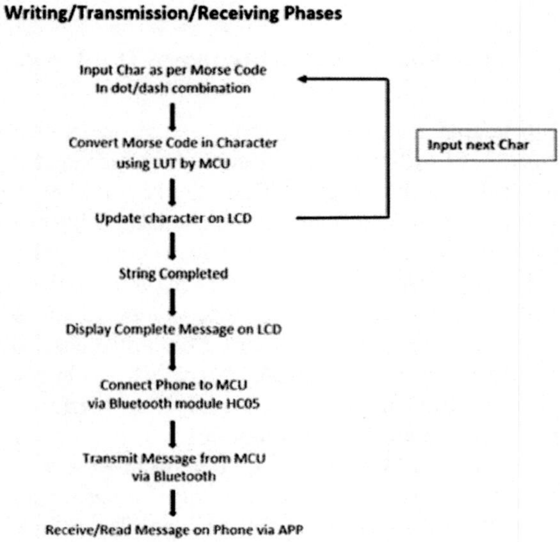

Fig. 3 Functioning of device

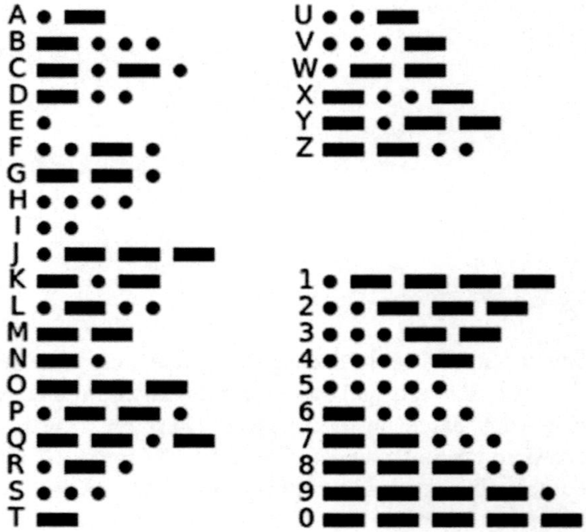

Fig. 4 International Morse code table. *Source* Wikipedia

3 Organization of the Solution

The rest of the paper discusses the proposed solution, its implementation, and the conclusions drawn from the system. Possible application scenarios of this technology and an outlook into further research are briefly discussed towards the end.

4 Implementation: Hardware Implementation

Morse Tool is a wired, hand-held, and, hence, portable device which runs on Morse code. It is a writing tool which consists of five keys for operation that are programmed to function at the key press. These five keys are dot, dash, space, reset, and sent. A tip key press is dot, and a side key press is dash. Other keys have been added to add/delete a space, to reset the device and send the text from LCD screen to Smartphone. Advantage of Morse Tool is that the implementation is made independent of the time intervals between the dot and dash combinations. So, it means a novice user does not have to wait for a time interval to implement dots and dashes or *dits* and *dahs*.

The device uses ATMEGA 328p, which is a single chip 8-bit microcontroller based on AVR RISC architecture. It is used for interfacing of LCD, Bluetooth, switches and to implement Morse algorithms.

A 16 × 2 LCD display is used to display the message entered by the user. Using a Bluetooth module HC-05, the device connects to an android application named "Unwired lite," installed in Smartphone. The received data on the app could be used further for various purposes like messaging, WhatsApp, etc. (Table 1).

Table 1 Hardware requirements

S. No.	Hardware requirements	Specifications
1	Microcontroller	Microcontroller ATMEGA328p 8-bit microcontroller based on AVR RISC architecture Flash program memory: 32 KB SRAM data memory: 2 KB MSSP: SPI and I^2C master and slave support
2	16 × 2 LCD display	Alphanumeric 16 × 2 backlit LCD display module operating voltage 4.7–5.3 V current consumption 1 mA
3	Bluetooth module	HC-05 Bluetooth module Speed: Asynchronous: 2.1 Mbps (Max)/160 kbps, Synchronous: 1 Mbps/1 Mbps Frequency: 2.4 GHz ISM band Sensitivity: ≤ −84 dBm at 0.1% BER
4	USB AVR programmer	USB ASP AVR programmer to program Atmel AVR controllers
5	Switches	Tactile push button switches operating current: 50 mA operating voltage (VDC): 12 V
6	Power supply	9 V HW battery
7	Miscellaneous	Resistance, capacitance, 50 MHz crystal oscillator, LED

5 Implementation: Software Implementation

Software requirements include Proteus design suite: EDA software tool for simulation of hardware design, ARES: A PCB design software, Atmel Studio IDE: Integrated development platform used for developing and debugging AVR microcontroller applications, Android application for data transfer using Bluetooth connectivity. The software part covers the outline of the code which has to be fed into the Atmega328p microcontroller. The software code is fed to the microcontroller using USB AVR ISP programmer. The skeleton of the code:

1. IDLE/RESET/INITIAL-STATE
Loop 1: 2. READ STATE

Input value Morse code (DOT-DASH)

3. CONVERSION STATE

Convert Morse code input value into CHAR

4. DISPLAY STATE

Update CHAR/String/Message on LCD and update flag value

5. COMPARISON STATE

```
While (! RESET)
If (Flag>0 && space Pin==LOW && Transmission Pin == LOW)
    BACK LOOP 1
else if (Flag>0 && Space Pin == HIGH && Transmission pin == LOW)
    UPDATE SPACE CHAR AND BACK TO LOOP 1
else if (Flag>0 && Space Pin == LOW && Transmission Pin == HIGH)
    MOVE TO TRANSMISSION STATE
else
    BACK TO IDEAL STATE
```

6. TRANSMISSION STATE

Send data to receiver mobile phone using Bluetooth communication and update LCD with "Sending Data and Morse code detected"

7. MOVE TO IDEAL/RESET/INITIAL STATE

The Morse code encodes the English alphabets, numeric, and the punctuations. Each Morse code is a sequence of dots and dashes. There is no distinction between upper and lower case letters. Apart from the standard alphabets and numeric, three extra command sequences have been added in the Morse pen—space, reset, and send. They have been added to write a space, to clear the LCD screen, and send the text from LCD screen to Smartphone. Figure 4 shows the international Morse code.

PCB designing "Proteus" tool is used for electronic design automation and to create schematics for manufacturing printed circuit boards. Figure 5 shows the circuit designing using Proteus software.

6 Result

The Morse Tool is developed after integration of hardware and software implementations. The device has been tested for various input conditions to work seamlessly well.

The contents below show the step-by-step working of the device. The word "YMCA" is typed using the tool and sent to display on the LCD and also to the Smartphone app via Bluetooth. The data in the app could be utilized further for various purposes like messaging, WhatsApp, etc.

The following figure shows the working of the Morse pen or Morse Tool:

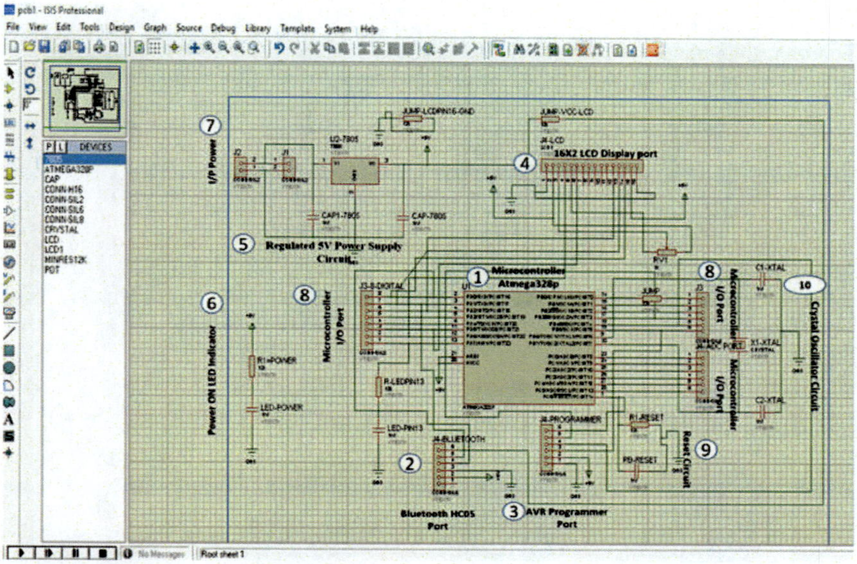

Fig. 5 Circuit designing using Proteus software

6.1 Power on State

The device is powered ON. Figure 6 represents the power ON stage and display "Morse Code Generator" on LCD display.

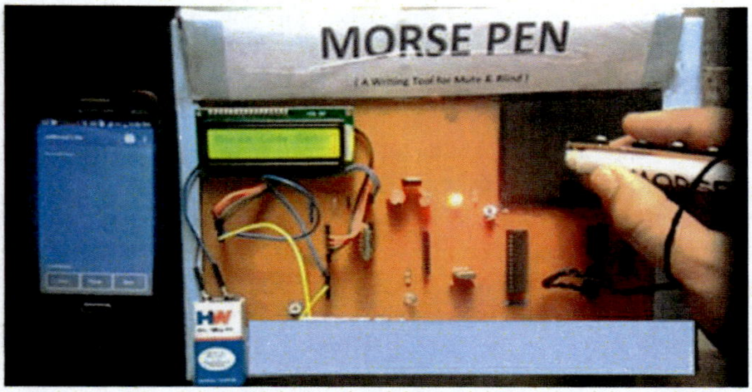

Fig. 6 Power on state of the device

Fig. 7 Morse code of character Y decoded

Fig. 8 String Y created and displayed on LCD display

6.2 Writing Y Character

Using Morse pen user inputs Y character in combination of dots and dashes as per International Morse code conventions.

To input Y character user inputs " - . - - ". Figure 7 shows Morse code of character Y is decoded by the device. Once after character is decoded, data string is created with Y character and displayed on LCD as shown in Fig. 8.

6.3 Writing M Character

Similar to Y character, user inputs M character in combination of dots and dashes as per International Morse code conventions. To input M character, user inputs " - - " using Morse tool. Figure 9 shows Morse code of character M is decoded by

Fig. 9 Morse code of character M decoded

Fig. 10 String YM created and displayed on LCD display

the device. After character is decoded, character M is appended to data string and displayed on LCD. Figure 10 shows the updated string "YM" on LCD display.

6.4 Writing C Character

Similar to Y and M character, user inputs C character in combination of dots and dashes as per International Morse code conventions. To input C character, user inputs " - . - . ". Figure 11 shows Morse code of character C is decoded by the device. After character is decoded, data string is updated, and character C is appended to it and displayed on LCD. Figure 12 shows the updated string "YMC" being displayed on LCD screen.

Fig. 11 Morse code of character C decoded

Fig. 12 String "YMC" is created and updated on LCD display

6.5 Writing A Character:

User now inputs next character A in combination of dots and dashes. To input A character, user inputs " . - ". Figure 13 shows Morse code of character A is decoded by the device. After character is decoded, character A is appended to the data string and displayed on LCD. Figure 14 shows the updated string "YMCA" being displayed on LCD screen.

6.6 Final Data "YMCA" Received on Phone

After entering required message, user presses send key on the Morse pen device. The message string is now transmitted to the Smartphone via Bluetooth connectivity between device and phone. Figure 15 shows the transmitted data on the phone.

Fig. 13 Morse code of character A decoded

Fig. 14 String "YMCA" created and updated on LCD display

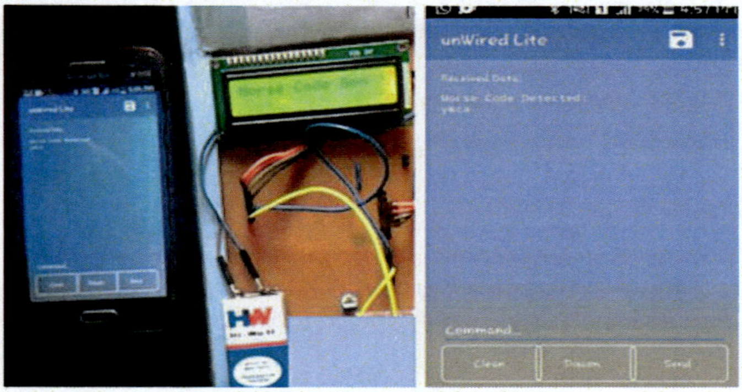

Fig. 15 Data is transmitted to the app

7 Conclusion

Morse pen, a writing tool for blind is ready to be put in use and withstands completely to meet its defined objective. The result and analysis section shows how the data is decoded character by character using the International Morse code. The string of characters is made and finally sent to a Smartphone via Bluetooth using a connectivity app between the device and Smartphone.

The other technologies such as Perkins Brailler, Braille Note, Pac Mate, and Orbit Reader are too costly to be employed in personal use. On top of that, all of these devices are electronically operated and quite cumbersome in size. It becomes very difficult for the blind people to handle and operate them. So, Morse Tool is a low-cost solution to the problem of communication for the blind. The cost of the pen or tool at a non-commercial level is merely INR 700 which certainly would prove to be a personally affordable device for the blind. It aims to help leverage the technology to enhance the quality of their communication.

8 Future Scope

There are a few shortcomings of this system. Firstly, the Morse Tool consists of a wired writing tool. So, the work can be done to modify the writing tool to adapt wireless technology. Secondly, the size of the device is quite large to carry around easily. The circuitry using VLSI technologies will miniature the device to become pocket size, which can be utilized by the user easily. Thirdly, a speech technology can be used in the device. This will make the device do text to voice conversion and could be played using a speaker. This text to speech conversion task can be done using the app. Also, cryptography methodology can be used to secure transmission of data between device and android application. The cryptography seals the information present inside a message [12]. With these improvements, the device will become more efficient and versatile.

References

1. N. Ezaki, K. Kiyota, S. Yamamoto, A pen-based Japanese character input system for the blind person. Proc. Int. Conf. Pattern Recogn. **15**(4), 372–375 (2000). https://doi.org/10.1109/icpr.2000.902936
2. S.A. Sabab, M.H. Ashmafee, BLIND READER: an intelligent assistant for blind, in *19th International Conference on Computer and Information Technology, ICCIT 2016* (2017), pp. 229–234. http://doi.org/10.1109/ICCITECHN.2016.7860200
3. V. Govardanam, T.N.V. Babu, N.S.H. Kavin, Automated read-write kit for blind using hidden Markov model and optical character recognition, in *Proceedings 2015 International Conference on Applied and Theoretical Computing and Communication Technology, iCATccT 2015* (2016), pp. 828–831. http://doi.org/10.1109/ICATCCT.2015.7456997

4. T. Choudhary, S. Kulkarni, P. Reddy, A Braille-based mobile communication and translation glove for deaf-blind people, in *2015 International Conference on Pervasive Computing: Advance Communication Technology and Application for Society, ICPC 2015* (2015), pp. 1–4. http://doi.org/10.1109/PERVASIVE.2015.7087033
5. R. Sarkar, S. Das, D. Rudrapal, A low cost microelectromechanical Braille for blind people to communicate with blind or deaf blind people through SMS subsystem, in *Proceedings of 2013 3rd IEEE International Advance Computing Conference, IACC 2013* (2013), pp. 1529–1532. http://doi.org/10.1109/IAdCC.2013.6514454
6. S. Manoharan, A smart image processing algorithm for text recognition information extraction and vocalization for the visually challenged. J. Innov. Image Process. (JIIP) **01**(01), 31–38 (2019). http://doi.org/10.36548/jiip.2019.1.004
7. K. Mukherjee, D. Chatterjee, Augmentative and alternative communication device based on eye-blink detection and conversion to Morse-code to aid paralyzed individuals, in *Proceedings of 2015 International Conference on Communication, Information and Computing Technology, ICCICT 2015* (2015), pp. 0–4. http://doi.org/10.1109/ICCICT.2015.7045754
8. P.S. Luna, E. Osorio, E. Cardiel, P.R. Hedz, Communication aid for speech disabled people using Morse codification, in *Annual International Conference of the IEEE Engineering in Medicine and Biology—Proceedings*, vol 3 (2002), pp. 2434–2435. http://doi.org/10.1109/iembs.2002.1053361
9. R. Li, M. Nguyen, W.Q. Yan, Morse codes enter using finger gesture recognition, in *DICTA 2017—2017 International Conference on Digital Image Computing: Techniques and Applications*, vol 2017 (2017), pp. 1–8. http://doi.org/10.1109/DICTA.2017.8227464
10. C.T. Lee, T.C. Shen, W. Der Lee, K.W. Weng, A novel electronic lock using optical Morse code based on the internet of things, in *Proceedings of IEEE International Conference on Advanced Materials for Science and Engineering, IEEE-ICAMSE 2016* (2017), pp. 585–588. http://doi.org/10.1109/ICAMSE.2016.7840206
11. C.P. Ravikumar, M. Dathi, A fuzzy-logic based Morse code entry system with a touch-pad interface for physically disabled persons, in *2016 IEEE Annual India Conference, INDICON 2016* (2017), pp. 1–5. http://doi.org/10.1109/INDICON.2016.7838961
12. M.R. Vinothkanna, A secure steganography creation algorithm for multiple file formats. J. Innov. Image Process. (JIIP) **01**(01), 20–30 (2019). http://doi.org/10.36548/jiip.2019.1.003

Software Effort Estimation Using Genetic Algorithms with the Variance-Accounted-For (VAF) and the Manhattan Distance

K. P. Mohamed Shabeer, S. I. Unni Krishnan, and G. Deepa

Abstract The cost and effort for developing software projects gain a growing interest in recent years. Defining these parameters is considered as a valuable goal in achieving efficiency in developing the projects. Implementing the COCOMO model in effort estimation helps the project developers to allocate the resources efficiently. But there lies a main problem to optimize the constants in the COCOMO model. In this study, we present a way to optimize these constants using genetic algorithm by comparing two different methods in calculating the fitness function. Identifying the efficient method among them will help the optimization of COCOMO parameters and makes the effort estimation more efficient.

1 Introduction

Developing large-scale software projects in a cost and time efficient way always seems to be a challenging task. The options such as identifying the cost estimate to evaluate the project progress and utilization needed to be considered. Constructive cost model [1] (COCOMO) developed by Boehm, Barry W is a famous model for estimating software effort. This model helps to define the mathematical relationship between software development time, man-months, and maintenance effort.

There are several effort estimation methods such as algorithmic methods and analogy-based methods have been proposed in the past. There have been several difficulties with the implementation of these techniques to overcome the calculation of software effort. Various heuristic optimization methods such as the genetic algorithm (GA) [2], the particle swarm optimization algorithm [3], the differential evolution algorithm [4], and others are used in optimization problems.

K. P. Mohamed Shabeer (✉) · S. I. Unni Krishnan · G. Deepa
Computer Science Department, Amrita School of Arts and Sciences, Kochi, Edappally North, Kochi, Kerala 682024, India

G. Deepa
e-mail: deepa@asas.kh.amrita.edu

P. Karuppusamy et al. (eds.), *Ubiquitous Intelligent Systems*, Smart Innovation, Systems and Technologies 243, https://doi.org/10.1007/978-981-16-3675-2_32

This research article focuses on how GA can be used to optimize the constant values (*A*, *B*) in the COCOMO model with two different evaluation criteria for calculating the fitness function. Once it is optimized, then these values can be used to estimate the effort and cost efficiently. The proposed system is evaluated with the help of NASA dataset [5].

1.1 The Constructive Cost Model (COCOMO)

Constructive cost model or COCOMO model is a cost estimation model which considers different parameters such as cost, effort, size, quality, and time in developing various projects.

The COCOMO model has three types—the basic, the intermediate, and the detailed [6]. One of the most important parameters in SEE is the project size.

The equation for a particular COCOMO model (basic) is defined as:

$$E = A \times \mathrm{KLOC}^{B} \quad (1)$$

1	E	Software effort in person-month
2	KLOC	Kilo-lines of code
3	A, B	COCOMO parameters

A project model depends on the value of *A*, and depending on the project size, COCOMO model has three types. These models are organic, semi-detached, and embedded models. Table 1 shows the basic COCOMO model.

When the size of a software project increases by two times, then the development time for the software project does not double instead it rises moderately. Thus, the development time for a software project is a sub-linear function of the size of the software project and is depicted in Fig. 1.

Table 1 Basic COCOMO models

Model name	Project size	Effort
Organic model	Less than 50 KLOC	$E = 2.4$ (KLOC)1.05
Semi-detached model	50–300 KLOC	$E = 2.4$ (KLOC)1.12
Embedded model	Over 300 KLOC	$E = 2.4$ (KLOC)1.20

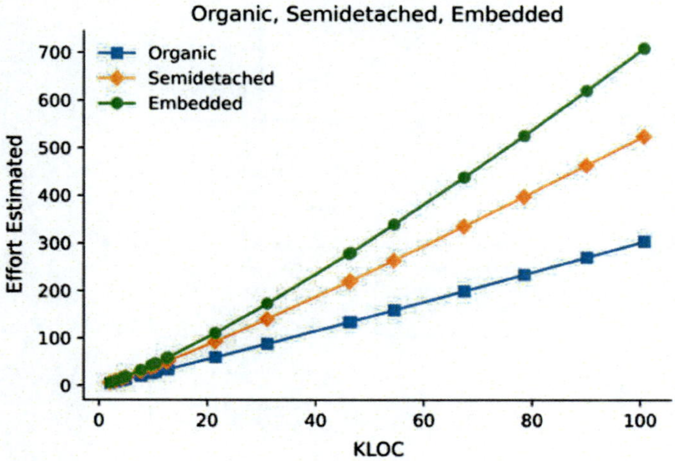

Fig. 1 Estimated effort versus software project size

1.2 Genetic Algorithm (GA)

The genetic algorithm is one of the algorithms among the area of nature-inspired algorithms and is a heuristic search algorithm. During the 1970s, the genetic algorithm was proposed by John Holland and his collaborators. This algorithm draws ideas from Darwinian evolution in which the fittest individual (unit) is selected for producing future generations. The result is an optimal or near to optimal solution which otherwise can take a lifetime to find the exact solution.

The main components of GA are gene, chromosome, initial population, fitness function, selection, crossover, mutation. Figure 2 depicts the steps in GA.

Gene. Genes are the basic unit of genetic algorithm. A bit in a chromosome is called gene. Thus, gene has a crucial role in this algorithm.

Chromosome. A chromosome is a representation of a collection of encoded genes, and the encoding can be integer encoding, binary encoding, etc.

Initial Population. The first generation is generated randomly from the available sample space of genes. The main goal of this step is to generate variety of chromosomes to find optimal solution in next steps.

Fitness Function. Fitness function acts as an evaluation method to find the fittest chromosome among the generation.

Selection. The fittest chromosomes are selected in this step for reproducing new generations, and the selected chromosomes act as a parent chromosome.

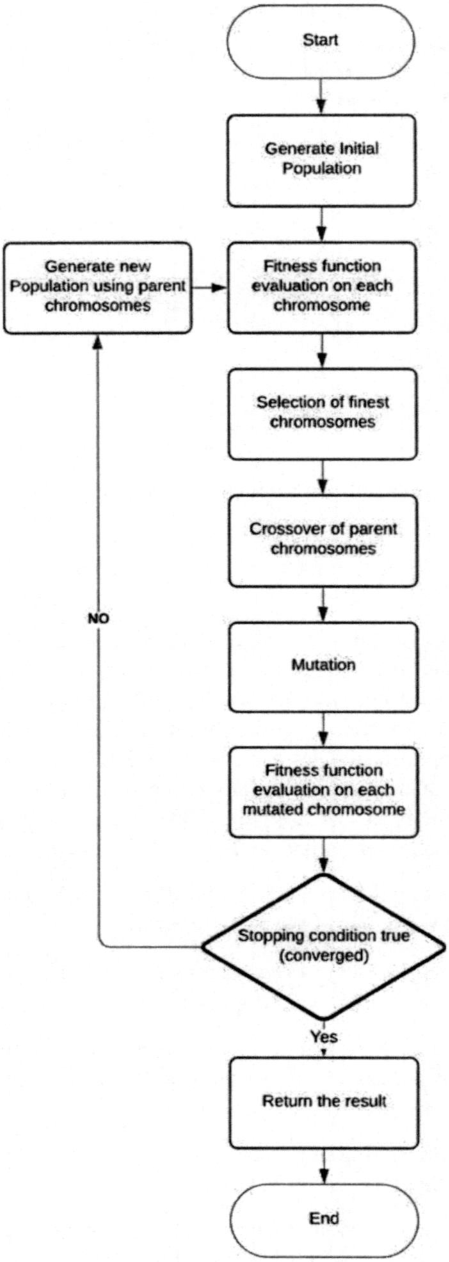

Fig. 2 Genetic algorithms (GA) steps

Crossover. The parent chromosomes are separated and crossed each other for generating varieties of chromosomes with mixed properties. There exist several crossover methods such as single point crossover, uniform crossover, arithXover crossover, etc.

Mutation. The chromosomes are then mutated in a very small ratio (usually 1%) to prevent premature convergence.

2 Literature Review

Benediktsson et al. [7] proposed a detailed study about how COCOMO effort estimation is used in incremental and iterative software development. This detailed study includes the benefits and challenges faced in software development while adopting the COCOMO method of effort estimation.

Galinina et al. [8] proposed that when comparing organic and semi-detached COCOMO models, the coefficient that are optimized with the help of GA (organic model) gives better results when it is compared to the results of current COCOMO model coefficient.

Sachan et al. [9] used a simple genetic algorithm for optimizing the basic COCOMO values. In the experiment, they chose Manhattan distance (MD) for calculating fitness function. The values a and b are then optimized using GA. The overall performance of the above system is analyzed using NASA dataset in the promise repository.

The results of their work show that the Manhattan distance of simplified GA is much better than basic COCOMO and shows that the parameter tuned with MD in GA is better.

Rahimunnisa [10] used a multi-population genetic algorithm to divide the whole population into subpopulations and then apply the genetic algorithm. This paper proposed a hybridized method which includes simulated annealing and a genetic algorithm to find the shortest path of transmission between the nodes. The study shows how much the genetic algorithm technique is capable of finding the optimal solution in the current scenario.

Hari and Reddy [3] introduced a technique of generalizing the COCOMO model with particle swarm optimization in 20 different projects. However, Aljahdali, Sultan, and Alaa F. Sheta present a way of using differential evolution to estimate the COCOMO parameters which provide a better result in effort estimation.

Sheta [11] chose Variance-Accounted-For (VAF) for calculating fitness function. The performance of the system is analyzed using a dataset provided by Bailey and Basili. Two models (Model 1 and Model 2) were implemented. With respect to VAF, the implemented models improved the estimation of effort.

Chhabra and Singh [12] took a fuzzy model approach on optimizing software cost estimation, and it was designed to reduce the imprecision of input range of cost drivers. The model is tested using the COCOMO NASA dataset, and the evaluation criteria used was mean magnitude of relative error (MMRE).

Saeed et al. [13] conducted a survey on estimation models, and most of these models used public dataset. Each model has its own limitations and advantages, most of them were efficient in some way.

Mukesh Mahadev and Gowrishankar [14] implemented genetic programming using Pred25 and MMRE as evaluation criteria for the prediction model. Various datasets with different size were used for testing the prediction model, and they concluded model build using GP shows better results.

3 Proposed System

In this research work, the constant values (*A*, *B*) of the COCOMO model were optimized by genetic algorithms. This optimization helps in calculating the effort and cost of developing the projects.

For optimizing the COCOMO model coefficients, we create/generate the initial population in which individuals are randomly created. After generating population, we can calculate the development time (predicted). Once the calculation of development time is done, then the individual project fitness is evaluated. For evaluating the individuals, we use the fitness value. The fitness function value should be minimized. We use the Manhattan distance (MD) and the Variance-Accounted-For (VAF) evaluation criteria to compute the fitness.

The next task is to check the break off condition. The break off condition defines when the algorithm is to be ceased. Selection is done to form a set of individuals. Crossover is done using the individuals from the previous process in which new individuals are generated. During the mutation process, a small unit of individuals was randomly selected from a pool of individuals, and a random pattern of mutant genes is defined for each individual unit. In the process of generating a new population, the best individuals among them are selected using the same selection method used previously.

We repeat the same until the fitness function is converged, and the values obtained are considered to be the optimal values for constants in the COCOMO model, and these constants are used to estimate software efforts and compare which among the computed effort values are better in comparison with the actual effort.

4 Experiment and Result Analysis

The experiment applies genetic algorithm for optimizing the value of constants *A* and *B* in Eq. (1) of COCOMO model. The fitness function for the proposed genetic algorithm is the absolute difference between the actual effort and the estimated effort of each software project. We chose two evaluation criteria for evaluating the performance. The Manhattan distance [15] and Variance-Accounted-For (VAF) [16] are the two evaluation criteria.

The Variance-Accounted-For (VAF) distance is calculated using Eq. 2.

$$\mathrm{VAF} = \left[1 - \frac{\mathrm{var}(\text{Actual Effort} - \text{Estimated Effort})}{\mathrm{var}(\text{Actual Effort})}\right] \times 100\% \quad (2)$$

The Manhattan distance is calculated using Eq. 3.

$$\mathrm{MD} = \sum_{i=1}^{n} |\text{Actual Effort}_i - \text{Estimated Effort}_i| \quad (3)$$

Dataset: Dataset presented in 1981 by Bailey and Basili [5] on the NASA project have been used for conducting this experiment. The dataset consists of two variables and a measured value—the developed line-of-code (KLOC), the actual effort, and the methodology (ME). KLOC is described as the kilo line-of-code (KLOC) of development; it acts as a measurement for calculating the size of a software project, and the effort is represented in man-months (man per months). The NASA project dataset is shown in Table 2.

The NASA project dataset consists of data regarding 18 software projects developed in NASA. This dataset is well known and public. We took this dataset from well-known PROMISE software engineering repository. From this dataset, we have

Table 2 Nasa software project dataset

Project No.	KLOC	Methodology (ME)	Actual effort
1	90.2	30	115.8
2	46.2	20	96
3	46.5	19	79
4	54.5	20	90.8
5	31.1	35	39.6
6	67.5	29	98.4
7	12.8	36	18.9
8	10.5	34	10.3
9	21.5	31	28.5
10	3.1	26	7
11	4.2	19	9
12	7.8	31	7.3
13	2.1	28	5
14	5	29	8.4
15	78.6	35	98.7
16	9.7	27	15.6
17	12.5	27	23.9
18	100.8	34	138.3

used the first 13 project data for estimating COCOMO parameters and other 5 were used for performance testing.

Tuning Parameters: In genetic algorithms (GA), users tune the parameters [17] of the design of a genetic algorithm (selection, crossover, mutation probability, number of generations and population size) manually. When genetic algorithm applications are being developed, it is essential to understand which parameters have the greatest influence on the performance of a genetic algorithm.

Method 1: The Variance-Accounted-For (VAF) Evaluation Criteria. The tuning parameters used for genetic algorithms for effort estimation using VAF are shown in Table 3. Selection mechanism used is the normalized geometric selection (normGeomSelect) which is the primary selection process used in this method. Crossover type is the Arith crossover (arithXover), performs an interpolation along the line formed by the P1 and P2 parents. The mutation parameter operator selected is a mutation type with non-uniform distribution (nonUnifMutation) which handles multiple variables well.

Method 2: The Manhattan Distance Evaluation Criteria. The tuning parameters used for genetic algorithms for effort estimation using MD are shown in Table 4. Selection mechanism used is the elitism selection in which few individuals with best fitness are selected and are passed to the next generation; thus, mutation operators are avoided. The arbitrary destruction of individuals with good fitness by mutation operators is prevented by elitism. Thus, mutation type is not applicable in this method. Crossover type is the two-point binary crossover; in this crossover, the points are selected from parent chromosomes in a random way. The bits in between the two-points are interchanged in between the parent organisms.

The other tuning parameters used for both evaluation criteria are shown in Table 5.

Table 6 shows the actual software effort, and software effort measured using basic COCOMO model of NASA 18 projects.

Figure 3 shows the actual software effort of the NASA 18 projects and also the calculated software effort using the basic COCOMO model.

Table 3 Parameter settings for GA with VAF evaluation criteria

Operator	Type
Selection mechanism	normGeomSelect
Crossover type	arithXover
Mutation type	nonUnifMutation

Table 4 Parameter settings for GA with MD evaluation criteria

Operator	Type
Selection mechanism	Elitism selection
Crossover type	Two point binary crossover
Mutation type	NA

Table 5 Common parameter settings for both evaluation criteria

Operator	Type
Population size	10
Maximum generation	100
Domain of search for A	1:2
Domain of search for B	0.3:2

Table 6 Actual effort and basic COCOMO effort

Project No.	KLOC	Actual effort	Basic COCOMO effort
1	90.2	115.8	271.1308
2	46.2	96	134.3028
3	46.5	79	135.2187
4	54.5	90.8	159.745
5	31.1	39.6	88.6358
6	67.5	98.4	199.977
7	12.8	18.9	34.8964
8	10.5	10.3	28.3439
9	21.5	28.5	60.1549
10	3.1	7	7.873
11	4.2	9	10.8298
12	7.8	7.3	20.7448
13	2.1	5	5.2304
14	5	8.4	13.0055
15	78.6	98.7	234.6414
16	9.7	15.6	26.0808
17	12.5	23.9	34.0382
18	100.8	138.3	304.6805

Table 7 shows the computed values of software effort using GA with VAF evaluation criteria and GA with MD evaluation criteria.

Figure 4 shows the software effort calculated using GA with Variance-Accounted-For (VAF) and GA with Manhattan distance.

Figure 5 shows the actual software effort, basic COCOMO model effort, effort values computed using GA with VAF, and effort values computed using GA with MD.

Figure 6 shows the actual software effort and effort computed using GA with VAF.

Figure 7 shows the actual software effort and effort computed using GA with MD.

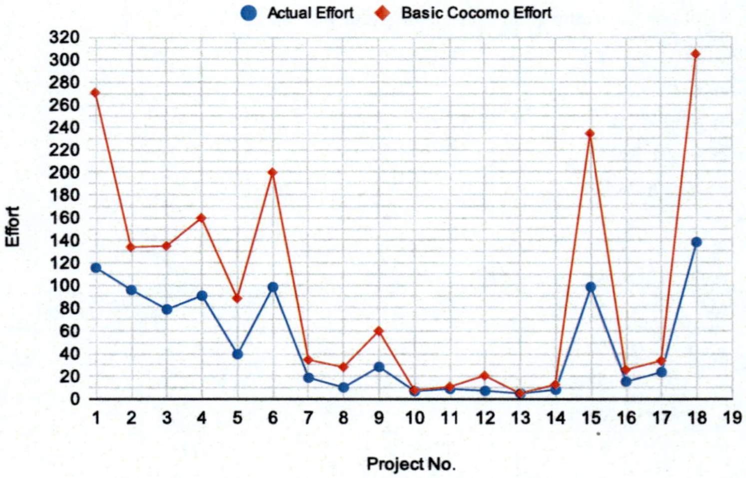

Fig. 3 Effort graph for actual effort and basic COCOMO effort

Table 7 Effort graph for calculated software effort using GA with VAF evaluation criteria and GA with MD evaluation criteria

Project No.	KLOC	GA with VAF (Effort)	GA with MD (Effort)
1	90.2	131.9154	114.851
2	46.2	80.8827	63.0152
3	46.5	81.2663	63.3821
4	54.5	91.2677	73.0839
5	31.1	60.5603	44.181
6	67.5	106.7196	88.5476
7	12.8	31.6447	19.9217
8	10.5	27.3785	16.6782
9	21.5	46.2352	31.7247
10	3.1	11.2212	5.5821
11	4.2	14.0108	7.3303
12	7.8	22.0305	12.774
13	2.1	8.4406	3.9359
14	5	15.9157	8.5715
15	78.6	119.285	101.5074
16	9.7	25.8372	15.5325
17	12.5	31.1008	19.5022
18	100.8	143.0788	126.89

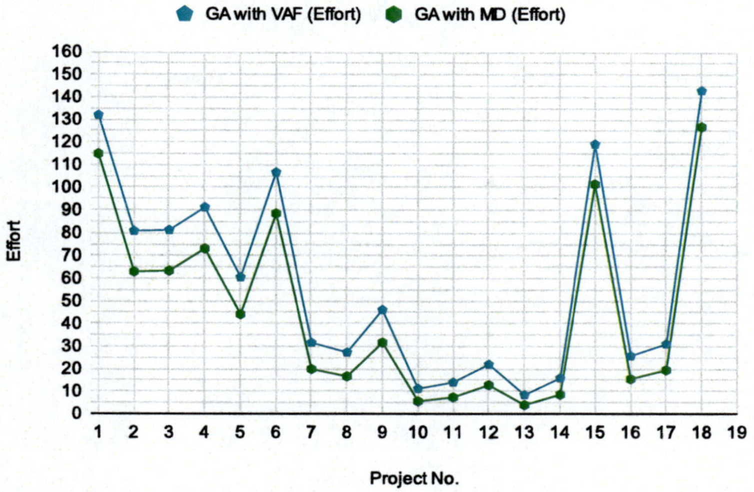

Fig. 4 Effort graph for GA with VAF evaluation and GA with MD evaluation

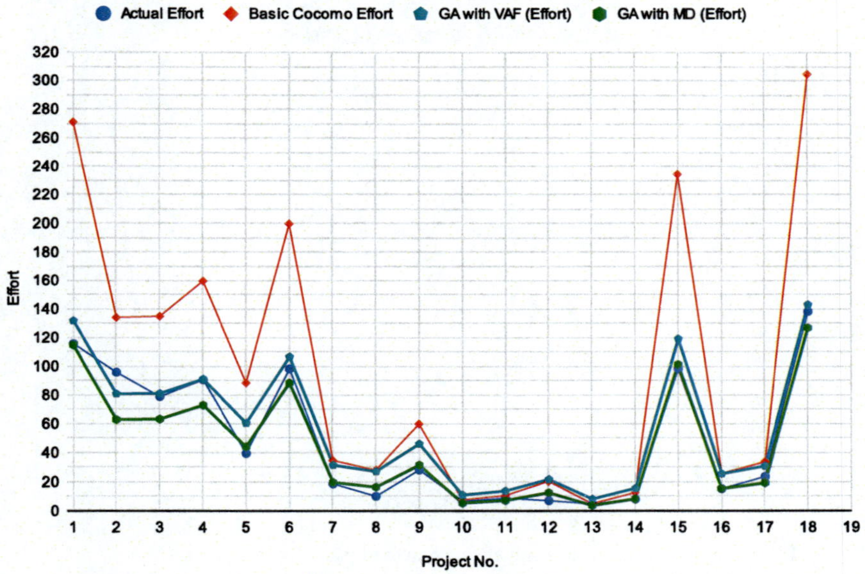

Fig. 5 Effort graph for actual software effort, basic COCOMO effort, effort computed using GA with VAF and effort computed using GA with MD

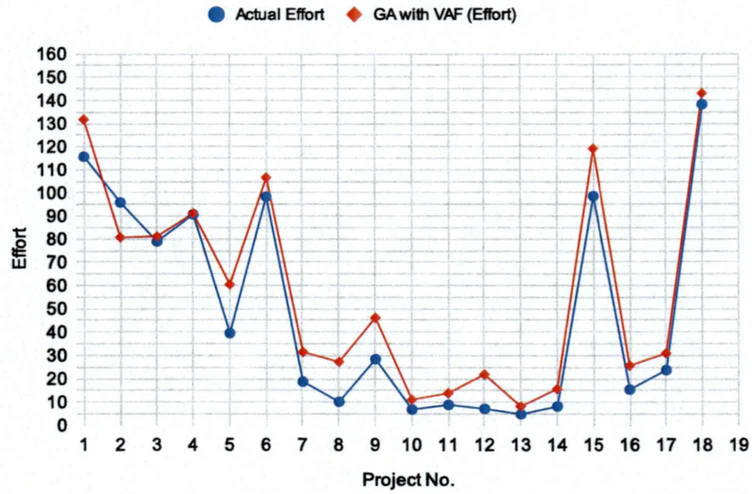

Fig. 6 Effort graph for actual effort and effort computed using GA with VAF

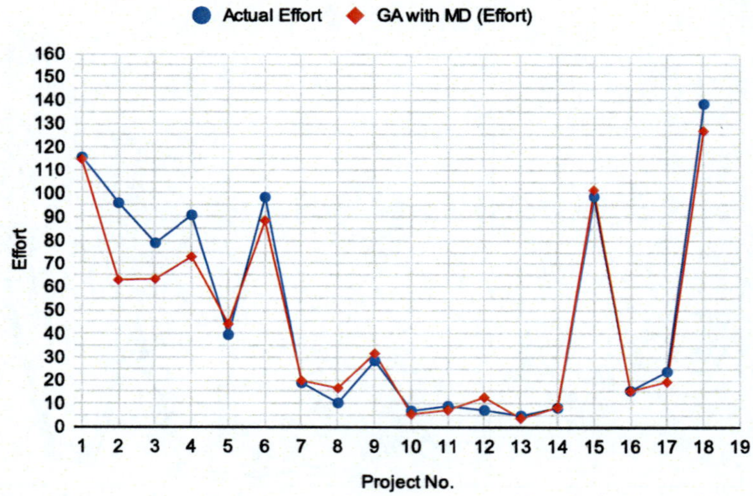

Fig. 7 Effort graph for actual effort and effort computed using GA with MD

5 Conclusion

This paper presents a SEE model that uses the genetic algorithm and compares two evaluation criteria in computing the fitness. We used GA for refining parameters in the basic COCOMO model using Variance-Accounted-For (VAF) and Manhattan distance. The analysis of the overall performance of the proposed models was evaluated with the help of NASA dataset in the promise repository. The comparison between VAF and MD models is shown in Fig. 2.

The result shows that effort estimated using Manhattan distance as evaluation criteria gives better results as compared to actual NASA effort and effort estimated by VAF as evaluation criteria. This makes the estimation done based on Manhattan distance very much reliable in software effort estimation. In each project, the Manhattan distance seems to be more accurate than VAF, and thus, the high deviation rate of actual effort and calculated effort seems to be a limitation of VAF in optimizing software effort estimation.

Undesired changes like risk in software project development might give different results as expected this is a limitation of this optimization model. In future, we will be using other evaluation criteria for comparing the fitness, and also, we will extend our study on optimizing the parameters of other COCOMO models.

Acknowledgements We are highly thankful to Head of our Department Dr. E. R. Vimina for her active guidance throughout the research process. We are also thankful to our learned faculty Asst. Professor G. Deepa for her active guidance throughout the research process. Last but not least, we would also want to extend our appreciation to those who could not be mentioned here but have well played in their role to inspire us.

References

1. B.W. Boehm, An experiment in small-scale application software engineering. IEEE Trans. Softw. Eng. **5**, 482–493 (1981)
2. J.H. Holland, An introductory analysis with applications to biology, control, and artificial intelligence, in *Adaptation in Natural and Artificial Systems*, 1st edn. (The University of Michigan, USA, 1975)
3. C.H.V.M.K. Hari, P.V.G.D. Reddy, A fine parameter tuning for COCOMO 81 software effort estimation using particle swarm optimization. J. Softw. Eng. **5**(1), 38–48 (2011)
4. S. Aljahdali, A.F. Sheta, Software effort estimation by tuning COCOMO model parameters using differential evolution, in *ACS/IEEE International Conference on Computer Systems and Applications-AICCSA* (IEEE, 2010)
5. J.W. Bailey, V.R. Basili, A meta-model for software development resource expenditures, in *ICSE*, vol 81 (1981)
6. B. Clark, S. Devnani-Chulani, B. Boehm, Calibrating the COCOMO II post-architecture model, in *Proceedings of the 20th International Conference on Software Engineering* (IEEE, 1998)
7. O. Benediktsson et al., COCOMO-based effort estimation for iterative and incremental software development. Softw. Qual. J. **11**(4), 265–281 (2003)
8. A. Galinina, O. Burceva, S. Parshutin, The optimization of COCOMO model coefficients using genetic algorithms. Inf. Technol. Manag. Sci. **15**(1), 45–51 (2012)
9. R.K. Sachan, et al., Optimizing basic COCOMO model using simplified genetic algorithm. Procedia Comput. Sci. **89**, 492–498 (2016)
10. K. Rahimunnisa, Hybridized genetic-simulated annealing algorithm for performance optimization in wireless adhoc network. J. Soft Comput. Paradigm (JSCP) **1**(01), 1–13 (2019)
11. A.F. Sheta, Estimation of the COCOMO model parameters using genetic algorithms for NASA software projects. J. Comput. Sci. **2**(2), 118–123 (2006)

12. S. Chhabra, H. Singh, Optimizing design parameters of fuzzy model based COCOMO using genetic algorithms. Int. J. Inf. Technol. 1–11 (2019)
13. A. Saeed, et al., Survey of software development effort estimation techniques, in *Proceedings of the 2018 7th International Conference on Software and Computer Applications* (2018)
14. K. Mukesh Mahadev, G. Gowrishankar, Estimation of effort in software projects using genetic programming. Int. J. Eng. Res. Technol. (IJERT) **09**(07) (2020)
15. A. Ardiansyah, M.M. Mardhia, S. Handayaningsih, Analogy-based model for software project effort estimation. Int. J. Adv. Intell. Inf. **4**(3), 251–260 (2018)
16. Z. Prokopova, P. Šilhavý, R. Šilhavý, VAF factor influence on the accuracy of the effort estimation provided by modified function points methods, in *Annals of DAAAM and Proceedings of the International DAAAM Symposium*, Danube Adria Association for Automation and Manufacturing, DAAAM (2018)
17. M. Angelova, T. Pencheva, Tuning genetic algorithm parameters to improve convergence time. Int. J. Chem. Eng. **2011** (2011)

High-Performance ANFIS-Based Controller for BLDC Motor Drive

R. Shanmugasundaram, C. Ganesh, A. Singaravelan, B. Gunapriya, and B. Adhavan

Abstract The BLDC motors are extensively used in aerospace, electric vehicles, medical equipment, etc., owing to their outstanding speed–torque characteristics. However, BLDC motors require controllers to control the speed, torque and output power based on the application. The PID, fuzzy and ANN-based controllers that are used for the control of BLDC motor have limitations due to their design complexity and implementation. In this paper, an adaptive neuro-fuzzy inference system (ANFIS) has been developed to control the speed of BLDC motor drive and the simulation results are investigated and compared with existing control techniques under the specified operating conditions.

1 Introduction

In recent years, the performance improvement is achieved by incorporating fuzzy controllers in motion control applications. However, the drawbacks of conventional fuzzy inference system [1–5, 7–9] are: (i) Only fixed membership functions can be used, (ii) chose arbitrarily membership functions, (iii) based on the interpretation of the user-defined variable characteristics, structure of the rule is created, and (iv) tuning the system by adjusting the limits of the membership functions. Therefore,

R. Shanmugasundaram (✉)
Sri Ramakrishna Engineering College, Coimbatore 641022, India
e-mail: rsseee@srec.ac.in

C. Ganesh
Rajalakshmi Institute of Technology, Chennai 600124, India

A. Singaravelan · B. Gunapriya
New Horizon College of Engineering, Bangalore 560103, India
e-mail: singaravelan@newhorizonindia.edu

B. Gunapriya
e-mail: gunapriya@newhorizonindia.edu

B. Adhavan
PSG Institute of Technology and Applied Research, Coimbatore 641062, India

P. Karuppusamy et al. (eds.), *Ubiquitous Intelligent Systems*, Smart Innovation, Systems and Technologies 243, https://doi.org/10.1007/978-981-16-3675-2_33

there is a need for an adaptive fuzzy inference system that can generate the desired input–output pairs from by constructing if–then rules with suitable membership functions. In this article, the performance improvement of BLDC motor drive is achieved by constructing an ANFIS for input to output mapping based on the input–output training data pairs. The input to output training data has been generated from the PID controller-based BLDC motor drive with controller gains that can produce optimum response for different parameter combinations.

The fuzzy and neural networks are combined to form an ANFIS hybrid intelligent system that improves the learning capability and adapt. Researchers are using intelligent systems for the modeling and predictions in the engineering applications. The neuro-adaptive learning techniques are used to develop the fuzzy model by learning from the training data set and alter the membership function parameters of fuzzy inference system by the least squares approximation and back-propagation (BPN) algorithm to relate the input to output.

The fundamental concept behind these neuro-adaptive learning techniques is to develop the fuzzy modeling procedure to learn from the training data set and automatically compute the membership function parameters that best match with the input–output training data provided to the FIS. The least squares technique and BPN algorithm are used to alter the membership function parameters of the FIS. During the learning process, these membership function parameters of the FIS are adjusted.

In order to eliminate errors between the real and desired outputs, these parameters are calibrated. This helps the FIS to create a suitable model and learn from the knowledge while constructing the model. ANFIS's benefit over the standard fuzzy model is that the human operator does not need to tune the membership functions.

In [1, 15], ANFIS implementation and performance review for BLDC motor speed control are discussed. The BLDC motor drive dynamic performance has been improved by incorporating ANFIS controller. In [2], methodology for tuning the parameters of adaptive controller of BLDC motor drive has been discussed. Simulation is carried out to show the superiority of this method in terms of dynamic performance and reliability over conventional methods. In [3], comparative study of various controllers such as PID and ANFIS speed-controlled BLDC motor drives has been discussed and their performance has been analyzed. In [4], a neuro-fuzzy adaptive inference system with supervisory learning method is developed for tracking and controlling the speed of BLDC motor drive. Thus, structure of the controller is simplified and dynamic performance improvement is achieved.

In [5], ANFIS structure is used to create a nonlinear model, identify online nonlinear parameters and construct a chaotic time series. Design and stability aspects of adaptive fuzzy systems are discussed in [6, 7]. In [8], implementation of ANFIS through GUI in MATLAB is illustrated. In [9], implementation of intelligent control of intelligent stepper motor drive using ANFIS is discussed and the performance of the drive is analyzed. In [10], the performance of ANN-based controllers is compared with fuzzy-controlled BLDCM drive subjected to variations in system parameters and load. The stability analysis of second-order system is discussed in [11]. In [12], the performance of fuzzy controllers is compared with conventional PID-controlled

BLDCM drive in the specified conditions. In [13, 14], modeling and effect of variation in the parameters of BLDC motor on its performance are analyzed. In [12], design of non-iterative compensator to improve the performance of higher-order systems is presented. In [16–18], development and performance analysis of PI and fuzzy PI speed-controlled BLDCM drive are presented. In [19–23], the efficacy of BLDC motor drives based on adaptive controllers is discussed.

In [24, 27], adaptive control techniques are employed to control the BLDCM drive system and improve the performance. The ANFIS controller design for controlling the solar-powered BLDCM-based wire feeder is discussed in [25]. In [26], hybrid PSO and least square estimation technique are used to develop and analyze the performance of ANFIS for BLDCM drive. In [27], an ANFIS-based control algorithm for controlling the BLDCM pump has been presented and the performance of the overall system has been analyzed.

This paper explores the effectiveness of BLDC motor drive controlled by ANFIS in dealing with nonlinearities occurring during the operating conditions in particular variation in parameters of the system and load. The ANFIS controller is constructed such that it can learn from the input–output data of the conventional PID controller-based BLDCM drive under the specified operating conditions and determine the optimum parameters of the membership function related to the fuzzy inference system. Utilizing the blend of least square approximation and BPN algorithm, the membership function parameters are modified to minimize errors between the real and desired outputs. Simulation results are provided to analyze the drive output under variations in parameters and loads.

2 Development of Adaptive Neuro-Fuzzy Inference Systems

The fuzzy inference system (FIS) [12] maps: (1) the input to membership functions of the input, (2) membership function of the input to rules, (3) rules to a set of output, (4) output to membership functions of the output and (5) the membership function output to a crisp output or a decision related to the output. In this fuzzy inference system, choose only fixed membership functions arbitrarily and predetermine the rules as per the user defined variables and their characteristics in the model. Using the training data set, the BPN alone or in combination with a minimum square method is used in the ANFIS to alter the parameters of the membership function. This alteration helps the FIS to build its model. The FIS does not require a predetermined structure for its model with variable characteristics defined in the system.

2.1 ANFIS Architecture

The architecture of ANFIS is shown in Fig. 1. ANFIS is an algorithm for automatically adjusting Sugeno fuzzy inference system by learning from the training data

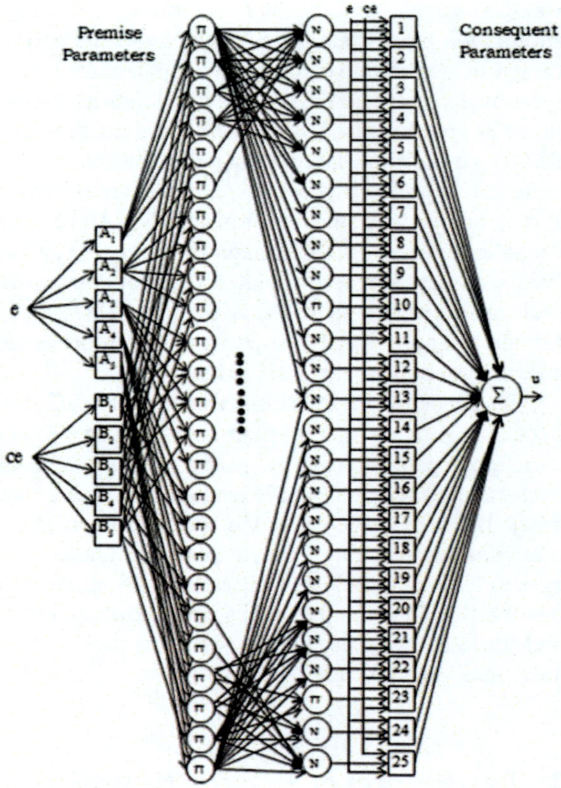

Fig. 1 ANFIS structure

set. It has two inputs with five membership functions each, twenty-five rules and one output. The "error (*e*)" and "rate of change of error (ce)" are the two inputs, and the "control signal (*u*)" is the output. There are five triangular membership functions for each input and a constant or linear function for the output. The ANFIS is a five-layer feed-forward fuzzy neural network. The function is same for all nodes in the same layer. The nodes in the first and fourth layer represented by square node are adaptive. The function of node in each layer is as follows [23]:

Layer 1: Every node i in this layer is a square node with a node function,

$$O_i^1 = \mu_{A_i}(x),\ i = 1, 2 \tag{1}$$

where x is the input and $\mu_{A_i}(x)$ is the membership value of the associated linguistic variable.

Layer 2: Every node in this layer is represented as $\prod$, and its output is firing strength of a rule as given below.

$$O_i^2 = w_i = \mu_{A_i}(x)\mu_{B_i}(y),\ i = 1, 2 \tag{2}$$

where x and y are inputs, and $\mu_{A_i}(x)$ and $\mu_{B_i}(y)$ are the membership values of their associated linguistic variable.

Layer 3: Every node in this layer is represented by a circle marked with N and gives output as normalized firing strength as given below.

$$O_i^3 = \overline{w_i} = \frac{w_i}{w_1 + w_2} i = 1, 2 \tag{3}$$

Layer 4: Every node i is an adaptive node and gives the node output as given below.

$$O_i^4 = \overline{w_i} f_i = \overline{w_i}(p_i x + q_i y + r_i) i = 1, 2 \tag{4}$$

where $\overline{w_i}$ is the output of layer 3 and $\{p_i, q_i, r_i\}$ are consequent parameters of Sugeno-type membership function.

Layer 5: Every node in this layer is represented by a circle and computes the overall output as the summation of all incoming signals as given below.

$$O_i^5 = \sum_{i=1}^{2} \overline{w_i} f_i = \frac{\sum_{i=1}^{2} w_i f_i}{\sum_{i=1}^{2} w_i} \tag{5}$$

The goal of training is to reduce the deviation between the real and predicted values by altering the anticipated (layer 1) and consequent parameters (layer 4). The output of the ANFIS is a linear combination of adjustable consequent parameters as given below.

$$f = (\overline{w_1}x)p_1 + (\overline{w_1}y)q_1 + (\overline{w_1})r_1 + (\overline{w_2}x)p_2 + (\overline{w_2}y)q_2 + (\overline{w_2})r_2 \tag{6}$$

In order to achieve fast learning, hybrid learning algorithm which is the combination of least squares and back-propagation learning algorithm is used. In this approach, optimal values of consequent parameters are obtained by least squares method and the premise parameters are altered by gradient descent method. The consequent parameters are used to compute the final output of ANFIS.

By presenting input–output data produced from the PID controller-based BLDC motor drive, the ANFIS network is trained by the hybrid learning process. The error between the real and desired output is reduced during learning by changing the premise and consequent fuzzy inference method parameters. Finally, the trained ANFIS is used as a BLDC motor drive controller.

3 Control Structure of ANFIS Controller-Based BLDC Motor Drive

Figure 2 shows the structure of BLDC drive controller with ANFIS controller. The inputs to the controller are "error (*e*)" and "rate of change of error (ce)." Initially, ANFIS is trained to learn input–output relationship from the training data. Therefore, ANFIS generates control signal (*u*) corresponding to its inputs (*e* and ce). Thus, the control signal (*u*) in turn drives the BLDC motor closer to reference speed. When the actual speed output deviates from the reference speed due to either variation in parameters of the system or load disturbances, then the error value increases. The ANFIS controller always tracks the change in inputs (*e* and ce) and adjusts the actuating signal (*u*) accordingly so that error is minimized.

3.1 Generation of Training Data

The generation of input to output training data for the ANFIS is one of the important steps in the development of ANFIS. In this research work, input–output training data is generated from the BLDC motor drive controlled by PID controller subjected to different operating conditions. In order to produce better response, the PID controller is tuned for different drive parameter combinations as discussed in [10]. The same controller gains are used to simulate the response and collect training data for different parameter combinations. The training data of 30,000 samples is collected for different parameter combinations of the drive and from which 10,000 samples are used as training data for the ANFIS.

3.2 Development of BLDC Motor Drive with ANFIS Controller

The steps followed in developing ANFIS are: (i) generation of input–output training data, (ii) generation or loading the initial FIS structure, (iii) loading of training data, checking and test data, (iv) training the FIS and (v) authenticating the trained FIS. In the first step, generate input–output data set to train the ANFIS as discussed in

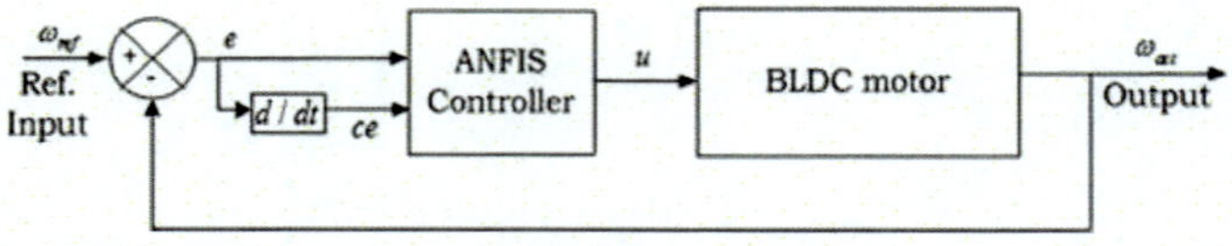

Fig. 2 Structure of BLDC drive controller with ANFIS controller

Sect. 3.1. In the second step, "anfisedit" command is used to open ANFIS editor. Now, load or create initial fuzzy inference system (FIS) with 2 inputs, each with 5 triangular membership functions and linear or constant membership functions for the output, and choose the defuzzification method as weighted average method. After loading or creating initial FIS, its model structure is automatically created with 25 fuzzy rules. Figure 3 shows the structure of ANFIS with 2 inputs, 1 output and Sugeno-type fuzzy inference. Figure 4 shows the membership functions of the inputs, and Fig. 5 shows the output functions.

The ANFIS model structure created through FIS is shown in Fig. 6. In this model structure, the input variables are "error (e)" and "rate of change of error (ce)," and the output is "control signal (u)." The input to output mapping data of ANFIS is given in Table 1. In the third step, load the training and test data. Now, FIS model is trained to emulate the training data by altering the membership function parameters based on the chosen error criterion and display the error plots. After training FIS, validate the model using the test data. The error plot and the validation of result after training are shown in Fig. 7. The error reduces to 3.398 after 1000 epochs of training. It is found that trained data and test data are closely matching with each training.

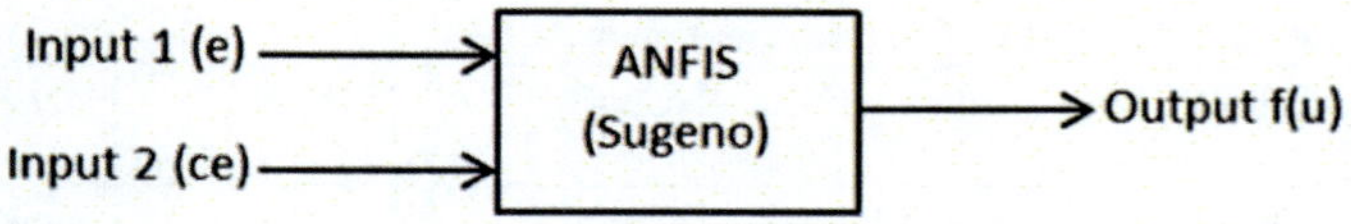

Fig. 3 Structure of ANFIS with inputs (e, ce) and output, $f(u)$

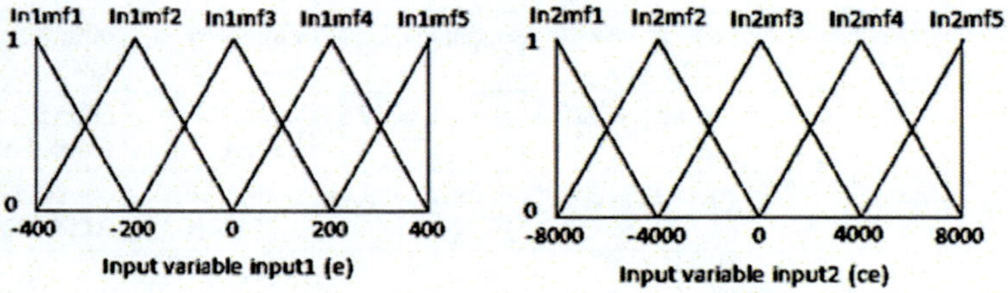

(a) (b)

Fig. 4 Membership functions "error" (input 1) and "rate of change in error" (input 2)

Fig. 5 Output functions

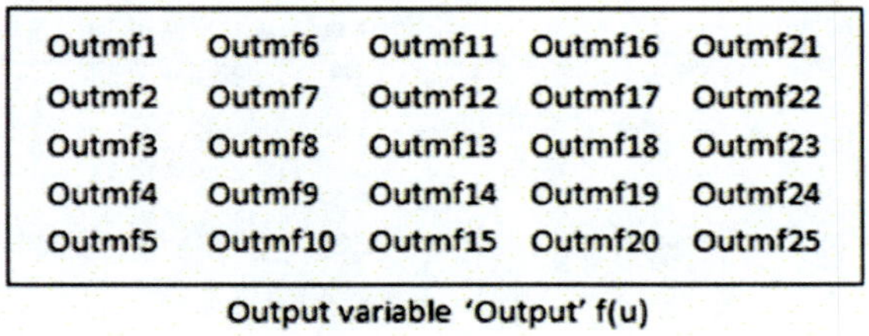

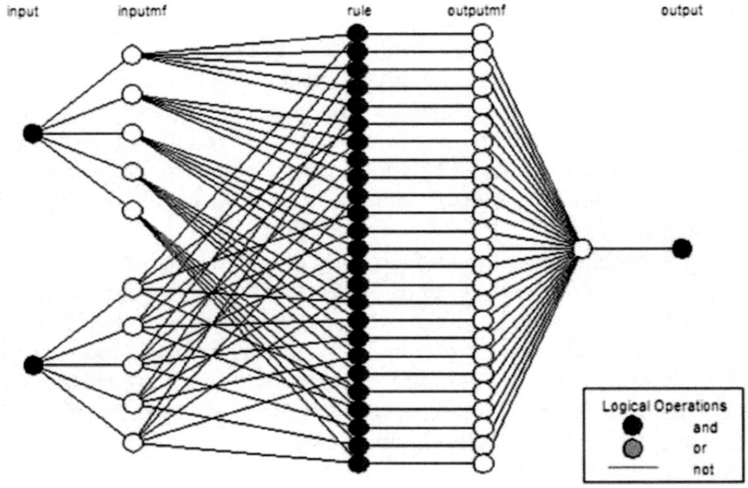

Fig. 6 ANFIS model structure

Table 1 Input–output mapping of ANFIS after training

ce *e*	in2mf1	in2mf2	in2mf3	in2mf4	in2mf5
in1mf1	outmf1 = −4.703e–5	outmf2 = 184.2	outmf3 = −2.129	outmf4 = 9.53	outmf5 = −5.383
in1mf2	outmf6 = −2.1e–09	outmf7 = 3424	outmf8 = −1134	outmf9 = −521.7	outmf10 = 11.38
in1mf3	outmf1 = −152.8	outmf1 = −105.3	outmf1 = 11.23	outmf1 = 574.7	outmf15 = 1246
in1mf4	outmf1 = 42.74	outmf1 = 220.8	outmf1 = 50.6	outmf1 = −2.65e4	outmf20 = 6.405e−06
in1mf5	outmf2 = 39.84	outmf2 = 34.66	outmf2 = 39.86	outmf2 = 3.751e–10	outmf25 = 0.0004006

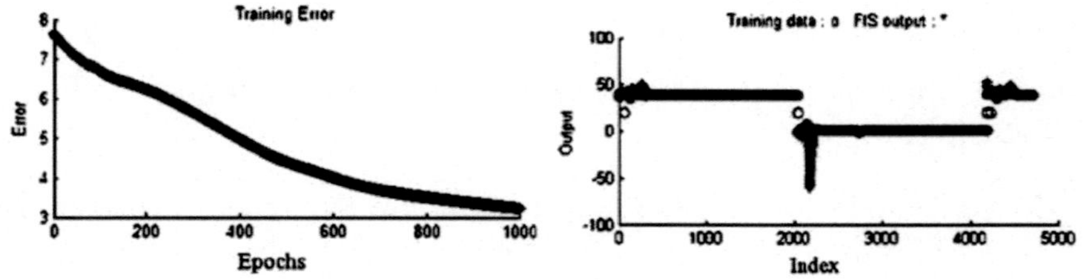

(a) (b)

Fig. 7 **a** Training error and **b** validation of ANFIS after training

4 MATLAB Simulation of ANFIS Controller-Based BLDC Motor Drive

The Simulink model of BLDC motor drive controlled by ANFIS controller is shown in Fig. 8. The trained FIS model is used as ANFIS controller. The ANFIS controller inputs are "error (*e*)" and "rate of change of error (ce)," and "control signal (*u*)" is the output. This control signal is used to drive the system output close to reference input. The load is modeled in terms of inertia and friction component of the load. The reference speed block does the function of step change in reference speed. Due to change in reference speed changes or load disturbances, there will be a deviation in actual speed from the reference speed. The "error (*e*)" and "rate of change of error (ce)" are computed, and these are applied as inputs to the ANFIS controller, which in turn produce an actuating signal in order to bring the output speed close to reference speed, thus minimizing the speed error.

The response of speed is obtained for different parameters of the system (i.e., inertia, J, and resistance, *R*) [12, 14] at full load by applying step change in input set speed, and the outputs are analyzed in the following cases.

Case 1: Output response for the system parameters ($R_1 = 0.57\Omega$, $J_1 = 350 \times 10^{-6}$ kg m^2). Figure 9 shows the BLDC motor drive output response with terminal resistance of the motor, R_1, and total inertia of the motor, J_1. The total inertia of the drive (J_1) is the sum of inertia of motor (J_M) and inertia of load (J_L). The load friction component (*B*) for this drive is 1 × 10^{-3} Nm/(rad/s). The performance evaluation parameters of the output response are settling time of 70 ms, time taken to reach 90% of the final value is 50 ms, error in the steady state is ± 10 rpm, and de-acceleration time is 100 ms. The ANFIS controller is able to track the change in set speed and maintain output speed close to set speed.

Case 2: Output response for the system parameters ($R_2 = 1.14\Omega$, $J_1 = 350 \times 10^{-6}$ kg m^2). Figure 10 shows the BLDC motor drive output response with terminal resistance of the motor, R$_2$, and total inertia of the motor, J_1 ($J_1 = J_M + J_L$). The

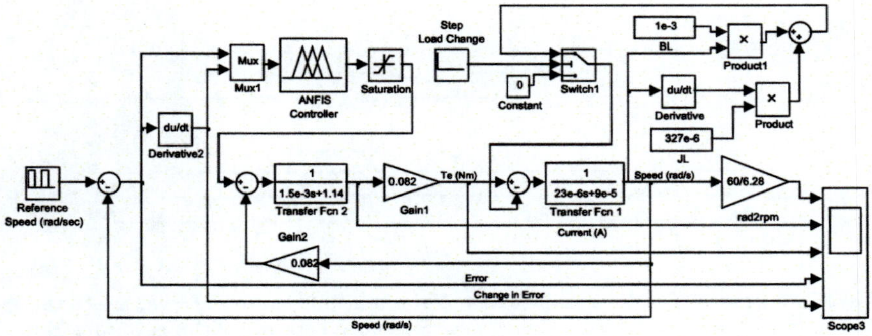

Fig. 8 Simulation model of ANFIS controller-based BLDC motor drive

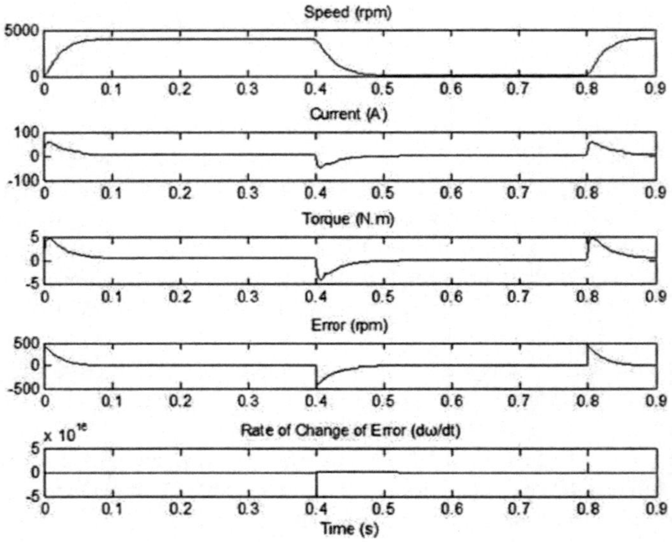

Fig. 9 Response of BLDC motor drive controlled with ANFIS controller (J_1, R_1 and maximum load) [*top to bottom* actual speed, DC supply current, torque, error, rate of change of error (ce)]

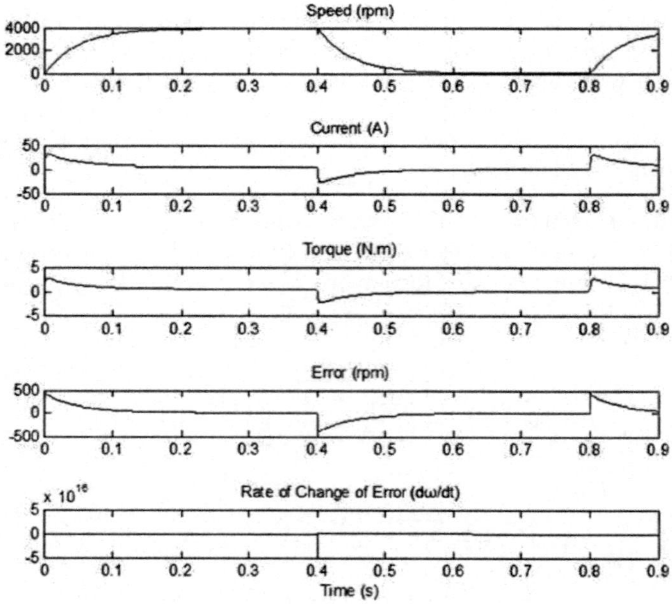

Fig. 10 Response of BLDC motor drive controlled with ANFIS controller (J_1, R_2 and maximum load) [*top to bottom* actual speed, DC supply current, torque, error, rate of change of error (ce)]

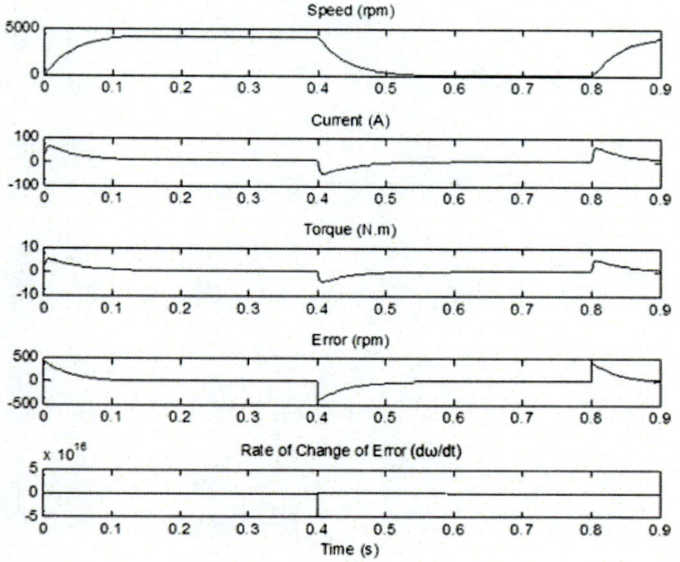

Fig. 11 Response of BLDC motor drive controlled with ANFIS controller (J_2, R_1 and maximum load) [*top to bottom* actual speed, DC supply current, torque, error, rate of change of error (ce)]

performance evaluation parameters of the output response are settling time of 128 ms, time taken to reach 90% of the final value is 120 ms, error in the steady state is ± 60 rpm, and de-acceleration time is 200 ms.

Case 3: Output response for the system parameters ($R_1 = 0.57\Omega$, $J_2 = 550 \times 10^{-6}$ kg m^2). Figure 11 shows the BLDC motor drive output response with terminal resistance of the motor, R_1, and total inertia of the motor, J_2. The performance evaluation parameters of the output response are settling time of 110 ms, time taken to reach 90% of the final value is 75 ms, error in the steady state is ± 10 rpm, and de-acceleration time is 120 ms.

Case 4: Output response for the system parameters ($R_2 = 1.14\Omega$, $J_2 = 550 \times 10^{-6}$ kg m^2). Figure 12 shows the BLDC motor drive output response with terminal resistance of the motor, R_2 and total inertia of the motor, J_2. The performance evaluation parameters of the output response are settling time of 240 ms, time taken to reach 90% of the final value is 130 ms, error in the steady state is ± 60 rpm, and de-acceleration time of 220 ms. The simulation results are tabulated in Table 2. The output speed response for all different combination of phase resistance and total inertia of the drive is shown in Fig. 13. It is found that the tracking performance of ANFIS controller is better as compared to conventional controllers under the specified operaitng conditions.

The results of PID [12], fuzzy [12], ANN [10] and ANFIS-based controllers are compared and analyzed. Compared to PID, fuzzy and ANN-based controllers,

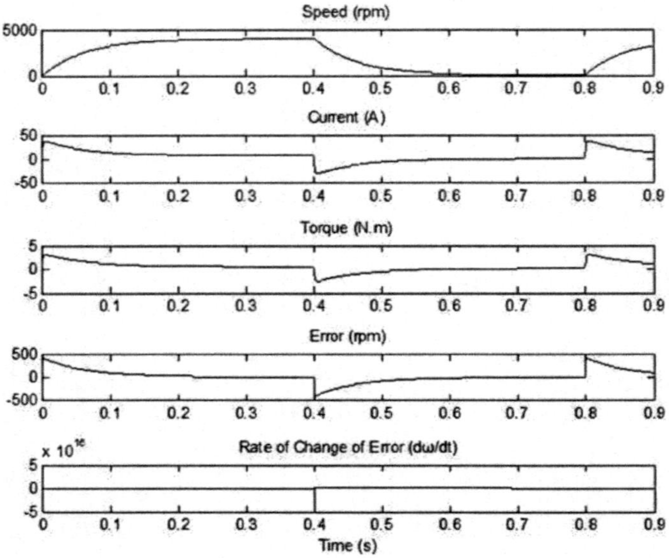

Fig. 12 Response of BLDC motor drive controlled with ANFIS controller (J_2, R_2 and maximum load) [*top to bottom* "Actual speed, DC supply current, Torque, Error, Rate of Change of Error (ce)"]

Table 2 Matlab Simulink output of ANFIS controller-based BLDCM drive

Inertia and resistance of the drive at Max. Load	Rise time t_r (ms)	Settling time t_s (ms)	Deceleration time t_d (ms)	Error
R_1, J_1	50	70	100	± 10 rpm
R_2, J_1	120	128	200	± 60 rpm
R_1, J_2	75	110	120	± 10 rpm
R_2, J_2	130	240	220	± 60 rpm

it is found that the settling time and tracking performance are more enhanced for ANFIS-based controller.

5 Conclusion

In this paper, an ANFIS-based BLDC motor drive speed controller has been developed and simulated to examine the performance of the drive exposed to variations in parameters. Various control parameters are collected, evaluated and compared with other controllers, such as rise time, setting time and steady-state error. The proposed ANFIS controller has a simple structure and high-performance monitoring

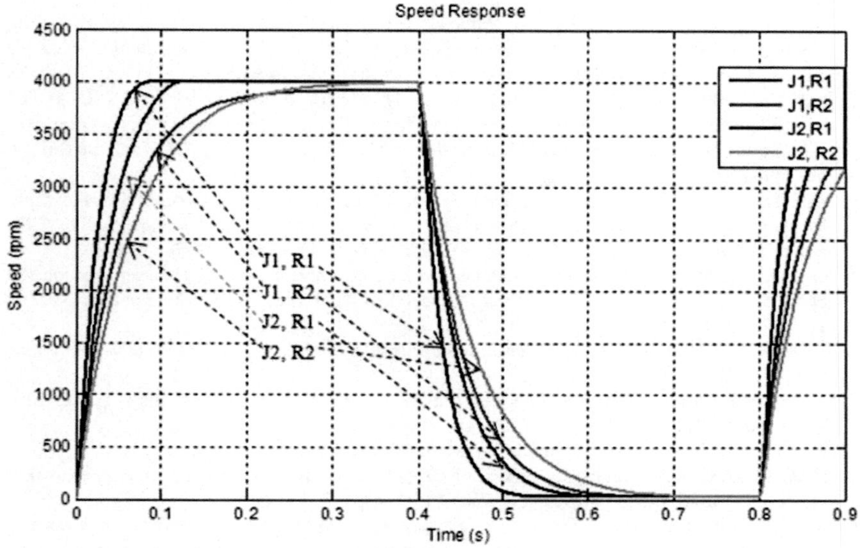

Fig. 13 Speed output for different combination of parameters of the drive

with learning capabilities. It is obvious from the findings that, as compared to other controllers, the ANFIS controller has many benefits as compared to other controllers such as: (i) Mathematical model is not required, (ii) fast and robust method is available to generate the suitable membership functions and rule base, (iii) membership tuning is not required, (iv) fast dynamic response and (v) less computation time.

Compared to PID, fuzzy [12] and ANN [10]-based controllers, the ANFIS controller is found to have better performance in terms of settling time and tracking performance. Since the results show that the ANFIS controller in all respects outperforms other controllers, it is suitable for real-time applications. Hence, ANFIS controller-based BLDC motor drive may be preferred for speed control applications to achieve better response under parameter variations and load disturbances. The future scope of this work is to create an experimental setup to implement the proposed ANFIS controller-based BLDCM drive and validate the experimental results with simulation results. In order to further improve the performance of the BLDCM drive, a hybrid controller may be developed to combine the futures of conventional and adaptive control techniques.

References

1. P.H. Sasongko, S. Sarjiya, Performance analysis of adaptive neuro fuzzy inference systems (ANFIS) for speed control of brushless DC motor, in *Proceedings of International Conference on Electrical Engineering and Informatics (ICEEI)*, Bandung (2011), pp. 1–6

2. V.M. Varatharaju, B.L. Mathur, K. Udhayakumar, ANFIS based controllers and modeling simulation of PMBLDC motor and drive system, in *Proceedings of International Conference on Sustainable Energy and Intelligent Systems (SEISCON 2011)*, Chennai (2011), pp. 518–522
3. P.H. Sasongko, S. Sarjiya, A comparative study of PID, ANFIS and hybrid PID-ANFIS controllers for speed control of Brushless DC Motor drive, in *Proceedings on International Conference on Computer, Control, Informatics and its Applications (IC3INA)*, Jakarta (2013), pp. 117–122
4. A.H. Niasar, A. Vahedi, H. Moghbelli, ANFIS-based controller with fuzzy supervisory learning for speed control of 4-switch inverter brushless DC motor drive, in *Proceedings on 37th IEEE Power Electronics Specialists Conference (PESC '06)*, South Korea (2006), pp. 1–5
5. G. Yanling, M.E.A. Mohamed, Study on the extent of the impact of data set type on the performance of ANFIS for controlling the speed of DC motor. J. Eng. Technol. Sci. **51**(1), 83–102 (2019)
6. L.-X. Wang, *Adaptive Fuzzy Systems and Control: Design and Stability Analysis* (Prentice Hall, NJ, 1994)
7. T. Johansen, Fuzzy model based control: Stability, robustness, and performance issues. IEEE Trans. Fuzzy Syst. **2**(3), 221–234 (1994)
8. Fuzzy Logic Toolbox User's Guide. The MathWorks Inc., Natick, MA (2011)
9. P. Melin, O. Castillo, Intelligent control of stepper motor drive using an adaptive neuro-fuzzy inference system. Int. J. Inf. Sci. (Elsevier) **170**(2–4), 133–151 (2005)
10. R. Shanmugasundaram, C. Ganesh, A. Singaravelan, ANN-based controllers for improved performance of BLDC motor drives, in *Proceedings of International Conference on Advances in Electrical Control and Signal Systems 2019, LNEE*, vol. 665 (Springer, Singapore, 2020), pp. 73–87s
11. C. Ganesh, R. Shanmugasundaram, Design of non-iterative first order compensator for type-1 higher order systems, in *Proceedings of the 2nd International Conference on Communication, Devices and Computing 2020, LNEE*, vol. 602 (Springer, Singapore, 2020), pp. 355–367
12. R. Shanmugasundram, K.M. Zakariah, N. Yadaiah, Implementation and performance analysis of digital controllers for brushless DC motor drives. IEEE/ASME Trans. Mechatron. **19**(1), 213–224 (2012)
13. R. Shanmugasundram, K.M. Zakariah, N. Yadaiah, Modeling, simulation and analysis of controllers for brushless direct current motor drives. J. Vib. Control **19**(8), 1250–1264 (2012)
14. R. Shanmugasundram, K.M. Zakaraiah, N. Yadaiah, Effect of parameter variations on the performance of direct current (DC) servomotor drives. J. Vibr. Control **19**(10), 1575–1586 (2012)
15. K. Premkumar, B.V. Manikandan, Adaptive neuro-fuzzy inference system based speed controller for brushless DC motor. Neuro Comput. **138**, 260–270 (2014)
16. A. Shyam, F.J.L. Daya, A comparative study on the speed response of BLDC Motor using conventional PI controller, anti-windup PI controller and fuzzy controller, in *Proceedings on International Conference on Control Communication and Computing (ICCC)*, Kerala, India (2013), pp. 68–73
17. R. Arulmozhiyal, R. Kandiban, Design of Fuzzy PID controller for Brushless DC motor, in *Proceedings on International Conference on Computer Communication and Informatics (ICCCI)*, India (2012), pp. 1–7
18. M.V. Ramesh, J. Amarnath, S. Kamakshaiah, G.S. Rao, Speed control of brushless DC motor by using fuzzy logic PI controller. ARPNJ. Eng. Appl. Sci. **06**(9), 55–62 (2011)
19. P. Devendra, G. Rajetesh, K.A. Mary, C. Saibabu, Sensorless control of brushless DC motor using adaptive neuro-fuzzy inference algorithm, in *Proceedings of International Conference on Energy, Automation, and Signal*, India (2011), pp. 28–30
20. Y. Guo, M.E.A. Mohamed, Speed control of direct current motor using ANFIS based hybrid P-I-D configuration controller. IEEE Access **8**, 125638–125647 (2020)
21. M. Gokbulut, B. Dandil, C. Bal, A hybrid neuro-fuzzy controller for brushless DC Motors, in *International Turkish Symposium on Artificial Intelligence and Neural Networks 2005, LNCS*, vol. 3949 (Springer, Berlin/Heidelberg, 2006), pp. 125–132

22. Q.C. Zhang, M. Jiang, Adaptive neuro-fuzzy control of BLDCM based on back-EMF. J. Comput. Inf. Syst. **07**(12), 4560–4567 (2011)
23. V.M. Varatharaju, B. Mathur, Udhayakumar, Adaptive controllers for permanent magnet brushless DC motor drive system using adaptive-network-based fuzzy interference system. Am. J. Appl. Sci. **08**(08), 810–815 (2011)
24. B. Rajani, K. Bapayya Naidu, Renewable source DC microgrid connected BLDC water pumping system with adaptive control techniques, in *Proceeding of 4th International Conference on Electronics, Communication and Aerospace Technology (ICECA)*, Coimbatore, India (2020), pp. 216–222
25. N. Hamouda, B. Babes, S. Kahla, A. Boutaghane, A. Beddar, O. Aissa, ANFIS controller design using PSO algorithm for MPPT of solar PV system powered brushless DC motor based wire feeder unit, in *Proceeding of International Conference on Electrical Engineering (ICEE)*, Istanbul, Turkey (2020), pp. 1–6
26. H. Suryoatmojo, M. Ridwan, D.C. Riawan, E. Setijadi, R. Mardiyanto, Hybrid particle swarm optimization and recursive least square estimation based ANFIS multi-output for BLDC motor speed controller. Int. J. Innov. Comput. Inf. Control 15(3), 939–954 (2019)
27. A.A. Hepzibah, K. Premkumar, ANFIS current-voltage controlled MPPT algorithm for solar powered brushless DC motor based water pump. Elect. Eng. **102**, 421–435 (2019)

Latency Aware Resource Scheduling and Queuing

Sharmila S. Patil and S. H. Brahmananda

Abstract In this digital world, clouds are one of the valuable resources which are widely available everywhere. In fulfilling services, it plays a vital role. Cloud is an optimum service provider in all areas. However, to satisfy the emergency requirement of resource allocation, clouds face many execution and design issues. Increasing demands of services in all areas affects the emergency service requirements due to bandwidth bottleneck, network performance problem, network size, and communication. The main issue is reducing latency produced by the processing time for queues by machines and other network intermediate processes. In the process of delay minimization, the initial stage is admission of the request, which will be scheduled and lined up it in an emergency queue. This process reduces the key time of handling requests. To fulfill this emergency resource requirement, a resource scheduling and queuing method is proposed. The algorithm is organized in two stages, where the priorities of a service will be considered for queuing and scheduling. The resource or service allocation will be carried out in a faster way with the combination of the gray wolf and the multidimensional queuing algorithms.

1 Introduction

The Internet of things (IOT) and application software are new tools in the healthcare industry which simplify the patient management and treatment processes [1, 2] and improve the total healthcare system. However, these IOT and supportive technologies services infrastructure development and maintenance are new challenges to this medical industry. The expenses of maintaining IT infrastructure, computational resources, and human resources are very high. IT infrastructure and services can easily be made available with a cloud [3]. Cloud is a central system to handle large services. Data servers provide services to clients that handle large communication services. All the healthcare services and applications can be accessed through the cloud [4–5]. Cloud structure reduces the operating cost of maintaining data centers

S. S. Patil (✉) · S. H. Brahmananda
Department of Computer Science and Engineering, GITAM School of Technology, (GITAM deemed university), Bangalore, India

P. Karuppusamy et al. (eds.), *Ubiquitous Intelligent Systems*, Smart Innovation, Systems and Technologies 243, https://doi.org/10.1007/978-981-16-3675-2_34

for the online-based services and resource allocations. Resource management with a cloud is always efficient and fast due to the central management and global availability through the Internet. To handle difficult situations efficiently, the cloud performs data backup and data recovery.

In cloud computing, clients are provided different services like infrastructure, software, as well as application platform and are charged based on service usage. Low bandwidth is a major problem while working with clouds at high traffic time [6]. In a cloud, data access, data storage, and application sharing work on the technique of virtualization. Resource scheduling and load balancing are pivotal for efficient cloud service provisioning [7, 8].

In medical fields, instant response for real-time requests is very essential. These requests have a time limit. In the medical field, many emergencies occur and need to be handled promptly. In case of heart attack or any accident prompt resources are required to be accessible. In such conditions before handling the demands and responding to them with the services, there is a need to line up the requests. The requests will have priorities at the entry point on the basis of the deadline. The highest priorities or lowest deadline jobs need to be handled primarily. The concept of this paper is to schedule and queuing of these jobs before they commence for the next low latency implementation. So that fuzzy resource scheduling and queuing (FRSQ) algorithm is proposed. The algorithm includes managing the process of workload traffic that exists over the Internet and reduces latency by enhancing the response time.

2 Related Work

Agarwal et al. [9] approached a three-layer algorithm for resource allocation in a fog computing environment, and this technique facilitates to overcome over-provisioning and under-provisioning. To solve fault tolerance, resource overflow, and underflow allocations of the resources, this algorithm is introduced. Proper and even distribution is the advantage of this algorithm. The author compared efficient resource allocation algorithm with existing other algorithms. This algorithm improves allocation of resources in low latency. The disadvantage of this method is that it is not able to handle resource allocation request at execution time. Lina et al. [10] introduced a resource allocation during execution. This strategy is implemented in a Fog computing environment, based on time, resource utilization, and requirement satisfaction. The result of the resource allocation algorithms was to predict the completion time of tasks and credibility of resource. Mahmud et al. [11] approached a management module in Fog computing environment. The module ensures the quality of service (QoS) in application. The application takes care of deadline and proper use of resources. The policy is a combination of two algorithms. The first was the application module and second module for simplification of constraint-based optimization problem in forwarding modules which helps the low latency implementation. Mahmud et al. [12] introduced a fuzzy-based approach for assigning position to request. Before assigning any request, a user's expectation is checked. The advantage of the policy was to improve processing time of data, congestion in a network, affordability of these resources, and quality of service.

Verma et al. [13] introduced a three-layer architecture and RTES pseudo code algorithm. This algorithm is implemented to overcome congestion of network bandwidth utilization issue with security. The result of the RTES algorithm shows efficient resource allocation, reduces response time, and increases throughput. Finally, the author's approach to architecture and algorithm was 90% efficient, with the remaining 10% as a future work with security factors. Mirjalili, S et.al approached a gray wolf optimizer optimization algorithm (GWO) [14], and the algorithm is an imitator of the natural wolf hunting method. In this algorithm, different levels are maintained for hunting rules. Gray wolf method arranges a preference wise 4 levels, so that emergency can be handled very promptly while attacking prey. The GWO is a very efficient optimization algorithm which satisfies the requested low latency demands. Priya et al. [15] proposed an algorithm multidimensional resource scheduling model for dynamic requests to access the available resources in a cloud environment. This algorithm arranges the demanded service requests efficiently with the balancing load approach, increasing machine utilization. Yujun ma [16] introducing health resource distribution problem, healthcare systems architecture, and technology challenges can be handled using information technologies. Author has discussed detailed key technologies and challenges. Chen, Joy Iong Zong et al. [17] proposed a method which performs the evaluation of delay, energy, and cost in the service provision in a cloud computing model. This method enhances the interoperability in the IOT devices.

Very less number of papers proposed work on latency improvement. This paper will try to contribute work on the latency improvement in the emergency resource allocation areas.

3 Proposed Work

In the healthcare department, requests from the users should be allowed to access resources all the time through the Internet. To achieve high user satisfaction and proper resource utilization, the proposed work reduces time at initial stage with priority between the connected nodes. It also ensures that every computing resource is distributed efficiently with a better response time. Highly efficient resource utilization and proper handling process help to minimize resource accessing and response time. The setup of the fuzzy resource scheduling and queuing (FRSQ) algorithm acquires less resource consumption and response time shown in Fig. 1 with the reference of GWO following resources are accessible to the highest emergency request as following way

$$Dp = |C \cdot X\mathrm{p}(t) - X(t)| \quad (1)$$

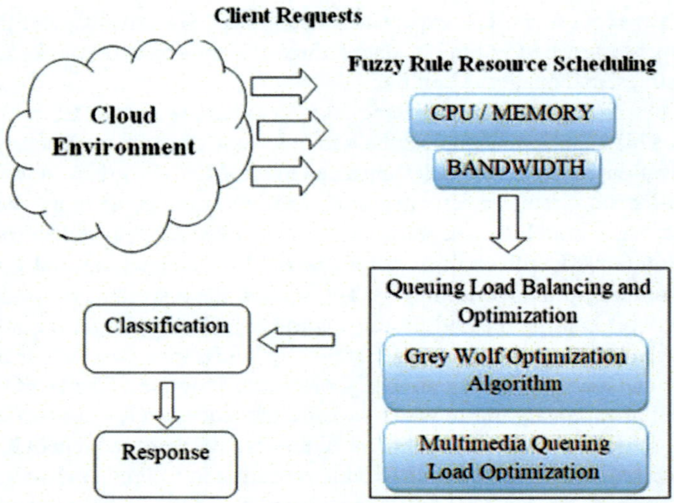

Fig. 1 Fuzzy resource scheduling and queuing (FRSQ) method

$$X(t + 1) = X\mathrm{p}(t) - A \cdot Dp \quad (2)$$

where t is the number of iteration, $X(t)$ is highest priority request, X(t + 1) is the next highest priority request it arrives, and $X\mathrm{p}(t)$ specifically refers to one of the priority levels of the request to the specific resource. Where A and C are coefficient vectors, expressed as follows.

$$A = 2ar1 - a \quad (3)$$

$$C = 2r2 \quad (4)$$

where $r1$, $r2$ are random vectors in [0,1], a is a decreasing value in [0, 2], typically $a = 2 - 2t/I$ (I is the maximum number of iterations). After queue arrangement of requests, the resources will be allocated dynamically.

The fuzzy resource scheduling and queuing (FRSQ) algorithm manages scheduling of emergency services and sets priorities to user request after evolution. It differentiates the resource requests. Resources for the requested services are generally computing machines, processors, or storage space like memory.

To reduce load on cloud infrastructure, the resource allocation algorithm is used for scheduling the resources, and the load is organized using a queuing optimization algorithm. It consists of two algorithms as gray wolf algorithm [14] and the multidimensional queuing load optimization algorithm [15]. The gray wolf optimization (GWO) algorithm is a population-based evolutionary algorithm inspired by the hunting behavior of gray wolves. In the GWO algorithm, different levels of the animal to manage to identify the hunting and attacking rules are decided. First stage using GWO finalizes the service or task set ($t1$, $t2$, $t3\ldots$ tn) and then maintain the queues of resources considering cloud user and computational time.

The multidimensional queuing algorithm considers memory, bandwidth, and CPU classes of requests. This algorithm dynamically selects the request from the queue and balances the load, avoiding underutilization and overutilization of the recourses that finally result in reduction of the latency.

The proposed method is initially the delay incurred in the services is optimized in the proposed model; the majority of the requests is executed in the device as in Eq. 1.

$$\text{Delay} = \Sigma(\text{time taken to Schedule}) + \Sigma\text{time spend in queue}$$

The prime purpose of the proposed method is the utilization of effective implementation of the scheduling algorithm incorporated with one more stage of queuing. The combination of both algorithms reduces the required process time and improves the average rate of success of requested resource allocation.

4 Algorithm

Fuzzy Resource Scheduling and Queuing (FRSQ)

Step 1 Bandwidth, memory and CPU, cloud user, cloud server, output variable "t" used to represent the efficient resource scheduling.

Step 2 Trapezoidal fuzzification function is obtained for scheduling multi-resources in cloud computing

Step 3 Mapping is checked for input and data center

Step 4 If mapped, perform centroid defuzzification for efficient resource scheduling

Step 5 Else, resource scheduling is unable to perform

Step 6 Using the gray wolf algorithm initializes the population for cloud user, server, resource, and time for queuing algorithm

Step 8 Finding the fitness value using iteration based on the resource threshold factor 'RTF value which should be less than RTF value

Step 9 Cloud user with balanced load and the resource to be assigned.

5 Results

The simulator result shows that the mentioned FRSQ improves performance in areas such as application delay, placement time, network usage, and power consumption.

The proposed two corporate methods improve performance a lot. The best of both results in proposed two objectives:

(a) To prepare categorized task queue considering their priorities for resources with GWO
(b) To successfully allocate resources to all lined up tasks with the help of multidimensional queuing load optimization to reduce latency

The metrics calculated are average success rate, load balancing ratio, throughput, computational time, and efficiency. Calculated result comparison of the FRS-QLO algorithm is performed with K-means [18], McMaster grid scheduling testing (MGST), [19] and sliding window daily profile (SWDP) [20] as in Table 1.

To analyze the effectiveness of resource allocation algorithm of FRSQ, the comparisons are shown in above Figs. 2, 3, 4, 5, and 6 with different algorithms. Figure 2 shows the task scheduling ratio enhancement time using this strategy using four virtual machines in Fig. 3 of FRSQ improves the task allocation average success rate, and the execution time is shorter than the time. However, the tasks kept fixed, which can vary in future work. The execution time effectively shortened due to effective balance of queuing.

Table 1 Comparison results

Parameters	Algorithms			
	FRS-QLO	MGST	SWDP	K-means
Scheduling length ratio	0.3125	0.956	1.1354	1.27407
Average success rate	27.2	25.6	23.87	21.3488
Load balancing ratio	1.55172	1.76	1.57	1.98248
Throughput	8.25	6.56	6.34	6.68889
Computational time	13	12	11	47
Efficiency	85.0897	82.4589	80.1268	63.7209

Fig. 2 Length ratio

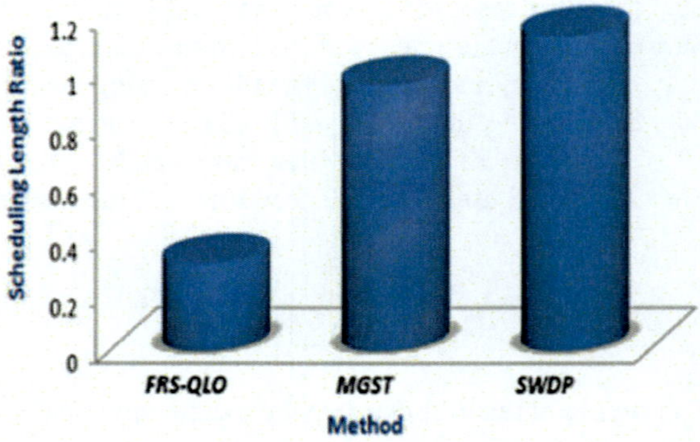

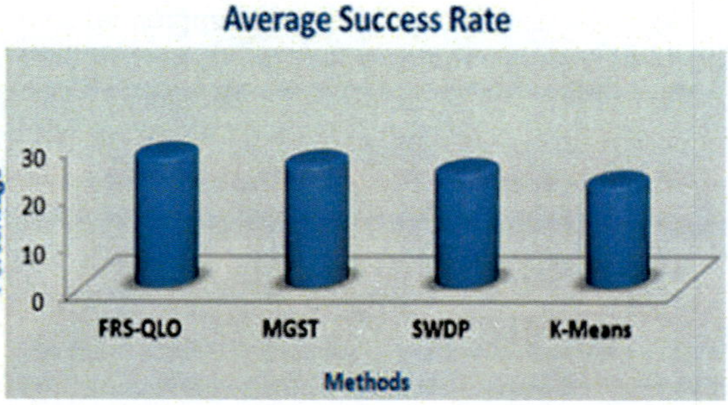

Fig. 3 Average success rate

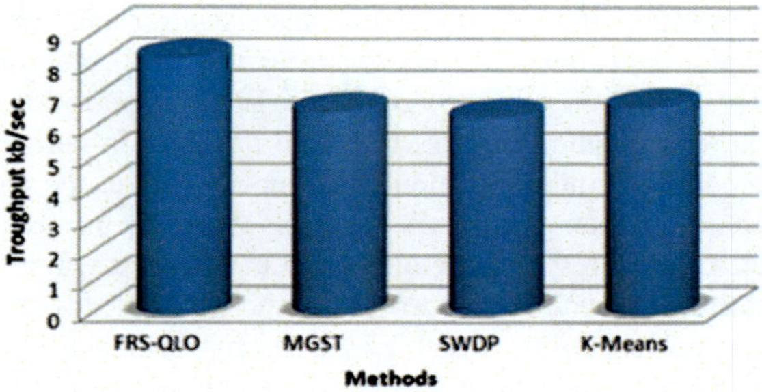

Fig. 4 Throughput

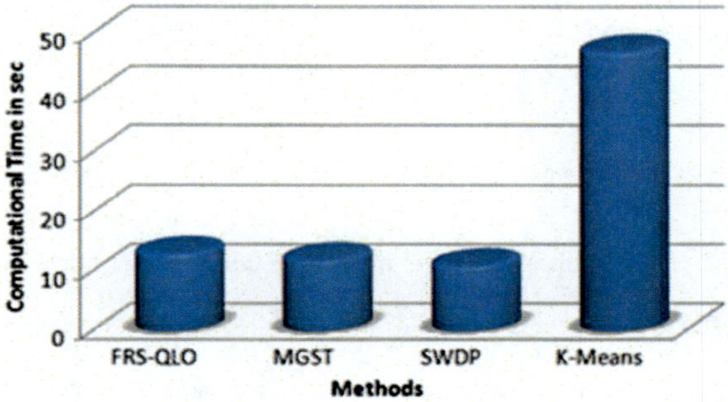

Fig. 5 Computational time

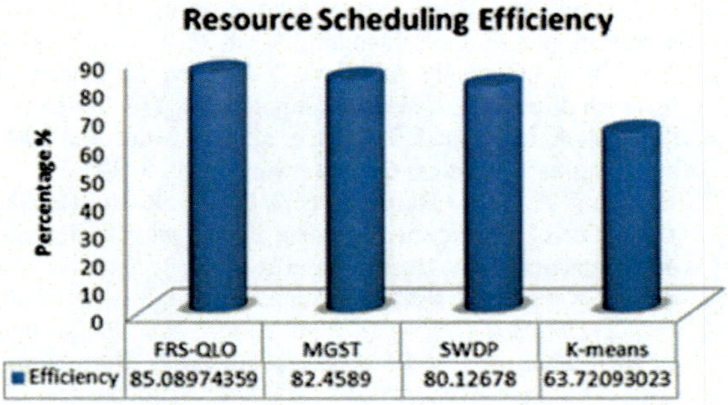

Fig. 6 Efficiency

Figures 4 and 5 show very slight improvement in throughput and computational time which can be more focused in the next future phase of implementation.

Task allocation efficiency is the percentage of allocated resources to the total task requested in Fig. 6 shows efficiency improvement due to the effective scheduling algorithm in second the stage followed by the load balancing stage.

6 Conclusion

The fuzzy resource scheduling and queuing (FRSO) algorithm is proposed to hold the processing of workload traffic that exists over the Internet and reduce latency. In this initial stage, this paper is explaining only about the scheduling and queuing process. This shows very little change in the latency improvement. The next stage will handle the optimization process for more progress in reducing the latency in emergency requests. Simulator in cloud data centers and results shows that the proposed method achieves better performance in terms of average success rate, resource scheduling efficiency, and response time. The major improvement reflects in the scheduling length ratio. The evaluation shows that the FRSQ reduces the processing time required for the scheduling which results in improvement latency time. The method shows success in the initial handling of the resource requests. Further reduction of latency process using an optimized method for resource allocation will be carried out in future work.

References

1. D.V. Dimitrov, Medical internet of things and big data in healthcare. Healthc. Inform. Res. **22**(3), 156–163 (2016)
2. T. Vijayakumar, Classification of brain cancer type using machine learning. J. Artif. Intell. **1**(02), 105–113 (2019)
3. N. Sultan, Making use of cloud computing for healthcare provision: Opportunities and challenges. Int. J. Inf. Manag. **34**(2), 177–184. ISSN 026 8-4012 (2014)
4. S.K. Sood, K.D. Singh, SNA based resource optimization in optical network using fog and cloud computing. Opt. Switching Netw. **33**, 114–121 (2019)
5. T.H. Noor, S. Zeadally, A. Alfazi, Q.Z. Sheng, Mobile cloud computing: Challenges and future research directions. J. Netw. Comput. Appl. **115**, 70–85 (1 Aug 2018)
6. H. Khattak, H. Arshad, S. Islam et al., Utilization and load balancing in fog servers for health applications. J Wireless Com Network **2019**, 91 (2019)
7. H. Gupta, A. Vahid Dastjerdi, S.K. Ghosh, R. Buyya, iFogSim: A toolkit for modeling and simulation of resource management techniques in the internet of things, edge and fog computing environments. Softw. Pract. Experience **47**(9), 1275–1296 (2017)
8. S. Patil-Karpe, S.H. Brahmananda, S. Karpe, Review of resource allocation in fog computing, in *Smart Intelligent Computing and Applications. Smart Innovation, Systems and Technologies*, ed by S. Satapathy, V. Bhateja, J. Mohanty, S. Udgata, vol. 159 (Springer, Singapore, 2020). https://doi.org/10.1007/978-981-13-9282-5_30
9. S. Agarwal, S. Yadav, A.K. Yadav, An efficient architecture and algorithm for resource provisioning in fog computing. Int. J. Inf. Eng. Electron. Bus. **8**(1), 48 (2016)

10. L. Ni et al., Resource allocation strategy in fog computing based on priced timed petri nets. IEEE Internet Things J. **4**(5), 1216–1228 (2017)
11. R. Mahmud, K. Ramamohanarao, R. Buyya, Latency-aware application module management for fog computing environments. ACM Trans. Internet Technol. (TOIT) **19**(1), 1–21 (2018)
12. R. Mahmud et al., Quality of experience (QoE)-aware placement of applications in Fog computing environments. J. Parallel Distrib. Comput. **132**, 190–203 (2019)
13. M. Verma, N. Bhardwaj, A.K. Yadav, Real time efficient scheduling algorithm for load balancing in fog computing environment. Int. J. Inf. Technol. Comput. Sci **8**(4) (2016), 1–10
14. S. Mirjalili, S.M. Mirjalili, A. Lewis, Grey wolf optimizer. Adv. Eng. Softw. **69**, 46–61 (2014)
15. V. Priya, C. Sathiya Kumar, R. Kannan, Resource scheduling algorithm with load balancing for cloud service provisioning. Appl. Soft Comput. **76**, 416–424 (2019)
16. Y. Ma, Y. Wang, J. Yang, Y. Miao, W. Li, Big health application system based on health internet of things and big data. IEEE Access **5**, 7885–7897 (2017). https://doi.org/10.1109/ACCESS.2016.2638449
17. J.I.Z. Chen, S. Smys, Interoperability improvement in internet of things using fog assisted semantic frame work. J. Trends Comput. Sci. Smart Technol. (TCSST) **2**(01), 56–68 (2020)
18. A.A. Mousa, M.A. El-Shorbagy, M.A. Farag, K-means-clustering based evolutionary algorithm for multi-objective resource allocation problems. Appl. Math **11**(6), 1681–1692 (2017)
19. M. Kokaly, I. Al-Azzoni, D.G. Down,MGST: A framework for performance evaluation of desktop grids, in *2009 IEEE International Symposium on Parallel and Distributed Processing*, Rome (2009), pp. 1–8. https://doi.org/10.1109/IPDPS.2009.5161133
20. D. Alberg, M. Last, Short-term load forecasting in smart meters with sliding window-based ARIMA algorithms. Vietnam J. Comput. Sci. **5**(3–4), 241–249 (2018)
21. J. Shi, J. Luo, F. Dong, J. Jin, J. Shen, Fast multi-resource allocation with patterns in large scale cloud data center. J. Comput. Sci. **26**, 389–401 (2018)

Smart Irrigation Monitoring System for Multipurpose Solutions

Vipina Valsan, Krishna Rajesh, Nikhila M. Santhoshlal, and Vykha Pradeep

Abstract The last two decades of the Information Age have been characterized by widespread proliferation of the Internet of things (IoT) technology. Besides diverse applications in various consumer, industrial, agriculture, and health care, IoT has enabled solutions for better management of natural resources. This paper illustrates the productive use of the IoT concept to automate the irrigation of vermicompost, encompassing three prime domains—waste management, smart irrigation, and (mobile) app. The soil moisture content and temperature of the compost bed are critical factors that delimit the earthworms' life expectancy. This paper includes an intelligent monitoring system for effective irrigation of the compost bed, at precise time intervals. The irrigation status and the compost bed's moisture content can be monitored ubiquitously through the Amrita Sparsham mobile application software, minimizing human intervention, facilitating water conservation. Thus, the multipurpose solution is the convergence of Amrita waste management through vermicompost and the automated smart irrigation system monitored using Amrita Sparsham mobile application.

1 Introduction

Vermicompost is a powerful crop nutrient in sustainable agriculture; it is an organic manure produced by earthworms that live in soil, eat biomass, digestion of which is excreted as compost. Vermicomposting is a cost-effective technology which converts organic wastes into organic enrichers, commonly known as vermicompost or compost, through the collaborative interaction of earthworms and mesophilic microorganisms [1]. Deployment of the innovative art of earthworm breeding and

V. Valsan · K. Rajesh
Department of Electrical and Electronics Engineering, Amrita Vishwa Vidyapeetham, Amritapuri, Kollam, India
e-mail: deepthi@am.amrita.edu

N. M. Santhoshlal · V. Pradeep (✉)
Department of Electronics and Communication Engineering, Amrita Vishwa Vidyapeetham, Amritapuri, Kollam, India

P. Karuppusamy et al. (eds.), *Ubiquitous Intelligent Systems*, Smart Innovation, Systems and Technologies 243, https://doi.org/10.1007/978-981-16-3675-2_35

propagation has proved beneficial for waste recycling. The relevance and benign effects of the adding compost to the soil, to improve the soil properties, have been well recognized and established by technologists and scientists [2]. Waste management in urban areas calls for heavy labor resources on a daily basis. Vermicompost, ingrained with easily ingestible nutrients—potassium (K), nitrogen (N), calcium (Ca), and phosphorus (P)—has been shown to be indispensable for plant growth and development [3]. The outmoded vermicompost harvest process mandates time-intensive extra manual labor, for irrigation of the compost. Nonchalant manual irrigation of compost fields is toxic to the worms, as they breathe through their mucus-coated skin, and less than 50% of moisture level in their skin can be fatal to them [4]. Temperature and moisture content of the compost bed are two key parameters that need to be regulated in a vermicomposting system. Optimal water consumption obligates persistent monitoring and control of these parameters, thereby fostering an ideal environment for development of worm population in the compost bed [5].

Rau et al. [6] developed an automated irrigation system using Raspberry Pi, temperature, and humidity sensors. The farmers can monitor and override the system if required, using the Android application. Disadvantages of this system include use of limited sensors, and for better interaction with more external hardware devices, the system can use an advanced microcontroller.

Singh and Saikia [7] implemented a controlled irrigation system, based on Arduino for agriculture. It is a fully automated user-friendly system, processing several environmental factors such as temperature and quantity of water required by the crops. The operators of this system can analyze the sensor data and control the irrigation pumps remotely through the Web site developed. Downside of this method includes high cost of installation and implementation hazards in marshy terrains or over large areas.

A. Gulati and S. Thakur [8] conceived an IoT-based automated system for agricultural irrigation, connected to the Internet via an ESP-8266 Wi-Fi chip module [9]. The time precise supply of water to the agriculture soil by the sensor-based system can be remotely observed by the farmers through the Android application developed. The main limitations of this study are higher installation and Opex costs as well as installation hardships in farmlands of a large area.

Internet of things (IoT) is characterized by a network of interconnected devices, implanted with sensors, software, or other technologies, interfaced with the Internet. Physical objects are connected to each other so that data can be transferred without human-to-human interaction [9]. Smart irrigation, an application of IoT, has been studied by many researchers, stimulated by the numerous opportunities to develop innovative systems in this field. Since the manual irrigation process of vermicompost is a cumbersome and time-consuming practice, an IoT-based automated system is a necessity to save time, water, and money. Concurrently, vermiculture processes require maintenance of temperature conditions of the compost bed, to increase its efficiency. Besides, water wastage and inaccuracy in maintaining the manure can affect the production process of manure. This study aims to develop an automated vermicompost irrigation system that monitors the compost bed soil characteristics like moisture and temperature. Based on the acquired sensor data, Arduino microcontroller, which is the cognitive segment and an open-source IoT analytics platform, manages the water pump of the smart system. Arduino receives and transmits data wirelessly, via a ESP8266 Wi-Fi module. The automation of vermicompost irrigation activities can enhance Amrita Vishwa Vidyapeetham's Amrita Waste Management System from manual to smart. The analyzed sensor data is also made available to the user remotely through an Android software application "Amrita Sparsham".

2 Amrita Waste Management Network (AWMN)—Deployment and Socioeconomic Impact

Vermicomposting is an efficient technique which recycles and reuses organic waste. Amrita Vishwa Vidyapeetham, Kochi campus, in Kerala produces almost 3000 kg of vermicompost monthly, which is a tremendous organic nourishment for agriculture [10]. This method is used to recycle tree litter, backyard wastes and harvest residues into vermicast, which is utilized as an organic manure for crop agriculture. Major nutrients like nitrogen (N), phosphorus (P), potassium (K), and micronutrients such as copper (Cu), zinc (Zn), and iron (Fe) are present more in vermicompost than garden compost. Advantages of vermicompost over other fertilizers are that they are user-friendly, eco-friendly and not costly, and a profitable resource too. There are two types of earthworms, surface feeders and deep feeders, of which surface feeders are suitable for vermicomposting [3]. Some important species of surface feeders are *Eisenia foetida, Eudrilus eugeniae, Perionyx excavatus, Lumbricus rubellus,* etc. Surface-feeding earthworms require a moist soil for survival. So, to sustain an ideal number of earthworms in the field for increased crop production, efficient use of irrigation water is mandatory [11]. *Eisenia foetida* is the most common worm used for vermicomposting. The criteria of reduce, reuse, and recycle which used in the waste management system are efficiently adopted in the vermicomposting process, thereby protecting our nature is a major conservative part of the project.

The various steps involved to prepare vermicompost at Amrita Vishwa Vidyapeetham is explained below. The waste materials like tree litter, backyard wastes and harvest residues are regularly gathered and fetched to the recycling unit for treatment (Fig. 1a). In the first phase of waste management, separation of the collected waste into different sections based on its biodegradability is done at the collection points itself. Since 60% of the total collected waste from AWMN is organic waste, a distinctive composting unit is built for vermicomposting. The collected waste is converted to a decomposed state in around 60 days by regularly sprinkling cow dung slurry and water and is later moved to an open area for fermentation. (Fig. 1b) *Eisenia fetida* worm is used to convert the fermented waste into vermicompost in the next 30–45 days. (Fig. 1c) Due to the extreme sensitivity of the worms to light, the vermicompost bins are covered with shade nets [10].

Fig. 1 Deployment of AWMN at Kochi Campus **a** Collection of waste materials. **b** Decomposition of collected waste. **c** Conversion to vermicompost

3 Problem Analysis

A wide observation was made on the AWMN's vermicomposting field. Detailed talks with the workers on the field made it possible to analyze the difficulty in handling the composting process. During the vermicomposting section of AWMN, the vermicomposting procedures were discussed with the laborers. Usually, the major task which requires heavy labor, during vermicomposting operation, is the setup of the compost bed. Once the compost bed is prepared, the laborers need to invest time in irrigating the compost bed, to maintain the soil moisture level of the bedding. This is required to maintain the limited range of temperature tolerance of red worms. The moisture content and temperature of the compost soil are important factors for the development of red worm population in the vermicomposting operation. The worms keep the temperature down by aerating the waste-based mix, but the bed needs to be occasionally watered. This is the problem faced by the laborers. They have to invest time in periodic watering of the compost, as the worms require moisture to breed fast. In order to help the laborers, the smart irrigation monitoring system has been proposed. Table 1 represents a comparison between the automated and manual irrigation system. The owners need to invest more time and effort, as well-timed irrigation of compost is crucial otherwise, lest the worms die prematurely due to overheating. Traditionally, laborers manually irrigate the soil, when the temperature goes above a threshold where the worms are grown. Manual irrigation can be automated and economically viable by the deployment of an IoT-based drip irrigation system. A well-planned introduction of such an IoT system can help solve the dual problems of manual labor and over or under-watering in the process of vermiculture.

Table 1 Comparison between automated and manual irrigation system

Irrigation system	Time	Monitoring
Manual	More time for delivering water at specific intervals of time	Uneven monitoring of the compost
Automated	Accurate delivery of water in specific intervals of time as per requirement	Persistent monitoring of the compost

4 Proposed Hardware Implementation of the Smart Multipurpose Irrigation Monitoring System

The hardware subsystem of the smart multipurpose irrigation monitoring system consists of soil moisture sensor, soil temperature sensor, Arduino Uno, DC water pump, and ESP8266 module. ESP8266 Wi-Fi module transmits the digital signal to turn ON/OFF the pump from the Arduino Uno to the Internet. The information from sensors and water pump status is also presented on an application named Amrita Sparsham, developed from the ThingSpeak Cloud Server.

Soil Moisture Sensor: The soil moisture sensor senses the moisture content present in the compost mixed with black soil particles. It measures the electrical resistance between the soil particles. The YL-69 probes of the soil sensor, when immersed in the soil, measure the electrical resistance to the flow of electricity in the soil between the probes. As the moisture content in the soil increases, the conductivity of electricity also increases due to an increase in ions present in the soil which in turn reduces the resistance. This condition gives a low output voltage reading from the sensor. In the case of dry soil, the resistance reading is high which gives a high output voltage reading to the probes.

Arduino Uno and Arduino Integrated Development Environment (IDE): Arduino Uno refers to a 28 pin ATmega328 microcontroller board. Arduino Uno works on advanced reduced instruction set computer (RISC) architecture and consists of transmit (TX) and receive (RX) pins for serial communications, pulse width modulation (PWM) pins, digital input/output pins, analog input pins, 16 MHz crystal oscillator, and one USB plug. The Arduino Uno is interfaced with the ESP 8266 Wi-Fi module. Arduino IDE is an open-source hardware and software development platform. Arduino Uno hardware board is programmed using the Arduino integrated development environment [12].

Soil temperature sensor: Soil temperature is absolutely critical to the growth and health of *Eisenia fetida* worms, in the compost. THERM200 temperature sensor along with a YL-69 probes of the soil sensor enables precise control of watering the compost.

DC Water Pump: DC water pump operates on direct current supply from a battery or any other power source, making it more convenient and portable. Hence, the pump is easier to operate and control. The pump is triggered to ON/OFF condition by the Arduino based on the data regarding the moisture and temperature collected from the sensor.

Espressif (ESP) Wi-Fi Module 8266: The ESP Wi-Fi module 8266 has dual functionalities. It can be added to any microcontroller-based design as a Wi-Fi adapter. It also has the ability to act as a self-contained Wi-Fi networking solution, which can carry and drive an entire application [13]. It can be easily connected to serial connect interface (SCI) /secure digital input output (SDIO) interface. This interface is embedded with a 32-bit microcontroller and works in the power range of (3.0 to 3.6)

Volts. It is a system on a chip, with capabilities of 2.4 GHz Wi-Fi communications. It also has the TCP/IP communication protocol stack written into the firmware.

Cloud for Data Aggregation: Nowadays, most electronic devices have sensors which can collect various information like temperature, pH, and moisture. The sensed information is transmitted to the receiving end in the form of electrical signals, numerical value. In this project, the received soil moisture value is collected and transmitted to the ThingSpeak Cloud via Espressif (ESP) 8266 module. The information about temperature and moisture collected can be viewed easily by logging into the ThingSpeak Cloud platform. The soil parameters details collected through the sensors. Here, we have updated all the data received to the system through the ThingSpeak Cloud and the Amrita Sparsham application.

5 Vermicompost Irrigation Monitoring Algorithm

Automated irrigation is a time-triggered system with zero manual intervention. This irrigation scheme is pre-programmed, unlike conventional irrigation. We are developing a smart vermicompost irrigation monitoring system which is an automated irrigation with a sensor network with remote access.

Figure 2 illustrates the working principle of the smart vermicompost irrigation monitoring system. When the moisture sensor response is obtained, it is tested if it ranges from 40 to 60% as it is the optimum moisture content required for the breeding of *Eisenia fetida* worm in vermicompost [5]. The water pump status stays OFF if the optimal moisture level is preserved, otherwise the soil temperature will be monitored. If the soil temperature level is not sustained between the ideal temperature range of 55–77 degrees Fahrenheit required by the red worm [5], the water pump is turned ON to rebalance the water content of the compost bed thus maintaining the temperature. In the case of sustained maintenance of optimum condition, the water pump remains OFF and the current readings of moisture sensor is collected thus repeating the cycle.

6 System Architecture

IoT is leading the world to an efficient, responsive, and smarter future. This is done by an intelligent blend of the physical and digital worlds. IoT helps to interconnect many sensors in a network, thus making it easier to learn and track the processes, which helps make better decisions.

The proposed irrigation monitoring system (Fig. 3) is an IoT based device, capable of automating the irrigation process by analyzing the moisture content and temperature of the soil. The soil moisture content and temperature are collected using YL-69 soil moisture sensor and temperature sensor and is transmitted to the Arduino UNO.

Fig. 2 Algorithm flowchart of the smart vermicompost irrigation monitoring system

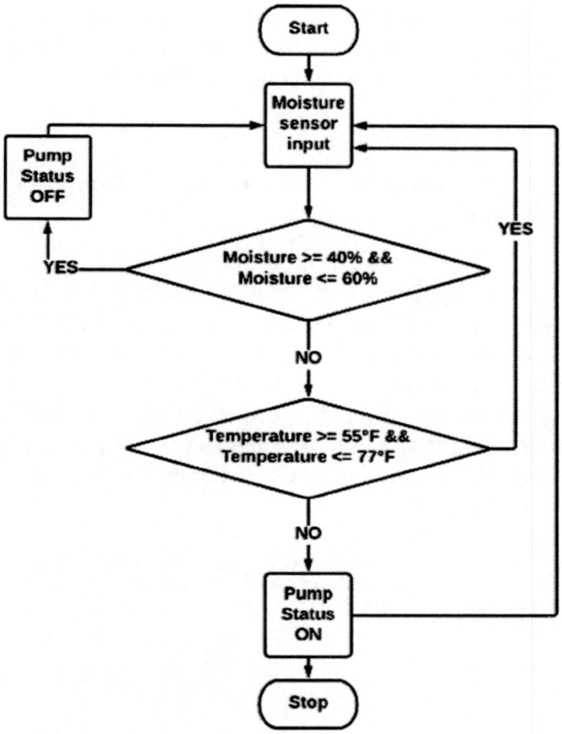

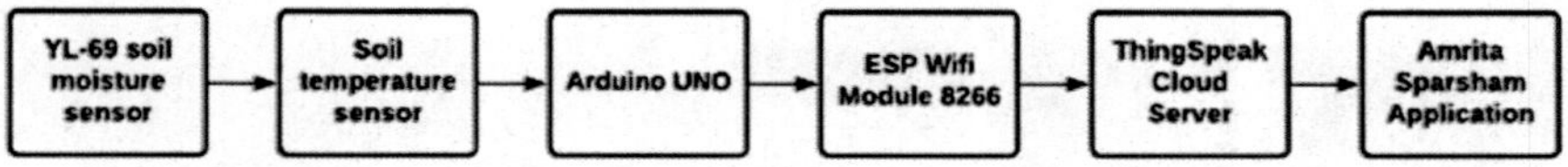

Fig. 3 Framework of the vermicompost irrigation monitoring system

The Arduino microcontroller controls the irrigation process based on the vermicompost irrigation monitoring algorithm. The DC water pump runs at full functional speed, when the moisture content of the vermicompost falls below the threshold values required by the *Eisenia fetida* worms, which helps maintain the optimum conditions for vermicomposting process. The sensor reading of the compost bed is transmitted to ThingSpeak Cloud Server via ESP Wi-Fi module. The transmitted information is monitored through ThingSpeak Cloud Server. ThingSpeak Cloud Server is an open data platform and API for the Internet of Things that allows us to gather, store, examine, visualize, and use information from sensors or microprocessors like Arduino. The analyzed data is also made available to the user remotely through an Android software application "Amrita Sparsham". Amrita Sparsham is made using Figma, which is a codeless user interface design tool and Bravo Studio, which converts the prototype to a functioning app. The app mainly has two pages which shows the soil moisture and pump status, respectively, and it can be easily accessed through a secure Internet connection.

6.1 Integration of IoT with a Multipurpose Smart Irrigation Monitoring System

A mobile application software "Amrita Sparsham" is developed using the user interface designer Figma [14] and prototype designer Bravo Studio [15]. The user interface/user experience (UI/UX) design is done using Figma. Frames are made for each page, viz. Splash Screen, Soil Moisture, Pump Status, FAQ, About Us, and Navigation Bar (Fig. 4). A frame allows one to combine different layers or components together to convert it into a single layer. The prototyping is done with the help of the free codeless prototyping tool that is available in the Figma. The lines in Fig. 4 show

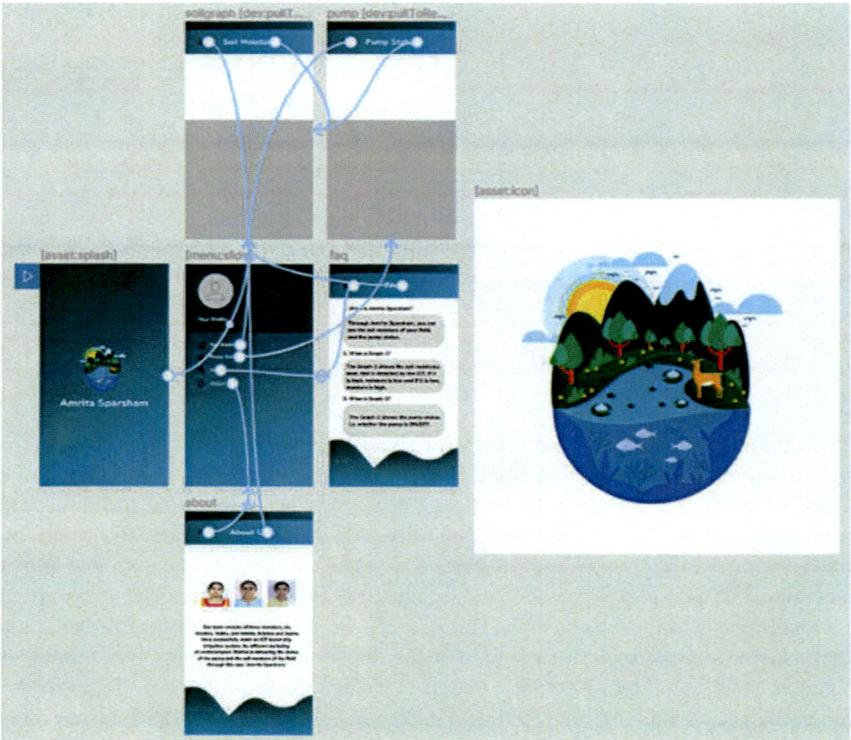

Fig. 4 Designing and prototyping in Figma app

the prototyping of the Amrita Sparsham app, it connects one frame to the navigation bar, and the titles of each frame are connected to the same frame itself to make it responsive. The vector image in Fig. 4 will act as the icon of the app. A color palette of within the range #00574C to #0CBEFF is used to design the app.

In the soil moisture and pump status page, a rectangle element (800 × 348.89 px) (Fig. 5) has been included to retrieve the soil moisture and pump status graph from the ThingSpeak channel of ThingSpeak Cloud Server.

Further, using the help of Bravo Studio, it is made usable for any Android device. To make the prototype that is designed in the Figma reactive, appropriate bravo tags are used, which is available on their Web site as "Bravo Tags Masterlist" [16].

Tried to integrate the Amrita Sparsham app with application programming interface (API), but the data fields were not displayed in the app. So, as an alternative method, Chart IFrame (Fig. 6) of the graphs from the ThingSpeak channel was used to display the soil moisture and pump status graphs to the "Amrita Sparsham" app, so it can be easily accessible by anyone. An <iframe> is a hypertext markup language (HTML) tag that gives an inline frame, and it is used to embed an existing document into the written HTML document.

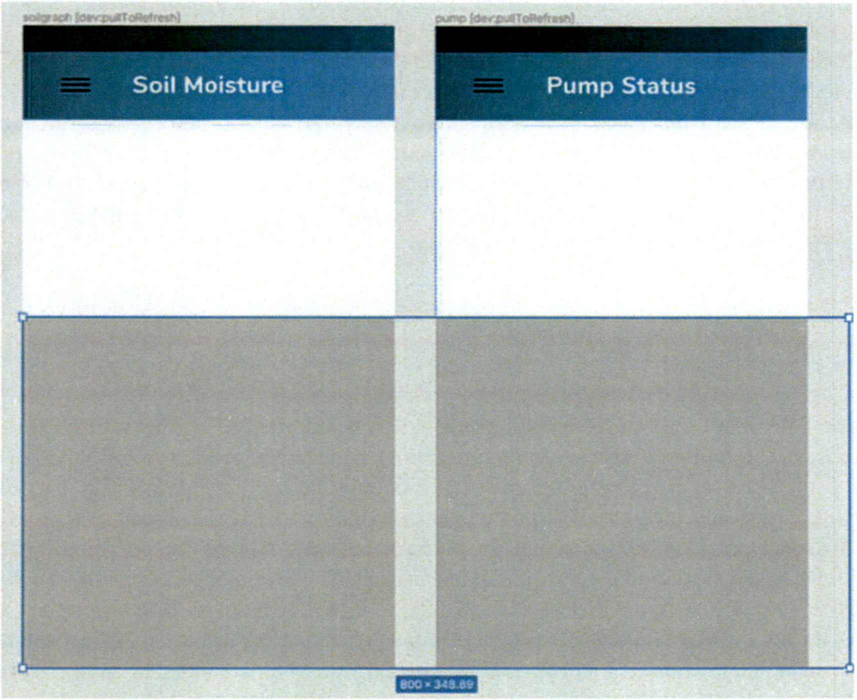

Fig. 5 Rectangle element which is added to the pages to retrieve the graph from ThinkSpeak Channel

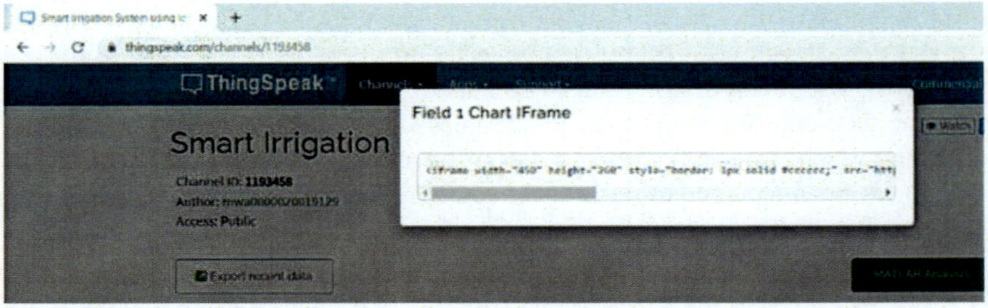

Fig. 6 Chart IFrame in ThingSpeak channel

The Amrita Sparsham app currently does not have the log-in page, as it is designed considering the fact that the project is now in small scale, so the APK file can be directly shared with the vermicompost farmers that are using the proposed system. As a future work, when the number of users is increasing, the app will be secured using log-in, OTP, and corresponding pages.

In the ThingSpeak Cloud Server, a channel is where you store the data that is sent from the IoT. The ThingSpeak channel continuously collects the moisture level sensed by the IoT device and gives the output as a graph in the public channel for the respective users of the proposed system. Using the private read and write API key, one can export the data to any web pages or app. It is available in the "My Channel" section of the ThinkSpeak Web site. Alternatively, one can use the Chart IFrame to export the graph to the designed app. It is particularly useful for projects which need an Internet connection, but the maintenance of the server is not necessary. There are many cloud storages available for exporting the data, but in this project, a free open-source cloud, i.e., ThingSpeak is used.

7 Results and Discussion

The data about the moisture level in the soil was recorded using the soil moisture sensor. The moisture values are recorded in standard units of Ohms, and when the soil moisture level is low, a high resistance is outputted and vice versa. The data of temperature and moisture was taken between 18:10 and 18:30 IST on October 26, 2020. Values of the soil sensor and the pump status were exported to the ThingSpeak Cloud-based system to the Amrita Sparsham app. The levels of soil moisture varied from 2 to 68 Ω with the lowest 1 Ω at 18:28:33 GMT + 05:30 and the highest at 68 Ω at 18:18:32 GMT + 05:30 (Fig. 7). From Fig. 8, we can see that the resistance is maintained, between 1–68 Ω, i.e., the soil moisture is maintained at an optimum level.

The Amrita Sparsham app has four pages with easy navigation. With a minimum requirement of having an Internet connection and the easy access of the graph, the

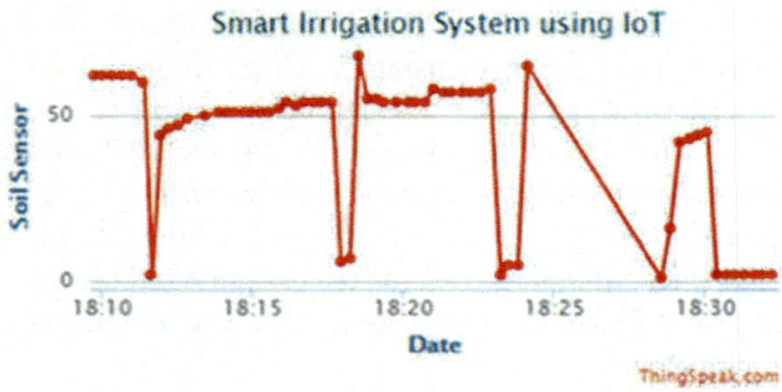

Fig. 7 Graph representation of the soil moisture content of soil

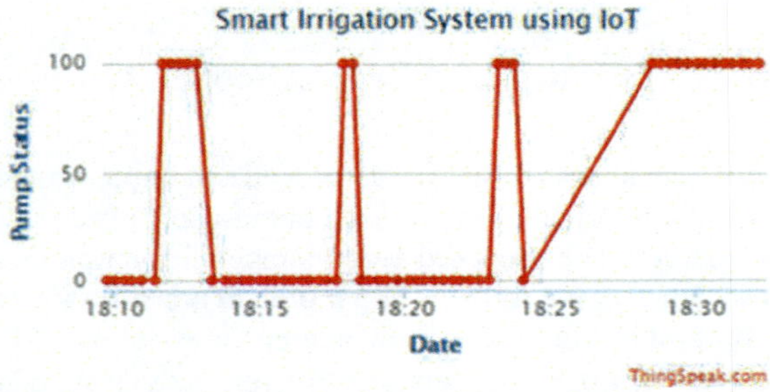

Fig. 8 Graph representation of the ON/OFF status of the water pump, where 0 indicates OFF mode, and 100 indicates ON mode of the water pump

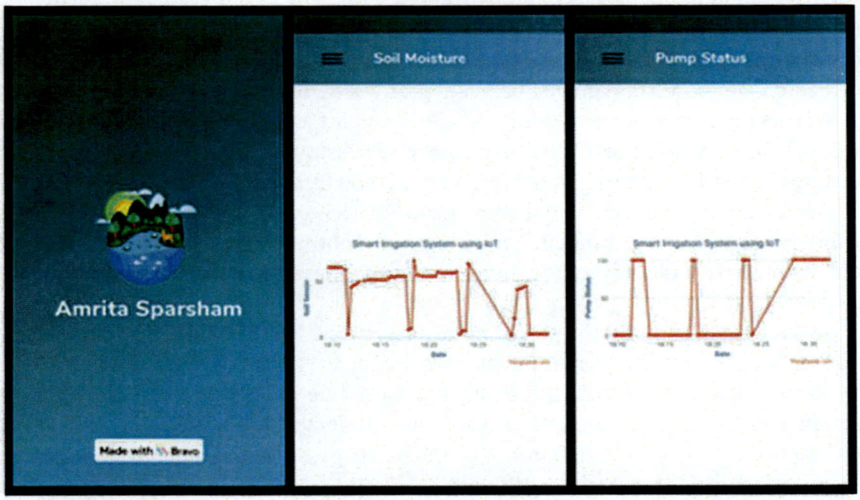

Fig. 9 User interface of Amrita Sparsham application: launch screen and two different pages of the app displaying the graphs

app is made user-friendly. The soil moisture graph (Fig. 9b) and pump status graph (Fig. 9c) are displayed using the Chart IFrame.

The implementation of the proposed vermicompost irrigation monitoring system framework builds the productivity of the AWMN, according to the moisture and temperature requisite of the vermicompost bed via consequently turning the water pump ON/OFF as required. This smart irrigation system has helped in reducing the manual labor required for the harvesting of vermicompost. The life expectancy of the worms utilized for vermicompost bedding is additionally sustained by the framework of smart irrigation system.

8 Conclusion and Future Scope

A sustainable solution was developed and implemented to solve the existing problem of manual irrigation process of compost bed, resulting in efficient utilization of water resources. The proposed vermicompost irrigation monitoring system increases the efficiency of the AWMN, as per the soil moisture and temperature requirement of the compost by automatically switching the water pump ON [start]/OFF [stop]. The life expectancy of the worms used for vermicompost bedding is also maintained by the smart vermicompost irrigation monitoring system. The system efficiently irrigates the vermicompost, by using the data collected from the soil sensors, thus preventing over-irrigation or under-irrigation of the compost bed. It can be concluded that we can automate the vermicompost harvesting of waste management field through IoT as well as reduce the water wastage. Thus, the smart irrigation system has achieved its multipurpose solution by the emergence of waste management through vermicomposting and monitoring of the irrigation system through Amrita Sparsham application for the soil moisture and temperature requirement of *Eisenia fetida* red worm. However, this system is limited to monitoring the optimum bedding requirements of red worms only. This prototype concept needs to be developed further, for large-scale implementation.

In future, we should incorporate more sensor data like pH, humidity, which can be combined into a sensor module, thus increasing the accuracy of the data provided by the system. The Amrita Sparsham app should be enhanced for additional soil parameters like temperature and pH for future implementation. We can also develop the application by providing control of the water pump to the labor so that he can remotely control the system, as required.

Acknowledgements We wish to express our sincere gratitude to the Chancellor of Amrita Vishwa Vidyapeetham, Mata Amritanandamayi Devi, our guiding light, the inspiration to undertake eco-friendly related works, providing all needed infrastructure facilities and guide, Vipina Valsan, for providing us with the enthusiasm and opportunity to develop this worth work. This endeavor would never have been successful without the team's coordination and blessings of parents and God Almighty.

References

1. S.A. Bhat, S. Singh, J. Singh, S. Kumar, A.P. Vig, Bioremediation and detoxification of industrial wastes by earthworms: Vermicompost as a powerful crop nutrient in sustainable agriculture. Biores. Technol. **252**, 172–179 (2018)
2. A. Lemma, Multiplication of red worms (Eiseniafetida) using different feeding materials and its effect on yield and quality of vermicompost. Int. J. Ecotoxicol. Ecobiology **5**(4), 48–53 (2020). https://doi.org/10.11648/j.ijee.20200504.12
3. R. Joshi, J. Singh, A.P. Vig, Vermicompost as an effective organic fertilizer and biocontrol agent: effect on growth, yield and quality of plants. Rev. Environ. Sci. Biotechnol. **14**, 137–159 (2015). https://doi.org/10.1007/s11157-014-9347-1
4. A.U. Aquino et al., Development of a solar-powered closed-loop vermicomposting system with automatic monitoring and correction via IoT and Raspberry Pi module, in *2019 IEEE 11th International Conference on Humanoid, Nanotechnology, Information Technology, Communication and Control, Environment, and Management (HNICEM)*, Laoag, Philippines, pp. 1–5 (2019) https://doi.org/10.1109/HNICEM48295.2019.9073372
5. G. Tripathi, P. Bhardwaj, Comparative studies on biomass production, life cycles and composting efficiency of *Eisenia fetida* (Savigny) and *Lampito mauritii* (Kinberg). Bioresour. Technol. **92**(3), 275–283 (2004). https://doi.org/10.1016/j.biortech.2003.09.005 (PMID: 14766161)
6. A.J. Rau, J. Sankar, A.R. Mohan, D. Das Krishna, J. Mathew, IoT based smart irrigation system and nutrient detection with disease analysis, in *2017 IEEE Region 10 Symposium (TENSYMP)*, Cochin, pp. 1–4 (2017). https://doi.org/10.1109/TENCONSpring.2017.8070100
7. P. Singh, S. Saikia, Arduino-based smart irrigation using water flow sensor, soil moisture sensor, temperature sensor and ESP8266 WiFi module, in *2016 IEEE Region 10 Humanitarian Technology Conference (R10-HTC)*, Agra, pp. 1–4 (2016). https://doi.org/10.1109/R10-HTC.2016.7906792
8. A. Gulati, S. Thakur, Smart irrigation using internet of things, in *2018 8th International Conference on Cloud Computing, Data Science & Engineering (Confluence)*, Noida, pp. 819–823 (2018). https://doi.org/10.1109/CONFLUENCE.2018.8442928
9. K. Goyal, A. Garg, A. Rastogi, S. Ankur, S. Singhal, A Literature survey on internet of things (IoT). Int. J. Adv. Manufact. Technol. **9**, 3663–3668 (2018)
10. Integrated Waste Management at Health Sciences Campus. https://www.amrita.edu/news/integrated-waste-management-health-sciences-campus
11. V. Valsan, G. Sreekumar, V. Chekkichalil, A. Kumar, Effects of service-learning education among engineering undergraduates: a scientific perspective on sustainable waste management. Procedia Comput. Sci. **172**, 770–776 (2020). https://doi.org/10.1016/j.procs.2020.05.110
12. K. Raveendran, R. Sai Sachin, A. Christy, A.G. Pillai, V. Valsan, T.S. Angel, Intelligent monitoring system for submersible motor protection, in *2020 Fourth World Conference on Smart Trends in Systems, Security and Sustainability (WorldS4)*, London, United Kingdom, pp. 677–680 (2020). https://doi.org/10.1109/WorldS450073.2020.9210391
13. P. Srivastava, M. Bajaj, A.S. Rana, Overview of ESP8266 Wi-Fi module based Smart Irrigation System using IOT, in *2018 Fourth International Conference on Advances in Electrical, Electronics, Information, Communication and Bioinformatics (AEEICB)*, Chennai, pp. 1–5 (2018). https://doi.org/10.1109/AEEICB.2018.8480949
14. Figma. https://www.figma.com/
15. Bravo Studio. https://www.bravostudio.app/
16. Bravo Tags Master List. https://bravostudio.help/145bec845f0b4afaa9e3bb8321b218a8

A Study on Data Compression Algorithms for Its Efficiency Analysis

Calvin Rodrigues, E. M. Jishnu, Chandu R. Nair, and M. Soumya Krishnan

Abstract For many computerized applications, data compression is a standard requirement. In order to minimize the capacity needed for that data, it decreases the redundancy in data representation and thus therefore reduces the connectivity cost by efficiently utilizing the available bandwidth. There are a range of algorithms for data compression that are used to compress various formats of data. There are also sets of different compression algorithms for a single data form, which use different approaches. This paper assesses lossless algorithms for compression of data and compares their performance. To assess the performance of compressing text data, images, audio, a set of selected algorithms are used. Experimental findings and comparisons of algorithms for lossless compression using methods for compression. It gives contrasting conclusions on the size and time ratios toward existing research. Different information that can be compressed not only text but it can be Audio, pictures and video information. In this article, there are variety of various forms of lossless compression algorithms discussed.

1 Introduction

Compression is the ability to represent the data instead of its initial or uncompressed form in a compact form. In other words, the size of a particular file can be minimized by the use of data compression. If you want to save storage space, this process may be useful. It is very convenient to share compressed files over the Internet, since they can be uploaded or downloaded much faster. In the area of file storage and distributed system compression, data compression is essential [4]. This is particularly helpful when a large file that takes a lot of resources is processed, stored or transferred. If the algorithms used for encryption work correctly, it is important to distinguish between the original and the compressed file. Data compression is used in a data transfer application, and speed is the key objective. The transmission speed depends on the number of bits sent, the time required to trigger the coded message by the

C. Rodrigues (✉) · E. M. Jishnu · C. R. Nair · M. Soumya Krishnan
Department of Computer Science and IT, Amrita School of Arts and Sciences Kochi, Amrita Vishwa Vidyapeetham, Kochi, India

P. Karuppusamy et al. (eds.), *Ubiquitous Intelligent Systems*, Smart Innovation, Systems and Technologies 243, https://doi.org/10.1007/978-981-16-3675-2_36

encoder and the time required to recover the original message group by the decoder [8]. It is possible to define compression as lossy or lossless. Without any information loss, lossless compression methods regenerate the original information from the compressed disk. Thus, during the compression and decompression processes, the data does not alter [2]. Only when it is possible to correctly recreate the original data from the compressed version will the data compression be lossless. Such a lossless approach is used when a source's original data is so important that we cannot afford to lose any information [3]. Data compression is the method of converting information to less bits than the original representation, so that when transmitting over a network, it takes less storage space and less transmission time [4]. Lossless compression approaches are used to compress medical images, text and images held for official purposes, executable data files and so on. With the loss of some details, lossy compression techniques regenerate the original message. It is not necessary to use the encoding method to regenerate the original message, and it is called irreversible compression. Approximate regeneration occurs in the decompression process. It could be desirable to disregard data from such ranges that could not be accepted by the human brain [2, 3]. To achieve more lightweight data compression, such techniques could be used for multimedia images, video and audio.

2 Literature Review

This study mainly focused on the efficiency comparison of different data compression algorithms. Based on these studies done, efficient algorithms for different data are compared. And also, the main focus was to find out the efficient algorithm for each data. For this, many papers are reviewed. This research paper gave different techniques for compression of data and compared their output. Data compression is a method that reduces the size of the information from which unnecessary information is removed [1]. Huffman coding and arithmetic encoding are used in this work. Huffman coding is a more sophisticated and powerful lossless data compression process, and in Huffman coding, the words in the text document are translated to binary code [2]. Test results and compression of the lossless compression algorithms were performed on text data [3]. This reduces the storage needed for information. A binary code tree is created for a given input in Huffman coding. For every symbol of input data, a code length is constructed by recurring of letters [4]. For big data, the research paper explains data compression techniques. Big data means vast volumes of information that can be in structured or unstructured format. Different techniques are used for image data compression in GIF and TIFF can be used [5]. This article includes a survey on the evaluation of various comparison algorithms. And this mainly focuses on text, doc, TIFF and GIF. And this paper suggests an efficient algorithm based on the compression ratio and compression time [6]. In this research paper, it is explained how to improve the Huffman coding using recursive splitting. Lossless compression of a sequence of symbols is an important part of data and signal compression [7]. This research paper shows the different data compression techniques that are widely used. This paper is written on the basis of different algorithms used for achieving the data compression. The paper contains different algorithms like Shannon–Fano coding testing which compresses the data based on probability of characters in the message, Huffman coding which is based on the probabilities of Shannon–Fano coding trying variable length codes, adaptive Huffman coding developed to overcome the problems faced in Huffman Coding, Arithmetic Coding Technique which uses code word to replace each bit [8]. This research paper focuses on data compression using logical truth tables where two bits of data can be represented using a single bit in both wired and wireless networks. The paper gives a brief introduction on what is data compression and mentions the conventional data compression method, lossless and lossy data compression techniques [9]. This paper focuses mainly on text data compression where data is reduced to a smaller size which can be handled easily. Here, they provide a brief description on data compression, how it takes place and the major two forms, namely lossless and lossy compression technique [10]. A comparison between the compression algorithms for lossless and lossy image compression is shown in the paper. First, two methods are compared such as RLE and Huffman, then two methods of lossy image compression are compared: discrete cosine transform and wavelets [11]. In this paper, they addressed the compression of images, the need for compression, its principles, compression types and various algorithms used for image compression [12]. This paper presents various techniques of image format compression and comparative study of Huffman encoding, arithmetic coding, RLE, transform coding and wavelet coding and ends with the best image compression techniques [13]. Compression ratio, signal-to-noise ratio, quality, compression time and decompression time are included in this paper to evaluate the output of image compression techniques [14]. This paper uses lossless compression techniques such as dynamic Huffman encoding and RLE to display audio compression. In order to find the sampling frequency, audio files are preprocessed, and dynamic Huffman and RLE are applied on the encoded data bits. The paper is to provide a thorough overview of compression techniques and then identify the better multimedia data compression technique [15]. This paper proposes a new approach that uses sampled data control theory to compress audio. The approach is easy than traditionally used [16].

3 Materials and Methods

The following approaches are used to measure the efficiency of lossless data compression algorithms.

3.1 Run Length Encoding Algorithm

The most basic type of data compression algorithms is run length encoding or simply RLE.

The series of continuous symbols are known as runs, and the others are defined as non-runs. In this with redundancies of some kind. This measures whether or not there are any recurring symbols and is focused on such redundancies and their lengths [2]. Consecutive repeating patterns are called loops, and non-runs are deemed to be all the other series. The text 'ABABBBBC' is used, for example, to compact, so the first three characters are examined as a non-run with a period of 3, and the next four characters are tested as a run with a period of 4 as symbol B is repeated. This algorithm's major objective is to understand the running of the source file and to record the symbol and the time of each run.

3.2 Huffman Coding

Huffman encoding algorithms use the probabilistic model of source alphabet letters to create coded language for symbols. To evaluate the distribution of probability, the frequency distribution of all the reference words is determined. Depending on the odds, the code words are distributed. For higher chances, simpler code words are assigned, and longer code words for lower probabilistic are assigned. For this purpose, a binary tree is generated to use the symbols as leaves with their chances in accordance, and then their paths are taken as code words [2]. Huffman encoding has been proposed in two forms: static Huffman algorithms and adaptive Huffman algorithms. Initially, static Huffman algorithms find the frequencies and then build a tree for both the compression and decompression phases. The compressed file should be sent with the details of this tree. While finding the frequencies, the adaptive Huffman algorithms grow the tree, and in both processes, there would be two trees. In this approach, a tree with both the flag symbols at the beginning is created and is modified when the next symbol is read [2, 3].

3.3 Lempel Ziv Welch Algorithm

It is a dictionary-based algorithm for compression based on a dictionary. A dictionary is a list of a language's possible terms and is stored in a table-like layout and describes larger and repeated dictionary terms using entry indexes. In this method, to index the previously seen string patterns, a dictionary is used. Other than repeated string patterns, these index values are used in the compression process. The dictionary is created in the compression process, and it does not need to be moved for decompression with the encoded letter. The same dictionary with original knowledge is generated in the process of decompression. This algorithm is, therefore, an algorithm for adaptive compression [2, 3].

3.4 Shannon–Fano coding

For lossless multimedia data compression, the Shannon–Fano algorithm is considered a word occurrence predictor. Named after Claude Shannon and Robert Fano, it gives a code word to each symbol, based on their probability. It is a variable length encoding program, i.e., the codes that are given to the letters will have different lengths. It is shorter than the symbol in the message that already exists [2]. The higher the chance, the smaller the code word would become. It is possible to calculate the length of each code word from the likelihood of each symbol and represent it with the code word. A code word that does not have the same length is generated by Shannon–Fano coding, so the code is special and can be encoded [2, 3].

4 Experiment and Result Analysis

The Huffman encoding algorithm, run length encoding algorithm, Shannon–Fano algorithm and Lempel Ziv Welch algorithm are used with a collection of files of different data, and then, efficiency of lossless compression algorithms should be calculated.

Measuring the Performance of Huffman Algorithm

Measuring Huffman Approaches Efficiency Huffman Encoding introduced and performed. File sizes, compression and decompression times are dependent on all of these.

Measuring the Performance of LZW Algorithm

Based on the output of each algorithm, compression time, decompression time, original sizes, compressed size, compression ratio and saving percentages are calculated.

Measuring the Performance of RLE Algorithm

Run length encoding algorithm for the compression process, the compression time, decompression times, original sizes and compression ratios are determined.

Measuring the Performance of Shannon–Fano Algorithm

For this algorithm, the compression and decompression times, compression ratio and file sizes are computed (Fig. 1).

Fig. 1 .

$$Compression\ ratio = \frac{size\ after\ compression}{size\ before\ compression} x\ 100$$

$$Compression\ factor = \frac{size\ before\ compression}{size\ after\ compression} x\ 100$$

$$Compression\ speed = \frac{uncompressed\ bits}{second\ to\ compress} x\ 100$$

$$Compression\ speed = \frac{compressed\ bits}{second\ to\ compress} x\ 100$$

4.1 Text Compression

Experimental result of text compression (Table 1).

Huffman encoding compression ratios are comparatively less than other algorithms. And also there is bit change in the compression time and decompression time. The probability distribution of the alphabet letters of the source is used by Huffman encoding algorithms to build code words for symbols (Table 2).

This algorithm provides better compression ratios for the given text files used. This algorithm offers a stronger compression ratio that ranges from 31 to 59%. When it is contrasted to the other algorithms, this is a sensible value. And also, in less time, all text files are compressed (Table 3).

For the RLE algorithm, compression and decompression times are comparatively less. This algorithm produces compressed files larger than the original files for the seventh and ninth files, however. This occurs because of the smaller quantities of the source runs. Both documents are compressed, but there are very high values in the compression ratios (Table 4).

Table 1 Huffman algorithm

Original file		Huffman encoding			
SL No	Original file size	Compressed file size	Compression ratio	Compression time (ms)	Decompression time (ms)
1	27,594	17,926	64.9633978	19,146	19,579
2	51,385	34,387	66.9203074	59,714	27,656
3	13,252	8584	64.7751282	4765	7756
4	15,654	8978	57.3527532	5943	9710
5	80,146	46,377	57.8656452	166,867	234,189
6	35,494	19,242	54.2119794	11,675	12,356
7	120,223	90,876	75.7762907	134,764	99,985
8	190,985	121,563	63.6505484	378,765	295,487
9	250,000	167,761	67.1044000	685,652	50,765
10	75,541	48,641	64.3901987	46,886	36,345

Table 2 LZW algorithm

Original file			LZW encoding			
File	File size	Number of characters	Compressed file size	Compression ratio	Compression time (ms)	Decompression time (ms)
1	27,594	23,765	14,875	53.90	3996	3756
2	51,385	44,654	16,436	31.98	7581	7834
3	13,252	11,654	7875	59.42	9686	7565
4	15,654	13,975	6765	43.21	6488	5421
5	80,146	75,678	38,876	48.50	7641	7132
6	35,494	38,654	18,876	53.18	4160	5512
7	120,223	114,532	51,786	43.07	9732	9423
8	190,085	70,897	78,858	41.29	10,670	9761
9	250,000	245,671	116,345	46.538	12,582	11,527
10	75,541	66,787	28,769	38.08	7943	6734

Table 3 Run length encoding

	Original file			Run length encoding		
File	File size	Number of characters	Compressed file size	Compression ratio	Compression time (ms)	Decompression time (ms)
1	27,594	23,765	18,651	67.590	4568	3211
2	51,385	44,654	30,569	59.4901	7869	3212
3	13,252	11,654	9,260	69.8762	5667	3123
4	15,654	13,975	9,890	63.17	11,034	3012
5	80,146	75,678	66,898	83.47	13,214	18,329
6	35,494	38,654	29,654	83.54	15,974	3455
7	120,223	114,532	120,492	100.223	45,395	1543
8	190,985	70,897	189,455	99.19	37,664	2342
9	280,679	245,671	280,784	100.037	39,334	1987
10	75,541	66,787	65,234	86.35	45,637	2687

The compression ratios are in the range of 57–72% for the Shannon–Fano method, which is average by comparing all other algorithms. Shannon–Fano coding based on variable lengthword, means that some of the symbols in the information which will be encoded is represented with a code word.

Comparison of the results

The compression times and file sizes are used to analyze in order to compare the performance of all algorithms.

Table 4 Shannon–Fano encoding

Original file		Shannon–Fano encoding			
S. No.	File size	Compression file size	Compression ratio	Compression time (ms)	Decompression time (ms)
1	27,594	19,167	69.46	16,259	20,623
2	51,385	33,445	65.08	59,708	72,016
3	13,252	9,672	72.98	4762	9731
4	15,654	9,234	58.98	6576	8747
5	80,146	55,298	68.99	172,609	259,545
6	35,494	20,514	57.79	14,632	13,024
7	120,223	78,765	65.51	173,869	135,787
8	190,985	114,376	59.88	330,686	267,833
9	280,679	178,786	63.69	569,523	467,953
10	75,541	51,345	67.96	48,997	33,459

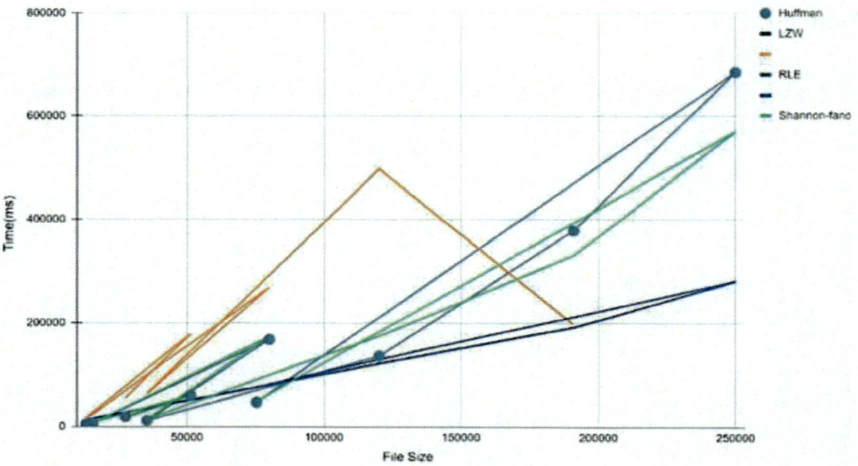

4.2 *Image Compression*

Experimental results of image compression

In this lossless compression, we use bitmap format images (*.bmp), and it compressed using two techniques, namely run length encoding and Shannon–Fano encoding. Then, it is possible to decompress the final result of the compressed file containing compressed information and then return it. Output from the original decompressed file is contained in the *.bmp extension (Table 5).

Table 5 Run length encoding

S. No.	Resolution	Original file (kb)	Compressed file	Compression ratio	Compression time (ms)	Space saving
1	100 × 150	210.98	297.54	141.02	5.654	−41.42
2	100 × 180	425.76	610.86	143.47	9.765	−43.52
3	200 × 200	500.32	980.12	195.89	9.983	−96.28
4	300 × 300	1232	1987.76	161.28	12.871	−61.28
5	400 × 400	2377	3123.72	131.38	15.704	−31.38
6	380 × 200	3976	4676.98	117.60	16.543	−17.60
7	250 × 200	1342	2234.45	166.46	11.452	−66.46
8	100 × 100	2412	2986.23	123.79	14.761	−23.79
9	200 × 200	4587	5243.70	114.30	17.753	−14.30
10	300 × 200	4876	5543.61	113.67	17.873	−13.67

Ten different sized images are compressed using run length encoding. It can be shown from this table that the average compression ratio is 140.886%, the average space saving is 40.97%, and the average compression time is 13.2359 percent. Using run length encoding, ten different sized images are compressed (Table 6).

Using Shannon–Fano encoding, ten differently sized images are compressed. It can be shown from this table that the average compression ratio is 77.54%, the average space saving is 22.67%, and the average compression time is 10.61%. Shannon–Fano variable length word encoding means that any of the symbols in the information to be encoded are represented by a code word.

Comparison of the result

Table 6 Shannon–Fano encoding

S. No.	Resolution	Original file (kb)	Compressed file	Compressed ratio	Compression time (ms)	Space saving
1	100 × 150	210.98	150.7	71.42	4.65	28.57
2	100 × 180	425.76	340.87	80.87	5.72	20.2
3	200 × 200	500.32	410.32	82.41	5.8	18.76
4	300 × 300	1232.8	823.43	66.83	11.65	33.20
5	400 × 400	2377.65	1604.43	67.49	13.76	32.52
6	380 × 200	3976.12	3193.45	80.31	13.98	19.68
7	250 × 200	1342.45	948.54	70.67	7.32	29.34
8	100 × 100	2412.32	2045.39	84.78	13.56	15.21
9	200 × 200	4587.23	3912.17	85.28	14.76	14.71
10	300 × 200	4876.21	4165.38	85.42	14.98	14.57

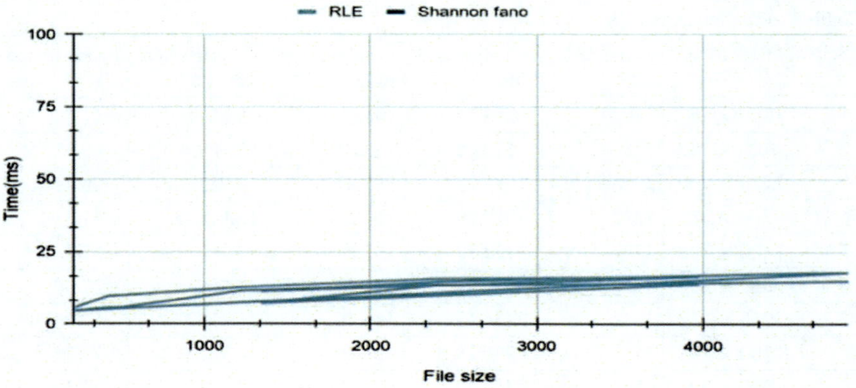

File size and compression time are used to compare the algorithm.

4.3 *Audio Compression*

Experimental results of audio compression

In this lossless compression, we use the WAV 2 channel audio format. The expected benefit of this research is to determine the optimal algorithm in the compression process of WAV 2 channel audio data so as to minimize memory or bandwidth usage and speed up the data transmission process. To calculate the efficiency of a compression algorithm, there are different requirements. The main problem, however, has always been the effectiveness of space and time efficiency. Comparing the results of a lossless compression especially on WAV audio data is very difficult to get supposition which algorithm is better, because various researches with audio data objects do not have the same data (Table 7).

The compression ratio for the Huffman approach is in the range of 13% to 76% which is an average value. And also compression ratio decreases when the original file size increases. Decompression speed is a bit high when compared to other algorithms. And the compression speed is average. The probability distribution of the alphabet letters of the source is used by Huffman encoding algorithms to create code words for symbols (Table 8).

The Shannon–Fano method compression ratio is between 17 and 77%, which is better than the Huffman algorithm. It takes only less time to compress all the files. Shannon–Fano coding based on variable length word, means that some of the symbols in the message (which will be encoded) is represented with a code word.

Table 7 Huffman encoding

Original file		Huffman coding				
SL No	File size	Compressed file	Compression ratio	Compression factor	Compression time	Decompression time
1	9356.0039	1684.5608	18.0051	81.9949	2.5962	329.374
2	5678.0027	780.0879	13.7387	727.8670	1.1123	143.658
3	15,432.765	7654.723	49.6029	201.611	3.123	565.876
4	4567.67	980.76	21.471	465.72	0.876	95
5	9434.533	1745.654	18.50	540.45	2.456	332.98
6	25,654.789	17,875.786	69.67	143.51	5.675	987.54
7	20,345.651	11,543.679	56.73	176.25	4.897	765.87
8	23,543.651	16,543.511	70.26	142.31	5.123	812.87
9	28,954.556	21,954.65	75.82	131.88	7.432	995.87
10	27,634.761	21,232.483	76.83	130.15	6.985	894.62

Comparison of the result

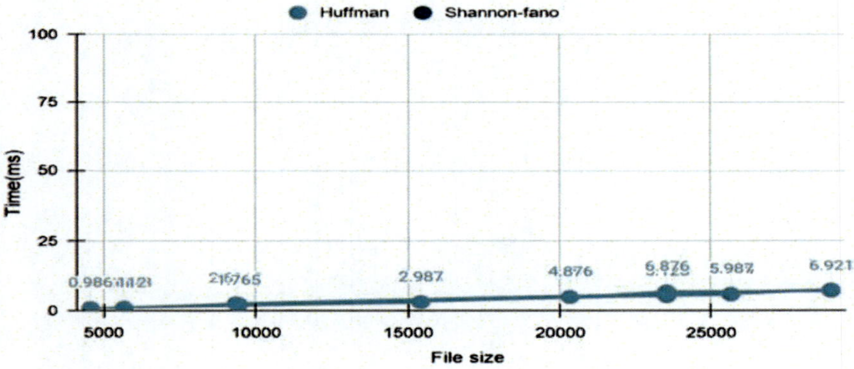

File size and compression time are used to compare the performance of the algorithms.

4.4 *Discussion*

By compressing different data types, different results are obtained. In text compression, Huffman encoding compression ratios are comparatively less than other algorithms. And also, there is bit change in the compression time and decompression time. And LZW provides better compression ratios for the given text files used. This algorithm offers a stronger compression ratio that ranges from 31 to 59%. For the RLE algorithm, compression and decompression times are comparatively less. The

Table 8 Shannon–Fano encoding

Original file		Shannon–Fano				
S. No.	File size	compressed file	Compression ratio	Compression factor	Compression time	Decompression time
1	9356.0039	3571.165	38.1698	61.8302	2.6212	241.0252
2	5678.0027	980.566	17.26	579.05	1.112	126.987
3	15,432.765	8434.532	54.65	182.97	2.987	390.876
4	4567.67	813.34	17.80	561.59	0.986	102.756
5	9434.533	3677.56	38.97	256.54	1.765	251.541
6	25,654.78	18,943.51	73.84	135.42	5.987	563.985
7	20,345.651	14,532.88	71.42	139.99	4.876	437.858
8	23,543.651	17,932.45	76.16	131.29	6.876	496.856
9	28,954.556	22,343.59	77.16	129.59	6.921	678.715
10	27,634.761	21,165.67	76.59	130.56	6.564	765.732

compression ratios are in the range of 57–72% for the Shannon–Fano method, which is average by comparing all other algorithms. In image compression, using RLE, the average compression ratio is 140.886%, the average space saving is 40.97%, and the average compression time is 13.2359 percent. Using Shannon–Fano encoding, ten differently sized images are compressed. It can be shown from this table that the average compression ratio is 77.54%, the average space saving is 22.67%, and the average compression time is 10.61 percent. And in audio compression, the compression ratio for the Huffman approach is in the range of 13% to 76% which is an average value. The Shannon–Fano method compression ratio is between 17 and 77%, which is better than the Huffman algorithm. It takes only less time to compress all the files. And in future, there is a chance to improve the compression ratio in all data type by improving the code efficiency, and also, compression and decompression time can be improved.

5 Conclusion

In this document, we compress different data types such as text, image and audio. Experimental comparison is used to compare different lossless compression algorithms. In order to compare their efficacy, many new lossless compression methods are used on various kinds of data. For text compression, four types of compression algorithms are used that are Huffman, LZW, run length encoding and Shannon–Fano. The LZW can be considered as the best algorithm for text compression by considering the compression ratio, compression time and decompression time of all compression algorithms. Two compression algorithms, run length and Shannon–Fano, are used in image compression, and the Shannon–Fano can be considered as the effective compression algorithm by considering the compression ratio, compression time and space saving. Two compression algorithms, including Huffman and Shannon–Fano, are used in audio compression. The Shannon–Fano can be considered as an effective algorithm by considering the compression ratio, compression time and decompression time of these algorithms.

References

1. P. Singh, Assistant Professor, Vadodara Institute of Engineering, Gujarat, India: "Lossless Data Compression and Comparison of the Performance"
2. S. Porwal, Y. Chaudhary, J. Joshi, M. Jain et al., *Data Compression Methodologies for Lossless Data and Comparison Between Different Algorithms* (2013)
3. S. Shanmugasundaram, R. Lourdusamy et al., *A Comparative Study of Text Compression Algorithms* (2011)
4. Prof. D. Mathpal, Prof. M. Darji, Prof. S. Mehta, *A Research Paper on Lossless Data Compression Techniques* (2017)
5. Ms. P. Bonde, Mr. S. Barahate, *Data Compression Techniques for Big Data* (2017)

6. M. Al-laham, M. Ibrahiem, M. El Emary, *Comparative Study between Various Algorithms of Data Compression Techniques*
7. K. Skretting, J.H. Husoy, S.O. Aase, *Improved Huffman Coding Using Recursive Splitting.*
8. S. Mishra, S. Sing, *Survey Paper on Different Data Compression Technique* (May 2016)
9. S. Mahmud, *An Improved Data Compression Method for General Data*
10. A.S. Sidhu [M.Tech], Er. M. Garg [M.Tech], *Research Paper on Text Data Compression Algorithm using Hybrid Approach compression system for multimedia IoT products*
11. J. Al-Shweiki, N. Hamdan, K. Alkaabneh, *Comparative Study between Different Image Compression*
12. P.B. Pokle, Dr. N.G. Bawane, *Comparative Study of Various Image Compression Techniques*
13. J.S. Kulchandani, S.H. Pal, K.J. Dangarwala, *Image Compression: Review and Comparative Analysis*
14. Md. A. Rahman, M. Hamada, *Lossless Image Compression Techniques: A State-of-the-Art Survey*
15. R.B. Patil, K.D. Kulat, *Audio compression using Dynamic Huffman and RLE Coding*
16. S. Ashida, H. Kakemizu, M. Nagahara, Y. Yamamoto, *Sampled-Data Audio Signal Compression with Huffman Coding*

Comparative Analysis of Apriori and ECLAT Algorithm for Frequent Itemset Data Mining

M. Soumya Krishnan, Aswin S. Nair, and Joel Sebastian

Abstract Frequent itemset data mining is the generation of association rules from a transactional dataset. For example, in a supermarket if bread and butter is placed nearby then the customer who chooses bread will also take butter which will increase the sales. There are many algorithms that are used for frequent itemset. Among those algorithms, the two most commonly used algorithms are Apriori and ECLAT algorithm. In this research, our aim is to prove that ECLAT is far better than Apriori for the frequent set mining in experimental and execution time-wise. We evaluate these two algorithms through experiment and also execute the code to find the time. With the dataset that we obtained, we have done problem solving of the two algorithms and the result obtained is Apriori to provide 11 and ECLAT to provide 9 combinations. When the problem solving was done, only a slight difference is detected so this research work is focused on the code of these algorithms. So, the code of these two algorithms is executed with the same dataset and final result of ECLAT and taken 0.2 s faster than Apriori taken 0.4 s.

1 Introduction

We live in the world where large amount of data is collected each day. Analyzing such data is an important need. The technique of analyzing and finding the required data from the large data is known as data mining. Data mining is the technique of finding the patterns and knowledge for decision making and also for other purpose from a large amount of data. Data mining can be conducted on any data where the

M. S. Krishnan (✉) · A. S. Nair · J. Sebastian
Department of Computer Science and IT, Amrita School of Arts and Sciences, Amrita Vishwa Vidyapeetham, Kochi, India

A. S. Nair
e-mail: aswinsnair@asas.kh.students.amrita.edu

J. Sebastian
e-mail: joelsebastian@asas.kh.students.amrita.edu

P. Karuppusamy et al. (eds.), *Ubiquitous Intelligent Systems*, Smart Innovation, Systems and Technologies 243, https://doi.org/10.1007/978-981-16-3675-2_37

data should be meaningful. The major dimensions of data mining are data knowledge, technologies and applications. The main objective of the methods used for data mining is to generate information knowledge. Dataset are amend it to advance use in the human understandable form. For research studies, these are analyzed by using statistical techniques available. Temporal data mining, association rule mining and frequent itemset mining are among these study designs. In this paper, we are focusing on frequent itemset data mining. An approach for market basket analysis is frequent itemset mining. It helps to determine regularities in customers' shopping conduct, supermarkets, mail-order shops, internet stores, etc. More generally, find collections of things that are often bought together. In the frequent itemset data mining, there are three categories of algorithm. They are as follows: join-based algorithm, tree-based algorithm and pattern-based algorithm. From these three categories, the most commonly used algorithms are as follows: Apriori and ECLAT, respectively. Apriori is a join-based algorithm, and ECLAT is a tree-based algorithm. Apriori work is in horizontal data format, and ECLAT work is in vertical data format. Apriori algorithm is the most commonly used algorithm because it is very easily understandable compared to ECLAT algorithm.

By using of this type of algorithm, companies can arrange these items in a combination of two or more and also providing some discount to them. The consumer can very easily purchase these items so that the companies will get there profit by increase in sales.

2 Literature Review

2.1 Data Mining

In computer science, data mining is one of the multi-disciplinary fields [2]. Data mining which is also known as knowledge discovery (KDD) is the technique where we can dig out relevant data from a huge set of raw data. In business field, the purpose of data mining is that they can understand the nature of their customer with provided data. These provided data are obtained with the help of data mining techniques. With the help of data mining techniques, it is easy to understand that how can we attract the customers. It implies the study of data trends in large batches of data using one or more tools. Data mining has many applications, including research and analysis. The knowledge obtained from data mining is new and highly profitable for the future purpose [2]. This data mining technique will also help in finding the best way to achieve the target and also to make strategic decisions. Effective data collection and storage as well as computer processing are necessary for data mining. By analyzing algorithm, data can be segmented and also the possibility of an event to occur can also be determined so easily.

2.2 *Apriori*

For frequent itemset mining and for mining association rules, the most commonly used algorithm is Apriori algorithm. The algorithm got the name as Apriori because the algorithm uses the previous knowledge of frequent itemset property [1]. Support, confidence and lift are the three important factors of Apriori algorithm. With the help of three factors, Apriori algorithm finds the best combination from the database. For calculating the efficiency of the itemset association, the algorithm uses breadth-first search which is an iterative process of finding the frequent itemset from dataset [1].

2.3 *ECLAT*

ECLAT is the short form of equivalence class clustering and bottom-up lattice which is another popular method for finding the association rule got evolved in 2001 [8]. Efficiency-wise and also scalability-wise ECLAT algorithm is better than the Apriori algorithm. The working of ECLAT algorithm is the depth-first search manner which makes the algorithm faster. The main factors that help in the working of ECLAT algorithm are support and confidence. With the help of the transaction id set which is also known as tidset, the support value is calculated. Here, also the function is recursive until no itemset can be combined [8]. For generating a new candidate, the function will verify each item with the rest of the pairs in the transaction. ECLAT algorithm will find the best combination of items with a short span of not more than three iteration. Since the algorithm follows DFS search, the usage of memory is less. For finding the item, ECLAT algorithm will not go for repeated scans. The main advantage of ECLAT algorithm over Apriori algorithm is the memory, computation and the speed. The main advantage of ECLAT algorithm over Apriori algorithm is the memory, computation and speed. ECLAT algorithm will scan the database to find the support count of the (k + 1)-itemset is not required.

3 Methodology

The phases in this study are based on the CRISP-DM method with the following steps.

3.1 Business Understanding Phase

The goal of this analysis is to search at relationships between products frequently bought together by consumers to make it easier to handle the layout of the merchandise. The main idea behind the business understanding phase is for a product named "X" which all other products show relationship so that it is easier to create a combo offer and also the merchandise can put offers and discount for the products having relationship. So, the main idea behind the business understanding phase is to understand the relationships between products. If we consider the product as bread, then the products that show the relationship with bread are butter, egg, milk, cheese and jams. This is the sample example showing the process that is being done in the business understanding phase.

3.2 Data Understanding Phase

Data understanding phase deals with the data that is used in the research for obtaining the final result. The data are used for solving the problem mathematically, and also the data are used for the execution of the code of the algorithms. The primary data within the sort of data on purchases of products collected directly from customer purchasing receipts over 30 days are that the data source utilized in this analysis.

3.3 Data Processing Phase

Data processing is carried out by picking the stored data, discretization of variables and tabular format transaction data on the transaction data collected over 30 days. Processed data are a transaction with more than 1 itemset where only "Transaction Number" and "Item Name" attributes are used. Although the above dataset of sales transactions has a relatively wide collection, it needs to be discretized. The laws of this range are often modified consistent with the desires that researchers wish to realize. The tabular format is the format in types 1 and 0 or the info format in binary format. The meeting point between the "Item Name" and "Transaction Number" variables is going to be binary 1, supported current sales transaction data, and 0 will become one that does not become the intersection.

Table 1 Products that have the minimum support value

Itemset	Frequent
Mineral water	131
Biscuits	210
Instant noodles	143
Bread	144
Snack	213
Milk	236
Tea	137

3.4 Evaluation Phase

Assessment phases are performed in order to ensure accuracy and effectiveness before they are distributed. They accomplished the original aims of this research, solved the issue and made decisions on the consequences of data mining.

4 Result

The rules developed in this research on transaction data are extracted from the two algorithms. The goal of the Apriori algorithm is to decide the association rule by taking into account the minimum support value (shows the combination of each item) and the minimum confidence value (shows the relationship between items) and the ECLAT algorithm by using the itemset pattern to determine the best frequent itemset. The frequent object collection can be derived directly with the aid of the data ECLAT algorithm pattern.

4.1 1-itemset Formation

The 1-itemset [6, 20] is formed by measuring the frequency of occurrence of the itemset with the minimum support value. The tables listed below display the products that are having the minimum support value (Table 1).

4.2 Combination of 2-itemset

The 2-itemset collection is achieved by comparing the two items in the contract with the minimum support value. The minimum support is 4%, and it can also be solved with the aid of the equation described below.

Table 2 Outcome of transaction value

Itemset	Frequent	Support (%)
Biscuit, snack	70	7.15
Biscuit, milk	63	6.44
Instant noodles, milk	39	3.98
Bread, milk	50	5.11
Snack, milk	66	6.74
Snack, tea	42	4.29

$$\text{Support}(A \cap B) = \frac{\Sigma \text{ Transactions containing } A \text{ and } B}{\Sigma \text{ Transactions}} * 100\%$$

The equation is formulated on the basis of two dimensions. The two dimensions are as follows: (1) sum of transaction containing A and B, which represent the number of transactions in which both the products A and B are present and (2) sum of the transaction which means the total number of transactions. So, for finding the support value the equation generated is sum of transactions containing A and B divided by sum of transaction * 100.

Since the minimum support value is 4%, the transaction data will be terminated by the itemset that has less support value or the itemset with equivalent support value. The outcome will be as shown in Table 2.

4.3 Combination 3-Itemset

The 3-itemset formation is based on the occurrence of three transaction data items along with the minimum support value as defined above. The 3-itemset combination can also be obtained from the equation described below.

$$\text{Support}(A \cap B) = \frac{\Sigma \text{ Transactions containing } A \text{ and } B}{\Sigma \text{ Transactions}} * 100\%$$

With the aid of the above equation, we were able to work out that the minimum confidence of this transaction data is as there is no 3-itemset event that satisfies the minimum support. The 2-itemset combination is then used for the formation of the association rules.

4.4 Association Rule Formation

The combination of the itemset with 19% as the minimum confidence is the formation of the association rule. With the support of the equation described below, minimum confidence can be evaluated

$$\text{Confidence}(A \rightarrow B) = \frac{\Sigma \text{ Transactions containing } A \text{ and } B}{\Sigma \text{ Transactions containing } A} * 100\%$$

The above equation is formed by two values, and the two values are as follows.

(i) Sum of transaction containing A and B
(ii) Sum of transactions containing A.

So, the confidence value can be obtained from the equation sum of transactions containing A and B divided by sum of transactions containing only A * 100%.

4.5 Formation of Final Association Rule

The overall formation of the association rule with the minimum confidence value follows. The 11 rules developed by the Apriori algorithm are listed below with the support, confidence, and the support and confidence products they have acquired (Table 3).

4.6 Comparison Result of Apriori Algorithm and ECLAT Algorithm

Association Rules Formed. The Apriori algorithm has developed 11 association rules where the ECLAT algorithm generates only 9 rules and nevertheless has the same final association value due to the variations in operation of each algorithm. Because the Apriori algorithm works on the basis of the combination between the items, the association rules generated by Apriori algorithm and ECLAT algorithm are different since the ECLAT algorithm works on finding the frequent itemset with the aid of the formed pattern.

Application Execution Time. ECLAT algorithm, results of manual analysis of the Apriori algorithm and ECLAT algorithm, is taken by 9 rules where Apriori algorithm generates 11 rules and ECLAT algorithm took just 0.2 s during execution time, but Apriori algorithm took 0.4 s.

5 Conclusion

From the research study, it can be concluded that ECLAT algorithm is better than then Apriori algorithm. The advantage of ECLAT algorithm is that the database is scanned for finding the support count of $(k + 1)$-itemset.

Table 3 11 rules developed by the apriori algorithm

Rule	Support (%)	Confidence (%)	Support × confidence (%)
If you buy **Biscuits**, then you will buy **Snacks**	7.15	33.33	2.38
If you buy a **Snack**, then you will buy **Biscuit**	7.15	32.86	2.35
If you buy a **Snack**, then you will buy **Milk**	6.74	30.99	2.09
If you buy **Biscuits**, then you will buy **Milk**	6.44	30.00	1.93
If you buy **Milk**, then you will buy **Snack**	6.74	27.97	1.89
If you buy a **Bread**, then you will buy **Milk**	5.11	34.72	1.77
If you buy **Milk**, then you will buy **Biscuits**	6.44	26.69	1.72
If you buy a **Tea**, then you will buy **Snack**	4.29	30.66	1.32
If you buy **Instant Noodles**, then you will buy **Milk**	3.98	27.27	1.09
If you buy **Milk**, then you will buy **Bread**	5.11	21.19	1.08
If you buy a **Snack**, then you will buy **Tea**	4.29	19.72	0.85

The disadvantage of ECLAT algorithm is that it requires more memory space. The advantage of Apriori algorithm is that it uses an iterative level-wise search technique to discover $(k + 1)$-itemsets from k-itemsets. The disadvantage of Apriori algorithm is to produce a lot of candidate sets if k-itemsets are more in numbers and also to scan the database repeatedly to determine the support count of the itemsets. The number of association rules received is different, but the final meaning of the association (support * confidence) is the same. The execution time needed for the 1846 item to run in the program indicates that the ECLAT algorithm is 0.2 s quicker than the 0.4-s Apriori algorithm. In scalability also ECLAT algorithm is better than Apriori algorithm. Apriori algorithm is not sufficient for large dataset.

References

1. Islamiyah, P.L. Ginting, N. Dengen, M. Taruk, Comparison of priori and FP-growth algorithms in determining association rules, in *2019 International Conference on Electrical, Electronics and Information Engineering (ICEEIE)* (2019)
2. S.I.T. Joseph, Survey of data mining algorithm's for intelligent computing system. J. Trends Comput. Sci. Smart Technol. (TCSST) **1**(01), 14–24 (2019)

3. A. Pitman, *Market Basket Synthetic Data Generator* (2011) [online] Available: http://mloss.org/software/view/294/.
4. G. Paulo, *Applied Data Mining: Statistical Methods for Business and Industry* (2003). ISBN 9812-53-178-5
5. C. Borgelt, Frequent item set mining. Wiley Interdisc. Rev. Data Min. Knowl. Discov. **2**(6), 437–456 (2012)
6. J. Han, H. Cheng, D. Xin, X. Yan, Frequent pattern mining: Current status and future directions. Data Min. Knowl. Discov. **15**(1), 55–86 (2007)
7. J. Leskovec, A. Rajaraman, J.D. Ullman, *Mining of Massive Datasets* (Cambridge University Press, 2014)
8. J. Heaton, *Comparing Dataset Characteristics that Favor the Apriori Eclat or FP-Growth Frequent Itemset Mining Algorithms* (SECON IEEE, 2016)
9. S. Kabir, S. Ripon, M. Rahman, T. Rahman, Knowledge-based data mining using the semantic web. Int. Conf. Appl. Comput. Computer Sci. Computer Eng. **7**, 113–119 (2014)
10. K. Khurana, S. Sharma, A comparative analysis of association rules mining algorithms. Int. J. Sci. Res. Publ. **3**(5), (May 2013) ISSN 2250-3153
11. B. Kamsu-Foguem, F. Rigal, F. Mauget, Mining association rules for the quality improvement of the production process. Expert. Syst. Appl. **40**(4), 1034–1045 (2013)
12. A. Kardan, M. Ebrahimi, A novel approach to hybrid recommendation systems based on association rules mining for content recommendation in asynchronous discussion groups. Inform. Sci. **219**(10), 93–110 (2013)
13. S. Kotsiantis, D. Kanellopoulos, Association rules mining: A recent overview. GESTS Int. Trans. Computer Sci. Eng. **32**(1), 71 (2006)
14. M. Hahsler, B. Gruen, K. Hornik, arules—A computational environment for mining association rules and frequent item sets. J. Stat. Softw. **14**(15), 1–25 (2005)
15. Maragatham, M. Lakshmi, G. Maragatham, A recent review on association rule mining. Indian J. Computer Sci. 2 (2012)
16. M. Sinthuja, N. Puviarasan, P. Aruna, Evaluating the performance of association rule mining algorithms. World Appl. Sci. J. 35(01), 43–53. ISSN 1818-4952 (2017)
17. P. Shendge, T. Gupta, Comparitive study of apriori & FP growth algorithms. Indian J. Res. 2(3) (Mar 2013)
18. P. Parikh, D. Waghela, Comparative study of association rule mining algorithms, in *UNIASCIT*, vol. 2, issue 1 (2012), 170–172. ISSN 2250-0987
19. R. Mishra, A. Choubey, Discovery of frequent patterns from web log data by using FP-growth algorithm for web usage mining. Int. J. Adv. Res. Computer Sci. Softw. Eng. **2**, 311–318 (2012)
20. S. Vijayarani, S. Sharmila, Comparative analysis of association rule mining algorithms, in *International Conference on Inventive Computation Technologies (ICICT)* (2016)
21. S. Kotsiantis, D. Kanellopoulos, Association rules mining: A recent overview. GESTS Int. Trans. Computer Sci. Eng. **32**(1), 71–82 (2006)
22. P.N. Tan, M. Steinbach, V. Kumar, "Introduction to Data Mining", Addison-Wesley, 769pp (2005); [22] Y. Yoon, G. Lee, *Two Scalable Algorithms for Associative Text C*

A Trusted User Integrity-Based Privilege Access Control (UIPAC) for Secured Clouds

S. Sweetlin Susilabai, D. S. Mahendran, and S. John Peter

Abstract The cloud is a virtual domain that offers a wide range of storage areas for domain users in the storage cloud. Clients will actually want to get to information and retrieve the assistance rapidly and without any problem. Protection and security are the main issues in the cloud platform. The main facets of security are availability, confidentiality, and integrity. An enormous number of clients keeps their information classified in the cloud for security reasons. Therefore, security plays an important role in cloud computing, which is used to protect user confidential data in the cloud. User information should not be hijacked by an unknown user, so client authentication is very important. In this work, we propose a new cryptographic RBAC model, called User Integrity-based Privilege Access Control (UIPAC) with master key algorithm (MKA) and privilege authentication policy algorithm (PAPA) to help different security features, including client identification, permission policy to the authorized client. The need for cloud storage security cannot be efficiently warranted; many users are unenthusiastic to transfer their vital information to the cloud for storage, making it really difficult to source cloud storage. At the moment, guaranteeing the secrecy of client data and forestalling unapproved access is the way to taking care of the security issues of cloud storage; a lot of encryption privilege access authentication has been conducted. The vast majority of existing authentication control schemes for enciphering data in cloud storage do not assist dynamic refreshing of access control strategies and computational overheads are high. The in-depth structure of our scheme including system setup, create and delete users and their roles, granting the privileges for the authorized user can read and write the file. This work mainly focus on increasing the integrity and authentication of confidential data over the cloud.

S. S. Susilabai (✉)
Manonmaniam Sundaranar University, Tirunelveli, Tamilnadu, India
e-mail: sweetlineric@gmail.com

D. S. Mahendran
Aditanar College of Arts and Science, Tiruchendur, Tamilnadu, India

S. J. Peter
St.Xavier's College, Tirunelveli, Tamilnadu, India

P. Karuppusamy et al. (eds.), *Ubiquitous Intelligent Systems*, Smart Innovation, Systems and Technologies 243, https://doi.org/10.1007/978-981-16-3675-2_38

1 Introduction

With the growth of the huge information period, cloud storage that evolved from cloud computing has become the most well-known external storage with its large limit, low cost, straightforward administration, resource sharing, and great adaptability [1–3]. Users and businesses can purchase storage services agility to suit their needs, which not only keep expensive investments in software and hardware infrastructure but also assure that storage resources are fully used. Therefore, cloud storage systems have received substantial attention from both industry and organization. Clouds [4, 5] can provide various services that can be provided on the Internet, such as Internet storage and software, servers, software development platforms commonly known as "clouds." Many organizations allow cloud resources such as Google, Amazon, and Cisco VMware.

1.1 Deployment Model of Cloud

Cloud implementation models are classified mainly in four ways: Public, private, community, and hybrid cloud [6, 7, 11–13].

1.1.1 Private Cloud

It is primarily designed by a single organization that contains many consumers. It can be managed and administrated by a third party, or it can be a combination of them and may be present inside or outside the structures. By using the private cloud, the user can create their own data center, e.g., Ubuntu Enterprise Cloud—UEC, Eucalyptus, Amazon.

1.1.2 Public Cloud

It is mainly designed for universal open use. It can be managed and administrated by a government, academic, or business organization, or it can be a combination of these elements. It can be present depending on the cloud provider, e.g., IBM Smart Cloud, Google App Engine, Microsoft Windows Azure, and Amazon EC2.

1.1.3 Community Cloud

It is mainly designed by a particular community of consumers in organizations that have shares. It can be managed and administrated by one or more of the specific community organizations, a third party, or it may be a combination of them and may be present inside or outside, e.g., Microsoft Government and Google Apps.

1.1.4 Hybrid Cloud

It is mainly of two or more different cloud communications only as "single" entities but standardized together that allows portability of data and applications, e.g., Windows Azure, VCloud, and VMware.

1.2 The Cloud Services

Cloud services are mainly classified by three services, such as Iaas, Saas, and Paas. Depending on the needs of the user application, adequate service will be provided. There are three cloud environment services [10, 14]:

Infrastructure as a service (IaaS): This level of service manages cloud storage and management activities, since it allows access to virtual machines, accessible storage network infrastructure components such as configuration services and firewalls.

Software as a service (SaaS): Cloud users can effectively coordinate with the application at this level, since a high-class service is offered at this level, eliminating the need to install software and hardware on end users. Furthermore, there is no need to pay attention to the management of services and infrastructure.

Platform as a service (PaaS): Applications and software are created at this next level, simplifying the distribution and management of the user application. The software and hardware tools offered by the service providers are also included by adapting the application frameworks that support the software as a service.

2 Related Work

Zhu et al. [1] discussed a role key hierarchy structure for large-scale systems. They introduced practical model with role-based security scheme to hold a signature, verification, and confidential data constructions on elliptic curves. They found that this method is very efficient and bendable enough to hold large-scale systems. Muthurajkumar et al. [2] presented another entrance control procedures specifically a smart transient job-based admittance control model. The proposed security system was attempted with real execution with a private cloud. The essential favored situation of the proposed model is the course of action of a two-level security system to ensure the security of data in cloud information bases. Zhou et al. [3] proposed to confide in issues in cryptographic RBAC frameworks for making sure about information storage in a cloud environment. They proposed trust models for data owners and roles in frameworks which are utilizing cryptographic RBAC plans to make sure about put away information. These secured models assist owners and roles to adaptable access arrangements, and cryptographic RBAC plans guarantee that these strategies are authorized in the cloud. They permit the information owners to utilize the trust assessment to choose whether or not to store their encoded information in the cloud for a specific role. They designed the architecture of a trust-based cloud storage system. Sridhar et al. [4] have brought up the downsides in the current confirmation convention. They proposed hybrid mechanisms to improve exhibition group registration strategy. It brings about lesser data transfer capacity utilization and decreases computational and communication cost. Su et al. [6] described that the advancement of distributed computing has advanced the procedure data. The individual clients can acquire information, programming, stage, or administration of security. Access control is one of the conventional innovations of data security. The fine-grained will control conditions are imported to the critical times of code text. Meanwhile, this arrangement doesn't grow the scrambling and key regulating work of individual customer. Geetha and Anbarasi [7] discussed that the utilization of XACML documentations is a simple method to speak to complex condition checking in access control components. This lets the utilization of number of strategy rules to be diminished and empowers a many access control approaches to be checked in a specified time. Thilakarathne et al. [8] portrayed an improved admittance control model that can be used in cloud and moreover that can be used for secure cloud data stockpiling. They demonstrated the exploratory outcomes of our executed admittance control model through the perspective of execution, data reliability, and security in veritable cloud condition. Xu et al. [9] proposed that an improved RBAC scheme depends on personality-based crypto method for cloud storage. Their method supports dynamic update of access control policies and also gets better encryption efficiency and system performance by using write time re encryption policy-based hybrid encryption policy. Ghazal et al. [10] have zeroed in on the requirement for a security instrument that can uphold the present progressively changing conditions and introduced our proposed I-RBAC model and structure which satisfies these prerequisites. The proposed I-RBAC structure is a general multi-specialist framework dependent on RBAC for furnishing admittance to authoritative resources with semantic jobs. These exercises are job creation, job mining, and job task contingent on the related errand data about the client.

3 Methodology

This document proposes a new framework of identity-role-based scheme to perform various cloud activities agreed upon by both cloud service providers and client users. A system manager is used to perform tasks on behalf of client users on virtual machines in cloud data centers. When client users request a query from the cloud, they must authenticate and verify with specific credential rules to achieve security in order to perform all the activities to create a reliable platform. This process helps both entities build mutual trust and use cloud services efficiently. The system manager and user of the cloud storage are designed with IBEM and HECC. The data owner wishes to share the file in the cloud, he must first convert the original file to encrypted file using a combination of IBEM and HECC algorithms. The client uploading the encryption file, he will get the master key for his secrecy. Therefore, only an authorized user can access confidential data. If a user wants to download data from the cloud, that user must specify the authentication policy with the master key, and if the data owner (DO) wants to share the credential file, then then only he will get the privilege authentication policy key. If the authentication information matches the database, the user decryption is performed through the same credentials but in the reverse order in which the two algorithms are applied.

The proposed framework can be run with the following proposed calculation.

Stage 1 All client users can make connection with the system.

Stage 2 They can communicate with the cloud to send and get to and oversee data.

Stage 3 Master key generation center (MKC), which resides on private cloud, who generates the master key of user depending on their roles, these secret key will not know by any unauthorized user.

Stage 4 After getting mater key, data owner is liable for the uploading confidential file signed with his own master key.

Stage 5 Privilege authentication policy center (PAPC) is responsible for provides access policy rights to the authorized user. An adversary is not able to build any factor authentication, which makes the server to verify his login request. This privilege access policy is generated only authorized users.

Stage 6 Furthermore, the first factor is only used once for each user's login request.

Stage 7 If final authentication of user to server is masked by using master key and privilege authentication policy key.

Stage 8 If the user masked value is valid, the requested confidential file will be decrypted with their three keys.

The overall proposed work layout is as follows (Fig. 1):

3.1 *Hierarchy of User Identity Role Cryptographic System (UIRCS)*

Access authority policies are allocated to both domain and their categories. UIRCS ties all users and their roles assigned and also it describes how a role restricts the authentication policies available to the user. Authorization determines whether a user is a person who claims to be. The authentication technique provides access to systems by verifying a user's credentials by matching the credentials in a database of authorized users. The user identity role-based hierarchy is defined as a tuple UIR

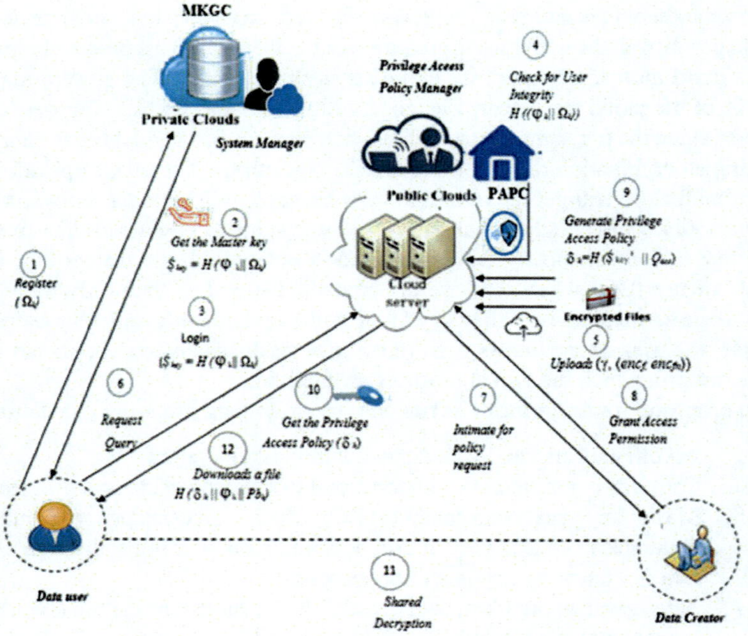

Fig. 1 Overall layout for UIPAC scheme

= US, RO, PER is a cryptographic partial order relation for the set of keys based on users (UO), roles (RO), and permissions (PER), satisfying the following conditions:

- <US, RO, PER>: This tuple represents users, roles, and authorized privilege permissions. Where $u_1, u_2, \ldots, u_n \varepsilon US, r_1, r_2, \ldots r_n \varepsilon$ RO and PER represents permission will be assigned by each user of the role who have request to access the file.
- $\mathrm{US_{Aas}} \subseteq \mathrm{US} \times \mathrm{RO}$, a M - to - M assignment of user role.
- $\mathrm{PER_{Ass}} \subseteq \mathrm{PER} \times \mathrm{US}$, a M-to-M authority for the role assignment.
- $K_{set} = \delta_k \cup \varphi_k \cup \mathrm{Pb_k}$, the key set K_{set} includes the user secret key k, the user privilege authentication policy key k, and the public key set $\mathrm{Pb_k}$;
- $\phi_k\mathbf{A} \subseteq U \times \phi_k$, a one-to-one user to key assignment relation, i.e., each user $u_{ij} \epsilon U$ is assigned to an user secret key $i, j \epsilon k$;
- $\delta_k\mathbf{A} \subseteq R \times \delta_k$, a one-to-one permission key with a particular role assignment relation which is prepared by PAPC cloud, i.e., each role $r_i \in R$ corresponds to a one-to-one permission key $k(\mathrm{PER}) \in \mathrm{k}$;

3.2 *Proposed User Integrity-Based Privilege Access Control (UIPAC)*

A proposed UIPAC scheme is based on the confidentiality of the user and their role-based privilege authentication control policy in the cloud storage space. In this segment, we consider the information proprietor trust models with the identity-based scheme and validation access control technique for RBAC structures.

Our proposed scheme is mainly composed of following entities:

- **Users** are the parties who want to access the data from the cloud and decrypt the file stored data. Each user must be authenticated by the SM and issued a credential, which is associated the identity of the user, after successful authentication.
- **Role** is the object that associates users with the access to data of owner
- System manager (SM): It is the administrator of the master key access control system, responsible for creating the secret key of user who has the register to access the resource.
- **Master key generation center** (MKC): Master key generation center generates the user master key and these key is for all users in the system.
- **Privilege authentication policy center** (PAPC): Privilege authentication policy center is deployed in the cloud and is responsible for coordinating authorized access to resources. **Privilege authentication policy administrator** (PAPA): The privileged authentication policy admin role is authorized to generate multi-factor authentication policy key for a user. He generates a key depends on role who can edit and access confidential file or private information.
- **Data creator** (DC) or user: He is a client of the cloud storage administration and is overseen by the entrance control monitor.

The RBAC scheme based on user identity privilege access cryptosystem (RBAC-UIPAC) in cloud storage can be represented by following RBAC−UIPAC tuple:

$$\langle \text{Setup, UEP, MKG, FM, PAP, PCS, DOF} \rangle$$

3.2.1 Setup

This is the principal layer in proposed structure, which represents the connection among users and their SM. The connection among user and system manager is straightforwardly relative to one another as user demand a service. To begin with, the user registers their data to neighborhood private cloud and SM is liable for giving login certifications to user. If any user $\hat{u}$ is already registered, he/she must enroll the next authentication stage to login with his UserID (Ω_k). Otherwise, he will be considered as a new user. When a new client wishes to enroll the system, first he needs to enroll with their information over the private cloud, which carries out the local cloud server operation and he will get identity of the User Ω_k. The user can create

Ω_k with their corresponding role r and all other credential information. The users in the list are $\hat{u}_1, \hat{u}_2, \ldots, u_n \, \varepsilon \tilde{U}$. Once he/she completed the registration process, then the client needs to enter the next authentication stage.

3.2.2 User Enrollment Process

In the second layer, SM and user must confirm and check each other so as to agree on certain assistance level agreements for accomplishing security to safely perform various administrations. User completed first verification with Ω_k function and he must able to get the identity-based signing key. If the first verification is failed, he must not able to get signing key. If the user credential is valid, then he precedes the next stage successfully.

Step 1 The user enters the framework with the assistance of his UserID (Ω_k).

Step 2 We allow users' secret credential matches by Ω_k as inquiry case; if it might be seamed as invalid secret password, he must not be able to continue following steps.

Step 3 In the setup and enrollment process, the user supplies her/his identity and his through a secure channel to data owner.

Step 4 Next, the user enters the first factor key generation of master key.

3.2.3 Master Key Generation (MKG)

Once the user completed the user enrollment process, then the user successfully gets the master key from the private cloud center. The master key generation center (MKGC) resides on private cloud, who generates the master key (ϕk) based on user identity Ωk, and depending on their roles γ, these secret key will not know by any unauthorized user. This master key is generated as onetime task. The first factor authentication process is as follows.

Step 1 The user connects the MKGC server with the help of $H(\Omega k \| \kappa)$.

Step 2 $U \rightarrow \dot{S} : \Omega_k, \kappa$ The user sends the first authentication factors $\langle \Omega k, k \rangle$ to $\dot{S}$.

Step 3 The server guarantees the legitimacy of hash function, in the event that it is legitimate matches, at that point it process the subsequent stage.

Step 4 $\dot{S} \rightarrow U : \varphi_k$, The $\dot{S}$ ensure the validity of hash query, then $\dot{S}$ generates the master key k and assigns a new hash signing key.

Step 5 $\dot{S}$ assigns a by a new hash signing key $\dot{S}$ key to the user as follows.

$$\dot{S}_{key} = H(\varphi_k \| \Omega_k)$$

The procedure of first factor key generation of master key (ϕ_k) is as follows:

```
Input   : User attribute set: υname, ρno, Ωi, user in list U;
Output  : Master Key(φk)
List the initialization of the parameters ⟨υname, ρno, Ωi⟩.2.
for each user Π υname, ρno, Ωi (U) are executed by the
System Manager, where u1,u2,……un ∈ U, υname is the name
of the user, ρno is the contact number of the user and Ωi is
the UserID. Get the user name and assign Ψ ← σ υname
(U) and Tokenize(Ψ), Find τ ← |Ψ|
For i= 0 to τ
  for each char in Ψ:
   Segregate Ψ by 2 by 2 order with two parts represented
  as Ψ1 and Ψ2
Get size of a part1 Γpart1 ← |Ψ1|, Start=0, Γend = Γpart1 -
1;
 For j= Start to Γpart1 -1
 If (Γpart1 -1 mod 2==0)
 Token1= Token1 + Asc (ψ1(j)) mod Asc (ψ1(Γend))
 Γend = Γend -1
 Else
 Γmid = (Γend +1)/2
 Check if Γmid ==j then
  Calculate
 Token1+= ψ1(j) mod (abs (Asc (ψ1(start))- Asc (ψ1(Γend
  )), Γend = Γend -1
7. Repeat the step6 for Ψ2

Let ϱ be a random variable, after the uniform discrete
appropriation over the set {2, 3, 4, 5, 6, 7, and 8}. So for r ϱ
∈ {2, …, 8}, The random variable ϱ can be defined as ϱ
∩[2,8]
Pick a two random numbers r1, r2 ∈ ϱ, Calculate Token=
Token1/r1 ∪ Token1/r2
To shuffle an array Token of n elements (indices 0..n-1)

For m =0 to |Token| do
      n ← random integer such that m ≤ n < 1
      m ← exchange Token [m] and Token [n]
      Assign S ← SHUFF(Token)
Take S' be the set of chosen indices and Pick a four
random seed (S) location from 0 to n- 1 size.
Get σ ρno (U) and find Plen= |ρno|
    For j= 0 to 4
         ρno'(j)+= ρno(Plen), Plen -= 1
Let Fstr be string array, Fstr is empty at the start.
For i=0 to |S| do
         If (S'(i)==i)) then perform Fstr(i) += ρno'
   (i) U S(i) otherwise perform Fstr (i) +=S(i)
Get the Ωi and assign ϴ ← σ Ωi (U) and
Tokenize(ϴ)
  For i= 0 to | ϴ|
       for each char in ϴ: calculate key=Asc (ϴ (i))
  If(| ϴ|>| Fstr|) then perform Rem=| ϴ|-| Fstr| and
  assign φk is a string array and starts with zero.
       For i=0 to | Fstr|
       for each char in Fstr
            φk +=Div (ϴi ; Fstri), k=| Fstr|
       For j= 0 to rem
            φk +=Div (ϴk ; Fstrj), K++

Else If(| ϴ|<| Fstr|) then Rem=| Fstr| -| ϴ| and assign φ
k is a string array and starts with zero.
For i=0 to | ϴ|
 for each char in ϴ
            φk +=Div (ϴi; Fstri), k=| ϴ|
  For j= 0 to rem
            φk +=Div (Fstrk; ϴj ), K++ otherwise
For i=0 to | ϴ|
 for each char in ϴ && Fstr
            φk +=Div (ϴi; Fstri)

 Return φk
```

This master key (ϕ_k) is personally secured by data creator as secret, and it will be maintained as confidential key of every client users of the cloud. After getting the k, the user needs signing the server for further process.

3.2.4 File Management

All the encrypted file can be proceed with the following ways to upload the file with cloud, getting permission to access the file from the data owner and the operations of the accessing the files.

Uploading a File and Granting Permission

For each encrypted file $£_f$ can be uploaded with their role γ and its authentication permission access is denoted as $(\gamma, \langle f_f, f_{fix})$. Where γ is the role name, $£_{fn}$ is the name of file $£_f$.For each encrypted file $£_f$ can be accessed by the role γ and its authentication permission access is denoted as

$$(\gamma, \langle £_{\mathrm{f}}, £_{\mathrm{fov}}, \theta_{n\sigma}, \varphi_{\mathrm{k}}\rangle)\epsilon \mathrm{Q}_{\mathrm{ass}})$$

where Q_{ass} describes the permission assignment relationship of user and role, Where γ is the role name, $£_{\mathrm{fn}}$ is the name of file $£_{\mathrm{f}}$, θrw represents the access authority to the file read or write file and ϕ_{k} is the master key. The policy key pair of tuple is generated as $\left(\mathrm{PO}_{\mathrm{kp}}, \langle\varphi_k, \delta_{\mathrm{k}}\rangle\right)$. The administrator generates and uploads a $\mathrm{PER}_{\mathrm{ass}}$ tuple of following form to the cloud:

$$|\left\langle Q_{\mathrm{ass},} \mathrm{PO}_{\mathrm{lop}} \right\rangle \epsilon \mathrm{PER}_{\mathrm{g}}\rangle$$

In this tuple, PER_g is a mark representing that this tuple describes the authorization task relationship. $\mathrm{PO}_{\mathrm{kp}}$ represents that the access permission signing pair key of the signer.

Procedure of File Access

Every uploaded files can be accessing by using the read or write permissions. The process of read and write procedure describes the following steps.

Steps for file reading:

If any user u is valid to read the file file $£_{fx}$, $u_i \exists\gamma$, $\langle u, \gamma\rangle\epsilon$USRO and $\left(F_{rp}, \left(£_{ff_n}, \lambda_{\mathrm{read}}\, \mathrm{PO}_{\mathrm{k}}\right)\right)\epsilon\, Q_{\mathrm{ass}}$, then the user must process the following four steps:

Step 1 Download the USRO tuples of u and γ, the Q_{ass} tuples of $F_{\mathrm{rp}}, £_{\mathrm{fn}}, \lambda_{\mathrm{read}}$, *and* $\mathrm{PO}_{\mathrm{kp}}$.

Step 2 Check the tag of the two tuples, continue to the following stage if the confirmation is passed, in any case a mistake error will be accounted for.

Step 3 The client decodes the decryption key $\mathrm{Pb}_{\mathrm{k}}(u^{\gamma})$ of the role γ of from the USRO tuple.

Step 4 Use $Pb_k(u^{\nu})$ to decrypt the file from the Q_{ass} tuple.

Steps for file writing:

The way of writing a file is same as reading file. If any user u is valid to write access to file $£_{fn}, u_i\gamma$, $\langle u, \gamma\rangle\epsilon$USRO$\cap$ $\left(F_{\mathrm{rp}}\left(£_{fn}\lambda_{\mathrm{write}}, PR_{kp}\right)\right)\epsilon\, Q_{\mathrm{ass}}$, then he can extract and decode the document as indicated by the over four stages and subsequent to adjusting the file transfers another Q_{ass} tuple to the cloud:

$$\left(F_{wp}\gamma, \left(£_{fn}, \quad \lambda_{\mathrm{write}}\right), £'_f, \mathrm{PO}_{kp}\right)$$

After the cloud gets the new Q_{ass} tuple, the arrangement of the tuple and the tag of the tuple will be confirmed. At that point, confirms regardless of whether the role γ has consent to write the file, replaces the old Q_{ass} tuple with the new Q_{ass} tuple if the confirmation is passed.

3.2.5 Privilege Authentication Policy (PAP)

If the user wants to get a file for read or write, then the user first send a request for file access to the PAPC. The cloud indicates the policy permission request to data owner. If the data creator wants to share the file for particular data user, then only he grants the permission to access file for further process. Otherwise, the file will not be shared. Once the cloud getting permission from the data owner, then the cloud is generating the privilege authentication policy key and this authorized key is generated only if the user is authorized.

After getting the permission for accessing the file, then server generates the PAP by using the following steps:

Step 1 $U \rightarrow \dot{S} : \dot{S}_{key}, \varphi_k, \Omega_k, \kappa$. The DC enters with first factor tuples $\langle \dot{S}_{\text{key}}, \varphi_k, \Omega_k, \mathbf{K} \rangle$ to cloud server.

Step 2 Upon receiving the first factor authentication details, $\dot{S}$ performs

$$\dot{S}'_{\text{key}} = H(\varphi_{\text{k}} \| \Omega_k \| \text{K})$$

Step 3 The DC ensures identity and first factor verification algorithm; if he is a trustee, then the server agrees $\dot{S}'_{\text{key}} = \text{H}(\varphi_{\text{k}} \| \Omega_{\text{k}} \| \kappa)$, if so, the server acknowledges the login query.

Step 4 Initially, DC is the one who uploads the file into the cloud storage and has full control over it. The DO uploads the credential file $£_{\text{f}}$ with their $\gamma, \theta rw$ to authentication server PAPC. Each file is identified by its file name.

Step 5 Upon receiving the first factor verification and once the credential file are uploaded, the PAPC generates the privilege access policy only which is applicable for the data creator who wish to share their credential file to those he wants to be share.

Step 6 Thus, the second factor authentication by depending on privilege authentication policy (PAP) mechanism.

The privilege authentication policy key (δ_{k}) is prepared by using following algorithm.

Input : φ_s, $£_f$, $£_{fn}$ **Output** : Privilege Authentication Policy Key(δ_k) 1. Get the $£_f$ and φ_s 2. Assign $q \leftarrow \sigma_{£_{fn}}(£_f)$ 3. Tokenize (q) 4. For i = 0 to $\|q\|$ for each char in q : Segregate q by with two parts represented as q_L and q_R 5. Get size of a part1 size $\leftarrow \|q_L\|$ Calculate $q_1 = \sum_{j=0}^{size} \mathrm{Asc}(q_L(j))$ 6. Get size of a part2 size $\leftarrow \|q_R\|$ Calculate $q_2 = \sum_{j=0}^{size} \mathrm{Asc}(q_R(j))$	7. Get the value of φ_s 8. Divide the φ_s into two halves φ_{left} and φ_{right} 9. φ_{left} and $\varphi_{right} \leftarrow split(\varphi_s)$ Where φ_{left} and $\varphi_{right} \in \varphi_s$ 10. Find $\varphi_L = \|\varphi_{left}\|$, $\varphi_R = \|\varphi_{right}\|$, $q_{left} = \|q_1\|$ and $q_{right} = \|q_2\|$ 11. Check if $\varphi_L \neq q_{right}$ 12. If so, then check if $\varphi_L < q_{right}$ Diff = $q_{right} - \varphi_L$ For i = Diff to $\|q_{right}\|$ Pad zero with $\varphi_L(i)$ 13. Scan each every φ_L and q_{right} tokens one by one $n = \|\varphi_L\|$ Calculate $\bigcup_{i=0}^{n} \delta_k(i) = \bigcup_{i=0}^{n} \varphi_L(i) - q_{right}(i)$
Output: δ_k	

The cloud storage server generates the privilege authentication policy key (δ_k) for the data creator and it will be maintained in the cloud. Setup and user enrollment stages are performed only once, and the authentication stage is executed whenever a user wishes to login. In the setup and enrollment phase, the user supplies her/his identity and his through a secure channel to DO.

3.2.6 Permission Control Strategy

The permission control strategy vector can be defined by using parameters a $\langle Q_{ass} : \theta_{rw}, £_{fn}, \delta k, \widetilde{U}, \gamma \rangle$ The input of permission assignment Q_{ass} of the user list are $\hat{u}_1, \hat{u}_2, \ldots, \hat{u}_n \mathcal{E} \tilde{U}$. by using the permission key *k*, the file name $£_{fn}$ and permission name op, which assigns the permission operations of the file $£_{fn}$ for the user on a particular role γ; the input of permission deletion involves a user act on a role γ; the file $£_{fn}$ and permission θrw,which deletes the permission operations of the $£_{fn}$ from the user in particular role.

3.2.7 Decryption of File

Step 1 The client sends a request to the cloud.
Step 2 The cloud service provider verifies their credential.
Step 3 The client's credentials are valid, and then CSP acknowledges if the client trust the CSP and decides to continue the connection.
Step 4 The cloud service provider crates a unique hash-encrypted key with client user master key and privilege key. The hash (key) is sent back to the client.
Step 5 The client user will decode the hash key; this process implies that a particular client can only read the hash key sent b CSP.
Step 6 Clients and CSPs can now exchange information securely.

The following table shows the proposed notation of using proposed User Integrity-based Privilege Access Control (UIPAC) (Table 1).

Table 1 Defined notations in UIPAC

Notations	Description
Ω_k	User ID
φ_k	Master key
γ_{name}	Name of the user
ρ_{no}	Phone number of the user
Γ	User Role
$£_f$, $£_{fn}$	Encrypted File &ftle name
θ_{rw}	Kile operations
Q_{ass}	Permission assignment relationship
PO_{kp}	Policy Key pair
λ_{read}, λ_{write}	Granting permission for file access permissions to read/write
Pb_k	Decryption key
δ_k	Privilege access policy key

4 Security Analysis

We characterize security objective and clarify the security of our proposed UIPAC against different attacks analyzed below.

Theorem 1 *Our proposed UIPAC method can uphold correctness.*

Proof of Correctness: This effectively confirms the valid pair of signatures, legitimate labels of individuals, in general; if valid, request sent to DO to obtain file sharing privileges. If DO wants to grant permission to access the particular file, then only the cloud prepares the privilege key. For all users, it can be checked by a pair of catch tags to catch the door. Pb_k is mutually shared by the DO to DU to decrypt the file, and if the invalid tag of signatures should fail in the verification, then access is denied.

Primal: Initially, the user enters the first level setup verification.
$U \rightarrow S:$ Token $=$ Verify(Hash($\Omega_k||\kappa$) $\rightarrow$ Valid/Invalid..

Sign: According to primal verification the master key generation algorithm to be used in UIPAC system for generating secret key of each individual user. This algorithm is executed by SM. Let $\mathbf{V} = \{U, M_{\dot{k}}\}$.Where $u_i \epsilon U$ and $M_{ki} \epsilon \varphi_k, i = 1 \ldots n$,. Each user has unique key $u_1 \leftarrow Mk_1, u_2 \leftarrow Mk_2, \ldots ..u_k \leftarrow Mk_N$. This secret key is not known by anyone. Gets the user private key (ϕ_k) from SM.

$$\sigma : \varphi_k \leftarrow \text{Sign (Mask(Token))}$$

Verify: Let us take the master key ϕ_k and confirm login credential token. It restores the approval result which would be either substantial or invalid. The latter acknowledgment can mean either that the user who generated has been revoked, or σ is not a valid signature, (in a bunch of invalid users):

∂ : Verify(Hash $(\delta_k, \sigma)) \rightarrow$ valid/invalid ;

Privilege Authentication Trace: Let us take a user key ϕ_k then this key is used to generate the privilege policy key using PAP algorithm. This is the second level factor to get credentials. It can trace a signature ∂ to legitimate user with particular one role member u_i who gets it:

$$\text{PATrace } (\text{Hash}(\partial, \sigma)) \rightarrow \text{ valid/invalid and Verifu}(\Omega_k, \text{Sign}(\varphi_k \delta_k), \text{Pb}_k) = \text{valid.}$$

Theorem 2 *Our proposed UIPAC method can uphold unforgeability.*

Proof of Unforgeability Just individuals from a role can make legitimate tags with the role. If the attacker wants to impersonate a valid user, he should create a substantial client's login request.

$$\dot{S}_{key} \leftarrow (\text{Hash}(\Omega_k, \text{Sign}(\varphi_k, \delta_k))$$

Because having no way $\dot{S}_{key} \leftarrow (\text{Hash}(\Omega_k, \text{Sign}(\varphi_k, \delta_k))$, the intruder cannot spoof a valid user's login request, and if so, but he could not able to guess any one of the key. Because the highly secure UIPAC system with the two levels of algorithms with the first level master key and the second level policy access key does not launch an impersonation attack.

Theorem 3 *The proposed UIPAC method can attack against known-key security.*

Proof of known-key security: When the user wishes to login the server, he yields $\Omega_k : \{\text{Uname}; U_{pwd}\} \rightarrow$ K Depending on their K, he receives master key. An adversary can access cloud; by using the previously generated master key, he is able to login the server (K; M_K). So an adversary cannot guess and attacks the user key.

Theorem 4 *The proposed UIPAC scheme can support anonymity.*

Proof of Anonymity: The integrity of the individual signer cannot be resolved without the secret key (ϕ_k) and privilege authentication policy key(δ_k).

Definition 4.1 From an owner's perspective, he is only able to make a policy for transaction, whereby a DO wants to share data to a particular role, and then only the credential is applicable for qualified users. In UIPAC system, successful credentials interaction is a legitimate interaction where only eligible users in the particular role to which data is decrypted by verifying the integrity of qualified user with matching the valid tags.

$$q_{\text{Intaritun}} : \text{Check } (\Omega_k, \ \text{Sign } (\varphi_k; \delta_k), \text{Pb}_k) \rightarrow \text{ valid/invalid}$$

An unsuccessful credentials interaction is an invalid interaction where an unqualified user could not able to access the credentials. We can mention two types of unsuccessful credentials interactions.

User Key Failure (UKF): User key failure is an unsuccessful credentials interaction caused by an intruder who did not give the correct user key; that is, the role with the particular user has not a qualified user for the role, so PAPM could not able to grant the permission key or does not sent the request to DO for unqualified users.

User Privilege Policy Failure (UPPF): User privilege policy failure is an unsuccessful credentials interaction; if the user is the qualified, but he is not eligible to access the particular role credentials, then DO does not provide the policy access to qualified user. When DO detects that user is eligible or else DO wants to be share the credentials, only he is trustee. Here, we can define user privilege policy failure as such an unsuccessful credentials interaction where the PAPM does not have the authority to generate the policy, only DO having the authority to which user has qualified. If DO knows who has not eligible access credentials, then he is considered as unqualified one. We consider this unsuccessful credential interaction as user privilege policy failure.

Definition 4.2 (Integrity trust vector). We can define an integrity trust vector to describe the user integrity for particular role as:

$$\tau = (\Delta, SUKF\gamma, SUPPF\gamma)$$

In this integrity trust vector, τ is the value related to successful credentials interactions $SUKF\gamma$ is the value related to the user key failure of the role, and $SUPPF\gamma$ is the value related to user privilege policy failure. Thus, UIPAC system defines the trust vector τ that represents the user integrity trust with anonymity.

Definition 4.3 (Policy trust vector). We can define a policy trust vector to describe the user integrity for particular file as:

$$\tau = (\Delta, SUKF\gamma, SUPPF\gamma)$$

In this policy trust vector, τ is the value related to successful policy key generation $SUKF\gamma$ is the value related to the user key failure of the role, and $SUPPF\gamma$ is the value related to user privilege policy Failure. Thus, UIPAC system defines the trust vector τ that represents the user integrity trust with anonymity.

Theorem 5 *The proposed scheme can support confidentiality*:

Proof of Confidentiality: The credentials are stored in the cloud, which prevents the cloud service provider from knowing the contents of the file. In addition, it is not easy to decrypt the encrypted content during transmission. When a snooper needs to break the secret, he must obtain the master key and the authorization access key.

We are currently checking the confidentiality of the proposed method. In the event that the client cannot comply with the data entry strategy characterized by the owner, the client will not be able to recover the correctly decrypted data. Users who are not authorized, i.e., those who do not have the required attributes, should be discouraged from accessing data stored on the cloud server. This data confidentiality is simply guaranteed within the proposed framework. If attributes owned by a user cannot meet the information access policy set by the owner, the user cannot retrieve the correctly decrypted information.

Theorem 6 *The proposed UIPAC scheme can support against collusion resistance*:

Proof of Collusion resistance: The proposed framework is very resistant to collusion. There is very little chance of collusion between the generated keys for the algorithm to include randomness in the key generation. It is difficult to find two pairs of keys that have identical results, i.e., it is not possible to break with two pairs of keys.

$$\mathrm{Hash}(\mathrm{Sign}\ (\varphi_{\mathrm{k}}, \delta_{\mathrm{k}}), \mathrm{Pb}_{\mathrm{k}}))$$

It is difficult to find two input attributes that have the same output as the hash, i.e., with two inputs F and S, Hash (F) = Hash (S) and F $\neq$ S. If the intruder hacks the $\phi_{k,}$ there is no chance to finding the $\delta_{\mathrm{k.}}$

Theorem 7 *The proposed UIPAC scheme can support reliability.*

Proof of Reliability: Since the DO controls the uploading and sharing of the confidential document according to its role and only grants permission to test the authorization policy, the risks of loss and loss of information are very low. Since the entire process is cloud dependent, it is deeply flawed, open-minded, and trustworthy. Since the owner is in control of the entire storage and sharing process, there is very little chance that any information will be lost or disclosed. Since the whole process depends on the cloud, it is highly fault tolerant and reliable.

5 Result and Discussion

5.1 Platform and Evaluation

The self-contained data protection is achieved by developing a cloud environment that utilizes UIPAC model for access control mechanism and identity-based encryption to protect user data. This interface is developed using Python 3.7 framework, and this framework needs user details and associated roles. This framework is compared with Amazon simple storage services, hosted by Amazon Web Services. The cloud framework provides different user interfaces that can be used by the administrator

Table 2 Key generation time

key generation time	
User name length	Time in seconds
Short length	0.07563
Mid length	0.0801
Long length	0.0894

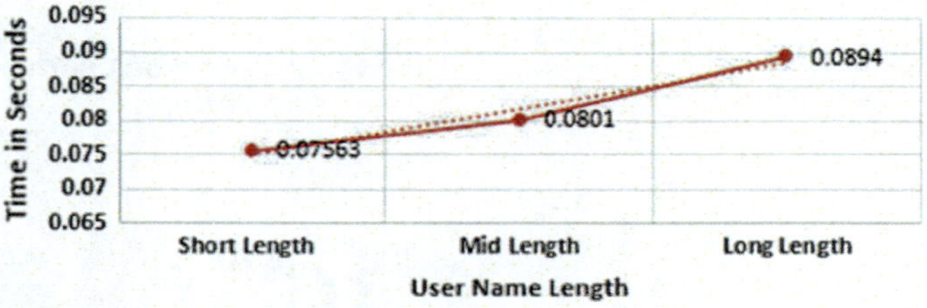

Fig. 2 Key generation time

or the user based on their roles. We make a comparison between AWS cloud scheme and our UIPAC scheme with regard to uploading time along with encryption, key generation time, downloading time along with decryption. We make an analysis on the scheme and AWS cloud. We have implemented a proposed UIPAC with the help of Python 3.7 based on boto3 and twinker library and also which uses the sqlite3 library. We performed the simulation on an Intel (R) with a Windows 7 framework,Core (M) i7-855 0U @ 1.80 GHz processor with 8 GB of RAM.

5.1.1 Key Generation Time

In this analysis, the key generation time is calculated for each user with their user name length which may be defined as short, mid, and long length, respectively. This key generation time can be measured with seconds. The result of key generation time is shown in Table 2 and Fig. 2.

5.1.2 Uploading Time Along with Encryption

One of the performance parameter is calculated by using the amount of time required for uploading an encrypted file. In this scenario, uploading time along with encryption is calculated for each file which is uploaded including encryption and it can be measured with seconds. The result of uploading time for AWS and proposed UIPAC scheme along with encryption is shown in Table 3 and Fig. 3

It is clear that the proposed UIPAC model algorithm encryption time is little bit more than AWS; it is negligible one compared to the security we attained with

Table 3 Comparison result of uploading time for AWS and proposed UIPAC

File size	Proposed UIPAC (Time in seconds)	AWS (Time in seconds)
50 Kb	1.005913496	0.90066
100 Kb	1.008733	0.910722
150 Kb	1.01053	0.9322
200 Kb	1.10009	0.9642

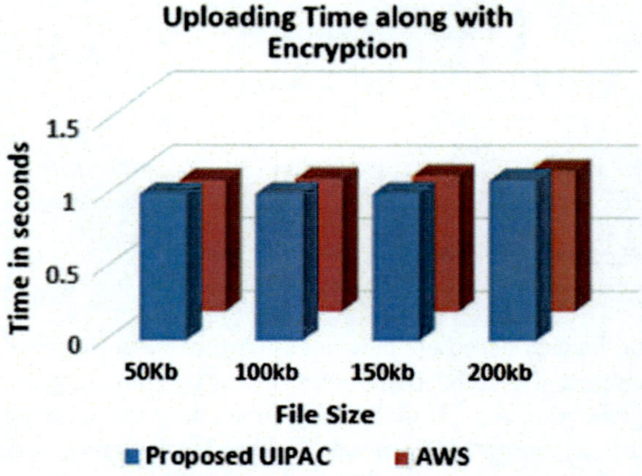

Fig. 3 Comparison result of uploading time for AWS and proposed UIPAC

double layer encryption algorithm which makes significant contribution to improving the highest security with double layer IBEM and HECC. In this way, the proposed model allows two layers of encryption algorithm (multi-model IBEM and HECC) for original input file with different sectors based on dynamic permutations. This dynamic permutation scheme allows to improve weak security model into more strong security model with encryption scheme but AWS uses AES 256 with single layer encryption.

5.1.3 Downloading Time Along with Decryption

One of the performance parameter is calculated by using the amount of time required for downloading a decrypted file. In this scenario, decrypted time is calculated for each file which is downloaded. This downloading time can be measured with seconds. The comparison result of downloading time for AWS and proposed UIPAC scheme along with decryption is shown in Table 4 (Fig. 4).

It is clearly shows that the proposed UIPAC model decryption time is little bit more than AWS. Even though while performing decryption in proposed UIPAC scheme, it provides double privilege authentication $(\varphi_k; \delta_k)$ Ωk, Pb_k. Here, the decryption key is not same for all the user and all files also.

Table 4 Comparison result of downloading time for AWS and proposed UIPAC

File size	Proposed UIPAC model (Time in seconds)	AWS (Time in seconds)
50 Kb	1.00323	0.90021
100 Kb	1.008211	0.9083
150 Kb	1.0107443	0.9297
200 Kb	1.10014	0.9639

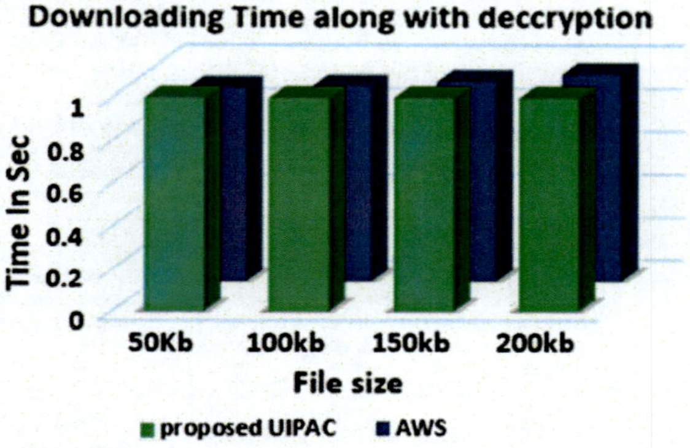

Fig. 4 Comparison result of downloading for AWS and proposed UIPAC

5.2 Comparative Analysis of AWS and UIPAC Scheme

The strength and properties comparisons between existing AWS scheme over UIPAC model have been analyzed. According to the study, current work has focused on the security study and provides a model that includes double layer security for the processing of encrypted data. However, the proposed UIPAC scheme has following strengths:

AWS	*Proposed UIPAC*
✓ Permissions can be directly assigned to users or groups.	✓ Permissions can be granted to users or groups depending on their request and this permission key is shared only applicable user.
✓ With the increase in IT infrastructure outsourcing to IaaS, the number of users and permissions can be quite dynamic and huge in scale. The cost of handling the mapping between users and permissions can be extremely high	✓ With the increase in IT infrastructure outsourcing to IaaS, the number of users and permissions can be quite dynamic and small in scale. The cost of handling the mapping between users and permissions can be extremely low.
✓ It assist groups to organize users and allow users to explicitly get the permissions assigned to the groups they belong to.	✓ Even though if the user is in the corresponding group and allow them to get the permission granted by the Data owner by implicit manner.
✓ Session management is absent from AWS, so all user permissions are in effective all the time, which clashes with the principle of least privilege.	✓ All user permissions are not in effective all the time, which is applicable with principle of least privilege.
✓ It arranges and deals with numerous entrance control strategy models for same role cloud clients' security necessities, and increments enforceable security approaches at public cloud providers.	✓ It arranges and deals with numerous entrance control strategy models for variation cloud clients' security necessities, and increments enforceable security approaches at public cloud providers.
✓ AWS offers a form of policy-based access control based on policy files attached to entities such as users, groups, roles and certificates. For all these entities, there are numerous policies defined. AWS may experience a problem with policy -explosion problem.	✓ UIPAC offers a form of policy-based access control based on requisition of user who are applicable. It avoids policy -explosion problem.
✓ It is secured with an access key with list of policies.	✓ It is well secured with Master key as well as privilege access key with two tier protection.
✓ It does not support customer based policy management tool.	✓ It supports customer-based policy management tool.
✓ The policy key is assigned for each role associated with low priority clients and high priority clients. It increases the operational and computation cost to a certain degree	✓ The policy key is granted based on low priority clients and high priority clients depending on data owner wish. It reduces the operational and computation cost to a certain degree

6 Conclusion

AWS was a challenge to incorporate all aspects and abilities of cloud storage into our UIPAC model. We focused in in fundamentally on access control and authorizations on a genuine cloud-empowered AWS platform. We accept that our model would fill in as an underlying arrangement to build up an overall access control model for cloud-viable system that can be dynamically improved to coordinate new advantage access control abilities. We additionally proposed a few upgrades to the UIPAC model dependent on our involvement with twofold layer advantage strategy. The proposed UIPAC model takes out the shortcomings of AWS and satisfies the necessities, especially in the quantity of clients and consents can be very unique and little in scale. The cost of taking care of the planning among clients and authorizations can be incredibly low and disseminated cloud computing situations. It gives most noteworthy security and improves the exhibition of the current advantage access control model in AWS cloud by joining worked on need-based advantage validation and approval framework. Thus, UIPAC classifies client privilege that reduces inefficiency, ineffectuality, and supports variant enforcement of policies to individuals. We play out the acknowledgment of the UIPAC model with the assessment showed how the proposed model successfully deals with a powerful customer. Our proposed model zeroed in on joining consideration of setting-based advantage access, i.e., need prior to giving advantage admittance to the administrations. Here, something more we need to presume that the advantage access strategy is allowed for record is not the equivalent for each role, contingent upon the master key premise, these can be changed and conceded. So we got most elevated security level with various arrangement advantage with various document.

References

1. Y. Zhu, G.-J. Ahn, Hu. Hongxin, Di. Ma, S. Wang, Role-based cryptosystem: A new cryptographic RBAC system based on role-key hierarchy. IEEE Trans. Inf. Forensics Secur. **8**(12), 2138–2153 (2013)
2. S. Muthurajkumar, M. Vijayalakshmi, A. Kannan, Intelligent temporal role based access control for data storage in cloud database, in *2014 Sixth International Conference on Advanced Computing (ICoAC)* (IEEE, 2014), pp. 184–188
3. L. Zhou, V. Varadharajan, M. Hitchens, Trust enhanced cryptographic role-based access control for secure cloud data storage. IEEE Trans. Inf. Forensics Secur. **10**(11), 2381–2395 (2015)
4. S. Sridhar, S. Smys, A hybrid multilevel authentication scheme for private cloud environment, in *2016 10th International Conference on Intelligent Systems and Control (ISCO)* (IEEE, 2016), pp. 1–5
5. L. Huang, Z. Xiong, G. Wang, A trust-role access control model facing cloud computing, in *2016 35th Chinese Control Conference (CCC)* (IEEE, 2016), pp. 5239–5242
6. M. Su, L. Wang, Fu. Anmin, Yu. Yan, Proxy re-encryption based multi-factor access control for ciphertext in cloud. J. Shanghai Jiaotong Univ. (Sci.) **23**(5), 666–670 (2018)
7. S. Selvi, M. Gobi, Hyper elliptic curve based homomorphic encryption scheme for cloud data security, in *International Conference on Intelligent Data Communication Technologies and Internet of Things* (Springer, Cham, 2018)
8. N. Geetha, M.S. Anbarasi, Role and attribute based access control model for web service composition in cloud environment, in *2017 International Conference on Computational Intelligence in Data Science (ICCIDS)* (IEEE, 2017), pp. 1–4
9. N. Thilakarathne, Improved hierarchical role based access control model for cloud. Int. J. Computer Sci. Netw. **8**(5), 2277–5420 (October 2019). ISSN (Online) (2019)
10. J. Zhao, J. Sun, Research on access control model based on RBAC model in microservice environment, in *Journal of Physics: Conference Series*, vol. 1437, no. 1 (IOP Publishing, 2020), p. 0120312020
11. R. Ghazal, A.K. Malik, N. Q.B. Raza, A.R. Shahid, H. Alquhayz, Intelligent role-based access control model and framework using semantic business roles in multi-domain environments. IEEE Access **8**, 12253–12267 (2020)
12. A. Majumder, S. Roy, S. Biswas, Data security issues and solutions in cloud computing, in *Handbook of Research on Security Considerations in Cloud Computing*, pp. 212–231 (n.d.). https://doi.org/10.4018/978-1-4666-8387-7.ch010

13. Z. Ali, A. Ghani, I. Khan, S.A. Chaudhry, S.K. Hafizul Islam, D. Giri. A robust authentication and access control protocol for securing wireless healthcare sensor networks. J. Inf. Secur. Appl. **52**, 102502 (2020)
14. R. El Sibai, N. Gemayel, J.B. Abdo, J. Demerjian, A survey on access control mechanisms for cloud computing. Trans. Emerg. Telecommun. Technol. **31**(2), e3720 (2020)

Securing Big Data in Hadoop Using Hybrid Encryption

Aswathi Sunder, Neetha Shabu, and T. Remya Nair

Abstract As the amount of data is increasing rapidly, its storage and security issues also increase. Hadoop is currently a popular framework for big data storage and processing. Since, security is not a primary consideration in Hadoop when it was initially developed makes the data stored in Hadoop distributed file system (HDFS) vulnerable to attacks. For protecting these data, numerous encryption techniques has been developed. This paper analyses and presents a comparison on advanced encryption standard (AES), a symmetric cipher and Rivest-Shamir-Adleman (RSA), an asymmetric cipher cryptographic algorithms. Finally, this paper proposes a hybrid method that combines both RSA and AES technique to improve the performance of encryption-decryption on HDFS files and thereby increases the security of big data in Hadoop framework. The proposed technique is found to be feasible and is able to enhance the confidentiality of HDFS files stored in Hadoop framework and is also able to overcome the limitations when using the AES encryption technique alone.

1 Introduction

Today, in this digital world the amount of data is increasing rapidly. The information transferred through digital applications are increasing day-by-day which requires extra storage space and processing assets. Every business organizations in order to become successful takes the aid of emerging technologies. A common model for handling such large data is MapReduce [1]. For processing big data set, Apache Hadoop [2] provides MapReduce model. Handling large datasets and the primary data storage system of Hadoop framework are Hadoop distributed file system (HDFS) [3]. Most of the organization now uses Hadoop to store vulnerable data hence a need for highly secure mechanism is required to protect these data.

A. Sunder (✉) · N. Shabu · T. Remya Nair
Department of Computer Science and IT, Amrita School of Arts and Sciences, Amrita Vishwa Vidyapeetham, Kochi, India

T. Remya Nair
e-mail: remyat@asas.kh.amrita.edu

P. Karuppusamy et al. (eds.), *Ubiquitous Intelligent Systems*, Smart Innovation, Systems and Technologies 243, https://doi.org/10.1007/978-981-16-3675-2_39

When Hadoop was initially established, security was not a consideration. The initial usage of Hadoop involves managing large amount of public web data, hence confidentiality was not an issue. Basic security was only implemented which are not much effective. Then, Hadoop cluster changed on to much more restricted networks where data stored in the cluster can be accessed by any users with equal rights [4, 5]. But giving equal access rights to all users is dangerous since it provides threat of confidentiality to the data [4, 5]. Researchers had experimented many approaches for providing security in HDFS files stored in Hadoop. Various encryption-decryption techniques have been developed and implemented in Hadoop clusters to enhance the data security. Certain techniques had showed some improvements in its data security feature. But there is no such technique exists that guarantee 100% data security, efficiency and confidentiality.

This paper deals with data security issues in Hadoop. The paper analyses the performance of AES (Advanced Encryption Standard) algorithm and RSA (Rivest-Shamir-Adleman) algorithm, compare its performance, analyse its merits and demerits and finally proposes a hybrid method that combines both AES and RSA cryptography algorithms to enhance the data security in Hadoop framework.

2 Literature Review

In 2008, Hadoop was developed with the aim of handling only a large amount of data limited to a particular situation. Security concerns were therefore not the top preference, hence there were no security encryption in Hadoop. Hadoop uses the name of the user for any data storage. There exist an absence of secured model between Hadoop and HDFS. Optimum security of such data has gained utmost importance and ensures that it remains secure from any invasion or tampering.

The initial versions of Hadoop lacks security features which affected several sectors like business fields, medical and healthcare department, national security and many more. Hence, it becomes very necessary that there should exist a mechanism that ensures Hadoop security. So security of HDFS files stored in Hadoop is the major leap that Hadoop Framework needs to take.

The four primary security concerns of Hadoop lies in the following domains:

1. Authentication: Identifying and validating the user as whom they claim to be is involved in the process of authentication. The most common ways of authentication are passwords, OTP and Biometrics [6].
2. Authorization: Giving the users certain privileges or permission to access specific resources or functionalities is defined as authorization [6].
3. Auditing: The process of tracking what an authenticated or authorized user did after they gets the access to specific resources is called auditing. It records all the activities of the user from the time they login to the cluster.
4. Data Protection: It is the process of preventing confidential data being accessed by unauthorized users through the technique like encryption.

As Hadoop becomes a common distributed programming platform with its distributed file system (HDFS) to process large data, demands for secure computing and file storage are growing rapidly. The current Hadoop does not, however, allow the encryption of HDFS block storage, which is a simple solution for stable Hadoop. Therefore, by adding encryption and decryption, this paper proposes a safe Hadoop architecture [7].

This work proposes a token-based authentication scheme to prevent security threats such as compromise and impersonation attacks from HDFS stored data. This work adds another security layer by performing HDFS client authentication on data node also hash chain of key approach is used [8].

In this work, in order to give maximum security to the encrypted data use of advanced encryption standard with message digest (MD5) and data encryption and the digital signature algorithm (DSA) was used to authenticate the data. The data from the user is encrypted and stored for further use, and this encrypted data is shared between different users [9].

This research work chose cipher block chaining to handle HDFS blocks with the electronic code book (ECB) AES algorithm mode, and the OTP algorithm for state cipher. For decrypting the data private key, residing with the user is required. Multiple encoding can be performed by using random key that has been developed by server, which is divided into dual key in order to pass a record to file [10].

In order to increase the security of big data, the proposed research work uses an ARIA encryption scheme, which is the encryption technique developed by South Korean researchers. The result of implementing this encryption technique shows that it offers a good level of data security compared with the advanced encryption standard (AES) algorithm [11].

2.1 AES Algorithm

Advanced encryption standard (AES), a substitution-permutation cipher, is used majorly for the encryption of electronic data thereby aiming to increase its protection. The steps of AES algorithm that converts plain text into cipher text as follows [12]:

(a) It includes a 128-bit plain text data block as the length of the input and keys can be 128, 192 or 256-bit and produce a 128-bit block output.
(b) The 128-bit plain text undergoes an initial round in which each byte of the state is combined with the round key using bitwise XOR.
(c) After the initial round, plain text goes through 10 rounds if the key length is 128-bit, 12 rounds if the key length is 192-bit, or 14 rounds if the key length is 256-bit. For each round, there are the following steps:

    (i) Sub-Bytes transformation
    (ii) Shift-Rows transformation
    (iii) Mix Column transformation

(iv) Add Round key.

2.2 RSA Algorithm

Rivest-Shamir-Adleman, one of the most commonly used asymmetric cryptographic methods is majorly used to secure confidential data in an insecure network. The principle of RSA algorithm is that "It is easy to multiply prime numbers, but hard to factor them" and hence it uses prime numbers to produce private key and public key [13].The steps of RSA algorithm are as follows [13]:

(a) Choose two large prime numbers P and Q such that P not equal to Q.
(b) Calculate N as: $N = P * Q$.
(c) Calculate S as: $S = (P - 1)\ (Q - 1)$.
(d) Select a public key e in such a way that e is not the S variable.
(e) Now select the private key d such that $(d * e) \bmod S = 1$
(f) Compute cipher text (C): $C = \text{Me} \bmod N$.
(g) Compute plain text (M): $M = \text{Cd} \bmod N$.

3 RSA Versus AES Analysis

The performance analysis of both RSA and AES algorithm is specified in Table 1. From this comparison table, it can be concluded that the AES algorithm performs better than the RSA algorithm. It also states that the symmetric cipher performs better than the asymmetric cipher. But experiments also shows that each algorithm has its own strength.

Considering the performance aspects like encryption speed, decryption speed, power consumption and throughput the AES cipher performs better than the RSA cipher. This symmetric cryptographic technique is faster as the key used here are much smaller than the key used in asymmetric cryptography. Additionally, the fact that a single key is used for encryption-decryption in AES also makes the entire

Table 1 RSA versus AES comparison

Parameter	RSA	AES
Encryption speed	Slower	Faster
Throughput	Low	High
Power consumption	High	Low
Security attacks	Low security	Highly secured
Scheme	Prime number factoring	Substitution-permutation
Type of encryption	Asymmetric	Symmetric

process faster. The slower speed of RSA cipher result in slow processes, issues with memory capacity and faster degradation of batteries.

Now considering the security aspects, the use of AES cipher is risky as the same key used to encrypt the data should be shared with anyone who wants to decrypt the data. RSA offers better security as it uses two different keys— a public key that is used to encrypt the messages and a private key to decrypt the messages that will not be shared. Also, AES has 128-, 192- or 256-bits key length only, whereas RSA has over 1024 bits. The encryption-decryption process of RSA is more complicated than DES. Therefore, comparing to AES it is more difficult to decode data that implements RSA algorithm which increase its security. So, in terms of security RSA is better, but it lacks in terms of performance.

4 Proposed System

The proposed system is to improve the data security in Hadoop by combining the strength of AES and RSA algorithms. The analysis of both RSA and AES algorithm shows that AES is better in terms of performance whereas RSA is better in terms of security.

The RSA technique is not able to implement Big Data. Since, RSA encrypts messages of limited size, a maximum of 245 bytes it cannot be used individually to provide security to large datasets in Hadoop framework. On the other hand, AES is able to encrypt large HDFS files stored in Hadoop. But it is not sufficient to provide security to HDFS files since AES is a symmetric encryption technique which uses a shared secret key for both encryption-decryption process, so key exchange is a major problem.

To overcome the limitation of AES technique, we propose a new hybrid method. The proposed encryption scheme combines the HDFS file encryption using advanced encryption standard (AES) algorithm and then it performs the AES key encryption using Rivest Shamir Adleman (RSA) algorithm. To strengthen the data protection, Hadoop distributed file system is encrypted using combination of AES and RSA algorithm. The key used to encrypt a HDFS file is then encrypted using RSA algorithm using user's public key (Fig. 1).

4.1 Proposed Algorithm

The encryption-decryption process for the proposed hybrid method is as follows:

HDFS file encryption process:

1. When a user upload a file it is stored in file buffer. Client request key management system to generate key using AES technique to encrypt data/file.
2. Key management system then generates AES key.

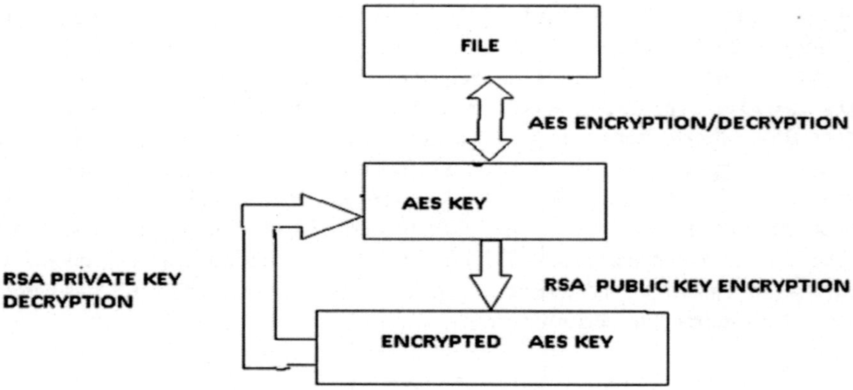

Fig. 1 Encryption—decryption process of proposed hybrid method

3. Find the user's public key from the database based on user ID and use it to encrypt (RSA encryption) AES key. If the user is new then the keys (public key and private key) are generated by the system and store it in database and also provide to the user.
4. Now the encrypted AES key is used to encrypt the user uploaded file stored in buffer.
5. The encrypted file is uploaded to HDFS.

HDFS file decryption process:

1. When a user needs to download a file, it provides its private key (RSA) and file information to find the file and its corresponding AES key.
2. Get the file from HDFS to the server.
3. Based on file information provided the server finds the encrypted AES key.
4. The encrypted AES key is decrypted using user's private key.
5. Use the decrypted AES key to decrypt the file.
6. Return the decrypted file to the requested user.

5 Results

5.1 Advantages of Hybrid Encryption Algorithm

- In AES, key exchange is a major problem because a same shared key is used for both encryption-decryption. However, using combination of AES and RSA algorithm can solve this since RSA asymmetric technique can solve key exchange problem.

- In RSA, memory usage and time consumption for encryption-decryption of files are a major problem, which can be solved using proposed method.
- Since RSA is used only for data key encryption the overall HDFS file encryption-decryption speed of proposed algorithm is almost same as that of AES which is faster than using RSA alone.
- In AES based encryption, the AES key needs to be sent to the client before file decryption. But in the proposed algorithm the key need not to be sent beforehand as it has been sent in the encrypted form along with the data.

5.2 *Security Analysis of Hybrid Encryption Algorithm*

Hadoop HDFS file protection using combination of AES and RSA algorithm is based on the security and application of AES and RSA algorithms. The data remains safe in AES algorithm due to the strength of all key length. In this proposed system, the data in Hadoop remains safe until the encryption key remains secret. It is difficult for the attacker to understand which part of message contain AES encryption key and cipher text which makes the algorithm better. Private key of receiver is required to decode the advanced encrypted standard key making the data in HDFS safe.

5.3 *Encryption and Decryption Time Analysis*

Figures 2 and 3 show the comparison of encryption-decryption time of the proposed hybrid method along with the RSA and AES standalone algorithm [14]. Experiment shows that RSA could not perform well over files with larger size. Figure 4 and 5 compare the behavior of proposed hybrid algorithm with existing AES algorithm on Hadoop [15] (Tables 2 and 3).

Fig. 2 Encryption time in sec (*y*-axis) and file size in kb

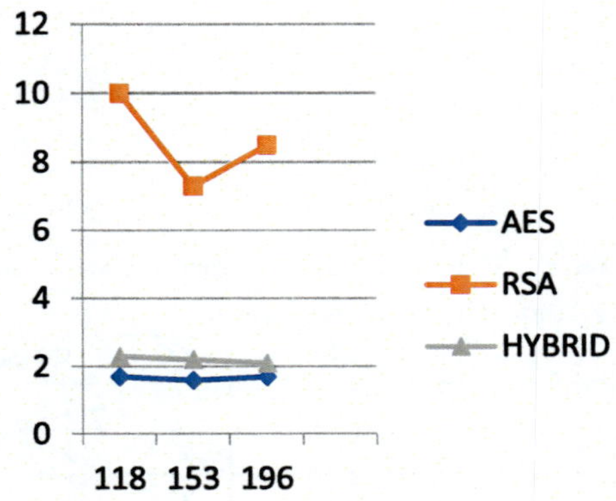

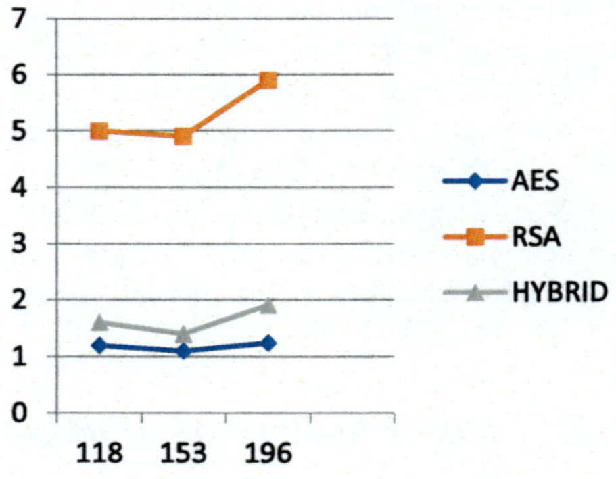

Fig. 3 Decryption time in sec (*y*-axis) and file size in kb

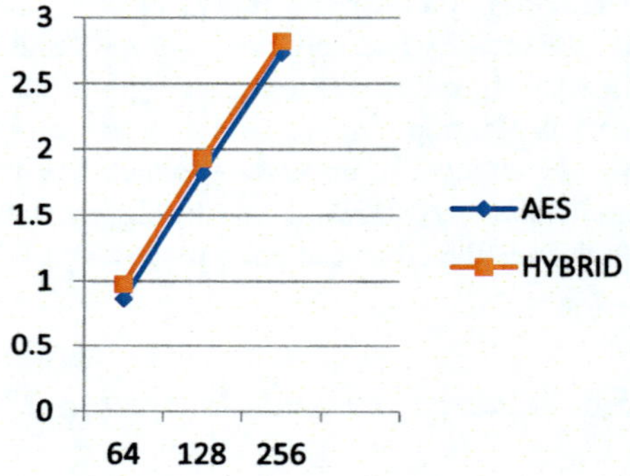

Fig. 4 Encryption time in min (*y*-axis) and file size in mb

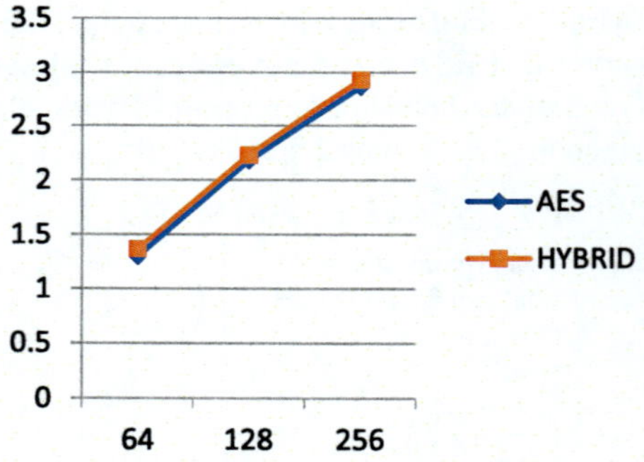

Fig. 5 Decryption time in min (*y*-axis) and file size in mb

Table 2 Encryption-decryption time analysis in sec

File size (kb)	Encryption (s)			Decryption (s)		
	AES	RSA	Hybrid	AES	RSA	Hybrid
118	1.7	10	2.3	1.2	5	1.6
153	1.6	7.3	2.2	1.1	4.9	1.4
196	1.7	8.5	2.1	1.24	5.9	1.9

Table 3 Encryption-decryption time analysis in min

File size (mb)	Encryption (min)		Decryption (min)	
	AES	Hybrid	AES	Hybrid
64	0.87	0.98	1.3	1.37
128	1.82	1.93	2.8	2.23
256	2.73	2.82	2.86	2.93

All these observation concludes that there is only a small variation in encryption-decryption time of the proposed method comparing with the existing AES method implemented in Hadoop storage. It is due to the use of asymmetric RSA algorithm which is time consuming, but it enhances the data security as well as manages the key sharing problem faced by the AES algorithm well.

6 Conclusion

This paper compares AES and RSA cryptographic techniques. It is found that both technique has its own strength and limits. But in terms of encryption, throughput and power consumption AES algorithm are better than RSA. AES is better in terms of performance and efficiency whereas RSA is better in terms of security. The proposed hybrid encryption approach (combination of AES and RSA) mainly concentrated on protection of HDFS files by implementing the proposed algorithm in Hadoop environment. The proposed technique is implemented in the Hadoop framework, analysed its encryption-decryption time and compared it with the existing encryption techniques. Experiments show that the proposed hybrid encryption is feasible and can improve the confidentiality of HDFS files.

The limitation of this proposed hybrid method is, it slightly consumes more encryption-decryption time compared to the existing AES encryption technique. This is because of the implementation of RSA algorithm to enhance the data security and overcome the limitation of AES algorithm. Future works can be performed to decrease the time consumed for encryption-decryption by the proposed hybrid encryption method and thereby increases its efficiency and performance without compromising the data security.

References

1. S. Jin, A. Yang, H. Yin, Design of a trusted file system based on hadoop, in *Trustworthy Computing and Services* (Springer, Berlin, 2013), pp. 673–680
2. Apache™ Hadoop®! Available: http://Hadoop.apache.org/
3. H. Zhou, Q. Wen, Data security accessing for HDFS based on attribute-group in cloud computing,in *International Conference on Logistics Engineering, Management and Computer*

Science (2014)
4. N. Somu, A. Gangaa, V.S. Sriram, Authentication service in Hadoop using one time pad.Indian J. Sci. Technol. **7**, 56–62 (2014)
5. E.B. Fernandez, Security in data intensive computing systems, in *Handbook of Data Intensive Computing* (Springer, Berlin, 2011), pp. 447–466
6. https://www.okta.com/identity-101/authentication-vs-authorization
7. S. Park, Y. Lee, *Secure Hadoop with Encrypted HDFS* (2013)
8. Y.S. Jeong, Y.T. Kim, A token-based authentication security scheme for Hadoop distributed file system using elliptic curve cryptography. J. Comput. Virol. Hack. Tech. **11**, 137–142 (2015)
9. M.I. Maheswari, S. Revathy, R. Tamilarasai, Secure data transmission for multi sharing in big data storage. Indian J. Sci. Technol. **9**(21) (2016)
10. H. Mahmoud, A. Hegazy, M.H. Khafagy, An approach for big data security based on Hadoop distributed file system, in *International Conference on Innovative Trends in Computer Engineering (ITCE)* (2018)
11. Y. Song, Y.S. Shin, M. Jang, J.W. Chang, Design and implementation of HDFS data encryption scheme using ARIA algorithm on Hadoop, in *International Conference on Big Data and Smart Computing (BIGCOMP)* (IEEE, 2017), pp. 2375–9356
12. W. Stallings, *Cryptography and Network Security* (Prentice Hall, 1995).
13. S. Shaili, S. Niraj, A comparative analysis of AES and RSA algorithms. Int. J. Sci. Eng. Res. **7** (2016)
14. B. Padmavathi, S. RanjithaKumari, A survey on performance analysis of DES, AES and RSA algorithm along with LSB substitution technique. Int. J. Sci. Res. **2** (2013)
15. S.R. Varsha, S. Shailaja, A novel technique for secure data transmission using cryptography and steganography. Int. J. Innov. Sci. Res. Technol. **2** (2017)

Handwriting Analysis Using Deep Learning Approach for the Detection of Personality Traits

Gayathry H. Nair, V. Rekha, and M. Soumya Krishnan

Abstract Handwriting is one of many that makes a person distinctive in their identity, and it is the method of identifying the writer's physical characteristics. The computerization is becoming more advanced in different fields and thus, handwriting analysis is gaining more importance these days. For recognizing handwriting, deep learning has been commonly used. In this paper, we have contrived Resnet50 comparison with CNN to predict personality traits. Hence, handwriting is one of the easiest ways to collect information on physical characteristics, such as mental health, emotional state, physical strength and many more.

1 Introduction

Handwriting reflects a person's mental state, and handwriting analysis is a method that is useful in areas of interpersonal skills, cogitation, fulfillments or work routine [1], and anyone can reliably and effectively predict the personality of a person by using handwriting analysis. Every individual is distinctive and that can be revealed in different ways. A method is proposed in this paper to analyze the person's personality traits from features derived from his handwriting. The characteristics taken into consideration are slant of word (slant ascending, slant descending, Balanced), margins (Left margin, Right margin) which are used to predict the human behavior. In the existing papers the prediction is carried out by using CNN algorithm and in proposed system, a comparison between two algorithms CNN and Resnet50 to find out which algorithm is best in predicting personality traits.

G. H. Nair (✉) · V. Rekha · M. Soumya Krishnan
Department of Computer Science and IT, Amrita School of Arts and Sciences, Amrita Vishwa Vidyapeetham, Kochi, India

P. Karuppusamy et al. (eds.), *Ubiquitous Intelligent Systems*, Smart Innovation, Systems and Technologies 243, https://doi.org/10.1007/978-981-16-3675-2_40

2 Literature Survey

Graphology can be expounded as an analysis of physical traits and patterns of an individual's handwriting to recognize conscious and unconscious mind at the time of writing. A particular personality trait is revealed with any written movements [2]. It works on the principle that our hand is controlled by our subconscious mind while writing [3].

As of now, many papers were published based on handwriting. The paper [4] explicates by Prachi Joshi and 4 others focused on the idea of machine learning with SVM method to predict personality characteristics, using template method, thresholding algorithm, and it also examines the personality characteristics revealed by the penmanship of a person by pattern (baseline), edge (margin), inclination of the words and height of *t* [5]. It emphasizes the concept of analyzing the numbers 0–9 and alphabets, such as s, u, d, o, y, t, p, b, from an individual's handwriting. It examines a person's personality from the baseline, pen pressure and strokes by dividing the works into 2 parts as Handwriting Recognition and Character Prediction for predicting the person's handwriting, and they were able to attain a reasonable accuracy [6] which has implicitly aimed on the method of handwriting analysis by considering nine features, such as pressure, inclination of words, space between words, pattern, break, size, edge, circle of 'e' and distance between the title(dot) and furthermore the stem of ' I.' The estimation of accuracy ranged from 90 to 95%.

In Guan et al. [7], the conclusions are determined based on regular activities, that can be compared using handwriting on the time series data can be converted into a profound network model by performing movement division utilizing a sliding window calculation and highlights brought into SoftMax classifier for expectation of 6 practices, for example, strolling, sitting, going higher up, going down, standing and resting.

Punetha et al.'s paper [8] is designed to develop techniques for predicting human mood and behavior, by following the procedure of Improved Susan Method and Principal Component Analysis (PCA) using signature analysis. In [9], authors focused on recognition of handwritten text with TensorFlow using Deep Learning to build a neural network using CNN, RNN, CTC layer steps with an accuracy of 90.3% [1]. A strategy has been proposed to foresee the doings of an individual from alphabets, such as I, f, a, d, e, g, and these boundaries are contribution to ANN which predicts the mannerisms of people. Paper [10] refers to an android application using deep learning for handwriting recognition. Paper [3] introduces a new method for the prediction of handwritten alphabets using neural networks. The method used for this purpose is diagonal-based feature extraction by extracting features from handwritten samples. The system is very useful in identifying handwritten names and also in converting written text into structural text. In [11] paper, the author has put forward a technique to predict whether the dragon fruit is fully grown and also to find out its time to harvest using RESNET152. The model was trained by collecting pictures of dragon fruit during different stages of development and testing is done with 100 new data using area of convergence and confusion matrix [12]. The paper reflects off-line handwritten signature regarding profound networks and it is a verification framework for disconnected signature acknowledgment put forward a two-venture hybrid classifier framework with better execution on GPDS information base, which had the option to extricate a high-portrayal of the signature pictures through multi-layers in a profound progressive construction will permits non-local generalization and fathom ability in this particular space. For extracting features, they used k-means, Histogram Frequencies, Discrete Cosine Transform Frequencies (DCT) Discrete Wavelet Transform Frequencies (DWT) and they conclude using extracting high-level representations, and future work can include Graphics Processing Units (GPU) [13]. The paper conveyed by utilizing Neural Network and Tensor Flow to obtain Character Recognition of handwriting. The disconnected manually written alphabets acknowledgment will definitely finished utilizing Convolutional neural organization and TensorFlow. The Technique SoftMax Regression is utilized in relegating the possibilities to transcribed alphabets being of the few alphabets that gives the qualities somewhere in the range of 0 and 1 summarizing to 1 and its presumed that is included extraction scheme like slanting and direction strategies were routed better in producing high precision output contrasted with large numbers of the conventional vertical and horizontal strategies.

3 Proposed System

The primary objective of the analysis is to obtain the characteristics of the person with the aid of their handwriting. A comparison is made between CNN and Resnet50 algorithms to decide which algorithm is better for predicting human behavior. The dataset is developed by collecting samples of handwriting from individuals in various age groups. Dataset is randomly divided into two groups in that 70% is for the training part and 30% is for the testing part.

3.1 Algorithms Used for classification

CNN

The most common neural network model that is used for the problem of image classification is Convolutional Neural Networks (CNNs) and it is often called as ConvNet. It is designed to imitate human visual processing and is highly optimized for the processing of 2D images [9]. CNN has the potential to generalize effectively than the networks with completely linked layers [14]. CNN is widely used in many research areas, such as face recognition, signature-based behavior prediction and many more. Feature extraction is the role of convolutional layer. In the convolution process, input data is processed for convolution operation by multiple trainable convolution kernels after input data then a convolution layer is acquired [7]. General architecture of CNN is depicted in Fig. 1.

Resnet50

Residual neural network (ResNet) is an artificial neural network. It has a convolutional neural network that is 50 layers deep. Network depth is important for neural networks, but it is harder to train deeper networks. By going deeper and deeper it will improve efficiency in various tasks. ResNet50 has a reliable initialization for image recognition and decreases training time [15]. To leap over some layers, it uses skip connections or shortcuts. ResNet uses Batch Normalization at its core, to improve the network efficiency. The Batch Normalization changes the issue of covariate is

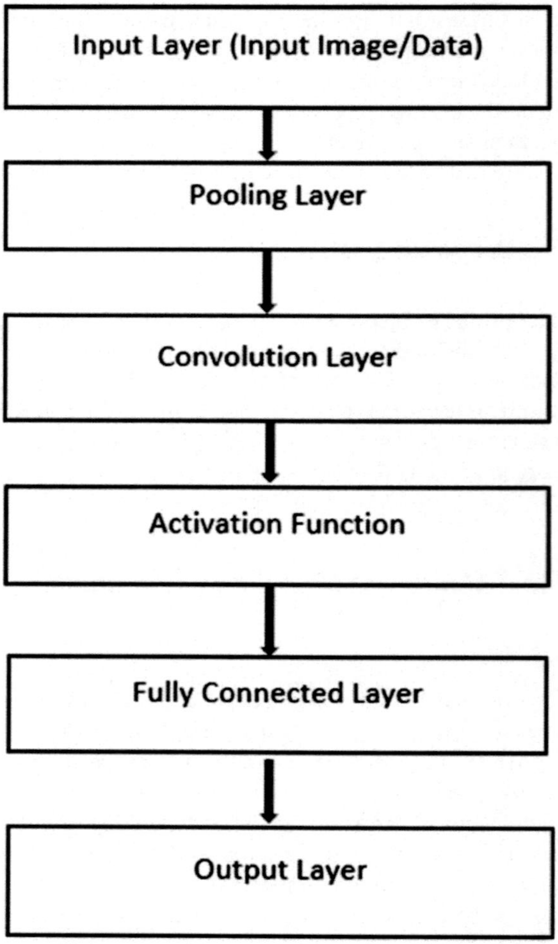

Fig. 1 General architecture of CNN [14]

mitigated. The Identity connection is used by ResNet to protect the network from vanishing gradient problem.

3.2 *Set of Characteristics*

This paper would be focusing on the following three parameters: Baseline, Letter slant, and Width of Margins. These are the most common parameters that help in identifying the personality traits of an individual through handwriting analysis and their corresponding traits are explained in Table 1.

CNN (Proposed Framework)

The images are scanned and are loaded into dataset folder. On Image pre-processing, peruse the image documents (stored in information envelope (data folder)). Decode the JPEG substance to RGB frameworks of pixels with channels. Convert these into floating-point tensors for contribution toward neural nets. Resize the pixel esteems (somewhere in the range of 0 and 255) to the [0,1], and then applying threshold and followed by erosion and dilation it will return the image. By the next step it is defining [CNN algorithm], and on training part 70% of dataset is used for training and we achieved an accuracy of 99.80% and thus saving the best model, testing part is done with 30% dataset and finally personality traits are predicted. Proposed framework is as given in Fig. 2.

Resnet50 (Proposed Framework)

The proposed approach uses ResNet50 which has 5 phases each with convolution and identity block. We are using a model trained on another problem as a first step rather than constructing from scratch. Transfer learning, it is the use of pre-trained images and one advantage is that training time can be reduced. The training phase

Table 1 Attributes and their corresponding traits

Attribute	Type	Personality traits
Slant	Left	A person who is afraid of future, difficulty in expressing emotions, self-sufficient, uncaring
	Right	Ultramodern, eloquent, philomath, a person with strong belief, talkative
Baselines	Ascending (rising upward) optimistic	Sanguine (optimistic or positive person)
	Descending (slanting downward)	A person who thinks worst in all situation (Pessimistic)
	Balanced level (straight)	A person who is sensible and do not have emotional problems (balanced)

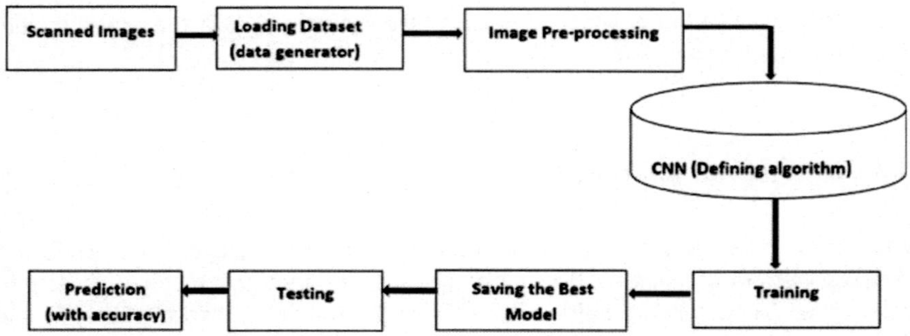

Fig. 2 Proposed Framework of CNN

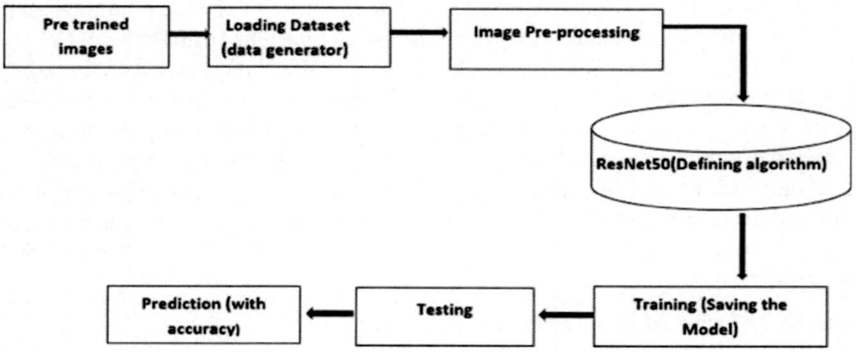

Fig. 3 Proposed framework of ResNet50

is carried out, and after the completion of the testing process the model predicts the personality traits and finally acquired an accuracy of 99.75% (Fig. 3).

3.3 Comparison Result

Accuracy of CNN

We achieved an accuracy of 99.80% by using CNN algorithm, and its validation accuracy is 99.75%, validation loss is 30% (as shown in Fig. 4).

Accuracy of ResNet50

We achieved an accuracy of 99.75% by using Resnet50 algorithm, and its validation accuracy is 75%, validation loss is 64% (as shown in Fig. 5).

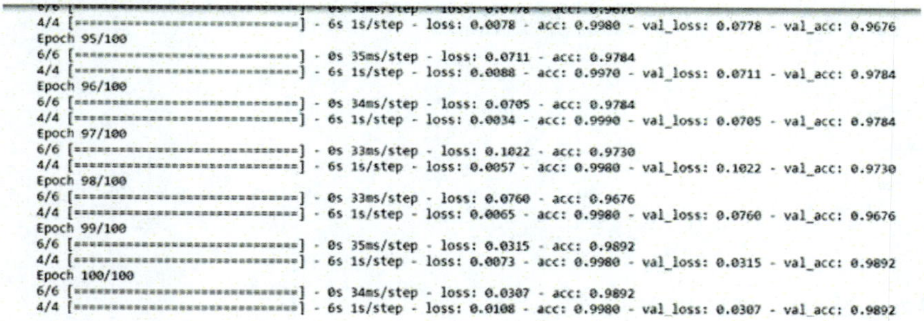

```
4/4 [==============================] - 6s 1s/step - loss: 0.0078 - acc: 0.9980 - val_loss: 0.0778 - val_acc: 0.9676
Epoch 95/100
6/6 [==============================] - 0s 35ms/step - loss: 0.0711 - acc: 0.9784
4/4 [==============================] - 6s 1s/step - loss: 0.0088 - acc: 0.9970 - val_loss: 0.0711 - val_acc: 0.9784
Epoch 96/100
6/6 [==============================] - 0s 34ms/step - loss: 0.0705 - acc: 0.9784
4/4 [==============================] - 6s 1s/step - loss: 0.0034 - acc: 0.9990 - val_loss: 0.0705 - val_acc: 0.9784
Epoch 97/100
6/6 [==============================] - 0s 33ms/step - loss: 0.1022 - acc: 0.9730
4/4 [==============================] - 6s 1s/step - loss: 0.0057 - acc: 0.9980 - val_loss: 0.1022 - val_acc: 0.9730
Epoch 98/100
6/6 [==============================] - 0s 33ms/step - loss: 0.0760 - acc: 0.9676
4/4 [==============================] - 6s 1s/step - loss: 0.0065 - acc: 0.9980 - val_loss: 0.0760 - val_acc: 0.9676
Epoch 99/100
6/6 [==============================] - 0s 35ms/step - loss: 0.0315 - acc: 0.9892
4/4 [==============================] - 6s 1s/step - loss: 0.0073 - acc: 0.9980 - val_loss: 0.0315 - val_acc: 0.9892
Epoch 100/100
6/6 [==============================] - 0s 34ms/step - loss: 0.0307 - acc: 0.9892
4/4 [==============================] - 6s 1s/step - loss: 0.0108 - acc: 0.9980 - val_loss: 0.0307 - val_acc: 0.9892
```

Fig. 4 Accuracy of CNN

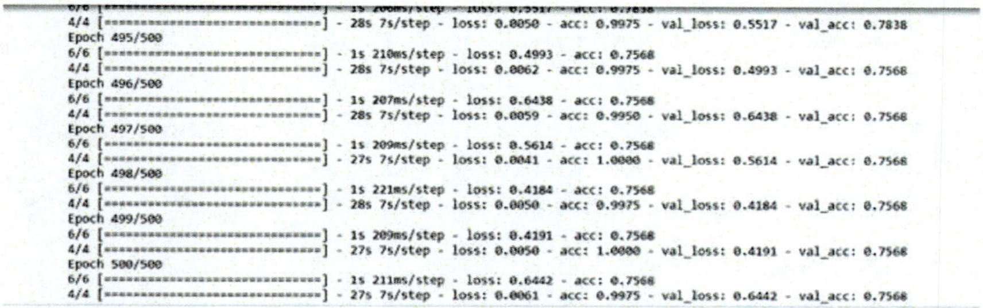

```
4/4 [==============================] - 28s 7s/step - loss: 0.0050 - acc: 0.9975 - val_loss: 0.5517 - val_acc: 0.7838
Epoch 495/500
6/6 [==============================] - 1s 210ms/step - loss: 0.4993 - acc: 0.7568
4/4 [==============================] - 28s 7s/step - loss: 0.0062 - acc: 0.9975 - val_loss: 0.4993 - val_acc: 0.7568
Epoch 496/500
6/6 [==============================] - 1s 207ms/step - loss: 0.6438 - acc: 0.7568
4/4 [==============================] - 28s 7s/step - loss: 0.0059 - acc: 0.9950 - val_loss: 0.6438 - val_acc: 0.7568
Epoch 497/500
6/6 [==============================] - 1s 209ms/step - loss: 0.5614 - acc: 0.7568
4/4 [==============================] - 27s 7s/step - loss: 0.0041 - acc: 1.0000 - val_loss: 0.5614 - val_acc: 0.7568
Epoch 498/500
6/6 [==============================] - 1s 221ms/step - loss: 0.4184 - acc: 0.7568
4/4 [==============================] - 28s 7s/step - loss: 0.0050 - acc: 0.9975 - val_loss: 0.4184 - val_acc: 0.7568
Epoch 499/500
6/6 [==============================] - 1s 209ms/step - loss: 0.4191 - acc: 0.7568
4/4 [==============================] - 27s 7s/step - loss: 0.0050 - acc: 1.0000 - val_loss: 0.4191 - val_acc: 0.7568
Epoch 500/500
6/6 [==============================] - 1s 211ms/step - loss: 0.6442 - acc: 0.7568
4/4 [==============================] - 27s 7s/step - loss: 0.0061 - acc: 0.9975 - val_loss: 0.6442 - val_acc: 0.7568
```

Fig. 5 Accuracy of ResNet50

Accuracy can be improved by increasing the number of steps. The validation loss in CNN is much lesser compared to Resnet50. Based on our analysis CNN performs better in predicting behavior more accurately and efficiently.

4 Performance Evaluation Result and Analyses

CNN (Accuracy and loss are shown in the graph in Fig. 6)

Resnet50 (Accuracy and loss are shown in the graph in Fig. 7)

The result is followed in two phases: Training and Testing, using both CNN as well as ResNet algorithms. The data set was randomly divided into two groups, 70% of which were training sets and 30% of which were test sets. The classification of handwriting is defined in various forms like left margin, right margin, slant ascending, slant descending and balanced. We achieved an accuracy of 99.80% by using CNN algorithm (as given in Fig. 6) and 99.75% accuracy using Resnet50 (as given in Fig. 7). Based on the classified forms matching is displayed on the output with accuracy. Accuracy can be improved by increasing the number of steps. By comparing both

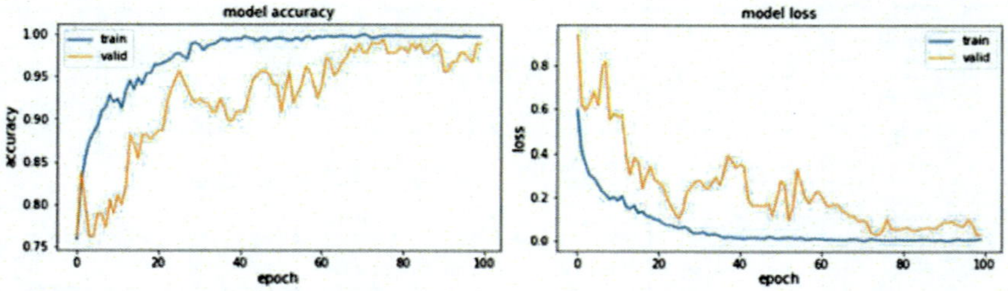

Fig. 6 Model accuracy and loss

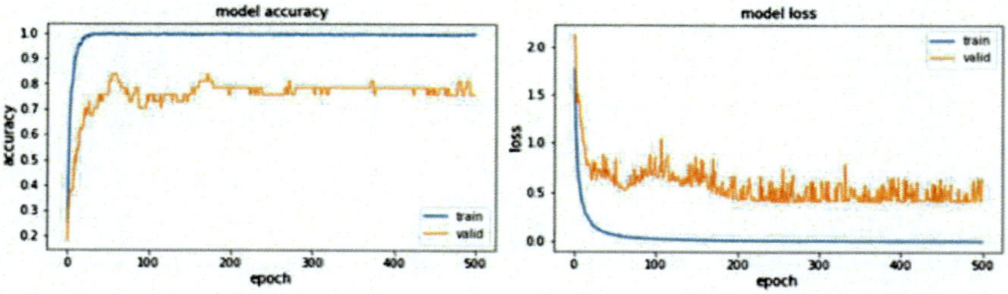

Fig. 7 Model accuracy and loss

the algorithms we concluded that CNN performs better in predicting the personality trait of a person.

5 Conclusion

The paper explores the personality traits, revealed by baseline, margin, slant (words) of a person's handwriting. Software used to test the functionality is anaconda and the latest version of python. Modern Techniques like neural networks are used to implement deep learning algorithm. In this paper we are comparing two algorithm Resnet50 and CNN algorithm to find out which algorithm is more accurate and efficient based on their performance and accuracy, that is done by comparing both algorithms CNN and Resnet50. The accuracy rate of CNN algorithm is 99.80 and 99.75% accuracy using Resnet50 and the validation loss of CNN is much lesser than compared to Resnet50 and from our analysis we concluded that CNN perform better in predicting the personality trait of a person. Future work is feasible and can include additional features and different algorithms.

Acknowledgements We would like to extend our deepest gratitude to all those who have directly or indirectly helped us in completing this paper. We want to thank our guide M. Soumya Krishnan for providing adequate information with constant guidance throughout our research. It helped us a lot to complete our research work successfully.

References

1. P. Joshi, A. Agarwal, A. Dhavale, R. Suryavanshi, S. Kodolikar. Article: handwriting analysis for detection of personality traits using machine learning approach. Int. J. Comput. Appl. **130**(15), 40–45, November 2015. Published by Foundation of Computer Science (FCS), NY, USA.
2. S. Sharma, S. Vohra, P. Mishra, S. Koli, Handwriting recognition and character prediction using neural networks, Int. J. Sci. Res. Deve. **6**(2) (2018)
3. S. Manimala, G. Megasree, P.G. Gokhale & Sindhu Chandrashekhar, Automated handwriting analysis for human behavior prediction. Department of Computer Science and Engineering, Sri Jayachamarajendra College of Engineering, Mysuru, Karnataka, India 2,3,4Student, Department of Computer Science and Engineering, Sri Jayachamarajendra College of Engineering, Mysuru, Karnataka, India (2016)
4. Y. Manchala , J. Kinthali, K. Kotha, K. Santosh Kumar, J. Jayalaxmi, Handwritten text recognition using deep learning with tensorflow. Int. J. Eng. Res. Tech. (IJERT) **09**(05) (2020)
5. Chanchlani, Predicting human behavior through handwriting. Int. J. Res. Appl. Sci. Eng. Technol. 624–628 (2017). https://doi.org/10.22214/ijraset.2017.10093.
6. S.S. Mor, S. Solanki, S. Gupta, S. Dhingra, M. Jain, Handwritten text recognition: with deep learning and android. Int. J. Eng. Adv. (2019)
7. N. Punetha, A.K. Pal, G. Singh, Kushwaha Characteristics and mood prediction of human by signature and facial expression analysis (2019)
8. S.G. Yinong, Z.Z. Tian, Research on human behavior recognition based on deep neural network, in *Proceedings of the 3rd International Conference on Mechatronics Engineering and Information Technology (ICMEIT 2019)* (Atlantis Press, 2019), pp. 777–781 2352–538X Guan2019/04
9. H.N. Champa, K.R. AnandaKumar, Article: artificial neural network for human behavior prediction through handwriting analysis. Int. J. Comput. Appl. **2**(2), 36–41 (2010)
10. S. Indolia, A. Goswami, S.P. Mishra, P. Asopa, Conceptual understanding of convolutional neural network- a deep learning approach. Proc. Comp. Sci. **132** (2018), 679–688
11. Y. Chu, X. Yue, L. Yu, M. Sergei, Z. Wang, Automatic image captioning based on ResNet50 and LSTM with soft attention. Wireless Commun. Mob. Comput. **2020**(7), 2020. , Article ID 8909458
12. D.T. Vijayakumar, M.R. Vinothkanna, Mellowness detection of dragon fruit using deep learning strategy. J. Innovat. Image Proc. **2**(1), 35-43 (2020). https://doi.org/10.36548/jiip.2020.1.004
13. P. Jayabala, E.Srinivasan, S.Himavathi, Diagonal based feature extraction for handwritten alphabets recognition system using neural network. Int. J. Comput. Sci. Inf. Technol. **3** (2011)
14. B. Ribeiro, I. Gonçalves, S. Santos, A. Kovacec, Deep learning networks for off-line handwritten signature recognition, in *Progress in Pattern Recognition, Image Analysis, Computer Vision, and Applications* ed by C. San Martin, S.W. Kim (eds) CIARP 2011. Lecture Notes in Computer Science, vol 7042 (Springer, Berlin, Heidelberg, 2011)
15. M. Agarwal, V.T. Shalika, P. Gupta, Handwritten character recognition using neural network and tensor flow. Int. J. Innovat. Tech. Expl. Eng. (IJITEE) **8**(6S4), 1445–1448 (2019)

Real-Time Emotion Recognition from Facial Expressions Using Convolutional Neural Network with Fer2013 Dataset

V. S. Amal, Sanjay Suresh, and G. Deepa

Abstract Over the most recent couple of years, facial detection and emotion recognition research has been widely considered. In many human–computer interactions, face-communication methods are used, such as facial expressions, eye movement and gestures, which are commonly used among them because they convey the individual emotional states and sentiments. The ability to recognize facial expressions in real time is automatically enabling fictitious applications in human and computer interactivity and different other regions like security, customer satisfaction recognition, etc. This paper provides a brisk measure to coordinate whether the CNN engineering model performs better if it uses only the raw pixels of image data for training or whether it is better to give the CNN some additional details, such as face landmarks and HOG characteristics. The faces are first detected using the LBP classifier; at that point, we extricate the face landmarks utilizing dlib. We additionally extricated the HOG features, and finally, we trained the CNN model with the raw pixel image data with the additional information. In support of this work, we used the FER-2013 Dataset collected from Kaggle. We obtained an accuracy of 75.1 and 59.1% when we trained the model with and without feature extraction, respectively. The results show that the additional information helps CNN to carry out better.

1 Introduction

Till late back, there were quite bigger challenges in computer vision problems but evolution in real-time technologies has trivially upgraded from problems like light, age, hair, etc. [1]. Methods of facial communication such as facial expressions, eye

V. S. Amal (✉) · S. Suresh · G. Deepa
Computer Science Department, Amrita School of Arts and Sciences, Edappally North, Kochi, Kerala 682024, India

G. Deepa
e-mail: deepa@asas.kh.amrita.edu

P. Karuppusamy et al. (eds.), *Ubiquitous Intelligent Systems*, Smart Innovation, Systems and Technologies 243, https://doi.org/10.1007/978-981-16-3675-2_41

movement and movements are used in many human–computer interaction applications, including facial emotions, which are commonly used because they communicate people's emotional states and feelings [2]. Face and emotion detection applications are used to recognize and validate facial characteristics of the individual or emotion, so describing the facial characteristics and their gesture is very important, as these characteristics and expressions uphold the study of the emotion in the human face [3]. With the evolution of artificial intelligent systems as the late technologies, these systems can sense and realize emotion through facial features. The versatility to perform facial expression recognition (FER) automatically through computer vision allows a variety of new applications in fields such as human–computer interaction and data analytics because of the significant part of facial expression in human communication consequently [4], FER has been generally examined in this area and important advancement has been made. A solved problem can now be considered to understand simple expressions under controlled conditions [5]. A fundamental term is many expressions that convey universal emotions, generally anger, displeasure, fear, joy, sorrow and surprise. It is, however, more difficult to accept such terms under naturalistic conditions. This is also due to shifts in the position of head and light, occlusions, and the way unexposed words, as seen in Fig. 1, are always subtle. Within the above-mentioned applications, reliable FER is mandatory under naturalistic conditions, but still an unresolved issue.

Convolutional neural networks (CNNs) can overcome these challenges. As per our paper, we trained CNN without information like face landmarks and HOG features and with additional pieces of information (face landmarks and HOG features). Histogram of oriented gradients (HOG) is a feature descriptor that frequently extracts features from image data. It is commonly used for object detection in computer vision tasks. It is used for the detection of objects in computer vision and image processing. The technique counts gradient orientation instances in localized portions of an image. In the first model, we developed a 3 $\times$ 3 layer network model where 1024 nodes with Softmax classification are available for the connected layers. Then, we use this

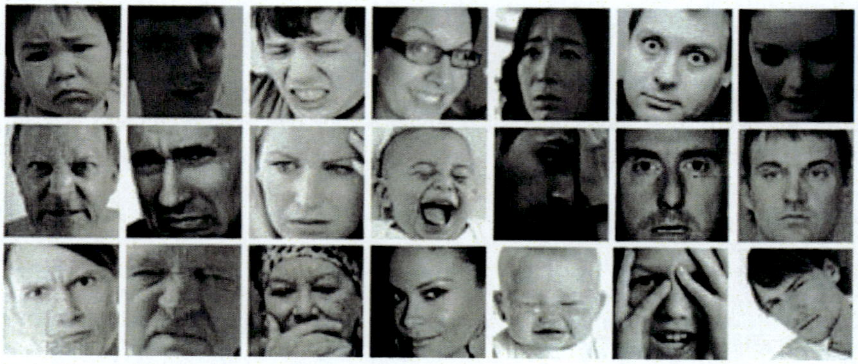

Fig. 1 Eg: of data from the Fer2013 dataset shows a lack of consistency in expression intensity, lighting, age and poses that occur under realistic conditions [4]

network to train the CNN model (we do not give any additional information) and after successful completion of training, we get the model and its weight data. Then, we use this model and weight data for prediction emotion in real time. We created a network model to improve this CNN model training that provides CNN with additional details such as face landmarks and HoG characteristics. By improving current CNN-based FER methods with additional data such as face landmarks and HoG features, this paper aims to reveal insight into this matter. A group of such CNNs has a 75.1% FER2013 test accuracy, surpassing existing CNN-based FER techniques.

2 Literature Review

There are several methods for detecting emotions from the face. The development across the range of impact recognition was evaluated by Sariyanidi et al. [5] to shed light on the essential signs for reading facial expressions and how to encode them. The gradient histogram (HoG) method describes images according to the directions of the edges they include. In Hog, local features are extracted in terms of gradient magnitude and angle by applying gradient operators across the image where the output is encoded. Once the local magnitude-angle histograms are removed from the cells, the dimensionality increases as the blocks overlap and are merged across large entities.

Hussain et al.'s [3] work had required three sequential methods, like detection, recognition and classification. A camera is used to record the face at first and detect the exact location of the face in real time with bounding box coordinates. This phase includes detecting the face using an OpenCV library and Haar cascade algorithm. The CNN model with VGG 16 is used to match the face in the database and to classify the name of the face. The image's properties in the dlib database and other libraries. The primary is then established using the train and test database of the CNN model with database and matching features. Finally, the recognized face is graded as rage, fear, disgust, happiness, neutrality and surprise as supporting the term in real time. Using the VGG16 and KDEF datasets, seven feelings are established and 88 per cent precision is obtained.

Xia et al. [6] used the CK+ dataset. The dataset comprises one hundred twenty-three elements, five hundred ninety-three-picture sequences and three hundred twenty-seven picture sequences and emotional labels. The dataset for countenance recognition may be common. The initial model v3 can be a very complex network, and it will cost you a lot of time to train the model directly, and thus the CK+ dataset data are comparatively limited, the training data are inadequate. To maintain the inception-v3 model, they used transfer learning technology. Here, the last network layer is deleted, and the output nodes number is set to seven, and then the network model is retained. The algorithm of rear propagation forms the last part of the model, and a cross-entropy cost function is also used to monitor the charging parameter by calculating the failure of the sample category mark vector between the softmax layer outputs. The model's classification precision is 97%.

Liliana et al. [7] proposed the extraction of facial features is kind of difficult to be manually crafted by humans since it is an important part of all stages. Using deep learning convolution neural networks to locate the occurrence of action units, automatic feature extractions have been developed. The CK+ dataset was used here as a contribution and thus separates it from other works that used SEMAINE and BP4D. The CK Database is one of the initial databases with extensive data on facial expression and behaviour units and ground reality.

The Haar cascades method was used by Yang et al. [8] to spot if a face exists inside the images or not. When the face is present, eyes and mouth have to be positioned, so that eye and mouth regions can be cropped out. Using the Sobel edge detection process, filter and edge detection are doled out, followed by feature extraction. Train the feature extraction using the neural network approach. Twenty pictures of six neural network emotion classes for training.

Meng et al. [9] used the video frames for spotting the face and positioning within the Dlib toolbox, with a ratio of 25%, they increased the face bounding box and then changed the cropped faces to a scale of 224–224. ResNet18 is already trained to face recall datasets and FER Plus expression datasets on MSCeleb-1M. For training, they set a batch to have 48 instances with K frames in each instance on both CK+ and AFEW 8.0. They initially split the video into K segments for frame sampling during a very video, then randomly choose one frame from each segment. We will set K to three by default.

Bartlett et al. [10], proposed that because of their high dimensionality of the Gabor representation O(105) and does not affect the training flow, first inspected face expression classification based on support vector machines. The machine carried out seven forced choices between distinct emotions. The forced seven-way option was performed in two steps. SVM initially conducted binary decision tasks using one-against-all data partitioning, where each SVM discriminated against one emotion from all else. The representation generated in the first phase is transformed into a probability distribution over the seven categories of expression in the second phase. With an overall precision of 94.8 per cent, the new system organizes 17 action groups, whether they take place individually or in conjunction with other actions. In Saransh et al. [11], the model has been categorized into two parts where the first is regarding the removal of the tuples whose attributes are least important and the second part deals with the addition of other algorithms for improving the model. The proposed model used random forest which is applied to those models generated by CNN. After removing the least important tuples from the FER dataset, the model has improved as compared to other models. The algorithm uses the combination of several trees, this project used the C4.5 algorithms to generate the decision tree, since it is necessary to compare and get the best accuracy.

3 Proposed System

The proposed method presents a brisk measure to coordinate whether the CNN engineering model performs better if it uses only the raw pixels of image data for training or whether it is better to give the CNN some additional details, such as face landmarks and HoG characteristics. First, we use the webcam to capture the live images. In the first model (CNN model without additional information), we use the Haar cascade classifier to detect the face part from the input image and predict the expression using the trained CNN model. In the second model (CNN model with additional information), we use an LBP cascade classifier to detect the face part and if the width and height of the detected image are less than 30, it will ignore the image considering it as a false detection. Further, the image is resized into 48 * 48 pixels and converted into a grayscale image. Then, we pass this converted grayscale image into the trained model for emotion prediction. This model is the improved and advanced version of the first model as we use the LBP cascade classifier, trained the CNN model with additional features and neglect the false detected face. Both the models are explained in the below session.

3.1 Real-Time Emotion Detection Using CNN Without Additional Information

Face detection and emotion classification are two sequential steps in this work. Within the initiative, a camera is used to capture the face portion and track the exact location of a face in real time by bounding the box coordinates. This sequence includes the detection of the face using the OpenCV library Haar cascade detection (frontal face) and Haar cascade features are integrated to detect part of the face. The built network model has 3×3 layers where there are 1024 nodes with Softmax classification for the linked layers. Then, we use this network to train the CNN model and we get the model after successful completion of training and then we use this model in real time for emotion prediction.

3.2 Real-Time Emotion Detection Using CNN with Additional Information (Face Landmark and HOG Features)

In this approach, each method is broken down into four components: (1) preprocessing, (2) CNN engineering, (3) training and inference for CNN, and (4) detection of real-time emotions.

Preprocessing: Usually, this involves face recognition, face registration to compensate for differences in poses, and means to correct variations in lighting.

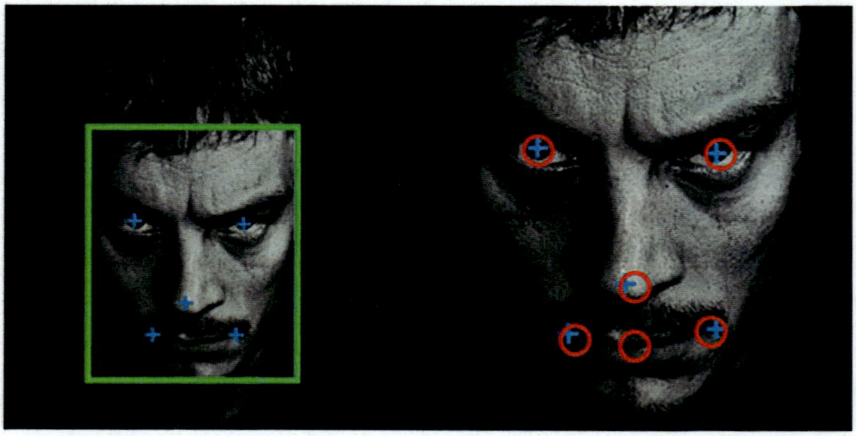

Fig. 2 The figure of a standard preprocessing pipeline, the green square indicates the face detection, blue crosses are the face landmark detection, red circles are the registration to reference landmarks, and illumination correction [12]

These measures are shown in Fig. 2. Using the pre-trained facial landmark detector within the dlib library, which is used to estimate the position of 68 x and y coordinates that map to facial structures on the face and HoG features from the FER-2013 Dataset, we extract the face landmark in preprocessing.

CNN engineering: We crafted 48 by 48-pixel single channelized input images (the size of images within the FER2013 dataset). CNN looks after images in different sizes in some of its works. After each convolution and completely connected (FC) layer, we add batch normalization layers to sub-optimal network initialization for sturdiness. Besides, after the third layer, we add two dropout layers and we use an ensemble of CNNs inside the first FC layer with various receptive fields and neuron numbers. The configuration is used for 3 to 3 receptive fields and 1024 $\times$ 1024 neurons.

Training and inference for CNN: To compensate for the FER image dataset's comparatively limited availability, training and inference for CNN are performed. So, to determine from face patches, we use additional features such as a vector of histogram of oriented gradients (Hog) features and also give face landmark features. Then, we use this additional data and then process it by CNN's first fully linked layer. This additional information is trained to perform FER on this basis. And we trained this enhanced CNN architecture by turning hyperparameter and setting the hyperparameter in 20 epochs. For the optimization of the cross-entropy loss, we use stochastic descent with a 0.95 momentum optimizer and set the batch size as 128, learning rate as 0.016, learning rate decay as 0.864, step decay as 50, and keep_prob as 0.956.

Detection of real-time emotions: The faces are first detected using OpenCV and LBP cascade classifiers (An LBP cascade can be trained to perform better than the

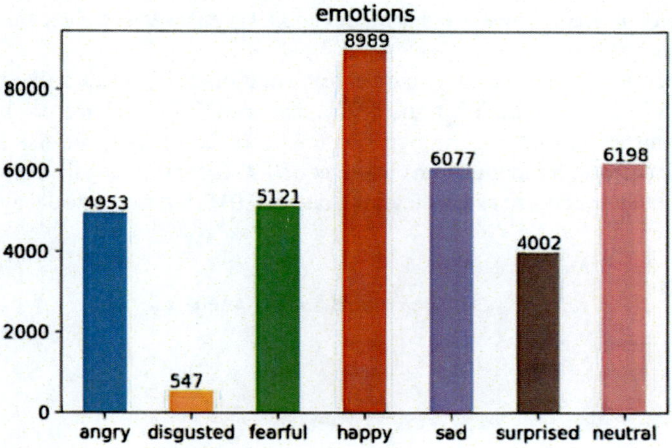

Fig. 3 FER2013 expression distribution [13]

Haar cascade classifier and the Haar cascade is about three times slower). If the width and height of the detected image are less than 30, it will ignore the image considering it as a false detection. Further, the image is resized into 48 * 48 pixels and converted into a grayscale image. Then, we pass this converted grayscale image into the trained model for emotion prediction. Finally, the resulting emotion will be displayed.

3.3 Dataset

Here, we selected the Fer-2013 dataset for emotion classification. 35,887 images with all basic expressions such as happy, surprise, rage, sad, fear, disgust, and neutral are in the dataset. The distribution of expressions is shown in Fig. 3. The FER-2013 dataset on the ICML 2013 Challenges was introduced in representation learning.

4 Experiment and Result Analysis

This section outlines the applications of our models and experiments conducted. We carried out a brisk measure to coordinate whether the CNN engineering model performs better if it uses only the raw pixels of image data for training or whether it is better to give the CNN some additional details, such as face landmarks and HOG characteristics. The dataset contained 35,884 images. In this analysis, we experimented with a CNN model without any additional information (face landmarks and HOG features). For this model, we achieved a 59.1% test accuracy shown in the following Figs. 4 and 5. We use this model and Haar cascade classifier for detecting emotion in real time.

We improved the above CNN model with additional information like face landmarks and HOG features. We trained this enhanced CNN architecture by turning hyperparameter and setting the hyperparameter in 20 epochs, we use stochastic gradient descent with a momentum optimizer of 0.95 for optimizing the cross-entropy loss, and set the batch size as 128, learning rate as 0.016, learning rate decay as 0.864,

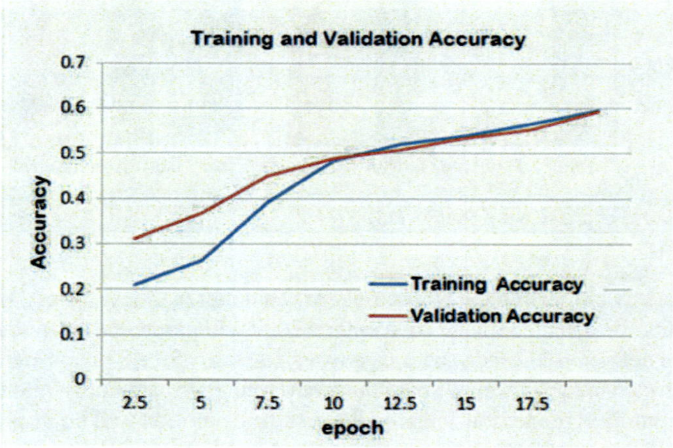

Fig. 4 CNN model without additional information training and validation accuracy

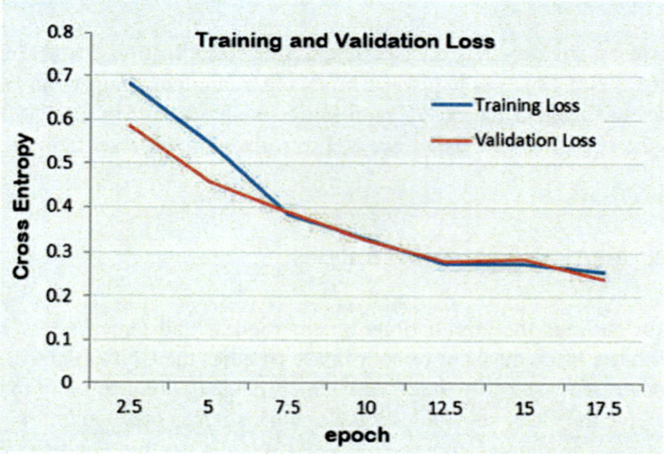

Fig. 5 CNN model without additional information training and validation loss

step decay as 50, and keep_prob as 0.956. This model achieved an accuracy of 75.1% as shown Figs. 6 and 7.

Table 1 shows the comparative study of different models with the FER2013 dataset conducted by Saravanan et al. [13]. Our proposed model with additional information like face landmarks and HOG features was found to be better and more accurate. In this improved system, we used the LBP cascade classifier because the LBP cascade classifier can be trained to perform better than the Haar cascade classifier and it is three times faster than the Haar cascade classifier and they also avoid small and inaccurate face detection leading to false detection.

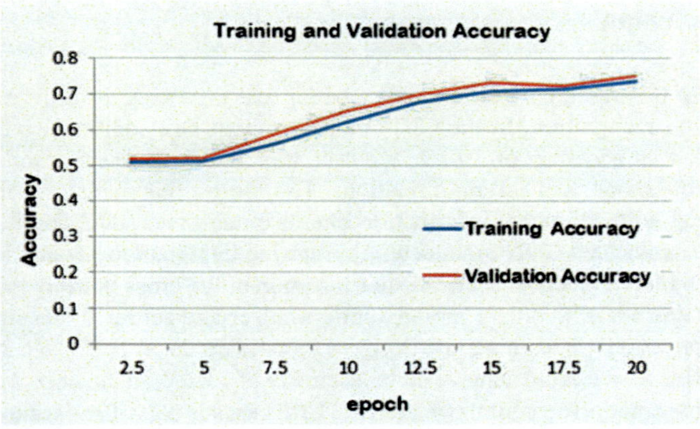

Fig. 6 CNN model with additional information training and validation accuracy

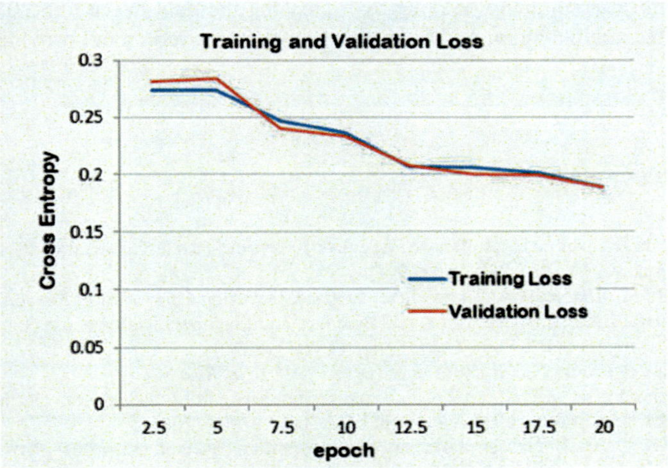

Fig. 7 CNN model with additional information training and validation loss

Table 1 Comparison of different models with FER2013 dataset [13]

Model	Accuracy (%)
Decision tree	30.84
Feed forward NN	17.38
Simple CNN	24.72
Saravanan et al. proposed CNN [13]	55.61
CNN without additional information	59.1
CNN with additional information (face landmark + HoG)	75.1

5 Conclusion

The goal of this research was to create a better real-time system for face detection and emotion recognition. In the CNN model without any additional information, we acquired an accuracy of 59.1%, we used this model to detect emotion in real time. But the model had certain limitations. To improve this model, we trained the CNN model with additional information like face mark and HOG features and we trained this enhanced CNN architecture by turning hyperparameter and setting the hyperparameter in 20 epochs. For the optimization of the cross-entropy loss, we use stochastic descent with a 0.95 momentum optimizer and set the batch size as 128, learning rate as 0.016, learning rate decay as 0.864, step decay as 50, and keep_prob as 0.956. Hence, we could achieve better accuracy of 75.1%. In the improved system for real-time emotion detection, we used the LBP cascade classifier because the LBP cascade classifier can be trained to perform better than the Haar cascade classifier and they also avoid small and inaccurate face detection leading to false detection, hence improving the overall performance of a real-time emotion detection system. Subsequent research could be done by increasing the data in the FER2013 dataset through data augmentation and training with our improved model provides a better result.

References

1. K. Nozaki, H. Ishibuchi, H. Tanaka, Adaptive fuzzy rule-based classification systems. IEEE Trans. Fuzzy Syst. **4**(3), 238–250 (1996)
2. M.F. Ali, M. Khatun, N.A. Turzo,*Facial Emotion Detection Using Neural Network*
3. S.A. Hussain, A.S.A. Al Balushi.A real time face emotion classification and recognition using deep learning model. J. Phys. Conf. Series **1432**(1) (2020)
4. I. Azizan, F. Khalid,*Facial Emotion Recognition: A Brief Review*
5. C. Pramerdorfer, M. Kampel,Facial expression recognition using convolutional neural networks: state of the art. arXiv preprint arXiv:1612.02903 (2016)
6. X.-L. Xia, C. Xu, B. Nan, Facial expression recognition based on tensorflow platform, in *ITM Web of Conferences*, vol. 12 (EDP Sciences, 2017)
7. D.Y. Liliana, Emotion recognition from facial expression using deep convolutional neural networks. J. Phys. Conf. Series **1193**(1) (2019)

8. D. Yang et al., An emotion recognition model based on facial recognition in virtual learning environment. Procedia Comput. Sci. **125**, 2–10 (2018)
9. D. Meng et al.,Frame attention networks for facial expression recognition in videos, in *2019 IEEE International Conference on Image Processing (ICIP)* (IEEE, 2019)
10. M.S. Bartlett et al.,Recognizing facial expression: machine learning and application to spontaneous behavior, in *2005 IEEE Computer Society Conference on Computer Vision and Pattern Recognition (CVPR'05)*, vol. 2 (IEEE, 2005)
11. S. Chiwande et al.,*Review on Deep Learning Based Stress Detection Using Facial Expressions*
12. http://securaworld.com/solutions/non-rule-based-artificial-intelligence/
13. A. Saravanan, G. Perichetla, K.S. Gayathri,Facial emotion recognition using convolutional neural networks. arXiv preprint arXiv:1910.05602 (2019)

New Era of Vernacular Voice Assistant

Jayant Agarwal, Nikhil Gulati, and Vishal Tyagi

Abstract Voice assistant is a type of digital assistant that integrates the voice and natural language processing procedures. To add a new innovation to this, the proposed system could change the way of communication among the end-users using artificial intelligence [AI], which is a perfect pivot between humans and their knowledge capacities. The proposed model is designed in such a way that all services provided by the electronic device are within the reach of the user commanding the assistant. Dialectal nature of the voice assistance can be used to sum up the interaction with people in a much more comfortable way. This paper describes the functionality for text-to-speech and vernacular language assistance to provide an enterprise-based solution and a better solution to support clients. The proposed application is not limited to different generations and functions, where it can be used in most of the real-time industrial applications.

1 Introduction

In recent times, everything is moving towards automation, be it your place or car. There is an increased chance for technology over the last few decades. As for nowadays, world you see can be connected with the machines. So here comes a question, what is Machine Interaction? Obviously to give input, but what if the input is not the regular typing method, rather it is your own speech wonder how? What if you can talk to machines, give them orders and want your machine to exclaim thoughts as if you hired a personal assistant. What if the machine is not able to give you desired results, but alternatively shows you possible results and suggests you on better solutions as well? Easy access to a machine with voice commands is a way to transform the communication in the near the future with the system [1]. To attain this, text-to-speech API is used to understand various inputs given by the user. Many business tycoon companies like Google, Apple and Amazon are trying hard to achieve this type of communication as standard. It is quite amazing that reminders should be set

J. Agarwal (✉) · N. Gulati · V. Tyagi
Galgotias University, Greater Noida, Uttar Pradesh, India

P. Karuppusamy et al. (eds.), *Ubiquitous Intelligent Systems*, Smart Innovation, Systems and Technologies 243, https://doi.org/10.1007/978-981-16-3675-2_42

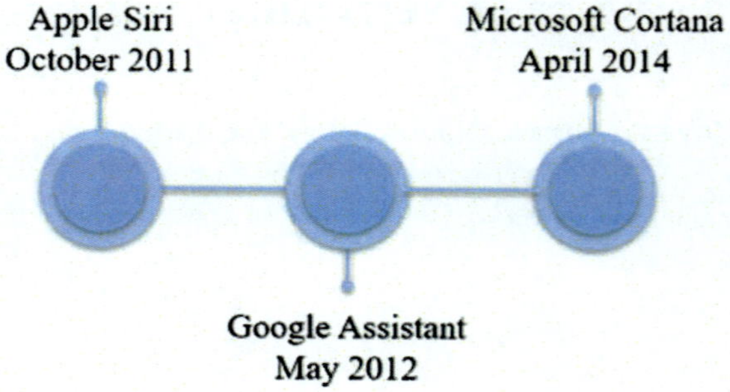

Fig. 1 Timeline of voice assistant

to remind you or set up alarm to wake you up. To absorb the importance of this, it has been decided to make a plan that can be placed anywhere nearby and you can ask it to help you do anything by simply interacting with it. In addition, you connect two or more such devices via Wi-Fi. This software would be convenient for colloquial use, and it can help you to execute better by perpetually providing the reminders and updates [2]. It is called as RICK with voice recognition intelligence, which takes the user input in the form of voice or text format and processes it and finally returns with an output in various ways such as actions or graphs, which are performed by the orders of the EU (refer Table 1). It would be definitely a major success in the business process outsourcing (BPO) companies to effectively manage their client and provide them the best solution. Automated calls are highly effective in the scenarios like where companies need to be more strategic towards the management of the people and overall efficiency towards client handling. Thus, our vernacular AI comes into this picture where it provides great solution approach to be more advanced in handling of business problem and providing better solution with the calculation and preprocessing done by the AI (Fig. 1).

1.1 Motivation

The main inducement behind selecting this project as our final year project predominantly lies in the large investments made by the substantial companies in promotion and testing but failed to get an equal response to their investment. They come in small packages and can do a multiplicity of things after hearing a name or command. The voice assistant is a project that aims to enhance the product and service performance by extracting commands form a user and provide the best solution to it over voice itself.

2 Related Work

See Fig. 2.

Google Now

Kepuska and Bohouta [3] Google Assistant, Google Now is preinstalled on all Android devices out there. It was launched way back in 2012, Google Now is used to schedule appointments, send text messages, search for references, etc. With the upcoming Android platforms, new advancements are being done over VA (refer Table 1) to provide better experience to EU (refer Table 1).

Siri

Developed by Apple Inc, Siri is a smart digital assistant that allows iPhone users to send messages, create schedules, makes calls, play music and videos, etc. A smart digital VA (refer Table 1) was first installed on the iPhone via iOS 5 showcased on 2011; gradually, it become available other apple products for the seamless connectivity [3].

Cortana

This is a leading digital VA (refer Table 1) founded by Microsoft and showcased with the launch of windows mobile devices 8.1 series back in 2014. In addition to humour and the ability to make jokes, Cortana can be used to set reminders, find files on the windows OS, etc. [3].

2.1 Literature Review

See Fig. 3.

1. **Graphical Interface**: It allows users to interconnect with the electronic devices by using animations, graphical icons and audio bit to enhance the overall experience.
2. **Speech Recognizer**: It is the tool or component that can translate the actual phrases spoken by the EU (refer Table 1) and syntactically convert them into readable text. It automatically stops when there is no audio bit present. It works using algorithms based on auditory and language designing.

Fig. 2 Voice assistants

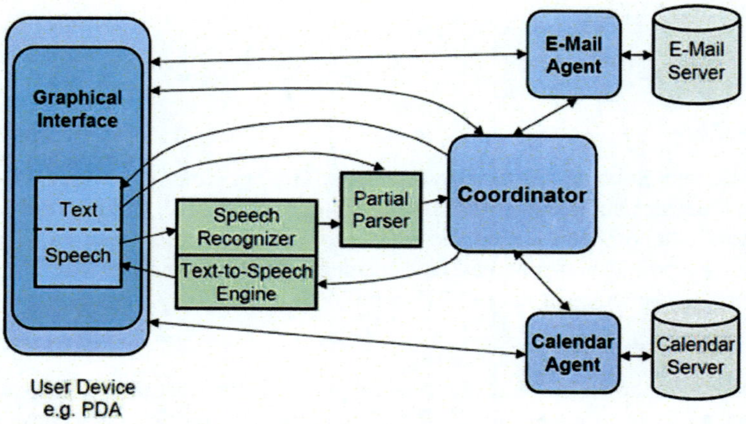

Fig. 3 Architecture of the basic voice-controlled personal assistant

3. **Text-to-Speech**: A screen reader program designed by Google works only on Android phones that enables specific apps (refer Table 1) to read out text which is written with multilingual support present.
4. **Partial Parsing**: Techniques which are desired to retrieve the syntactic fragments of the text instead of providing the necessary data already contained in the traditional semantic analyses.
5. **Agent**: An Agent (IA) (refer Table 1) is the term used in artificial intelligence; thus, its means that learning is the utmost important characteristics of term "intelligence".

3 Proposed Work

The work started with scrutinizing the audio bit by EU (refer Table 1) delivered through a microphone present on the electronic device. Test cases are made by programming according to books and online resources, with the explicit goal to get best practices and more advancement in usage of the VA (refer Table 1).

The basic workflow of a smart voice assistant is given at a glance in Fig. 4.

Speech recognition is required to derive the speech input to text format. The text is then transferred to central processor that determines the type of commands, checks the semantical errors and then calls the appropriate routines to further execute. Hundreds and hundreds of inputs, some things can play a major role whether the software would relate to what you said or what you aim for. The most important factor which needs to be filter out is the background noise or surrounding noise because it is mandatory process as the system does not able to distinguish between the actual useful audio bit and surrounding sound for reference people talking behind or car honking around.

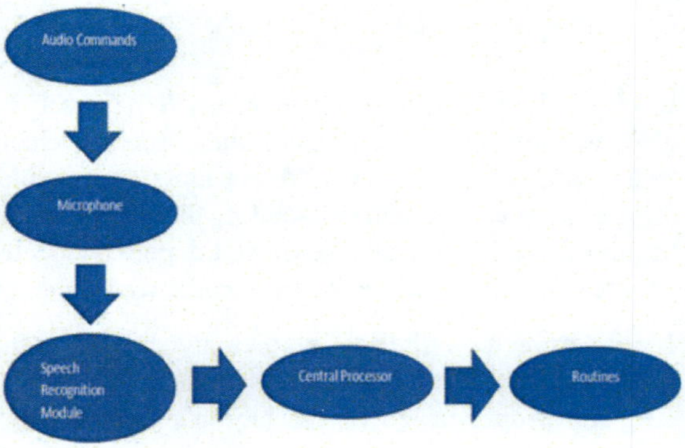

Fig. 4 Workflow

We now get that VA (refer Table 1) actually has to polarize the surrounding noise and critically it has to get tone of a person (low pitch and high pitch) which usually makes a greater impact in the sentence framing. Speech recognition can be sensitive to these changes.

The overall system design consists of the following phases:

1. Data collection in the form of speech.
2. Voice analysis and conversion to text.
3. Data storage and processing.
4. Generating speech from the processed text output.

Now let us understand the use of routines as we have discussed earlier.

Python Backend: It retrieves the output which it gets from speech recognition modules and performs search whether the commands received are a system call or context extraction or an API call. Thus, the output which is given is handed over to the EU (refer Table 1) waiting for the desired result.

API Call: The API is abbreviated for application programming interface. An API is a software interface that allows to achieve interaction between two or more programs. For reference, an API is a messenger that brings your requests made and delivers it to the result provider from which you are expecting results from and finally returns a particular response to it.

Context Extraction: It is the function that according to its own extracts out the structured data also known as the labelled data from random machine-readable texts (unstructured). Through this, the software can make useful informed decision and thus performs better at recommendations. Generally it involves the processing of indigenous language texts through natural language processing (NLP).

System Call: Commonly known as Syscall. It is a program in which a computer program requests a service from the app's kernel (core component of the OS [refer Table 1]) where the execution already started. It plays an important role between process and application.

3.1 Methodology

In vernacular language acceptance, it will make very easy for every person across the globe to interact with the assistant and different dialects would not be resisting people to interact. Understanding different languages is one thing but replying back with different language and understanding their accents and speech would possibly be the best modification to improve overall experience of the user. The main contribution towards the vernacular in the tech world would include the following [4].

1. Learning new dialects and dialogue delivery is the ability will be learned during the participation in a conversation.
2. The proper usage of the jargon's human-like responses will provide more human–machine comfort.
3. Learnable task is important as a self-feeding process, thus resulting in the best approach based on the uncertainty of the model.

The model is categorized into three user task, namely:

1.1. TASK 1: Dialogue

The primary task for the agent is to observe the pattern in the language delivery and build a connection with words from the data set like a coherence nature. The training of the agent is figured out in the form (x: y) pairs where

x- > Context of the conversation.
y- > Appropriate response from the human.

1.2. TASK 2: Satisfaction

The objective behind this is to judge whether the user is actually been satisfied with the result shown and the quality of the speech produced by the system in the current interaction. The training of the agent is figured out in the form (x: s) pairs where

x- > Context of the conversation
s- > Ranging from [0,1].

1.3. TASK 3: Feedback

Feedback acceptance is necessary as the agent never stops learning so to accelerate the performance of learning the agent must learn new dialects which it never came across.

Fig. 5 Experiment flow

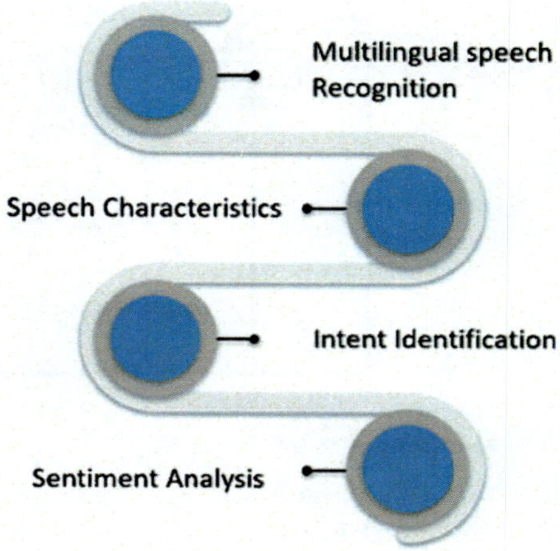

3.2 *Experiment*

Idiolect Engine—Idiolect the word itself is self-explanatory which means "person's unique usage of lingual, whether it may be vocabulary or may it be pronunciation". Idiolect Engine understands the regular semantics and the lexical components required for the spoken language understanding (SLU). The SLU contributes to generating responses that are more human-like. The outcome of having an idiolect engine is to provide better rich conservation with an end customer (Fig. 5).

Contextual Conversation Clustering—The concept behind this is to provide a cluster [5] of problems raised by different users at a different time with different queries, clustering of these problems provide high performance in satisfying business needs as its groups down the similar problem into cases and following automated result are opted to fulfil the requirements. This saves a lot of time instead of auditing the query raised. This C3 technology ensures our AI to learn over time and become future proof to solve similar issues on its own.

3.3 *Feasibility Analysis*

1. **Model Performance**: Our project uses algorithms which are good at both time and space complexities and aimed to produce best results.
2. **Technological Considerations**: The analysis will be performed on various instruction given to the software, and based on the results given, necessary amendments would be done.

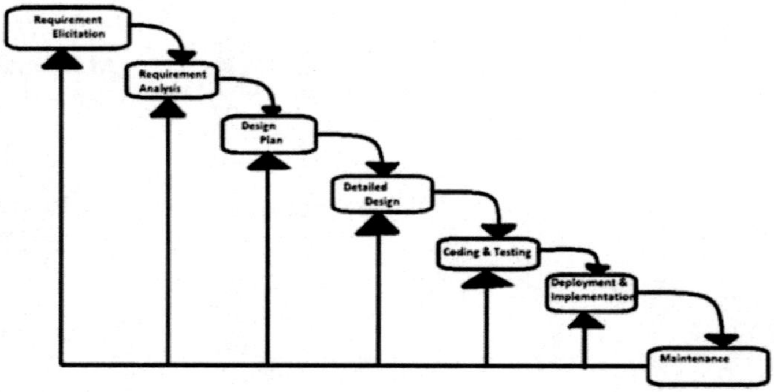

Fig. 6 Software development life cycle

3. **Financial Feasibility**: Our project is not expensive as we have gathered our data from government sites and manpower for building this project required for execution is minimal.
4. **Resource Feasibility**: Our project is primarily dependent on large data set as it makes recommendation. So nearly achieving good fit training data sets, thus providing best results.

3.4 Development Analysis

The steps we followed for developing this project:

1. Analysis of the problem.
2. System requirement specification was filed.
3. Feasibility Study of the project.
4. Appropriate algorithms were decided.
5. Studied about the pros and cons.
6. Initialize the development phase.
7. Installation of software like Web Browser, Pysttsx3, PyCharm, Python3.9, Visual Studio Code.
8. Discussed the algorithms with the guide.
9. Coded as per the algorithm opted (Fig. 6).

3.5 Implementation

Modules required [6]:

1. **Subprocess**: Used to allow subsystem details used in various commands, namely Shutdown, Sleep, etc. This module comes with a built-in Python function.

2. **Tkinter**: Used to build a GUI and comes with Python. This module will be built internally with Python.
3. **Wolframalpha**: Used to calculate responses at a professional level using Wolfram's expertise, Knowledge base and AI domain.

```
pip install wolframaplha
```

4. **Pyttsx3**: Used to convert text into speech in an offline application.

```
pip install pyttsx3
```

5. **Wikipedia**: Wikipedia provides us ample information about anything, and Wikipedia module is needed to search for information on Wikipedia.

```
pip install Wikipedia
```

6. **Speech Recognition**: VA [refer Table 1] application, one of the most important things, in this case, is that your assistant should be able to know your voice.

```
pip install SpeechRecognition
```

7. **Web Browser**: Used for performing Web Search. This module will be built in with the Python function.
8. **ES Capture**: Used for capturing your photos from your camera.

```
pip install escapture
```

9. **Pyjokes**: Used to collect Python jokes via the Internet.

```
pip install pyjokes
```

10. **Date Time**: Used for displaying date and time by this module. This module will be built in with the Python function.
11. **Twilio**: Used to make calls and messages.

```
pip install twilio
```

12. **Requests**: Used for making GET and POST requests.

```
pip install requests
```

For Windows User:

The **Pyttsx3** module is stored in a variable name engine.

Microsoft text-to-speech engine (**Sapi5**) will be used for voice recognition.

The voice ID value will be ranging 0 or 1.

1. 0 refers for male Voice.
2. 1 refers for female Voice.

4 Results

Cost Reduction

Many companies hire staffs for the purpose of attending the client queries and being MNC and have a global expansion so queries will not have specific time to attend as we have different time zones so automated agent is very helpful for attending these queries without using that much man power.

Customer Satisfaction

Connected to the Internet is the soul of the VA (refer Table 1) and providing correct results in their idiolect speech thus results in higher customer retention to which we can see in the business growth.

Advancement in Automation

Leading technology implementation in every aspects of the business and in life is very fruitful for day-to-day problems, and we can totally rely on the statement as Necessity is the Mother of Invention. Until and unless we will not implement the new solutions, the forthcoming technology will not have an essence in life.

Precise Results

Precise output generated by the strong analytics running behind the system is very important to get to the exact solution that user wants. In order to provide correct solution, AI requires some prior information to come up with factual result (Fig. 7).

5 Recommendation for Further Work

No software is well equipped at the first place, and every modification is done to increase the useability of the product at every iteration of software development life cycle. However, this program is working quite well and is always hungry for new changes. For the future updates are concerned, we are working on adding more functions to software so as to provide more comprehensive nature, practicality and need of the software is taken care. The increasing date volume and fitting must be checked at utmost priority to make it more optimizable as never before [7].

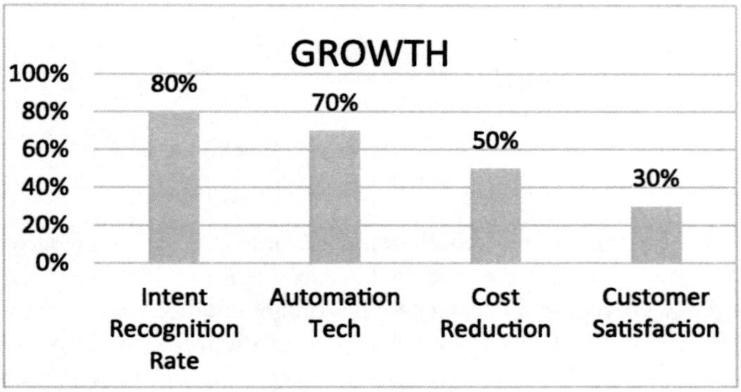

Fig. 7 Growth opportunity

Ambiguity would be another factor a VA (refer Table 1) that may lag behind as word suggests a sentence which has more than one meanings so adding more volume and rationalizing the word power used by our artificial intelligence, the system has to presume notion to make it work with the corresponding commands. The program is restricted to certain number of words adding possible sentences and meaning of words which can make it more reliable for further use.

Voice remembering capabilities can also be useful for a VA (refer Table 1) situated at home or even at work place greeting people with their names based on deciding who is speaking to the assistant. This can also be helpful for people who are working for a cause and virtual assistant assisting every single person depending upon their queries concurrently.

Be it with anything, improved visual experience always strikes the eyes new interface of the VA and is always welcomed with more capabilities assigned to it giving user more satisfied result. User interface will always be configured and optimized.

Dependency over the Internet is pulling back the capabilities of the VA; therefore, some places which do not have reliable Internet connection make them difficult to use. To overcome this, we would require offline speech recognition system; as VA is linked to cyberspace, it will be gigantic progress in the automation technology industry.

6 Conclusion

This project is built using open-source software components with PyCharm community backing up, which can accommodate any updates in the near future. The innovative nature of this project makes it more reliable for long use and thus makes it flexible to adapt new features without affecting current performance. Judging the commands

Table 1 Abbreviation

VA	Voice assistant
EU	End-user
OS	Operating system
App	Application
IA	Intelligent agent

by pitch wave be it high or low so as it gets the commands as interrogative, exclamatory, neutral, etc. Greeting the user feels more comfortable to interact. VA uses natural language processing integrated with artificial intelligence to achieve human-like smartness. Technology is always designed to minimize the human efforts so does the voice assistant. Business processing outsourcing would be able enjoy more customer retention; as we have discussed in Results section, we totally say customer satisfaction and cost reduction are the two major components that a company desire, so with this technology in mind we are one step closer to achieve this goal.

Acknowledgements In the present world of competition, there is a race of existence in which those who are willing to succeed will only come forward as better person. A project is like a bridge, connecting theoretical learning with practical working. Firstly we would to thank the Almighty for guiding us on the right path. Next to him are our parents whom we greatly indebted for the love and courage they imbibed us till this day. We are feeling obliged in taking this opportunity and sincerely thank our Professor Mr. Pratyush Kumar Deka, for helping us and guiding us through different stages of the project. We are thankful to our all teachers for imparting the knowledge to us that we are capable now of conducting this project on our own. We have no valuable words to express our gratitude and thanks, but hearts are still overflowing with the favours received from every person.

References

1. https://en.wikipedia.org/wiki/Virtual_assistant
2. A. Dekate, C. Kulkarni, R. Killedar, Study of voice controlled personal assistance device, in *Proceedings of IJCTT* vol. 42, no. 1 (2016)
3. V. Kepuska, G. Bohouta, Next-generation of virtual personal assistants (Microsoft Cortana, Apple Siri,Amazon Alexa,Google Home), IEEE
4. B. Hancock, A. Bordes, P.-E.M.J. Weston, *Learning from Dialogue after Deployement Feed Yourself, Chatbot!*
5. J. Wu, X. Wang, W.Y. Wang, Self-supervised dialogue learning, in *Proceedings of 57th Annual Meeting of the Association for Computational Linguistics*, Florence, Italy (2019)
6. https://towardsdatascience.com/how-to-build-your-own-ai-personal-assistant-using-python-f57247b4494b
7. S. Hui, Q. Song, *Intelligent Voice Assistant*, Bachelor Thesis, School of Health and Society, Spring (2012)

An Effective Classification Algorithm for Rainfall Prediction Using Time Series Data

G. Rahul, S. Vinayak, and L. Nitha

Abstract Rainfall is the most important factor that affects Kerala's economy. Kerala is a highly populated state when compared to its availability for water. Therefore, the need for a large amount of water can be compromised by proper management of rainfall. The main objective of this paper is to compare the three classification algorithms—SMOreg, Linear Regression and Multilayer Perceptron for Time series forecasting feature using WEKA tool. For the analysis, a data set containing 116 samples of Kerala's annual rainfall from the year 1901–2017 that has been collected from a government website. The performance of the algorithms is checked on the basis of Root-Mean-Squared Error. The model shows that the least RMSE seems the better algorithm and found that Multilayer Perceptron is least with it. So, in the three algorithms Multilayer Perceptron is the best followed by Linear Regression and SMOreg.

1 Introduction

Rainfall forecasting is one among the greatest problems faced by the meteorology department of Kerala. Rainfall prediction is a much-needed factor in a wide range of areas. Rainfall forecasting can be helpful in many sectors, such as flood prevention, disaster management, coastal area warning. Rainfall prediction can be very crucial for states like Kerala which mainly depend on agriculture. The factors like accuracy and precision have a very important role in the field of rainfall forecasting.

Most of the prediction techniques made by forecasters nowadays need different variables like climate variables like temperature, speed of the wind, direction of the wind, humidity, etc. Using many variables increases the complexity of the model and it does not produce more accuracy always.

Rainfall prediction will help to protect many resources like human power, natural resources which are essential for the existence of life. So it is mandatory to make use of rainwater appropriately.

G. Rahul (✉) · S. Vinayak · L. Nitha
Department of Computer Science and IT, Amrita School of Arts and Sciences, Amrita Vishwa Vidyapeetham, Kochi, India

P. Karuppusamy et al. (eds.), *Ubiquitous Intelligent Systems*, Smart Innovation, Systems and Technologies 243, https://doi.org/10.1007/978-981-16-3675-2_43

In this paper the possibility of doing prediction of rainfall is discussed using Time Series Forecasting feature in WEKA tool. Here, comparison of three different algorithms which are SMOreg, Linear Regression and Multilayer Perceptron are done on the basis of Direction Accuracy and RMSE; the algorithm that consists of higher Direction Accuracy and least RMSE is the best. Thus, the classification of above-mentioned algorithms (SMOreg, Linear Regression and Multilayered Perceptron) is done.

2 Related Works

The rainfall can be managed quite well and make it useful for the living beings. The most important attribute to satisfy the same is rainfall prediction. Chaitanya Damle introduced prediction of flood using TDSM to forecast the occurrence of time delayed events using space reconstruction and data mining to dig out the patterns of future occurrence in nonlinear and non-stationary TDSM. The current methodologies for the flood forecasting are nonlinear time series way like HMM, ANN, NLP are used to estimate the prediction; it may be accurate for one day but after the period of prediction increases accuracy decreases. Due to the above-mentioned reason for overcoming that introduction of time series into the prediction, TDSM can offer more accuracy in the matter of prediction. The prediction output from TDSM can be used as a management tool for decision making, and estimating the action plan. The effect of flood can be calculated by the area that is the severity of the flood can be easily foreseen [1].

A project for building a weather prediction application launched by Dhore et al. [2] which can be used in local area weather. These types of models are mostly sophisticated and it requires more amounts of dataset and resources; therefore the estimation of weather can be done for future and now. The methodology will extract the useful data from the data set. In this paper a genetic algorithm is used and by using this impact of the flood can be calculated and the methods for the prevention can be selected. For the partition of the data like temperature or weather condition Clustering kmean and FP Growth algorithms are used. The findings of this paper will be really useful to the people to save their life, property and other belongings from destruction [2].

A comparative study of the different methodologies used to foresee weather has been put forward and it is mainly focused on an advanced system which can forecast the upcoming weather. The sudden extreme changes in weather causes large numbers of loss in the world. Barde et al. [3] found some models used for real time, monthly, yearly. In this system weather is predicted using the previous data (Historical data), such as temperature, humidity, etc. Multilayer Perceptron, K-Nearest Neighbor and naive Bayes are some methods used to forecast. These three are compared with their evaluation parameters and identify which is more accurate for the prediction [3].

The flood forecasting will be really useful for the people and officials, where the areas in which chance of flood is high and the disaster management can be done appropriately, the officials will get enough time to tackle the situation. Wu et al. [4] bring in different techniques like the Novel Hybrid Model which is based on AI, GA. The paper mainly focused on the flood forecast of Yangtze River, Republic of China. The forecasting is based on the current water level of the river. Linear regression is compared with Genetic algorithm and Conventional ANN. On the basis of performance Genetic algorithm-based ANN gives better results [4].

Rohit is considering the state of Maharashtra and the data set is collected from Yavatmal which is a rainfall monitoring group of agencies. In this paper, acute tool for time series prediction of rainfall using Multilayer Perceptron is used. The rainfall can be higher or lower. The cycle of rainfall may be high or low. This method is selected on the basis of its performance like RMSE and NMSE. The Multilayer Perceptron Neural Network can predict the effect and amount of rainfall accurately as proven [5].

Due to drastic human empowerment the weather condition of the land is drastically exploited which causes climate changes. Flora and fauna are necessary for the life in earth. Climate is changing according to the exploitation done in nature. So we are responsible for the protection of our wildlife. Jayakumar et al. [6] introduce a FUZZY ANFIS methodology to foresee the wildfire. The combination of FUZZY and ANN is very useful to analyze the chances for wildfire. FCM and ANFIS are the different algorithms to foresee the wildfire [6].

3 Materials and Methods

3.1 Dataset

The rainfall data from the year 1901–2017 in Kerala are taken for analysis. For training, a total of 116 records are used. History often repeats itself, so whatever events happened in the past, they are likely to happen again in the future [7, 8]. Whenever data or observations or something is recorded at regular time intervals it's called a time series data. The figure shown below is the sample image of dataset that was used in this work (Fig. 1).

3.2 WEKA 3.8.4

WEKA is software which provides data preprocessing, implementation tools in different algorithms of Machine Learning, and helps in picturing it; these help to create new machine learning techniques and use them into data mining problems [7]. WEKA provides implementation of several algorithms of your choice, and dataset can be run using desired parameters which will provide with the statistical output of model processing using WEKA. Many models can be used in the same dataset.

1	SUBDIVISI	YEAR	JAN	FEB	MAR	APR	MAY	JUN	JUL	AUG	SEP	OCT	NOV	DEC	ANNUAL	JF	MAM	JJAS	OND
2	Kerala	1901	28.7	44.7	51.6	160	174.7	824.6	743	357.5	197.7	266.9	350.8	48.4	3248.6	73.4	386.2	2122.8	666.1
3	Kerala	1902	6.7	2.6	57.3	83.9	134.5	390.9	1205	315.8	491.6	358.4	158.3	121.5	3326.6	9.3	275.7	2403.4	638.2
4	Kerala	1903	3.2	18.6	3.1	83.6	249.7	558.6	1022.5	420.2	341.8	354.1	157	59	3271.2	21.7	336.3	2343	570.1
5	Kerala	1904	23.7	3	32.2	71.5	235.7	1098.2	725.5	351.8	222.7	328.1	33.9	3.3	3129.7	26.7	339.4	2398.2	365.3
6	Kerala	1905	1.2	22.3	9.4	105.9	263.3	850.2	520.5	293.6	217.2	383.5	74.4	0.2	2741.6	23.4	378.5	1881.5	458.1
7	Kerala	1906	26.7	7.4	9.9	59.4	160.8	414.9	954.2	442.8	131.2	251.7	163.1	86	2708	34.1	230	1943.1	500.8
8	Kerala	1907	18.8	4.8	55.7	170.8	101.4	770.9	760.4	981.5	225	309.7	219.1	52.8	3671.1	23.7	328	2737.8	581.7
9	Kerala	1908	8	20.8	38.2	102.9	142.6	592.6	902.2	352.9	175.9	253.3	47.9	11	2648.3	28.8	283.7	2023.6	312.2
10	Kerala	1909	54.1	11.8	61.3	93.8	473.2	704.7	782.3	258	195.4	212.1	171.1	32.3	3050.2	65.9	628.3	1940.4	415.5
11	Kerala	1910	2.7	25.7	23.3	124.5	148.8	680	484.1	473.8	248.6	356.6	280.4	0.1	2848.6	28.4	296.7	1886.5	637
12	Kerala	1911	3	4.3	18.2	51	180.6	990	705.3	178.6	60.2	302.3	145.7	87.6	2726.7	7.3	249.7	1934	535.7
13	Kerala	1912	1.9	15	11.2	122.7	217.3	948.2	833.6	534.4	136.8	469.5	138.7	22	3451.3	16.9	351.1	2453.1	630.2
14	Kerala	1913	3.1	5.2	20.7	75.7	198.8	541.7	763.2	247.2	176.9	422.5	109.9	45.8	2610.8	8.3	295.2	1729	578.3
15	Kerala	1914	0.7	6.8	18.1	32.7	164.2	565.3	857.7	402.2	241	374.4	100.9	135.2	2899.1	7.6	215	2066.1	610.5
16	Kerala	1915	16.9	23.5	42.7	106	154.5	696.1	775.6	298.8	396.6	196.6	302.5	14.9	3024.5	40.4	303.1	2167	514
17	Kerala	1916	0	7.8	22	82.4	199	920.2	513.9	396.9	339.3	320.7	134.3	8.9	2945.3	7.8	303.4	2170.2	463.9
18	Kerala	1917	2.9	47.6	79.4	38.1	122.9	703.7	342.7	335.1	470.3	264.1	256.4	41.6	2704.8	50.5	240.4	1851.7	562.1
19	Kerala	1918	42.9	5	32.8	51.3	683	464.3	167.5	376	96.4	233.2	295.4	54.1	2501.9	47.9	767	1104.3	582.6
20	Kerala	1919	43	6.1	33.9	65.9	247	636.8	648	484.2	255.9	249.2	280.1	53	3003.3	49.2	346.8	2025	582.3
21	Kerala	1920	35.2	5.5	24.1	172	87.7	964.3	940.8	235	178	350.1	302.3	8.2	3303.1	40.6	283.7	2318.2	660.6
22	Kerala	1921	43	4.7	15	171.3	104.1	489.1	639.8	641.9	156.7	302.4	136.2	15.8	2719.9	47.8	290.3	1927.5	454.3
23	Kerala	1922	30.5	21.4	16.3	89.6	293.6	663.1	1025.1	320.6	222.4	266.3	293.7	25.1	3267.6	51.9	399.4	2231.2	585.1
24	Kerala	1923	24.7	0.7	78.9	43.5	80	722.5	1008.7	943	254.3	203.1	83.9	41.6	3484.7	25.3	202.3	2928.4	328.6
25	Kerala	1924	19.3	2.9	66.6	111	185.4	1011.7	1526.5	624	289.1	176.5	162.9	50.4	4226.4	22.2	363	3451.3	389.9
26	Kerala	1925	4.1	16.5	76.9	93.4	258.2	688.8	593.5	554.1	158.8	295.4	223.7	98.8	3062.1	20.5	428.5	1995.2	617.9
27	Kerala	1926	28.6	5.8	23.1	55.8	222.6	563.9	885.2	536	322.7	216.7	88.8	16.2	2965.4	34.4	301.5	2307.8	321.7
28	Kerala	1927	18.8	35.3	49.6	86.5	265.4	720.2	888.2	315	335.6	135.8	137.6	6.8	2994.7	54.1	401.4	2258.9	280.2
29	Kerala	1928	12.7	65.9	51.3	121.1	81.9	590.7	420.6	553.2	75.9	321.5	155.2	52.7	2502.8	78.6	254.3	1640.4	529.4
30	Kerala	1929	12.8	29.8	58.9	210.7	148	946.6	844	293.9	268.9	350.4	158.2	39.4	3361.6	42.6	417.6	2353.5	548
31	Kerala	1930	10.8	10.8	39	102.7	404.9	633.1	401.7	273.4	411.5	433.9	207	89.2	3018	21.6	546.5	1719.7	730.2

Fig. 1 A glimpse on the dataset

The different outputs of models which are used can be compared and find the apt one for the purpose. Here, WEKA is used for the dataset preprocessing, training and for the performance check of the three algorithms (SMOreg, Linear Regression, and Multilayer Perceptron). Direction accuracy and Root-Mean-Square is calculated by WEKA for finding the best suite algorithm for the purpose.

3.3 Root-Mean-Squared Error [RMSE]

Dutta and Gouthaman [7] Root-Mean-Square Error (RMSE) can be said as the measure of difference between the sample and populated values that is predicted by any model. It is represented by the standard deviation of the difference between the actual values and predicted values. Lesser the value of Root-Mean-Squared Error means more fit the model. RMSE is calculated using the equation given below

$$\mathrm{RMSE_{fo}} = \left[\sum_{i=1}^{n} (Z_{\mathrm{fi}} - z_{0i})^2 / N\right]^{1/2}$$

Σ = summation
$(Z_{\mathrm{fi}} - Z_{\mathrm{oi}})^2$ = differences between sample and populated values, squared
N = sample size.

Here, RMSE is used to compare the three algorithms to find which one has the least value and is best for the purpose.

3.4 Direction Accuracy

Usha and Appavu Alias Balamurugan [9], Usha and Balamurugan [10] the direction accuracy can be said as the number of times the movement of the predicted values matches the movement of actual values. It shows the percentage of the number of the values predicted. Direction accuracy is also considered with RMSE to find the best algorithm among the three algorithms (Linear Regression, SMOreg, Multilayer perceptron). Direction Accuracy can be defined by following equation [10]

$$\frac{1}{Z}\sum_{t}^{1} \text{sign}(x_t - x_t - 1) = \text{sign}(y_t - y_t - 1)$$

X_t = real value at the point of time
Y_t = predicted value at the point of time
Z = the number of uniform time interval of forecasting.

3.5 Data Preprocessing

Usha and Appavu Alias Balamurugan [9], Kumar and Shobika [11] the data preprocessing is an important step in creating the models. The dataset taken for analysis may not be perfect; it contains many unwanted rows. We need to clean the dataset and a proper filtering should be done. The first step is to remove the unwanted fields. In this work the required data field is just the Annual Rainfall data while the dataset contains monthly rainfall data too. The monthly field containing rainfall data from January to December is removed, thus making the dataset feasible for further process.

In this paper, three different classification algorithms are used, like Multilayer Perceptron, Linear Regression and SMOreg to compare RMSE and propose the best fit for the given dataset.

1. Al-Muqrashi and Soosaimanickam [12] linear regression consists of two variables. One is an independent variable and another is a dependent variable. It might be a numerical value. It is used to find the periodical (time series) values of the given dataset. Time series values are independent variables and the final outcome is of course a dependent variable. We have to find an interconnection between the variables.
2. Agrawal et al. [13], Joseph and Ratheesh [14] multilayer Perceptron with one hidden layer are capable of approximating any continuous function. Multilayer Perceptron are commonly used for supervised learning problems. The group of inputs and outputs are trained and learn to model the dependencies with respect to those inputs and outputs.
3. Al-Muqrashi and Soosaimanickam [12] SMOReg algorithm is used for numerical input variables. It will convert nominal values to numerical values. SMOreg

implements the support vector machine (SVM) for regression. The parameters can be learned using various algorithms. The first versions of the SVMs were created for binary classification problems, while the later versions, such as SMOReg or SVR (Support Vector for Regression) have been used to support regression issues and multi-class classification (Fig. 2).

Once the preprocessing is done, the classification algorithms can be applied to the datasets. WEKA tool provides an option for time series forecasting. The data are taken into the forecasting option in WEKA. The data set contains the rainfall data for the period 1901–2017, using this dataset the model will predict the rainfall data of 2018. A total of 116 instances are taken. First SMOreg classifier is used to predict the future rainfall, following with Linear regression and Multilayer Perceptron, respectively. In the evaluation part, the Root-Mean-Squared Error (RMSE) and Direction accuracy of the three algorithms are checked. The one with less RMSE and higher Direction accuracy is the best algorithm for the given dataset.

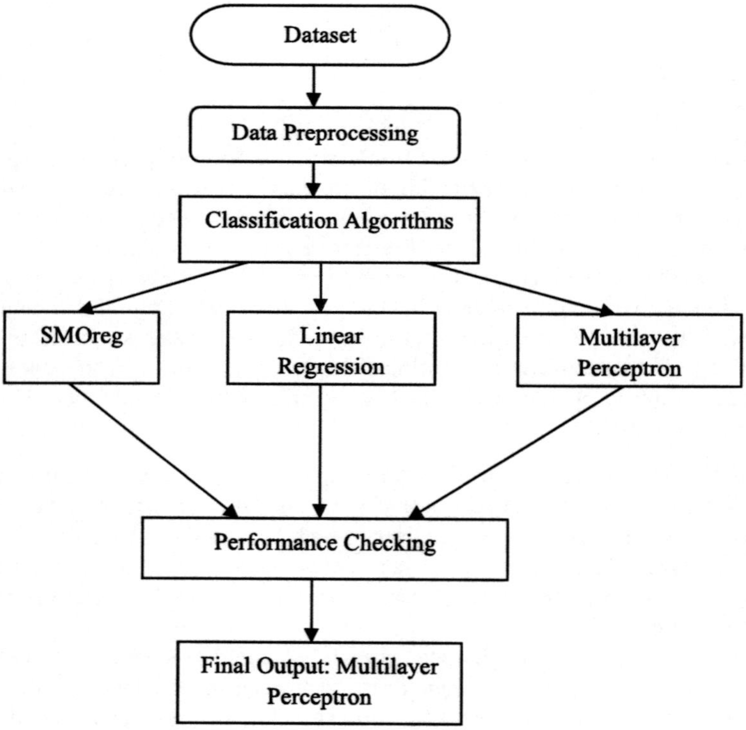

Fig. 2 Process workflow

4 Results and Observations

Table 1 shows the values acquired by doing evaluation on the three algorithms. From the performance check, the Direction Accuracies of SMOreg, Linear Regression, Multilayer Perceptron are 53.1532, 45.9459, 71.1712, respectively, and their RMSE are 406.1387, 405.8745, and 313.4686, respectively. From the observations, it shows Multilayer Perceptron has the least Root-Mean-Squared Error and highest Direction Accuracy making it the best model among these three algorithms.

By doing evaluation in WEKA, several graphs are acquired showing how close the model predicts rainfall by each classification algorithms. Using each classification algorithms a prediction is done every year in the training set and the graphs are plotted comparing the actual rainfall value and the value predicted by the model.

In the below graph the *x*-axis consists of instances (year from 1901 to 2017, respectively) and the *y*-axis consist of the annual rainfall in (mm). This graph shows the prediction done by the model using Linear regression algorithm. The blue nodes are the predicted rainfall and the red nodes are the actual rainfall of the respective years (Fig. 3).

In the below graph the *x*-axis consists of instances (year from 1901 to 2017, respectively) and the *y*-axis consist of the annual rainfall in (mm). This graph shows the prediction done by the model using Multilayer Perceptron algorithm. The blue nodes are the predicted rainfall and the red nodes are the actual rainfall of the respective years (Fig. 4).

In the below graph the *x*-axis consists of instances (year from 1901 to 2017, respectively) and the *y*-axis consist of the annual rainfall in (mm). This graph shows the prediction done by the model using SMOreg algorithm. The blue nodes are the

Table 1 Values obtained during evaluation of the three algorithms

	SMOreg	Linear regression	Multilayer perceptron
Direction accuracy	53.1532	45.9459	71.1712
Root-mean-squared error	406.1387	405.8745	313.4686

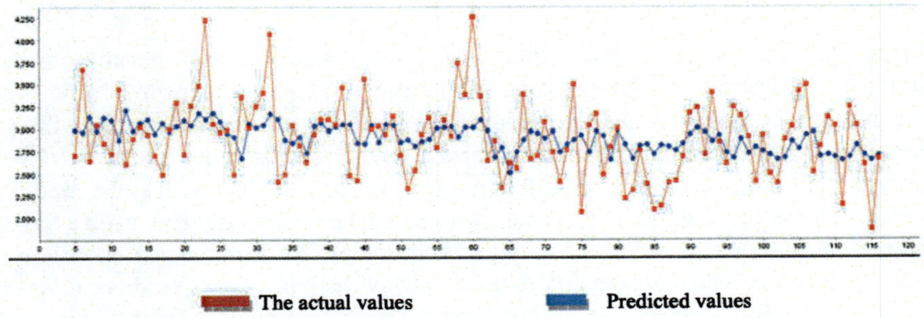

Fig. 3 Graph showing one step-ahead prediction of annual rainfall (linear regression)

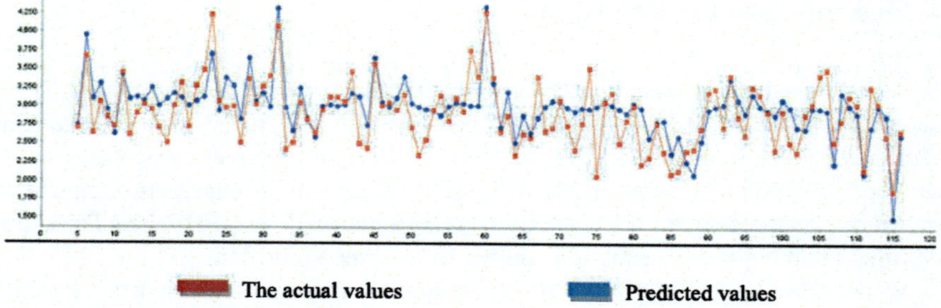

Fig. 4 Graph showing one step-ahead prediction of annual rainfall (multilayer Perceptron)

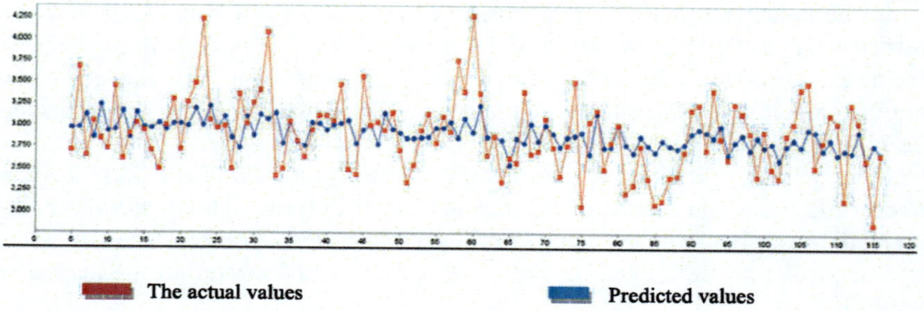

Fig. 5 Graph showing one step-ahead prediction of annual rainfall (SMOreg)

predicted rainfall and the red nodes are the actual rainfall of the respective years (Fig. 5).

As it sees from the graphs, the closest predictions are done by the model using the Multilayer Perceptron algorithm.

5 Conclusion

In the case of rainfall many management issues are present nowadays, because sometimes the lack of correct management of rainfall will cause another time. It may be the most important factor that affects India's economy. It is the most important thing for everyone; all our living conditions depend on the average rainfall received. The current system of predicting rainfall is not as good as it is. Nowadays, the prediction is not correct in some tough cases. Most of the rainfall predictions need a large amount of data. Also, it uses many types of attributes like Temperature, Humidity, Wind speed, etc. And more data types do not always help in a better prediction. As it all fails to predict in some cases, it is compelled to find a new way. Historic data are only used for prediction. By using the previous year rainfall data, the future prediction is done. Time series forecasting with Multilayer perceptron algorithm, SMOreg algorithm and Linear Regression is compared; all of these algorithms were compared on the basis of accuracy, such as Root-Mean-Squared Error (RMSE) and Direction Accuracy were taken into consideration. The one with the least RMSE is the best model; therefore, the process with less RMSE is taken for making predictions.

Here, the prediction is done by using the dataset of a large area; therefore it may show some variations between predicted values and actual values, in future this work can be improved by predicting the rainfalls from using historic data collected from smaller regions. Thus, combining the rainfall predictions from smaller regions altogether, the prediction accuracy can be improved.

References

1. C. Damle, *Flood Forecasting Using Time Series Data Mining*. Graduate Theses and Dissertations (2005)
2. A. Dhore, A. Byakude, B. Sonar, M. Waste, Weather prediction using data mining techniques. Int. Res. J. Eng. Technol. (IRJET) **04**(05) (2017). e-ISSN: 2395-0056
3. N.C. Barde, M. Patole, *Classification and Forecasting of Weather using ANN, k-NN and Naive Bayes Algorithms*
4. C.L. Wu, K.W. Chau, A flood forecasting neural network model with genetic algorithm. Int. J. Environ. Pollut. **28**(3/4) (2006)
5. R.R. Deshpande, On the rainfall time series prediction using multilayer perceptron artificial neural network. Int. J. Emerg. Technol. Adv. Eng. **2**(1) (2012). www.ijetae.com. ISSN 2250-2459
6. A. Jayakumar, A. Shaji, L. Nitha, Wildfire forecast within the districts of Kerala using Fuzzy and ANFIS, in *Proceedings of the 4th International Conference on Computing Methodologies and Communication, ICCMC 2020*, art. no. 9076432, (2020), pp. 666–669
7. K. Dutta, P. Gouthaman., Rainfall prediction using machine learning and neural network. IJRTE **9**(1) (2020). ISSN: 2277-3878
8. A. Graham, E.P. Mishra, Time series analysis model to forecast rainfall for Allahabad region. J. Pharma. Phytochem. **6**(5), 1418–1421 (2017)
9. T.M. Usha, S. Appavu Alias Balamurugan, Seasonal based electricity demand forecasting using time series analysis. Circ. Syst. **7**, 3320–3328 (2016). Published Online August 2016 in SciRes
10. T.M. Usha, S.A.A. Balamurugan, Seasonal based electricity demand forecasting using time series analysis. Circ. Syst. **7**, 3320–3328 (2016)
11. S. Kumar, S. Shobika, Monthly rainfall forecasting using 1-D deep convolutional neural network. Int. Res. J. Eng. Technol. (IRJET) **07**(12) (2020). e-ISSN: 2395-0056
12. A. Al-Muqrashi, A. Soosaimanickam, A comparative study of the efficient data mining algorithm for forecasting least prices in Oman fish markets. Int. J. Appl. Eng. Res. **13**(11) (2018). ISSN 0973-4562
13. A. Agrawal, V. Kumar, A. Pandey, I. Khan, An application of time series analysis for weather forecasting. Eng. Res. Appl. (IJERA) **2**(2) (2012). ISSN: 2248-9622 www.ijera.com
14. J. Joseph, T.K. Ratheesh, Rainfall prediction using data mining techniques. Int. J. Comput. Appl. **83**(8), 0975–8887 (2013)

Analysis of MQTT-Based Mesh Networks for Industry 4.o Applications

K. Ramamoorthy, S. Karthikeyan, and T. Chelladurai

Abstract Nowadays, the need for energy conserving in industries is in big rise. To meet the demand, various technologies have been adopted to fit industrial standards. This paper relies on the modern standard 4.o conventions used in critical industrial applications at each point. In Internet of Things (IoT), the use of short packets is expected to meet the stringent latency requirement in ultra-reliable low-latency communications network. In view of the reaction of ping production, idleness and throughput and analyzing their upgraded exhibition in companies such as oil and gas, energy utility, media transmission and some more industries are carried out. A number of application layer protocols have been considered in order to accommodate a wide range of applications. HTTP, a well-known, fundamental client–server protocol and the protocol that is most compatible with current network infrastructure, is one of the leading candidate solutions in this paper. In this paper, the Trial outcomes are acquired through demo industrial setup and recognized through Message Queuing Telemetry Transport (MQTT) and resultant checked through Wire shark and LCD in real-time.

1 Introduction

Industry 4.0–the term tossed over the bench of German industrialists and researchers has become an eye of correspondence layer in Cyber Physical Space (CPS). The physical entity is completely shaped over digital physical space by the machine, assessed, checked, adjusted dependent on client necessity, and re-done back to the physical

K. Ramamoorthy (✉) · S. Karthikeyan · T. Chelladurai
Department of ECE, PSNA College of Engineering and Technology, Dindigul, Tamil Nadu, India
e-mail: krmoorthy@psnacet.edu.in

S. Karthikeyan
e-mail: skarthik@psnacet.edu.in

T. Chelladurai
e-mail: chelladurai@psnacet.edu.in

P. Karuppusamy et al. (eds.), *Ubiquitous Intelligent Systems*, Smart Innovation, Systems and Technologies 243, https://doi.org/10.1007/978-981-16-3675-2_44

element (gadget machine-gadget circle). This paper will chip away at CPS-IoT coordination for critical things based on MQTT protocol. Machine to machine (M2M) instruction fetch and execution is dealt by Industry 4.0. This involves lightweight protocols like Constrained Application Protocol (CoAP). In [1] the security standards for long-distance transmission, inter-water communication, different layers and stacks, medium access control ecosystem have been discussed. Evaluations of protocols like XMPP (Extensible Messaging and Presence Protocol); MQTT; CoAP; RESTFUL; DSS (Decision Support System); AMQP (Advanced Message Queuing Protocol); and Web Socket were performed in [2]. Reem Abdul Rahman, Babar Shah have been described the details over security standards in [3] and list out the redefined HTTP (Hypertext Transfer Protocol) to enable data layer security through CoAP.

Application of CoAP for class 1 and 2 machines serves a limitation. Important constraints like power consumption, performance, reliability, predictability, and efficiency were major highlights in [4]. Throwing light on MQTT-reliability, operating layers table, and data distribution system to power, grid communication devices. Though the paper holds scope for 5 g, its compatibility cannot be matched during consequent updates. Industrial Internet of Things hierarchy, models of IIOT, and its functionality stand as design pattern [5]. This paper's scope relies on heterogeneous model implementations.

In [6], inter-operability illustrates heterogeneous approach on exchange of data between devices, networks, protocols, physical connectors that works on platform even after upgrade. All protocols do not overlap functions and stay different for different devices and their upgrades. This holds as the paper's limitation. Organizing huge database by efficient resource management and intelligent broker is the main aspect in [7].

Segregation of data into meta-modules enables the intelligent broker to index and prioritize the request and acknowledgment. Paper [7] can be implemented over interactive and real-time updates can be done in MQTT communication. Challenge of MODBUS/TCP reliability and wired embedded design of recent trend is sorted in [8]. Its opportunity may hold good for one-to-many client-subscriber paradigms.

Utilization of short packets for low-latency communications is altered by finite block-length coding. By using actuator and eaves-antenna dropper, resulted in low latency [9]. Here, short packets provide high quality data. The antenna dropper instead of crypto technology for physical security layer is considered. Research may follow relaying nodes addition, inclusion of more antenna droppers, aiming over Wi-Fi security stacking ways to minimize block-length code. In [9], limitation consists of

1. Addition of nodes can cause more error
2. More users or more signal-to-noise ratio (SNR), the total concept will collapse
3. Much of a theoretical approach.

Considering mission critical approach design models are discussed in [10]. This is followed by the communication technology for industry4.0 [11] where it illustrates about Production Performance Management Protocol (PPMP), Reference Architecture Model Industry (RAMI) model shown in Fig. 1. In [12], uneven network node energy consumption and local optimum are reached by the algorithm protocol due to

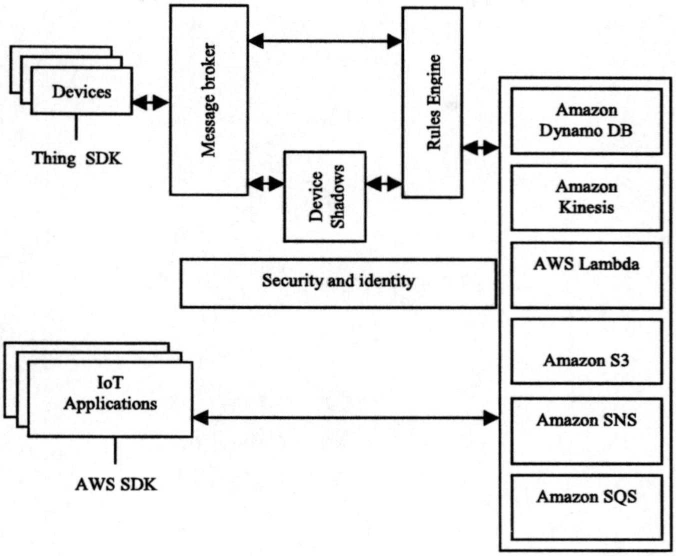

Fig. 1 Architecture-device and broker implementation

the high energy consumption issues relating to the routing strategy. The survey work in [13] mainly provides an insight about deep learning through an intensive analysis of deep learning architectures and its characteristics along with its limitations.

2 Methodology

The Wi-Fi interestingly with the Zigbee has higher data transfer capacity and bolsters IP-based exchanging. The Zigbee router configuration is illustrated in Fig. 2.

The industrial IoT conventions like Wireless HART, MQTT, COAP, and REST can be actualized effectively with our Wi-Fi Network. In industries particularly for the mission critical applications emerge a requirement for the quicker control and speedier observing, which cannot be provided by the different remote advances like ZIGBEE, Bluetooth, and so forth. The bandwidth utilization for different remote devices is shown in Fig. 3.

The protocol of Wi-Fi mesh networks for real-time applications is analyzed. It involves comparison of available internet protocols by their ping response, data length, latency, and throughput. An efficient protocol was identified for mission critical application. An industrial real-time model was built, and resultant efficient protocol was applied, tested, and verified.

Heavy data transfer is routed with low latency and high security. Disadvantages of Wired system in industry can be totally converted to the standards of Industry 4.0 in

Fig. 2 ZigBee router configuration

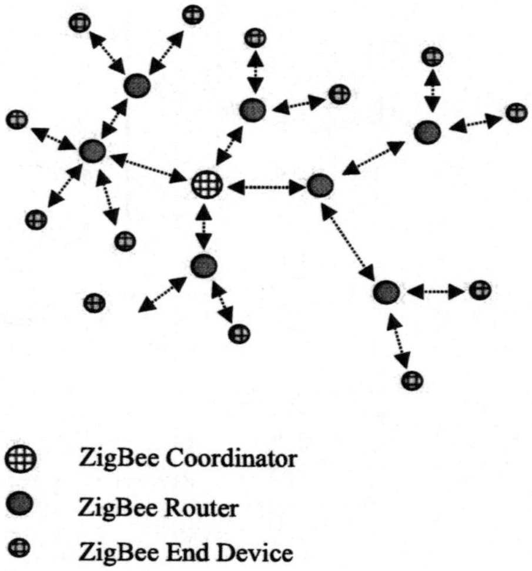

Fig. 3 Bandwidth utilization

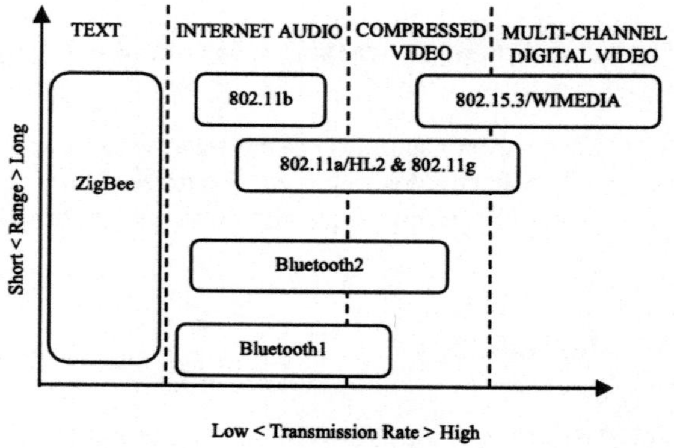

real time. In order to meet the demands of the energy-saving requirements in industry, a novel method of MQTT-based self-organizing and self- healing mesh network which allows numerous devices spread over a large physical area (both indoors and outdoors) to be interconnected under a single WLAN (Wireless Local-Area Network) through Wi-Fi is proposed. The work features:

a. True Ad hoc networking
b. JSON based
c. Wi-Fi-based networking over single routing device
d. Deep sleep mode can be enabled to save more power
e. Reconfigurable.

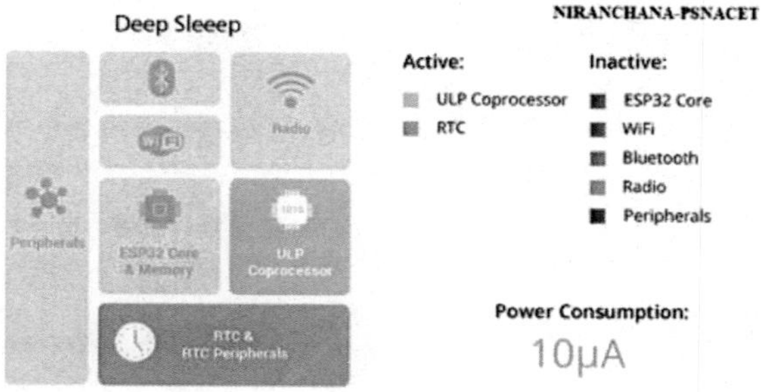

Fig. 4 Power saving-deep sleep mode

2.1 True Ad Hoc Networking

Network is a genuine specially appointed system, implying that no-arranging, focal controller, or switch are required. Any arrangement of at least 1 hub or more will self-compose into completely useful work. The most extreme size of the work is constrained by the measure of memory. But in the pile that can be allotted to the sub-associations support is actually very high. Each hub will have a novel number. Messages can either be communicated to the entirety of the hubs on the mesh or sent explicitly to an individual hub which is recognized by its 'nodeld.'

2.2 Deep Sleep Mode Versus Active Mode

The power consumption objective is observed in epic way of efficiency and battery longevity shown in Fig. 4.

3 Working Principle-MQTT Protocol

MQTT is a binary-based protocol with binary bytes rather than text strings as control elements. A command and command acknowledgment format is used by MQTT. This implies that each instruction has a corresponding acknowledgment. UTF-8 strings are used to encrypt topic names, client IDs, user names, and passwords. The payload is binary data, except MQTT protocol information such as Client ID, and the content and format are application specific. A 2 byte fixed header (always present) + variable-header (not always present) + payload make up the MQTT packet or message format (not always present). The control field and the variable length packet length field make up the fixed header field. For messages with a total length of less than 127 bytes, the packet length area has a minimum size of 1 byte. (Not including the duration and control fields) The maximum size of a packet is 256 MB. A 1 byte packet length

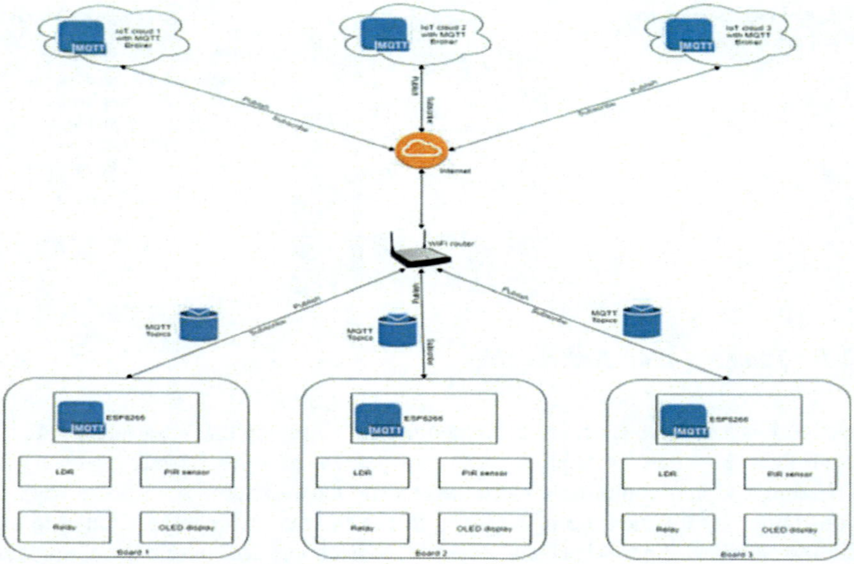

Fig. 5 Block diagram-MQTT protocol

area is used for small packets of less than 127 bytes. 2 bytes will be used for packets greater than 127 and smaller than 16,383.

For the demo, three sets of customized board are made to communicate with the IoT cloud loaded with different firmwares for communicating with IoT cloud to justify the capabilities of the MQTT protocol shown in Fig. 5.

Each customized contains the following:

- Node MCU-based ESP8266 board with ESP12E module.
- LDR circuit.
- Relay circuit.
- I2C-based OLED display.

First application deals with the control of the relay with the PIR sensor using MQTT where the values of the PIR sensor are sent to IoT cloud by publishing via MQTT broker under a MQTT topic. If the values of PIR sensor are turned positive, then the IoT cloud publishes back with a value which is in-turn subscribed by the NodeMCU board. Depending on the char value received via MQTT topic, the relay is triggered, whose status is again published in the cloud. So in the MQTT dashboard app, we can able to see the status of the PIR, relay. To display the messages that has been published or subscribed to IoT cloud locally an OLED display is used.

Second application deals with the control of the relay with the LDR sensor using MQTT where the values of the LDR sensor are sent to IoT cloud by publishing via MQTT broker publishes back with a value which is in-turn subscribed by the NodeMCU board. Depending on the char value received via MQTT topic, the relay is triggered, whose status is again published in the cloud. So in the MQTT dashboard app we can able to see the status of the LDR, relay. To display the messages that has been published or subscribed to IoT cloud locally an OLED display is used.

Third application deals with the control of the relay with the manual MQTT topic using MQTT where from the MQTT dashboard app user publishes a value under a MQTT topic to the customized. If the values subscribed by the customized board turn high the relay are triggered, whose status is again published in the cloud. So, in the MQTT dashboard app, we can able to see the status of the LDR, PIR, and relay. To display the messages that has been published or subscribed to IoT cloud locally an OLED display is used.

4 Results and Discussion

The proposed work was simulated, and the synthesis report was obtained by using Wire shark network analyzer. The internet protocol MQTT was derived as the most effective, one-to-many systems with publish-subscribe paradigm. The detailed MQTT working is studied and was tested using chrome MQTT open-source broker shown in Fig. 6. MQTT packets are explained in case of mosquito MQTT broker. Google chrome MQTT server enables the task easier and researcher friendly with inbuilt packet updates.

The initially determined efficient protocol without load plus interference was studied in real-time with load and interference. Proposed work comprises the analysis of the optimal internet protocol that was estimated and implemented for mission critical applications in Industry4.0. Model of 4.0 was built, and testing was completed.

Now the whole process of MQTT communication is done, we must determine the Wi-Fi ping request and response by estimating (i) actual data length of this MQTT connection (ii) ping timing response shown in Fig. 7. Wire shark, an open-source platform, is used for determining the same. Thus, latency of MQTT shall be calculated

Fig. 6 MQTT broker

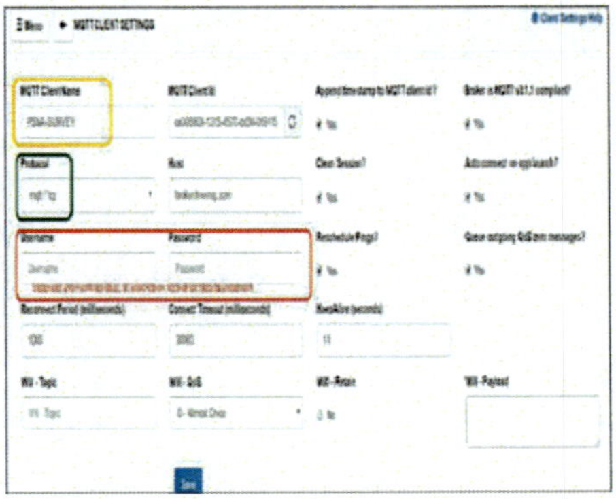

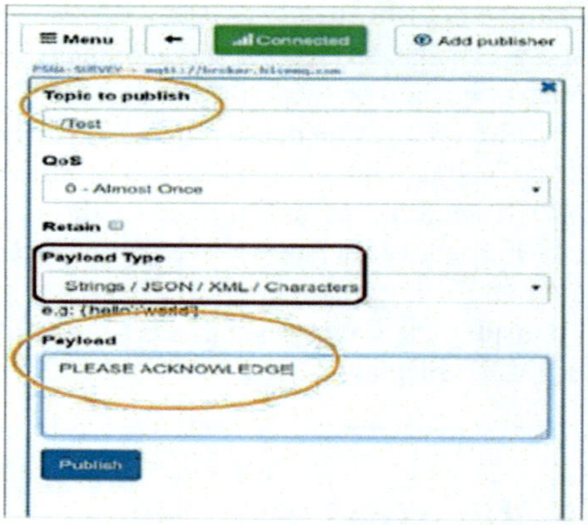

Fig. 7 Publish and acknowledgment

from the derived framework. The various readings based on interference and load were calculated and shown in Fig. 10, while connected to different MQTT brokers.

The variant result was monitored through the OLED as below; and the resultant Wire shark reading is noted in Fig. 8 for acute analysis of ping response and latency as with load plus interference and without both shown in Fig. 11. Here, LDR unit of the topics was subscribed on a remote client observation (mobile). The ping and latency were observed, and it does not cause much of a delay even in real time. Thus, the wire shark reading for the transmission of data through MQTT is shown in Fig. 9 and the MQTT data format is shown in Fig. 10 which provides a best data transmission.

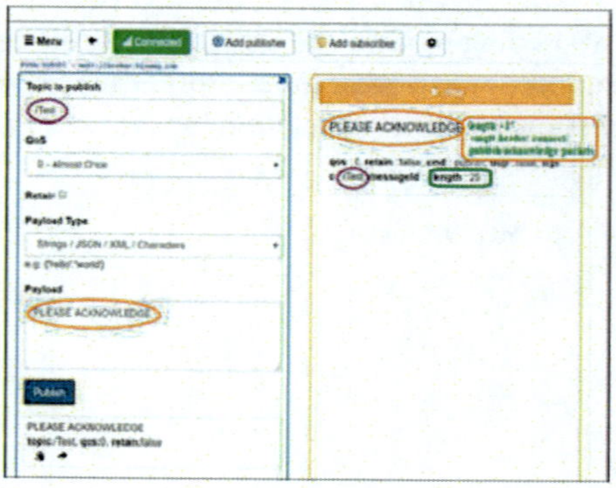

Fig. 8 Topic-subscribe length and ping test

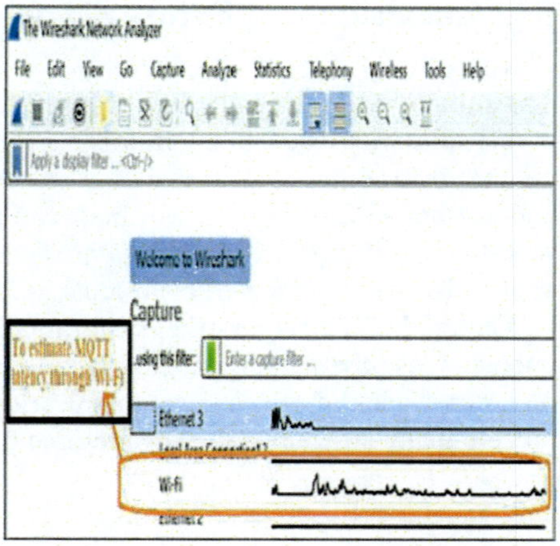

Fig. 9 Wire shark reading

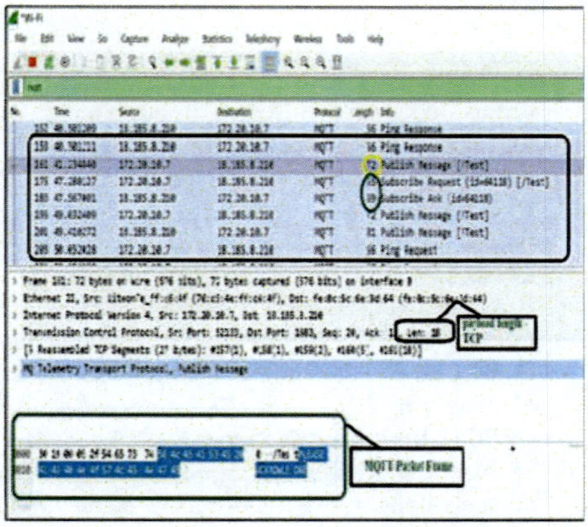

Fig. 10 MQTT frame formats for data

Fig. 11 Load and interference readings

5 Conclusion and Future Scope

The initially determined efficient protocol without load plus interference was studied in real-time with load and interference. Proposed work comprises the analysis of the optimal internet protocol that was estimated and implemented for mission critical applications in Industry 4.0 was built and testing was completed. As aimed, there was slighter difference in ping response and latency when load and interference were applied but still held the efficient protocol.

Though MQTT serves efficient for multi-unit industrial communication through publish-subscribe paradigm. CoAp ping latency impresses researchers but holds a major drawback of one–one communication like a wired connection. Working over one too many coAp models may increase the resultant in higher efficiency.

References

1. F. Ciccozzi, I. Crnkovic, D. Di Ruscio, P. Pelliccione, Model-driven engineering for mission-critical IoT systems. IEEE Comput. Soc. (2017)
2. M. Iglesias-Urkia, A. Orive, M. Barcelo, A. Moran, J. Bilbao, A. Urbieta, Towards a lightweight protocol for Industry 4.0: an implementation based benchmark, in *IEEE ISSNIP* (2014)
3. M.B. Yassein, D. Al-zoubi, M.Q. Shatnawi, Application layer protocols for the internet of things: a survey, in *2016 International Conference on Engineering and MIS (ICEMIS)* (IEEE, 2016)
4. R.A. Rahman, B. Shah, Security analysis of IoT protocols: a focus in CoAP, in *3rd MEC International Conference on Big Data and Smart City* (2016)
5. G. Bloom, B. Alsulami, E. Nwafor, I.C. Bertolotti, Design patterns for the industrial internet of things, in *2018 14th IEEE International Workshop on Factory Communication Systems (WFCS)*, Imperia, (2018), pp. 1–10. https://doi.org/10.1109/WFCS.2018.8402353
6. M. Noura, M. Atiquzzaman, M. Gaedke, Interoperability in internet of things: taxonomies and open challenges. J. Sens. Actuator Netw. (2019)
7. M. Saqlain, M. Piao, Y. Shim, J.Y. Lee, Framework of an IoT-based industrial data management for smart manufacturing. J. Actuator Netw. Sens. Actuators **8**(2) (2019)
8. S. Jaloudi, Communication protocols of an industrial internet of things environment: a comparative study. Future Internet (2019)
9. H.-M. Wang, Q. Yang, Z. Ding, H. Vincent Poor, Secure short-packet communications for mission-critical IoT applications. IEEE Trans. Wireless Commun. arXiv:1903.01433 (2019)
10. P. Marcon, F. Zezulka, I. Vesely, Z. Szabo, Z. Roubal, O. Sajdl, E. Gescheidtova, P. Dohnal, Communication technology for industry 4.0, in *Progress In Electromagnetics Research Symposium*, Spring (PIERS) (2017)
11. B. Safaei, A.M.H. Monazzah, M.B. Bafroei, A. Ejlali, Reliability side-effects in internet of things application layer protocols, in *2nd International Conference on System Reliability and Safety* (2017)
12. J.I.Z. Chen, K.-L. Lai, Machine learning based energy management at internet of things network nodes. J. Trends Comput. Sci. Smart Technol. (TCSST) **2**(03), 127–133 (2020)
13. S. Smys, J.I.Z. Chen, S. Shakya,Survey on neural network architectures with deep learning. J. Soft Comput. Paradigm (JSCP) **2**(03), 186–194 (2020)

An Improved Dehazing and De-raining Technique for Haze and Rain Streaks Removal

Anjana Anand, Aparna Suresh, P. R. Meera, and L. Nitha

Abstract De-raining and dehazing is a challenging problem. To overcome the challenge, this research work develops a novel de-raining and dehazing technique with a combination of dark channel prior technique for dehazing and L_0 gradient minimization for de-raining. Almost all the works done in this field have performed either dehazing or de-raining, but this paper addresses both dehazing and de-raining. This method efficiently eliminates haze and rain from the images. The proposed method eliminates the haze and rain effectively than any other existing methods. The minimization approach can throughout control the gradients (non-zero), which are upshot in image. This technique is self-sufficient for local characteristics. DCP method is established to restore the haze-affected picture. The outcome shows the efficiency obtained by combining DCP and L_0 gradient minimization in eliminating the haze and rain from images, and also this research work results in an acceptable method to achieve a better clarity loss and color loss.

1 Introduction

Weather conditions like rain or fog cause distortion in image as shown in Fig. 1 (image shown is covered with rain and haze, which reduces the image quality). Rain streaks cause poor human vision and not only that, it also dramatically degrades the efficacy of computer vision algorithms utilized for object recognition, object monitoring, object restoration, and so on. The rainwater and haze droplets reduce the picture quality but haze is also responsible for the occurrence of many road accidents, so it is very important to eliminate the haze and rain from images. The clearing of rain or snow has acquired a lot of attention, particularly the clearing of rain. Hazed pictures cause the quality of the image to degrade. The degraded images have low color contrast and fidelity. These images are affected due to fog, smoke, dust, etc. These factors are nothing, but haze and rain affect the images. Such images

A. Anand (✉) · A. Suresh · P. R. Meera · L. Nitha
Department of Computer Science & IT, Amrita School of Arts & Sciences, Amrita Vishwa Vidyapeetham, Kochi, India

P. Karuppusamy et al. (eds.), *Ubiquitous Intelligent Systems*, Smart Innovation, Systems and Technologies 243, https://doi.org/10.1007/978-981-16-3675-2_45

Fig. 1 Image covered with haze and rain

are reduced in quality and which in turn reduces contrast and visibility. So to remove these factors, in-depth knowledge of the scene is required.

This paper addresses both the dehazing and de-raining of a single image. Inborn occurrences create unclear scenes, lessen clearness, and alter colors, which is known as haze. It is an infuriating issue for photographers as it degrades picture standards. It also warns against the accuracy of numerous applications, such as exterior surveillance, object diagnosis and aerial imaging. So detaching haze from images is dominant in computer graphics. The rain significantly degrades visibility and is responsible for many computer vision problems. Generally, rain initiates hardly any kind of visibility degradation. Raindrops block, deform and unfocus scenes, especially background. Rain streaks stockpile causes outcomes comparable to smog, which seriously diminish visibility by the dispersion of light out and into the line of vision. The nearby rainfall trails show strong specular high points which block the backdrop scenes. These rain trails can have various shapes and regulations, especially in heavy rains, resulting in drastic degradation of visibility. All the works which have done in the field were either haze or rain is done. This paper addresses the elimination of both rain and haze effectively.

In this paper, we introduce a combination of dark channel prior [1] for dehazing and L_0 gradient minimization [2] for de-raining. The minimization procedure can over domination of how gradients (non-zero) are out-turned in the picture. This method is self-sufficient in local characteristics and alternatively detects major edges overall. We have also put in place better comprehensible groundwork on the properties of the clear background edges.

The primary borders are conserved and unimportant features are declined. Rain components can be separated by this method. The outcome depicts that the algorithm which is put forward is effective as it also separates and maintains the features of the image while it is raining heavily. DCP is an unquestioned and strong dehazing method. We introduce an effective and efficient method of dehazing and de-raining to enhance the properties and attributes of the image using the MATLAB tool.

2 Literature Review

Fu et al. [3] suggested a rain elimination framework based on a single framework by formulating rainfall elimination as a problem of image corruption resolved by running sparse dictionary algorithms [3]. The outcome proves that this proposed method can efficiently remove rain from the original image without blurring. The implementation can furthermore be improved in future works.

Zhang et al. [4] suggested that in order to eliminate the rain from an image, a GAN algorithm can be used. To boost training steadiness and reduce the artifacts instituted by GAN in resultant images, they established usage of loss operation in the GAN optimization framework [4] in collation with various novel approaches which shows that their method makes substantial advancement in terms of various measures. Furthermore, a thorough ablation survey is performed to distinctly demonstrate improvements gained due to various modules in the suggested technique.

Huang et al. [5] suggested a framework that learns context details in an unmonitored setting, while the rain patterns can be inevitably determined and eliminated from dictionaries learnt for each context classification. Distinct from prior rain suppressing techniques, they tackle this problem as the diagnosis of rain patterns from the input image. Their objective is to design an image and situation-dependent rain elimination framework without any training image data or presumptions on any image priors.

Guo et al. [6] have proposed a method to solve the decomposition problem. This method uses simple patches of Gaussian mixture models [6]. On the basis of their fact-finding, they have shown that this is more efficient than techniques on low-ranking constraints and learning dictionaries.

Schaul et al. [7] proposed an approach that relies on combining multi-resolution images. They combined an edge-conserving smoothening filter with a multiresolution decomposition method. Their method provides better degradation in scattering. Compared to local enhancement techniques, their procedure does not demand heuristics to track down fog because NIR images are basically fog-free.

Zheng et al. [8] have projected a brand new technique for de-raining and dehazing image. They use the contrast between background edges with rain and haze. The various properties are specified by the low frequency, so a rain or haze elimination technique is projected. The elimination half is generally created from a target-hunting filter. The findings show that their technique is effective and economical for dehazing and de-raining.

Gaofeng Meng graphist [9] introduced an effective formalized approach to eliminate haze from an image. This technique uses built-in constraints on the imparting operation. This is united with contextual adjustment on the L1 standard and is modeled as an expansion problem to evaluate the transmittal of unknown scenes. A fairly systematic bust variable algorithm is also present to fix the issue, and results show that several dry fog images indicate the effectiveness of the suggested technique.

Zhang et al. [10] introduced an easy method called change of detail prior [10] also called cod prior. This technique solves unsatisfactory dehaze in native regions. The change of detail prior established on native content. Their technique is a lot meatier than the present color primarily based ways, and as a result shows, it effectively eliminates the haze from images.

Makarau et al. [11] proposed an algorithmic rule separate spatially differing haze supported the HTM [11]. Associate in estimation of haze removal technology employing a scene from constant space reveals that haze removal is unit spectrally trusted.

3 Proposed System

There are many techniques for removing fog and rain trails, but all this eliminates rain or fog trails. Here we introduce a combination of dark channel prior [1] for dehazing and L_0 gradient minimization [2] for de-raining. Figure 2 illustrates how dispersion of air particles due to rain and haze causes distortion to an image and how it can be solved. The primary goal of combining both is for the retention of picture clarity and color. There are two steps for this.

(a) Dark Channel Prior
(b) L_0 Gradient Minimization.

3.1 Dark Channel Prior

Dark channel prior (DCP) is one of the notable dehazing techniques based on the observation of the key features of the haze-free images. The steps of dark channel prior are shown in Fig. 3.

Expression for the hazy image (input image) is got from the atmospheric scattering model and it is

$$I(x) = t(x) \cdot J(x) + (1 - t(x)) \cdot A \tag{1}$$

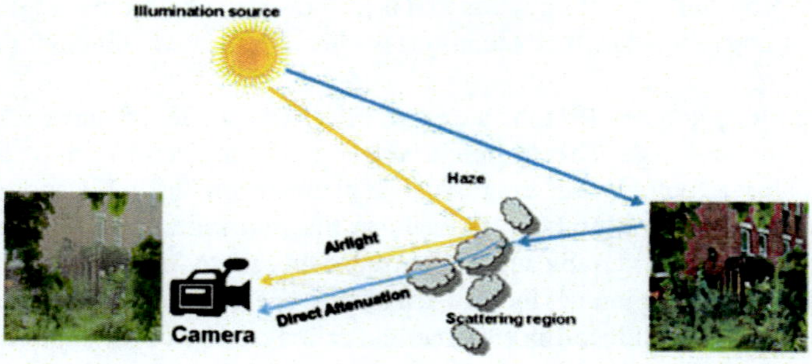

Fig. 2 Climatic scattering of air particles

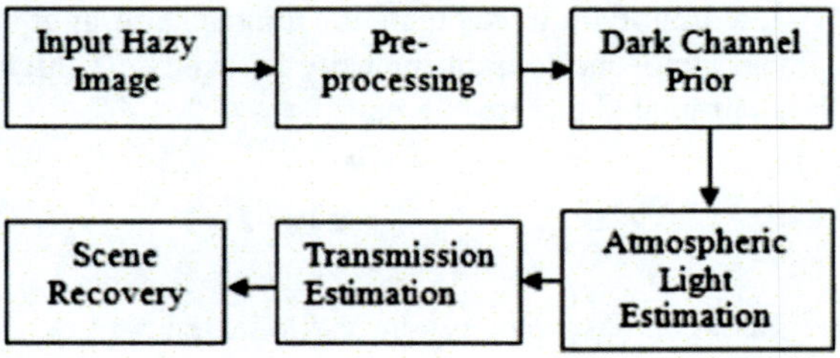

Fig. 3 Steps done in dark channel prior

where I = Input image intensity, J = Recovery image intensity, t = Transmission, A = Atmospheric light

$$J^{\text{dark}}(x) = \min_{c \in \{r,g,b\}} \left(\min J^c(x) \right) \tag{2}$$

J^c = Recovery image (J) color channel. J^{dark} = darkest pixel in picture. Assume, $J^c = 0$, J^c is referred to as J^{dark}.

$$J^{\text{dark}} = 0 \tag{3}$$

Hence, Eq. (1)

$$\frac{I^c(x)}{A^c} = t(x) \cdot \frac{J^c(x)}{A^c} + (1 - t(x)) \tag{4}$$

Apply min function on both sides:

$$\min\left(\min \frac{I^c(x)}{A^c} \right) = t(x) \cdot \min\left(\min \frac{J^c(x)}{A^c} \right) + (1 - t(x)) \tag{5}$$

Equation (2)

$$\min\left(\min J^c(y) \right) = 0 \tag{6}$$

To estimate transmission, apply Eq. (6) in Eq. (5),

$$t(x) = 1 - \min\left(\min \frac{I^c(y)}{A} \right) \tag{7}$$

During estimation transmission of light, weight coefficient is added in order to process the small amount of reserved haze. The light transmission

$$t(x) = 1 - \omega \cdot \min\left(\min \frac{I^c(y)}{A^c} \right) \tag{8}$$

It is important to calculate the atmospheric light of images with thick haze. The global atmospheric light intensity is taken as I (highest intensity of input image). Fog eliminated picture can be

$$J(x) = I(x) - \frac{A}{t(x)} + A \tag{9}$$

Equation (9) provides a haze eliminated picture.

3.2 L_0 Gradient Minimization

Performed by a procedure called L_0 gradient smoothening, Xu et al. [12] recognizes high-contrast borders using limited number non-zero gradients. It is independently of the attributes of rain. L_0 smoothing has already been used efficiently for [12] artifacts suppression, Edge boosting, extraction image abstraction and pencil sketching.

Smoothing Operation: There are two types of smoothing operations. ID Smoothing and 2D smoothing.

ID Smoothing [12] denotes input as '*g*' and smoothened output as '*f*' *the counting function is*

$$c(f) = \#\{p | |f_p - f_{p+1}| \neq 0\} \tag{10}$$

p and $p + 1$ Adjacent pixel index $|f_p - f_{p+1}|$ is gradient with respect to p.

{} is counting operator that outputs the number of p that satisfies

$$|f_p - f_{p+1}| \neq 0$$

Resultant 'f' must be structurally similar to g

$$\min_f \sum_p (f_p - g_p)^2 \tag{11}$$

$C(f)$ = non-zero gradient (k) exist. Then equation (11) is

$$\min_f \sum\nolimits_p (f_p - g_p)^2 + \lambda \cdot c(f) \tag{12}$$

where λ = smoothing specification that controls value k and finds a balance between result similarity with input signal.

In 2D Smoothing [12], input image = 'I' and output = 'S.' The gradient $\nabla S_p = (\partial_x S_p, \partial_y S_p)$, p can be estimated in x and y directions and is stated as:

$$C(S) = \#\{p | |\partial_x S_p| + |\partial_y S_p| \neq 0\} \tag{13}$$

The cost function will be

$$\min_{S}\left\{\sum_{p}(S_p - I_p)^2 + \lambda \cdot c(S)\right\} \tag{14}$$

Equation (14) is solved using half quadratic splitting technique. Wang et al. [13] established supplementary variables *hp* and *vp* to expand the initial terms and upgrade them repeatedly. Then Eq. (14) evolve into:

$$\min_{s,h,v}\left\{\sum_{p}(S_p - I_p)^2 + \beta\left(\left(\delta_x S_p - h_p\right)^2 + \left(\delta_y S_p - v_p\right)^2\right)\right\} \tag{15}$$

where β is the criteria, to monitor *(h, v)* and its correspondent gradients $\partial_x S_p$ and $\partial_y S_p$.

Subproblem 1: Calculate *S* by minimizing given equation:

$$\left\{\sum_{p}(S_p - I_p)^2 + \beta((\delta_x S_p - h_p)^2 + \left(\delta_y S_p - v_p)^2\right)\right\} \tag{16}$$

Subproblem 2: Estimate value for *(h,v):*

$$\min_{h,v}\left\{\sum_{p}(\partial_x S_p - h_p)^2 + (\delta_x S_p - v_p)^2 + \frac{\lambda}{\beta}C(h, v)\right\} \tag{17}$$

Algorithm 1: L_0 Gradient Minimization. **Input**: image I, smoothing weight λ, smoothing rateκ, parameters β and β_{max}.

Output: smoothed image S

Initialization: S ← I, β_{max} ← β, κ = 2.0. λ = 2E − 2, β_{max} = IE5

Read input image I

Evaluate FFT on I

Introduce auxiliary variables h and v

β = 2 * λ

While β < βmax

  Solve (h, v) subproblem

  Solve S subproblem

  Evaluate IFFT

  β = β*κ

End

Output smoothed image S

Enhance S

In almost all techniques based on this idea, either de-raining or dehazing is done. This has a consequence of clarity loss or color loss of image. The image also gets dark when attempting to reinstate color. This paper stresses more on elimination of the fog and rain in the picture without reducing clarity. It is attained by the method dark channel prior along with the procedure of evaluating airlight (global), evaluation of dark channel and depth estimation, we come to an acceptable way to get the better of the issue color and clarity loss. The L_0 gradient minimization method is efficient and can be used for pictures with heavy showers. The technique is self-sufficient of properties (local) like space and chromatics, it also overall preserves the prominent areas. The lack of local filtering and averaging functions helps in not blurring the edges. The improvement of the smoothened pictures produces quality images. The variety of still pictures taken in illumination to heavy atmospheric conditions have substantiated that the suggested technique eliminates both rain streaks and haze efficiently.

Algorithm of Proposed Method Input: image I, smoothing weight λ, smoothing rate K, parameters β and βmax

Output: output image S
Initialization: S ← I, βmax ← β, K = 2.0, λ = 2E-2, βmax = IE_5
Read input image I
Evaluate FFT on I
Introduce auxiliary variables h and v
B = 2*λ
While β < βmax
Solve (h,v) subproblem
  Solve S subproblem
  Evaluate IFFT
  β = β * K
End
Output smoothed image S
Enhance S
Consider the patch(mask) of 15*15
Initiate the padding size ass half of the patch size i.e., 7*7
Initiate a blank matrix of size similar to input image(S)
For h = 1; height of image(s)
For w = 1; width of image(s)

$$\text{Patch} = \text{imgh} : (\text{h} + \text{patchsize} - 1, (\text{w}; \text{w} + \text{patchsize} - 1))$$

Calculate minimum value within the patch.

$$\text{J(Dark(h, w)} = \min(\text{patch})$$

Repeat the step until all pixel in the image are processed
End

Fig. 4 **a** Hazed picture, **b** dark channel picture, **c** depth estimation, **d** de-hazed image

4 Results Analysis

4.1 Haze Elimination

Figure 4 shows images obtained in MATLAB for eliminating haze from a picture. The processing of pictures in MATLAB is shown in Fig. 4. Hazed picture is given as input here, and after applying the algorithms, we obtain the output of the dark channel prior to the image, the value is above 0 which shows a hazed image.

The transmission map is also called as depth estimation [14] and the comparison of both hazed and de-hazed pictures. Dehazed picture acquired will preserve its color and eliminate haze without any clarity loss.

4.2 Rain Elimination

Elimination of rain streaks from still images utilizing the L_0 gradient reduction [2] method is efficient and can be used on all pictures with heavy showers. The method is self-sufficient of local attributes globally. It also preserves the prominent areas. Because of the lack of local filters and average functions edges are not blurred. The improvement of the smoothened pictures produces quality pictures, and it also gives a substantial simulation outcome on several still images taken in heavy rain conditions. The processing steps of eliminating rain in MATLAB are shown in Fig. 5.

Connected Components Analysis

$A1 =$ number of rain streaks in input image.
$A2 =$ number of rain streaks in restored image

Fig. 5 **a** Input image. **b**, **c** obtained from the proposed method, **d** de-rained image

$$\text{Percentage of Rain Removal} = (1 - (A2/A1)) * 100)$$

By using the combination of both, haze and rain streaks from the images are completely removed and thus provides a clear dazed and de-rained image. The steps that are processed for dehazing and de-raining by dark channel prior [1] and L_0 gradient minimization [2] are shown in Fig. 6. Contrast in rain and haze condition of the restored image are shown in Fig. 7.

Fig. 6 **a** Input rain and haze image. **b, c** obtained from proposed technique, **d** final dehazed and de-rained image

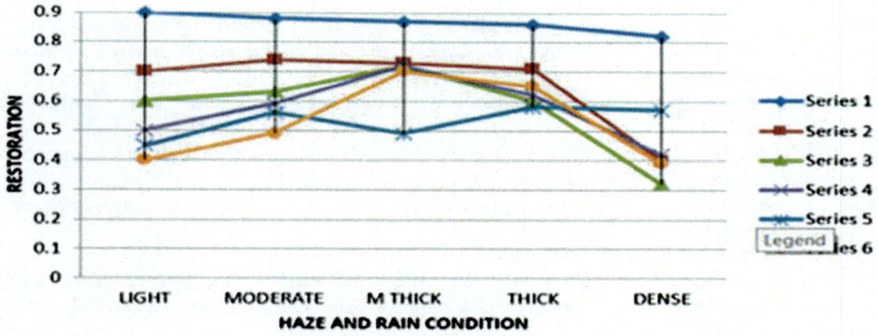

Fig. 7 Graph of contrast of rain and haze in the image

5 Conclusion

De-hazing and de-raining techniques use a combination of dark channel prior method [1] and L_0 gradient minimization technique [2]. Our execution reduces the color loss and clarity loss restored pictures. The method of haze removal and rain removal can be used at any time to misty and rainy pictures—outdoor pictures, sky images, undersea pictures and so forth. Aside from fog and rain decrement, the purity of the picture is improved without unfavorable color loss. The advantage of our method is that it is independent of the attributes (local) and also maintains and improves the prominent areas. More research works could be done in this field in order to get improved results. One of the demerits is the processing time of our technique is very slow when compared with other methods. Due to the lack of local filters and average functions, edges are not blurred. Improvement of smoothened images provides higher quality de-hazed and de-rained images.

References

1. P. Gurav, A. Patil, S. Londhe, S. Waghmare, Haze removal using dark channel prior
2. B.N. Manu, Rain removal from still images using L0 gradient minimization technique
3. Y.-H. Fu, L.-W. Kang, C.-W. Lin, C.-T. Hsu, Single-frame-based rain removal via image decomposition
4. H. Zhang, V. Sindagi, V.M. Patel, Image de-raining using a conditional generative adversarial network
5. D.-A. Huang, L.-W. Kang, M.-C. Yang, C.-W. Lin, Y.-C. Frank Wang, Context-aware single image rain removal
6. Y. Li, R.T. Tan, X. Guo, J. Lu, M.S. Brown, Rain streak removal using layer priors
7. L. Schaul, C. Fredembach, S. Süsstrunk, Color image dehazing using the near-infrared
8. X. Zheng, Y. Liao, W. Guo, X. Fu, X. Ding, Single-image-based rain and snow removal using multi-guided filter
9. G. Meng, Y. Wang, J. Duan, S. Xiang, C. Pan, Efficient image dehazing with boundary constraint and contextual regularization

10. J. Li, H. Zhang, D. Yuan, M. Sun, Single image dehazing using the change of detail prior
11. A. Makarau, R. Richter, R. Müller, P. Reinartz, Haze detection and removal in remotely sensed multispectral imagery
12. X. Li, et al., Image smoothing via L0 gradient minimization. ACM Trans. Graph. (TOG) **30**(6) (2011)
13. Y. Wang, J. Yang, W. Yin, Y. Zhang, A new alternating minimization algorithm for total variation image reconstruction. SIAM J. Imag. Sci. **1**(3), 248–272 (2008)
14. N. Aiswarya Menon, K.S. Anusree, A. Jerome, An enhanced digital image processing based dehazing techniques for haze removal
15. https://github.com/vaibzz/Fog-Removal-using-dark-channel-Prior
16. J.S. Manoharan, G. Jayaseelan, Single image dehazing using deep belief NN to reduce computational complexity, in *New trends in computational vision and bio-inspired computing*, ed. by S. Smys, et al. (2018). https://doi.org/10.1007/978-3-030-41862-5_151.

Minimized Error Rate with Improved Prediction Accuracy Using Pre-processing Models

K. Saravana Kumar and N. Shenbagavadivu

Abstract Over the last few decades, heart-related diseases or cardiovascular diseases (CVDs) are considered to be the deadly disease which affects both men and women not only in India, as well as across the world. Diagnosing heart disease in the health care field can be a challenging task and many researchers devoting their time and knowledge to develop intelligent clinical decision support systems to improve the ability of the clinicians. Out of many machine learning techniques, classification is the most powerful technique which is commonly used for prediction. The rapid growth of complex data about patient medical records is the main source of discovering hidden knowledge for error-free decision-making. The main purpose of this study is to assist non-specialists in diagnosing heart disease risk limits and in diagnosing heart disease early. This paper introduces pre-processing models that have been applied on the University of California, Irvine (UCI) repository—Cleveland heart disease database for examining the accuracy and the error rate on predictions. The results obtained are compared and then verified using the Jupyter Notebook interface with the Scikit-Learn Library in PYTHON. In order to validate the efficacy of the proposed model, 75% training–25% testing is performed on the data set. Our analysis over few classification algorithms shows the efficacy of our proposed pre-processing models.

1 Introduction

In recent years, CVD is the major cause of human death [1, 2]. CVDs are the leading cause of death worldwide. More people die from CVDs each year than any other cause. In high- and low-income countries, more than 80% of CVD deaths occur equally among men and women. According to WHO figures, 17.9 million people died of CVDs in 2016, accounting for 31% of global deaths. Of these deaths, an

K. Saravana Kumar (✉)
IT Department, UCE (BIT CAMPUS), Anna University, Trichy, Tamil Nadu, India

N. Shenbagavadivu
MCA Department, UCE (BIT CAMPUS), Anna University, Trichy, Tamil Nadu, India

P. Karuppusamy et al. (eds.), *Ubiquitous Intelligent Systems*, Smart Innovation, Systems and Technologies 243, https://doi.org/10.1007/978-981-16-3675-2_46

estimated 7.3 million were due to coronary heart disease and 6.2 million were due to stroke (85% were due to heart attack and stroke). Approximately, 23.6 million people, primarily those with heart disease and stroke, will die from CVDs by 2030 [2]. According to the Global Partner of Disease Report released on September 15, 2017, heart disease killed 1.7 million Indians in 2016. The World Health Organization (WHO) estimates that India has lost $237 billion since 2005–2015, due to heart-related or cardiovascular diseases [3]. Therefore, the accurate prediction of heart-related diseases is very important. The common risk factors related to heart disease are high level of blood pressure, obesity, alcohol intake, physical inactivity, cholesterol, diabetes, pulse rate, and few other hereditary risk factors [4–7]. Due to the lack of an early diagnosis system, the prediction of heart failure will be a challenging task. Formerly, the heart diseases are diagnosed by medical tests like electrocardiogram (ECG), nuclear scan, angiography, and echocardiogram prescribed by physicians, and the recommendations are made under the symptoms analysed. Out of those, angiography is the promising key tool for heart failure diagnosis [8]. However, this method of diagnosis has some limitations like high cost, side effects, and further required high-level technical experts for the examinations [8, 9]. Machine learning-based expert systems will lower the barriers in the health care industry and further help to diagnose heart disease earlier with improved accuracy. The rapid growth of complex data about patient medical records is the main source of discovering hidden knowledge for error-free decision-making. This paper provides a detailed survey on various machine learning algorithms which has been applied on the UCI repository—Cleveland heart disease database [10] for the heart disease prediction. Further, we introduce pre-processing models on the data set to examine the accuracy and the error rate on prediction. Finally, a comparison has been made with few other machine learning techniques, and the results analysed the efficiency of the pre-processing method.

The rest of the paper is organized as follows: Sect. 2 provides the detailed literary survey of the work carried out earlier. Section 3 describes the database description and proposed data pre-processing techniques over the UCI repository—Cleveland heart disease database. Section 4 outlines the result obtained with a comparative analysis. Finally, Sect. 5 concludes the paper with future work direction.

2 Literature Survey

To date, various decision-making support systems have been proposed for the prediction of heart failure (HF) disease in the literature with regard to decision tree (DT), logistic regression (LR), Naïve Bayes (NBs), random forest (RF), neural network (NN), k-nearest neighbour (KNN), support vector machine (SVM), genetic algorithms (Gas), fuzzy logic-based algorithms, and artificial neural network (ANN) assemblies [4, 11–24]. Various research contributions were made over the UCI online repository—Cleveland database for heart disease prediction. Few benchmarked contributions are considered for this literature study based on their approaches, algorithms analysed, validation techniques, feature selection, tools utilized, accuracy, and error rate which is shown in Table 1.

3 Database Description and Pre-processing

The database was taken from an online machine learning repository of the UCI [10]. Out of the available databases, Cleveland database suits for heart disease prediction which was obtained from V.A. Medical Center, Long each, and Cleveland Clinic Foundation from Dr. Robert Detrano. This database has a total of 303 samples and 14 raw attributes out of those 297 are with complete samples and 6 are with missing values. Most of the researchers took only 14 of 76 attributes (including num attributes) and out of those chosen 14 attributes, 8 are categorical attributes and 5 are numerical attributes and the 14th attribute is a class attribute that is used for prediction analysis (angiographic disease status) and those all are listed in Table 2.

As early mentioned out of 303 patient records, six records are with missing cell values. Many researchers considered only the 297 patient records for their research experimentation. For our research analysis, we considered the entire 303 patient records with missing cell values. Those missing cell values are filled in three different ways first by the lower cell value, next by the upper cell value, and finally by the mean value. The classification accuracy and error rate were analysed, and the results were examined by validation methods over this pre-processed data on various machine learning algorithms (Fig. 1).

4 Results and Discussions

Jupyter Notebook interface with Scikit-Learn Library in PYTHON has been utilized to perform a detailed comparative analysis between machine learning algorithms. This analysis is made by training 75% of samples and tested over the remaining 25% of samples.

First, by removing the missing cell values from 303 patient records, 297 patient records have been taken for analysing the prediction accuracy and their error rate. Out of the results, the prediction accuracy of the decision tree, neural networks, and random forest algorithms is relatively high. The error rate of logistic regression, Naïve Bayes, and neural networks is less when compared to other algorithms. Figure 2 depicts the overall results obtained during computation. While considering accuracy as the only metric for evaluation, the decision tree is superlative on prediction accuracy. Next, by considering the error rate as an evaluation metric, logistic regression performance is on the top when compared to other algorithms.

By removing null values, several machine learning algorithms are examined, those missing null values are filled using three different ways first by the lower cell value,

Table 1 Classification accuracies of existing methods in the literature that used the UCI online repository—Cleveland heart disease data set

S. No.	Study (year)	Approach	Algorithm	Validation technique	No. of attributes	Tool/software	Accuracy (%)	Error rate (%)
1	Latha and Jeeva (2019) [11]	Ensemble classification (bagging, boosting, stacking, and majority voting)—combine multiple classifiers—a comparative analytical approach	NBs, BN, C4.5, multilayer perception, PART, RF	Tenfold cross-validation test	Input—13, Output—1 (num—class attribute) FS1–FS6	WEKA	85.48 (Max increase of 7% accuracy for weak classifiers)	Not handling
2	Ali et al. (2019) [12]	χ^2 statistical model and Gaussian Naive Bayes (X^2– GNB)	NIL	K-fold train-test hold-out validation	Input—13 attributes (297 samples used)	NIL	93.33	Not handling
3	Makumba et al. (2019) [13]	GUI-based decision support system developed	DT, NBs, KNN	40% training (296 samples) and 60% testing (444 samples)	Cleveland database and public data. (Total 740 samples) Input—13 and Input—15 attributes	WEKA	13 attributes. DT—90.76, NBs 97.07, KNN-9.28	13 Attributes. DT—9.24, NBs—2.93, KNN—20.72
							15 Attributes. DT -94.59, NBs–99.77, KNN–82.43	15 Attributes. DT-5.41, NB – 0.23, KNN-17.57

(continued)

Table 1 (continued)

S. No.	Study (year)	Approach	Algorithm	Validation technique	No. of attributes	Tool/software	Accuracy (%)	Error rate (%)
4	Maji et al. (2019) [14]	Hybridization techniques	ANN and DT classifiers are hybridized	Tenfold cross-validation test	Input -13, Output—1 (num—class attribute)	WEKA	Accuracy—78.14, Sensitivity—78, and Specificity—22.9	Not handling
5	Bashir et al. (2019) [15]	Feature selection techniques and algorithms Minimum redundancy maximum relevance feature selection (MRMR)	DT, LR, LR SVM, NBs, and RF	Data validation test	Input—13, Output—1 (num—class attribute)	Rapid miner	DT-82.22, LR-82.56, LR (SVM)-84.85, NB-84.24, RF-84.17	Not handling
6	Ali et al. (2019) [16]	Hybridization model named Statistical Model χ^2—DNN (deep neural network)	Conventional ANN and DNN	K-fold test validation	Input—13 attributes (297 samples used)	Python software package	91.57	Not handling
				Train-test hold-out validation			93.33%	
7	Ali et al. (2019) [17]	Hybrid grid search algorithm (HGSA)—stacks two SVM models	L1 Linear SVM + L2 Linear and RBF SVM	Train-test hold-out validation	Input—13 attributes (297 samples used)	Python software package	92.22	Not handling

(continued)

Table 1 (continued)

S. No.	Study (year)	Approach	Algorithm	Validation technique	No. of attributes	Tool/software	Accuracy (%)	Error rate (%)
8	Mohan et al. (2019) [18]	Hybridization techniques named random forest with a linear model (HRFLM)	DT, deep learning (DL), generalized linear model (GLM), gradient boosted trees (GBT), LR, NB, RF, SVM, VOTE, HRFLM (proposed model)	85% training and 15% testing. 252-training remaining for testing	Input—13 attributes (297 samples used)	R studio rattle	DT—85, DL—87.4, GLM—85.1, GBT—78.3, LR—82.9, NB—75.8, RF—86.1, SVM—86.1, VOTE—87.41, HRFLM (proposed model)—88.4	Not handling
9	Rathnayake et al. (2018) [19]	Combining the algorithms with increased number of attributes	NBs, NN, KNN, DT, LR	NIL	14/15 attributes. 13 attributes identical. 14—depression and exercise. 15—smoking	NIL	NBs—83, NN—78, KNN—75, DT—77, LR—77	Not handling

(continued)

Table 1 (continued)

S. No.	Study (year)	Approach	Algorithm	Validation technique	No. of attributes	Tool/software	Accuracy (%)	Error rate (%)
10	Purushottama et al. (2016) [20]	Covering rules model for classification (taking into account decision trees) as C4.5 Rules. (1) Original Rules. (2) Pruned rules. (3) Rules without duplicates. (4) Classified rules. (5) Polish	SVM, C4.5, 1-NN, PART, MLP(multilayer perception), RBF(radial basis function), TSEAFS and proposed framework	Tenfold cross-validation test	Input—13, Output—1 (num—class attribute)	WEKA	Proposed framework—86.7%. 86.3% in testing phase and 87.3% in training phase SVM—70.59, C4.5—73.53, 1-NN—76.47, PART—73.53, MLP—74.85, RBF—78.53, TSEAFS—77.45	Not handling
11	El-Bialy et al. (2015) [21]	Handling different data sets. To avoid incorrect, missing, and inconsistent problems in the data set	Pruned C4.5 tree, fast decision tree (FDT)	Tenfold cross-validation test	Input—13, Output—1 (num—class attribute)	WEKA	C4.5—78.54 FDT—77.55	Not handling
12	Chaurasia and Pal (2013) [22]	Survey on machine learning techniques	Bagging algorithms, J48 DT, and NBs	Tenfold cross-validation test	Input—13 attributes	WEKA	NBs-82.31, J48 DT-84.35 and bagging algorithm—85.03%	Not handling

(continued)

Table 1 (continued)

S. No.	Study (year)	Approach	Algorithm	Validation technique	No. of attributes	Tool/software	Accuracy (%)	Error rate (%)
13	Das et al. (2009) [4]	Neural networks ensemble model (SAS-based software) feedforward neural networks called backpropagation networks	Neural network ensembles	Train-test hold-out validation	Input—13 attributes (297 samples used)	SAS enterprise minor 5.2	Accuracy—89.01% Sensitivity—80.95% Specificity—95.91%	Not handling
14	Otoom et al. (2015) [23]	Mobile application real-time monitoring component (intelligent classifier)	Bayes Net, SVM, FT (functional trees)	Tenfold cross-validation test	Input—13 attributes (297 samples used)	Nokia lumia 520 mobile phone	Bayes Net—84.5, SVM—85.1, FT—84.5	Not handling
15	Chaurasia et al. (2013) [22]	Developed prediction models using a large data set	CART, ID3 and decision table extracted from a decision tree or rule-based classifier	Tenfold cross-validation test	Input—11 attributes (297 samples used)	WEKA	CART—83.49%, ID3—72.93%, Decision Table—82.50%	CART classifier has the lowest average error at 0.3 compared to others
16	Tan et al. (2009) [24]	Based on a wrapper approach a new hybridization model	GAs and SVMs	Tenfold cross-validation test	Input—13 attributes (297 samples used)	WEKA and LIBSVM	GA + SVM—84.07	Not handling

Table 2 UCI—cleveland heart disease data base attributes detailed information

S. No.	Attributes	Values and ranges
1	Age	29–79
2	Sex	0–1
3	Chest pain types (cp)	1–4
4	Resting blood pressures in mm hg (trestbps)	94–200
5	Serum cholesterol in mg/dl (chol)	126–564
6	Fasting blood sugar (fbs) > 120 mg/dl	0–1
7	Resting electrocardiographic results	0–2
8	Maximum heart rate achieved (thalach)	71–202
9	Exercise induced angina (exang)	0–1
10	Old peak = ST depression induced by exercise relative to rest	1–3
11	The slope of the peak exercise ST segment	1–3
12	Number of major vessels coloured by fluoroscopy (ca)	0–3
13	Status of heart illustrated through three distinctly numbered values (thal)	3, 6, 7
14	Num (class attribute)	0–1

next by the upper cell value, and finally by the mean value. Now, the total samples for examining will be 303 samples. We got analogous results while evaluating the performances. Figure 3 depicts the performance of algorithms by filling the missing field by lower cell value. Figure 4 depicts the performance of algorithms by filling the missing field by upper cell value. Figure 5 depicts the performance of algorithms by filling the missing field by mean.

Figure 3 depicts the overall results obtained during computation by filling the missing cell value by lower value. It has been observed that the accuracy and the error rate of both the neural network and random forest are relatively the same, but on the other hand, prediction accuracy of the decision tree seems to be higher with less error rate. The error rate of decision tree, logistic regression, and support vector machine is less when compared to other algorithms. Next, by considering the error rate as an evaluation metric, support vector machine performance is on the top.

Figure 4 depicts the overall results obtained during computation by filling the missing cell value by upper cell value. From the observation prediction accuracy of neural networks, random forest and decision tree algorithms are relatively high. While considering accuracy as the only metric for evaluation, the decision tree is superlative on prediction accuracy. Next, by considering the error rate as an evaluation metric, both Naïve Bayes and support vector machine results are considered the same. Finally, on considering both error rate and accuracy as the evaluation metrics, the performance of the support vector machine is on the top when compared with the Naïve Bayes algorithm.

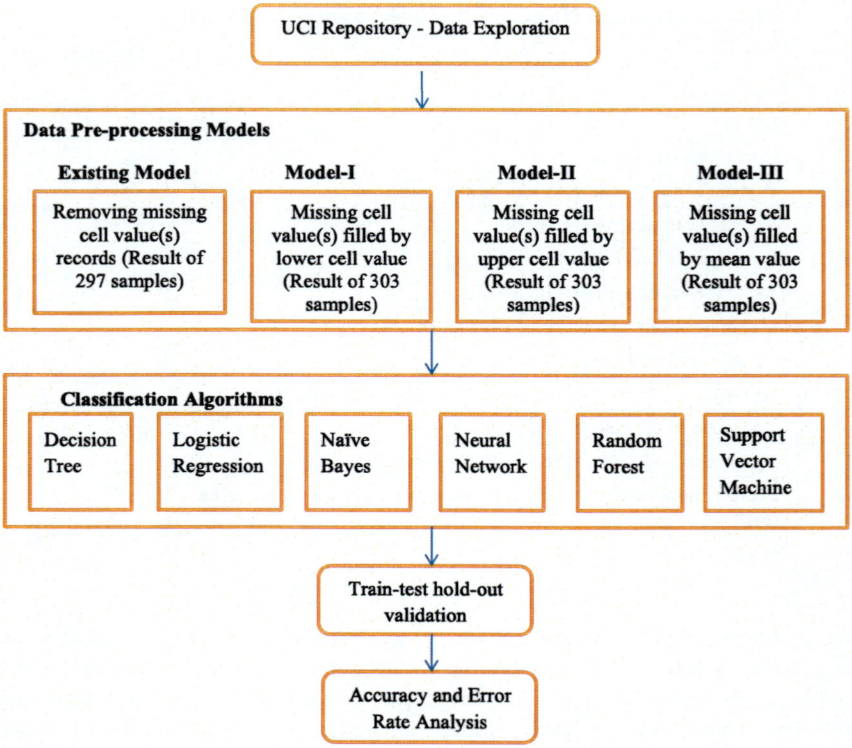

Fig. 1 Block diagram of the proposed data pre-processing models with classification algorithms

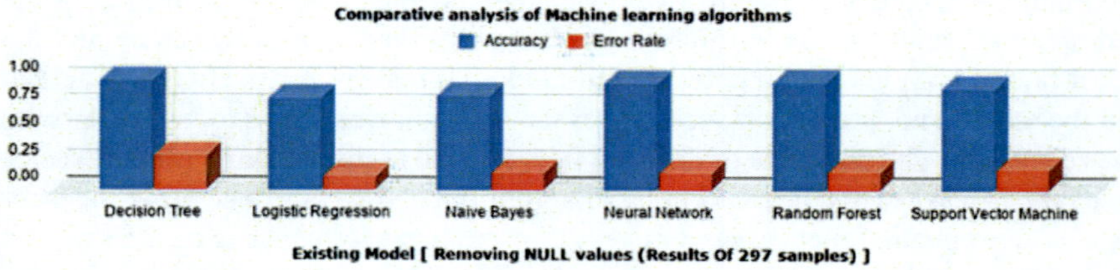

Fig. 2 Existing model (removing NULL values—results of 297 samples)

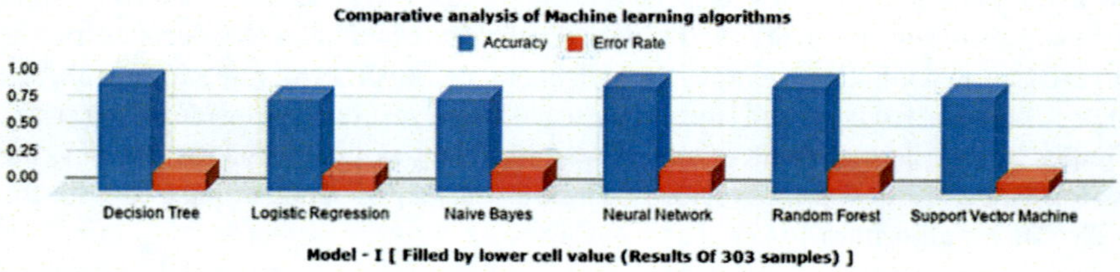

Fig. 3 Model-I (filled by lower cell value—results of 303 samples)

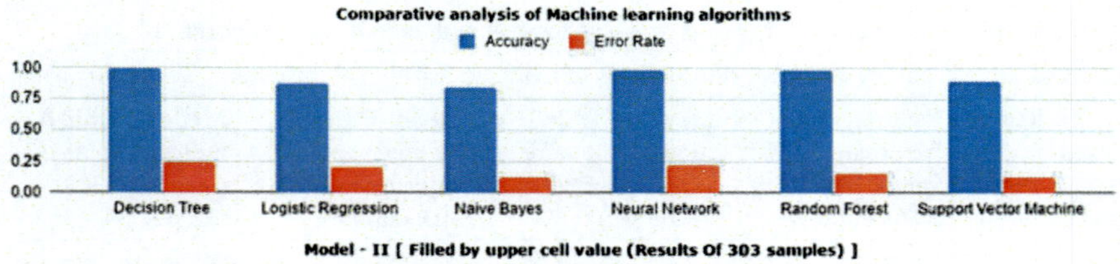

Fig. 4 Model-II (filled by upper cell value—results of 303 samples)

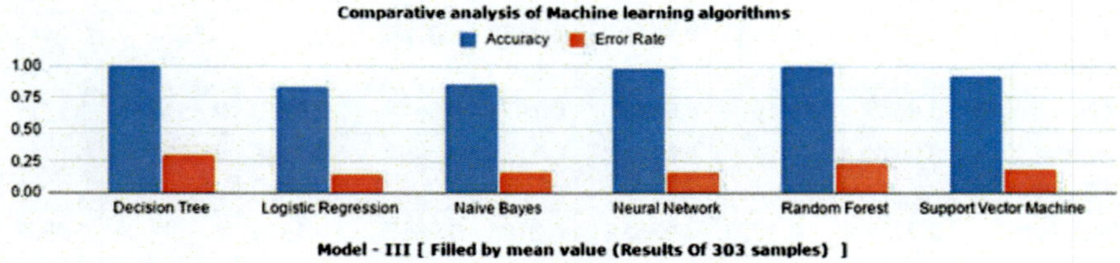

Fig. 5 Model-III (filled by mean value—results of 303 samples)

Figure 5 depicts the overall results obtained during computation by filling the missing cell using the mean. From the observation, prediction accuracy of the decision tree and random forest algorithms is alike but the random forest is superlative while considering error rate as an additional metric when compared with the decision tree algorithm. Next by considering the error rate as an evaluation metric, logistic regression performance is on the top when compared to other algorithms. Table 3 shows the comparative analysis made with the machine learning techniques with respect to error rate and accuracy. This analogous result proves the efficacy of our pre-processing method.

Figure 6 depicts the overall comparative analysis of accuracy on various algorithms with our pre-processing models. Figure 7 depicts the overall comparative analysis of the error rate on various algorithms with our pre-processing models.

5 Conclusion

The introduced data pre-processing models of filling the missing cell values in the data set of UCI repository—Cleveland heart disease database are examined over various machine learning algorithms like DT, LR, NBs, NN, RF, and SVM. The different types of experiments were conducted and to diagnose heart disease in a fully automatic manner. The existing model with removing NULL values (results of 297 samples), filled by upper cell value (results of 303 samples), filled by lower cell value (results of 303 samples), and filled by mean (results of 303 samples) were employed for experiments. The results obtained during computation prove the

Table 3 Results of various data pre-processing models with classification algorithms

Comparative accuracy and error rate analysis of machine learning algorithms								
Algorithms used	Existing model [removing NULL values (results of 297 samples)]		Model-I [filled by lower cell value (results of 303 samples)]		Model-II [filled by upper cell value (results of 303 samples)]		Model-III [filled by mean value (results of 303 samples)]	
	Accuracy	Error rate	Accuracy	Error rate	Accuracy	Error rate	Accuracy	Error rate
Decision tree	1	0.32	1	**0.171**	1	**0.25**	1	**0.3026**
Logistic regression	0.8423	0.1333	**0.8546**	0.1578	**0.8766**	0.2105	0.8414	0.1447
Naive Bayes	0.8603	0.16	**0.8678**	0.1973	0.8458	**0.1315**	0.8546	0.171
Neural network	0.9774	0.16	**0.9955**	0.2105	**0.9867**	0.2236	**0.9823**	0.171
Random forest	0.9954	0.1733	**0.9955**	0.2105	0.9867	**0.1578**	**1**	0.2368
Support vector machine	0.9324	0.2	0.903	**0.1184**	0.8942	**0.1315**	0.9251	**0.1842**

Bolded values represents the improved results obtained during the evaluation process of our proposed method when compared with the existing model [Removing NULL values (Results of 297 samples)]

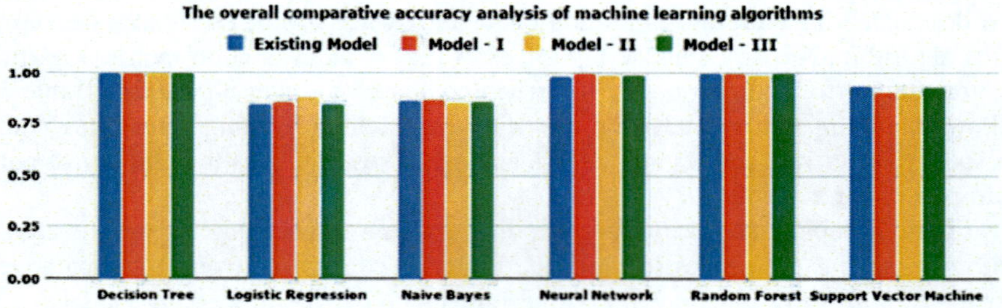

Fig. 6 Overall comparative analysis of accuracy on various algorithms

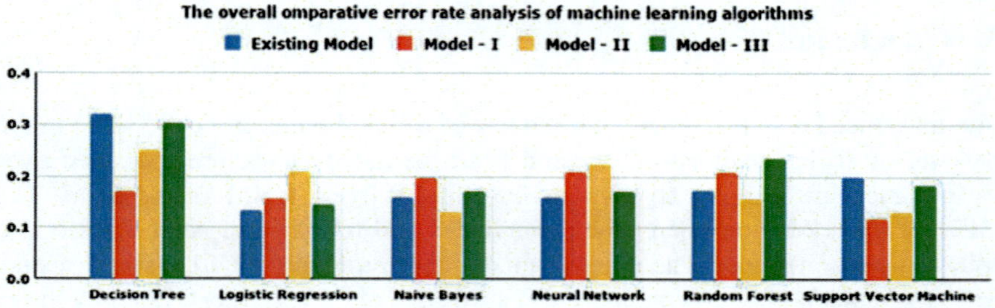

Fig. 7 Overall comparative analysis of the error rate on various algorithms

efficacy of our pre-processing models with improved prediction classification accuracy with less error rate. As for future work directions we planned to apply various machine learning techniques to fill those missing cell values and that result will be analysis for fixative the better pre-processing techniques for improved predictive analysis and more complex and fast algorithms, such as genetic algorithms, will be used.

References

1. Coronary Artery Disease: MedlinePlus: 2020. https://medlineplus.gov/coronaryarterydisease.html
2. World Health Organization. Available from: https://www.who.int/cardiovascular_diseases/about_cvd/en/
3. S.B. Patel, P.K. Yadav, D.P. Shukla, Predict the diagnosis of heart disease patients using classification mining techniques. IOSR J. Agric. Veterin. Sci. (IOSR-JAVS) **4**(2), 61–64 (2013)
4. R. Das, I. Turkoglu, A. Sengur, Effective diagnosis of heart disease through neural networks ensembles. Expert Syst. Appl. **36**(4), 7675–7680 (2009)
5. H.G. Lee, K.Y. Noh, K.H. Ryu, Mining biosignal data: coronary artery disease diagnosis using linear and nonlinear features of HRV, in *Pacific-Asia Conference on Knowledge Discovery and Data Mining* (Springer, Berlin, 2007), pp. 218–228
6. H.D. Masethe, M.A. Masethe, Prediction of heart disease using classification algorithms, in *Proceedings of the world Congress on Engineering and computer Science*, vol. 2, pp. 22–24 (2014)
7. J. Nahar, T. Imam, K.S. Tickle, Y.P.P. Chen, Computational intelligence for heart disease diagnosis: a medical knowledge driven approach. Expert Syst. Appl. **40**(1), 96–104 (2013)
8. L. Verma, S. Srivastava, P.C. Negi, A hybrid data mining model to predict coronary artery disease cases using non-invasive clinical data. J. Med. Syst. **40**(7), 178 (2016)
9. U.R. Acharya, O. Faust, V. Sree, G. Swapna, R.J. Martis, N.A. Kadri, J.S. Suri, Linear and nonlinear analysis of normal and CAD-affected heart rate signals. Comput. Methods Programs Biomed. **113**(1), 55–68 (2014)
10. UCI—Heart disease dataset from http://archive.ics.uci.edu/ml/datasets/Heart+Disease
11. C.B.C. Latha, S.C. Jeeva, Improving the accuracy of prediction of heart disease risk based on ensemble classification techniques. Inf. Med. Unlock. **16**, 100203 (2019)
12. L. Ali, S.U. Khan, N.A. Golilarz, I. Yakubu, I. Qasim, A. Noor, R. Nour, A feature-driven decision support system for heart failure prediction based on statistical model and Gaussian Naive Bayes. Comput. Math. Methods Med. (2019)
13. D.O. Makumba, W. Cheruiyot, K. Ogada, A model for coronary heart disease prediction using data mining classification techniques. Asian J. Res. Comput. Sci. 1–19 (2019)
14. S. Maji, S. Arora, Decision tree algorithms for prediction of heart disease, in *Information and Communication Technology for Competitive Strategies* (Springer, Singapore, 2019), pp. 447–454
15. S. Bashir, Z.S. Khan, F.H. Khan, A. Anjum, K. Bashir, Improving heart disease prediction using feature selection approaches, in *2019 16th International Bhurban Conference on Applied Sciences and Technology (IBCAST)* (IEEE, 2019), pp. 619–623
16. L. Ali, A. Rahman, A. Khan, M. Zhou, A. Javeed, J.A. Khan, An Automated diagnostic system for heart disease prediction based on ${\chi^{2}} $ statistical model and optimally configured deep neural network. IEEE Access **7**, 34938–34945 (2019)
17. L. Ali, A. Niamat, J.A. Khan, N.A. Golilarz, X. Xingzhong, A. Noor, R. Nour, S.A.C. Bukhari, An optimized stacked support vector machines based expert system for the effective prediction of heart failure. IEEE Access **7**, 54007–54014 (2019)

18. S. Mohan, C. Thirumalai, G. Srivastava, Effective heart disease prediction using hybrid machine learning techniques. IEEE Access **7**, 81542–81554 (2019)
19. B.S.S. Rathnayakc, G.U. Ganegoda, Heart diseases prediction with data mining and neural network techniques, in *2018 3rd International Conference for Convergence in Technology (I2CT)* (IEEE, 2018), pp. 1–6
20. K. Saxena, R. Sharma, Efficient heart disease prediction system. Procedia Comput. Sci. **85**, 962–969 (2016)
21. R. El-Bialy, M.A. Salamay, O.H. Karam, M.E. Khalifa, Feature analysis of coronary artery heart disease data sets. Procedia Comput. Sci. **65**, 459–468 (2015)
22. V. Chaurasia, S. Pal, Early prediction of heart diseases using data mining techniques. Caribbean J. Sci. Technol. **1**, 208–217 (2013)
23. A.F. Otoom, E.E. Abdallah, Y. Kilani, A. Kefaye, M. Ashour, Effective diagnosis and monitoring of heart disease. Int. J. Softw. Eng. Appl. **9**(1), 143–156 (2015)
24. K.C. Tan, E.J. Teoh, Q. Yu, K.C. Goh, A hybrid evolutionary algorithm for attribute selection in data mining. Expert Syst. Appl. **36**(4), 8616–8630 (2009)

Ensemble Based-Cross Project Defect Prediction

Rajni Jindal, Adil Ahmad, and Anshuman Aditya

Abstract In Software Testing, there are typically two ways to predict defects in the software—within-project defect prediction (WPDP) and cross project defect prediction (CPDP). In this research, we are using a hybrid model for cross project defect prediction. It is a two-phase model consisting of ensemble learning (EL) and genetic algorithm (GA) phase. For our research, we used datasets from the PROMISE repository and created clusters after normalization using k-means clustering algorithm. This further helped us improve the accuracy of the model. Our dataset consists of 22 attributes and were labeled defective or not. Our results show that our hybrid model after implementing k-means clustering achieved an F1 score of 0.666. CPDP is a newer and faster approach for software defect prediction but is often error prone. This method can change the software industry as it will lead to improved software development and faster software delivery.

1 Introduction

Software defect prediction is one the prominent techniques used in the testing phase of the software development life cycle (SDLC). It helps in allocating test resources (i.e., files, modules, classes) by predicting whether a resource is defective or not. This helps in performing a more efficient testing during the testing phase. Numerous defect prediction algorithms have been trained and tested using the data from some particular project. Most of these approaches are trained and applied within the same project. This approach has been in practice for a long time but is often time consuming leading to delay in software development. The reasons vary from having insufficient dataset to delay in dataset collection. On almost all occasions any new project had limited amounts of training data. Cross project defect prediction enables the training of a machine learning model from instances of data from certain projects which are

R. Jindal · A. Ahmad (✉) · A. Aditya
Delhi Technological University, Delhi 110042, India

R. Jindal
e-mail: rajnijindal@dce.ac.in

P. Karuppusamy et al. (eds.), *Ubiquitous Intelligent Systems*, Smart Innovation, Systems and Technologies 243, https://doi.org/10.1007/978-981-16-3675-2_47

utilized in order to anticipate the presence of defects in some other projects, thereby providing a new outlook to the prediction of defects.

Currently, features like—LOC, Cyclomatic Complexity, Essential Complexity, Design Complexity, Volume, Program Length, Effort, Time Estimator, etc., are used to successfully predict and classify software defects. In reality, it is hard to obtain enough training data from within a project for the testing phase. However, one can find an abundance of data using different relatable projects. PROMISE repository is one such collection of publicly released software defect prediction datasets.

Cross project defect prediction faces challenges due to the fact a prediction model that's trained on one project may not generalize well on the other projects. By proposing an ensembled genetic model consisting two phases—genetic algorithm and ensemble learning. We are basically first identifying optimal project with most correlation with current project using genetic algorithm and after that be application of an ensemble-based classifier which utilizes the results of previous genetic algorithm as well, for that we used random forest classifier. We tried with various other ensemble-based classifiers, however, random forest provided better results in comparisons because of the nature of data orientation in the random forest model. We normalize our datasets and form clusters by performing K-means clustering after analyzing the inertia curve for the dataset and choosing a suitable cluster number.

2 Motivation

Since the significance of programming imperfection expectation in quality affirmation and programming support, lately it has bit by bit become one of the most dynamic exploration fields in software engineering. To the best of our knowledge, cross project defect prediction (CPDP) and within-project defect prediction (WPDP) are the two mainstream techniques for software defect prediction. However, there are some approaches that propose a unified solution [1, 2] for SDA [3]. In our work, we opted to continue with cross project defect prediction approach.

2.1 *Why Cross Project?*

WPDP has a major drawback that its training needs sufficient amounts of data from the same project, which does not happen on a regular basis, especially for smaller organizations. Due to sudden increase in startup culture, this has become a widespread need to better detect and identify software defects without compromising any historical needs. Since CPDP uses training data from different projects, hence it reduces the time usually lost in gathering enough data from within the project. Thus, this method can potentially change the software industry as it will lead to improved software development time and faster software delivery.

CPDP is considered to be the most effective when it comes to predicting flaws present in a software, however, there is a lot of criticism regarding its performance when compared with WPDP, but it's been improving ever since with constant research and effort being poured into this domain.

3 Research Methodology

The job of cross project defect prediction is an uphill one as a forecast model which has been prepared on various bunch of tasks probably won't sum up well to different undertakings. The major task here is to build a model which will understand the generalizable properties of defective instances in a more lucid manner, and in consideration of the target project, it shall work perfectly fine. Moreover, the model shall (fully or partly) ignore the properties which cannot be generalized for the target project. In this section, we present various techniques, concepts, and the methodologies used in order to build the model, train the model, and formulate the results. A comparison of cross project defect prediction approaches is covered in other works [4].

3.1 Model Architecture

We opted for a 2-phase compositional model because not even one state of the art single model has been able to perform very well across a wide range of projects. There are instances where a single model has been able to perform good for a few projects but not for a wide variety of projects [5]. That is why there has been a need for a compositional model.

The architecture of the compositional model is depicted in Fig. 1. In phase 1, i.e.,

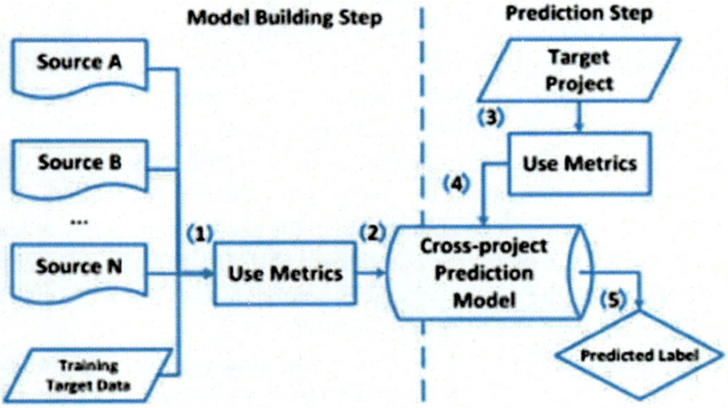

Fig. 1 Model architecture—representing the 2 phase model

Table 1 Attributes used in the dataset

S. No.	Attributes	Description
1	loc	Count of lines of code
2	v(g)	Cyclomatic complexity
3	ev(g)	Essential complexity
4	iv(g)	Design complexity
5	IOComment	Halstead's count of lines of comments
6	IOCode	Halstead's count of lines
7	IOBlank	Halstead's count of blank lines
8	IOCodeAndComment	Count of lines of code and comments
9	b	Halstead
10	t	Halstead's time estimator
11	n	Total count of operators and operands
12	v	Volume
13	1	Program length
14	d	Difficulty
15	i	Intelligence
16	e	Effort
17	uniq_Op	Unique number of operators
18	uniq_Opnd	Unique number of operands
19	total_Op	Total number of operators
20	total_Opnd	Total number of operands
21	branchCount	Branch count of the flow graph
22	defects	Number of reported defects

model building phase, we obtain training target data from multiple sources and build a model learned from these instances. In phase 2, the prediction phase, we apply this model to predict whether any module is clean or defective.

Since source component shift does occur, if a source project is chosen without proper care, we choose to cluster every component based on available features into 15 clusters. A method which will be able to utilize the available source projects and also mitigate the consequence of source component shift [6] will be really helpful. And in pursuit of achieving the same a massive compositional model having 2 phases, namely the GA phase and the EL phase, is built.

All the 22 attributes [7] used by our model and selected as the features for performing clustering on the dataset are listed in Table 1, explaining the significance of each of the attributes.

4 Proposed Approach

We considered P source projects $\{S_1, S_2, \ldots, S_p\}$ along with a target project T. Every individual project had a large number of instances and corresponding to every instance, there was a module. For every instance, there is a set of attributes x and an output label y ($y = 1$ means defective and $y = 0$ means defect free). Our proposed model consists of 2 phases, namely the genetic algorithm phase and the ensemble learning phase.

4.1 Genetic Algorithm Phase

In genetic algorithm phase, $(N + 1)$ classifiers are created for any given source project S_i; $\{1 \le i \le N\}$, we combine it with T_t, i.e., $S_i \bigcup T_t$, and using it we create classifier M_i over the combined data. $(N + 1)$th classifier is created for M_{N+1} for the target training data T_t by training it over the training data itself. By calculating the F1-score of the classifier on the target training data T_t, we measure the performance of M_i. By default, in order to create $(N + 1)$ classifiers, logistic regression was used as the underlying classifier.

$$\text{Comp}(j) = \frac{\sum_{i=1}^{N+1} a_i \times \text{Score}_i(j)}{\text{LOC}(j)} \tag{1}$$

A GA classifier is a weighted composition of $(N + 1)$ classifiers. For any instance j, M_i will return the likelihood of j being defective, represented by $\text{Score}_i(j)$, ranging between 0 and 1. This prediction is made by comparing the likelihood scores on the instance j using preset threshold score against their weighted sum of $(N + 1)$ classifiers following settings from previous studies [5, 8].

4.2 Ensemble Learning Phase

In ensemble learning phase, the genetic algorithm phase is iterated multiple times giving us an ensemble of GA classifiers. In order to achieve it, we use AdaBoost [9]. We are adjusting the weights of instances distinctively for source projects and target training data. While performing these iterations, the objective of ensemble learning phase is minimizing the errors of instance prediction in target training data. AdaBoost is responsible for minimizing errors in prediction of every training instance.

In order to prevent overfitting of the model, we created an ensemble of GA classifiers instead of using only the best performing GA classifier [10]. There are further studies covering imbalanced learning [11, 12].

4.3 K-Means Clustering

To further improve the accuracy of our phase 1 model, we are forming clusters of initial features like—LOC, Cyclomatic Complexity, Essential Complexity, Design Complexity, Volume, Program Length, Effort, Time Estimator, etc., listed in Table 1.

What is clustering? It is an unsupervised technique which tries to learn patterns from input data. The aim is to group similar data into one group and in such a way we can divide complete data into different clusters or groups.

In Fig. 2 provides us with an idea about how clustering works on the datapoints. Before applying K-Means, data points are available vaguely to us. Here in Fig. 2, K-means help us to group similar data into groups (clusters) on the basis of the initial features.

Our Approach. We initially performed data cleaning and pre-processing steps on the dataset. These include filling NaN values with the mean value of their corresponding columns, normalizing the data with standard scaling and removal of duplicate rows.

In order to obtain some useful insights from this data, we performed K-means clustering while minimizing error sum of squares (SSE) [13] and we plotted the inertia values (Fig. 3) with respect to the number of clusters we obtained.

Inertia of a clustering job is the sum of the distances of each point in each cluster to their centroid. It indicates how "tight" the clusters are so the smaller this value the better. We don't want too many clusters, because this will slow the model down on large datasets. For this purpose, we plotted an inertia against the number of clusters (Fig. 3).

That is why we chose the number of clusters to be 15 since the curve started flattening out after that. For every new data point, we first identify the cluster in which our data point lies, next we determine the cohesiveness within features by plotting a heat map.

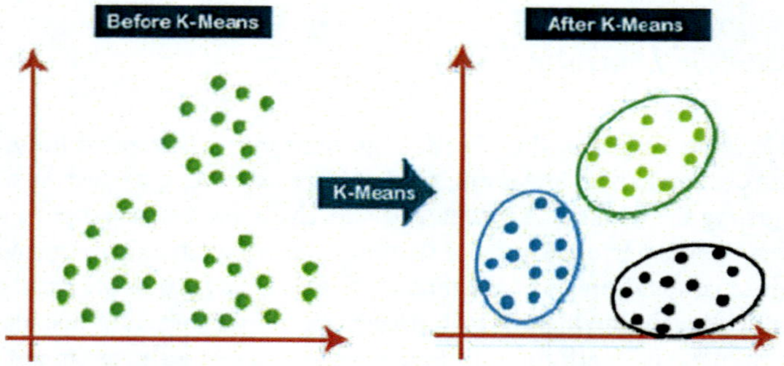

Fig. 2 Example of forming clusters on the dataset using K-means clustering

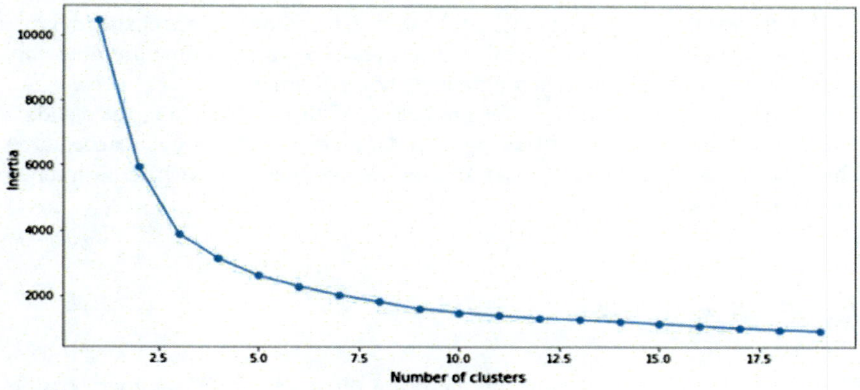

Fig. 3 Plot of inertia versus number of clusters

4.4 Experimental Setup

In the model building step, a cross project forecast model is made which gains information from examples present in numerous source ventures and also the preparation target information.

For prediction, this model is applied to anticipate whether there is another class or record or module in the objective project that has defects or not.

Next, in each iteration, new classifiers are being learnt and the combination of these are able to capture the generalizable properties in a more refined way and also in every new iteration the relevant feature selection is done using the Chi-Square method [14] and recursive feature elimination [15].

$$x^2 = \sum \frac{(O_i - E_i)^2}{E_i} \tag{2}$$

We are not only learning a small bunch of classifiers but also tuning a 2-layer graded composition of numerous other classifiers. The whole process of tuning comprises many iterations, utilizing genetic algorithm (GA), and ensemble learning (EL), which step by step enhances the model to inculcate the generalizable properties in a proper manner.

GA fit works by gathering a bunch of basic classifiers which individually comprise to become chromosomes and then list of genes. Number of genes is determined by the number of classifiers + Threshold.

First, the initial population is gathered and then the value range of the threshold in the last column is adjusted with which optimal solution in the initial population is found. Next breeding within the population occurs.

This process repeats in each iteration. With each breeding, a slight chance of cross mutation is always there which can lead to mutation which is done by randomly generating mutation positions and producing new offspring.

After that the model is ready for prediction. With each iteration, the standard deviation will lower down until the optimal flattening of the curve is reached after which the computation capacity will take over the accuracy. Due to which the training has to be terminated.

4.5 Evaluation Metric

We used F1-score as our evaluation metric in this research. It has four possible outcomes. Using the confusion matrix, these values can be calculated. Hence, the F1-score can be defined as follows:

$$\text{Recall}(R) = \frac{\text{TP}}{(\text{TP} + \text{FN})} \tag{3}$$

$$\text{Precision}(P) = \frac{\text{TP}}{(\text{TP} + \text{FP})} \tag{4}$$

$$\text{F1 score} = \frac{(2 \times P \times R)}{(P + R)} \tag{5}$$

We chose F1-score to avoid the exchange between recall and precision. It is the harmonic mean of the two metrics and has been used by previous studies [5].

5 Result and Discussion

In order to facilitate our work, we ignored NaN values to avoid underfitting our model. Another approach we tried was putting running average in place of NaN values, however, this approach didn't result in better accuracy. Overall, with our approach, the F1 score came out to be 0.666. One downside of our approach is that it cannot be used against models with non-deterministic scenarios, however, nowadays most software predetermines risk and cost, so the non-deterministic aspect can be ignored.

Table 2 F1-scores achieved

Iterations	F1-score
1	0.6666
2	0.6666
3	0.6666

6 Conclusion

With our research, we were able to achieve a high accuracy for a cross project defect prediction approach. We did this by, firstly, the dataset is classified under a cluster only that cluster of data is being selected from the historical set followed by two-phase processes including—genetic algorithm phase and ensemble learning phase. First, for each project within the cluster classifier is built. Next a composition classifier which is made by ensembling the model from the genetic algorithm phase. In each iteration, a genetic classifier is built and weights are assigned. Overall, we achieved an F1 score of 0.666 with our approach (Table 2).

7 Future Work

In the future work, we would like to study more and come up with a solution in case of non-deterministic scenario where risk associate is higher. Coming up with a solution in case of NaN values presence will improve the overall confidence of our approach. We would like to test the model with different datasets and compare its results with the other pre-existing models out there. Our research shows that there is scope for improvement in the field of software defect prediction and more emphasis should be given toward advancement in cross project defect prediction.

References

1. F. Wu et al., Cross-project and within-project semisupervised software defect prediction: a unified approach. IEEE Trans. Reliab. **67**(2), 581–597 (2018). https://doi.org/10.1109/TR.2018.2804922
2. F. Wu et al., Cross-project and within-project semi-supervised software defect prediction problems study using a unified solution, in *2017 IEEE/ACM 39th International Conference on Software Engineering Companion (ICSE-C)*, Buenos Aires, 2017, pp. 195–197. https://doi.org/10.1109/ICSE-C.2017.72
3. X. Jing, F. Wu, X. Dong, B. Xu, An improved SDA based defect prediction framework for both within-project and cross-project class-imbalance problems. IEEE Trans. Softw. Eng. **43**(4), 321–339 (2017). https://doi.org/10.1109/TSE.2016.2597849
4. S. Herbold, A. Trautsch, J. Grabowski, [Journal First] A comparative study to benchmark cross-project defect prediction approaches, in *2018 IEEE/ACM 40th International Conference*

on Software Engineering (ICSE), Gothenburg, 2018, pp. 1063–1063. https://doi.org/10.1145/3180155.3182542
5. X. Xia, D. Lo, S.J. Pan, N. Nagappan, X. Wang, HYDRA: massively compositional model for cross-project defect prediction. IEEE Trans. Softw. Eng. **42**(10), 977–998 (2016). https://doi.org/10.1109/TSE.2016.2543218
6. B. Turhan, On the dataset shift problem in software engineering prediction models. Empir. Softw. Eng. **17**, 62–74 (2012). https://doi.org/10.1007/s10664-011-9182-8
7. M. Cetiner, O.K. Sahingoz, A comparative analysis for machine learning based software defect prediction systems, in *2020 11th International Conference on Computing, Communication and Networking Technologies (ICCCNT)*, Kharagpur, India, 2020, pp. 1–7. https://doi.org/10.1109/ICCCNT49239.2020.9225352
8. F. Rahman, D. Posnett, I. Herraiz, P. Devanbu, Sample size vs. bias in defect prediction, in *Proceedings of the 9th Joint Meeting Foundations Software Engineering*, 2013, pp. 147–157
9. Y. Freund, R.E. Schapire, A decision-theoretic generalization of on-line learning and an application to boosting, in *Proceedings of the 2nd European Conference on Computational Learning Theory*, 1995, pp. 23–37. J. Nam, S.J. Pan, S. Kim, Transfer defect learning, in *Proceedings—International Conference on Software Engineering*, 2013, pp. 382–391
10. Y. Zhang, D. Lo, X. Xia, J. Sun, An empirical study of classifier combination for cross-project defect prediction, in *2015 IEEE 39th Annual Computer Software and Applications Conference*, Taichung, 2015, pp. 264–269. https://doi.org/10.1109/COMPSAC.2015.58
11. L. Gong, S. Jiang, L. Bo, L. Jiang, J. Qian, A novel class-imbalance learning approach for both within-project and cross-project defect prediction. IEEE Trans. Reliab. **69**(1), 40–54 (2020). https://doi.org/10.1109/TR.2019.2895462
12. M.F. Sohan, M.I. Jabiullah, S.S.M.M. Rahman, S.M.H. Mahmud, Assessing the effect of imbalanced learning on cross-project software defect prediction, in *2019 10th International Conference on Computing, Communication and Networking Technologies (ICCCNT)*, Kanpur, India, 2019, pp. 1–6. https://doi.org/10.1109/ICCCNT45670.2019.8944622
13. K. Kohara, T. Kawaoka, Well-balanced learning for reducing the variance of summed squared errors, in *Proceedings 1993 The First New Zealand International Two-Stream Conference on Artificial Neural Networks and Expert Systems*, Dunedin, New Zealand, 1993, pp. 29–33. https://doi.org/10.1109/ANNES.1993.323089
14. M. Göl, A. Abur, A modified Chi-Squares test for improved bad data detection, in *2015 IEEE Eindhoven PowerTech*, Eindhoven, 2015, pp. 1–5. https://doi.org/10.1109/PTC.2015.7232283
15. J.D. Souza, B. Parvathavarthini, Machine learning based intrusion detection framework using recursive feature elimination method, in *2020 International Conference on System, Computation, Automation and Networking (ICSCAN)*, Pondicherry, India, 2020, pp. 1–4. https://doi.org/10.1109/ICSCAN49426.2020.9262282
16. G. Canfora, A. De Lucia, M. Di Penta, R. Oliveto, A. Panichella, S. Panichella, Multi-objective cross-project defect prediction, in *2013 IEEE Sixth International Conference on Software Testing, Verification and Validation, Luxembourg*, 2013, pp. 252–261. https://doi.org/10.1109/ICST.2013.38
17. Q. Wu et al., Online transfer learning with multiple homogeneous or heterogeneous sources. IEEE Trans. Knowl. Data Eng. **29**(7), 1494–1507 (2017). https://doi.org/10.1109/TKDE.2017.2685597
18. J. Sun, X. Jing, X. Dong, Manifold learning for cross-project software defect prediction, in *2018 5th IEEE International Conference on Cloud Computing and Intelligence Systems (CCIS)*, Nanjing, China, 2018, pp. 567–571. https://doi.org/10.1109/CCIS.2018.8691373

Effective Plant Discrimination Using Deep Learning

Advyth Ashok, M. S. Devadeth, and E. R. Vimina

Abstract Effective plant differentiation plays a vital role in agriculture, wherein this includes identifying crops and weeds in the farmland. The common method used for weed control is herbicides. However, the excessive use of herbicides will result in herbicides-resistance in weeds. An effective plant differentiation can decrease the expense of agriculture and increase crop quality, yield, and weed control. Henceforth, this research work has proposed a deep learning-based convolutional neural network to efficiently classify the plant leaves of three crops like canola, corn, and radish. In the experiment, the dataset used here is "bccr-segset," which includes four categories like background, radish, corn, and canola. The dataset contains 30,000 images of these subclasses in four different growth stages.

1 Introduction

Food security is one of the basic needs for human population. Any advancements in this field directly result in our daily lives. In farming effectively, identifying crops and weeds has always been a problem, weed infestation constrains the yield and the productivity of the crops. The capability to precisely differentiate plants in real time will improve crop production and weed management, which results in preventing the weeds in the field from competing for the resources like nutrients, water, and light required by the crops. The commonly followed method for weed management is blanket spraying of herbicides, but this brings down the range of agriculture chemicals available and their longevity as weeds develop resistance for them. Figure 1 [1] illustrates the aggressive spread of weed infestation in spring and winter crops.

Le et al. [2] used a combination of local binary pattern (LBP) operators and support vector machine (SVM) method, for the extraction of crop leaf textural features and

A. Ashok (✉) · M. S. Devadeth · E. R. Vimina
Computer Science Department, Amrita School of Arts and Sciences, Kochi, Edappally North, Kochi, Kerala P.O. 682024, India

E. R. Vimina
e-mail: viminaer@asas.amrita.edu

P. Karuppusamy et al. (eds.), *Ubiquitous Intelligent Systems*, Smart Innovation, Systems and Technologies 243, https://doi.org/10.1007/978-981-16-3675-2_48

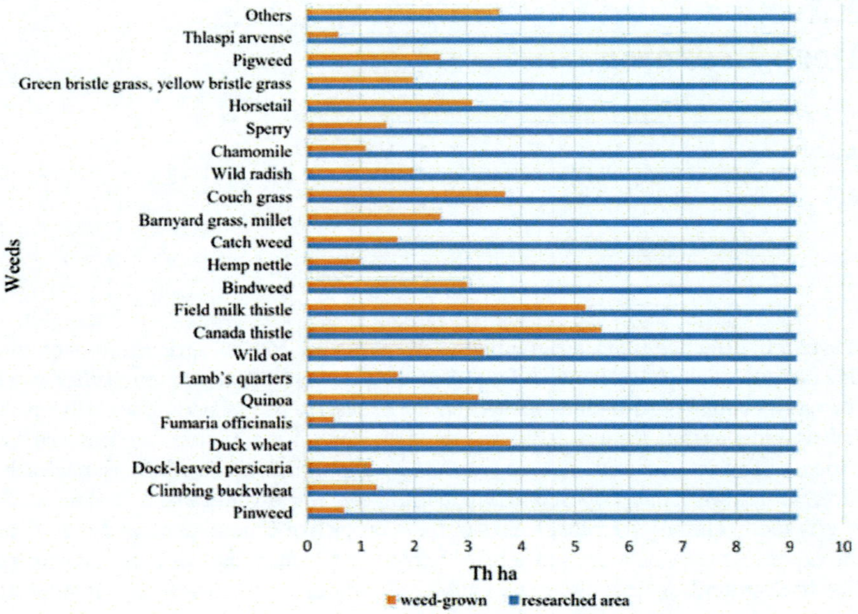

Fig. 1 Correlation between weeds in spring wheat culture of Tomsk oblast in 2015 [1]

multi-class plant classification. Le et al. [3] in their work say that wild radish is an invasive weed of dominant nature.

Even though there have been a lot of attempts to identify leaf feature set and perform classification of plants using sophisticated computer vision algorithms, plant discrimination is still regarded as a very difficult issue to solve. For classification-based on computer vision plant leaves in real conditions many factors come into play, this includes overlapping, sub-optimal lighting conditions, occlusion, and leaves with defects and damages. Models utilizing deep learning architecture have been frequently used for the classification of plants using images. We can achieve high accuracies with deep learning models due to their ability to easily learn from spatial information present in images.

2 Literature Review

There are several methods to discriminate plants. In Le et al. [2] approach Excess Green minus Excess Red (ExG-ExR) method is used to segment the green plant region in images. It is a color index-based method that exhibits high accuracy and adequate robustness. Here, LBP is used for edge detection. LBP considers 3 × 3 matrix, it compares the center element with the surrounding elements, and if the outer element is greater than the center element, it is assigned 1 otherwise 0. This gives a binary chain, which we can change into a decimal. It is then used to build a histogram, which is the texture representation of the image. This gives a binary chain, which we can change into a decimal. It is then used to build a histogram, which is the texture representation of the image. Because SVM can perform classification more accurately in applications involving high dimensional data. This makes SVM an optimal combination for LBP. Le et al. [3] improved on their previous work by using a combination of contour masks and LBP on "bccr-segset" dataset. In this, they used morphological operations like erosion and dilation on grayscale images to reduce the noise caused by the ExG-ExR method.

For segmenting and classifying seven plants like Jamun, Guava, Apple, Tomato, and Grapes. Kour et al. [4] used particle swarm optimization-based support vector machine (P-SVM). In their work, they partition the whole process into different phases. In the first phase, preprocessing is done for the input images for performing noise removal, contrast enhancement, and resizing. In the second phase, feature extraction from the images is based on color and texture. For segmentation of images, the K-means algorithm is used, this constitutes the third phase and in the fourth phase, and SVM is used for training. Finally, from both the segmentation and classification particle, swarm optimization algorithm is used to select the best suitable values of the initialization parameter. With this approach, they achieved an accuracy of 95.23%.

Hu et al. [5] proposed a local-global scheme to classify weeds that produce proper weed representations. For the proposed graph weeds net (GWN) architecture consists of major three components multi-scale graph convolutional layer, graph pooling layer, and recurrent neural network (RNN). The convolutional layer is used for obtaining the associated graph representations for each scale from the input images. The pooling layer is used over the vertices of each graph to highlight the components of the target weed and discard the irrelevant information, and finally, RNN is used to sum up the multi-class graph and obtain predictions.

Kattenborn et al. [6] study CNN and remote sensing technology to reveal both spatial and temporal vegetation patterns. In the studies, it is also stated that CNNs are very effective in extracting a wide range of properties from vegetation and detecting different individual parts of the plants or the pixel-wise segmentation of vegetation classes.

Sharma et al. [7] in their paper which improved on their previous work [8] on plant disease detection, and they were able to increase the accuracy of the modal to 98.6% from 93%. This was achieved by feeding the modal with segmented images that only contain regions of interest with disease symptoms. To detect leaf blight and rust in corn, Mishra et al. [9] used CNN. The system architecture of the CNN consists of convolutional layers, max-pooling layers, activation layers, and dropout layers. An accuracy of 88.48% was achieved by the modal which showcases the feasibility of such a system in real time. The hyperparameter like learning rate and number of epochs were adjusted in the training stage to attain this accuracy.

Ciocca et al. [10] in their work demonstrate that we need a large image database for effective feature extraction. In the paper, they compared a fine-tuned CNN-based residual network (ResNet-50) for feature extraction with the following databases like Food-50, UECFOOD-256, Food-101, VIREO, Food-524, and Food-475, and each of these databases has different characteristics like a different number of food classes, etc. The results show that the Food-475 food database outperforms all the other databases. Food-475 includes 475 different classes and 247,636 images, and it is the largest publicly available food database.

Olsen et al. [11] in their work they also used Keras and TensorFlow but rather than a custom-made model they used Inception-v3 [12] and ResNet-50 [13] to classify a dataset called DeepWeed. Dataset consists of eight different weed species relevant to Australia. In this, ResNet-50 was able to achieve an accuracy of 95.7% and Inception-v3 was able to achieve an accuracy of 95.1%.

Sharpe et al. [14] used deep learning to detect goosegrass in tomato and strawberry fields. The model they chose for the experiment was YOLOv3-tiny [15]. For feature extraction, Darknet-19 was used, which is derived from YOLOv2. The model consists of 19 convolution layers with 3×3 filters and 5 max-pooling layers with 1×1 filters. This model was able to achieve an accuracy of 82% in the strawberry field and an accuracy of 38% in the tomato field.

3 Proposed System

In the proposed system, we are using convolutional neural network (CNN) for image classification. Image classifications using CNN accept an input image, processes it, and classifies it under specific class labels. For training and testing in CNN models, each input image must go through a sequence of layers such as convolution layers, pooling layers, fully connected layers, and apply an activation function that classifies an object with a probability value which indicates what class an image belongs to. The first layer of the CNN model is the convolution layer which extracts features from the given input images. After feature extraction pooling layers significantly reduces the number of parameters if the dimensions of the image are large. Pooling is also called subsampling or downsampling which reduces the resolution of each map but keeps the important information. For building CNN, we are using Keras with TensorFlow backend. Keras is a python library that is open-source and free to use. Keras makes building a deep learning model easy because it already contains implementation for the commonly used layers, activation function, and optimizers.

3.1 Dataset

The dataset "bccr-segset" [5] used for the experiments was developed by Electron Science Research Institute (ESRI), Edith Cowan University, Australia. The images were captured by using the Xilinx Zynq ZC702 development platform that captures HD images using an On-Semi VITA 2000 camera sensor. The images were captured on a custom-build test-bed on which the Xilinx Zynq ZC702 development board and camera are attached on a movable trolley above the test-bed. The trolley had the camera optics mounted perpendicular to the ground and moves at a maximum allowed speed of 5 m/s above the test-bed. The captured images were of size 228 × 228 pixels. The dataset consists of 30,000 images. The dataset is made up of four categories of images, canola, radish, corn, and background. Each category contains four growth stages of the crops.

3.2 Data Augmentation

To prevent overfitting and lower accuracy, we are using data augmentation with the help of Keras preprocessing. We are applying random flip, random rotation, and random zoom to the images in the database. This results in double the number of images for training. Random flip augmentation reverses the rows and columns of pixels. The random rotation rotates the images from 0 to 360 in the clockwise direction. A random zoom augmentation zooms in on the image or adds new pixel values or interpolates the pixel values around the image.

4 Experimental Results and Analysis

We experimented using Keras with TensorFlow as the backend on a computer with 8 cores and 16 threads along with an Nvidia GTX 1650ti which has 1024 CUDA cores. With an epoch count of 25, it took about 30 min to perform training and validation on 60,000 images. We split the images in the ratio of 8:2 splitting it into training and validation tests, respectively. Figure 2 shows the architecture of the CNN model we used for training and validation.

First, the model has an input layer that accepts images, and then it is passed to the Sequential layer which is used to create a model layer-by-layer. Then, the model has a Rescaling layer which is Keras preprocessing layer that rescales and offsets the value of images. Then, the model has a Conv2D layer which is used to generate a

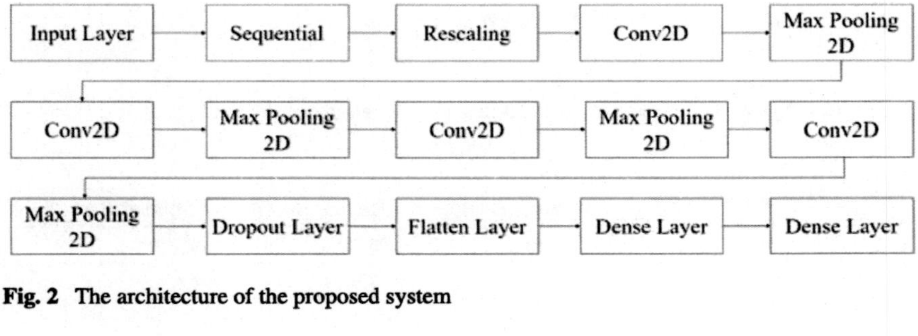

Fig. 2 The architecture of the proposed system

tensor, it is a kernel or mask that is convolved over the input image. Then, the model has a max-pooling layer that downsamples the input representation according to a pool size by taking the maximum value. Then, the input passed through a series of Conv2d and max-pooling layers. To avoid overfitting, the model has a dropout layer that sets the input unit to 0 randomly during the training with a certain frequency. Then, the model has a Flatten layer that flattens the multi-dimensional tensor to a single-dimensional array which does not affect the batch size. Then, the model has two dense layers that apply an activation function on the input and gives an output (Table 1).

From Fig. 3, it is clear that the model is overfitted training accuracy is greater than the validation accuracy. This is caused by the dataset having a lower number of images. To fix this, we add data augmentation to the images which double the image count. In the data augmentation, we applied random flip, random zoom, and random rotation what is explained in Sect. 3.2. Table 2 shows the accuracy after performing data augmentation and dropout (Fig. 4).

With data augmentation, we were able to avoid overfitting and achieve better accuracy from the model, but to increase the accuracy we added a convolution and pooling layer. Table 3 shows the accuracy and loss achieved by adding a convolution and pooling layer (Fig. 5).

Table 1 Accuracy and loss without data augmentation or dropout

Evaluation	Training set (%)	Validation set (%)
Accuracy	99.75	92.13
Loss	0.73	56

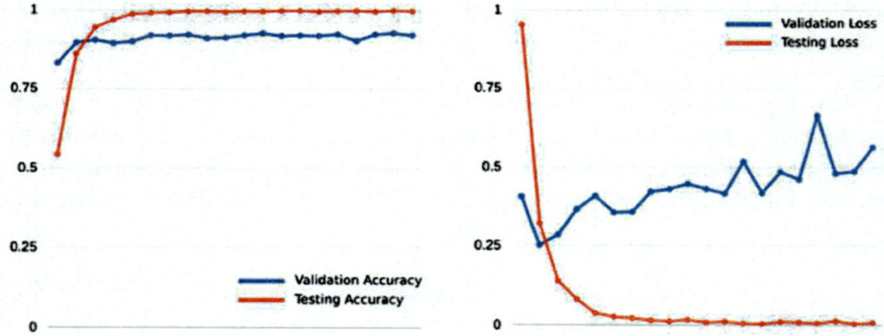

Fig. 3 Training/validation accuracy without data augmentation and dropout

Table 2 Accuracy and loss with data augmentation or dropout

Set type/value type	Training set (%)	Validation set (%)
Accuracy	96.89	95.30
Loss	7.92	12.7

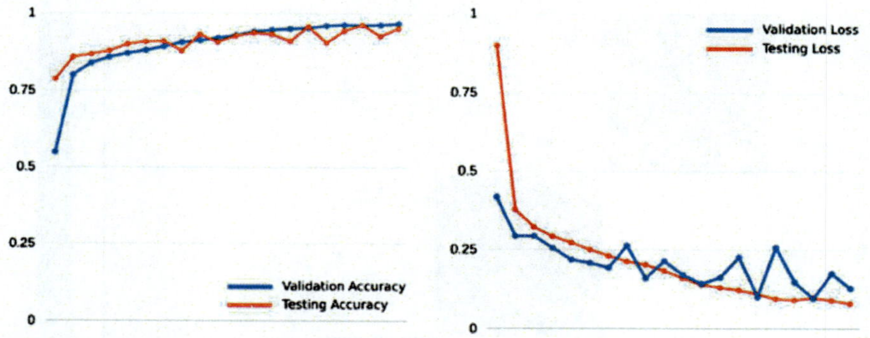

Fig. 4 Training/validation accuracy with data augmentation and dropout

Table 3 Accuracy and loss with tuning

Set type/value type	Training set (%)	Validation set (%)
Accuracy	98.65	98.58
Loss	3.70	4.78

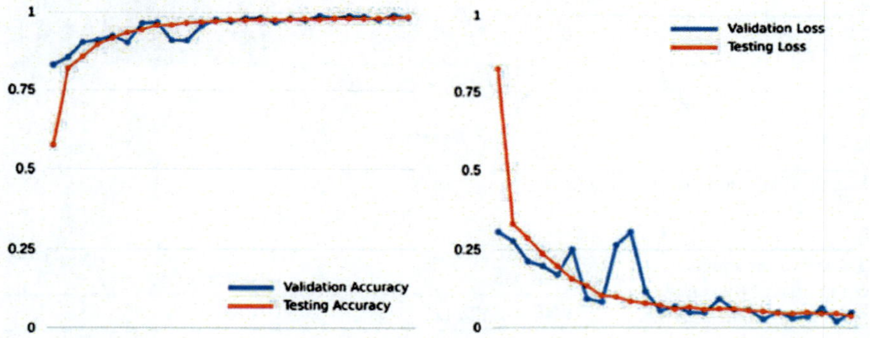

Fig. 5 Training/validation accuracy achieved by tuning

The confusion matrix below shows the performance of the CNN model with all four classes. The confusion matrix as plotted here indicates to some degree the actual accuracy or performance of the model. The values along the Y-axis of the matrix indicate the true label and the values along the X-axis refers to the predicted label. In our case, we can see that the actual and predicted labels are always the same which is due to the high accuracy of our model (Fig. 6).

Table 4 shows the comparison of the accuracy of the proposed system and other methods using the "bccr-segset" dataset.

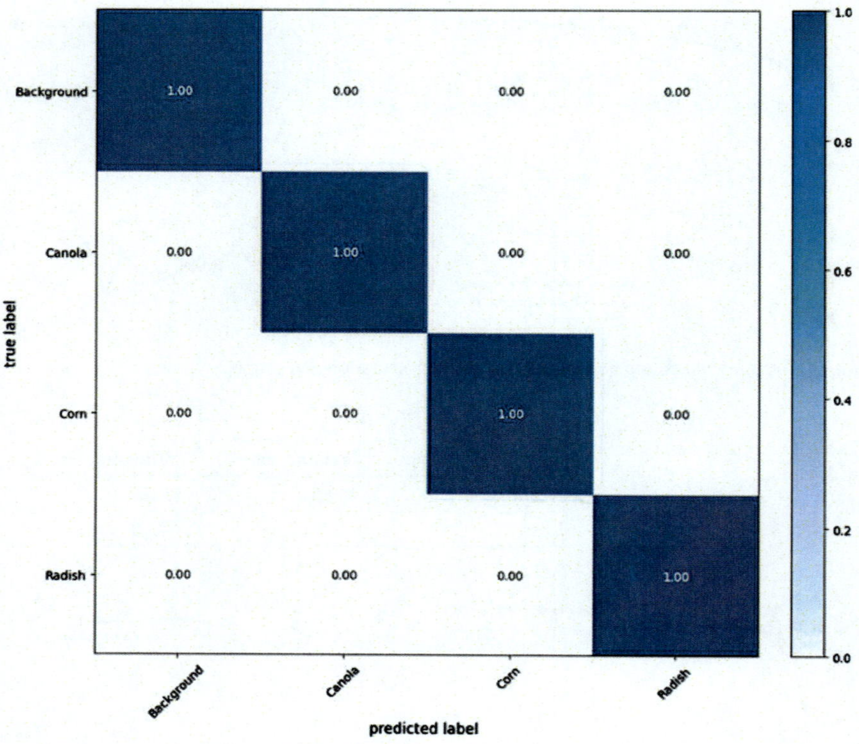

Fig. 6 Confusion matrix of the model

Table 4 Performance comparison between the proposed system with other methods

Methods	Accuracy (%)
LBP + SVM [2]	91.85
Proposed system	98.58

5 Conclusions and Scope for Future Work

The base paper [2] which we referenced showed an experimental accuracy of 91.85%. Compared to this, our experiments indicated an accuracy of 98.58%. This goes on to show that CNN's are a superior way of classifying images because of higher accuracy and lesser effort for feature extraction compared to using LBP and SVM. Our experiment and consequent analysis go on to conclude that CNN's are an optimal way to classify large image data sets. We can further improve on this experiment by changing different aspects of this model such as the loss function, optimizer, batch size, and epoch count to either increase the accuracy or decrease the training time. Future work based on this experiment can be extended to using color images of plants instead of greyscale as given in the data set for training. This could lower the effort for data collection and significantly cut down on time taken for data preprocessing.

References

1. V.L. Bogdanov, et al., The issues of weed infestation with environmentally hazardous plants and methods of their control, in *IOP Conference Series. Earth and Environmental Science*, 2016
2. V.N.T. Le, B. Apopei, K. Alameh, Effective plant discrimination based on the combination of local binary pattern operators and multiclass support vector machine methods. Inf. Process. Agric. **6**(1), 116–131 (2019)
3. V.N.T. Le, S. Ahderom, B. Apopei, K. Alameh, A novel method for detecting morphologically similar crops and weeds based on the combination of contour masks and filtered Local Binary Pattern operators. GigaScience **9**(3), giaa017 (2020)
4. V.P. Kour, S. Arora, Particle swarm optimization based support vector machine (P-SVM) for the segmentation and classification of plants. IEEE Access **7**, 29374–29385 (2019)
5. K. Hu, et al., Graph weeds net: a graph-based deep learning method for weed recognition. Comput. Electron. Agric. **174**, 105520 (2020)
6. T. Kattenborn et al., Review on Convolutional Neural Networks (CNN) in vegetation remote sensing. ISPRS J. Photogramm. Remote. Sens. **173**, 24–49 (2021)
7. P. Sharma, Y.P.S. Berwal, W. Ghai, Performance analysis of deep learning CNN models for disease detection in plants using image segmentation. Inf. Process. Agric. **7**(4), 566–574 (2020)
8. P. Sharma, Y.P.S. Berwal, W. Ghai, KrishiMitr (Farmer's Friend): using machine learning to identify diseases in plants, in *2018 IEEE International Conference on Internet of Things and Intelligence System (IOTAIS)* (IEEE, 2018)
9. S. Mishra, R. Sachan, D. Rajpal, Deep convolutional neural network based detection system for real-time corn plant disease recognition. Procedia Comput. Sci. **167**, 2003–2010 (2020)
10. G. Ciocca, P. Napoletano, R. Schettini, CNN-based features for retrieval and classification of food images. Comput. Vis. Image Underst. **176**, 70–77 (2018)
11. A. Olsen et al., DeepWeeds: a multiclass weed species image dataset for deep learning. Sci. Rep. **9**(1), 1–12 (2019)
12. C. Szegedy et al., Rethinking the inception architecture for computer vision, in *Proceedings of the IEEE Conference on Computer Vision and Pattern Recognition*, 2016
13. K. He et al., Deep residual learning for image recognition, in *Proceedings of the IEEE Conference on Computer Vision and Pattern Recognition*, 2016
14. S.M. Sharpe, A.W. Schumann, N.S. Boyd, Goosegrass detection in strawberry and tomato using a convolutional neural network. Sci. Rep. **10**(1), 1–8 (2020)
15. J. Redmon, A. Farhadi, Yolov3: An Incremental Improvement. arXiv preprint arXiv:1804.02767 (2018)

Efficient Iterative Linear Precoding Scheme for Downlink Massive MIMO Systems

A. Augusta, C. Manikandan, S. Rakesh Kumar, and K. Narasimhan

Abstract Massive multiple-input multiple-output (MIMO) is the crucial technology to increase the 5G wireless communication system's reliability and throughput. Massive MIMO uses a combination of a precoder and massive antennas at the base station (BS). A simple beamforming strategy such as zero-forcing (ZF) can be exploited in massive MIMO. ZF is a linear precoding technique usually adopted in low complexity massive MIMO systems. However, ZF precoding techniques involve matrix inversion, whose size increased with the user equipment. It increases the system's overall computational complexity. In this paper, a modified weighted two-stage (WTS) algorithm is proposed to minimize that effect. The existing WTS algorithm used two symmetric half iterations and combined the iterations for faster convergence. It demands computation in the forward and the reverse order in each iteration. However, the proposed modification considers the present and past iterations, eliminating the reverse iterations and minimizing the complexity. The proposed change reduces the computational complexity by 17%. Simulation results show that modified WTS achieves the near-optimal capacity and similar bit error rate (BER) performance as ZF precoding.

1 Introduction

MIMO is a multi-antenna system [1, 2]. A massive MIMO system is an extension of a MIMO system. It has large antennas at BS, simultaneously serving multiple user equipment, using the same spectrum and time resources, thereby increasing the spectral efficiency and minimizing the BER [3, 4]. This is achieved by forming directional beams towards the intended users and nullifying the inter-user interference. This type of beamforming is done by precoding at the base station. Precoding can be classified as linear and non-linear based on its computation method. Linear precoders are widely preferred due to their low complexity and near-optimal performance. The linear precoding scheme, matched filter (MF) [5], has low complexity than ZF,

A. Augusta · C. Manikandan (✉) · S. R. Kumar · K. Narasimhan
School of EEE, SASTRA Deemed To Be University, Tamil Nadu, Thanjavur 613401, India
e-mail: manikandan@ece.sastra.edu

P. Karuppusamy et al. (eds.), *Ubiquitous Intelligent Systems*, Smart Innovation, Systems and Technologies 243, https://doi.org/10.1007/978-981-16-3675-2_49

but ZF precoding performs better than MF. However, the ZF precoding involves complex matrix inversion, which has more computational complexity. Two different methods, namely, (i) Approximation technique and (ii) Iterative technique, are available to reduce the computational complexity of pseudoinverse matrix. Neumann series approximation [6] and truncated polynomial expansion [7] are proposed under the approximation technique. These two methods are based on truncating the series expansion, but the reduction in complexity is only marginal. The second method is the iterative method. It is iteratively finding the transmitted signal using low complexity algorithms [8–17] without calculating the complex matrix inversion. Iterative algorithms such as Jacobi algorithm (JA), Gauss–Seidel (GS), conjugate gradient (CG), symmetric successive over-relaxation (SSOR), accelerated over relaxation (AOR) and successive over-relaxation (SOR) have been proposed in the literature. Predominant of the existing algorithms considerably reduce the complexity but still require more than three iterations to reach the near-optimal performance of ZF precoding. The weighted two-stage (WTS) method [18] has little more complexity than the Gauss–Seidel approach. However, it achieves faster convergence and gives better performance close to ZF precoding.

This paper proposed a new methodology to modify the existing WTS to achieve low complexity and better performance with less iteration. The following are the significant contribution of this work, (i) Mathematical model of precoder is formulated with modified WTS, which reduces the two half iterations (forward and reverse iteration) to single iterations (forward iteration), (ii) Optimal initial solution is introduced in the precoder design, (iii) Simulation studies are carried, and performance of the proposed method is evaluated in terms of complexity, bit error rate (BER) and capacity.

The remaining sections are arranged as follows: Sect. 2 illustrates the system model used for precoding in massive MIMO systems and describes the proposed modified WTS precoding technique. Section 3 provides the simulation results. Section 4 gives the conclusion.

2 System Model

The massive MIMO downlink system is illustrated in Fig. 1. In a massive MIMO downlink, the BS has multiple antennas (M), simultaneously serving various single-antenna user equipment (K). Usually, the number of transmitting antennas at BS is larger than user equipment ($M \gg K$). BS antennas are going to use the same frequency and time resources to serve K users simultaneously. This is achieved by forming multiple beams towards the user equipment, as shown in Fig. 1. To mitigate the multiuser interference, precoding is done at the BS.

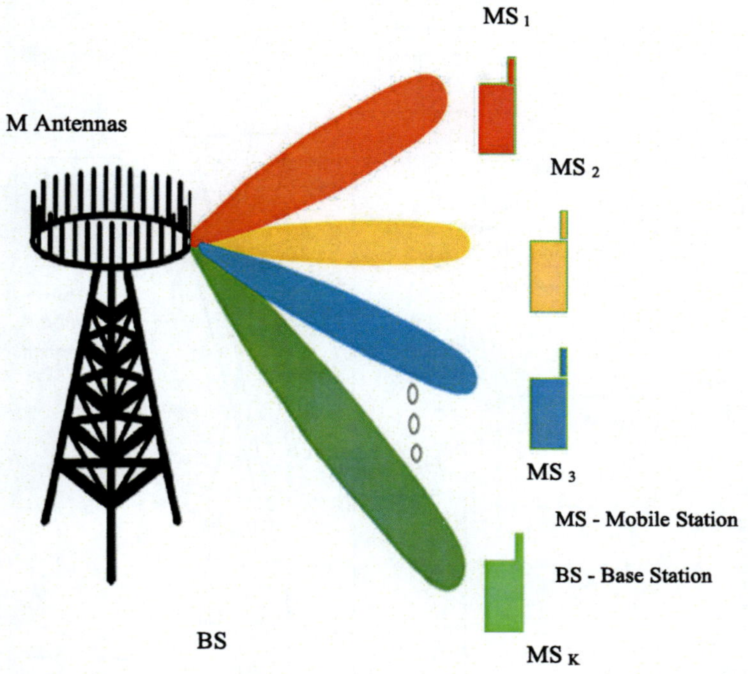

Fig. 1 Downlink massive MIMO system

2.1 Problem Formulation

Formulation of a mathematical model for the downlink-massive MIMO is presented in this section. Figure 2 represents the proposed work's block diagram. The binary data for K users are modulated using 64-QAM. The data vector $d = [d_1, \ldots, d_k]^{\mathrm{T}}$ containing the QAM modulated data symbols for K users. The data vector d is precoded using the precoding matrix P. The precoded vector t is transmitted by M transmitting antennas at BS. The size of massive MIMO used for the simulation is $M \times \mathrm{K} = 128 \times 16$. The received data vector $y = [y_1, \ldots, y_k]^{\mathrm{T}}$ can be represented as

$$y = \sqrt{\rho} H t + n \tag{1}$$

where $n = [n_1, \ldots, n_k]^{\mathrm{T}}$ is the additive white Gaussian noise (AWGN), ρ is the signal-to-noise ratio (SNR), $H \in \mathbb{C}^{K \times M}$ is the Rayleigh flat fading channel matrix, and $t = [t_1, \ldots, t_m]^{\mathrm{T}}$ is the precoded vector which contains the symbols to be transmitted can be represented as,

$$t = Pd \tag{2}$$

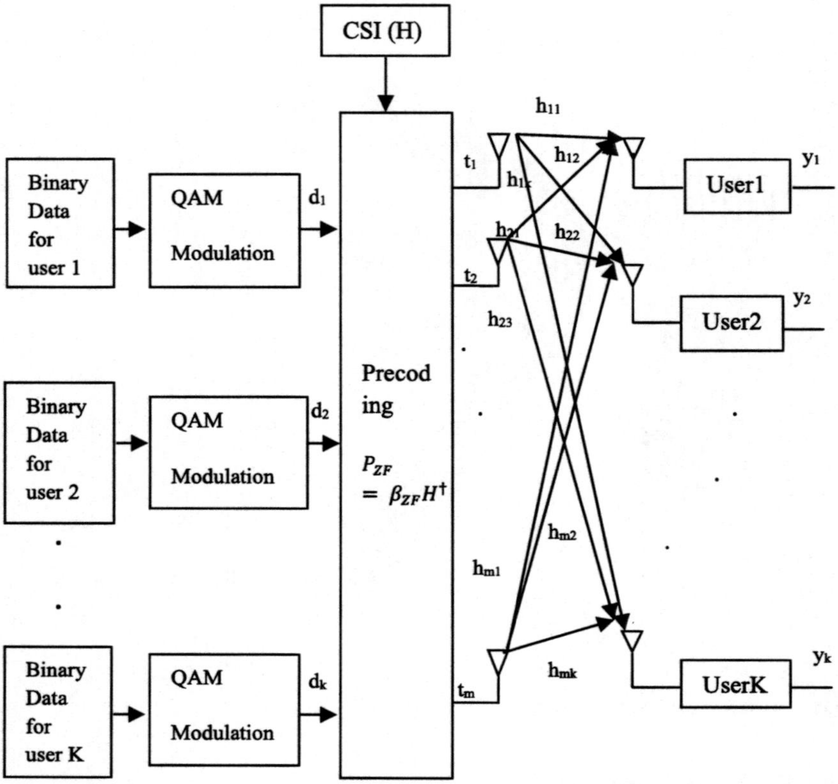

Fig. 2 Block diagram of the proposed system

where $P \in \mathbb{C}^{M \times K}$ is the precoding matrix. Precoding is done to mitigate multiuser interference. The Zero Forcing (ZF) precoding is expressed as,

$$P_{\text{ZF}} = \beta_{\text{ZF}} H^{\dagger} \tag{3}$$

where $H^{\dagger}$ is the pseudoinverse of the channel matrix.

The pseudoinverse is written as,

$$H^{\dagger} = H^{H}\left(HH^{H}\right)^{-1} \tag{4}$$

$$H^{\dagger} = H^{H} W^{-1} \tag{5}$$

where $W = HH^{H}$. Then the precoding matrix is written as,

$$P_{\text{ZF}} = \beta_{\text{ZF}} H^{H} W^{-1} \tag{6}$$

where $\beta_{ZF} = \sqrt{K/tr(W^{-1})}$ [5], is power normalization factor.

The precoded vector t can be expressed as,

$$t = \beta_{ZF} H^H W^{-1} d \tag{7}$$

However, the computational complexity of W^{-1} is more. The computational complexity of W^{-1} is expressed as $O(K^3)$.

2.2 Proposed Modified WTS Precoding

A new algorithm, modified WTS precoding, is proposed. The computational complexity to calculate W^{-1} is considerably reduced. The precoded vector can be written as,

$$t = \beta_{ZF} H^H x \tag{8}$$

where $x = W^{-1}d$. Where x is the unknown, it needs to be calculated. Equation of x can be written as,

$$Wx = d \tag{9}$$

Equation (9) resembles the linear equation of the form $Ax = b$. The linear equations can be solved using the iterative method with reduced complexity. The algorithm has the following steps,

Decompose W matrix as below,

$$W = U + L + D \tag{10}$$

L is a strictly lower triangular matrix, D is a strictly diagonal matrix and U is a strictly upper triangular matrix. Two half iterations are proposed in WTS [18] as below,

$$(D + L)x^{(j+1/2)} = d - Ux^{(j)} \tag{11}$$

$$(D + U)x^{(j+1)} = d - Lx^{(j+1/2)} \tag{12}$$

where j is the iteration number, when $j = 0$, the vector x is set to zero, the two half iterations are used in WTS, which calculates x in forward and reverse directions. A modification is proposed in this work which removes the reverse calculation of x, and an optimal initial solution is proposed for faster convergence. Hence, the below equation is used to calculate x iteratively,

$$(D + L)x^{(j+1)} = d - Ux^{(j)} \tag{13}$$

Equation (13) can be written as,

$$x^{(j+1)} = (D + L)^{-1}\left(d - Ux^{(j)}\right) \tag{14}$$

An optimal initial solution from [19] is taken as mentioned below,

$$x^{(0)} = D^{-1}d \tag{15}$$

The computational complexity of $D^{-1}d$ is low, and the convergence is faster compared to previously discussed algorithms. To further speed up the convergence, two levels of combining of current and the previous iteration value is done as below

$$x^{(j+1)} = (1 - \alpha)x^{(j+1)} + \alpha\, x^{(j)} \tag{16}$$

where $\alpha = (K/M)^2$

$$x^{(j+1)} = (1 - \omega)x^{(j)} + \omega x^{(j+1)} \tag{17}$$

where $\omega = 1 + (K/M)$.

After several iterations, the obtained value of x is used to calculate the precoded vector $t = \beta_{\mathrm{ZF}} H^H x$.

The proposed modified WTS method has low complexity than the WTS scheme since the two half iterations are reduced to one. The initial solution for x is given in the modified WTS scheme, which has very low complexity and speeds up the convergence.

3 Results and Discussion

The proposed work aims to decrease the complexity of computations in the precoding of massive MIMO. A modified WTS method proposed for this purpose is simulated using MATLAB R2020b with parameters as mentioned in Tables 2, 3 and 4. The simulation results and the complexity comparison are provided in the following sections. In Sect. 3.1, Table 1 gives the complexity comparison of the proposed and the existing algorithms. In Sect. 3.2, capacity analysis is carried out. In Sect. 3.3, BER performance analyses are carried out. The size of massive MIMO used for this purpose is $M \times K = 128 \times 16$. The modulation technique used is 64 QAM. The binary data for each user is modulated using 64 QAM, and the QAM symbols are precoded using the existing and proposed methods. The precoded data is passed through the channel, and noise is added based on the SNR value. The received data is demodulated and compared against the transmitted data, and the BER is calculated for each precoding technique. In Sect. 3.4, BER performance is analysed for the variable number of user equipment and transmitting antennas. It is found that the proposed

Table 1 Comparison of computational complexity

Precoding technique	Computational complexity
ZF method [18]	$M + MK + K^3$
NS method [6]	$M + MK + (j-2)K^3$
SSOR method [12]	$M + MK + (j-2)K^2 + 3K$
WTS method [18]	$M + MK + 2jK^2 + 2jK$
Modified WTS method	$M + MK + jK^2 + 4jK$

Table 2 Simulation parameters

Simulation parameters	Values
Modulation	64QAM
Base station	1
Number of users	16
Number of antennas at each user	1
Number of transmitting antennas at the base station	128

Table 3 Simulation parameters for calculating BER performance for variable users

Simulation parameters	Values
Modulation	64QAM
Number of transmitting antennas	128
SNR(dB)	26
Range of users	6–22

Table 4 Simulation parameters for calculating BER performance for variable transmitting antennas

Simulation parameters	Value
Modulation	64QAM
Transmitting antennas range	64–192
Number of users	16
SNR(dB)	26

modified WTS scheme has low complexity, and a 17% reduction in complexity is observed as compared to the WTS scheme.

3.1 Computational Complexity Analysis

The proposed modified WTS method has low computational complexity, and the computation of the precoded vector has three parts. The first part is solving x using Eq. (14). Equation (14) for calculating x is written as

$$x_m^{(j+1)} = \frac{1}{w_{mm}}\left(d_m - \sum_{n<m} w_{mn}x_n^{(j+1)} - \sum_{n>m} w_{mn}x_n^{(j)}\right)$$
$$m, n = 1, 2, \ldots, K. \quad (18)$$

$x_m^{(j+1)}$ and d_m denote mth elements of x and d vectors, respectively. There are K elements in x and d, respectively. The number of complex multiplications required is K^2 for each iteration. So solving x requires jK^2 times of complex multiplications. For the combining process, the number of complex multiplications required is $4jK$. The second part is the computation of H^Hx, the complex multiplications required are MK. The third element is the determination of $\beta_{ZF}H^Hx$ the number of complex multiplications required is M. So, the proposed scheme has a computational complexity of $M + MK + jK^2 + 4jK$ which is having less complexity than the WTS, SSOR and Neumann method.

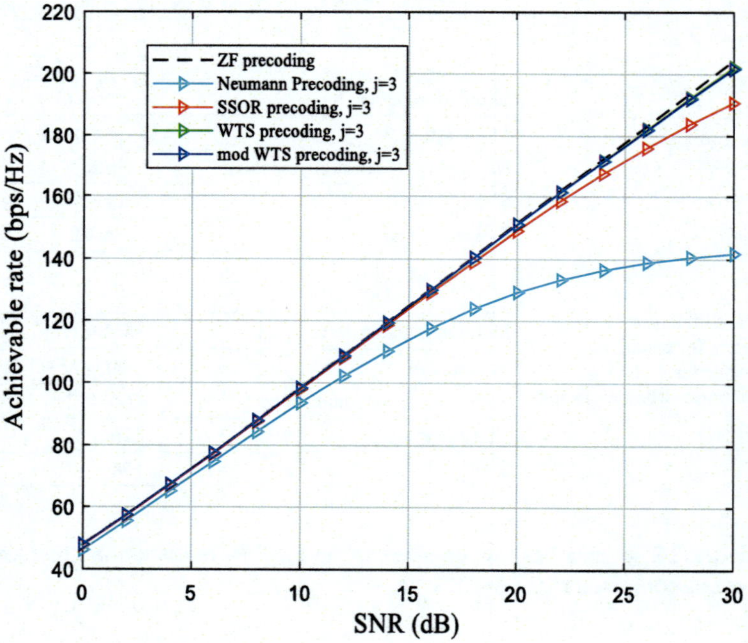

Fig. 3 Capacity comparison of low complexity algorithms

3.2 Capacity Analysis

Figure 3 gives the capacity comparison of ZF, Neumann, SSOR, WTS and modified WTS scheme. It shows that the WTS and modified WTS approach's sum-rate is closer as ZF precoding within three iterations. But the modified WTS achieves the capacity with reduced complexity. The sum rate is calculated using (25). The received signal to interference noise ratio (SINR) can be defined for each user k as [13],

$$\gamma_k = \frac{\frac{\rho}{K}|g_{kk}|^2}{\frac{\rho}{K}\sum_{n\neq k}^{K}|g_{nk}|^2 + 1} \tag{19}$$

For ZF precoding, when $n \neq k$, $|g_{nk}|^2 = 0$ and $G = \text{HP}$ [13], using Eq. (6) for P, G becomes,

$$G = H\beta_{\text{ZF}}H^H W^{-1} \tag{20}$$

Substituting $\beta_{\text{ZF}} = \sqrt{K/tr(W^{-1})}$ in Eq. (20)

$$G = \sqrt{K/tr(W^{-1})}HH^H W^{-1} \tag{21}$$

Since $W = HH^H$, Eq. (21) becomes,

$$G = \sqrt{K/tr(W^{-1})}WW^{-1} \tag{22}$$

$$G = \sqrt{K/tr(W^{-1})}I_K \tag{23}$$

where I_K is the identity matrix.
Substitute Eq. (23) in Eq. (19), we obtain

$$\gamma_k = \frac{\rho}{tr(w^{-1})} \tag{24}$$

The sum rate of ZF precoding is expressed as [13]

$$C_{\text{ZF}} = \sum_{i=1}^{K}\log_2(1+\gamma_i) = K\log_2\left(1 + \frac{\rho}{tr(w^{-1})}\right) \tag{25}$$

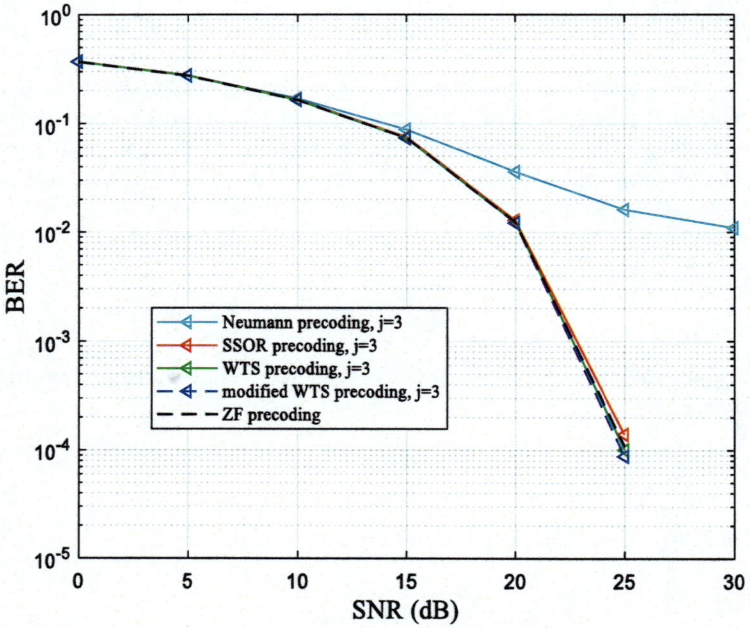

Fig. 4 BER performance comparison of low complexity algorithms

3.3 BER Performance Analysis

Figure 4 gives the BER performance comparison of SSOR, Neumann series, WTS and the proposed modified WTS algorithm. When $j = 3$, modified WTS reaches a similar performance as the WTS algorithm but with reduced complexity. It is shown that the SSOR method needs an SNR of 23.5 dB to reach 10^{-3} BER, and Neumann precoding has poor BER performance, and it reaches 10^{-2} BER at 30 dB. But the SNR required to achieve the 10^{-3} BER is 23 dB for both WTS and modified WTS, but the computational complexity for the proposed method is 17%, which is less than WTS.

3.4 BER Performance Analysis for Variable Number of Users and Transmitting Antennas

Figure 5 shows the BER performance of the existing and proposed methods by varying the number of users from 6 to 22 by keeping the number of transmitting antennas (M = 128) and SNR (26 dB) constant. The BER performance of Neumann and SSOR precoding reduces as the number of user equipment increases. But it is

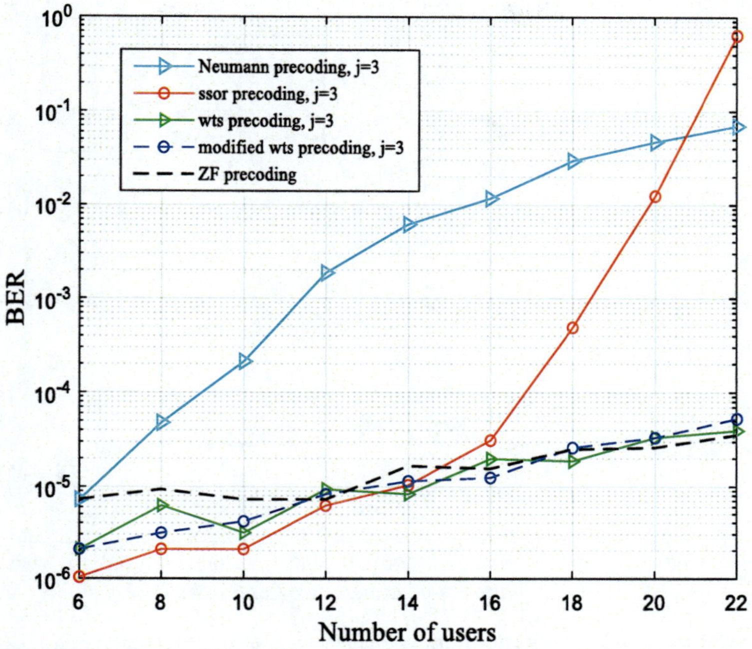

Fig. 5 BER comparison of low complexity algorithms for a range of users from 6 to 22

shown that WTS and modified WTS enjoy good BER performance of less than 10^{-3} up to the number of users 22. But the proposed scheme achieves this performance with reduced complexity.

In Fig. 6, BER performance is compared by varying the number of transmitting antennas by keeping the user equipment ($K = 16$) and SNR (26 dB) constant. Neumann and SSOR precoding's BER performance is low as the number of transmitting antennas varies from 64 to 128. But it is shown that WTS and modified WTS enjoy good BER performance of less than 10^{-3} for transmitting antennas range from 64 to 192. But the proposed scheme achieves this performance with reduced complexity.

4 Conclusion

A low complexity precoding method called modified WTS precoding is proposed in this paper. Neumann method is giving a poor performance with less iteration. SSOR method and WTS method are giving improved performance, but still, the complexity is more. It is found that modified WTS-based precoding has similar sum-rate and BER performance as that of WTS with reduced complexity (by 17%) and

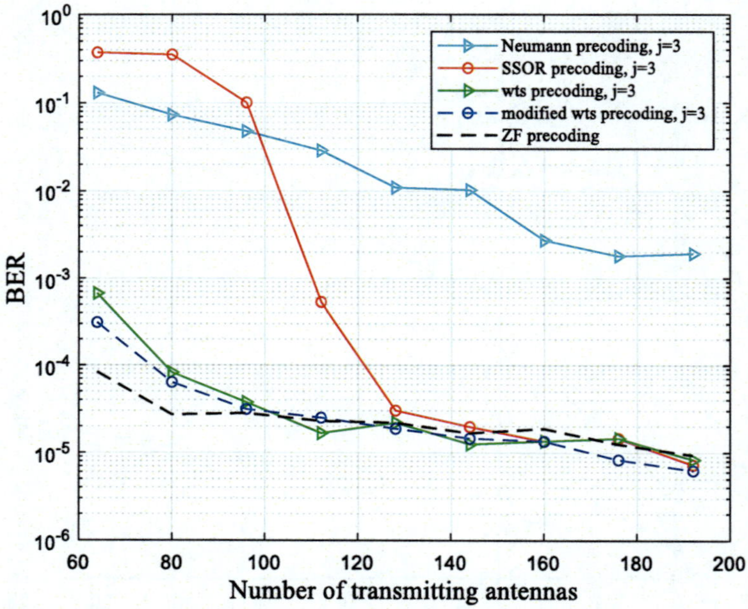

Fig. 6 BER comparison of low complexity algorithms for a range of transmitting antennas from 64 to 192

having near-optimal performance close to classical ZF precoder. The experiment is carried out by varying the user count and the transmitting antenna count. It is shown that the proposed algorithm provides better BER performance. One limitation of the proposed linear scheme is that suboptimal in performance compared to other non-linear techniques. Low complexity precoders have potential research scope in millimetre Wave (mmWave) massive MIMO communication systems. The digital precoding section of the mmWave massive MIMO system can be designed using the proposed technique.

References

1. A. Bashar, Artificial intelligence based LTE MIMO antenna for 5th generation mobile networks. J. Artif. Intell. **2**, 155–162 (2020)
2. C. Manikandan, P. Neelamegam, A. Srivishnu, K.G. Raj, A survey of MIMO transceiver designs in wireless communication systems. Int. J. Appl. Eng. Res. **10**, 12073–12078 (2015)
3. L. Lu, G.Y. Li, A.L. Swindlehurst, A. Ashikhmin, R. Zhang, An overview of massive MIMO: Benefits and challenges. IEEE J-STSP. **8**, 742–758 (2014)
4. N. Hassan, X. Fernando, Massive MIMO wireless networks: an overview. Electronics **6**, 1–29 (2017)

5. F. Rusek, D. Persson, B.K. Lau, E.G. Larsson, T.L. Marzetta, O. Edfors, F. Tufvesson, Scaling up MIMO: Opportunities and challenges with very large arrays. IEEE Signal Process. Mag. **30**, 40–60 (2012)
6. H. Prabhu, J. Rodrigues, O. Edfors, F. Rusek, Approximative matrix inverse computations for very-large MIMO and applications to linear precoding systems, in *IEEE Wireless Communications and Networking Conference* (IEEE, Shanghai, China, 2013), pp. 2710–2715
7. A. Mueller, A. Kammoun, E. Björnson, M. Debbah, Linear precoding based on polynomial expansion: reducing complexity in massive MIMO. EURASIP J. Wirel. Commun. Netw. **63** (2016)
8. J. Wu, Y. Hu, Y. Wang, An improved AOR-based precoding for massive MIMO systems, in *4th International Conference on Communication and Information Processing* (2018), pp. 251–255
9. S. Berra, M.A. Albreem, M.S. Abed, A low complexity linear precoding method for massive MIMO, in *International Conference on UK-China Emerging Technologies* (IEEE, Glasgow, United Kingdom, 2020), pp. 1–4
10. X. Qiang, Y. Liu, Q. Feng, J. Liu, X. Ren, M. Jin, Approximative matrix ınversion based linear precoding for massive MIMO systems, in *International Conference on Computing, Networking, and Communications* (IEEE, Big Island, HI, USA, 2020), pp. 950–955
11. D. Subitha, J.M. Mathana, J.S. Leena Jasmine, R. Vani, Modified conjugate gradient algorithms for gram matrix ınversion of massive MIMO downlink linear precoding. Int. J. Recent Technol. Eng. **8** (2019)
12. T. Xie, L. Dai, X. Gao, X. Dai, Y. Zhao, Low-complexity SSOR-based precoding for massive MIMO systems. IEEE Commun. Lett. **20**, 744–747 (2016)
13. X. Gao, L. Dai, J. Zhang, S. Han, I. Chih-Lin, Capacity-approaching linear precoding with low-complexity for large-scale MIMO systems, in *IEEE İnternational Conference on Communications* (IEEE, London, UK, 2015)
14. S. Hashima, O. Muta, Fast matrix inversion methods based on Chebyshev and Newton iterations for zero-forcing precoding in massive MIMO systems. EURASIP J. Wirel. Commun. Netw. **34**, 1–12 (2020)
15. M.N. Boroujerdi, S. Haghighatshoar, G. Caire, Low-complexity statistically robust precoder/detector computation for massive MIMO systems. IEEE Trans. Wirel. Commun. **17**, 6516–6530 (2018)
16. C. Zhang, Y. Jing, Y. Huang, L. Yang, Performance analysis for massive MIMO downlink with low complexity approximate zero-forcing precoding. IEEE Trans. Commun. **66**, 3848–3864 (2018)
17. Q. Xie, H. Han, Z. Xu, Qi, W. Shen, A low-complexity linear precoding scheme based on SOR method for massive MIMO systems, in *IEEE 81st Vehicular Technology Conference* (IEEE, Glasgow, UK, 2015), pp. 1–5
18. Y. Liu, J. Liu, Q. Wu, Y. Zhang, M. Jin, A near-optimal iterative linear precoding with low complexity for massive MIMO systems. IEEE Commun. Lett. **23**, 1105–1108 (2019)
19. L. Dai, X. Gao, X. Su, S. Han, I. Chih-Lin, Z. Wang, Low-complexity soft-output signal detection based on Gauss-Seidel method for uplink multiuser large-scale MIMO systems. IEEE Trans. Veh. Technol. **64**, 4839–4845 (2014)
20. E.G. Larsson, O. Edfors, F. Tufvesson, T.L. Marzetta, Massive MIMO for next-generation wireless systems. IEEE Commun. Mag. **52**, 186–195 (2014)
21. M. Costa, Writing on dirty paper (corresp.). IEEE Trans. Inf. Theory. **29**, 439–441 (1983)
22. A. Razi, D.J. Ryan, I.B. Collings, J. Yuan, Sum rates, rate allocation, and user scheduling for multiuser MIMO vector perturbation precoding. IEEE Trans. Wirel. Commun. **9**, 356–365 (2010)
23. J.H. Lee, Lattice precoding and pre-distorted constellation in a degraded broadcast channel with finite input alphabets. IEEE Trans. Commun. **58**, 1315–1320 (2010)

New Topologies of 9 Level CHMLI Based on DVR Using FLC for Compensate the Harmonics

N. Eashwaramma, J. Praveen, and M. VijayaKumar

Abstract The objective of this research work is to improve the power quality in a three phase distribution field using fuzzy logic controller (FLC)-based dynamic voltage controller along with œ, ß, o transformation functions. To improve the power quality in the supply process, power systems must limit the harmonics like voltage sag and swell. Custom dynamic voltage controller devices are widely used to reduce power quality issues. Proposed work presents a three phase three legs 9 levels CHMLI topology based on dynamic voltage controller along with switching technique SVPWM. The proposed topology design includes 36 and 24 IGBTs which are connected in 9 levels cascaded H-bridge MLI. The proposed inverter circuit produces less harmonics in the load output voltage and reduces the requirements of external DC sources and switches. Simulation results of the proposed design are compared with FLC in terms of THD and power factors using MATLAB/SIMULINK tool.

1 Introduction

The most important issue in the present power systems is its power quality requirements. Reduced power quality in customer premises or distribution areas affects the customers in terms of cost and time. So, it is essential to meet the demand with quality power, and new technologies are evolved to handle these power quality issues. Sensitive devices like IGBTs and programmable logic controller (PLC), power electronic circuits are widely used to solve these power quality issues. The poor quality power leads into harmonics like voltage sag and swell, and it will create severe disturbances in the distribution system [1]. Voltage sag is known as "decrease in RMS AC voltage between "0.1 pu" and 0.9 pu frequency for durations from 0.5 cycles to one minute

N. Eashwaramma (✉)
JNTUA, Ananthapuramu, India

J. Praveen
Department of EEE, GRIET, Affiliation JNTUH Hyderabad, Hyderabad, India

M. VijayaKumar
Department of EEE, JNTUA, Ananthapuramu, India

P. Karuppusamy et al. (eds.), *Ubiquitous Intelligent Systems*, Smart Innovation, Systems and Technologies 243, https://doi.org/10.1007/978-981-16-3675-2_50

[2], whereas voltage swell is an increase in RMS AC voltage greater than 1.1 pu and 1.8 pu frequency for a duration longer than in [3, 4].

DVR is modified as per customer requirements, i.e. called custom power device, which provides an efficient and fast dynamic response or solution to the power quality issues. It is used to compensate the voltage harmonics like sag and swell [5]. The control unit (CU) in the DVR is used to find the power quality issues in the system by calculating the required injecting voltage and generating a reference voltage. To switch IGBTs of an inverter, PWM is generated [6].

The FLC has given good applications, and it has advantages over topology controllers. However, FLCs does not provide accurate values as the time duration is very less. It works on inputs that handle nonlinearity and nonlinear controllers. Also FLCs require many membership functions. Various research works are evolved to decrease the number of nonlinear functions and membership functions. An FLC controller is introduced in [7] enhance the effectiveness and robustness in compensating the missing voltage. Based on the analysis, this research work presents a DVR with control technique which is suitable to handle linear load considering issues in power systems.

2 Conventional Multilevel Inverter

Working of MLI

Figure 1 depicts the cascaded multilevel inverter, the full bridge configuration with a separate DC source, which may be used solar cells are connected in series. "Each full bridge inverter unit can generate a three level output: $+V_{dc}$, 0, $-V_{dc}$ by connecting the DC source to the AC load side by different combinations of the four switches S1, S2, S3 and S4. Using the top level as the example, turning on S1 and S4 yields $+V_{dc}$ output. Turning on S2 and S3 yields $-V_{dc}$ output. Turning off all switches yields 0 V output. The AC output voltage at other levels can be obtained in the same manner. The number of voltage levels at the load generally defines the number of full bridge inverters in cascade. If N is the number of DC sources, the number of output voltage levels is $m = 2N + 1$. The number of full bridge inverters units N is $(m - 1)/2$ where m is the sum of the number of positive, negative and zero levels in multilevel inverter output. The number of converters N also depends on: the injected voltage and current harmonic distortion requirements the magnitude of the injected voltage required and the available power switch voltage ratings". The AC load rating and, therefore, the DC source rating depend upon the total compensation voltage required". The main "features of cascaded multilevel inverters are:

- For real-power conversions from DC to AC, the cascaded inverters need separate DC sources, and it is separate well suited for various renewable energy sources such as fuel cell, photovoltaic and biomass.
- A least number of components are required to achieve the same number of voltage levels.

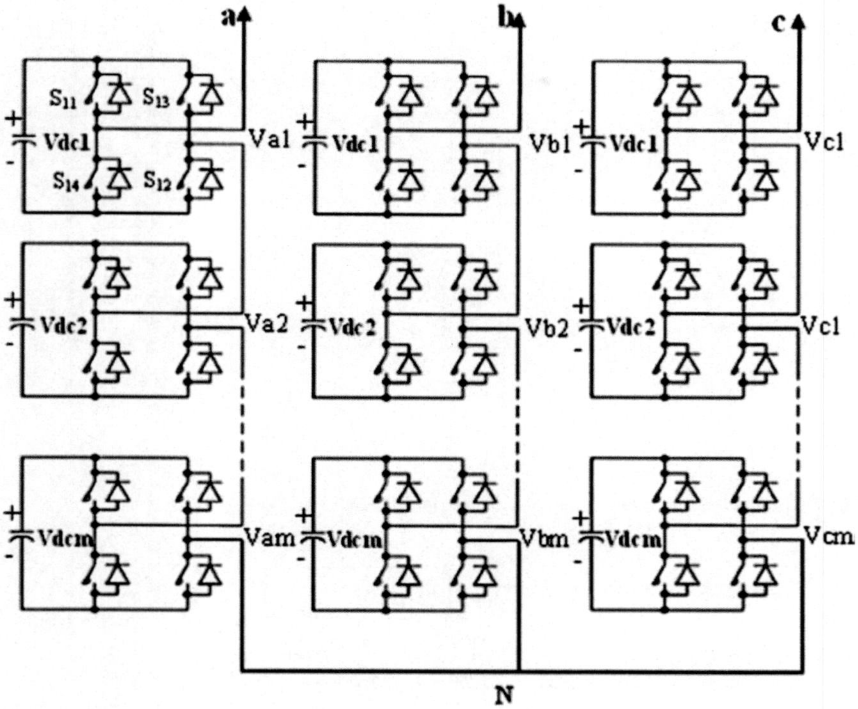

Fig. 1 Conventional type of cascaded MLI m level inverter (using 48 IGBTs)

- Optimized circuit layout and packaging are possible
- Soft-switching techniques can be used to reduce switching losses and device stresses" (Fig. 2).

3 New Topologies of Multilevel Inverter

Multilevel Inverter (MLI) is made with power electronic devices which its converts DC power to AC power at desired output voltage and frequency at different multilevel. The proposed circuit was developed based on a new topology type of MLI. The proposed method is a known reduction technique also. It is designed with 36 IGBTS and 24 IGBTS (Figs. 3 and 4).

Three phase three legs 9 level CHMLI designed based on this equation

$$\text{No.of switches} = \text{No.of legs of MLI} * \text{No.of voltage sources}$$
$$\text{No.of switches} = 3 * 12$$
$$\Rightarrow 36$$

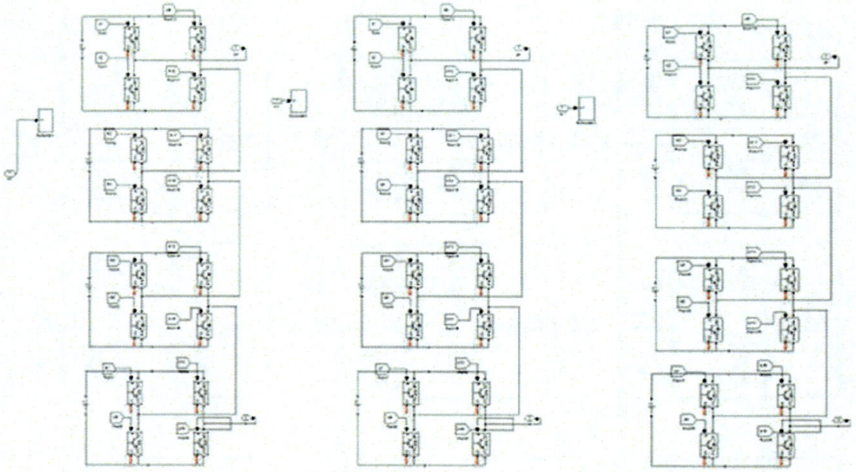

Fig. 2 Simulation diagram of three phase three level CHMLI using 48 IGBTS

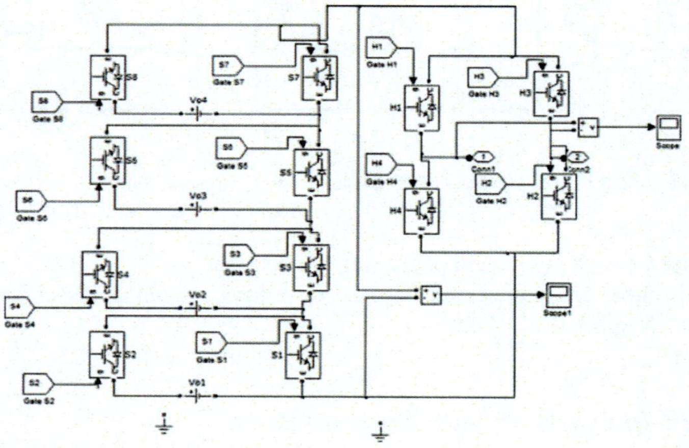

Fig. 3 New topology (first method) of CHMLI for one leg (reduction type)

Above switching status for one leg. 36 IGBT circuit diagram working as per Table 1 (Fig. 5).

Three DC sources and capacitors for using self-balancing purpose. This inverter circuit generated low harmonics compared to the 36 IGBTS topology method one conventional type of inverters.

Generally existing circuit that is the conventional type like 9 level CHMLI 48 IGBTS inverter circuit which operates at high frequency, high harmonic distortion and high voltage stress from these disadvantages are compensated by introducing a

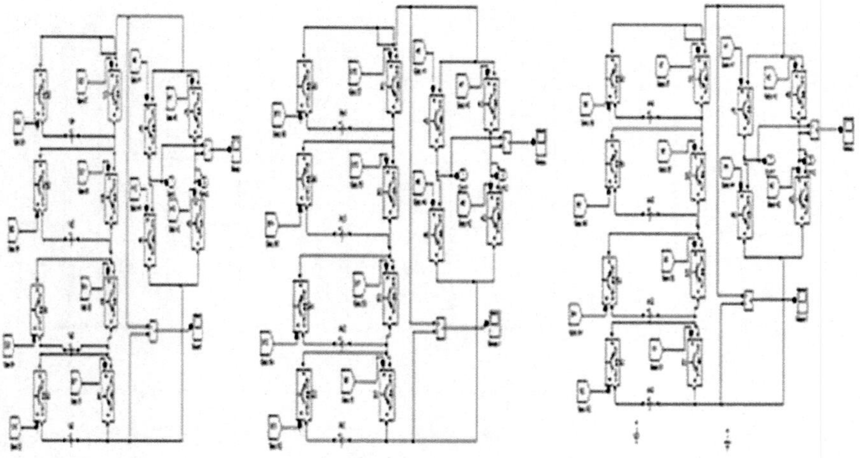

Fig. 4 New topology first method (Reduction type) I (36 IGBITS, 12 DC sources)

Table 1 Switching states of 9 level CHMLI proposed new topology first method

S	S	S	S	S	S	S	S	H	H	H	H	Voltage level at V_o load	Voltage level at V_o load
1	2	3	4	5	6	7	8	1	2	3	4		
1	0	1	0	1	0	0	1	1	1	0	0	(V_{dcl})	$+V_{dc}$
0	1	1	0	1	0	0	1	1	1	0	0	$(V_{dcl} + V_{dc4})$	$+2V_{dc}$
0	1	0	1	1	1	1	0	1	1	0	0	$(V_{dc2} + V_{dc3} + V_{dc4})$	$+3V_{dc}$
0	1	0	1	0	1	0	1	1	1	0	0	$(V_{dcl} + V_{dc2} + V_{dc3} + V_{dc4})$	$+4V_{dc}$
0	0	0	0	0	0	0	0	1	0	1	0	(0)	0
1	0	1	0	1	0	1	0	0	0	1	1	$-(V_{dc4})$	$-V_{dc}$
0	1	1	0	1	0	0	1	0	0	1	1	$-(V_{dc4} + V_{dcl})$	$-2V_{dc}$
0	1	0	1	0	1	1	0	0	0	1	1	$-(V_{dc2} + V_{dc3} + V_{dc4})$	$-3V_{dc}$
0	1	0	1	0	1	0	1	0	0	1	1	$-(V_{dcl} + V_{dc2} + V_{dc3} + V_{dc4})$	$-4V_{dc}$

new topology, i.e. called 9 levels cascaded H-bridge multilevel inverter(“Reduction Techniques”). By these new topologies are used to overcome the disadvantages by increasing the power rating of the device. The advantages of reduction techniques or new topology of cascaded the multilevel inverters are as follows:” (a) Inverter output, increase in the number of voltage levels. (b) Decrease the harmonics in Injecting voltage, (c) Switching losses are less as well as voltage stress also decreases, (d) Output waveform of MLI in the stair case model, (e) It is used for high power applications”.

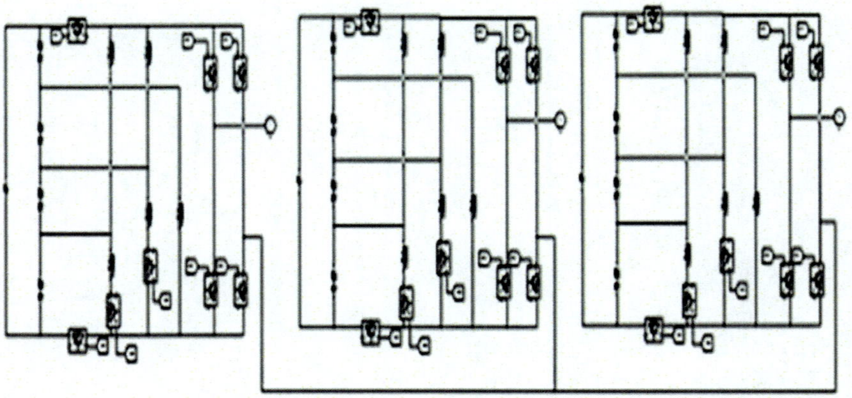

Fig. 5 Topology second method "Three phase three legs 24 IGBTS Topology (CHMLI) and using 3 DC sources and capacitors"

Total harmonic distortions (THD) and power factor (PF) results of conventional type three phase three legs 9 levels CHMLI, 48 IGBTs and proposed new topologies of nine level CHMLI ("three phase three legs 9 levels CHMLI, 36 and 24 IGBTs") were compared. As per these results, the proposed new topologies are best methods. From these topology methods, second method is good because as per these advantages are like 3 DC source, less number of IGBTS, low switching losses, capacitors are used for self-balanced voltage purpose and less harmonic content in compensating voltage (Fig. 6) (Table 2).

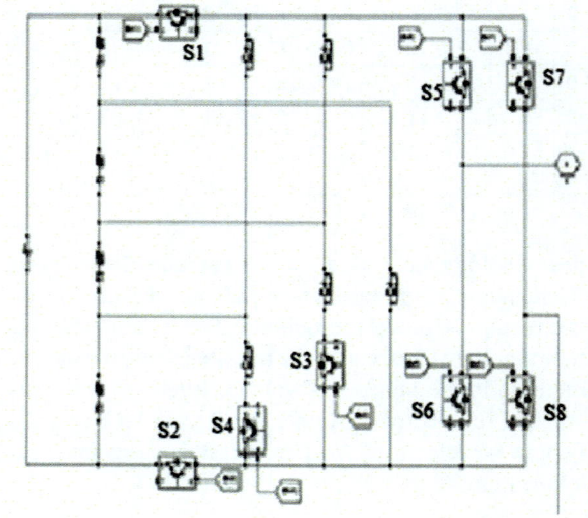

Fig. 6 Second method topology "diagram of one leg of CHMLI (used 8 switches)"

Table 2 Switching table for proposed "9 levels CHMLI level inverter"

Switches turn on	Voltage level
S1, S5, S8	V_{dc}
S2, S5, S8	$V_{dc}/4$
S3, S5, S8	$2V_{dc}/4$
S4, S5, S8	$3V_{dc}/4$
S7, S8	0
S1, S6, S7	$-V_{dc}$
S2, S6, S7	$-V_{dc}/4$
S3, S6, S7	$-2V_{dc}/4$
S4, S6, S7	$-3V_{dc}/4$

4 Dynamic Voltage Restorer ("DVR")

Figure 7 depicts the DVR block diagram, and it consists of an injection transformer, low pass filter ("LPF"), voltage source converter ("VSC"), control unit ("CU"). The two main aim of injection transformer of the DVR, first one connected to the distribution lines and second one generated Injection voltage from VSC, i.e. "injected in to the distribution lines". The purpose of low pass filter is eliminating the "harmonic" in the Injecting voltage. The use of energy storage devices is to supply the necessary energy to the voltage source converter. A VSC is a power electronic device consists of an "Energy storage device" and IGBT triggering or switching devices. This device

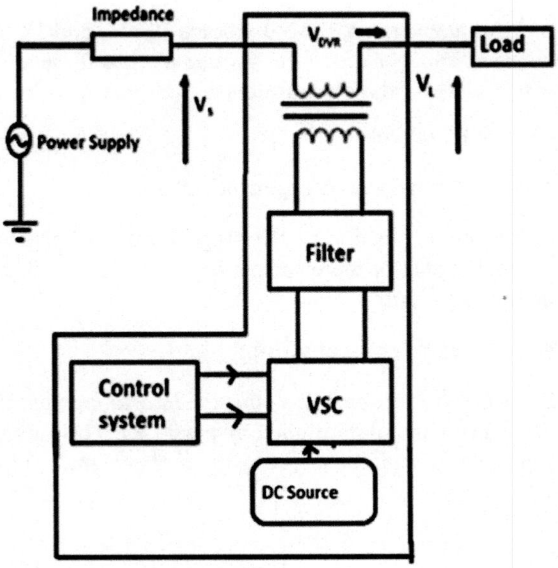

Fig. 7 Schematic diagram of DVR

produces the sinusoidal with the required magnitude of "frequency and phase angle". The dynamic voltage restorer first important task compensation of "voltage Sag and voltage Swell". VSC is developing reduction techniques or new topology methods of three phase three legs nine level CHMLI.

5 Control Technique

The controller rules are DVR to detect the harmonics occurs in the distribution lines. These controllers generate the signals that will be sent to the pulse width modulation generator, which will generate the triggering pulses. These pulses are giving to the Inverter circuit of "VSI". The DVR produced the compensating voltage by injecting the distribution lines for reducing the harmonics. In this research work, the "d–q–o" park transformation is used for voltage calculation. The "œ-ß-o" is a transformation of co-ordinate from these 3 phase constant co-ordinate [8, 9]. This œ ß o method gives the information of the direct axes and shift of quadrature axes. This "œ- ß –o" method is to give the information, the direct axes "œ" and quadrature axes "ß".

$$\begin{aligned}
V_0 &= \tfrac{1}{3}(V_a + V_b + V_c)\\
V_{oe} &= \tfrac{2}{3}(V_a \sin(\omega t) + V_b \sin(-2\pi/3)\\
&\quad + V_c \sin(\omega t + 2\pi\ 3)\\
V_\beta &= \tfrac{2}{3}(V_a \cos(\omega t) + V_b \cos(\omega t - 2\pi/3)\\
&\quad + V_c \cos(\omega t + 2\pi\ 3)\omega t
\end{aligned}$$

After conversion, the 3Ø voltage V_a, V_b and V_c, which are converted into both constant voltages are of the $V_{œ}$ and $V_{ß}$ and these are controlled easily [10]. In this research work, three controllers are proposed, which are PI, FLC.

(A) **PI Controller**

- **P. "Proportional compensation"**

The main function of the proportional compensator is to introduce a gain that is proportional to the error reading which is produced by comparing the system's output and input".

- **I "Integral compensation"**

It is a unitary feedback system, "the integral compensator will introduce the integral of the error signal multiplied by a gain KI. This means that the area under the error signal's curve will be affecting the output signal" (Fig. 8).

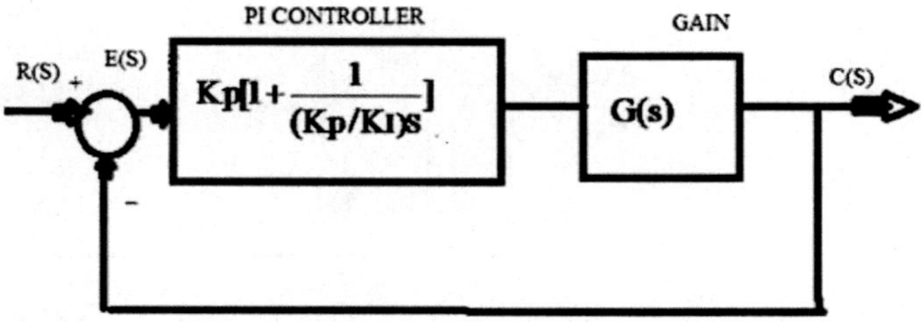

Fig. 8 Block diagram of PI controller

$$\text{output} = K_p\, e(t) + \int_0^t K_I e(t)\text{d}t$$
$$u(t) = K_p e(t) + K_I \int e(t)\text{d}(t)$$
Where $K_p, K_I =$ proportional constant

By applying laplace transformation
$$u(s) = k_p\, E(s) + k_I \frac{E(s)}{S}$$
$$u(s) = E(s)\left[Kp + \frac{K_I}{S}\right]$$
$$\frac{u(s)}{E(s)} = Kp\left[1 + \frac{1}{(Kp/K_I)S}\right]$$
Where$(Kp/Kt) =$ Transfer function
$$\frac{u(s)}{E(s)} = K_p\left[1 + \frac{1}{(Tt)S}\right]$$
PI Controller

(B) Fuzzy Logic Controller" (FLC)

The FLC is "the most important operations of fuzzy set theory, and it is used major features of linguistic variables rather than numerical variables. This control technique is understanding the systems behaviour and based on quality control rules. FLC provides definite conclusion based upon vague, ambiguous, imprecise, noisy or missing input information [11]. Figure 9 is a basic structure of FLC.

(1) "A fuzzyfication inference: It converts input data into suitable different variables of linguistic values or membership functions". (2) "A knowledge base: which consists of a database with the necessary linguistic definitions and control rule

Fig. 9 Basic block diagram of FLC

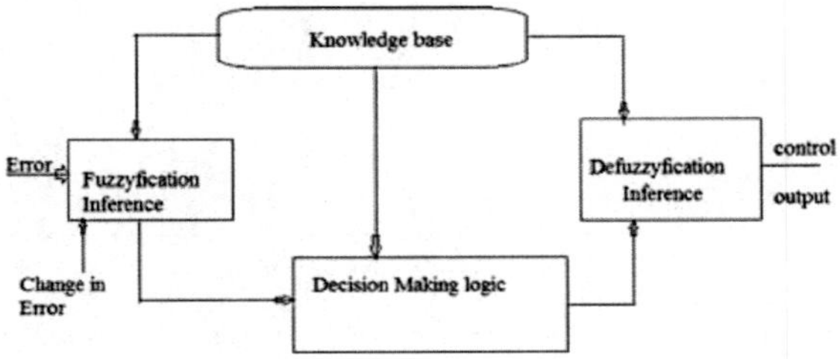

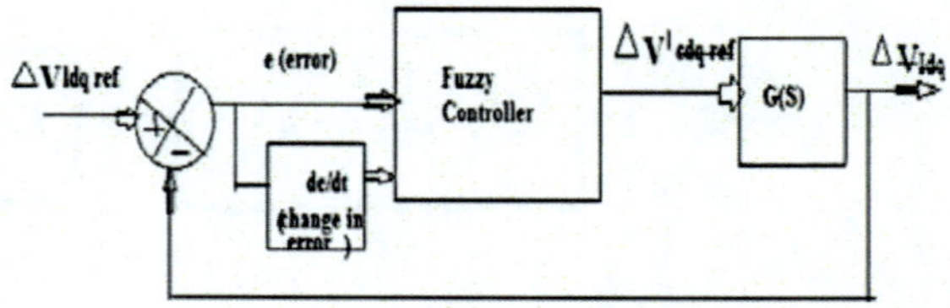

Fig. 10 Feedback control loop of DVR

set". (3) "A decision making logic: which simulating a human decision process infers the fuzzy control system from the knowledge of the control rules and linguistic variables definitions and (4) A defuzzyfication interface: which yields a non-fuzzy control action from an inferred fuzzy control action [12].

This research work has two fuzzy logic controller blocks are used for error signal—œ and error signal—ß". Error and change in error are the inputs to the fuzzy controller are shown in Fig. 10 (Figs. 11 and 12).

Table 3 shown member functions "LP = low positive, MP = Medium positive, SP = small positives = small, SN = Small negative MN = medium negative and LN = low negative. Above membership functions are 8 * 8 matrix, there are 49 rules".

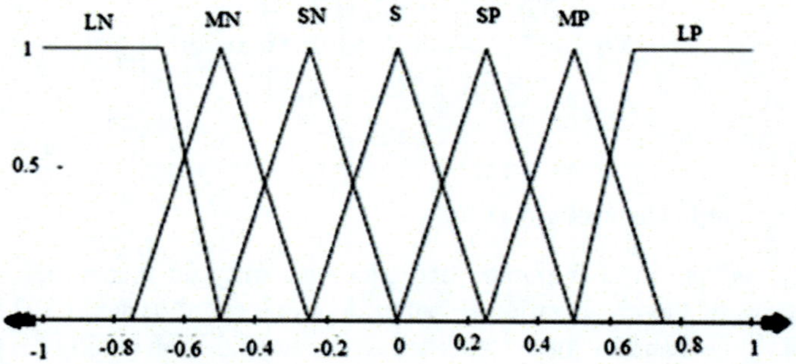

Fig. 11 "Membership" function input variables "Error"

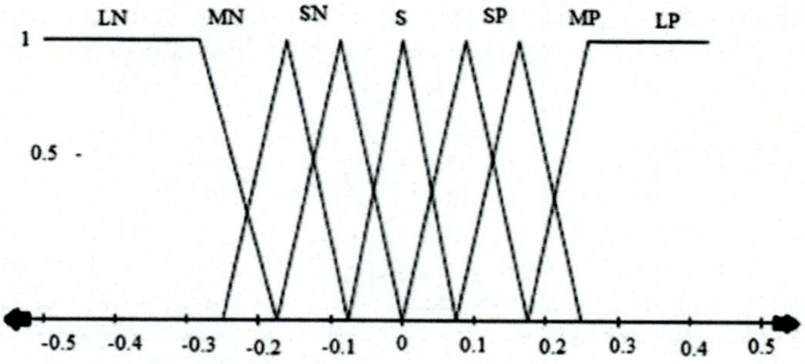

Fig. 12 "Membership" function input variables "Change in Error"

Table 3 Decision table for logic control (Member functions)

Error	Error rate						
	LP	MP	SP	S	SN	MN	LN
LP	PB	PB	PB	PM	PM	PS	Z
MP	PB	PB	PM	PM	PS	Z	NS
SP	PB	PM	PM	PS	Z	NS	NM
S	PM	PM	PS	Z	NS	NM	NM
SN	PM	PS	Z	NS	NM	NM	NB
MN	PS	Z	NS	NM	NM	NB	NB
LN	Z	NS	NM	NM	NB	NB	NB

There are two types of methods of fuzzy logic controller, which are Mamdani and Sugenio method. In this research work, Mamdani method is used. This method is suitable for nonlinear loads.

6 Explanation of Simulation Diagrams Working

This research work designed 3Ø 3 legs 9 levels of CHMLI for compensated voltage sag and swell based on DVR using "SVPWM" switching technique. This SVPWM switching techniques result more "efficient" compared to "SPWM" switching technique. The conventional type of CHMLI used 48 power switches and 12 DC sources. This type MLI, disadvantages are (1) cost is high, (2) occupied space area more and (3) more number of harmonics are generated. These disadvantages are reduced by designed or developed new technique that is a "Reduction Technique or New Topology". It is developed with three phase three legs nine CHMLI and designed with "36 IGBTS, 12DC sources", i.e. method one, and second method is "24 IGBTS, 3 DC sources, 12 capacitors" are connected in series for "self-balanced voltage purpose". This second method new topology advantages are reduced "number of power switches, DC sources, reduced switching losses, the voltage stresses of power devices and occupied space area and cost". The harmonic contents are in the output voltage is very less by using reduction technique, when compared both topology using PI controller and new topology of FLC method has given excellent THD and power factor results. The proposed topology is validated using the software "MATLAB/Simulink." FLC is used in the second method's new topology to calculate THD from FFT and PF results. Table 6 compares FLC results to PI controller results.

7 Simulation Results

See Figs. 13, 14 , 15 and 16.

THD and PF results of New Topology of 9 level CHMLI Using FLC

Results III			
Reduction technique by using	SVPWM technique compensated	sag and swell	with FLC (24 switches)
Type of control technique	No. of switches in CHMLI	THD%	Power factor
SVPWM with FLC	24	0.8	0.976

Figure 17 diagram PI controllers replaced by figure FLC (Fig. 18).

These harmonics occurs in three phase distribution lines because of three phase faults and large load additions. Voltage sag and swell harmonics are very harmful in distribution lines. Reduction method is generated very few or low harmonics. Low harmonics were observed in the load voltage. This type of voltage or standard voltage is supplied to near-consumer locations or loads, extending the life of sensitive equipment and reducing damage (Fig. 19).

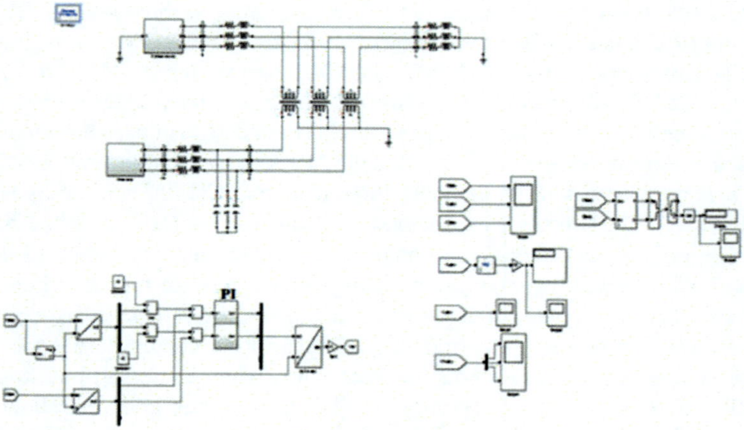

Fig. 13 Simulation diagram of 9 level reduction technique of CHMLI based on DVR using SVPWM with PI controller for compensated voltage sag and swell (36 IGBTS and 24 IGBTS)

Fig. 14 PI controller

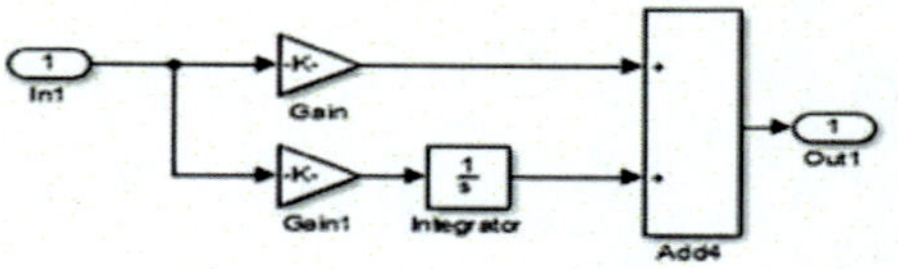

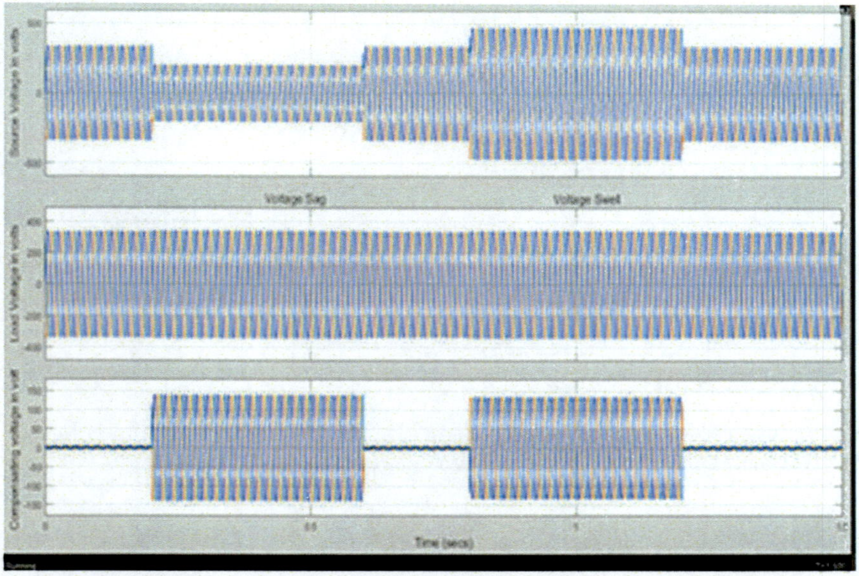

Fig. 15 Using PI controllers in main simulation circuit diagram, this figure shows voltage sag, swell, injecting voltage and constant load voltage

Table 4 THD and PF results of new topology of 9 level CHMLI using SVPWM

Results I					
Reduction technique by using SVPWM technique compensated sag and swell with PI controller (36 Snitches)					
Compensation technique	Compensated harmonics	Type of CHMLI	No. of switches	THD%	power factor
SVPWM	Sag, swell	Reduction technique	36	1.33	0.968

Table 5 THD% and power factor for 24 switches proposed 9 level inverter of PI Controller

Results II			
Reduction technique by using SVPWM technique Compensated Voltage sag and swell with PI Controller (24 switches)			
Type of control technique	No. of switches	THD%	Power factor
SVPWM with PI controller of 9 levels CHMLI	24	1.03	0.9701

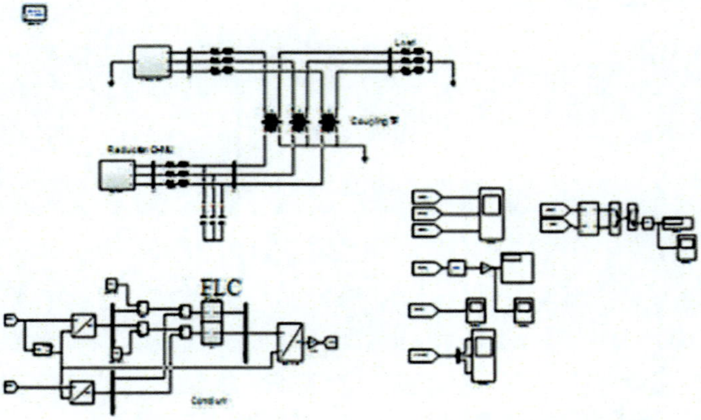

Fig. 16 Simulation diagram of "9 level reduction technique of CHMLI based on DVR using SVPWM with FLC controllers for compensated voltage sag and swell"

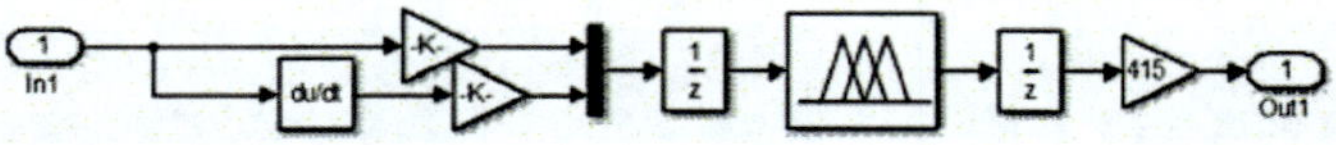

Fig. 17 Using "Fuzzy Logic Controller"

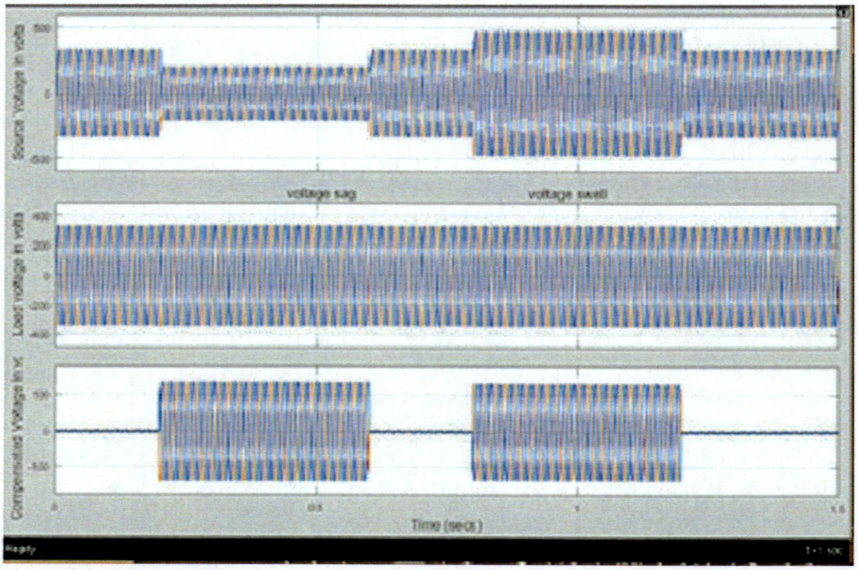

Fig. 18 Using FLC controllers in main simulation circuit diagram, this figure shows voltage sag, swell, injecting voltage and constant load voltage

For Fig. 14, main simulation diagram of 9 level CHMLI is based on DVR using SVPWM with FLC controllers for compensated voltage sag and swell. This above figure created voltage sag, swell because of three phase fault, large load additions and power line switching. DVR consists of control unit, VSC, low pass filter and coupling T/F. FLC and SVPWM methods are designed in control unit. These techniques are generated to control signals to give PWM methods for generating triggering pulses for switching to the power switches of VSI. In VSI, a reduction technique for 9 levels of CHMLIAQ13 was designed. This method of reduction produced compensating voltage with low harmonics, switching losses are low. 9 level CHMLI reduction model is used 24 IGBTS, its gives more advantages are low cost, occupied space is very less and fast dynamic solution. Fuzzy logic technique is used in main diagram for improving the performance of the DVR. Combination of FLC with SVPWM method is suitable for generating low harmonics. This type of method generated less harmonics, and sampling time is less compared to PI controllers, where above diagram generated THD% = 0.8 and power factor = 0.976. This figure is called power quality circuit. Finally, this circuit is generated low harmonics, less occupied space area, less cost and simple circuit (Figs. 20 and 21).

Figure 19 is known as space vector for 9 level CHMLI. "X" axis is called "direct axis", it represents alpha, and "Y axis" is called "quadrature axis" it represents beta. This space vector is designed as per nine levels. It shows "vectors and sectors". Vector is known as "magnitude and phase", and these are applied to source voltage, whereas harmonics occurs (Figs. 22 and 23).

From Tables 4, 5 and 6 (simulation results) draw barcode diagram.

Reduction technique with "PI and FLC" from the above "Barcode" diagram. Using FLC controller results generated THD% = 0.8 and power factor = 0.976 compared to reduction technique of PI controllers, its generated THD% = 1.33 and power factor = 0.96. FLC technique is excellent technique compared to PI controller method.

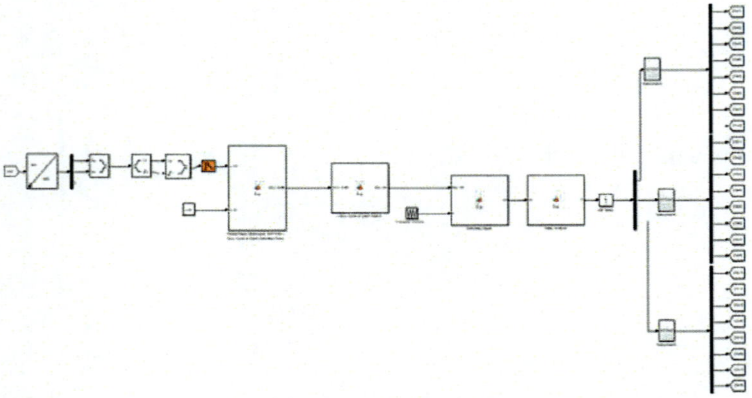

Fig. 19 SVPWM control technique for reduction type 9 level CHMLI

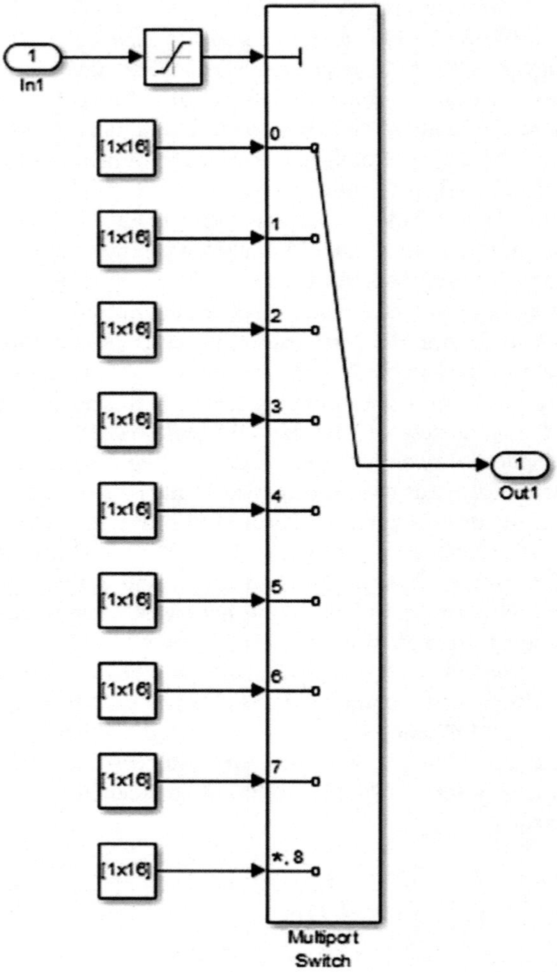

Fig. 20 Switch selector (SVPWM)

Total Harmonic Distortion (in %)

$$V_{\text{THD}} = \frac{\sqrt{\sum_{n=2}^{\infty} V_n^2}}{V_1}$$

where

V_n nth harmonic rms voltage.
V_1 fundamental frequency of rms voltage.

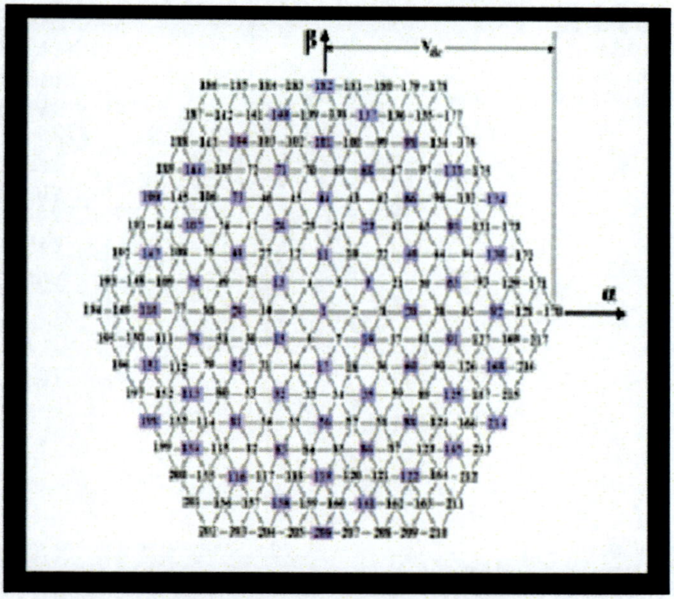

Fig. 21 Vector diagram of SVPWM for 9 level CHMLI

Fig. 22 Output wave forms of 9 levels CHMLI

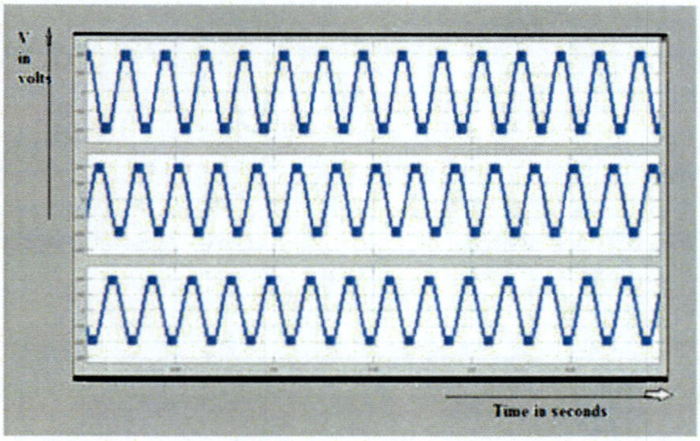

Figures 24 and 25 observed that the signal curve of "FLC using SVPWM technique is pure compared to output of curve PI". From above two diagrams, FLC technique gives less harmonics, and it is a best method compared to PI controller method.

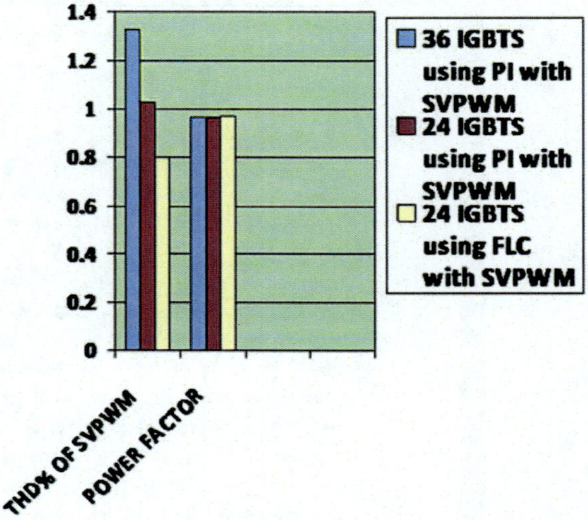

Fig. 23 "THD and PF comparison of 36 and 24 IGBTS used in reduction method" of 9 level CHMLI

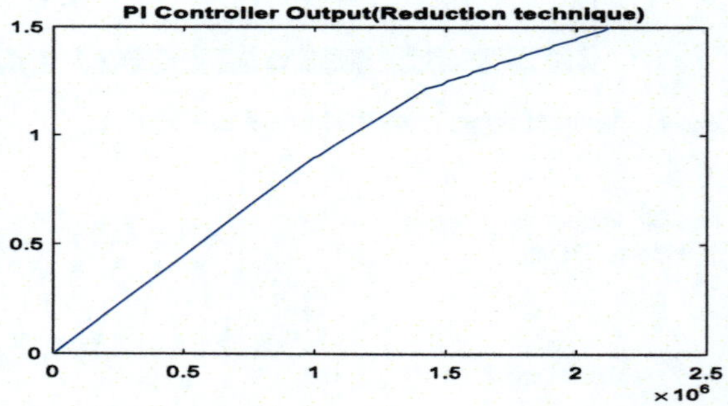

Fig. 24 Output signal of PI using SVPWM

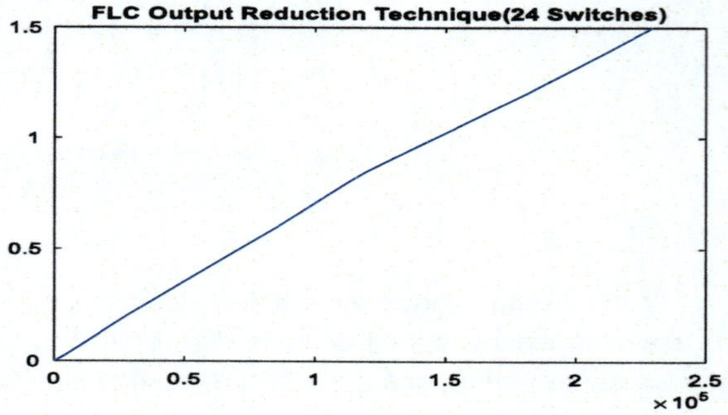

Fig. 25 Output signal of FLC using SVPWM

8 Conclusion

This research work presents a 9 level CHMLI based on DVR to reduce the power switches in the power systems to improve the quality of the power. Reduction techniques of MLI, IGBTs are reduced to 36 and 24 IGBTs. This new topology technique used 24 switches with FLC attains better performance compared to 36 switches with PI controller type of three phases three legs 9 level CHMLI. The proposed topology has merits of reduced space area, cost, harmonics and good power factor. The DVR has been simulated with SVPWM techniques. FLC results are compared to PI controller of CHMLI of reduction method. The proposed design reduces the power quality issues like voltage sag and swell in three phase electrical distribution lines and improves the quality of power.

References

1. M. Jayabalan, Thamizharasan, A new reduced switch count pulse width modulated multilevel inverter, in *IET, Power Electronics*, pp. 1–24
2. N.G. Hingorani, Introducing custom power, in *IEEE Spectrum*, pp. 41–48 (1995)
3. N. Nagalakshmi U. Ramesh Babu, S. M. Ali, K. Sudheer, A Single Phase Nine Level Inverter with reduced switches, in *IEEE International Conference on power, control, signals and Instrumentation Engineering (ICPCSI 2017)*
4. S. Chenthur Pandian, Mahalingam, R. Karthikeyan, An efficient multilevel inverter system for reducing THD with space vector modulation. Int. J. Comput. Appl. (0975 – 8887) **23**(2) (2011)
5. R. Omar, Modelling and simulation of a the phase MLI for harmonics reduction based on modified SVPWM. J. Theor. Appl. Inf. Technol. **77**(2) (2015)
6. E.A. Mohamad, N.D.D. Rao, Artificial Neural Network based fault diagnostic system for electric power distribution feeders. Electrical power systems research detailed explained in 1995
7. A. Gosh, G. Ledwich (2007 Flickers, voltage Flickers is mainly caused by rapid change of loads in power systems, pp. 417–421
8. R. Omar, N.A. Rahim, Modeling and simulation for voltage sags/swells mitigation using dynamic voltage restorer (DVR), in*Power Engineering Conference, 2008. AUPEC '08* (Australasian Universities, 2008), pp. 1, 5 (2008)
9. C. Fitzer, M. Barnes, *Voltage Sag Detection Technique for a Dynamic Voltage | Restorer* (IEEE, 2002), pp 917–924. |
10. A. Rai, A.K. Nadir, Modeling & Simulation of Dynamic Voltage Restorer (DVR) for Enhancing Voltage Sag, pp. 85–93 (2008)
11. C. Zhan, C. Fitzer, V.K. Ramachandaramurthy, A. Arulampalam, M. Barnes, N. Jenkins, Software phase-locked loop applied to Dynamic Voltage Restorer | (DVR), in *IEEE Proceedings of PES Winter Meeting*, pp. 1033–1038 (2001)
12. N.H. Woodley, L. Morgan, A. Sundaram, Experience with an inverter-based dynamic voltage restorer. IEEE Trans. Power Delivery **14**, 1181–1186 (1999)

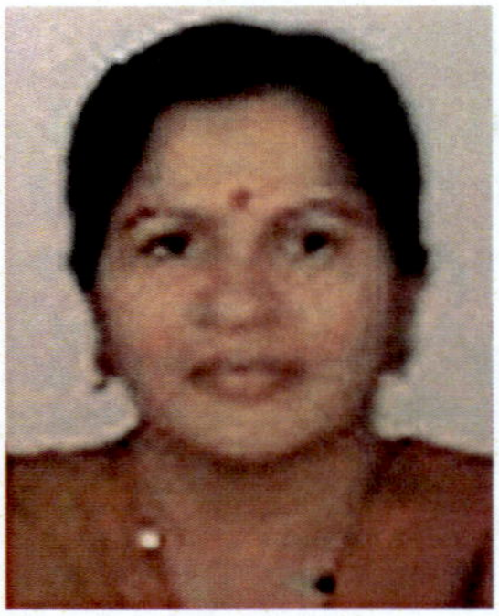

N. Eashwaramma B.Tech. (EEE),M.Tech (Ph.D.) I did B.Tech. from GNITS and M.Tech from JNTUH Hyderabad and I am pursuing Ph.D. in the field power Electronics from JNTUA, Anantapuram (AP).

J. Praveen B.E (EEE), M.Tech, Ph.D. Principal of GRIET Hyderabad and professor in EEE Dept. Graduate from Osmania University College of Engineering Autonomous) in EEE Hyderabad. He has done Masters from Jawaharlal Nehru Technological University Hyderabad, He has Doctrate in philosophy in Electrical Engineering from Osmania University in the field of power Electronics.

Dr. M. Vijay Kumar M.Tech., Ph.D. Professor & Registrar of JNTUA, professor in EEE Dept. 30 years teaching experience and 25 years research experience. Publications: Journals, international and national 60 and Conferences: international and national 63″. Research Areas: Power Electronics & Industries.

Developing Preeminent Model Based on Empirical Approach to Prognose Liver Metastasis

Shiva Shankar Reddy, Gadiraju Mahesh, V. V. R. Maheswara Rao, and N. Meghana Preethi

Abstract Liver metastasis is a type of cancer which is associated with diseases like diabetes, non-alcoholic fatty liver and obesity. The main aim of this paper is to develop efficient and accurate predictive model that helps in making strong and better decision by medical experts for liver metastasis. Machine learning techniques naïve Bayes and support vector machine (SVM) with radial basis function will be compared with deep learning techniques including ELM. ELM is the proposed technique. R programming is chosen for implementing these techniques, and few performance metrics are used for evaluation and comparison. From this work, a better algorithm will be recognized among ML and deep learning techniques.

1 Introduction

Drastic reduction in the quality of lifestyle and food habits is being observed in modern days. This leads to many hazardous diseases like diabetes and cancer. In a recent study done by national cancer registry program, in India had reported that about 1,392,179 people are estimated to be newly diagnosed with cancer in 2020 [1]. Cancer is the disease which most of the people are suffering with, and there is no proper treatment for it, which unfortunately leads to death.

Liver metastasis is a type of cancer also known as secondary liver cancer that is spread from another part of body. In human body, liver plays a major role like

S. S. Reddy (✉) · G. Mahesh · N. M. Preethi
Department of Computer Science and Engineering, SRKR Engineering College, Bhimavaram, Andhra Pradesh, India
e-mail: shiva@srkrec.ac.in

G. Mahesh
e-mail: gmahesh@srkrec.ac.in

V. V. R. M. Rao
Department of Computer Science and Engineering, Shri Vishnu Engineering College for Women (A), Kovvada, Andhra Pradesh, India

P. Karuppusamy et al. (eds.), *Ubiquitous Intelligent Systems*, Smart Innovation, Systems and Technologies 243, https://doi.org/10.1007/978-981-16-3675-2_51

production of enzymes, proteins, bile and storage of glucose. Apart from these functionalities, cleansing of blood is most important function performed by liver. Non-functioning of liver may lead to various severe side effects like loss of life. Usually, the cancerous tumor in liver metastasis will be spread from another body parts to liver and disrupts its functioning. Untreated liver metastasis will increase the severity of the disease by reaching to final stage. In this stage, the life expectancy is very less [2].

Diabetes correlated with few other medical conditions like obesity and non-alcoholic fatty liver (NAFLD, complication of diabetes) will indeed increase the risk for liver metastasis. Most common disease occurring to human irrespective of age group is diabetes, directly leads to major health complications. Correlation of diabetes, obesity and NAFLD eventually considered as the dangerous situation with highest liver cancer probability [3].

Colorectal cancer (CRC) is one of the main causes that promote the spread of cancerous tumor to liver and lead to liver metastasis. A study was performed to identify the exact association between diabetes and liver metastasis in which colorectal cancer is the source for tumor spreading to liver. In this study, various cancers related to liver, lung and peritoneum metastases were observed in colorectal cancer diabetic patients. In particular, nerve and vascular invasions are also frequently observed in those patients, which are related to malfunctioning of liver [4].

Every disease can be diagnosed in its early stage by monitoring changes in health and being aware of risk factors associated with it. Prevention or early detection of a disease is the best solution to treat a disease in its initial stage. However, liver metastasis (cancer) is a high rate mortal disease, also it was hard to provide treatment and increase survival rate if it is not detected in early stage. This can be done by monitoring few risk factors like chronic viral hepatitis, cirrhosis, non-alcoholic fatty liver, primary biliary cirrhosis, inherited metabolic diseases, type-2 diabetes, obesity, tobacco and alcohol usage. Type-2 diabetes correlated with chronic viral hepatitis and alcohol use will increase the risk of liver cancer [5].

Through few symptoms, an individual can doubt on having liver metastasis. They are weight loss, bloating in abdomen, jaundice, dark urine, sweating, fever, fatigue, loss of appetite, enlarged liver known by clinical tests, pain in upper right side of abdomen and right shoulder. However, these symptoms may vary from person to person. Upon having any of these symptoms in addition to risk factors, the medical specialist will conduct few clinical tests to diagnose the disease. These clinical tests include CT scan done on abdomen, ultrasound, laparoscopy, MRI scan, angiogram and few liver functioning tests. Based on the severity of the liver metastasis, various treatments like radiation therapy, chemotherapy and targeted therapy are given to patient. Even so, the life expectancy of liver metastasis is very less. By early detection, it may increase survival period [6].

Various new technologies are being introduced which can strengthen the decision made by the medical expert. Machine learning and deep learning are of the same sort. Among the machine learning is widely used for simple problems and deep learning is used for complex real-life problems to make better predictions. Strong and accurate diagnosis of life threat disease like liver metastasis is very essential. In this scenario, few machine learning and deep learning techniques are being compared for obtaining better predictions. In addition, the deep learning techniques are also compared among themselves to recognize best deep learning algorithm. The naïve bayes, ANN, SVM with radial basis function, DBN and ELM are selected in this proposed work. Entire implementation and evaluation process is performed in R programming which is a software platform. Comparison of obtained results will be needed in order to get the best working technique. Accordingly, implementing this efficient algorithm in real life will strengthen the medical experts' decision which indeed very helpful to increase survival period by providing treatment immediately.

2 Literature Survey

Reddy et al. [7] detected diabetes mellitus using voting strategy on Pima dataset. Employed k-fold cross validation with SMO, SVM, decision tree, Adaboost-M1 and naïve Bayes data mining techniques. Applying voting strategy on them has given 95% overall accuracy.

Pruthvi et al. [8] employed various ML techniques to perform analysis (survey) on liver cancer. Abdominal images are used for the purpose of survey. Various research papers are being considered which have contributed ML techniques like SVM, hidden Markov model and naïve bayes. They supported that SVM has performed better among all other techniques.

Reddy et al. [9] employed deep belief network on hospital readmission dataset for diabetic patients. In addition to it Adaboost, decision tree, logistic regression, random forest and gradient boosting techniques are also evaluated. Only DBN has obtained highest values of 0.6917, 0.6814, 0.6644 and 0.7032 for accuracy, precision, specificity and NPV, respectively.

Chen et al. [10] employed logistic regression to determine risk factors related to liver cancer in type-2 diabetic patients and predict the chances of affecting to it in the future. A model for comparing T2DM with and without liver cancer obtained 0.925—AUROC, 78.68%—sensitivity, 84.50%—accuracy and 90.12%—specificity. Another model for comparing T2DM with liver cancer and other type of cancer obtained 0.810—AUROC, 66.14%—sensitivity and 85.54%—specificity and 77.20%—accuracy.

Reddy et al. [11] reviewed various data mining techniques like neuro cognitive, C4.5, fuzzy, I-SVM, İmage Net and predicted ailments correlated to diabetes and diabetes itself. These are implemented using k-cross validation technique. Accordingly, image Net data mining technique was observed with highest accuracy.

Kumar and Agilan [12] predicted liver cancer in type-2 diabetic patients who are suffering from 6 years using ANN, random forest, logistic regression and Adaboost. Among them, random forest was observed with less error and better prediction results with 85% accuracy.

Reddy et al. [13] employed fuzzy logic along with k-cross validation to predict various ailments correlated to diabetes. It is being compared to various other schemes and recognized that their proposed model has contributed better. Observed values of performance metrics for their work are 97% overall accuracy and 80 ms of computation time.

Phan et al. [14] employed deep learning technique, namely convolution neural network (CNN) to predict liver cancer in patients who have history of viral hepatitis. From their study, the chance of affecting to liver cancer was less in young people. Accuracy of 98% was observed in their proposed approach of using CNN on image classification dataset.

Reddy et al. [15] proposed a predictive model for diabetes using gradient boosting. In addition to its decision tree, logistic regression, random forest and adaptive boosting are also employed. However, only gradient boosting was found to obtain highest accuracy (0.665) and f1 score (0.783).

Rau et al. [16] used ANN and logistic regression for developing a web application to predict liver cancer in diabetic patients. Data related to people suffering from type-2 diabetes from 6 years are considered for developing the model. ANN algorithm obtained 0.757—sensitivity, 0.873—AUROC and 0.755—specificity which overcomes the performance of logistic regression.

Sathurthi and Saruladha [17] employed random forest ensemble to predict liver cancer into different stages. Random forest has been considered with C4.5, J48 and naïve Bayes individually. J48 with random forest has obtained 43% accuracy which is recognized as best among all.

Schau et al. [18] proposed a deep learning technique, namely neural estimator of metastatic origin to detect the tumor of liver metastasis (secondary liver cancer). Image classification has been performed by this algorithm and obtained 90.2% accuracy while classifying liver metastasis into three categories.

Liang et al. [19] analyzed rectal cancer and performed prediction of metachronous liver metastasis based on that. Machine learning algorithms logistic regression and SVM with 5-cross validation are being used in their work. Logistic regression performed better with 0.80—accuracy, 0.83—sensitivity, 0.76—specificity and 0.87—AUROC.

Ramkumar et al. [20] contributed their work for predicting liver cancer by employing conditional probability Bayes theorem. The implementation is done in WEKA tool to predict the probability of having liver cancer. An overall accuracy of 50% was obtained for their proposed approach.

Spelt et al. [21] employed artificial neural network to predict the survival after undergoing the surgery for colorectal cancer metastasis which leads to liver metastasis. Five-cross validation is implemented on data followed by ANN (on the cross-validated data) gave a good value of 0.722 c-index for ANN.

Wen et al. [22] compared the performance of logistic regression and Adaboost in the context of predicting liver metastasis in patients with colorectal cancer. Feature selection is being performed with genetic algorithm and information gain. Adaboost obtained 78% accuracy and LR obtained 73% accuracy and they have concluded to use any one of the algorithms as there is no vast difference between accuracies.

Li et al. [23] developed prediction models based on non-invasive imaging to predict liver metastasis in patients with colon cancer. A model for clinical, radiomics and hybrid datasets individually is developed. Hybrid dataset model is employed with SVM and 5-cross validation, which has performed better with accuracy of 90.63% for training and 85.50% for validation datasets, respectively.

Lee et al. [24] used CT scan images to predict the period of survival for liver cancer patients. The prediction is limited to only 24 months. After extracting the images, SVM regression algorithm combined with CNN and 5-cross validation is employed on them. About 86.5% accuracy and 11.6 RMSE are observed for their proposed approach.

Chen et al. [25] focused on predicting survival period in liver cancer patients using CART decision tree and ANN algorithms. Among the 9 variables used in the dataset 6 of them are found to be more significant. ANN was found to be best performing with 0.915—AUROC, 0.87—accuracy, 0.88—sensitivity and 0.87—specificity.

Also, few other researchers contributed their work on various aspects of liver metastasis like predicting survival rate, liver transplantation, KRAS mutant of liver metastasis and relation between NAFLD, colorectal cancer. Margonis et al. [26] contributed their work on prediction of the survival period for colorectal liver metastasis patients. Chatterjee et al. [27] performed prediction analysis on liver transplantation based on artificial intelligence and machine learning. Chakraborty and Wang [28] done a research on relation between NAFLD and colorectal cancer. These are major risk factors of liver metastasis. Zou et al. [29] predicted KRAS mutation of liver metastasis based on MRI scan images obtained before treating for it. KRAS is a mutant of liver metastasis that spreads the cancerous tumor so that the disease reaches to advanced stage.

3 Research Approach

3.1 Objectives of Work

Cancer is the most hazardous disease that human race is suffering with now-a-days. Irrespective of age group, the population affecting to cancer is increasing drastically. Diabetes is also the disease of same sort, whereas cancer will give least life expectancy than diabetes. Unfortunately, these two diseases are correlated with each other. A diabetic person having other health complications like NAFLD has high chance of getting affected to secondary liver cancer (liver metastasis). Diagnosing it early may increase the survival time, so accurate prediction for this disease is the most needed which will help medical experts to take better decision. In this scenario, the problem of accurate prediction of liver metastasis was chosen and its objectives are as follows:

- Develop efficient predictive model for liver metastasis that help medical experts to take better decision;
- Collect the most appropriate dataset with all the required clinical test results;
- Implement ML and deep learning techniques and compare them using prominent performance metrics;

- Recognize and suggest algorithm with best performance.

3.2 Dataset

Dataset considered in this work is comprised of 38 attributes with 451 instances collected from figshare repository. This dataset can be classified into two categories one each for positively and negatively tested to liver metastasis given in Table 1. This scenario is said to be binary classification which the algorithms will perform. Attribute named as liver_metastasis is the target variable, rest all are various components, risk factors and clinical test results to identify the liver cancer. Few general attributes like gender, age, height and weight were also included. Detailed demonstration for all of them is provided in Table 1 where the attributes from 2 to 4 are called as fibrosis score used to identify liver damage. Attributes 13–20 are components of liver functioning test used to diagnose liver diseases. In the same way, attributes 23–37 are used to detect liver cancer whose functionality is provided in Table 1.

3.3 System Architecture

The flow of working given in Fig. 1 demonstrates the entire working process clearly. Loading the dataset and performing data preprocessing on it will resolve issues regarding missing data if any. Two partitions of pre-processed dataset obtained by performing percentage split of 70% are training data with 316 records and test data with 135 records. Considered SVM with RBF, ANN, NB, DBN and ELM is individually applied on the training data which produces its corresponding model. Evaluating each model obtained is performed by passing test data to it which in turn gives the prediction results in terms of performance metrics. Comparison process is entirely done based on these obtained results to recognize the model with better performance metrics which is suggested for other related works.

4 Algorithms

4.1 Naïve Bayes

It classifies the problem using the concept of probability. A probabilistic value calculated for all the possible groups in a classification problem is obtained from formula (1). This is said to be posterior probability denoted as $P(C/V)$. All the predictor variables are considered in V to calculate likelihood probability on target variable, and the same is denoted as $P(V/C)$. The chance for a record belonging to particular group is known as prior probability denoted as $P(C)$ [30].

Table 1 Description of dataset used

S. No.	Attribute	Description
1	Gender	Gender: 1—Female and 2—Male
2	APRI	AST to Platelet Ratio Index. Normal range is < 0.5
3	FIB-4	Fibrosis-4. Normal range is < 1.45. Intermediate range is ≥ 1.46 and ≤ 3.25
4	NFS	NAFLD Fibrosis Score. Normal range is < 0.676
5	Age	Patient's age (years)
6	Height	Patient's height (cm)
7	Weight	Patient's weight (kg)
8	BMI	Body mass index-BMI (kg/m^2). Normal range is between 18.5 and 24.9, if value is out of range then considered as obese
9	HBsAg	Hepatitis B: 0—Negatively tested, 1—Positively tested
10	Diabetes	Diabetes mellitus: 0—Negatively tested, 1—Positively tested
11	Hypertension	Hypertension or High blood pressure: 1—Has High BP, 0—No High BP
12	NAFLD	Non-Alcoholic Fatty Liver Disease: 0—Negatively tested, 1—Positively tested
13	ALT	Alanine amino transferase (IU/L): Normal range is ≥ 0 and ≤ 45 IU/L
14	AST	Aspartate amino transferase (IU/L): Normal range is ≥ 0 and ≤ 35 IU/L
15	GGT	Gamma-glutamyl transferase (IU/L). Normal range is ≥ 0 and ≤ 30 IU/L
16	ALP	Alkaline phosphate (IU/L): Normal range is ≥ 30 and ≤ 120 IU/L
17	ALB	Albumin (g/L): Normal range is ≥ 40 and ≤ 60 g/L
18	TBIL	Total Bilirubin (μmol/L): Normal range is ≥ 1.71 and ≤ 20.5 μmol/L
19	DBIL	Direct Bilirubin (μmol/L): Normal range is < 5.1 μmol/L
20	IBIL	Indirect Bilirubin (μmol/L). Formula: TBIL–DBIL. Normal range depends on TBIL and DBIL
21	TG	Triglycerides (mmol/L): Used to identify cholesterol. Normal range is < 1.7 mmol/L
22	PLT	Platelet count (109/L): Normal range is ≥ 150 * 109/L and ≤ 400 * 109/L
23	CEA	Carcinoembryonic antigen (ng/mL): It is used to identify various cancer cells. Normal range is ≤ 3 ng/mL. Value > 20 ng/mL is considered as very high
24	AFP	Alpha Fetoprotein (ng/mL). It is a protein which is stimulated in liver and is used to detect liver cancer at an early stage. Normal range is > 0 and < 8 ng/mL

(continued)

Table 1 (continued)

S. No.	Attribute	Description
25	CA19-9	Carbohydrate antigen 19–9 (U/mL). Normal range < 37 U/mL. Higher value indicates cancer
26	Tumor_site	Area of cancer tumor. 1—Colon, 2—Rectum
27	Tumor_type	Type of tumor. 1—Protuberant, 2—Ulcerative, 3—Infiltrative
28	Tumor_size	Size of tumor (cm)
29	Differentiation	Differentiable probability. 0—Well, 1—Moderate, 2—Poor
30	T stage	Describes number and size of tumor. Stages: 1—T1, 2—T2, 3—T3 and 4—T4
31	N stage	Identifies whether cancerous tumor is present near lymph nodes. Stages: 0—No (safe), 1—N1 and 2—N2
32	TNM	Four different stages of liver cancer 1, 2, 3 and 4 from stages 1 to 4, respectively.
33	Vascular invasion	Identifies the tumor cells in lumen of lymph vessels. 1—Present, 0—Not present
34	Nerve invasion	Identifies invasion of cancer cells to surroundings of nerves. 1—Yes, 0—No
35	KRAS	It is a mutation. 0—Missing data, 1—No mutation and 2—No mutation
36	NRAS	It's a mutation. 0—Missing data, 1—No mutation and 2—No mutation
37	BRAF	It's a mutation. 0—Missing data, 1—No mutation and 2—No mutation
38	Liver_metastasis	1—Positively tested for live metastasis. 0—Negatively tested for liver metastasis

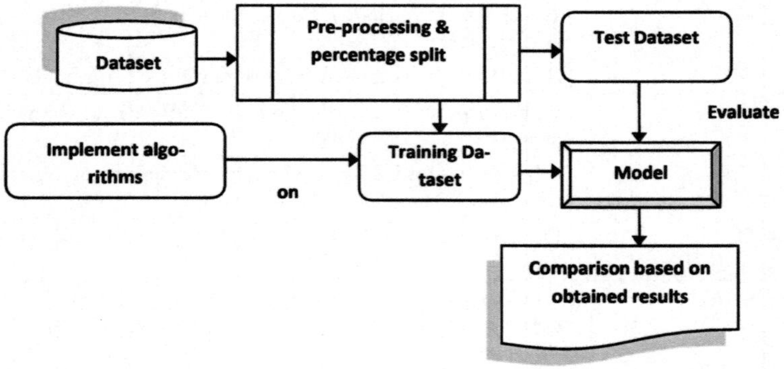

Fig. 1 Architecture for the proposed work

$$P\left(\frac{C}{V}\right) = P\left(\frac{V}{C}\right) * P(C);\ \text{where}$$
$$P\left(\frac{V}{C}\right) = P\left(\frac{v_1}{C}\right)P\left(\frac{v_2}{C}\right)\ldots P\left(\frac{v_n}{C}\right) \tag{1}$$

4.2 SVM with Radial Basis Function

It is one of the most efficient techniques for classification problems. Different kernel types are used to classify a problem by dividing its data into possible categories. Radial basis function or Gaussian kernel is one among such kernel types. A situation that can be arrived in case of complex problems with nonlinearly separable data is a feature space which is of low dimensional, due to which classification of data by using hyper plane is not effective. To overcome this, feature space was converted to high dimensional one which can produce effective results by making it possible for linear separability. The formula (2) said to be Gaussian function is the one used in this kernel type for conversion of feature space dimension. The parameters x_t, x_s are low dimensional feature space vectors, and σ is known as free parameter [31].

$$kf(x_t, x_s) = e^{-\frac{(x_t - x_s)^2}{2\sigma^2}} \tag{2}$$

4.3 Artificial Neural Network (ANN)

Neural networks are widely used now-a-days for complex real-life problems. It resembles the human nervous system. Prediction is made by this algorithm by passing the given input data through three layers. They are commonly called as input, hidden and output layers, where there can be one or more hidden layers but only one input and output layers individually. Neurons present in these layers are responsible for data processing and calculates the possible outcome in terms of probabilistic value. Some parameters like weights and biases are assigned to each hidden and output neuron which are used along with user input to predict the possible output [32].

4.4 Deep Belief Network (DBN)

It is also a type of neural network and is basically a deep learning technique. As a neural network, it will have one input, one or more hidden and one output layers. Unlike ANN, the neurons in the hidden layer are not connected, rather the hidden layer itself is connected with the input layer. Pre-training and fine tuning are the two phases in which a DBN model will get trained and adjusted accordingly to reduce error rate in predictions made. Restricted Boltzmann machine (RBM) which is an unsupervised network is implemented in the pre-training phase. To reduce the errors, biases and weights between layers will be adjusted [33].

4.5 Extreme Learning Machine (ELM)

It is basically a feed-forward neural network (FFNN) in which there will be only one hidden layer. As the number of hidden layers is limited, the training speed and computation time for ELM are far less than the FFNN. In addition, FFNN uses back propagation which in turn takes extra time in a worst case to reduce error. Unlike FFNN, ELM uses concept of matrix and stores the values in an inverse matrix which decides the final prediction values and doesn't need to back propagate [34].

As illustrated in Table 2, Assigning values to all neurons in the input layer is done in step-2 followed by random assignment of weights and biases between layers in step-3. Hidden layer's output matrix is obtained in step-4 using step-3 values, which is demonstrated in formulae (3, 4). The output function in formula (3) is the criteria used to calculate an element in matrix S of formula (4). Based on this matrix the weight matrix for output layer is obtained in step-5. Formula (5) demonstrates the formula to get the output matrix, and its representation in matrix form is shown in formula (6). These matrices from steps 4 and 5 are used to obtain the final predictions from the model. These predictions are stored in matrix shown in formula (8) obtained by applying formula (7) on S and α.

5 Results and Discussion

Obtained results for the considered algorithms ANN, deep belief network, Naïve bayes, SVM with radial basis function and ELM are provided in this section. These results are comprised of balanced accuracy, MCC, Youden's J index, threat score and accuracy. Parameters of confusion matrix, namely TP, TN, FP and FN obtained for ELM are 122, 5, 1 and 7, respectively. Utilizing these values various evaluation metrics were evaluated for ELM in Eqs. (9)–(15). Same approach is performed in the context of evaluating remaining algorithms as well.

Table 2 ELM algorithm

Algorithm: ELM
INPUT: Dataset
OUTPUT: Predicted target values for input dataset provided
ASSUMPTIONS: xi—input vector with instance, sr—output value for hidden neuron 'r' (r = 1, 2,….,q), br—bias of hidden neuron r, wr—weight vector for connections between input and hidden layer neurons (r), α—weight vector for connections of hidden and output layers, f()—activation function
STEPS:
1. Start
2. Assign values to the input neurons based on given input data
3. Assign weights between input-hidden and hidden-output layers randomly
4. Obtain hidden layer output matrix $s_r = f(w_r * x_i + b_r),\ \text{where } i = 1, 2, \ldots, n$ (3) $S = \begin{bmatrix} f(w_1 x_1 + b_1) & \cdots & f(w_q x_1 + b_q) \\ \vdots & \ddots & \vdots \\ f(w_1 x_1 + b_1) & \cdots & f(w_q x_n + b_q) \end{bmatrix}$ (4)
5. Obtain weight matrix for the output layer which is the pseudo inverse of hidden layer output matrix (S). The actual target values for given input is maintained in matrix D $\alpha = (S * S^T)^{-1} S * D$ (5) $\alpha = \begin{bmatrix} \alpha_1 \\ \vdots \\ \alpha_q \end{bmatrix}$ (6)
6. Obtain the output matrix that contains predicted values $T = S * \alpha$ (7) $T = \begin{bmatrix} t_1 \\ \vdots \\ t_q \end{bmatrix}$ (8)
7. Return predicted values
8. Stop

5.1 Performance Metrics

Accuracy. Measuring the correctness in the predictions made by the algorithm is calculated using accuracy formula (9). The value is closer to 1 points out the best prediction accuracy and closer to 0 points out the worst prediction accuracy.

$$\text{Accuracy} = \frac{TP + TN}{TP + TN + FP + FN} = \frac{122 + 5}{122 + 5 + 1 + 7} = .9407 = 94.07\% \tag{9}$$

MCC. The quality of classification is very much needed while developing a predictive model. MCC is one of such performance metric that measures binary classification quality using formula (10). Value in between − 1 and 1 is the possible range for MCC where negative value represents low quality and positive value represents high quality.

$$\begin{aligned}\text{MCC} &= \frac{(\text{TP} * \text{TN}) - (\text{FP} * \text{FN})}{\sqrt{(\text{TP} + \text{FP})(\text{TP} + \text{FN})(\text{TN} + \text{FP})(\text{TN} + \text{FN})}} \\ &= \frac{(122 * 5) - (1 * 7)}{\sqrt{(122 + 1)(122 + 7)(5 + 1)(5 + 7)}} = 0.5641\end{aligned} \tag{10}$$

Balanced Accuracy. In a binary classification context, there will two different possible values 0 and 1. Accordingly, in case of medical data, (i.e.) disease prediction the possible values are tested positive and tested negative. While predicting disease, the balance between these positively tested and negatively tested records is obtained by calculating the value of balanced accuracy in formula (13). Value closer to 1 indicates the best value that can be obtained for any prediction algorithm.

$$\text{Sensitivity} = \frac{\text{TP}}{\text{TP} + \text{FN}} = \frac{122}{122 + 7} = 0.9457 \tag{11}$$

$$\text{Specificity} = \frac{\text{TN}}{\text{TN} + \text{FP}} = \frac{5}{5 + 1} = 0.8333 \tag{12}$$

$$\begin{aligned}\text{Balanced accuracy} &= \frac{\text{Sensitivity} + \text{Specificity}}{2} \\ &= \frac{0.9457 + 0.8333}{2} = 0.8895\end{aligned} \tag{13}$$

Youden's J Index. The objective of this performance metric is to precise a diagnostic test's performance. When the value got by evaluating an algorithm using formula (14) is close to 0 that resembles no use of the model or least performance. Contrarily the value close to 1 resembles high performance.

$$\begin{aligned}\text{Youden's J index} &= \text{Sensitivity} + \text{Specificity} - 1 \\ &= 0.9457 + 0.8333 - 1 = 0.7791\end{aligned} \tag{14}$$

Threat Score (TS). It is also known as critical success index (CSI) used to define the percentage of correctly predicted positive instances to remaining instances. Formula for the same is given in Eq. (15). Usually, its value is bound between 0 and 1, where value which is near to 1 resembles better performance.

$$TS = \frac{TP}{TP + FN + FP} = \frac{122}{122 + 7 + 1} = 0.9384 \quad (15)$$

5.2 *Obtained Results*

Results obtained for all considered techniques are given in Table 3. The comparison table and its related comparison graphs from Figs. 2 and 3 are demonstrated here.

From this comparative analysis, it can be clearly observed that the proposed deep learning technique ELM has achieved highest accuracy of 94.07% and better values for remaining metrics also. It has obtained good values for most of the metrics that are being considered, this resemblances best predictive algorithm among all. Neural network techniques are the most prominent approaches for complex real-time problems. As three of them are considered in this proposed work ANN, DBN

Table 3 Results obtained

Algorithm name	Accuracy in %	Balanced accuracy	MCC	Youden's J index	Threat score
Naïve bayes	91.11	0.6504	0.3626	0.3008	0.9084
SVM (RBF)	92.59	0.5833	0.3926	0.1667	0.9248
ANN	90.37	0.6990	0.3827	0.6583	0.9
DBN	91.85	0.5417	0.2766	0.8333	0.9179
ELM	94.07	0.8895	0.5641	0.7791	0.9384

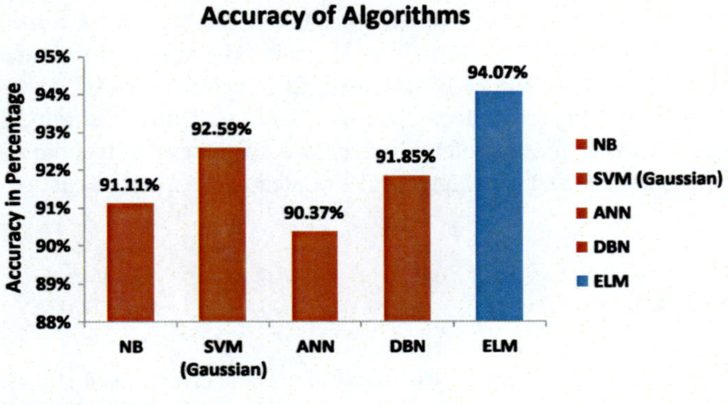

Fig. 2 Accuracy comparison plot

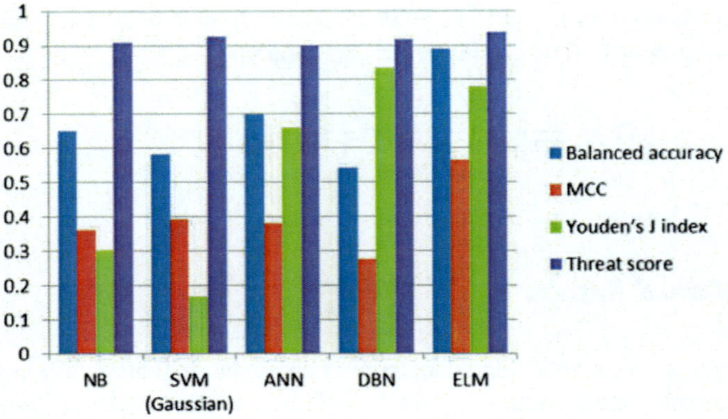

Fig. 3 Comparison plot including performance metrics

and ELM, comparison among them will finally prove the better performing one among neural networks as well. In this context, it can be observed that ANN has not performed well in any of the metrics and DBN only performed better in terms of Youden's J index. However, the difference between ELM and DBN in this metric is very minute. Apart from this metric all the remaining metrics including the most prominent accuracy measure obtained by ELM proves it as the best algorithm, even among neural network techniques.

The literature works related to liver metastasis (cancer) are compared with proposed work in Table 4. By comparing them, it was observed in Fig. 4 that proposed algorithm has obtained 94.07% accuracy which is better than all other best algorithm in the literature work. Every literature work in Table 3 used various machine learning techniques. However, they also performed better but the importance of using deep learning for complex problems was considered in this proposed work. So, ML techniques are compared with few deep learning techniques from which the actual potential of a deep learning technique can be recognized. From the result analysis, DBN has obtained best value only for Youden's J index. But as a prominent metric, accuracy of DBN is less than ELM. Moreover, there is very minute difference between Youden's J index of DBN and ELM. Even so, ELM achieved 94.07% which shows the effectiveness of it to predict accurately. As ELM technique was found with good performance even when compared with other two deep learning techniques and ML techniques, it must be used for future works related to liver metastasis.

6 Conclusion

Liver metastasis is also known as secondary liver cancer is one of such hazardous diseases that is associated with some other major health problem like diabetes and

Table 4 Literature survey and proposed work comparison

Authors	Techniques implemented	Findings in their work	Identified best algorithm	Results evaluation for best algorithm
This work	Naïve bayes, SVM with radial basis function, ANN, DBN and ELM	Predicted the liver metastasis. Compared ML techniques with deep learning techniques to recognize algorithm which can give efficient and accurate results	ELM	Accuracy—94.07%, balanced accuracy—0.8895, MCC—0.5641, Youden's J index—0.7791 and threat score—0.9384
Chen et al. [10]	logistic regression	Predicted the risk of affecting to liver cancer in diabetic patients Model 1: developed for comparing T2DM with and without liver cancer Model 2: developed for comparing T2DM with liver cancer and other type of cancer	logistic regression	Model 1: 0.925—AUROC, 78.68%—sensitivity, 84.50%—accuracy and 90.12%—specificity Model 2: 0.810—AUROC, 66.14%—sensitivity and 85.54%—specificity and 77.20%—accuracy
Kumar and Agilan [12]	ANN, random forest, logistic regression and Adaboost	Predicted liver cancer in type-2 diabetic patients who are suffering from 6 years	Random forest	85% accuracy
Rau et al. [16]	ANN and logistic regression	Developed a web application to predict liver cancer in type-2 diabetic patients who are suffering from 6 years	ANN	0.757—sensitivity, 0.873—AUROC and 0.755—specificity
Sathurthi and Saruladha [17]	Random forest ensemble with C4.5, J48 and naïve Bayes individually	Predicted liver cancer and classified into different stages	J48 with random forest	43% accuracy

(continued)

Table 4 (continued)

Authors	Techniques implemented	Findings in their work	Identified best algorithm	Results evaluation for best algorithm
Schau et al. [18]	Neural estimator of metastatic origin (NEMO)	Detected the tumor of liver metastasis (secondary liver cancer) using image classification	NEMO	90.2%—accuracy
Liang et al. [19]	Logistic regression and SVM	Prediction of metachronous liver metastasis based on rectal cancer. Used 5-cross validation	Logistic regression	0.80—accuracy, 0.83—sensitivity, 0.76—specificity, 0.87—AUROC
Ramkumar et al. [20]	Conditional probability Bayes theorem	Predicted liver cancer in terms of probability	Conditional probability Bayes theorem	50%—accuracy
Spelt et al. [21]	ANN	Predicted the survival after undergoing the surgery for colorectal cancer metastasis which leads to liver metastasis	ANN	0.722 C-index
Wen et al. [22]	Logistic regression and Adaboost	Prediction liver metastasis in patients with colorectal cancer. Feature selection is being performed with genetic algorithm and information gain	Adaboost	78%—accuracy
Li et al. [23]	SVM	Developed prediction models based on non-invasive imaging to predict liver metastasis in patients with colon cancer. A model for clinical, radiomics and hybrid datasets individually is developed. Used 5-cross validation	Hybrid dataset: SVM	90.63%—accuracy for training dataset and 85.50%—accuracy for test dataset

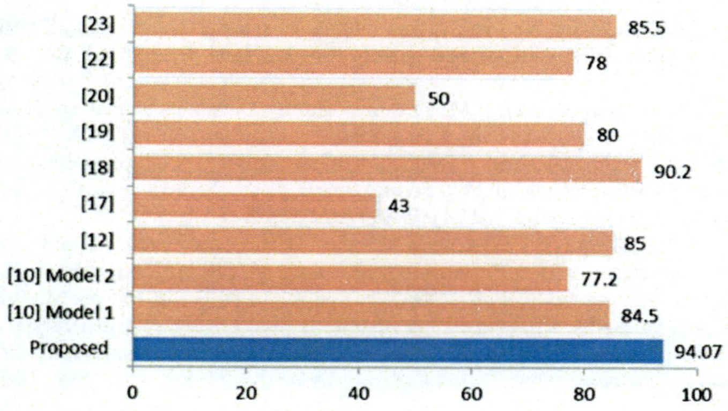

Fig. 4 Literature work versus proposed work in terms of accuracy

NAFLD. As liver is an important organ of human body, early diagnosis and treatment of liver metastasis could increase survival period of patients. In this work, an effective model is being developed and proposed to predict liver metastasis. Two ML techniques naïve Bayes and SVM with radial basis function along with three deep learning techniques ANN, DBN and ELM were applied to the dataset and compared with one another. In the comparisons, Youden's J index for DBN is better than the other algorithms. However, accuracy of DBN is less than ELM. The proposed deep learning algorithm ELM was found to be best based on most of the performance metrics. ELM obtained 94.07% of accuracy, 0.8895 of balanced accuracy, 0.5641 of MCC, 0.7791 of Youden's J index and 0.9384 of threat score. After comparative analysis, ELM outperformed all other ML and deep learning techniques. Hence, it is strongly recommended to consider ELM for effective prediction of liver metastasis.

References

1. P. Mathur, K. Sathishkumar, M. Chaturvedi, P. Das, K.L. Sudarshan, S. Santhappan, V. Nallasamy, A. John, S. Narasimhan, F.S. Roselind, ICMR-NCDIR-NCRP investigator group: cancer statistics, 2020: report from national cancer registry programme, India. JCO Glob. Oncol. **6**, 1063–1075 (2020)
2. Liver Metastasis: healthline. https://www.healthline.com/health/liver-metastases
3. Liver Cancer: Diabetes.co.uk. https://www.diabetes.co.uk/diabetes-complications
4. R. Fujiwara-Tani, T. Sasaki, K. Fujii, Y. Luo, T. Mori, S. Kishi, S. Mori, S. Matsushima-Otsuka, Y. Nishiguchi, K. Goto, I. Kawahara, Diabetes mellitus is associated with liver metastasis of colorectal cancer through production of biglycan-rich cancer stroma. Oncotarget **11**(31), 2982 (2020)
5. Liver Cancer Risk Factors: American Cancer Society. https://www.cancer.org/cancer/liver-cancer/causes-risks-prevention/risk-factors.html
6. C. Chun, J. Fletcher, What to know about liver metastases. Med. News Today (2019). https://www.medicalnewstoday.com/articles/325379

7. S.S. Reddy, R. Rajender, N. Sethi, A data mining scheme for detection and classification of diabetes mellitus using voting expert strategy. Int. J. Knowl. Based Intel. Eng. Syst. **23**(2), 103–8 (2019)
8. P.R. Pruthvi, B. Manjuprasad, B.M. Parashiva Murthy, Liver cancer analysis using machine learning techniques—a review. Int. J. Eng. Res. Technol. (Ijert) **5**(22) (2017)
9. S.S. Reddy, N. Sethi, R. Rajender, Evaluation of deep belief network to predict hospital readmission of diabetic patients, in *2020 Second International Conference on Inventive Research in Computing Applications (ICIRCA)* (IEEE, 2020), pp. 5–9
10. H. Chen, Y. Xin, Y. Yang, F. Li, G. Cheng, X. Zhang, Related factors and risk prediction of type 2 diabetes complicated with liver cancer, in *Proceedings of 2019 IEEE International Conference on Mechatronics and Automation (ICMA)*, Tianjin, China (IEEE, 2019), pp. 2138–2143
11. S.S. Reddy, N. Sethi, R. Rajender, A review of data mining schemes for prediction of diabetes mellitus and correlated ailments, in *Proceedings of 5th International Conference On Computing, Communication, Control And Automation (ICCUBEA)*, Pune, India (IEEE, 2019), pp. 1–5
12. J.K. Kumar, S. Agilan, Liver cancer prediction for type-II diabetes using classification algorithm. Int. J. Adv. Res. Comput. Sci. **9**(2), 472–477 (2018)
13. S.S. Reddy, N. Sethi, R. Rajender, Mining of multiple ailments correlated to diabetes mellitus. Evol. Intel. 1–8 (2020)
14. D.V. Phan, C.L. Chan, A.A. Li, T.Y. Chien, V.C. Nguyen, Liver cancer prediction in a viral hepatitis cohort: a deep learning approach. Int. J. Cancer **147**(10), 2871–2878 (2020)
15. S.S. Reddy, N. Sethi, R. Rajender, A comprehensive analysis of machine learning techniques for ıncessant prediction of diabetes mellitus. Int. J. Grid Distrib. Comput. **13**(1), 1–22 (2020)
16. H.H. Rau, C.Y. Hsu, Y.A. Lin, S. Atique, A. Fuad, L.M. Wei, M.H. Hsu, Development of a web-based liver cancer prediction model for type II diabetes patients by using an artificial neural network. Comput. Methods Prog. Biomed. **125**, 58–65 (2016)
17. S. Sathurthi, K. Saruladha, Prediction of liver cancer using random forest ensemble. Int. J. Pure Appl. Math. **116**(21), 267–273 (2017)
18. G.F. Schau, E.A. Burlingame, G. Thibault, T. Anekpuritanang, Y. Wang, J.W. Gray, C. Corless, Y.H. Chang, Predicting primary site of secondary liver cancer with a neural estimator of metastatic origin. J. Med. Imag. **7**(1), 012706 (2020)
19. M. Liang, Z. Cai, H. Zhang, C. Huang, Y. Meng, L. Zhao, D. Li, X. Ma, X. Zhao, Machine learning-based analysis of rectal cancer MRI radiomics for prediction of metachronous liver metastasis. Acad. Radiol. **26**(11), 1495–1504 (2019)
20. N. Ramkumar, S. Prakash, S.A. Kumar, K. Sangeetha, Prediction of liver cancer using conditional probability Bayes theorem, in *2017 International Conference on Computer Communication and Informatics (ICCCI)* (IEEE, 2017), pp. 1–5
21. L. Spelt, J. Nilsson, R. Andersson, B. Andersson, Artificial neural networks–a method for prediction of survival following liver resection for colorectal cancer metastases. Eur. J. Surg. Oncol. (EJSO) **39**(6), 648–654 (2013)
22. J. Wen, X. Zhang, Y. Xu, Z. Li, L. Liu, Comparison of AdaBoost and logistic regression for detecting colorectal cancer patients with synchronous liver metastasis, in *2009 International Conference on Biomedical and Pharmaceutical Engineering* (IEEE, 2010), pp. 1–6
23. Y. Li, A. Eresen, J. Shangguan, J. Yang, Y. Lu, D. Chen, J. Wang, Y. Velichko, V. Yaghmai, Z. Zhang, Establishment of a new non-invasive imaging prediction model for liver metastasis in colon cancer. Am. J. Cancer Res. **9**(11), 2482 (2019)
24. H. Lee, H. Hong, J. Seong, J.S. Kim, J. Kim, Survival prediction of liver cancer patients from CT images using deep learning and radiomic feature-based regression, in *Medical Imaging 2020: Computer-Aided Diagnosis, International Society for Optics and Photonics* (2020), pp. 849–854
25. C.M. Chen, C.Y., Hsu, H.W. Chiu, Rau, Prediction of survival in patients with liver cancer using artificial neural networks and classification and regression trees, in *2011 Seventh International Conference on Natural Computation* (IEEE, 2011), pp. 811–815

26. G.A. Margonis, N. Andreatos, M.F. Brennan, Predicting survival in colorectal liver metastasis: time for new approaches. Ann. Surg. Oncol. **27**(13), 4861–4863 (2020)
27. P. Chatterjee, O. Noceti, J. Menéndez, S. Gerona, M. Toribio, L.J. Cymberknop, R.L. Armentano, Machine learning in healthcare toward early risk prediction: a case study of liver transplantation. Data Anal. Biomed. Eng. Healthc. 57–72 (2021)
28. D. Chakraborty, J. Wang, Nonalcoholic fatty liver disease and colorectal cancer: correlation and missing links. Life Sci. **262**, 118507–118507 (2020)
29. W. Zou, J. Wang, Z. Zhang, Y. Wang, M. Zhou, KRAS mutation prediction based on pretreatment liver metastasis MRI images in mCRC patients. Int. J. Radiat. Oncol. Biol. Phys. **108**(3), e784 (2020)
30. S. Kapoor, R. Verma, S.N. Panda, Detecting kidney disease using Naive Bayes and decision tree in machine learning. Int. J. Innov. Technol. Explor. Eng. (IJITEE) **9**(1), 498–501 (2019)
31. V.A. Kumari, R. Chitra, Classification of diabetes disease using support vector machine. Int J. Eng. Res. Appl. **3**(2), 1797–1801 (2013)
32. N.S. El_Jerjawi, S.S. Abu-Naser, Diabetes prediction using artificial neural network. Int. J. Adv. Sci. Technol. **124**, 1–10 (2018)
33. P. Prabhu, S. Selvabharathi, Deep belief neural network model for prediction of diabetes mellitus, in *3rd International Conference on Imaging, Signal Processing and Communication* (IEEE, 2019), pp. 138–142
34. J.J. Pangaribuan, Diagnosis of diabetes mellitus using extreme learning machine, in *2014 International Conference on Information Technology Systems and Innovation (ICITSI)* (IEEE, 2014), pp. 33–38

Development and Assessment of Outdated Computers: A Technology Waste for Alternative Using Parallel Clustering

Jeffrey John R. Yasay

Abstract Technology is constantly evolving to the point that computers that are purchased then are inevitably outmoded in terms of speed and their ability to process new applications. The study aims to provide procedure and measurement in viewing the process of the parallel clustered computers via graphical representation. The idea of the development procedure has been conceptualized by the author to elevate obsolete computer for alternative use. Likert scale was used (experts and users) in assessing the system. It was found out that the development has a promising result as evident in the assessment of experts on the system's reliability (availability and stability) and the users' assessment of the system's accessibility (ease of use and flexibility). It is also noted that obsolete computers have alternative disposal technique of e-wastes. With this, the development of clustering (using the interconnectivity of a master node and slave nodes) that is reliable, accessible and with a minimal cost was conceptualized as an alternative for managing e-waste and addressing the demand of new technology in the public sectors.

1 Introduction

Nowadays, technology has become an essential part of our lives. New technology has paved the way for smartphones, faster and more powerful computers, more compact televisions and so much more. Technology has made our lives simpler, quicker, safer and more enjoyable.

Technology has truly revolutionized the way we live and the way we work. It has provided opportunities for productivity and development. It has made working more effective and efficient in general as companies continue to invest in cutting-edge technologies.

With all the promising outcomes of technology, companies have embraced it and enjoy all the profits it could give. It has played a crucial role in companies that

J. J. R. Yasay (✉)
Department of Computer Studies, College of Engineering and Technology, Tarlac Agricultural University, Camiling, Tarlac, Philippines
e-mail: jryasay@tau.edu.ph

P. Karuppusamy et al. (eds.), *Ubiquitous Intelligent Systems*, Smart Innovation, Systems and Technologies 243, https://doi.org/10.1007/978-981-16-3675-2_52

technology is no longer seen as cost but more of an investment. At present, various companies and industries have strategically advanced their technologies to cope with the ever-changing world.

However, as technology progresses, there were also setbacks created by them. So much of the wastes from various industries come from the technologies that are utilized in their gateways. E-wastes or the electronic products nearing the end of their "useful life" such as computers, televisions, copiers, and fax machines are some of the challenges in the fast-paced technology development.

E-waste, also known as "a wide and growing range of electronic devices ranging from large household appliances such as refrigerators, air conditioning, cell phones, personal stereos and consumer electronics to computers that have been discarded by their users" [1], has a major effect as technology progresses. Technology has developed and progressed so fast. Rapid application development has become challenging for developers to adapt although some are searching for alternatives that will potentially help urbanized communities develop those technologies.

As we live in a world that is geographically complex and unpredictable, new business forces are generated by the rush of mega-trends, including dramatic shifts in globalization and advances in technology. For any organization to survive and prosper in such an environment, innovation is imperative.

However, innovation is no longer just for creating value to benefit individuals, organizations or societies. Innovation's overall goal can be far more far-reaching, helping to build a smart world where people can achieve the highest possible quality of life [2].

Over the past decade, technical advances have accelerated the exponential use of multimedia tools by learners of all ages. These global trends also include the constant progression of the e-learning assessment. Evaluation is the practice of clarifying what needs to be done and relating it to what needs to be done, in order to promote the evaluation of performance and how it should be achieved [3]. In terms of speed and their ability to process new applications, computers which are then bought are ultimately outdated. When this happens, outdated computers are considered to be redundant. This also happens in sectors where computation plays a crucial role in development and achievement. As necessity dictates, there is a need to find a way in which these devices, considered redundant and worthless, can be useful in constructing computers that can meet the demands of whatever endeavors.

A cluster consists of a series of interconnected stand-alone computers operating together as a single consolidated computing resource and is a type of parallel or distributed computer system [4]. Clustering is commonly used in a network to reduce the energy consumption and thus increase the network longevity [5]. In other terms, cluster is a series of separate and inexpensive computers, used together to provide a solution as a supercomputer.

Cluster computing provides a single general approach for designing and implementing high-performance parallel systems independent of individual hardware manufacturers and their product preferences [6]. A typical application of cluster parallel computing is to load and disperse the demand for processes by the master

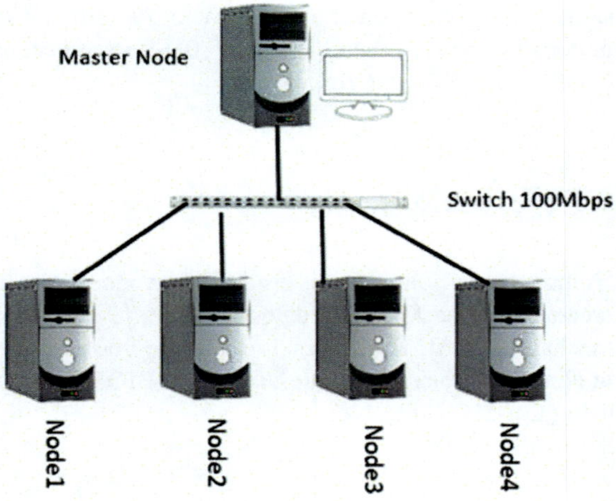

Fig. 1 indicates the cluster clustering connectivity. The development was based on computer architecture clustered in parallel

node to the slave nodes. The information is transmitted from the source to its respective cluster head and then to the base station in order for the selected head to bear all of the information that needs to be transmitted and route it to the intended target [7]. A commodity cluster is an array of entirely autonomous computer systems that are interconnected by an off-the-shelf networking network of commodity interconnections [8] and play a major role in redefining the supercomputing concept. As a result, high-performance, high-throughput and high-availability computing have arisen as parallel and distributed standard platforms.

With this, the development of clustering (using the interconnectivity of a master node and slave nodes) that is reliable, accessible and with a minimal cost was conceptualized as an alternative for managing e-waste in the public.

2 Build and Architecture

2.1 The Parallel Clustered Uniform SetUp

After the selection of obsolete system attachments on peripherals, cluster computers must be built.

2.2 Production Instruments

The design of the clustered computers was based on the hardware and software needed to meet the demand of cluster computers are (a) personal computers consist of the same basic components: a CPU, memory, circuit board, storage and input/output devices [9] (b) fast ethernet switch [10] (c) straight cable (T568A–T5668A) [11] and (d) Ubuntu ABC GNU/Linux [12].

2.3 Setup Clustering

Homogeneous computing is used to interconnect identical processor cores or units to create a high-performance device in order to use a homogeneous parallel clustering mechanism [13]. The nodes 1–4 and the master node all come in the same "Boot to Network" basic input-output system (BIOS) configuration connected via T568A using Cat-5E UTP cable.

2.4 Installation (Software)

The next move is to install the program after the computers have been assembled. ABC GNU Linux (Ubuntu 9.04) [12] was used with the default kernel as a basis. Upon the installation of ABC GNU Linux (Ubuntu 9.04), gathered the information about the hardware specifications.

2.5 Specification and Checking of Device

Step 1: Upon determining the master node and slave node, this will be the basis of heterogenousity of the system as the specification be processor: Intel Celeron M CPU with a CPU Speed: 2266 MHz.

Step 2: Setting up of ABC GNU Linux kernel ISOLINUX3.63 Debian to the master node. Boot from the CD-ROM then chooses an install mode and press enter then follow the directions on the screen. The default language of the distro is Spanish. Changing to your preferred language is necessary. After which select use entire disk to partition the hard disk, then create username and password and lastly install ABC GNU (Ubuntu 9.04).

Step 3: Setting up the slave nodes, first enter the configuration or setup of CMOS, choose halt on ALL ERROR and finally, set up to boot from the network.

Step 4: This procedure will check the master node via Command Line Interface (CLI), master@master-desktop: ~ $ cat cluster hosts 192.168.0.1. Upon checking proceed to connectivity check this will test the network connectivity of master node, node1, node 2, node 3 and node 4, master@master-desktop: ~ $ cat cluster hosts 192.168.0.1 192.168.0.13 192.168.0.3 192.168.0.10 192.168.0.8.

3 Monitoring

3.1 Cluster Interpretation of GANGLIA Monitoring Tool (GUI) [14]

Figure 2 shows the device view of the cluster. A series of small graphs display the master node, and processes are used for nodes 1–4. It also indicates that the master node and nodes 1–4 work with various processes.

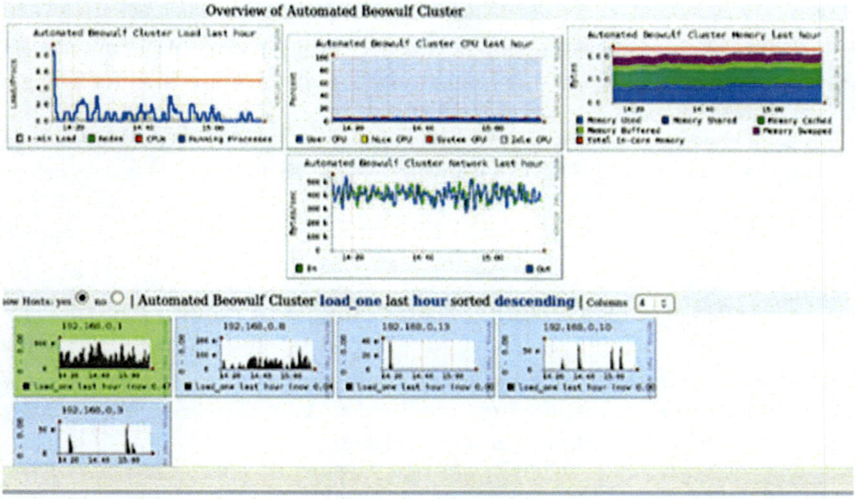

Fig. 2 Overview of automated Beowulf cluster using Ganglia

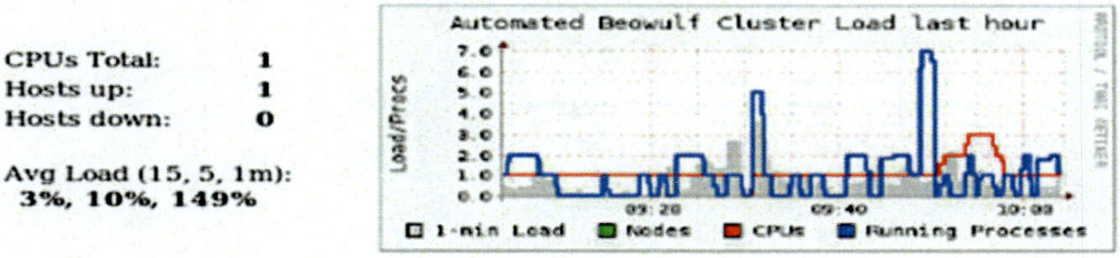

Fig. 3 Performance of total hosts (1 CPU)

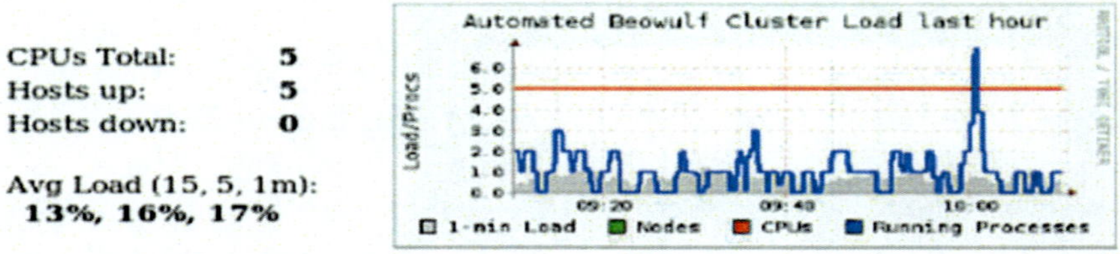

Fig. 4 Performance of Total hosts (5 CPU)

3.2 *Performance Differences of Machine Loaded*

Figure 3 displays performance representation from 1 host. It showed that the average capacity of a single CPU was 3, 10 and 149%, showing that it is hard for a single host to process.

Figure 4 shows the performance of 5 host computer. It indicates that the average load of performance is 13, 16 and 17% which reveals that a multiple hosts process smoothly.

3.3 *Network Flow by Graph (Master Node and Node 1)*

Figure 5 reveals the master node and node 1. It ensures that the master node process and node 1 process are distinct from one another. This also shows how process efficiency and relation identification are calculated.

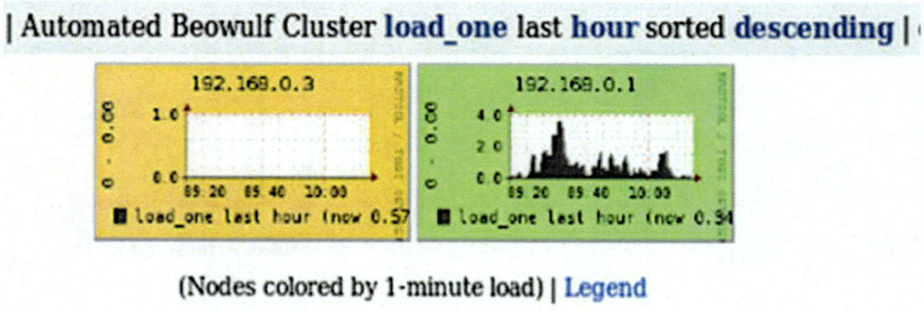

Fig. 5 Master node and node 1

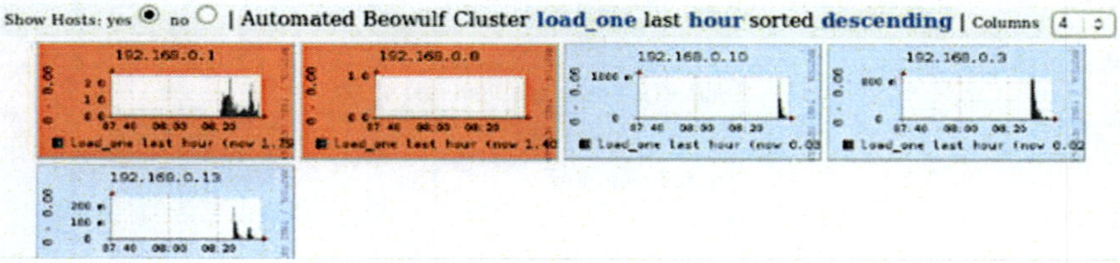

Fig. 6 Process identification of nodes

3.4 *Network Movement Process by Graphs (Master Node and Node 1–4)*

Figure 6 shows that the use of the CPU is 100%, it also shows that the master node and node 1 used their processing power in the process distribution. It also reveals that different nodes have distinct processes.

3.5 *Network Movement Process by Graphs in Shutting Down of Nodes*

Figure 7 indicates the nodes have been successfully shut down. In the image and graph, the master node and the remaining nodes used that homogeneous parallel clustering processes are established and used.

4 Evaluation and Results

Two approaches are applied to test the homogeneous parallel clustering of alternatives for success acceptance and creation: by IT experts and by the users. The IT experts assessed the system as to reliability with system availability and system stability [15] while the users rated the system as to accessibility with ease of use and flexibility of the system [16]. The questionnaire was based on the Likert scale suggested by ISO

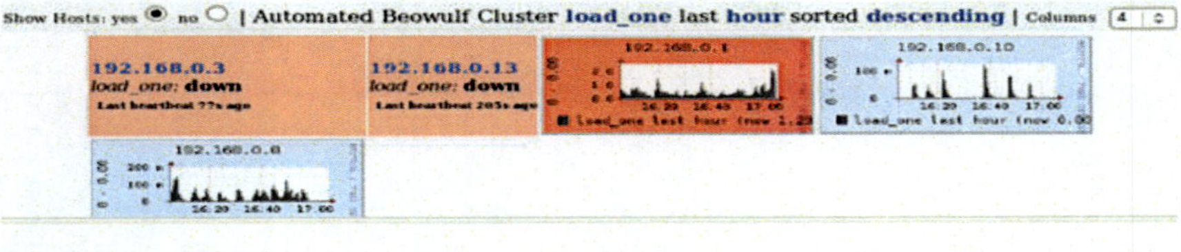

Fig. 7 Node process in shutting down

Table 1 Assessment of the system by IT experts

Assessment criteria	Mean	Descriptive rating
Reliability (composite mean: 4.08)		
System availability	4.50	Excellent
System stability	3.67	Very good

9126 [17] and used to analyze the results from scales 4.01–5.0 as excellent, 3.01–4.0 as very good, 2.01–3.0 as good, 1.01–2.0 as fair and 0–1.0 as poor with the following informative equivalents.

4.1 IT Experts

Table 1 shows the results of the evaluation based on the reliability of the system. It obtained a composite mean of 4.08.

The IT experts evaluated the reliability of the system based on the system availability with a 4.50 mean with a descriptive rating of excellent and system stability with a 3.67 mean with a descriptive rating of very good.

4.2 Assessment of Users

Table 2 shows the results of users' assessment using a homogeneous parallel cluster. The users of the system were the students, IT faculty and employees of Tarlac Agricultural University. To obtain the reliability of the evaluation, there were sixty (60) users who evaluated the system.

They evaluated the system accessibility based on ease of use with 4.67 as excellent and flexibility of the system with 4.72 as excellent. The system accessibility obtained a composite mean of 4.69 with a descriptive rating of excellent. The result indicates the uncomplicatedness of the system's operation.

Table 2 Assessment of the system by users

Assessment criteria	Mean	Descriptive rating
Accessibility (composite mean: 4.69)		
Ease of use	4.67	Excellent
Flexibility of the system	4.72	Excellent

5 Conclusion

The study found that the development and assessment result of the homogeneous parallel process clustering as alternatives is significant. Over the course of the review and testing, the performance of the machine was not damaged. Hence, the achievement of serviceable machines with low development costs has been established and guaranteed. Moreover, based on expert opinion and review, the use of a homogeneous parallel clustering method is strongly appropriate. The functionality of the framework was based on the efficiency of parallel clustering, and it also notes that operating is the master process and the nodes. Finally, because of the ease of service, the system's assessment is strongly appropriate to consumers in terms of usability.

In addition, the study findings have been established and could be introduced to other universities and schools in the area which will be used as an alternative computer to run application in today's technology demands. This will assist faculty, teachers and staff in researching other technical development and device efficiency.

Nevertheless, with the use of clustering strategies and encouraging e-waste management, universities and schools to alternatively develop outdated computers.

A future collection of machines with changed architectures will be selected for future work to enhance the analysis, in order to observe the effect of heterogeneity on the efficiency and growth of the clustering technique.

Lastly, to increase device reliability, an implementation can require additional measures, configurations and performance review. Additional testing methods are also recommended to determine the efficiency of the device being built.

References

1. D. Sinha-Khetriwal, *The Management of Electronic Waste: A Comparative Study on India and Switzerland, M.S.*
2. M. Sang, M. Leea, S. Trimi, Innovation for creating a smart future. J. Innov. Knowl. **3**(1), 1–8 (2016). ISSN: 2444-569X
3. D.D. Williams, C.R. Graham, *International Encyclopedia of Education*, 3rd edn. (2010)
4. C.S. Yeo, R. Buyya, H. Pourreza, R. Eskicioglu, P. Graham, F. Sommers, Cluster computing: high-performance, high-availability, and high-throughput processing on a network of computers, in *Handbook of Nature-Inspired and Innovative Computing*. ed. by A.Y. Zomaya (Springer, Boston, MA, 2006), p. 2006
5. S.R. Mugunthan, Novel cluster rotating and routing strategy for software defined wireless sensor networks. J. ISMAC **2**(02), 140–146 (2020)
6. T. Sterling, *Encyclopedia of Physical Science and Technology*, 3rd edn. (2003)
7. J.S. Raj, Machine learning based resourceful clustering with load optimization for wireless sensor networks. J. Ubiquit Comput. Commun. Technol. (UCCT) **2**(01), 29–38 (2020)
8. T. Sterling, et al., *High Performance Computing* (2018)
9. J. Casey, *Computer Hardware: Hardware Components and Internal PC Connections* (Technological University Dublin, Guide for Undergraduate Students, 2015)
10. S. Kangovi, *Peering Carrier Ethernet Networks* (2017)
11. N.J. Alpern, R.J. Shimonski, *Eleventh Hour Network+* (2010)

12. I. Castaos, I. Garrido, A. Garrido, M. Sevillano, *Design and İmplementation of an Easy-to-Use Automated System to Build Beowulf Parallel Computing Clusters* (2009), pp. 1–6. https://doi.org/10.1109/ICAT.2009.5348420
13. M. Van Steen, A.S. Tanenbaum, A brief introduction to distributed systems. Computing **98**(10), 967–1009 (2016). https://doi.org/10.1007/s00607-016-0508-7
14. M.L. Massie, B.N. Chun, D.E. Culler, The ganglia distributed monitoring system: design, implementation, and experience. Parallel Comput. **30**(7), 817–840 (2004)
15. D.T. O'Connor, A. Kleyner, *'Practical Reliability Engineering'*, 5th edn. (Wiley Ltd., Chichester, UK, 2012)
16. H. Petrie, N. Bevan, The evaluation of accessibility, usability, and user experience. C. Stepanidis (2009). https://doi.org/10.1201/9781420064995-c20
17. ISO/IEC 9126, Software Engineering—Product quality, Parts 1–4, 1999–2004

A Novel Approach to Detect Leaf Disease and Feature Extraction Using IoT

K. V. Prasad, S. Sri Harsha, Sudhakar Putheti, and Katragadda Raghu

Abstract The primary significance of this research work is to develop a prototype for detecting the paddy diseases like bacterial leaf spots, target spots, spectral spots, and leaf molding spots. The proposed research work mainly concentrates on the image processing technique used for enhancing the quality of image and neural networking methods for the categorization of paddy disease. The proposed method engages the image acquisition, pre-processing, segmentation, categorization, and analysis of paddy disease for image segmentation done by k-means clustering technique that are calculated by affecting disease clustering features such as contrast, correlation, energy, homogeneous, mean, standard deviation, and variance that are removed by features given as classified inputs of disease.

1 Introduction

An item quality control is in a general sense required so as to acquire esteem included products [1]. Numerous investigations shown by the nature of farming substance are reduced by various causes. One of the majority important essentials of such excellence is plant infection. Thus, preventive plant maladies permit generously humanizing scenery of the items. Rice known as Oryza Sativa (explicit name) is most used food plants and generally developed started in ASIA [2]. Rice is a significant harvest worldwide and over portion of the whole population depended on it for food. Abundant persons on the planet including Malaysia eat rice as staple food. Notwithstanding, there are abundant variables to construct paddy rice creation developed

K. V. Prasad (✉) · S. S. Harsha
Department of Computer Science and Engineering, Koneru Lakshmaiah Education Foundation, Vaddeswaram, Guntur 522502, India

S. S. Harsha · S. Putheti
Department of Computer Science and Engineering, Vasireddy Venkatadri Institute of Technology, Guntur, Andhra Padesh, India

K. Raghu
Department of Business Management, V R Siddhartha Engineering College, Vijayawada, India

P. Karuppusamy et al. (eds.), *Ubiquitous Intelligent Systems*, Smart Innovation, Systems and Technologies 243, https://doi.org/10.1007/978-981-16-3675-2_53

into reasonable and less beneficial. One of the principle factors is paddy ailment. A strange circumstance that evils the plant or lead it to work improperly is called as a malady. Ailments are punctually apparent by their side effects. Here are a great deal of paddy infection types, which are Bakanae, red sickness infection, earthy colored spot malady, and abundant [3]. Image handling and PC vision innovation are advantageous to the rural business. Here, it is more important to numerous territories present in farming innovation [3]. Paddy disease detection system is one of the advantageous frameworks. It can help the paddy rancher distinguish the sickness quicker. This examination means to build-up a model framework to consequently identify and group the paddy sicknesses by utilizing picture preparing strategy as another option or supplemental to the customary manual technique. India is rapid to create, and farming is the vertebrae for nation's advancement in the commencement phase. Since of industrialization and globalization thoughts, where the field is going up against hindrances and on top of that, the care and the need of the advancement ought to be instilled in the minds of the more energetic age. Developments are accepted as a key part in all the fields, yet till today, some old methodologies are utilized in agribusiness. Perceiving the plant disease wrongly prompts huge loss of yield, time, money, and nature of thing. Perceiving the condition of plant expect a noteworthy capacity for productive turn of events. The picture planning techniques are used for generating ID of plant infection. By taking all the sign of disease on leaves, stems, blooms, etc., into consideration, here leaves are utilized for performing unmistakable verification of ailment affected plants. The component extraction is done in RGB, HSV, YIQ, and dithered images. The component extraction from RGB picture has incorporated the proposed system. Another modified technique is used to achieve disease sign division for mechanized photographs of plant leaves. The illnesses of different plant species has also referred. This research work implements the disease affirmation for intended leaf picture. India is famous for agriculture, and the farming business occupies a huge part in the economic sectors.

2 Literature Review

This part briefly reviews, clarifies, and examines the existing expositive expression survey related to the current undertaking, which may be "Paddy infection identification framework utilizing picture processing." This part comprises three areas. This starts with area that depicts the overview of paddy [4]. The subsections are the definition, sort about paddy disease, paddy manifestation, and paddy oversaw economy. The second area may be those audit about a few existing framework that utilized same systems and strategies. The third segment examines those audit with respect to procedure and system utilized toward the framework. Those subsections require to aid picture acquisition, picture division, and simulated neural system.

There would huge numbers components that aggravate paddy rice processing turned moderate furthermore less profitable. A standout among those principle factors will be paddy infection. Those table beneath will indicate you the type of paddy disease along with the side effect about paddy illness and oversaw the economy of paddy illness [5]. These researches concentrate on three sorts for diseases, which need to aid paddy blast and tan. Gavhale, furthermore u. Gawande, Gavhale furthermore Gawande (2014) exhibited review. Furthermore, image transforming strategies summarize a few plant species used for distinguishing plant infections. Those real strategies to recognize plant illnesses are back propagation neural system (BPNN), support vector machine (SVM), K-nearest neighbor (KNN), and spatial graylevel reliance matrices (SGDM). Further, this strategy is used to analyze the concrete and poorly plants abandon. Canny diagnose framework of Wheat sicknesses in the view of bisexuality by Y. Q. Xia, Y. Li, Also c. Li. Previously, 2015, Xia Also li have suggested those bisexuality outline of canny wheat sicknesses diagnose framework. In this process, clients gather information from the pictures on wheat maladies by utilizing bisexuality phones and send the pictures to organize the server and detect the illness [6]. After obtaining the ailment images, the server performs picture division by converting the pictures starting with RGB shade space and HSI color space. The color and composition offer of the maladies are dictated by utilizing color minute grid and the gray level co-event grid. The favored characteristics have entered the backing vector machine to distinguish the ID number that comes about to nourish the back of the customer. The RGB and grayscale pictures are used for plant abandons and illness identification—a comparative investigation carried out by Padmavathi and Gadurai (2016) have provided the similar outcomes about RGB. Furthermore, analyzing grayscale pictures over leaf fat discovering procedure. To identify the contaminated leaves, shade turns into a critical characteristic to find that illness force. This needs to be viewed as grayscale. Also RGB pictures and utilized average filter for picture upgrade and division are to extract the ailing part, which is used to recognize the infection level. The plant malady distinguish model is dependent upon leaf beet picture classification. 13 sorts about ailments are eminent from those sound, and it intends to separate abandons from their surroundings.

Soil dampness determination is performed by utilizing remote sensing information for the property insurance. Furthermore, the expansion of agribusiness preparation provides acceptable geospatial information that empowers the era about sufficient data identified with floods and droughts; this research work connected the remote sensing strategy that depends on the utilization of soil-moisture list (SMI), where the information is obtained from satellite sensors. Similarly as exhibited by hunt, the list will be dependent upon the genuine content of water (), water ability, and wilting perspective. Multispectral satellite pictures from unmistakable (red band) and infrared groups (near infrared and warm bands) are vital to that count of the list. A short audit around plant sickness identification by utilizing the image transforming, where the advanced picture transform is the utilization for machine calculations to perform picture methodology once the advanced gets portrayed. It permits a wide margin that differ of calculations should make them connected to the machine document. Also, it might stay away from issues like the build-up of clamor and sign twisting for procedure. Advanced picture procedure needs unpleasantly essential part done in the farming field. It is generally acclimated to watch those crop malady for high correctness. An IoT-based dirt dampness checking with respect to Losant platform and those webs about things (IoT) may be converting the agribusiness industry and comprehending the enormous issues or the major tests confronted toward the farmers in the field. India stands as 13th nation on the planet for hosting the shortage of water assets [7]. Because of ever expanding globe population, the world confronts the challenges in the deficiency of water resources, constrained accessibility for land, and further, it was troublesome to deal with those expenses, and at the same time, it gathers the requests of expanding the utilization necessities of a worldwide populace that is expected to develop by 70% of the quite a while 2050. Keen Farming: IoT-based keen sensors agribusiness adhere for live temperature and dampness screening that utilizes Arduino and cloud registering and sun-oriented engineering organization. Those agribusiness stick constantly, and they are recommended by means of this paper to get coordinated for Arduino technology, breadboard blended for different sensors and carry on with information encourage might obtained from the web by starting with Things speak.com. The result has continuously recommended the live agribusiness fields provided for high exactness of over 98% of the information encourage.

3 Existing Technique

Figure 1 explains the entire cycle of the projected strategy by beginning with the picture procurement, i.e., gathering the pictures by using information base. These pictures get sectioned by utilizing K-implies bunching calculation, and highlights are extricated from the groups and that is specified to the ANN categorizer. Picture acquisition is the way toward gathering the pictures from information base. The

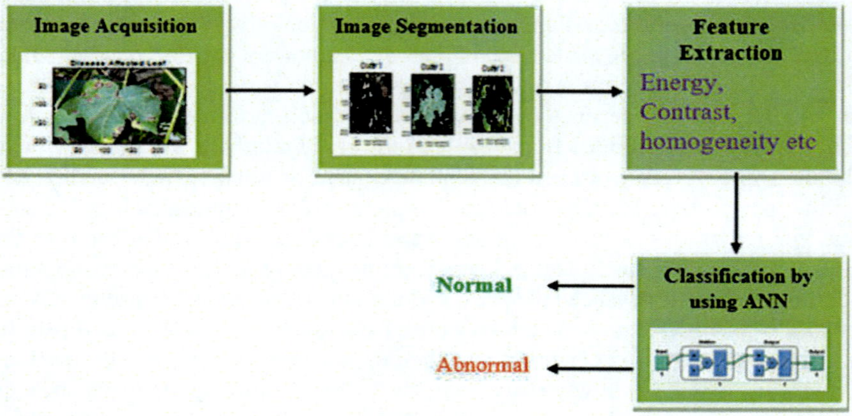

Fig. 1 Projected technique for disease recognition

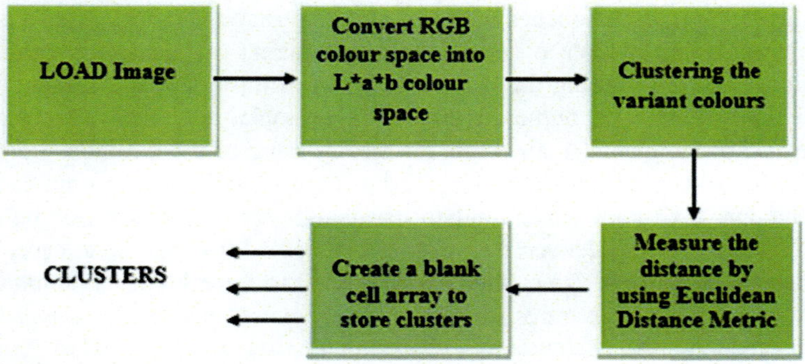

Fig. 2 K-means clustering technique

pictures are stacked from plant town information base. Stacked pictures are tomato solid, tomato leaf spot, and cotton leaf sound and leaf spot. Picture division is the way toward isolating the pictures into various parts, for example: bunches and that is finished by utilizing the k-implies clustering technique (Fig. 2).

K-means clustering calculation is utilized to isolate the recolored part and sound leaf area. In this, initial step is Load the picture into MATLAB by the information base, and at that point, it translates the RGB picture into L*a*b* shading space. L* refers to the delicacy, a* and b* refers to the chromaticity layers. The entire shading data is in the a* and b* layers, and next step is bunching the variation hues [8]. The image gets parceled into three locales by re-allocating every pixel to its closest bunches that diminishes the entirety of separation and recalculates the centroids of the groups. Every bunch comprises of various fragments of leaf picture. Three groups have record esteems, which are utilized to name each pixel in the picture by utilizing the results from K imply, and the subsequent stage is making a clear cell exhibit to accumulate the consequences of clustering.

4 K-Means Clustering

The k-implies bunching [1] is one of the traditional and very much contemplated solo learning calculations that take care of major grouping issues. It endeavors to find potential classes in the given information through a cycle of arranging objects into bunches, whose individuals are comparative on somehow or another. A group compares in this way to an variety of article that is "relative" to one another and are "single" to the substance that have to leave by singular bunches. The k-implies grouping can be viewed as the most significant solo learning approach. Area 2 gives more insights regarding the k-implies grouping. The k-implies grouping strategy has the accompanying potential points of interest [9]: (1) managing various kinds of qualities; (2) finding bunches with self-assertive shape; (3) negligible prerequisites for space information to decide input boundaries; (4) managing clamor and exceptions; and (5) limiting the difference between information. Subsequently, it discovers applications in numerous fields, for example, advertising, science, and picture acknowledgment. In any case, for a fruitful use of the k-implies bunching, we need to beat its inadequacies: (1) The method of introduction implies is not indicated. One well-known technique to start is to haphazardly pick k of the examples; (2) The outcomes delivered rely upon the method that we use at every level. (3) It can occur with the agreement of tests nearest to some bunch communities as vacant, so they cannot be refreshed. This is an inconvenience, and it must be dealt with an execution; (4) The outcomes rely upon the measurement used to gauge the uniqueness between a given example and a specific bunch of community. A well-known agreement helps to normalize each issue by its standard deviation and (5) the outcomes rely upon the value of k.

Classification: Present classification may be completed by utilizing the neural system device. Seven concentrated features, for example, such that contrast, correlation, energy, homogeneity, mean, standard deviation, and difference would be provided for information of the neural system and target information provided for of the neural system by population vector. In this, once more proliferation neural system might have been used to arrange that information. It provides for those execution plot, perplexity grid, and slip histogram plot then afterward fruition about preparing to organize.

Neural Networks: Neural networks need aid the workhorses of profound taking in. Furthermore, same time they might look like bootleg boxes, where it counts down (sorry, I will stop the repulsive puns), and they would attempt to fulfill the same relic similar to whatever viable model in order to make great predictions. Neural networks are the multi-layer networks about neurons (the blue and fuchsia hubs in the graph below) that are used to arrange things, settle on predictions, and so forth. The following will be the outlines of anticipated procedure.

Convolutional Neural Networks: It resonances similar to a weird mix from the claiming science. Also, math with a minimal CS sprinkled in, yet all these networks have been a percentage of the vast majority persuasive innovations in the field for machine dream. 2012 might have been the primary year, where quite a while that neural nets growth remains as unmistakable quality by concerning the illustration. Alex Krizhevsky utilized them to win that year's picture net rival (basically, the yearly Olympics from claiming workstation vision), dropping the order slip record starting with 26% on 15%, an amazing change at those long run. Since the time, a group of organizations have utilized it for their benefits [10]. Facebook utilizes neural nets for their programmed tagging algorithms, Google to their photograph search, Amazon to their item recommendations, Pinterest for their home bolster personalization, and Instagram for their scan foundation.

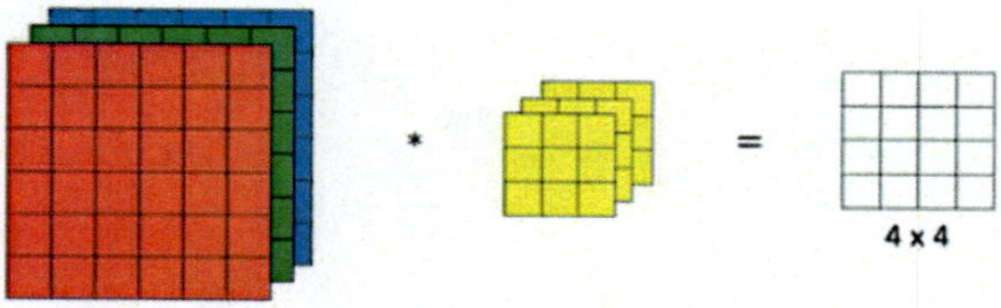

Fig. 3 Input and output filters

Convolutions over volume believe as an alternative of 2D image having a 3D input image of shape $6 \times 6 \times 3$. How to relate convolution once this image? A $3 \times 3 \times 3$ filter is used as opposed to a 3×3 filter. Let us take a gander at a sample.

Input: $6 \times 6 \times 3$
Filter: $3 \times 3 \times 3$

The dimensions over symbolize the height, width, and filters in the input and filter. Here that, the number of filters in the input and filter ought to be same. This will bring about an output of 4×4. Let us observe it outwardly (Fig. 3).

As here, there are three filters in the input and that filter will therefore additionally bring three filters. After difficulty, those yield state may be a 4×4 grid. So, those initially component of the yield is those entirety of the element-wise item of the main 27 values from those enter (9 values from each filter) and the 27 values starting with the filter [11]. Then afterward that we convolve through the whole picture. As opposed to utilizing simply an absolute filter, we could utilize various filters too. Though we utilize various filters, the yield measurement will progress. So, as opposed to hosting a 4×4 yield similarly as we would be a $4 \times 4 \times 2$ output (if we use 2 filters) (Fig. 4).

Comprehensive dimensions are specified as:

$$\text{Input: } n \times n \times nc$$

$$\text{Filter: } f \times f \times nc$$

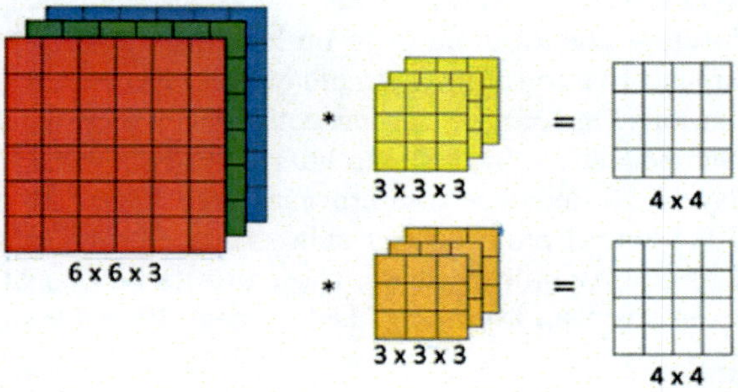

Fig. 4 Dimensions of input and output filters

$$\text{Padding: } p$$

$$\text{Stride: } s$$

$$\text{Output: } \left[n + 2p - \frac{f}{s} + 1\right] x \left[n + 2p - \frac{f0}{s} + 1\right] \times nc'$$

at this time, n_c is the integer of filters in the input and filter, even as n_c' is the number of filters.

Internet of things

Internet of things (IoT) is an superior mechanization and analytics framework which exploits networking, sensing, enormous data, and also counterfeit consciousness innovation organization to convey complete frameworks for an item alternately administration. These frameworks permit more stupendous transparency, control, and execution at connected on whatever industry alternately framework.

IoT frameworks need provisions crosswise over commercial enterprises through their interesting adaptability. Furthermore, capability with a chance to be suitableness done any earth. They upgrade information collection, automation, operations, and a great deal additional through keen units also capable empowering innovation organization.

IoT

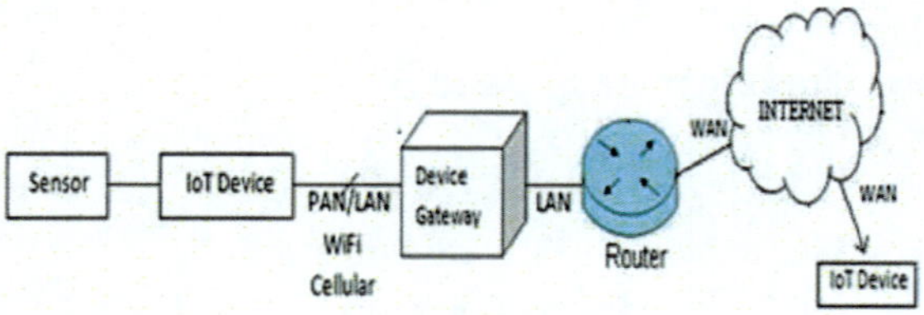

Wireless cellular companies are attempting toward giving collectivity. Also, the upgrade for existing remote gadgets rely on backing the rising IoT advertise.

IOT device and components: The IoT gadget principally comprises the battery to leverage control. It ought to further bolster for long existence and approximation around 10 a considerable length of time. Those parts incorporate the interfacing for sensors. Furthermore, the connectivity with remote makes it as a wired system [12]. Subsequently, it incorporates little and only physical layer. Also, the upper protocol layers will interface with provision layer. Units are ought to back both IPV4 and IPV6 built IP protocols. IoT units must need recipient affectability for at least 20 dB superior to non-IoT gadgets. Those IoT units are ought to provide a chance to be less expensive with something like less than $10. Elude IoT parts to additional data.

Result

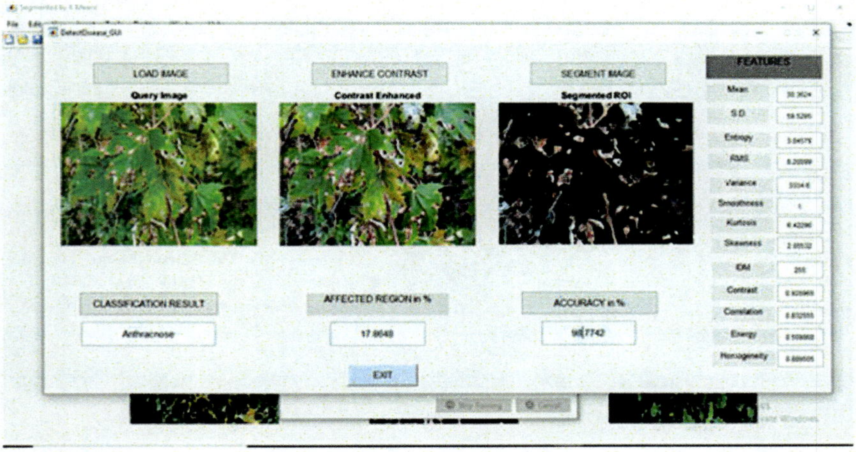
DetectDisease_GUI
LOAD IMAGE
ENHANCE CONTRAST
SEGMENT IMAGE
FEATURES
Query Image
Contrast Enhanced
Segmented ROI
Mean
S.D
Entropy
RMS
Variance
Smoothness
Kurtosis
Skewness
IDM
Contrast
Correlation
Energy
Homogeneity
CLASSIFICATION RESULT
AFFECTED REGION in %
ACCURACY in %
Anthracnose
17.8648
EXIT

Segmented by K Means
File Edit View Insert Tools Desktop Window Help
Original Image
Cluster 1
Cluster 2
Cluster 3

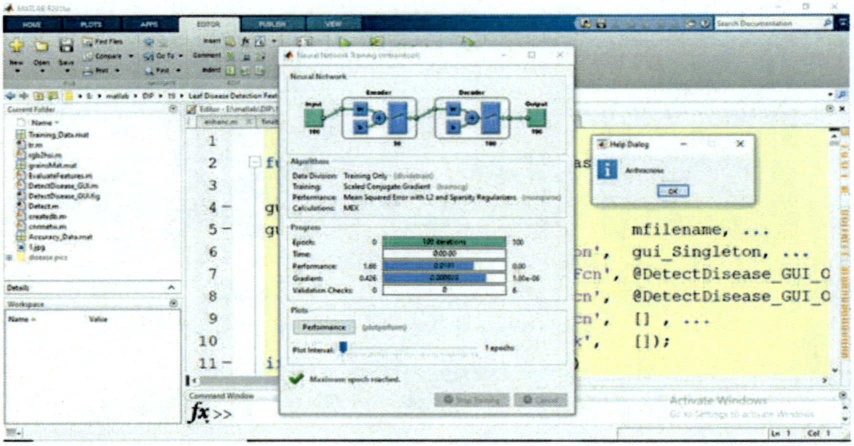

5 Conclusion

This project executes a creative plan to distinguish the influenced crops and give cure events to the agrarian business. Here, the utilization of k-mean grouping calculation, the contaminated locale of the leaf is divided and broke down. The pictures are taken care to make the requests recognizable for proof of illnesses. It gives a decent decision to farming the network especially in distant towns. It goes about as a productive framework as far as diminishing the bunching time and territory of the tainted locale. Highlight extraction strategy assists in extricating the contaminated leaf and furthermore to arrange the plant illness. CNN and IOT addition yields a fantastic effective solution to this idea for the development of exactness in the projected technique as an important methodology, which can essentially uphold a precise discovery of leaf infection in a little computational exertion.

References

1. S. Ananthi, S. Vishnu Varthini, Detection and classification of plant leaf diseases||. IJREAS **2**(2), 763–773 (2012)
2. S.E. Grigornescu, N. Petkov, P. Kruizinga, Comparison of texture features based on Gabor filters. IEEE Trans. Image Process. **11**(10), 1160–1167 (2002)
3. S. Naikwadi, www.mathworks.in
4. Pushparaj, Jagalingam; Malarvel, Muthukumaran "Panchromatic image denoising by a log-normal-distribution-based anisotropic diffusion model " published in JOURNAL OF APPLIED REMOTE SENSING volume 13, issue-1 Feb-2019
5. L.K. Rao, P. Rohini, L.P. Reddy, Local color oppugnant quantized extrema patterns for image retrieval. Multidim. Syst. Signal Process. **30**(3) (2019)

6. N. Sasikala, P.V.V. Kishore, D.A. Kumar, Ch.R. Prasad, Localized region based active contours with a weakly supervised shape image for inhomogeneous video segmentation of train bogie parts in building an automated train rolling examination. Multimed. Tools Appl. **78**(11) (2019)
7. K. Raveendra, R. Vinothkanna, Hybrid ant colony optimization model for image retrieval using scale-invariant feature transform local descriptor. Comput. Electr. Eng. **74** (2019)
8. R. Lenka, A. Khandual, S.R. Nayak, G. Palai, An improvement of visual perception of single image by fusion of DCP with patch processing. OPTIK **183** (2019)
9. S. Manoharan, Improved version of graph-cut algorithm for CT images of lung cancer with clinical property condition. J. Artif. Intell. **2**(04), 201–206 (2020)
10. R.B. Vallabhaneni, V. Rajesh, On the performance characteristics of embedded techniques for medical image compression. J. Sci. Ind. Res. **76**(10) (2017)
11. G. Suryanarayana, R. Dhuli, Super-resolution image reconstruction using dual-mode complex diffusion-based shock filter and singular value decomposition. Circ. Syst. Signal Process. **36**(8) (2017)
12. N.K. Gattim, V. Rajesh, R. Partheepan, S. Karunakaran, K.N. Reddy, Multimodal image fusion using curvelet and genetic algorithm. J. Sci. Ind. Res. **76**(11) (2017)

Improvement of Trade-Off Between Global and Local Search in Hybridization GA-PSO with Fuzzy Adaptive Acceleration Coefficients

Rodrigo Possidônio Noronha

Abstract In this paper, a new stochastic optimization methodology is proposed, resulting from the hybridization involving genetic algorithm (GA) and particle swarm optimization (PSO). This hybridization is performed with the goal of performing a search process with fast and non-premature convergence, through sharing and attenuating, respectively, the desirable and undesirable characteristics of GA and PSO. To avoid premature convergence and increase the speed of convergence, it is performed a parametric adaptation, through a fuzzy system, of the acceleration coefficients of the PSO; since through a correct selection of values, it is possible to perform an efficient trade-off between global and local search. Thus, through the proposed optimization methodology, it is possible to obtain an efficient optimization method which performs a search process with a fast and non-premature convergence.

1 Introduction

Evolutionary computation is one of the fastest growing areas within engineering and computer science, providing methods inspired by biological processes to solve complex problems, where traditional methods are unable or limited to solve them. In evolutionary computation theory, the population-based stochastic optimization methods have been used to solve problems in control theory [1], in system identification [2], in pattern classification [3], and in time series prediction [4]. Specifically, the use of GA and PSO, which are population-based stochastic optimization methods, has been widely performed in the development of methodologies in various fields due to the efficient and robust performance that these methods provide for the optimization of real-world problems.

PSO is an optimization method belonging to algorithms family of the swarm theory developed by Kennedy J. and Eberhart R. with the goal of simulating the social and cooperative behavior exhibited by various species, such as flocks of birds and schools of fish in search of food or reproduction [5]. In unimodal problems, the

R. P. Noronha (✉)
Federal Institute of Education, Science and Technology, Imperatriz, Maranhão, Brazil

P. Karuppusamy et al. (eds.), *Ubiquitous Intelligent Systems*, Smart Innovation, Systems and Technologies 243, https://doi.org/10.1007/978-981-16-3675-2_54

search process performed by the PSO converges quickly due to its unidirectional behavior. However, due to unidirectionality, the premature convergence can occur in multimodal problems. Premature convergence occurs due to the inefficient trade-off between global and local search. Several methodologies have already been published in the literature that propose strategies to avoid premature convergence of PSO. In [6], in order to maintain the diversity of positions, several swarm topologies were introduced in the search space. In [7], in order to avoid premature convergence, the mutation genetic operator was used, so that particles could explore regions in the search space that were still unexplored. In [8], two new versions of the PSO were proposed to prevent premature convergence. Both versions are memory-based, where in the first version is used the concepts of Ebbinghaus forgetting curve and in the second version is used the concept of splitting swarms into multiple subswarms. In this way, both versions use memory to store the best positions, which are used when particles are stuck in optimal locations and, thus, can avoid premature convergence.

GA, on the other hand, is an optimization method inspired by the theory of genetic evolution, developed by Holland [9]. In order for the principles of the theory of genetic evolution to be simulated by the GA, the genetic operators of parent selection, crossover and genetic mutation are used. Due to the great diversity of positions inserted in the search space by genetic operators, for GA, if a particle gets stuck in a local optimum, it will have an easier time jumping to better positions. However, due to the use of genetic operators, the search process is performed in a multidirectional manner, making convergence slow. Several attempts to increase the speed of convergence of GA have already been proposed in the literature. In [10], to improve the convergence speed, it was proposed to run GA in parallel utilizing both CPU and GP-GPU. In [11], multi-domain inversion was used to improve the convergence speed. In addition, individuals with worse fitness were eliminated through with a second fitness evaluation. In [12], the improvement in convergence speed was realized according to fine-tuning the positions of individuals through the binary space partition (BSP) tree topology, in each generation.

Although GA and PSO are simple to implement, robust optimization methods and with great applicability, both have undesirable characteristics that can be mitigated through hybridization. Hybridization of optimization methods is performed with the goal of sharing desirable features and mitigating undesirable features, as can be seen in [13–15]. The hybridization involving GA and PSO is originally done in order to solve optimization problems with fast convergence and high diversity of positions in the search space (aiming to avoid premature convergence). There are some applications of hybridization involving GA and PSO published in the literature, such as in [16] that GA-PSO based on two-dimensional parallel coding was proposed for resource optimization in the Jammers target penetration blocking process. In the paper [17], the application of adaptive GA-PSO in bearing fault detection was proposed. On the other hand, only inserting diversity of positions in the search space is not enough to avoid premature convergence, because particles can get stuck in local optimal due to excess or insufficient diversity [18]. Moreover, the large diversity inserted by GA causes slow convergence of the search process performed by hybridization. Since diversity is a measure of proximity between particles, such that a high diversity results in a global search and a small diversity results in a local search, through an efficient trade-off between global and local search, it is possible to obtain a hybrid optimization algorithm that provides both fast and non-premature convergence.

Thus, in this paper, a new stochastic optimization methodology called genetic algorithm–particle swarm optimization (GA-PSO) with fuzzy adaptive acceleration coefficients is proposed. Since, according to what proposed by Kennedy J. and Eberhat R. in the article [8], the acceleration coefficients C_1 and C_2 control, respectively, the sharing of individual and global knowledge about the search process, by adapting them, it is possible to realize an efficient trade-off between global and local search. For the adaptation of the acceleration coefficients C_1 and C_2, a Mamdani fuzzy system is used, where the inputs are the normalized diversity and the percentage of elapsed iterations. With the use of a Mamdani fuzzy system, it is possible to model the dynamic behavior of the acceleration coefficients in function of the expert's knowledge about how the trade-off between global and local search should be performed. Thus, through the proposed methodology, it is possible to perform the optimization of complex problems with a fast and non-premature convergence. In order to evaluate the performance of the proposed methodology, it was performed the optimization of unimodal, multimodal and multidimensional benchmarks.

1.1 Contribuitions

The main contributions of the proposed optimization methodology are described as follows:

- Structurally, the hybridization proposed in the optimization methodology is cascaded. In this way, the multidirectional search is refined into a unidirectional search, which contributes to increasing the speed of convergence and to avoiding premature convergence;
- The fuzzy adaptation of the acceleration coefficients C_1 and C_2 is performed as a function of the proximity between the particles and the time advance of the search process. By combining the inputs through fuzzy propositions, it is possible to obtain an efficient trade-off between global and local search;
- Due fuzzy adaptation of the acceleration coefficients C_1 and C_2, an optimization methodology is obtained that both circumvents premature convergence and quickly converges to the optimal solution.

This paper is organized as follows. Section 2 is presented the proposed stochastic optimization methodology. Section 2.1 is presented the parent selection genetic operator. Section 2.2 is presented the crossover genetic operator. Section 2.3 is presented the mutation genetic operator. Section 2.4 are presented the equations for updating the velocity and positions of the particles. In subsection fuzzy, the methodology for parametric adaptation of the acceleration coefficients C_1 and C_2 of the PSO, using a Mamdani fuzzy system, is presented. Section 2.6 is presented a table containing the computational complexity analysis of the proposed optimization methodology. Finally, Section 3 are presented the computational results obtained by optimization benchmarks.

2 Proposed Optimization Methodology

In this section, the proposed stochastic optimization methodology is presented. As can be seen in Fig. 1, the proposed stochastic optimization methodology is performed in a cascade structure. The genetic operators, the each iteration k, update the positions $\mathbf{x}_i$ of the particles in the search space. Following the cascade hybridization structure, the acceleration coefficients C_1 and C_2 are adapted by a Mamdani fuzzy system, in which the inputs are the normalized diversity entered by GA and the percentage of elapsed iterations up to instant k; the outputs are the acceleration coefficients C_1 and C_2. Right after that, the velocities $\mathbf{v}_i$ and positions $\mathbf{x}_i$ of the particles in the search space are updated. Finally, the best individual positions $\mathbf{p}_i$ and the best global position $\mathbf{p}_g$ are updated. The solution to the problem is obtained when the stopping condition is satisfied; otherwise, the search process is continued. The proposed optimization methodology aims to obtain the solution that maximizes or minimizes the following

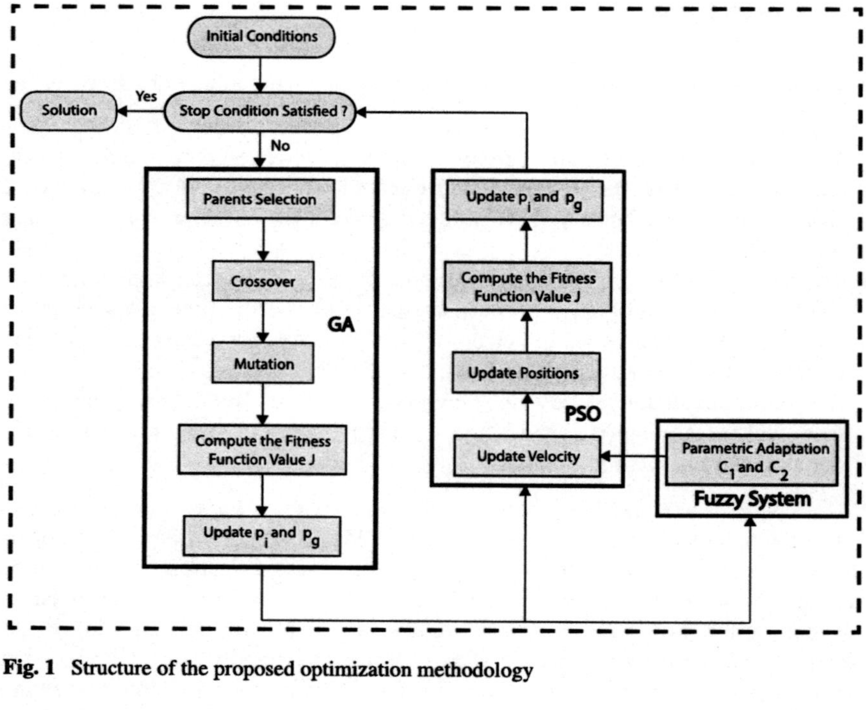

Fig. 1 Structure of the proposed optimization methodology

fitness function:

$$\begin{cases} J = J(x_{i,1}[k], x_{i,2}[k], \ldots, x_{i,n}[k]) \\ \text{subject to :} \\ \min(\mathbf{x}_i) \leq \mathbf{x}_i[k] \leq \max(\mathbf{x}_i) \\ \min(\mathbf{v}_i) \leq \mathbf{v}_i[k] \leq \max(\mathbf{v}_i) \\ k \in [1, T] \end{cases} \tag{1}$$

where $J : \mathbb{R}^n \rightarrow \mathbb{R}$ is the fitness function, $\mathbf{x}_i[k] \in \mathbb{R}^n$ is the position vector of the ith particle in the search space, $\mathbf{v}_i[k] \in \mathbb{R}^n$ is the velocity vector of the ith particle, N is the particle population size, and T is the maximum number of iterations. The position and velocity vectors are defined as $\mathbf{x}_i[k] = [x_{i,1}[k], x_{i,2}[k], \ldots, x_{i,n}[k]]$ and $\mathbf{v}_i[k] = [v_{i,1}[k], v_{i,2}[k], \ldots, v_{i,n}[k]]$, where n is the dimension of the search space. Aiming to perform maximization or minimization of a generic problem described according to Eq. (1), it is necessary to mathematically formulate the optimization goal of the proposed methodology described in Fig. 1. Let be a fitness function $J : \mathbb{R}^n \rightarrow \mathbb{R}$, the update of the best position $\mathbf{p}_i$ of the i-th particle, for a minimization problem, is given by:

$$\mathbf{p}_i[k+1] = \begin{cases} \mathbf{p}_i[k] \text{ if } J(\mathbf{x}_i[k+1]) \geq J(\mathbf{p}_i[k]) \\ \mathbf{x}_i[k+1] \text{ if } J(\mathbf{x}_i[k+1]) < J(\mathbf{p}_i[k]) \end{cases} \tag{2}$$

For a maximization problem, the update of the best position $\mathbf{p}_i$ of the ith particle is given by:

$$\mathbf{p}_i[k+1] = \begin{cases} \mathbf{p}_i[k] \text{ if } J(\mathbf{x}_i[k+1] \leq J(\mathbf{p}_i[k]) \\ \mathbf{x}_i[k+1] \text{ if } J(\mathbf{x}_i[k+1]) > J(\mathbf{p}_i[k]) \end{cases} \tag{3}$$

For a minimization problem, the update of the global best position $\mathbf{p}_g$ is given by:

$$\begin{aligned} \mathbf{p}_g[k+1] &\in \{\mathbf{p}_1[k+1], \mathbf{p}_2[k+1], \ldots, \mathbf{p}_N[k+1]\} \\ &= \arg\min_{1 \leq i \leq N} J(\mathbf{p}_i[k+1]) \end{aligned} \tag{4}$$

For a maximization problem, the update of the global best position $\mathbf{p}_g$ is given by:

$$\begin{aligned} \mathbf{p}_g[k+1] &\in \{\mathbf{p}_1[k+1], \mathbf{p}_2[k+1], \ldots, \mathbf{p}_N[k+1]\} \\ &= \arg\max_{1 \leq i \leq N} J(\mathbf{p}_i[k+1]) \end{aligned} \tag{5}$$

2.1 *Genetic Operator of Parent Selection*

The objective of the parent selection genetic operator is to select the fittest particles to be reproduced. The selection method used in this methodology is of the tournament type, which consists of holding a tournament between m randomly chosen particles, in each generation or iteration, where the winning particle will be the parent particle. This way, the next generation of particles will be more apt to be possible solutions to the optimization problem. One control parameter of the tournament type parent selection genetic operator is the number of particles that will participate in the tournament. The higher the value of m, better will be the fitness of the selected particles and, higher will be the computational complexity of the parents selection genetic operator. The pseudocode for the genetic operator of parent selection of the tournament type is presented below:

Algorithm 1 Pseudocode of the Tournament Method

Set the crossover rate μ_c;
for $g = 1$ to $N\mu_c$ **do**
  for $l = 1$ to 2 **do**
    Choose randomly m particles;
    $parent_l$ = The particle with the best fitness;
  end for
end for

2.2 *Genetic Operator of Crossover*

The crossover genetic operator is used, according to the crossover rate μ_c, to combine the characteristics of the selected parents, aiming to obtain the fittest particles to be possible solutions to the problem. This way, the new generation will contain a larger amount of characteristics possessed by the fittest particles, and with this, the characteristics considered "good" are distributed by the population of particles. It is important to note that the higher the value of μ_c, the greater will be the quantity of individuals submitted to the crossover. If the value of μ_c is too small, the next generation of particles will be practically the same as the past generation. Typical values of μ_c are in the range $[0.5,\ 1]$. The crossover method used in this methodology is of the uniform type, according to the following pseudocode:

Algorithm 2 Pseudocode of the Uniform Crossover

for $g = 1$ to $\frac{N\mu_c}{2}$ **do**
  Choose two particles at random;
  Obtain randomly $\alpha \in [0, 1]$;
  $child_1 = \alpha parent_1 + (1 - \alpha) parent_2$;
  $child_2 = \alpha parent_2 + (1 - \alpha) parent_1$;
end for

2.3 Genetic Operator of Mutation

The goal of the genetic operator mutation is to insert features not yet present or present in small quantity in the current population. The mutation genetic operator is important to insert diversity aiming to explore new regions in the search space, according to the mutation rate μ_m and, consequently, contribute to avoid premature convergence. It is important to note that if μ_m is high, many particles of the population will be mutated. If μ_m is small, few particles of the population will be mutated. Typical values of μ_m are in the range [0.005, 0.05]. The mutation method used in this methodology is of the random type, according to the following pseudocode:

Algorithm 3 Pseudocode of the Ramdom Mutation

1: Set the mutation rate μ_m;
2: **for** $i = 1$ to N **do**
3: Obtain randomly $\beta \in [0, 1]$;
4: **if** $\beta < \mu_m$ **then**
5: Obtain randomly $\gamma \in \{1, 2, \ldots, n\}$;
6: Mutate the gene position corresponding to the γ index, using the equation: $x_{i,\gamma}[k] = max(x_{i,\gamma}) + \beta(max(x_{i,\gamma}) - min(x_{i,\gamma}))$;
7: **end if**
8: **end for**

2.4 Velocity and Positions Update

The velocity update equation is composed of two terms that introduce the stochastic nature of the optimization method, that are $r_1 \in [0, 1]$ and $r_2 \in [0, 1]$. The velocity update equation of the ith particle is given by:

$$\begin{aligned} v_{i,j}[k+1] = v_{i,j}[k]\omega + C_1 r_1(p_{i,j}[k] - x_{i,j}[k]) \\ +C_2 r_2(p_{g,j}[k] - x_{i,j}[k]) \end{aligned} \tag{6}$$

where $0 < C_1, C_2 \leq 2$ is the acceleration coefficients and $\omega \in [0,\ 1.2]$ is the inertial weight. According to [8], the acceleration coefficients C_1 and C_2 control the knowledge sharing of individual and global about the search process, respectively, by weighting the terms $(p_{i,j}[k] - x_{i,j}[k])$ and $r_2(p_{g,j}[k] - x_{i,j}[k])$. Thus, through a correct selection of values for the acceleration coefficients C_1 and C_2, it is possible to perform efficient trade-off between global and local search and, consequently, avoid premature convergence. After the velocity vector of the ith particle has been updated, the update of the position vector is given by:

$$\mathbf{x}_i[k+1] = \mathbf{x}_i[k] + \mathbf{v}_i[k+1] \tag{7}$$

2.5 Fuzzy Adaptation of the Acceleration Coefficients

To model the dynamic behavior, through mathematical formulations, of the acceleration coefficients C_1 and C_2, aiming at an efficient trade-off between the global and local search, is not a simple task. Moreover, other factors influence the trade-off between global and local search, such as the size of the search space, the number of local maximums and minimums, initialization of initial positions, and others. A solution to get around problems, where a mathematical description is not possible, is through a Mamdani fuzzy system. Using a Mamdani fuzzy system, in function of the expert's knowledge of how the trade-off between global and local search should be performed, and it is possible to model the dynamic behavior of the acceleration coefficients C_1 and C_2 in order to obtain an efficient trade-off between global and local search.

For example, the dynamic behavior of the acceleration coefficients C_1 and C_2 can be modeled as follows. As described in [8], during the first interactions, the particles must acquire individual knowledge about the best regions, so that it is necessary to increase the diversity of positions in order to explore the search space. As the end of iterations, the diversity of positions must be decreased aiming at a greater contribution of global knowledge and, consequently, the convergence of the search process. Thus, during the first iterations, the acceleration coefficient C_1 should be greater than C_2; as the end of iterations, C_2 should be greater than C_1.

On the other hand, the dynamic behavior of the acceleration coefficients C_1 and C_2 should not be modeled only in function of the time advance of the search process. Furthermore, the dynamic behavior of the acceleration coefficients C_1 and C_2 should not be linear or piecewise linear. For an efficient trade-off between global and local search, it is necessary that the acceleration coefficients C_1 and C_2 are also adapted in function of the distance between the particles, that is, in function of the diversity of positions in the search space, since through diversity, it is possible to verify whether the particles are performing a global or local search. Within a context of fuzzy rules, the dynamics of the acceleration coefficients can be modeled, for example, as follows: if the diversity is small and the amount of elapsed iterations is small, then C_1 must be high and C_2 must be small, so that the particles explore the search space. If the diversity is high and the amount of iterations is high, then C_1 must be small and C_2 must be high. If the diversity is small and the amount of iterations is high, then C_1 should be medium high and C_2 should be medium small. It is important to note that the linguistic descriptions high, small, and medium are defined as a function of the expert's knowledge of how the trade-off between global and local search should be performed.

Figure 2 is presented, the Mamdani fuzzy system that describe the dynamic behavior of the acceleration coefficients C_1 and C_2, aiming to obtain an efficient trade-off between the global and local search and, consequently, to obtain a search process with a fast and not premature convergence. The inputs of the fuzzy system are the normalized diversity inserted by GA and the percentage of elapsed iterations. The outputs are the adapted acceleration coefficients C_1 and C_2. The antecedent and con-

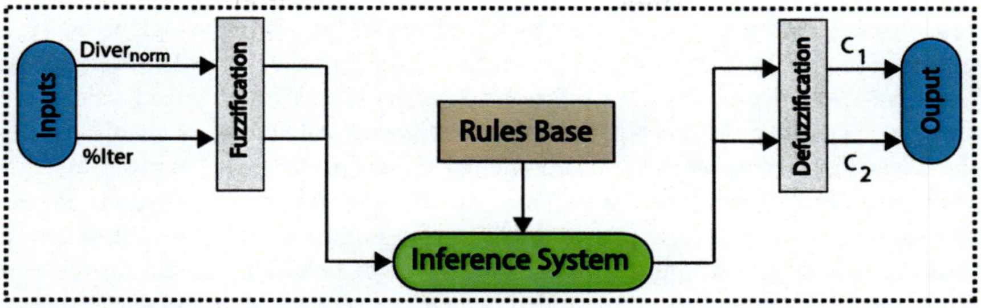

Fig. 2 Block structure of the fuzzy adaptation of acceleration coefficients

sequent of the fuzzy system are composed of fuzzy sets. The membership functions, defined based on the expert's knowledge, are of triangular type, the inference engine was defined as inference minimum, and the defuzzifier is of centroid type.

The dynamic behavior of the acceleration coefficients C_1 and C_2, aiming to obtain an efficient trade-off between global and local search, is described by the rth rule:

$$L_r : \text{If } X_1 \text{ is } A_j \text{ and } X_2 \text{ is } A_q \text{ Then } Y_1 \text{ is } B_f \text{ and } Y_2 \text{ is } B_g \tag{8}$$

where the variables $X_1, X_2 \in U_1 = U_2 = [0,\ 1]$ are the linguistic variables of the antecedent referring, respectively, to the percentage of elapsed iterations and the normalized diversity. The linguistic values A_j and A_q are the fuzzy sets that model the dynamic behavior, based on the expert's knowledge, of the input variables. The variables $Y_1, Y_2 \in H_1 = H_2 = [0,\ 3.0]$ are the linguistic variables of the consequent referring to the acceleration coefficients C_1 and C_2, respectively. The linguistic values B_f and B_g are the fuzzy sets that model the dynamic behavior, based on the expert's knowledge, of the output variables C_1 and C_2, respectively.

In order to model the dynamic of the acceleration coefficient as a function of information about the temporal advance of the search process, is used as the first fuzzy input the percentage of elapsed iterations, given, by Eq. (9). Three membership functions are defined, for the linguistic variable X_1, with the following linguistic values: small iteration, medium iteration, and high iteration. Thus, for example, when starting the search process, the linguistic variable X_1 will receive the linguistic value "low," with a certain degree of pertinence. At the end of the search process or near 100%, for example, the linguistic variable X_1 will receive the linguistic value "high," with a certain degree of pertinence. Each element belonging to the universe of discourse U_1 is associated to a linguistic value or fuzzy set with a given degree of pertinence $\mu_{A_j}(\%\text{Iter}) : \mathbb{R} \to [0,\ 1]$, where μ_{A_j} is the fuzzy set defining the jth linguistic value.

$$\%\text{Iter}(k) = \frac{k}{T}, \qquad k = 1, 2, \ldots, T. \tag{9}$$

So that the acceleration coefficients are also adapted as a function of the distance between the particles, and it is used as the second fuzzy input, the diversity of positions in the search space inserted by GA, given by Eq. (10). Through diversity, it is possible to verify if the particles are performing a global or a local search, that is, when particles are closer to each other, the diversity is small, and when particles are far from each other, the diversity is high. The diversity calculation is given by mean Euclidean distance between the position $\mathbf{x}_i[k]$ of each particle and the position $\mathbf{p}_g[k]$ of the particle with the best fitness. Three membership functions were used, for the linguistic variable of diversity, with the following linguistic values: small diversity, medium diversity, and high diversity. Each element belonging to the universe of discourse U_2 is associated to a linguistic value or fuzzy set with a given degree of pertinence $\mu_{A_q}(\text{Diver}_{\text{norm}}) : \mathbb{R} \rightarrow [0,\ 1]$, where μ_{A_q} is the fuzzy set defining the qth linguistic value.

$$\text{Diver}[k] = \frac{1}{N}\sum_{i=1}^{N}\sqrt{\sum_{j=1}^{n}(x_{i,j}[k] - p_{g,j}[k])^2} \tag{10}$$

With the goal of that the two fuzzy inputs have equal maximum and minimum values, the diversity normalization is performed, given by:

$$\text{Diver}_{\text{norm}}[k] = \frac{\text{Diver}[k] - d_{\min}[k]}{d_{\max}[k] - d_{\min}[k]} \tag{11}$$

where $d_{\min}$ and $d_{\max}$ are the minimum and maximum Euclidean distance, respectively, obtained between $\mathbf{x}_i[k]$ and $\mathbf{p}_g[k]$.

In Table 3, it is described the fuzzy rule base used to model the dynamics of the acceleration coefficients, aiming to obtain an efficient trade-off between global and local search. Still in Table 3, the linguistic values of the antecedent and consequent were represented by abbreviations, which are: S = Small, M = Medium, H = High, MS = Medium small and MH = Medium High. Tables 1 and 2 are presented, based on the expert's knowledge, the parameters of the triangular membership functions that describe the linguistic values of the fuzzy rule base. In Tables 1 and 2, for simplicity, the triangular functions are defined by three parameters, in the following order: $[a\ b\ c]$, where a, b, and c are the minimum, center, and maximum values of the membership function, respectively. Thus, a generic triangular membership function is defined as:

$$\mu(x) = \begin{cases} 0, & x \leq a \\ \dfrac{x-a}{b-a}, & a \leq x \leq b \\ \dfrac{c-x}{c-b}, & b \leq x \leq c \\ 0, & c \leq x \end{cases} \tag{12}$$

The defuzzified output, obtained using the centroid method, is given by:

Table 1 Intervals of the triangular functions of the antecedent

%Iter		$Diver_{norm}$	
Linguistic value	Interval	Linguistic value	Interval
Small	[0 0 0.1]	Small	[0 0 0.5]
Medium	[0 0.1 0.3]	Medium	[0 0.5 1]
High	[0.1 0.3 1.0]	High	[0.5 1.0 1.0]

Table 2 Intervals of the triangular functions of the consequent

C_1		C_2	
Linguistic value	Interval	Linguistic value	Interval
Small	[0 0.5 1.0]	Small	[0 0.5 1.0]
Medium small	[0.5 1.0 1.5]	Medium small	[0.5 1.0 1.5]
Medium	[1.0 1.5 2.0]	Medium	[1.0 1.5 2.0]
Medium high	[1.5 2.0 2.5]	Medium high	[1.5 2.0 2.5]
High	[2.0 2.5 3.0]	High	[2.0 2.5 3.0]

Table 3 Fuzzy rule base for parametric adaptation of the acceleration coefficients C_1 and C_2

Rules base
L_1: **If** (%Iter is S) **and** ($Diver_{norm}$ is S) **Then** (C_1 is H) **and** (C_2 is S)
L_2: **If** (%Iter is S) **and** ($Diver_{norm}$ is M) **Then** (C_1 is MH) **and** (C_2 is M)
L_3: **If** (%Iter is S) **and** ($Diver_{norm}$ is H) **Then** (C_1 is MH) **and** (C_2 is MS)
L_4: **If** (%Iter is M) **and** ($Diver_{norm}$ is S) **Then** (C_1 is MH) **and** (C_2 is MS)
L_5: **If** (%Iter is M) **and** ($Diver_{norm}$ is M) **Then** (C_1 is M) **and** (C_2 is M)
L_6: **If** (%Iter is M) **and** ($Diver_{norm}$ is H) **Then** (C_1 is MS) **and** (C_2 is MH)
L_7: **If** (%Iter is H) **and** ($Diver_{norm}$ is S) **Then** (C_1 is M) **and and** (C_2 is H)
L_8: **If** (%Iter is H) **and** ($Diver_{norm}$ is M) **Then** (C_1 is MS) **and** (C_2 is H)
L_9: **If** (%Iter is H) **and** ($Diver_{norm}$ is H) **Then** (C_1 is S) **and** (C_2 is H)

$$C_1[k] = \frac{\sum_{l=1}^{R} C_1 \mu_{B_f}(C_1)}{\sum_{l=1}^{R} \mu_{B_f}(C_1)}, \quad C_2[k] = \frac{\sum_{l=1}^{R} C_2 \mu_{B_g}(C_2)}{\sum_{l=1}^{R} \mu_{B_g}(C_2)} \tag{13}$$

Table 4 Computational complexity of proposed optimization methodology

Methodology step	Operations number
Parents selection genetic operator	$f(N, \mu_c, m) = 3N\mu_c + m + 7$
Crossover genetic operator	$f(N, \mu_c, n) = N\mu_c/2 + 6n + 9$
Mutation genetic operator	$f(N) = 2N + 9$
Velocity update	$f(N, n) = N + 7n + 3$
Position update	$f(N, n) = N + n + 3$
First input of the Mamdani fuzzy system	$f = 1$
Second input of the mamdani fuzzy system	$f(N, n) = n(N + 4) + N + 7$
Fuzzification	$f = 6$
Minimum inference	$f(R) = R + 5$
Defuzzification	$f(R) = 2R^2$

2.6 *Computational Complexity of the Proposed Optimization Methodology*

After presenting the proposed optimization methodology, it is necessary to evaluate the computational complexity. The computational complexity was analyzed through the number of operations performed by the proposed optimization methodology, presented in Table 4, calculated independently of the number of iterations for the worst case.

3 Computational Results

To validate the proposed optimization methodology, this section is proposed the minimization of three benchmarks, which are Sphere, Rastrigin, and Griewank functions. The descriptions of the benchmarks are presented in (14)–(16) and in Table 5.

1. **Sphere Function**:

$$f_1(\mathbf{x}) = \sum_{i=1}^{n}(\mathbf{x}_i{}^2) \tag{14}$$

2. **Rastrigin Function**:

$$f(\mathbf{x}) = 10n + \sum_{i=1}^{n}\left[\mathbf{x}_i^2 - 10\cos 2\pi\mathbf{x}_i\right] \tag{15}$$

Table 5 Description of benchmarks

Function	Dimension	Search space	Modes	Global Min.
f_1	20	[−5.12, 5.12]	Unimodal	0
f_2	20	[−5, 10]	Multimodal	0
f_3	20	[−5.12, 5.12]	Multimodal	0

3. **Função Griewank:**

$$f_3(\mathbf{x}) = 1 + \frac{1}{4000}\sum_{i=1}^{n}\mathbf{x}_i^2 - \prod_{i=1}^{n}\cos\frac{\mathbf{x}_i}{\sqrt{i}} \quad (16)$$

The set of parameters used to obtain the results are as follows: the crossover rate was set $\mu_c = 0.9$, and the probability rate was set $\mu_m = 0.1$; for the canonical PSO, the acceleration coefficients were set as $C_1 = C_2 = 2$. These parameters were set by trial and error. The number of particles randomly selected to participate, each generation, in the selection of parents by tournament was set to $m = 3$. The initial positions $\mathbf{x}_i[1]$ were defined randomly, belonging to the search spaces defined in Table 5. Furthermore, the best individual positions and particle velocities were initialized as $\mathbf{p}_i[1] = \mathbf{x}_i[1]$ and $\mathbf{v}_i[1] = \mathbf{0}$, respectively. The stopping condition was set when the number of iterations k is greater than T, where $T = 300$ iterations. The particle population was defined to contain $N = 100$ individuals. In order to maintain competitiveness between the methodologies, all positions $\mathbf{x}_i[1]$ were initialized equally. The results presented in this paper were obtained on a notebook computer with a 4th generation Intel Core i5 processor and 6 GB of RAM.

Since the proposed optimization methodology is stochastic, in order to obtain a greater statistical consistency of the results presented in Table 6, the optimization of each benchmark was performed during 100 realizations. Through the optimization results presented in Table 6, it can be seen that in the benchmark f_2, the worst results were obtained. Thus, the comparative analysis of the performance of the methodologies presented in Table 6 was performed only in relation to the results obtained in the optimization of benchmark f_2. Although the proposed optimization methodology has a higher computational complexity when compared to GA, PSO, and GA-PSO without adaptation of the acceleration coefficients, according to the optimization curves presented in Fig. 3, it can be seen that the proposed optimization methodology obtained the best results in convergence speed. Furthermore, as per the results presented in Table 6, it can be seen that the proposed methodology obtained the best optimization fitness results, for 300 iterations of the search process and 100 realizations of the optimization methodologies.

Through analysis of the optimization curve in green color, referring to the GA-PSO without adaptation of the acceleration coefficients, it can be seen that the search

Table 6 Results obtained from optimizing benchmarks for 100 realizations

Methodology	Results	f_1	f_2	f_3
PSO	Mean fitness	0.0213	163.0306	0.0017
	Standard deviation	0.0328	162.1489	0.0028
	Mean time (s)	0.9664	0.8270	1.0398
GA	Mean fitness	4.8759×10^{-4}	8.0351	0.0069
	Standard deviation	4.8759×10^{-4}	8.1492	0.0032
	Mean time (s)	3.5141	3.5261	3.7494
GA-PSO	Mean fitness	4.3499×10^{-12}	3.9898	0
	Standard deviation	1.37241×10^{-11}	4.9871	0
	Mean time (s)	119.9296	114.7730	117.2899
Proposed	Mean fitness	1.0936×10^{-21}	1.9899	0
	Standard deviation	3.4432×10^{-21}	1.8761	0
	Mean time (s)	7.5051	6.4968	7.7095

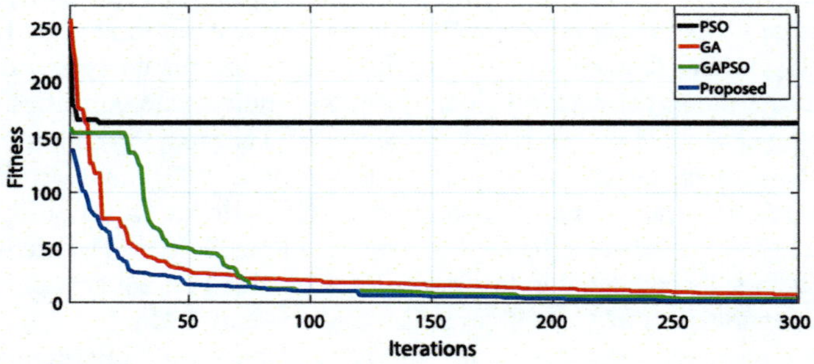

Fig. 3 Optimization curves

process converged slowly to the permanent regime, due to the large diversity inserted by GA, resulting in an inefficient trade-off between global and local search. The slow convergence did not occur when analyzing the optimization curve of the GA-PSO with adaptation of the acceleration coefficients in blue color, which is due to the efficient trade-off between the global and local search. In function of the fuzzy adaptation of the acceleration coefficients C_1 and C_2, it was possible to notice that when the particles were stuck in local optimal, due to the correct sharing of individual and global knowledge, the particles jumped quickly to other regions in the search space, thus avoiding premature convergence. This characteristic can be seen in Fig. 3, where it can be seen that the optimization curve of the proposed methodology presents greater uniformity, when compared to the other methods.

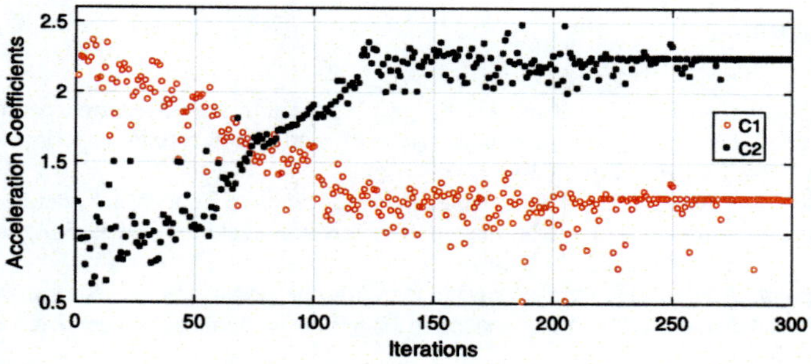

Fig. 4 Fuzzy adaptation of the acceleration coefficients

The parametric adaptation of the acceleration coefficients C_1 and C_2 is presented in Fig. 4. In the first iterations, in order for the particles to perform a global search, the acceleration coefficient C_1 was larger than C_2, favoring a global search. As iterations advance, so that the particles could perform a local search near the best global position, the acceleration coefficient C_2 was larger than C_1, favoring a local search and the convergence of the search process. For an efficient trade-off between global and local search, the adaptation of each acceleration coefficient should not be linear or piecewise linear. Since premature convergence occurs due to too much or not enough diversity in the search space, to avoid premature convergence, it is necessary that the adaptation of the acceleration coefficients C_1 and C_2 is also performed as a function of a measure of proximity between the particles. Thus, combining the percentage of elapsed iterations and the normalized diversity, based on the expert's knowledge, into a set of fuzzy rules, it was possible to consistently model the dynamic behavior of the acceleration coefficients C_1 and C_2, as can be seen in Fig. 4.

4 Conclusion

A new GA-PSO optimization methodology with fuzzy adaptation of the acceleration coefficients was proposed, in this paper. Through a Mamdani fuzzy system, the dynamic behavior of the acceleration coefficients was modeled, based on the expert's knowledge about how the trade-off between global and local search should be performed. Due to the dynamic behavior of the acceleration coefficients, it was noticed that the optimization of unimodal, multimodal, and multidimensional functions was performed with a fast and non-premature convergence, showing competitiveness when compared to GA, PSO, and GA-PSO without adaptation of acceleration coefficients. Furthermore, since the optimization methodology is stochastic, the optimization of the benchmarks was performed during 100 realizations, thus being able to obtain greater consistency in the statistical analysis of the results.

References

1. H. Su, Y. Hu, H.R. Karimi, A. Knoll, G. Ferrigno, E. Momi, Improved recurrent neural network-based manipulator control with remote center of motion constraints: experimental results. Neural Netw. **131**, 291–299 (2020)
2. N.N. Son, C.V. Kien, H.P.H. Anh, Parameters identification of Bouc-Wen hysteresis model for piezoelectric actuators using hybrid adaptive differential evolution and Jaya algorithm. Eng. Appl. Artif. Intell. **87**, 103317 (2020)
3. S. Bouzbita, A.E. Afia, R. Faizi, A new hidden Markov model approach for pheromone level exponent adaptation in ant colony system, in *Heuristics or Optimization and Learning* (2021), pp. 253–267
4. K. Khelil, F. Berrezzek, T. Gabased, Gabased design of optimal discretewavelet filters for efficient wind speed forecasting. Neural Comput. Appl. 1–14 (2020)
5. R. Eberhart, J. Kennedy, Particle swarm optimization, in *Proceedings of the IEEE International Conference on Neural Networks*, vol. 4 (1995), pp. 1942–1948
6. E.S. Peer, F.V. Van Den Bergh, A.P. Engelbrecht, Using neighbourhoods with the guaranteed convergence PSO, in *Proceedings of the 2003 IEEE Swarm Intelligence Symposium*, 2003, pp. 235–243
7. A.R. Jordehi, Enhanced leader PSO (ELPSO): a new PSO variant for solving global optimisation problems. Appl. Soft Comput. **26**, 401–417 (2015)
8. K. Chaitanya, D.V.L.N. Somayajulu, P. Radha Krishna, Memory-based approaches for eliminating premature convergence in particle swarm optimization. Appl. Intell. 1–34 (2021)
9. J.H. Holland, *Adaptation in Natural and Artificial Systems*, vol. 1(97) (Ann Arbor, MI, 1975), p. 5
10. A.K. Ghoshal, N. Das, S. Bhattachrjee, G. Chakraborty, A fast parallel genetic algorithm based approach for community detection in large networks, in *11th International Conference on Communication Systems & Networks*, 2019, pp. 95–101
11. J. Xin, J. Zhong, F. Yang, Y. Cui, J. Sheng, An improved genetic algorithm for path-planning of unmanned surface vehicle. Sensors **19**(11), 2640 (2019)
12. Y. Su, N. Guo, Y. Tian, X. Zhang, A non-revisiting genetic algorithm based on a novel binary space partition tree. Inf. Sci. **512**, 661–674 (2020)
13. K. Rahimunnisa, Hybridized genetic-simulated annealing algorithm for performance optimization in wireless adhoc network. J. Soft Comput. Paradigm (JSCP) **1**(1), 1–13 (2019)
14. D. Sayantan, A. Banerjee, Highly precise modified blue whale method framed by blending bat and local search algorithm for the optimality of image fusion algorithm. J. Soft Comput. Paradigm (JSCP) **2**(4), 195–208 (2020)
15. S. Manoharan, Population based meta heuristics algorithm for performance improvement of feed forward neural network. J. Soft Comput. Paradigm (JSCP) **2**(1), 36–46 (2020)
16. F. Liu, Y. Wang, J. Chen, Q. Wang, N. Yuan, Research on jamming resource allocation technology based on improved GAPSO algorithm. J. Phys.: Conf. Ser. **1738**(1), 012075 (2021)
17. X. Zhang, W. Zhang, Q. Guo, W. Lei, Optimization of hmm based on adaptive GAPSO and its application in fault diagnosis of rolling bearing, in *IEEE 2020 5th International Conference on Control and Robotics Engineering* (IEEE, 2020), pp. 53–57
18. Y. Tian, R. Cheng, X. Zhang, Y. Su, Y. Jin, A strengthened dominance relation considering convergence and diversity for evolutionary many-objective optimization. IEEE Trans. Evol. Comput. **23**(2), 331–345 (2018)

A Vision-Based Real-Time Driver Identity Recognition and Attention Monitoring System

Md. Khaliluzzaman, Siddique Ahmed, and Md. Jashim Uddin

Abstract In the current decade, many automobile crashes are occurred due to the non-professional drivers and the negligence of legal drivers during driving. To reduce the automobile crashes, illegal driver recognition and legal driver's attention estimation requires significant research attention in the field of computer vision. In this paper, a driver recognition and assistance system is proposed to monitor the driver's attention and drowsiness based on face and eyes tracking. For that, initially, the driver is recognized before starting the automobile through SVM, where features are extracted by using uniform LBP and Gabor filters. After recognize the driver, the driver is allowed to drive. Furthermore, from the recognized face, the eye's pupils are detected and tracked to estimate the attention of the driver. Here, color feature is utilized to track the face and eyes through the mean shift algorithm. The system generates the alarm for the illegal driver and awareness alarm of the driver in the case of driver face angle and eye's fatigue. The effectiveness of the proposed system is demonstrated through the real-time experiment. The proposed system is evaluated in the different lighting conditions and presented outcomes demonstrate the adequacy.

1 Introduction

In the current decade, vehicle accident remains as one of the most common issues. Different causes are associated with the occurrence of automobile accidents. However, it is due to driver's negligence and drowsiness, which leads to the occurrence of serious automobile accidents. Driver's fatigue and distraction can occur due to feeling sleepy or fall asleep of driver, side talking, and talking over the phone. As a result, it causes great damage to the automobiles, human lives, and properties. In order to reduce the automobile accidents, it is very essential to explore innovative

Md. Khaliluzzaman (✉) · S. Ahmed
Department of Computer Science and Engineering, International Islamic University Chittagong (IIUC), 4318 Chittagong, Bangladesh

Md. Jashim Uddin
Department of Electrical and Electronics Engineering, International Islamic University Chittagong (IIUC), 4318 Chittagong, Bangladesh

P. Karuppusamy et al. (eds.), *Ubiquitous Intelligent Systems*, Smart Innovation, Systems and Technologies 243, https://doi.org/10.1007/978-981-16-3675-2_55

technology for predicting the vehicle accident on the basis of drivers' drowsiness and distraction. One of the common ways to reduce the automobile accident is by monitoring the driver attention. Before that, illegal driver detection and recognition for specific automobile is also an important task.

This section describes resent state of the art of the face recognition and driver attention. Until today, many researchers have developed various techniques to face detection and recognitions. Nowadays, some effective and interesting methods such as AdaBoost face detection, robust face detection using Hausdorff distance, deep learning-based face detection are used to detect the face. Various efficient feature extractor methods such as Gabor filters [1–3], local binary pattern (LBP) [4], kernel base method [5] are used to recognize the face through renowned classifier such as support vector machine (SVM) [6] and artificial neural network (NN) [7]. Some researchers developed some methods to recognize the face by combining the LBP and Gabor filter features [5, 8]. In this paper, utilized both Gabor filters and LBP features to recognize the driver's face.

For monitoring driver attention, many real-time systems are developed based on the yawn frequency, percentage of eyelid closure over time (PERCLOS), and gaze direction [9], head pose [10], face orientation [11, 12], lip motion [11], eyes blink [13, 14], and pupil effect [12] monitoring algorithms. In this paper, face orientation and pupil status parameters used to estimate the attention of the driver.

For tracking the object, many researchers use different algorithms such as Kalman filer [11] and mean shift algorithm [15, 16]. In this work, mean shift algorithm for tracking the face and eye's pupils based on the color features is used. The non-parametric iterative mean shift algorithm is chosen in this work due to its advantage of low computational cost and easy implementation process.

The paper is organized as follows. In Sect. 2, the proposed method is explained. The experimental results are described in Sect. 3. The paper is concluded in Sect. 4.

2 Proposed Method

The main concern of the proposed method is to develop a framework to monitor the attention of the driver. Since there are lots of factors are involved in measuring the attention of drivers, this research work focuses on the face orientation and parameters of eye pupil. The face orientation is estimated by central axis point line describe in Sect. 2.4. Before estimating the drowsiness of the drive, the driver face has to be recognized to identify the illegal drivers. The proposed method is shown in Fig. 1.

According to Fig. 1, the proposed method is divided into six primary steps. Initially, the input video fame is preprocessed. From the preprocessed frame, the driver face is detected through Viola and Jones algorithm [17]. The detected face is recognized by SVM through uniform LBP and Gabor filters features to ensure the original driver is derived the vehicle.

After that, from the recognized driver face, the eye's pupils regions are detected through detecting the pupil of the eyes. Finally, the face and eyes, i.e., pupils, are

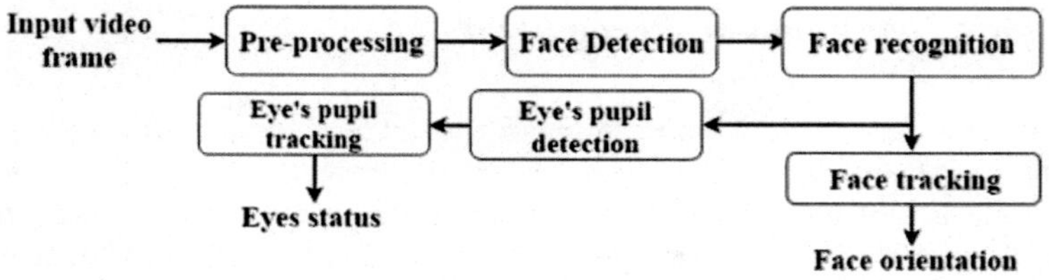

Fig. 1 Proposed method to driver recognition and driver's attention monitoring system

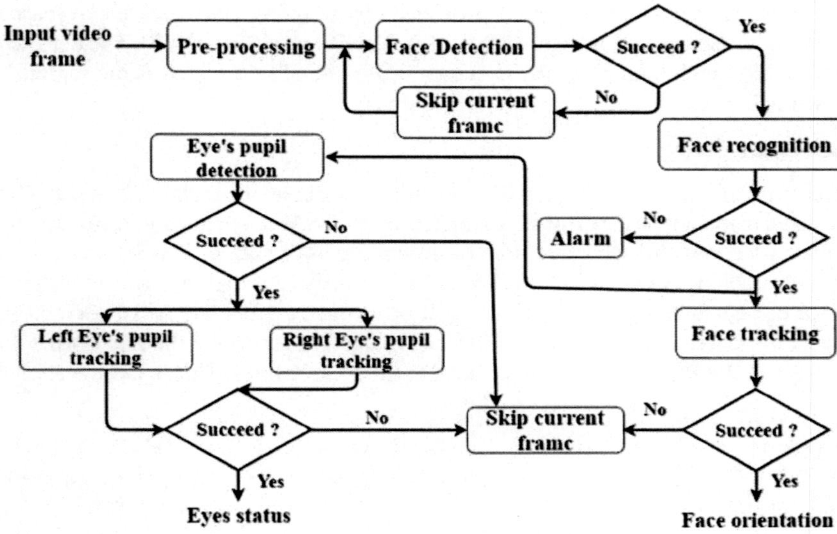

Fig. 2 Schematic working flow diagram of the proposed driver attention monitoring system

tracked by utilizing mean shift algorithm based on the color feature to monitor the driver attention. The schematic working flow diagram of the proposed system is presented in Fig. 2.

2.1 Preprocessing

The sample dataset that is used in this work is our own created dataset which is collected from different environmental conditions. Initially, the input RGB frame is converted to the gray scale image to optimize the computational complexity. The gray scale image is resized into 400 × 400.

2.2 Face Detection and Recognition

Face detection and recognition is an importance part in various research areas for computer vision technology, such as for security system. To recognize the legal automobile driver from input image frame, firstly, we have to detect the facial region of the driver from input image frame. The driver face detection and recognition procedure is presented in Fig. 3.

Face Detection

In this work, driver's facial region is detected by utilizing Haar like feature-based Voila–Jones face detection algorithm with the AdaBoost classifier [17]. The detected facial region with bounding box is shown in Fig. 4b. The cropped facial region is shown in Fig. 4c.

Skin Color Conversion

After extracting the facial region from input image frame, the facial features will be extracted to recognize the original driver. However, the Voila–Jones algorithm detects the facial region with some additional surrounding information. This information is created problem to recognize the driver appropriately. For that reason, the features that are used to recognize the driver will be extracted from the facial region. To extract the facial region from detected facial image, the skin color method is used. In this regard, the RGB image is represented to the YCbCr form. This is more effective

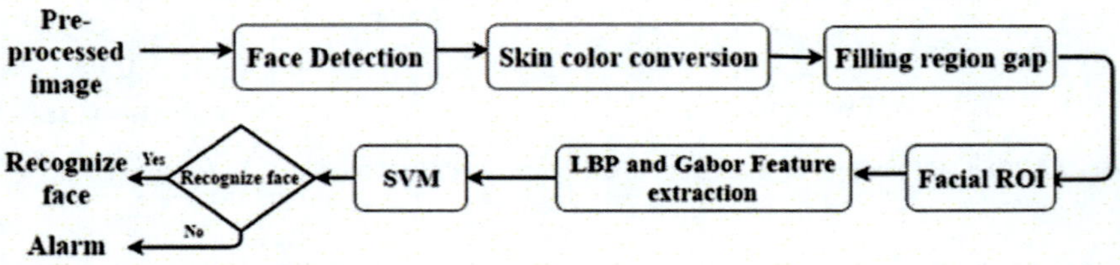

Fig. 3 Face detection and recognition procedure

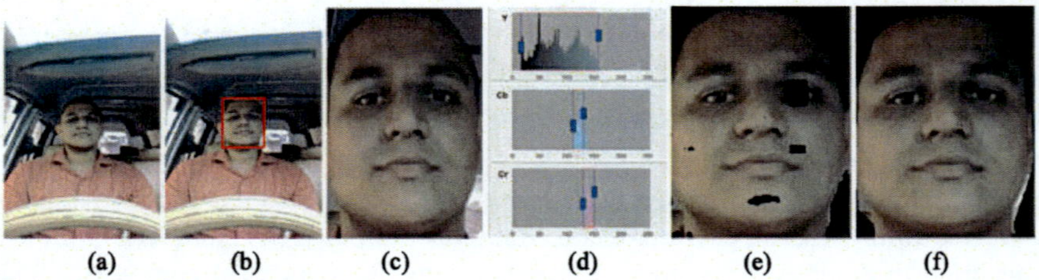

Fig. 4 Processing example of driver's face detection: **a** input image frame, **b** detected facial region with bounding box, **c** the cropped facial region, **d** YCbCr channel values, **e** facial color region detected by the YCbCr color model, and **d** final facial region

for skin color conversion. In this formation, the value of RGB is encoded, for which, the values of Y, Cb, and Cr range from 16 to 235 and 16 to 240, respectively. Finally, Eqs. (1), (2), and (3) are used to convert the RGB image to YCbCr color space.

$$Y = 16 + (65.481 * R + 128.553 * G + 24.966 * B) \tag{1}$$

$$Cb = 128 + (-37.797 * R - 74.203 * G + 112 * B) \tag{2}$$

$$Cr = 128 + (112 * R + 93.786 * G + 18.214 * B) \tag{3}$$

After transforming the RGB information to YCbCr color space, the skin color regions have to be extracted. For that, the threshold information of Cr as well as Cb is required. The estimated Cr and Cb information is shown in Fig. 4d. Figure 4e presents the extracted skin color image.

Facial Region of Interest (ROI)

In this section, the facial region of interest is extracted from the facial skin color image. For that, firstly, this facial skin color image is transform to binary image. In this binary image, the morphological operations, i.e., erosion and dilation, are performed to extract the appropriate facial image without any gap. After applying the morphological operations, the gaps in Fig. 4e are filled up. The resulting facial region of interest region is presented in Fig. 4f.

LBP and Gabor Filter Features Extraction

From extracted facial region of interest shown in Fig. 4f, the uniform local binary pattern (LBP) and Gabor filters features are extracted. For extracting the features from driver facial image, at first, the facial image is resized into 128 × 128 pixels. After that, uniform local binary pattern with rotational invariant and Gabor filters features is extracted from the facial resized image. The feature size for uniform LBP with rotational invariant is 59, i.e., $P \times (P - 1) + 3$, where, the value of P is eight. For Gabor features here, use 32 Gabor filters in 4 scales and 8 orientations. As the size of the facial image is 128 × 128 pixels, after employing the Gabor filter, the feature vector will be 128 × 128 × 32 = 524,288. To speed up the process, the feature vector dimensionality is reduced by down sampling process. The factor of down sample used in this work is 4. Hence, the final feature vector size will be 655,360/(4 × 4) = 32,768. By combining two features, total final feature vector size will be 59 + 32,768 = 32,827.

Face Recognition

The estimated facial features are sending to the support vector machine (SVM) to recognize the driver. If the face is recognized, then the driver is allowed to drive the vehicle; otherwise, alarm will be generated.

2.3 Face Tracking with Mean Shift Tracking Algorithm

This section explained the process of driver face tracking process after recognized the driver face. In this work, non-parametric and iterative mean shift tracking algorithm based on hue color feature [15] is used to tracking the driver face. The mean shift (MS) tracking algorithm is efficiently used due to its low computational cost and easy implementation. In MS tracking algorithm, the target object region is presented by the histogram, i.e., color histogram. A gradient ascent process is utilized to move the tracker to the location that maximizes a similarity score between the model and current tracked object region. The target model basically showed by the probability density function (PDF) and is regularized by spatial masking with an asymmetric kernel. The similarity of target and candidate model can be presented with a metric based on the Bhattacharyya coefficient.

The main concept of MS tracking algorithm is to track the specific region clustering points with the direction of the density growing, and the region clusters point of recursive iteration is drifting with the iterative direction to track the maximum values to the maximum growing points. The MS algorithm processing concept is shown in Fig. 5a. The driver face recognition with graphical user interface is shown in Fig. 5b. Face tracking result is shown in Fig. 5c.

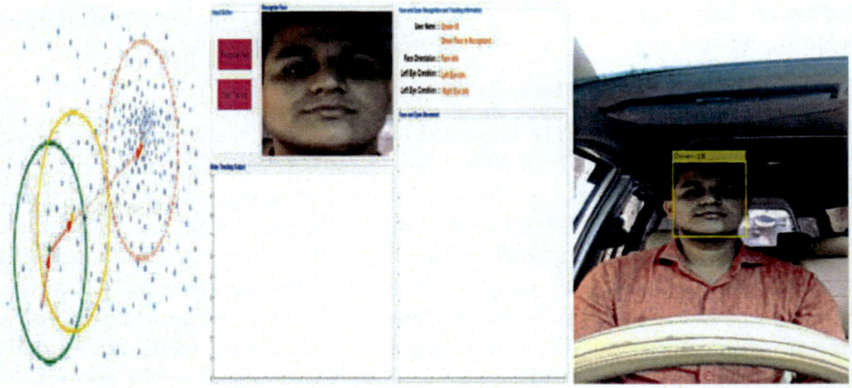

(a) (b) (c)

Fig. 5 Experimental example of face recognition and tracking: **a** concept of mean shift, **b** driver's face recognition in GUI, and **c** driver's face tracking

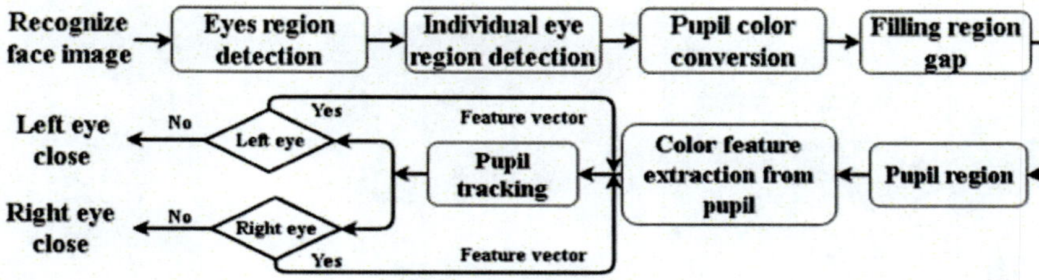

Fig. 6 Eyes detection and tracking procedure

2.4 *Eye's Pupil Detection and Tracking*

After recognized the driver's face, the driver eye's pupils are detected to estimate the driver attention based on the driver's eye's pupils tracking. The eye's pupil detection and tracking procedure is shown in Fig. 6.

In this section, again Voila–Jones algorithm [17] is utilized to extract the eyes region from the facial image. The extracted eyes region in face image is shown in Fig. 7a. Both eyes are separately detected by the further process shown in Fig. 7b, where upper one is left eye and bottom one is right eye.

To track the eyes, in this work, the pupil's color features are utilized. For that, both eyes pupil's area is detected by utilizing the color model. In this work, the YCbCr color space is utilized. The YCbCr color conversion binary image is presented in Fig. 7c. After color conversion, the morphological operation is used to extract the appropriate pupil shape presented in Fig. 7d. The extracted left and right pupils from image are presented in Fig. 7e. From these, both left and right eye's pupil's color features are extracted to track the eyes. A central axis point line is drawn vertically to estimate the face angle. The concept of the central axis point is explained in Fig. 9. The face and eyes tracking result with central axis point is presented in Fig. 8.

Fig. 7 Experimental example of eye's pupil detection: **a** eyes region with bounding box, **b** left (top) and right (bottom) eye region, **c** binary pupil region after color conversion, **d** pupil region after morphological operation, and **e** left (top) and right (bottom) eye with eye's pupil

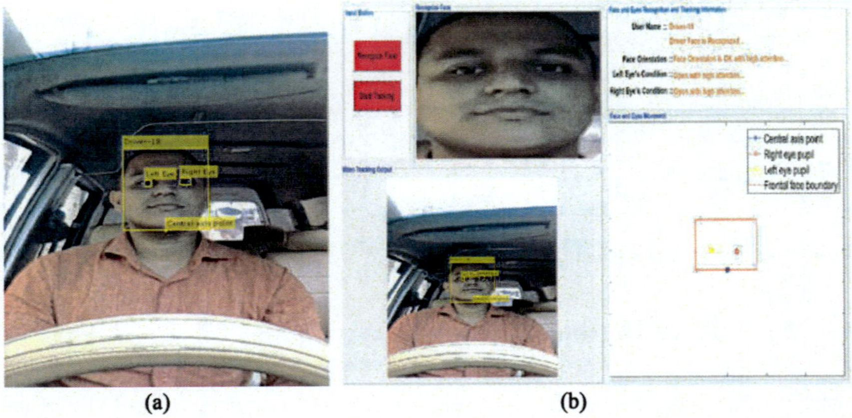

Fig. 8 Experimental example of face and eye's pupil tracking: **a** tracking result in normal window and **b** tracking result in the GUI

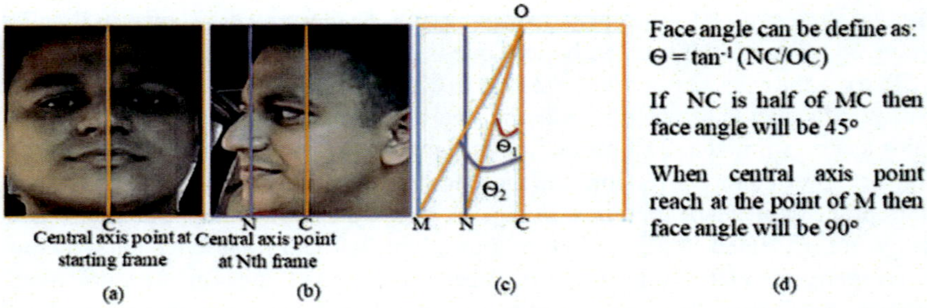

Fig. 9 Experimental example of face angle estimation: **a** face and central axis point at starting frame, **b** central axis point at Nth frame with face angle, **c** process to estimate the face angle, and **d** face angle estimation formula and some justification

3 Experimental Results

The experimental processing results of driver recognition and attention estimation are presented in this section. All experiments evaluation were performed on Intel(R) Core(TM) i3-3120 M@2.5GHZ processor with 8 GB RAM in the MATLAB environment. The video frame of the driver is captured by a camera which is placed in front of the driver. In this work, experimental results are divided into two parts. Firstly, the driver face is recognized, and on the other part, the driver attention is estimated. To recognize the driver face from the image 150 images with three orientations (O°, 45° left and 45° right), i.e., 450 images are used for train and 35 images are used for test. To monitor the driver attention, ten drivers' automobile driving experiments are considered.

In this work, the MATLAB GUI is developed to observe the driver's attention and the status of two parameters that are face orientation and eyes movement. The attention of the driver is shown in the GUI in real time, which is very important to awareness the driver through alarm.

During the experiment, the participants are asked to drive the automobile slowly and change the face angle randomly. The participants are also asked to close and open their left and right eyes randomly to ensure the functionality of eye's pupil detection and tracking activity. Every participant is involved in more than ten trials to measure the face orientation and eye's status. The face and eye's movements are shown in the visual form in the GUI as shown in Fig. 8 (bottom right).

If any point (x, y) in the face region is not satisfy the condition of *((y < 1 or y > I_Height-H) or (x < 1 or x > I_Width-W))*, then it is confirmed that the face is not on the frame boundary, where *W* and *H* are the width and height of the face image and *I_Width* and *I_Height* are the width and height of the video frame. The face angle can be estimated by tracking the central axis point shown in Fig. 9.

The experimental result of the driver face and eye's movement monitoring with alarm generation is shown in Fig. 10. In this figure, the experimental result is shown for the situation where face is moved to the left. The experimental result for the situations where face is moved to the right is shown in Fig. 11. The alarm will be different for the different level of attention, i.e., high or low. The driver's face movement is also classified in the different attention labels, i.e., high, low, or very low. The driver's attention is shown high if the central axis point of the respective frame is in the zero to 45 degree. In this situation, the left and right eyes focus is in the frame boundary. The driver's attention is shown in low if the central axis point is in the 45 to 90 degree; in that situation, the left eye during left movement or right eye during right movement will be in the no attention label. That means the focus of the left eye for left movement or right eye for right movement will be out of the frame boundary. If the central axis point moves to the 90 degree upper, in that situation, the focus of the left and right eyes will be out of the frame boundary. In that case, the attention status will be very low.

The performance of the face recognition method is compared with the some existing state of the arts. The performance comparison is shown in Table 1, where the proposed method shows higher performance compare to the present works. The performance evaluation of the driver attention based on the face orientation (FO) and eye's status (ES) parameter is shown in Table 2.

From Table 2, it is revealed that the eye's status is detected more accurately than faces orientation. The reason behind that is the status of eyes, i.e., close or open, is dependent on the face orientation. When the face orientation is changed to left or right, then the eye's pupil is not visible properly, so why the eyes status are detected appropriately. On the other hand, the face orientation is changed rapidly in either site at any time, so it is really difficult to detect face orientation appropriately all time.

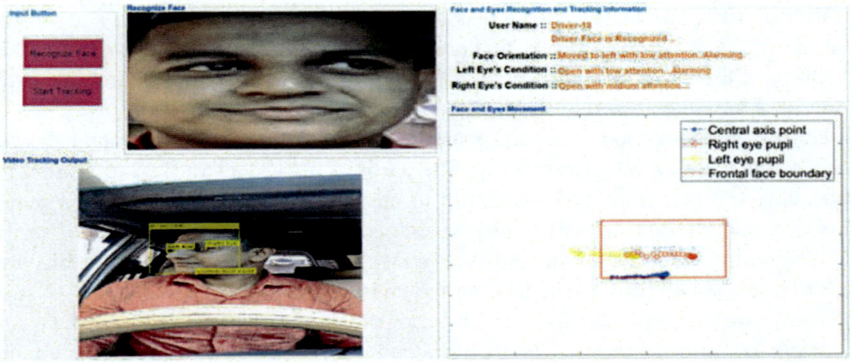

(a)

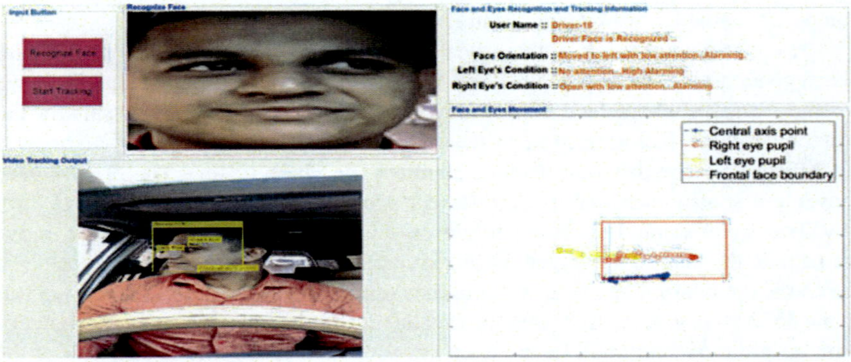

(b)

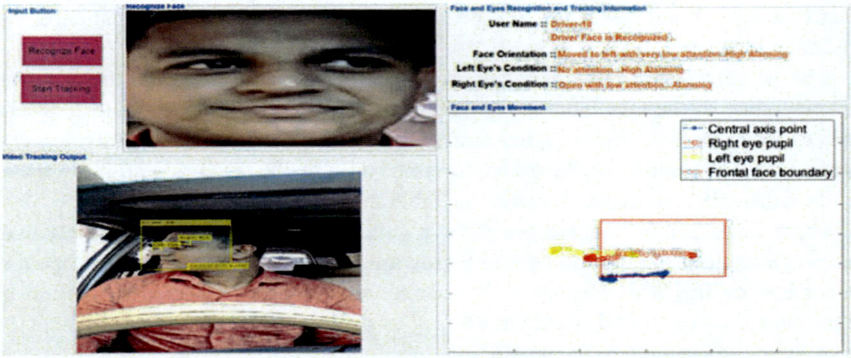

(b)

Fig. 10 Experimental result of different face orientation during moving to left side at different video frame: **a** face is moved to left with low attention, **b** face is moved to left with low attention where left eye is in no attention level, and **c** face is moved to left with very low attention where left eye is in no attention level

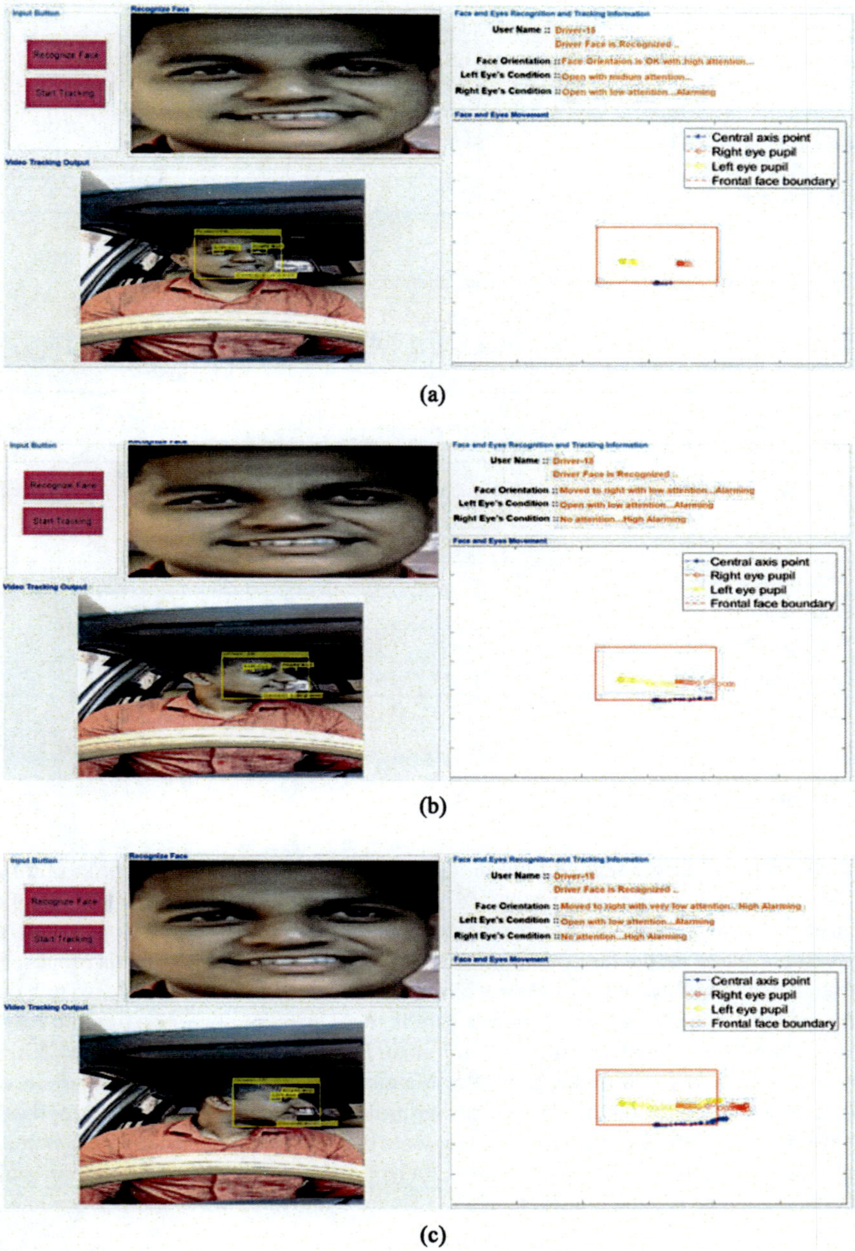

Fig. 11 Experimental result of different face orientation during moving to right side at different video frame: **a** face is moved to right with low attention, **b** face is moved to right with low attention where right eye is in no attention level, and **c** face is moved to right with very low attention where right eye is in no attention level

Table 1 Performance evaluation of driver's face recognition

Methods	LBP [7]	HGPP [14]	Gabor [13]	LBP + Gabor (Person Re – identification) [4]	PERCLOS, yawn frequency and gaze direction [9]	Proposed method (LBP + Gabor)
Success rate	81%	62.9%	93.1%	72%	81.5%	95.02%

Table 2 Performance evaluation of the driver attention based on FO and ES

	Actual FO and ES	Face orientation		Eyes status	
		FO detection	Accuracy (%)	ES detection	Accuracy (%)
Participant 1	13	11	84.62	12	92.31
Participant 2	12	10	83.33	12	100
Participant 3	15	14	93.33	14	93.33
Participant 4	11	10	90.91	9	81.82
Participant 5	14	12	85.57	13	92.86
Participant 6	10	10	100	10	100
Participant 7	15	13	86.67	14	93.33
Participant 8	12	12	100	11	91.67
Participant 9	13	12	92.31	12	92.31
Participant 10	11	11	100	11	100

4 Conclusion

A vision-based driver recognition and attention monitoring system is proposed in this research work. The proposed work is effective for reducing the accident rate by ensuring a next-generation safe driving environment. The driver monitoring system is developed based on the face orientation and parameters of eye pupil. In this proposed system, the unauthorized driver is not allowed to drive the vehicles in order to reduce the accident rate. For that, the driver's face is recognized by incorporating the LBP and Gabor filters. The face and pupil of human eyes are monitored through tracking system, which is based on the hue color feature. The proposed system has attained 95.02% accuracy for the driver face recognition. The effectiveness of the proposed real-time driver monitoring system is adequate by incorporating the real-time experiment in different environments. More efficient parameters such as lip motion and alcohol detector will be considered in the future work to improve the performance of the system.

References

1. Z. Chai, Z. Sun, H. Mendez-Vazquez, R. He, T. Tan, Gabor ordinal measures for face recognition. IEEE Trans. Inf. Forensics Secur. **9**(1), 14–26 (2014)
2. M. Yang, L. Zhang, Gabor feature based sparse representation for face recognition with Gabor occlusion dictionary, in *European Conference on Computer Vision* (Springer, Berlin, Heidelberg, 2010), pp. 448–461
3. T. D'Orazio, M. Leo, C. Guaragnella, A. Distante, A visual approach for driver inattention detection. Pattern Recogn. **40**(8), 2341–2355 (2007)
4. T. Ahonen, A. Hadid, M. Pietikainen, Face description with local binary patterns: application to face recognition. IEEE Trans. Pattern Anal. Machine Intell. **28**(12), 2037–2041 (2006)
5. X. Tan, B. Triggs, Fusing Gabor and LBP feature sets for kernel-based face recognition, in *International Workshop on Analysis and Modeling of Faces and Gestures* (Springer, Berlin, Heidelberg, 2007), pp. 235–249
6. P.J. Phillips, Support vector machines applied to face recognition, in *Advances in Neural Information Processing Systems*, pp. 803–809 (1999)
7. K. Kumar, Artificial neural network based face detection using gabor feature extraction. Int. J. Adv. Technol. Eng. Res. (IJATER) **2**, 220–225 (2012)
8. Y. Zhang, S. Li, Gabor-LBP based region covariance descriptor for person re-identification, in *2011 Sixth International Conference on Image and Graphics (ICIG)*(IEEE, 2011), pp. 368–371
9. L. Alam, M.M. Hoque, Vision-based driver's attention monitoring system for smart vehicles, in *Advances in Intelligent Systems and Computing*, vol. 86 (Springer, 2019), pp. 196–209
10. E. Murphy-Chutorian, M.M. Trivedi, Head pose estimation and augmented reality tracking: an integrated system and evaluation for monitoring driver awareness. IEEE Trans. Intell. Transp. Syst. **11**(2), 300–311 (2010)
11. P. Chowdhury, L. Alam, M.M. Hoque, Designing an empirical framework to estimate the driver's attention, in *2016 5th International Conference on Informatics, Electronics and Vision (ICIEV)* (IEEE, 2016), pp. 513–518
12. C.H. Morimoto, D. Koons, A. Amir, M. Flickner, Pupil detection and tracking using multiple light sources. Image Vis. Comput. **18**(4), 331–335 (2000)
13. S. Ghosh, T. Nandy, N. Manna, Real time eye detection and tracking method for driver assistance system, in *Advancements of Medical Electronics* (Springer, 2015), pp. 13–25
14. A. Rahman, M. Sirshar, A. Khan, Real time drowsiness detection using eye blink monitoring, in *2015 National Software Engineering Conference (NSEC)* (IEEE, 2015), pp. 1–7
15. J. Liu, X. Zhong, An object tracking method based on Mean Shift algorithm with HSV color space and texture features. Cluster Comput., 1–12 (2018)
16. B. Wang, B. Fan, Adoptive mean shift tracking algorithm based on the feature histogram of color and texture. J. Nanjing Univ. Posts Telecommunication. **33**(03), 18–25 (2013)
17. P. Viola, M.J. Jones, Robust real-time face detection. Int. J. Comput. Vision **57**(2), 137–154 (2004)

Precision of Product Reviews Using Naive Bayes and Linear Regression

M. R. Lakshmanan, Kashyap Kumar, Arjun G. Nair, and L. Nitha

Abstract As the digital information technology domain is growing unprecedently, the process of making the devices mimic human like actions has gained an increased research attention. One of the main features of human like system is the human emotion or sentiment detection. Online shopping has made it easy for everyone in the world to buy different varieties of products from a single place. It also gives the opportunity for them to try new products. While trying to buy new products, people check for the reviews and ratings of those products given by the other customers. Those reviews and ratings can be used to determine the sentimentality of users, who have bought those products by using text mining techniques and classify them either as positive or negative reviews. To analyze these data, the proposed research work has used both Naive Bayes and linear regression algorithms. In our analysis, the accuracy rate of both algorithms is checked.

1 Introduction

With the advances of technology, human population is becoming more dependent on the autonomous systems. Nowadays, online shopping has become a new trend for the younger generation to get all the desired products of different brands under one place. Also, the online platforms offer a lot of discounts while buying different products from those online platforms. Different machine learning techniques are applied for different purposes such as sentiment analysis and data analytics for business purposes. Here, the sentiment analysis approach of the data has been considered and the data can be obtained from online shopping platforms.

Machine learning has different approaches to analyze the data from the online platform for different purposes; one among them is sentiment analysis. Sentiment analysis determines the sentimentality of the user or the person who has written down the statement. This can be determined using different machine learning tools like support vector machine, linear regression, Naive Bayes, etc. Most of the customers

M. R. Lakshmanan (✉) · K. Kumar · A. G. Nair · L. Nitha
Department of Computer Science and IT, Amrita School of Arts and Sciences, Amrita Viswa Vidyapeetham, Kochi, India

P. Karuppusamy et al. (eds.), *Ubiquitous Intelligent Systems*, Smart Innovation, Systems and Technologies 243, https://doi.org/10.1007/978-981-16-3675-2_56

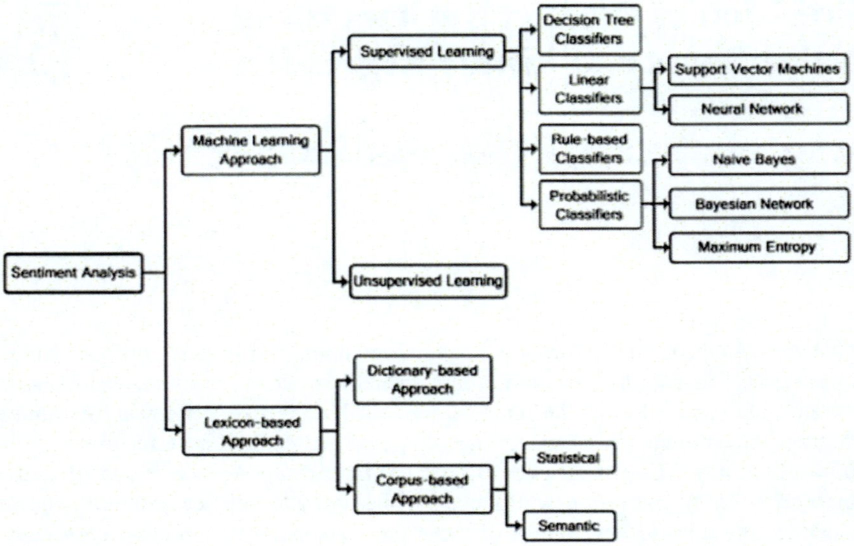

Fig. 1 Different approaches toward sentiment analysis

in online shopping platforms first go through the reviews and the ratings of the products before they buy them.

Many number of research and development works are conducted nowadays as for letting artificial intelligence has more knowledge about the human sentiments even through in the written form. Sentiment analysis can be used in the business to know the impact that has been made on product in people's minds.

According to this analysis, the businessman can incorporate changes or improve their product to have more impact on the customers.

Figure 1 shows various approaches through which the sentiment analysis can be performed. The machine learning approach is considered in this research paper. The Python language is used to implement the proposed algorithm. In the machine learning approach, the probabilistic classifier and linear classifier are used.

2 Literature Review

Yi et al. [1], in this paper, aim on the sentimental behavior of each person. Mostly, customers used to purchase more products through online mode. To know the quality and other details, all the purchaser looks at the reviews and rating of those products, so it is important to analyze the behavior of customers by analyzing ratings.

Guerreiro et al. [2], in this paper, cover the making of poor decisions that can be decreased by considering the opinions shared by the peer travelers. But at times, it also makes the whole process difficult as there are a lot of reviews based on each person's experience. When reviewers use expressions such as 'I highly recommend' or 'don't recommend,' it may affect the purchasing behavior of customers. The text mining method was applied in review and it was found out that negative attitudes from the provider triggered negative recommendations and positive feelings predicted positive recommendations.

Isah et al. [3], in this paper, cover the development of a machine learning framework which collects and analyzes the feedback from people who use drug and cosmetic items using natural language processing (a type of machine learning) techniques such as text mining and sentiment analysis. They also analyse the collection of data required for brand analysis (brand popularity, i.e., which brands are preferred, which are not) from comments, tweets, and other social media platforms such as Facebook and Twitter as well as finding the correct jargon for expressing safety hazards so that the layman can easily understand the possible implications of using the mentioned products, and data for determining brand popularity.

Chauhan et al. [4], in this paper, have used product reviews to get the sentiment analysis. They determined the reviews into positive, negative, and neutral reviews. They have used Naive Bayes and support vector machine algorithms to arrive at the result and plot them into a graph. The dataset was the reviews that were taken from Amazon.

Priyanka et al. [5], in this paper, have taken the social media post and comments as the dataset. As people have started to put their opinions, feedback, and reviews into their social media, those reviews can be a useful content to use for sentiment analysis. These content analysis can be useful for business, government, and individuals. They have used the artificial neural network for getting the output. They have improved the system by providing the motive that could have made the user post it in social media.

Xu et al. [6], in this paper, have done a continuous Naive Bayes learning framework for large and multi-domain ecommerce platform product review analysis. They have used the Amazon product review and movie reviews to showcase that the knowledge of the past domain can also be used for the different domain results. The old dataset results are used to get more accuracy rate by doing so.

Safrin et al. [7], in this paper, discussed the negation words that are used in the product reviews. They use the negation phrase identification algorithm to identify these negation phrases. The dataset was collected through a sample website that they have created for this purpose.

Yassir et al. [8], here in this paper, are trying to increase the knowledge finding from the multi-view text data. The system proposed by them was done by the use of X-clustering algorithm and hotspot algorithm of association rule.

Valsan et al. [9], in this paper, have taken reviews of two camera manufacturers. They have used the support vector machine (SVM) algorithm to get high accuracy of the sentiment analysis. They used the K-means algorithm for getting the most discussed word by the users from the review. By using this means of system, the user can get the most relevant category that is needed for them.

Gopinath et al. [10], in this paper, have used tweets to identify the sentimentality of those tweets. They have used the support vector machine (SVM) and K-means algorithm for this purpose. They have classified the tweets into positive, very positive, neutral, negative, and very negative feedback.

D'Andrea et al. [11], in this paper, mainly overview different sentimental approach and tools used in sentimental analysis. The approach means different techniques, features use in it. Mainly ever field it is being checked like business, politic, public actions and finance. In this, it explained about the sentimental analysis in social media how to figure the review of products, mainly knows by the people reaction which is emotional, positive, or negative. Also they explain about different approaches, each has its features to get the analysis. It not only focuses in the social media itself but also says that sentimental analysis is mainly in marketing, political, and sociological. It describes each field to categorise the analysis, the approaches that this paper focused on are (i) machine learning (ii) lexicon based and (iii) hybrid, for each of these approaches different algorithms are used that are mainly used are emotions that show the positive and negative reactions and also happy or sad, then Linguistic Inquiry and Word Count are also used.

Jagdale et al. [12], in this paper, mainly focused in sentimental analysis and opinion mining which is an easy way to analyze from various source like Facebook, Twitter, etc. Here, they say that sentimental analysis and opinion mining is the best way to gather information about data and product reviews so that it can be modified accordingly. Algorithm used in this research is Naive Bayes and support vector machine. They got a point difference in the result using these both algorithms. They also mention that not only the product but in every field, we can use the sentiment analysis and has a major role to analyze the emotions of people to make human like machine with emotions.

Anto et al. [13], in this paper, mainly focus on the feedback, review which has been taken from the Twitter which is filtered and analyzed using opinion mining. Feedback of a product is considered as the most important aspect for a company growth. The manufacturer will only get his product review from the customer side. Only through feedback, the manufacturer can make necessary changes to its product. Here, they use feedback which is obtained from mobile phones. This technique which they have used has shown about 80% of accuracy in sentimental analysis; this helps to provide valuable and effective feedback for the company.

Gan et al. [14], in the paper, mainly focus to find the architecture of online restaurant review and also to find impact of review attributes and sentiments on the rating (star) of the restaurant. The main four attributes which the reviews are focused are food, ambience, price, and service. This paper finds a fifth attribute which is also an important one for online review. Using sentimental analysis the result shows, customers sentiments in the major five attributes have a drastic change in star rating. Food, context, and service are the major three attributes that affect the star rating.

Cuizon et al. [15], in the paper, discuss the reviews and the star rating of different hotels based on five aspects like ambiance, cost, food, hygiene, and services. They provide a platform where different users and reviewers can share their experiences, strengths, and weakness of the hotel. They have used Stanford core NLP library and

AFINN library to determine the noun-adjectives and adjectives, respectively. They developed applications to get the unstructured text to get the relevant data like the user and reviews. At the end they say, we can use this over the existing reviews to get automatic ratings.

3 Proposed System

See Fig. 2.

3.1 Data Collection

The raw data is collected from ecommerce platforms. The data consists of star ratings and text reviews of portable water filters. We have collected 179 sample data from Amazon. We collected reviews by different users for the purpose of our research. These reviews were collected manually by us from Amazon. We stored these reviews and star ratings in an excel file in a table format.

3.2 Data Extraction

Data extraction is the process of collecting or retrieving a dataset from different sources. The star ratings and text reviews are the main dataset for our analysis as shown in Fig. 3. These datasets we collected are of one product category, that is

Fig. 2 Workflow of the proposed system

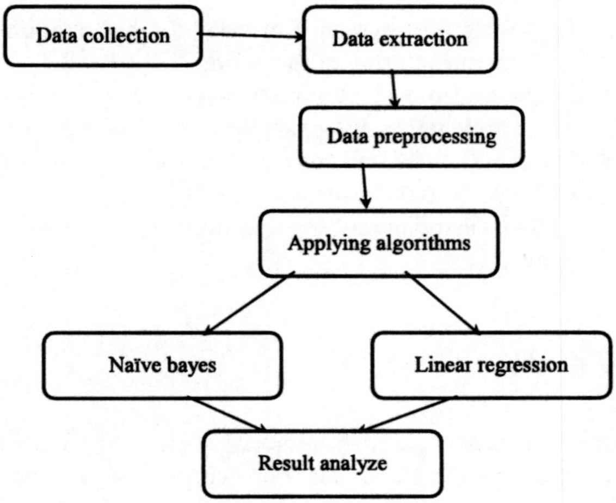

star ratings	reviews
5	Safe purified drinking water, now available without fire and electricity!!!
3	Great for treks even though it lacks in capacity
1	Think again!!
3	purifies the water well. But the only problem with this bottle is when the bottle is inverted water spills out
3	The valve leaks
5	Amazing Product! Just awesome!
5	Much needed utility for backpacking and Hiking
5	great product
5	Value for money
1	Filters dirt & bacteria, sends the chemicals and odours to your mouth.
5	TRUE LIFE FILTER
4	Great bottle for a decent cost, though bottle material quality needs to be improved.
3	Scratches bottle but lifestraw best brand for trekking
2	Design is not Leak proof but Filter is Excellent.
5	Good bottle for trip-goers but amazon delivery might have damaged my product
4	Sipper deteriorate

Fig. 3 Star ratings and reviews

portable water purifiers. We collected both the star ratings and review of the verified profile into a CSV file for the purpose of the analysis.

3.3 Data Preprocessing

Data preprocessing is the next stage after data extraction. It is the process of getting the text into the form that can be used in the program. This process is the first step in natural language process (NLP). In this process, we make the necessary changes in the review text file.

We have gone through the preprocessing steps that were used in the research paper [11, p. 4]. Valsan et al. [11, p. 4] proposed that the preprocessing steps followed are stop word removal, removing punctuations. In order to perform data preprocessing, we have added a few more steps.

- Removing punctuation: Removed the punctuation marks from the text reviews for better processing of the words.
- Stopword removal: Stopwords were removed as they may not make much meaning to the review text. For example words like 'a,' 'an,' 'and,' etc.
- Letter casing: We changed the case of every word to lowercase, since those words will belong to the same case.
- Tfidf—Vectorization: We transformed the words into the vectors to use as the inputs.

3.4 Naive Bayes

It is a machine learning approach which comes under the supervised learning. It is based on the Bayes theorem with an assumption of independence among the

Cassification report

	precision	recall	f1-score	support
1	1.00	0.25	0.40	4.00
2	0.00	0.00	0.00	1.00
3	0.00	0.00	0.00	1.00
4	0.00	0.00	0.00	2.00
5	0.59	1.00	0.74	10.00
accuracy rate			0.61	18.00
macro avg	0.32	0.25	0.23	18.00
weighted avg	0.55	0.61	0.50	18.00

accuracy : 61.11111111111114

Fig. 4 Result of Naive Bayes

predictors. It is a classifier technique which assumes that the presence of a feature is unrelated to any other feature.

$$P(c|x) = \frac{P(x|c)P(c)}{P(x)} \tag{1}$$

The above formula is the general formula that is used in Naive Bayes algorithm. It is one of the fastest classifier algorithms which is used. After the data preprocessing, we divided the dataset to training dataset and test dataset as shown below:

#Split the data into train and testing.

X_train, X_test, Y_train, Y_test = train_test_split(X_tfidf, y, test_size = 0.1, random_state = 0).

The above of statement is used for split the dataset into training and testing dataset. This was done as when we executed the algorithm without splitting we got lesser rate or accuracy than after the split.

We applied this partition as to increase the accuracy of the algorithm. As we tried executing the dataset with and without the partition, we saw a difference in the accuracy rate. Then we applied the Naive Bayes algorithm to get the accuracy rate. After using the Naive Bayes algorithm, we got the accuracy and classification report as shown in Fig. 4.

3.5 Linear Regression

Linear regression is also a supervised machine learning approach. It is a part of regression analysis, which is a predictive model that helps to find out the relationship between input and target variable. It is linear in the model that assumes a linear

slope	3.15575708 -1.92110164 1.19794134 -2.57123314 - 0.17512188 0.23817597 -2.9132895 -1.7564348 0.17474026 0.24944449 0.26216258 0.86581735 0.35999388 0.69924544 0.19574213 -1.07738419 - 1.92110164 -0.31523531 0.76552822 -3.02831962 - 0.56415136
intercept	3.45378204
root mean squared error	1.816650648
r2 score	0.338613496

Fig. 5 Result of linear regression

relationship between input variable and output variable. We have used the multi-linear regression model as the input variable, and it is not a one dimension array.

The general formula of linear regression is:

$$Y = mx + c \tag{2}$$

In the linear regression algorithm, the dataset is not split into training and testing dataset since it showed a lower score in r2 score. So we used the dataset as such and found the results like slope, intercept, rooted mean squared error, and r2 score as shown in Fig. 5.

4 Result Analysis

In result analysis, we compare the accuracy of both Naive Bayes and linear regression algorithms. After employing the Naive Bayes algorithm, we got the accuracy rate as 61.1 as shown in Fig. 4.

Whereas in the linear regression algorithm, we got the r2 score as 0.33. If we compare it with Naive Bayes, this score is even less accurate than the former.

5 Conclusion

We can conclude as Naive Bayes algorithm gives more accuracy than the linear regression algorithm. We have used 179 datasets from an ecommerce platform for analysis. So in future, if we use bulk amount of dataset, the rate of accuracy will also change. If sentiment analysis is retrieved, the data can be classified into positive, negative, and neutral. The business firms can obtain a clear image on how their product had an impact on the customers. Also, they can decide what all changes they have to do for staying in the business. These can be done as an improvement or add-on to this paper in the future.

References

1. S. Yi, X. Liu, Machine learning based customer sentiment analysis for recommending shoppers, shops based on customers' review. Complex Intell. Syst. **6**, 621–634 (2020)
2. J. Guerreiro, P. Rita, How to predict explicit recommendations in online reviews using text mining and sentiment analysis. J. Hospitality Tourism Manag. **43**, 269–272 (2020)
3. H. Isah, P. Trundle, D. Neagu, Social media analysis for product safety using text mining and sentiment analysis, in *2014 14th UK Workshop on Computational Intelligence (UKCI)*
4. M. Chauhan, D. Yadav, Sentimental analysis of product based reviews using machine learning approaches. J. Netw. Commun. Emerg. Technol. (JNCET) **5** (Special Issue 2) (Dec, 2015)
5. B. Priyanka, J.T. Thirukrishna, Sentimental data analysis on social media. IJSRD Int. J. Sci. Res. Dev. **3**(09). ISSN (online): 2321–0613 (2015)
6. F. Xu, Z. Pan, R. Xia, E-commerce product review sentiment classification based on a naïve Bayes continuous learning framework. Inf. Process. Manag. **57**(5), 102221 (Sept 2020)
7. R. Safrin, K.R. Sharmila, T.S. Shri Subangi, E.A. Vimal, Sentiment analysis on online product review. Int. Res. J. Eng. Technol. (IRJET), **04**(04) (Apr 2017)
8. A.H. Yassir, A.A. Mohammed, A.A.J. Alkhazraji, M.E. Hameed, M.S. Talib, M.F. Ali, Sentimental classification analysis of polarity multi-view textual data using data mining techniques. Int. J. Electr. Comput. Eng. (IJECE) **10**(5), 5526–5533 (Oct 2020)
9. A. Valsan, C.T. Sreepriya, L. Nitha, Social media sentiment polarity analysis: A novel approach to promote business performance and consumer decision making. Artif. Intell. Evol. Comput. Eng. Syst. 1–12
10. G.P. Gopinath, A. Raj, L. Nitha, Multi-class sentiment analysis. Int. J. Appl. Eng. Res. **10**(55) ISSN 0973-4562 (2015)
11. A. D'Andrea, F. Ferri, P. Grifoni, T. Guzzo, Approaches, tools and applications for sentiment analysis implementation. Int. J. Comput. Appl. (0975–8887) **125**(3) (Sept 2015)
12. R.S. Jagdale, V.S. Shirsat, S.N. Deshmukh, Sentiment analysis on product reviews using machine learning techniques. Cogn. Informat. Soft Comput. 639–647
13. M.P. Anto, M. Antony, K.M. Muhsina, N. Johny, V. James, A. Wilson, Product rating using sentiment analysis, in *2016 International Conference on Electrical, Electronics, and Optimization Techniques (ICEEOT)*
14. Q. Gan, Yu. Yang, L. Jin, A text mining and multidimensional sentiment analysis of online restaurant reviews. J. Qual. Assur. Hospitality Tourism **18**(4), 1–28 (2016)
15. J.C. Cuizon, J. Lopez, D. Rose Jones, Text mining customer reviews for aspect-based srestaurant rating. Int. J. Computer Sci. Inf. Technol. (IJCSIT) **10**(6) (Dec 2018)

Analysis of Bandwidth Consumption in VoIP

M. Sai Prasanthi, I. Yuva Krishna Kishore, G. Satyanarayana, Sai Venkata Reddy Vanga, and Pamulapati Nitheesh Prasad

Abstract In this modern era of advancement in computer networks, quality communication plays a major role in establishing an effective understanding among individuals. While interacting with the people around us, they act and acknowledge. In order to attain this kind of quality communication, one should invest a good amount to select the traditional calling methods. This type of communication costs more and ends up in having different kinds of bills to different places for obtaining the required communication. But in VoIP, there is no such price sheet for the same communication. VoIP is also cost-effective compared to traditional calling and requires very less time to set up and maintain. Still, VoIP is growing and achieving good quality communication. Though VoIP has good and quality communication in real time, it consumes more bandwidth than usually required, and this issue in VoIP affects a lot in terms of quality of service (QoS) in VoIP. These issues are variable for different protocols, and hence, they are discussed briefly in this paper.

1 Introduction

Voice over Internet Protocol (VoIP) is a technology that makes it possible for the users to make telephone calls over the internet or intranet networks. VoIP has great benefits for increased saving, video streaming, and several other added value services.

Now in this current generation, there is a tremendous competition among the VoIP suppliers in the market and further, the improvements in the quality of services are felt by subscribers. Businesses are now more attracted to use VoIP Systems. A larger number of organizations today use VoIP systems.

The extent of a VoIP usage can differ as per an association's necessities and wants, going from the moderately direct—utilizing VoIP for neighborhood and significant distance calls or to convey between an organization's different workplaces—to more unpredictable arrangements like call places. We can add different features to the VoIP communications like call forwarding, recording, encryption, etc., as an added

M. Sai Prasanthi (✉) · I. Yuva Krishna Kishore · G. Satyanarayana · S. V. R. Vanga · P. N. Prasad
K L University, Guntur, India
e-mail: prasanthi1216@kluniversity.in

P. Karuppusamy et al. (eds.), *Ubiquitous Intelligent Systems*, Smart Innovation, Systems and Technologies 243, https://doi.org/10.1007/978-981-16-3675-2_57

advantage to the communication. The VoIP can also provide an anonymous identity across the network which avoids the spam content. This creates a virtual identity in place of a real telephone number.

As discussed earlier, VoIP has a great potential to deliver a great quality of communication under a minimal cost. Keeping quality of service (QoS) as an aspect into consideration [1], we are trying to investigate at what rate VoIP is delivering the quality to its users. QoS consists of various aspects to measure a network, they are delay, jitter, packet loss, bandwidth, noise, error rate, etc. [2], and all these are used to determine the quality of a communication [3].

When you look into the VoIP, it manages as a good alternative to traditional calling in every aspect and even sometimes a better to the regular one. However, still VoIP has issues in some aspects while using different protocols to implement it. For example, when you consider TCP protocol for communication in network, then it provides some advantages like security [4], compatibility, audio quality, and at the same time it lacks in terms of jitter and latency [5]. For UDP case, it has broadcast and multicast transmission is available, but it doesn't provide reliability in transmission.

VoIP offers a great efficiency as it acts as your dedicated receptionist because the system is delivered online it can include advanced functionality, and it is also helpful in making efficiency saving in your business. As compared to the traditional PBX telephone systems, VoIP is far better and cheaper on a monthly basis because it doesn't need wires and hard-wares to install or maintain. Thus, VoIP is a lower service in cost.

2 VoIP Architecture

VoIP architecture is a network topology that upholds the real-time audio with a broadband/internet connection.

As VoIP changes over audio signals into digital signals, and it sends them over the web. VoIP is presently the highest standard source to provide staff with reliable business communications. The amount of your network bandwidth may affect your call quality as it is the foundation of VoIP architecture. With high network usage in the system, VoIP calling will work, but due to high usage of the network, the voice quality may decrease or experience latency issues [6] (Fig. 1).

Since most of the VoIP stages are cloud-based, the network configuration is the responsibility for your cloud service provider to guarantee that the service is dependable, functioning optimally, and specific security rules [4].

Codec is also known as coder and decoder or compression—decompression. It is mostly used in digital media, especially audio and video, which have traditionally consumed significant bandwidth [7]. If we concentrate on proper packet transfer, we need to face a huge delayed transmission which is not good to suggest an alternative to the existing mode of communication (Telephone calling) [8]. If we choose to avoid delays by eliminating the acknowledgment part, we may face issues with the packet loss [9]. One of the main drawbacks in VoIP is excess bandwidth consumption. If

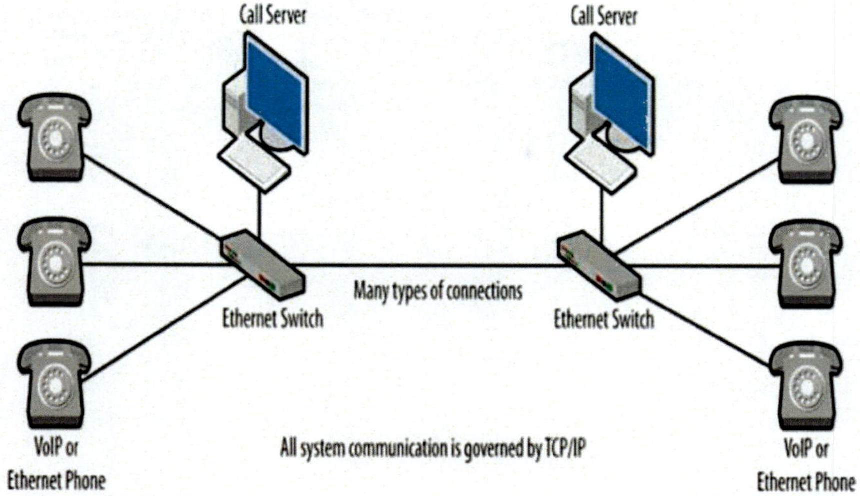

Fig. 1 VoIP architecture

you see the real-time transmission quality now-a-days we can hardly achieve the same quality under 5 KB/s, but the VoIP transmissions are demanding more than that every time due to various reasons. So, now we will investigate this issue in VoIP which might be a considerable drawback in terms of QoS [2].

3 Existing System

There are various implementations of VoIP in day-to-day life like WhatsApp, Facebook, Telegram Calling, etc. All these applications follow different protocols based on their features. Each has their own uncommon issue over using different protocols in them. They need to be rectified in the future. VoIP is a type of making voice over using data via the internet. It is commonly used by mainstream cellular networks to provide free similar network calling.

Every time you use your computer or phone to call someone using the internet, you are using VoIP. For instance, when you use Skype or Facebook Messenger, all these come under VoIP.

Codec & Bit Rate (kbps)	Codec Sample Size (Bytes)	Codec Sample Interval (ms)	Mean Opinion Score (MOS)	Voice Payload Size (Bytes)	Voice Payload Size (ms)	Payload Size (ms) Packets Per Second (PPS)	Bandwidth MP or FRF.12 (kbps)	Bandwidth w/cRTP MP or FRF.12 (kbps)	Bandwidth Ethernet (kbps)
G.711 (64 kbps)	80	10	4.1	160	20	50	82.8	67.6	87.2
G.729 (8 kbps)	10	10	3.92	20	20	50	26.8	11.6	31.2
G.723.1 (6.3 kbps)	24	30	3.9	24	30	33.3	18.9	8.8	21.9
G.723.1 (5.3 kbps)	20	30	3.8	20	30	33.3	17.9	7.7	20.8
G.726 (32 kbps)	20	5	3.85	80	20	50	50.8	35.6	55.2
G.726 (24 kbps)	15	5		60	20	50	42.8	27.6	47.2
G.728 (16 kbps)	10	5	3.61	60	30	33.3	28.5	18.4	31.5
G722_64k(64 kbps)	80	10	4.13	160	20	50	82.8	67.6	87.2
ilbc_mode_20(15.2 kbps)	38	20	NA	38	20	50	34.0	18.8	38.4
ilbc_mode_30(13.33 kbps)	50	30	NA	50	30	33.3	25.867	15.73	28.8

4 Procedure

4.1 Installing the Required Tools

To run this model, we need to install the required tool called Wireshark. Wireshark is a free and open-source packet analyzer. It is mainly used by the network security engineers and developers to examine network security problems. As Wireshark allows the users to observe all the traffic being passed over the network.

4.2 VoIP Bandwidth Requirements

The required bandwidth is determined by the codec and the transmission medium. Two events necessitate the use of bandwidth. The media stream needs at least 17–106 kbps of bandwidth, depending on the codec, header content, and Layer 2 and 3 headers. Call signaling isn't the only thing to think about. Although the bandwidth required by call signaling is much less, irregular interests can cause problems on a network.

5 Working Principle

The proposed model has calculated the packets sent over a network. A packet is a unit of data, which is transmitted over a network between the sender and receiver.

Networks packets are small, i.e., maximum 1.5 kb for Ethernet packets and 64 kB for IP packets.

Firstly, before making a call, we should need a tool known as Wireshark, a packet sniffing tool as it captures the data packets over the transmission medium. After installing the tool, start the call and also start capturing the data in the tool. The tool captures every packet in the transmission. After completing the call, stop capturing the packets and export the captured data to the CSV file, as it is flexible to process the data in parameter wise.

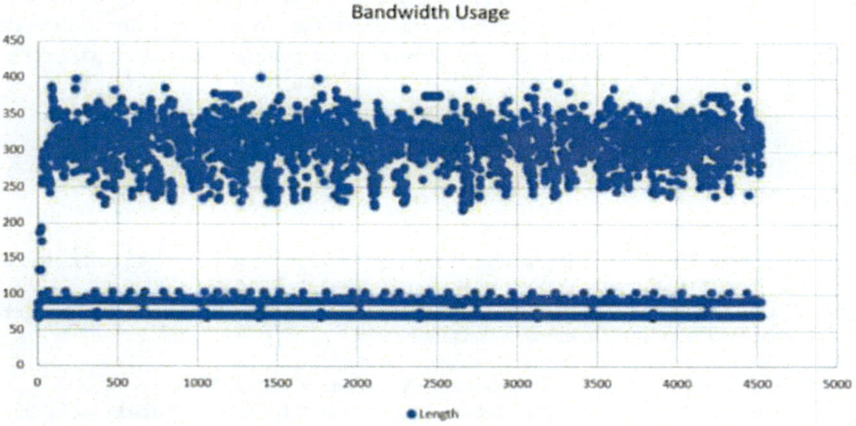

This data gives us the bandwidth consumption rate on Telegram app which uses the UDP protocol for data transmission [10], where it is consuming more bandwidth than required. When it is in TCP case, it uses too much bandwidth and may create huge delays in transmission, which again leads to packet discarding, in simple words, it finally comes to data loss. So, we need to use a good protocol which serves as best alternative by inheriting the all positive note in the both protocols (TCP and UDP). So, we can choose RTP as a good alternative protocol for the VoIP transmission.

When you get a doubt, why to use RTP instead of TCP and UDP. Real-time transmission protocol (RTP) has the mix-up nature of both TCP and UDP in functionality. It provides you the security factor present in TCP, and low delay transfers like UDP [11].

For example, consider the scenario of WhatsApp vs. Telegram in terms of VoIP calling. Here, WhatsApp follows RTP protocol and Telegram uses UDP protocol (most of the VoIP services are running on UDP-based protocol). Telegram uses MadelineProto API and WhatsApp uses Opus Codec. Telegram uses UDP protocol, whereas WhatsApp uses RTP protocol. As per communication, telegram uses peer-to-peer type and WhatsApp uses client–server type of communication. In consumption of bandwidth, telegram consumes more bandwidth of 71 kb/s, whereas WhatsApp only takes 30 kb/s [12]. Telegram has a low ping of 20-30 ms than WhatsApp of 30–40 ms. As consumption of bandwidth is more in telegram, Quality of Service in Telegram is far better than WhatsApp [10].

6 Conclusion

From this analysis, we came into conclusion that there is a bandwidth waste in UDP protocol using VoIP. There is an excess consumption of bandwidth during the packet transmission. UDP is still used in many mainstream applications that we use daily like video conferencing and streaming video. Our criteria to maintain a good quality of service and low packet loss, we see that TCP out-performs UDP and RTP [13]. Packet loss is minimal in TCP unless the background traffic is high. RTP performs extremely well but when compared to other protocols it provides slightly low quality calling [14], but it can be improved in the future but certain modifications to its code [15]. At present, RTP provides our desirable outcome near to our QoS constraints in VoIP compared to other protocols.

References

1. M.H. Miraz, S.A. Molvi, Simulation and analysis of quality of service (QoS) parameters of voice over IP (VoIP) traffic through heterogeneous networks. IJACSA
2. J.D. Gupta, S. Howard, A. Howard, Traffic behaviour of VoIP in a simulated access network, in *Proceedings of World Academy of Science* (2006)
3. U. Shaw, B. Sharma, A survey paper on VOIP. IJCA (2016)
4. M. Marjalaakso, Security Requirements and Constraints of VoIP (2001)
5. T.J. Walsh, D.R. Kuhn, Challenges in security Voice over IP. IEEE Security and Privacy (2005)
6. A. Leon-Garcia, I. Widjaja, *Communication Networks: Fundamental Concepts and Key Architectures*, 2nd edn. (McGraw-Hill, 2005)
7. D. Collins, *Carrier Grade Voice Over IP* (McGraw-Hill, 2002)
8. C. Demichelis, P. Chimento, *IP Packet Delay Variation Metric for IP Performance Metrics (IPPM)*, RFC 3393 (2002). http://www.ietf.org/rfc/rfc3393.txt. Accessed in Mar 2012
9. T. Uhl, QoS by VoIP under use different audio codecs, in *Joint Conference-Acoustics* (2018)
10. R.A. Manna, S. Ghosh, A comparative study between Telegram and Whatsapp in respect of library services. IJLIS (2018)
11. O. Hersent, *IP Telephony: Deploying VoIP Protocols and IMS Infrastructure* (Wiley, West Sussex, UK, 2011)
12. M.T. Meeran, P. Annus, Evaluation of VoIP QoS performance in wireless mesh networks. MDPI (2017)
13. B. Goode, Voice over internet protocol, in *Proceedings of the IEEE*, vol. 90, no. 9 (2002)
14. K. Tambe, R. Bhor, Study of VOIP services and its applications. *IJSER* (2013)
15. H. Schulzrinne, S. Casner, R. Frederick, V. Jacobson, RTP: A Transport Protocol for Real-Time Application. Network Working Group (2003)

Author Index

P. Karuppusamy et al. (eds.), *Ubiquitous Intelligent Systems*, Smart Innovation,
Systems and Technologies 243, https://doi.org/10.1007/978-981-16-3675-2